FUNDAMENTALS OF GAS DYNAMICS

BOARD OF EDITORS

I. Thermodynamics and Physics of Matter. Editor: F. D. Rossini
II. Combustion Processes. Editors: B. Lewis, R. N. Pease, H. S. Taylor
III. Fundamentals of Gas Dynamics. Editor: H. W. Emmons
IV. Theory of Laminar Flows. Editor: F. K. Moore
V. Turbulent Flows and Heat Transfer. Editor: C. C. Lin
VI. General Theory of High Speed Aerodynamics. Editor: W. R. Sears
VII. Aerodynamic Components of Aircraft at High Speeds. Editors: A. F. Donovan, H. R. Lawrence
VIII. High Speed Problems of Aircraft and Experimental Methods. Editors: A. F. Donovan, H. R. Lawrence, F. Goddard, R. R. Gilruth
IX. Physical Measurements in Gas Dynamics and Combustion. Editors: R. W. Ladenburg, B. Lewis, R. N. Pease, H. S. Taylor
X. Aerodynamics of Turbines and Compressors. Editor: W. R. Hawthorne
XI. Design and Performance of Gas Turbine Power Plants. Editors: W. R. Hawthorne, W. T. Olson
XII. Jet Propulsion Engines. Editor: O. E. Lancaster

VOLUME III
HIGH SPEED AERODYNAMICS
AND JET PROPULSION

FUNDAMENTALS OF GAS DYNAMICS

EDITOR: HOWARD W. EMMONS

PRINCETON, NEW JERSEY
PRINCETON UNIVERSITY PRESS
1958

London: Oxford University Press

L. C. Card 57-6331

Printed in the United States of America by
The Maple Press Company, Inc., York, Penna.

FOREWORD

On behalf of the Editorial Board, I would like to make an acknowledgement to those branches of our military establishment whose interest and whose financial support were instrumental in the initiation of this publication program. It is noteworthy that this assistance has included all three branches of our Services. The Department of the Air Force through the Air Research and Development Command, the Department of the Army through the Office of the Chief of Ordnance, and the Department of the Navy through the Bureau of Aeronautics, Bureau of Ships, Bureau of Ordnance, and the Office of Naval Research made significant contributions. In particular, the Power Branch of the Office of Naval Research has carried the burden of responsibilities of the contractual administration and processing of all manuscripts from a security standpoint. The administration, operation, and editorial functions of the program have been centered at Princeton University. In addition, the University has contributed financially to the support of the undertaking. It is appropriate that special appreciation be expressed to Princeton University for its important over-all role in this effort.

The Editorial Board is confident that the present series which this support has made possible will have far-reaching beneficial effects on the further development of the aeronautical sciences.

Theodore von Kármán

PREFACE

Rapid advances made during the past decade on problems associated with high speed flight have brought into ever sharper focus the need for a comprehensive and competent treatment of the fundamental aspects of the aerodynamic and propulsion problems of high speed flight, together with a survey of those aspects of the underlying basic sciences cognate to such problems. The need for a treatment of this type has been long felt in research institutions, universities, and private industry and its potential reflected importance in the advanced training of nascent aeronautical scientists has also been an important motivation in this undertaking.

The entire program is the cumulative work of over one hundred scientists and engineers, representing many different branches of engineering and fields of science both in this country and abroad.

The work consists of twelve volumes treating in sequence elements of the properties of gases, liquids, and solids; combustion processes and chemical kinetics; fundamentals of gas dynamics; viscous phenomena; turbulence; heat transfer; theoretical methods in high speed aerodynamics; applications to wings, bodies and complete aircraft; nonsteady aerodynamics; principles of physical measurements; experimental methods in high speed aerodynamics and combustion; aerodynamic problems of turbomachines; the combination of aerodynamic and combustion principles in combustor design; and finally, problems of complete power plants. The intent has been to emphasize the fundamental aspects of jet propulsion and high speed aerodynamics, to develop the theoretical tools for attack on these problems, and to seek to highlight the directions in which research may be potentially most fruitful.

Preliminary discussions, which ultimately led to the foundation of the present program, were held in 1947 and 1948 and, in large measure, by virtue of the enthusiasm, inspiration, and encouragement of Dr. Theodore von Kármán and later the invaluable assistance of Dr. Hugh L. Dryden and Dean Hugh Taylor as members of the Editorial Board, these discussions ultimately saw their fruition in the formal establishment of the Aeronautics Publication Program at Princeton University in the fall of 1949.

The contributing authors and, in particular, the volume editors, have sacrificed generously of their spare time under present-day emergency conditions where continuing demands on their energies have been great. The program is also indebted to the work of Dr. Martin Summerfield who guided the planning work as General Editor from 1949–1952. The cooperation and assistance of the personnel of Princeton University Press and of the staff of this office has been noteworthy. In particular, Mr. H. S.

Bailey, Jr., the Director of the Press, and Mr. R. S. Snedeker, who has supervised the project at the Press and drawn all the figures, have been of great help. Special mention is also due Mrs. E. W. Wetterau of this office who has handled the bulk of the detailed editorial work for the program.

Coleman duP. Donaldson
Joseph V. Charyk
General Editors

PREFACE TO VOLUME III

Gas dynamics as a branch of physics and applied mathematics has grown with the growth of high speed flight. In this volume the fundamentals of gas dynamics are developed and then applied to the flow through nozzles and passages, to shock phenomena of various kinds, to condensation effects, to flames and detonations, and to the flow of rarefied gases.

The enormous rate of development of gas dynamics has caused some delay in publication date. The outstanding contributors to this volume were so busy developing new areas that only little time was obtainable to write up the present state of knowledge.

Several authors have, besides writing the original manuscript, spent time, just prior to publication, in bringing their manuscripts up to date. The move of Professor Tsien from California Institute of Technology to China has made communication in the final stages rather slow. The move of Kantrowitz from Cornell to AVCO has so increased his work load that a manuscript revision since the 1952 version has not been possible.

Special thanks are due ONR London for providing the time for Professor Hayes to complete his assignment which had been enlarged as the work progressed. Although Section G, Chapter 3 on detonations was written in rough draft form by Professor G. I. Taylor, the press of other work prevented its completion and the considerable effort of Richard Tankin in completing this part warranted his inclusion as junior author.

Section H on rarefied gas flow while originally prepared for Volume IV was ready so much before the remainder of that volume and, in fact, forms such an obvious extension of gas dynamics into some of the latest problems that it was added as a most appropriate concluding section.

Initially Professor Crocco served as volume editor and thus launched the volume through its initial stages. I served as volume editor through its manuscript and proof stages. I know I speak for Professor Crocco as well as for myself as I express appreciation for the fine cooperation of all the authors, the General Editor, and the Princeton University Press.

Howard W. Emmons
Volume Editor

CONTENTS

FUNDAMENTALS OF GAS DYNAMICS

SECTION A

THE EQUATIONS OF GAS DYNAMICS

H. S. TSIEN

A,1. Introduction. The information needed by design engineers of either aircraft or flow machinery is the pressure, the shearing stress, the temperature, and the heat flux vector imposed by the moving fluid over the surface of a specified solid body or bodies in a fluid stream of specified conditions. To supply this information is the main purpose of the discipline of gas dynamics. The description of the state of fluid is generally considered to be complete if, in addition to the above information, the velocity vector and the density of the fluid are specified. The problem of gas dynamics is then completely solved if all these quantities are determined for every point of the specified spatial region and every time instant of the specified interval. It is the purpose of this section to formulate the general problem and to supply the equations for the analysis.

However, before going into the detailed analysis, it is appropriate to compute an all-important quantity in gas dynamics, namely the speed of propagation of small disturbances. This speed is often called the *velocity of sound*. Since the presence of a body in a stream of fluid creates disturbances in the fluid, it is to be expected that the characteristics of the flow field are determined to a large extent by the ratio of the average speed of motion to the speed of sound. In fact, flow problems are divided into four categories by this ratio alone: subsonic flows when the fluid speed is less than the sound speed, transonic flows when the fluid speed is comparable with the sound speed, supersonic flows when the fluid speed is larger than the sound speed, and finally hypersonic flows when the fluid speed is much larger than the sound speed.

To compute the speed of propagation of small disturbances, consider a uniform flow from left to right crossing a small discontinuity. The position of the discontinuity is held fixed, and the flow field is then steady, i.e. invariant with respect to time. By crossing the discontinuity, the fluid velocity is increased from u to $u + du$, the density from ρ to $\rho + d\rho$, and the pressure from p to $p + dp$. Consider now a stream tube of unit cross-sectional area. Then since no fluid mass is created by passing the discontinuity,

$$\rho u = (\rho + d\rho)(u + du) \tag{1-1}$$

The increase in momentum in crossing the discontinuity must be balanced by the increase in pressure acting on the fluid. Therefore

$$(\rho + d\rho)(u + du)^2 - \rho u^2 = -dp \tag{1-2}$$

From Eq. 1-1 and 1-2, one has by taking only first order differentials

$$u^2 = \frac{dp}{d\rho}$$

It is evident that for an observer moving with the fluid, the fluid is at rest but the disturbance moves with a speed u. Thus u is really the speed of propagation of small disturbances, or sound speed. Denote the sound speed by a, then

$$a^2 = \frac{dp}{d\rho} \tag{1-3}$$

Eq. 1-3 gives the sound speed whenever the relation between pressure and density is known. For incompressible fluids, the sound speed is infinite, because the pressure p can change but density ρ cannot. Since the propagation of disturbance cannot involve the creation or destruction of energy, the process must be adiabatic. In fact, the proper pressure-density relation is the isentropic relation. For a perfect gas, i.e. for gas in which interactions between its constituent molecules can be completely neglected, as is true for most gases under ordinary temperatures and pressures,

$$\frac{dp}{d\rho} = \gamma \frac{p}{\rho} \tag{1-4}$$

where γ is the ratio of specific heats. Therefore the calculation of velocity of sound is very simple, namely

$$a^2 = \gamma \frac{p}{\rho} \tag{1-5}$$

A,2. Basic Equations. Let the coordinates of a point in the Cartesian system be denoted by x_i where $i = 1, 2, 3$. The components of velocity of the fluid at x_i and the time instant t are u_i where $i = 1, 2, 3$. The density and the temperature are denoted by ρ and T. The temperature of the fluid is the temperature that would be measured by a thermometer of negligible time lag moving with the fluid. The heat flux vector, i.e. the quantity of heat flow through a unit area in one unit of time, is q_i ($i = 1, 2, 3$).

The stresses in the fluid can be represented as intensity of forces acting in various directions on different surfaces of an elementary volume. Fig. A,2a gives such a representation for an elementary cube of sides dx_1, dx_2, dx_3. In particular π_{11} is the tensile force per unit area or tensile stress in the direction of the x_1 axis acting on the face perpendicular to

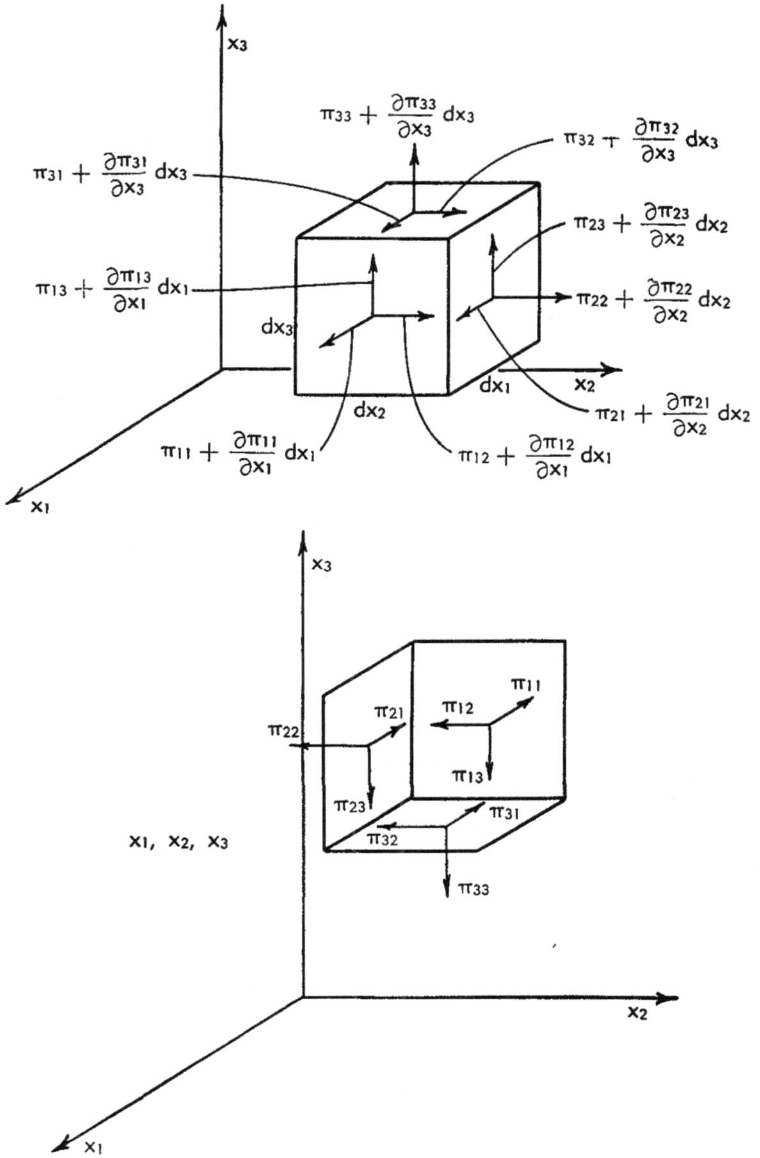

Fig. A,2a.

the x_1 axis. π_{12} is the shear force per unit area or shear stress in the direction of x_2 axis acting on the face perpendicular to the x_1 axis. In general then, π_{ij} is the stress in the direction of the x_j axis acting on a face perpendicular to the x_i axis, and the group of nine quantities is simply called the *stress tensor* (see IV,B). There are, of course, two faces of the elementary cube perpendicular to the same axis. For instance, perpendicular to the x_1 axis, there is a face at x_1 and another face at $x_1 + dx_1$. If the stresses are functions of position, at the left face they are π_{11}, π_{12}, π_{13}; but at the right face the stresses are $\pi_{11} + (\partial\pi_{11}/\partial x_1)dx_1$, $\pi_{12} + (\partial\pi_{12}/\partial x_1)dx_1$, $\pi_{13} + (\partial\pi_{13}/\partial x_1)dx_1$. A similar situation exists for other pairs of faces of the elementary cube.

Now compute the components of the resultant force due to these stress systems. The component in the x_1 direction is

$$\left[-\pi_{11} + \left(\pi_{11} + \frac{\partial\pi_{11}}{\partial x_1}dx_1\right)\right]dx_2dx_3 + \left[-\pi_{21} + \left(\pi_{21} + \frac{\partial\pi_{21}}{\partial x_2}dx_2\right)\right]dx_3dx_1$$
$$+\left[-\pi_{31} + \left(\pi_{31} + \frac{\partial\pi_{31}}{\partial x_3}dx_3\right)\right]dx_1dx_3 = \left(\frac{\partial\pi_{11}}{\partial x_1} + \frac{\partial\pi_{21}}{\partial x_2} + \frac{\partial\pi_{31}}{\partial x_3}\right)dx_1dx_2dx_3$$
$$= \left(\sum_{j=1,2,3}\frac{\partial\pi_{j1}}{\partial x_j}\right)dx_1dx_2dx_3 = \frac{\partial\pi_{j1}}{\partial x_j}dx_1dx_2dx_3$$

The last expression adopts the "summation convention," i.e. whenever an "index" or subscript is repeated in one term, the term is summed over that index. The i components of the resultant force due to stresses are then

$$\frac{\partial\pi_{ji}}{\partial x_j}dx_1dx_2dx_3$$

These components of the resultant force are balanced by the body forces and the inertia forces of the fluid mass. It is evident, however, that such body forces and inertia forces for an elementary cube must be proportional to the volume of the cube or $dx_1dx_2dx_3$; otherwise such forces cannot be in equilibrium with the resultant of the stresses. Actually, of course, this condition is fulfilled by the fact that body forces and the inertia forces are proportional to the mass in the elementary cube, and the mass is $\rho dx_1dx_2dx_3$. Before setting up this equilibrium equation, consider the equilibrium of the moment of forces. Take the elementary cube again, and compute the moment of forces about an axis parallel to the x_3 axis. The moments due to body forces and inertia forces will be proportional to the volume of the cube multiplied by dx_1 or dx_2, or a differential of fourth order. Fig. A,2a then shows that for equilibrium of moments about an axis parallel to the x_3 axis and passing through x_1, x_2, x_3,

$$0 = (\pi_{12}dx_2dx_3)dx_1 - (\pi_{21}dx_3dx_1)dx_2 + \text{terms of fourth order}$$

Therefore

$$\pi_{12} = \pi_{21}$$

or in general

$$\pi_{ij} = \pi_{ji} \tag{2-1}$$

This result, a consequence of equilibrium of moments, is universally true, and shows that the stress tensor now has only six independent components. Such a tensor is called a symmetric tensor.

It is customary to define a thermodynamic pressure p such that p, ρ, and T satisfy the equation of state of the fluid concerned.

$$p = p(T, \rho) \tag{2-2}$$

This equation of state is that determined under static conditions. The "viscous stress tensor" τ_{ij} is then defined as

$$\tau_{ij} = \pi_{ij} + \delta_{ij}p \tag{2-3}$$

where δ_{ij} are the Kronecker deltas defined by

$$\begin{aligned} \delta_{ij} &= 1, \qquad i = j \\ &= 0, \qquad i \neq j \end{aligned} \tag{2-4}$$

Since $\tau_{ij} = \pi_{ij}$ when $i \neq j$, Eq. 2-1 shows that $\tau_{ij} = \tau_{ji}$. Therefore τ_{ij} is also a symmetric tensor. A nonviscous fluid is a fluid for which τ_{ij} is identically zero. The underlying physical concept for this separation of hydrostatic pressure p and the viscous tensor τ_{ij} is discussed in the next article.

Consider the elementary cube with sides dx_1, dx_2, dx_3. The net flux of matter out of the cube per unit time is

$$\left(\frac{\partial \rho u_1}{\partial x_1} + \frac{\partial \rho u_2}{\partial x_2} + \frac{\partial \rho u_3}{\partial x_3}\right) dx_1 dx_2 dx_3$$

If the matter is conserved as should be the case, then the density of fluid in this cube must change with respect to time, i.e.

$$\left(\frac{\partial \rho u_1}{\partial x_1} + \frac{\partial \rho u_2}{\partial x_2} + \frac{\partial \rho u_3}{\partial x_3}\right) dx_1 dx_2 dx_3 = -\left(\frac{\partial \rho}{\partial t}\right) dx_1 dx_2 dx_3$$

Therefore the equation for the conservation of mass, or the *equation of continuity* is

$$\frac{\partial \rho}{\partial t} + \frac{\partial \rho u_i}{\partial x_i} = 0 \tag{2-5}$$

Here again the summation convention is adopted. Thus

$$\frac{\partial \rho u_i}{\partial x_i} = \sum_{i=1,2,3} \frac{\partial \rho u_i}{\partial x_i}$$

By using similar methods of computation, the net increase in the i component of the momentum per unit volume per unit time is

$$\frac{\partial \rho u_i}{\partial t} + \frac{\partial \rho u_i u_j}{\partial x_j}$$

This increase in momentum is made possible by the net forces acting on the matter contained in the elementary cube. These come from two sources. If X_i are the components of forces per unit mass due to external sources, then one part of the accelerating forces is X_i. The other part comes from the stress tensor π_{ij}. As computed previously the net i component of force per unit volume is

$$\frac{\partial}{\partial x_j}(\pi_{ji}) \quad \text{or} \quad \frac{\partial}{\partial x_j}(\pi_{ij})$$

due to symmetry of the tensor π_{ij}. Therefore the dynamic equations which state that the net increase in the i component of the momentum per unit volume per unit time must equal the i component of the force per unit volume, are

$$\frac{\partial \rho u_i}{\partial t} + \frac{\partial}{\partial x_j}(\rho u_i u_j) = \rho X_i + \frac{\partial \pi_{ij}}{\partial x_j} \tag{2-6}$$

By expanding the terms on the left and then using the equation of continuity (Eq. 2-5), Eq. 2-6 can also be written as

$$\frac{\partial u_i}{\partial t} + u_j \frac{\partial u_i}{\partial x_j} = X_i + \frac{1}{\rho}\frac{\partial \pi_{ij}}{\partial x_j} \tag{2-7}$$

If the viscous stress tensor τ_{ij} is introduced, then Eq. 2-6 and 2-7 become

$$\frac{\partial \rho u_i}{\partial t} + \frac{\partial}{\partial x_j}(\rho u_i u_j) = -\frac{\partial p}{\partial x_i} + \rho X_i + \frac{\partial \tau_{ij}}{\partial x_j} \tag{2-8}$$

and

$$\frac{\partial u_i}{\partial t} + u_j \frac{\partial u_i}{\partial x_j} = -\frac{1}{\rho}\frac{\partial p}{\partial x_i} + X_i + \frac{1}{\rho}\frac{\partial \tau_{ij}}{\partial x_j} \tag{2-9}$$

Eq. 2-6 to 2-9 are four forms of the *dynamic equations.*

In Eq. 2-7 and 2-9, the left sides are the acceleration of the fluid particles computed by following the fluid, i.e. the actual acceleration of the fluid. To show that this is correct, consider any quantity $f(t, x_i)$ at time instant t and space point x_i. For an observer who moves with the fluid, at the time instant $t + \delta t$, the coordinate of the space point will be $x_i + u_i\delta t$. For this observer then, the rate of change of the quantity f with respect to time is

$$\lim_{\delta t \to 0} \frac{[f(t + \delta t, x_i + u_i \delta t) - f(t, x_i)]}{\delta t} = \left(\frac{\partial}{\partial t} + u_i \frac{\partial}{\partial x_i}\right) f$$

Thus if the time differentiation by following the fluid particles is denoted by D/Dt, then

$$\frac{D}{Dt} = \frac{\partial}{\partial t} + u_i \frac{\partial}{\partial x_i} \tag{2-10}$$

If e is the internal energy of the fluid per unit mass, then increase of this energy per unit volume per unit time is

$$\rho \frac{De}{Dt}$$

This energy increase must come from three sources: Firstly, there can be heat addition by such means as the absorption of radiation, chemical reaction, and combustion. Let such heat addition be Q per unit mass per unit time. Secondly, the heat flux vector q_i due to heat conduction actually decreases the energy contained in the elementary cube. The increase per unit volume per unit time is then

$$-\frac{\partial q_i}{\partial x_i}$$

The stress tensor π_{ij} also does work on the fluid. This last source requires a detailed analysis:

Consider first the simple case of one-dimensional flow in the direction of x_i axis. Take a stream tube of unit cross section (see Fig. A,2b). At time instant t, a fluid element extends from x_1 to $x_1 + dx_1$. At a later time instant $t + dt$, the position of fluid originally at x_1 is $x_1 + u_1 dt$ while the

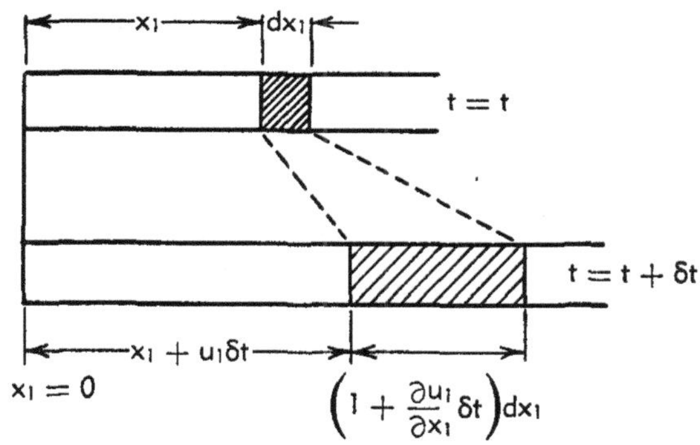

Fig. A,2b.

position of fluid originally at $x_1 + dx_1$ is $x_1 + dx_1 + [u_1 + (\partial u_1/\partial x_1)\, dx_1]\, dt$. Therefore while the volume of the original fluid element is $dx_1 \cdot 1$, the volume is now $[1 + (\partial u_1/\partial x_1)dt]dx_1 \cdot 1$. The change of volume is thus $(\partial u_1/\partial x_1)dx_1 dt$. The work done is the tensile stress π_{11} multiplied by the change in volume, or

$$\pi_{11}\frac{\partial u_1}{\partial x_1}dx_1 dt$$

The work done per unit of fluid volume per unit time is then

$$\pi_{11}\frac{\partial u_1}{\partial x_1}$$

For the general case of three-dimensional flow, there are additional terms similar to the one discussed above: $\pi_{22}(\partial u_2/\partial x_2)$, $\pi_{33}(\partial u_3/\partial x_3)$. Actually, $\partial u_1/\partial x_1$ is the time rate of the tensile strain of the fluid element in the x_1 direction, $\partial u_2/\partial x_2$ the tensile strain rate in the x_2 direction, and

$\partial u_3/\partial x_3$ in the x_3 direction. Therefore the work done per unit volume per unit time should be equal to the sum of the stresses multiplied by the corresponding *strain rates,* including the shear stresses and shear strain rates. The question is what are the rates of shear strain? To answer this question, consider first the tensor $\partial u_i/\partial x_j$ expressing the rate of deformation of a fluid element by its motion. This tensor can be broken into a symmetrical part and an antisymmetrical part as follows:

$$\frac{\partial u_i}{\partial x_j} = \begin{bmatrix} \frac{\partial u_1}{\partial x_1} & \frac{\partial u_1}{\partial x_2} & \frac{\partial u_1}{\partial x_3} \\ \frac{\partial u_2}{\partial x_1} & \frac{\partial u_2}{\partial x_2} & \frac{\partial u_2}{\partial x_3} \\ \frac{\partial u_3}{\partial x_1} & \frac{\partial u_3}{\partial x_2} & \frac{\partial u_3}{\partial x_3} \end{bmatrix}$$

$$= \begin{bmatrix} \frac{\partial u_1}{\partial x_1} & \frac{1}{2}\left(\frac{\partial u_1}{\partial x_2} + \frac{\partial u_2}{\partial x_1}\right) & \frac{1}{2}\left(\frac{\partial x_1}{\partial x_3} + \frac{\partial u_3}{\partial x_1}\right) \\ \frac{1}{2}\left(\frac{\partial u_1}{\partial x_2} + \frac{\partial u_2}{\partial x_1}\right) & \frac{\partial u_2}{\partial x_2} & \frac{1}{2}\left(\frac{\partial u_2}{\partial x_3} + \frac{\partial u_3}{\partial x_2}\right) \\ \frac{1}{2}\left(\frac{\partial u_1}{\partial x_3} + \frac{\partial u_3}{\partial x_1}\right) & \frac{1}{2}\left(\frac{\partial u_2}{\partial x_3} + \frac{\partial u_3}{\partial x_2}\right) & \frac{\partial u_3}{\partial x_3} \end{bmatrix}$$

$$- \begin{bmatrix} 0 & \frac{1}{2}\left(\frac{\partial u_2}{\partial x_1} - \frac{\partial u_1}{\partial x_2}\right) & -\frac{1}{2}\left(\frac{\partial u_1}{\partial x_3} - \frac{\partial u_3}{\partial x_1}\right) \\ -\frac{1}{2}\left(\frac{\partial u_2}{\partial x_1} - \frac{\partial u_1}{\partial x_2}\right) & 0 & \frac{1}{2}\left(\frac{\partial u_3}{\partial x_2} - \frac{\partial u_2}{\partial x_3}\right) \\ \frac{1}{2}\left(\frac{\partial u_1}{\partial x_3} - \frac{\partial u_3}{\partial x_1}\right) & -\frac{1}{2}\left(\frac{\partial u_3}{\partial x_2} - \frac{\partial u_2}{\partial x_3}\right) & 0 \end{bmatrix} \tag{2-11}$$

The physical significance of this separation of deformation tensor can be understood as follows: The antisymmetrical part has essentially three components,

$$\frac{1}{2}\left(\frac{\partial u_3}{\partial x_2} - \frac{\partial u_2}{\partial x_3}\right), \quad \frac{1}{2}\left(\frac{\partial u_1}{\partial x_3} - \frac{\partial u_3}{\partial x_1}\right), \quad \frac{1}{2}\left(\frac{\partial u_2}{\partial x_1} - \frac{\partial u_1}{\partial x_2}\right)$$

These can be easily shown to be the rate of angular rotation of the fluid element about axes through the center of the fluid element and parallel to the three coordinate axes. It is evident that such angular rotations do not strain the fluid element. Therefore all the strain must be expressed by the symmetrical part. Since

$$\frac{\partial u_1}{\partial x_1}, \quad \frac{\partial u_2}{\partial x_2}, \quad \frac{\partial u_3}{\partial x_3}$$

are shown to be the tensile strain, the remainder

$$\frac{1}{2}\left(\frac{\partial u_1}{\partial x_2} + \frac{\partial u_2}{\partial x_1}\right), \quad \frac{1}{2}\left(\frac{\partial u_2}{\partial x_3} + \frac{\partial u_3}{\partial x_2}\right), \quad \frac{1}{2}\left(\frac{\partial u_3}{\partial x_1} + \frac{\partial u_1}{\partial x_3}\right)$$

must be the rate of shear strain. This can be shown to be so by a detailed computation similar to that given above for the tensile strain $\partial u_1/\partial x_1$. Each shear strain has two corresponding shear stresses. For instance,

$$\frac{1}{2}\left(\frac{\partial u_1}{\partial x_2} + \frac{\partial u_2}{\partial x_1}\right)$$

has both π_{12} and π_{21}. Therefore the total work done per unit volume per unit time is

$$\pi_{ij} \frac{\partial u_i}{\partial x_j}$$

Finally then the equation for the *conservation of energy* is

$$\frac{De}{Dt} = Q - \frac{1}{\rho}\frac{\partial q_i}{\partial x_i} + \frac{1}{\rho}\pi_{ij}\frac{\partial u_i}{\partial x_j} \tag{2-12}$$

This equation can be modified by the dynamic equations (2-7). By multiplying each of the dynamic equations by the corresponding velocity component and summing the resultant equations,

$$\frac{D}{Dt}\left(\tfrac{1}{2}u_i u_i\right) = u_i X_i + \frac{1}{\rho} u_i \frac{\partial \pi_{ij}}{\partial x_j}$$

where $\frac{1}{2}u_i u_i$ is of course the kinetic energy of the fluid per unit mass. The sum of the above equation and Eq. 2-12 is

$$\frac{D}{Dt}\left(e + \tfrac{1}{2}u_i u_i\right) = Q + u_i X_i - \frac{1}{\rho}\frac{\partial q_i}{\partial x_i} + \frac{1}{\rho}\frac{\partial}{\partial x_i}(\pi_{ij}u_j) \tag{2-13}$$

This equation gives the change of the sum of internal energy and the kinetic energy of the fluid.

The last term of Eq. 2-13 can be written differently: By using the viscous stress tensor π_{ij} as defined by Eq. 2-3 and 2-4,

$$\frac{1}{\rho}\frac{\partial}{\partial x_i}(\pi_{ij}u_j) = \frac{1}{\rho}\frac{\partial}{\partial x_i}(\tau_{ij}u_j) - \frac{1}{\rho}\frac{\partial}{\partial x_i}(pu_i)$$

The second term on the right can be further converted by using the continuity equation, and is equal to

$$-\frac{D}{Dt}\left(\frac{p}{\rho}\right) + \frac{1}{\rho}\frac{\partial p}{\partial t}$$

Therefore another form of the energy equation is

$$\frac{D}{Dt}\left(h + \tfrac{1}{2}u_i u_i\right) = Q + u_i X_i + \frac{1}{\rho}\frac{\partial p}{\partial t} - \frac{1}{\rho}\frac{\partial q_i}{\partial x_i} + \frac{1}{\rho}\frac{\partial}{\partial x_i}(\tau_{ij}u_j) \tag{2-14}$$

where $h = e + (p/\rho)$ is the enthalpy per unit mass. Then Eq. 2-14 expresses the change of the sum of enthalpy and the kinetic energy. A further modification of the energy equation is possible if the external forces X_i have a potential $\mathcal{V}$, i.e.

$$X_i = -\frac{\partial \mathcal{V}}{\partial x_i} \tag{2-15}$$

Then Eq. 2-12 can be written as

$$\frac{D}{Dt}(h + \tfrac{1}{2}u_iu_i + \mathcal{V}) = Q + \frac{1}{\rho}\frac{\partial p}{\partial t} + \frac{\partial \mathcal{V}}{\partial t} + \frac{1}{\rho}\frac{\partial q_i}{\partial x_i} + \frac{1}{\rho}\frac{\partial}{\partial x_i}(\tau_{ij}u_j) \tag{2-16}$$

In this form of energy equation [*1*], the left side expresses the change of the sum of enthalpy, kinetic energy, and potential energy.

A,3. Viscous Stresses and Heat Flux. The unknowns in the problem of gas dynamics are the thermodynamic variables p, ρ, and T, the velocity vector u_i, the heat flux vector q_i, and the viscous stress tensor τ_{ij}. The internal energy e and the enthalpy h are known when the thermodynamic variables are specified. The basic equations developed thus far, the continuity equation, the dynamic equations, the energy equation, together with the equation of state of the fluid are thus not sufficient to determine the problem, because the number of unknowns exceeds the number of equations. To completely specify the problem, additional equations are needed.

Since any fluid, viewed microscopically, consists of a tremendously large number of molecules, the basic approach to the problem of gas dynamics is through the principles of statistical mechanics. Assume classical mechanics to be valid, and the total number of particles to be N, and the number of degrees of freedom of each particle to be n. Any specified instantaneous state of the N particles is represented by a point in the Gibbs phase space of nN coordinates of position and nN coordinates of momenta, a total of $2nN$ dimensions. The fundamental problem of statistical mechanics is to determine the probability of finding the system of the N particles at any point in the phase space for every time instant, or the probability distribution function of $2nN$ coordinates of the phase space and the time t. Once this is done, any macroscopic property of the fluid, such as flux and stresses, can be computed by averaging over the probability distribution function. This averaging process was carried out recently by Irving and Kirkwood [*2*].

The general problem of determining the probability distribution function is, of course, very difficult. Fortunately, for gases under ordinary and low pressures, the density of molecules is such that only binary encounters need to be considered. In other words, when two molecules approach each other to a distance so close that there is an appreciable mutual influence, other molecules in the gas can be considered so far away that they have no influence on the two interacting molecules. The kinetic theory of gases is based upon this concept of binary "collision." In this theory, there is a corresponding great simplification of the probability distribution function in that only the probability distribution function of a single molecule needs to be considered. This distribution function is determined by a single integro-differential equation called the Boltzmann equation. Chapman and Enskog [*3*, Chap. 3] assume that the space and time variations of the distribution function are small and compute the heat flux and the viscous stress tensor as

$$q_i = -k \frac{\partial T}{\partial x_i} \tag{3-1}$$

$$\tau_{ij} = \mu \left(\frac{\partial u_i}{\partial x_j} + \frac{\partial u_j}{\partial x_i} \right) + (\mu' - \tfrac{2}{3}\mu)\delta_{ij} \frac{\partial u_k}{\partial x_k} \tag{3-2}$$

where k is the coefficient of heat conduction, μ the *ordinary viscosity coefficient*, and μ' the dilational viscosity coefficient. μ is thus associated with both shear stress and tensile stress; but very often only the shear stress is of importance because the divergence $\partial u_k/\partial x_k$ is often small in comparison with the shear strain. However, μ' is associated only with the tensile stress and its effect vanishes when the divergence is zero. μ' is thus the viscosity coefficient for the rate of bulk expansion or compression. μ and μ' are determined by laws of interaction between molecules and molecular properties. For instance, μ' vanishes when the molecule has no internal degrees of freedom or when the internal motion is not excited. For polyatomic molecules, μ' is not zero. It is of interest to note that the symmetry of the stress tensor demonstrated previously by macroscopic consideration is validated also by the kinetic theory.

The two viscosity coefficients, μ and μ', are essentially functions of temperature, being only weakly dependent upon pressure. For gases, μ is found to increase with temperature, a property quite different from that of liquids. A detailed treatment on the viscosity coefficients and the heat conduction coefficient can be found in I,D of this Series. Eq. 3-1 and 3-2 are correct, however, only for gases under ordinary pressure. For low pressures, additional terms in the heat flux and stress tensor appear. These additional terms were computed by Burnett [*3*, Chap. 15) and were very complicated. (For a discussion of this topic see III,H.)

Eq. 3-1 and 3-2 show that the heat flux and the viscous stresses are absent if the gradients of temperature and the gradients of velocity vanish. They appear only when there are spatial variations which require the fluid to fit into new conditions as it flows from one point to another. It is quite evident that such adjustments cannot be instantaneous but, on the other hand, require many collisions of the molecules. It is, then, this process of trying to accommodate to different thermodynamic equilibriums at different points of the flow field that causes the heat flux and the viscous stresses. When there are only very small spatial variations of fluid conditions, then there is little necessity to adjust the states of the fluid, and thus the fluid must be very nearly at the thermodynamic equilibrium. Therefore at vanishingly-small heat flux and viscous stresses, the state of fluid and, in particular, the relation between pressure, density, and temperature must be that at thermodynamic equilibrium. This is the reason for the separation of the stress tensor π_{ij} into the pressure p and the viscous stress tensor τ_{ij} with the pressure defined as the pressure calculated from the equation of state for fluid at thermodynamic equilibrium.

Recently, an entirely different approach in solving the Boltzmann equation was discovered by Grad [*4*]. His results are appropriate for rapid space and time variations of the probability distribution function, such as the condition within the shock wave. In this analysis, the heat flux vector q_i and the stress tensor τ_{ij} cannot be expressed explicitly in terms of velocity and temperature derivatives. On the other hand, the additional equations involve all the unknown variables. Therefore, in this theory, the variables q_i and τ_{ij} are raised to equal footing with other unknowns p, ρ, T, and u_i. The application of Grad's theory to the dynamics of rarefied gases is of particular importance (III,H).

Of course the kinetic theory gives more than indicated by Eq. 3-1 and 3-2. It also determines the coefficient of heat conduction and the coefficients of viscosity in terms of molecular properties (I,D). If only the form of the relations between heat flux and stress tensor and other specified fluid field variables is desired, a straightforward analysis based upon dimensions and invariance under coordinate transformation generally suffices. For instance, if the stress tensor is specified to depend linearly upon the space derivatives of the velocity components, then Eq. 3-2 can be obtained uniquely with μ and μ' as unknown parameters, magnitudes undetermined. Truesdell [*5*] has carried this procedure to great length resulting in extremely complicated expressions. The initial terms are naturally given by Eq. 3-1 and 3-2.

The three principal forms of the energy of molecules are the translational, the rotational, and the vibrational energies. Of these three, the vibrational degrees of freedom are the most difficult to excite (I,H). Therefore if there is rapid change in the states of the gas, the vibrational energy of the molecules generally lags behind the equilibrium value. For gases at high temperatures, there is an appreciable amount of molecular vibrational energy. In such cases, the main contribution to the second viscosity coefficient μ' must then come from the lag of vibrational energy.

There is, however, another way of looking at the problem, introduced first by Kantrowitz [*6*]. This method is particularly appropriate when the time required to excite the molecular vibrations is very much longer than the time required to excite other forms of molecular motion. This is so because the viscosity effects appear as a derivative of velocity in the equations of motion, and thus are local effects, incapable of expressing any long-time integrated influence. Let $e(T)$ be the equilibrium value of the internal energy per unit mass at the thermodynamic temperature T, $e_{\text{vib}}(T)$ the equilibrium value of the vibrational energy per unit mass at T, and e'_{vib} the actual vibrational energy. Then assuming that all other forms of the internal energy are practically at equilibrium, the actual internal energy is

$$e = e(T) - e_{\text{vib}}(T) + e'_{\text{vib}} = e(T) + [e'_{\text{vib}} - e_{\text{vib}}(T)] \tag{3-3}$$

The quantity $e'_{\text{vib}} - e_{\text{vib}}(T)$ is the lag. Now the rate of approach to equilibrium of the vibrational energy for small deviations can be approximated as a linear function of the deviation, i.e.

$$\frac{De'_{\text{vib}}}{Dt} = -\beta[e'_{\text{vib}} - e_{\text{vib}}(T)] \tag{3-4}$$

where β is a constant, having the dimension of inverse time. By substituting Eq. 3-3 in the energy equations (Eq. 2-12 and 2-13), a new dependent variable e'_{vib} is introduced. But then Eq. 3-4 is the additional equation for this unknown. Needless to say, when this is done, the values of μ and μ' in the system of equations should not include the effects of the vibrational lag. Otherwise the same effect will be counted twice. The appropriate value of μ' in the Kantrowitz theory is thus zero, and the appropriate value of μ is that computed by the usual kinetic theory (I,D). The net effect of the vibrational lag, which is equivalent to viscosity, is of course an energy dissipation. For instance, Kantrowitz has estimated that a thin body in pure carbon dioxide might have a resistance twice as large as that due to ordinary viscosity alone.

For gases under ordinary conditions, the heat flux and the stress tensor as given by Eq. 3-1 and 3-2 are sufficiently accurate. When they are substituted into the dynamic equations and the energy equation, the system of basic equations is then complete in the sense that the number of unknowns is equal to the number of equations. This system is loosely called the Navier-Stokes equations. In this section, the discussion will henceforth be based upon the Navier-Stokes equations. In particular the energy equation (Eq. 2-12) becomes

$$\frac{De}{Dt} = Q + \frac{1}{\rho}\frac{\partial}{\partial x_i}\left(k\frac{\partial T}{\partial x_i}\right) - \frac{p}{\rho}\frac{\partial u_i}{\partial x_i} + \frac{\Phi}{\rho} \tag{3-5}$$

where Φ, called the *dissipation function*, is

$$\Phi = \tau_{ij}\frac{\partial u_j}{\partial x_i} = \tfrac{1}{2}\tau_{ij}\left(\frac{\partial u_i}{\partial x_j} + \frac{\partial u_j}{\partial x_i}\right) = \frac{\mu}{2}\left(\frac{\partial u_i}{\partial x_j} + \frac{\partial u_j}{\partial x_i}\right)^2 + (\mu' - \tfrac{2}{3}\mu)\left(\frac{\partial u_i}{\partial x_i}\right)^2 \tag{3-6}$$

Φ is then the heat generated by viscous dissipation. Another form of Eq. 3-5 is

$$\frac{Dh}{Dt} = Q + \frac{1}{\rho}\frac{\partial}{\partial x_i}\left(k\frac{\partial T}{\partial x_i}\right) + \frac{1}{\rho}\frac{Dp}{Dt} + \frac{\Phi}{\rho} \tag{3-7}$$

where h is of course the enthalpy per unit mass.

With the differential equations established, the only missing part in the complete formulation of the gas dynamic problem is the question of *boundary conditions*. This involves the specification of the velocity component u_i and the temperature T, or the heat flux vector q_i at the surface of the solid body. Historically, the conditions at the surface of contact of a fluid with a solid body have been a subject of considerable controversy [7, p. 676]. Fortunately for gas dynamics, the fluid can be treated microscopically by the kinetic theory. The result given by this theory is that under ordinary density, the velocity of the fluid at the surface must be equal to the velocity of the surface. In other words, the relative velocity between the fluid and the surface must vanish. This is the so-called no-slip condition. Similarly the fluid temperature T must be equal to the temperature of the surface. This is the condition of no "temperature jump" at the surface. Of course, at low densities, these surface conditions have to be modified (III,H), but at ordinary densities, the no-slip and the no temperature jump are the correct and sufficient conditions to determine the problem completely.

During the above discussion, it was tacitly assumed that the fluid could be considered as a single-component gas. If the fluid consists of several components nonuniformly mixed, or if there is chemical reaction between the components, then the situation is greatly complicated (II,F). Moreover, even if fluid consists of uniformly mixed gases initially, special effects such as thermal diffusion in regions of large temperature gradient will destroy the uniformity of the mixture. Fortunately for the main body of problems in gas dynamics for aircraft and flow machinery, these complications do not appear, and the simpler system of Navier-Stokes equations suffices.

A,4. Integral Forms of Basic Equations. It is often convenient for particular problems to use the integral forms of the basic equations instead of the differential forms given in the previous article. The integration extends over a *fixed* region of the space. The region whose volume is V may be simply connected or not simply connected and may be enclosed by one single surface or by several surfaces. The element of volume of the region is denoted by dV, the element of surface enclosing V by dA. Let n_i be the unit vector normal to the surface element dA (Fig. A,4a). The basic operation of transforming the differential form to integral form is defined by the Stokes theorem: If f_i is any vector, then

$$\int_V \frac{\partial f_i}{\partial x_i} dV = \int_A (f_i n_i) dA \tag{4-1}$$

Therefore by applying Eq. 4-1 to the continuity equation (2-5), we have

$$\frac{\partial}{\partial t}\int_V \rho dV + \int_A \rho(u_i n_i) dA = 0 \tag{4-2}$$

Physically, the meaning of this equation is quite clear: $\rho(u_i n_i)dA$ is the time rate of flow out of the region through the element of surface dA. The integral over A is thus the net outflow. Since it is assumed that no fluid

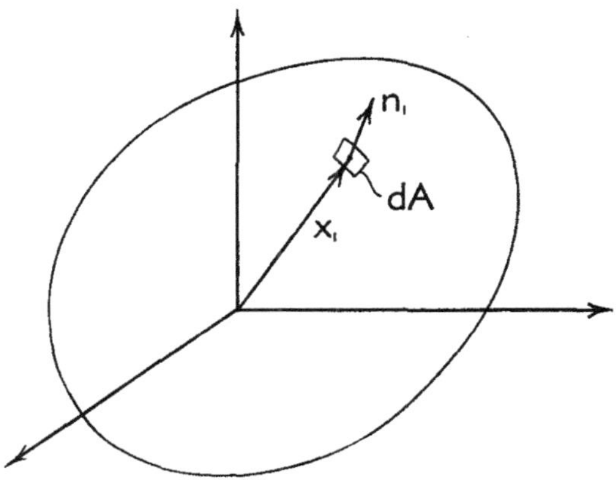

Fig. A,4a.

is introduced or removed from this region, this net outflow must be balanced by the rate of change of mass which remains in the region. This is the statement of Eq. 4-2. In the general case with introduction or removal of fluid, the right-hand side of Eq. 4-2 should be equal to the total algebraic strength of sources and sinks in the region.

Similarly, the dynamic equations (2-6) are transformed into

$$\frac{\partial}{\partial t}\int_V \rho u_i dV + \int_A \rho u_i(u_j n_j) dA = \int_V \rho X_i dV + \int_A (\pi_{ji} n_j) dA \tag{4-3}$$

The physical meaning of this equation is again very simple. The left-hand side expresses the increase in the momentum of the fluid in the region in the direction of the x_i axis, while the right-hand side expresses the cause for this increase, namely the resultant body force and the stresses on the surface. Eq. 4-3 is of course based upon the assumption that no fluid is introduced or withdrawn within the region V. When this is not the case, the right side of Eq. 4-3 should have, in addition, the time rate of momentum introduced by this means.

Instead of linear momentum, one can also compute the angular momentum about the three coordinate axes. For this purpose, it is convenient to introduce the notation δ_{ijk}. The meaning of this is:

$\delta_{ijk} = 1$, if $i \neq j \neq k$ and ijk is a cyclic permutation of 1, 2, 3

$\delta_{ijk} = -1$, if $i \neq j \neq k$ and ijk is not a cyclic permutation of 1, 2, 3

$\delta_{ijk} = 0$, otherwise.

Thus $\delta_{123} = 1$, but $\delta_{132} = -1$. Eq. 2-6 can be written as

$$\frac{\partial \rho u_k}{\partial t} + \frac{\partial}{\partial x_l}(\rho u_k u_l) = \rho X_k + \frac{\partial \pi_{kl}}{\partial x_l}$$

By multiplying this equation by $\delta_{ijk}x_j$ and integrating over the volume V,

$$\frac{\partial}{\partial t}\int_V \rho\delta_{ijk}x_j u_k dV + \int_A \rho\delta_{ijk}x_j u_k(u_l n_l)dA = \int_V \rho\delta_{ijk}x_j X_k dV + \int_A \delta_{ijk}x_j(\pi_{kl}n_l)dA \qquad (4\text{-}4)$$

This is the equation for the angular momentum about the x_i axis. The terms are summed over j and k according to the summation convention. Needless to say, when there are points within V where the fluid is introduced or removed, the effects of such sources or sinks should be properly included in Eq. 4-4.

To transform the energy equation (2-13) into integral form, one notes that by means of the continuity equation (2-5), it can be written as

$$\frac{\partial}{\partial t}\rho(e + \tfrac{1}{2}u_i u_i) + \frac{\partial}{\partial x_i}\rho u_i(e + \tfrac{1}{2}u_j u_j) = \rho Q + \rho u_i X_i - \frac{\partial q_i}{\partial x_i} + \frac{\partial}{\partial x_i}(\pi_{ij}u_j)$$

Hence by integrating this equation over a region V,

$$\frac{\partial}{\partial t}\int_V \rho(e + \tfrac{1}{2}u_i u_i)dV + \int_A \rho(e + \tfrac{1}{2}u_j u_j)u_i n_i dA = \int_V \rho Q dV + \int_V \rho u_i X_i dV - \int_A (q_i n_i)dA + \int_A (\pi_{ij}u_j n_i)dA \qquad (4\text{-}5)$$

The left side expresses the net rate of increase of the sum of internal energy and the kinetic energy of the fluid. The right side expresses the cause of this increase. The first term is the heat addition by such means as chemical reaction, the second term is the work done by the body force, the third term is the heat loss by conduction, and the fourth term is the work done by stresses on the surface A of the region. Another form of the integral energy equation can be obtained from Eq. 2-16. Thus

$$\frac{\partial}{\partial t}\int_V \rho(h + \tfrac{1}{2}u_iu_i + \mathcal{V})dV + \int_A \rho(h + \tfrac{1}{2}u_iu_i + \mathcal{V})u_jn_jdA = \int_V \rho Q dV$$

$$+ \frac{\partial}{\partial t}\int_V p dV + \frac{\partial}{\partial t}\int_V \rho\mathcal{V} dV - \int_A (q_in_i)dA + \int_A (\tau_{ij}u_jn_i)dA \quad (4\text{-}6)$$

The left side of the equation expresses the net rate of increase of the sum of enthalpy, kinetic energy, and potential energy. The right side gives the cause of this increase. The first term is again heat increase through Q, the second term is the rate of change of pressure energy, the third term

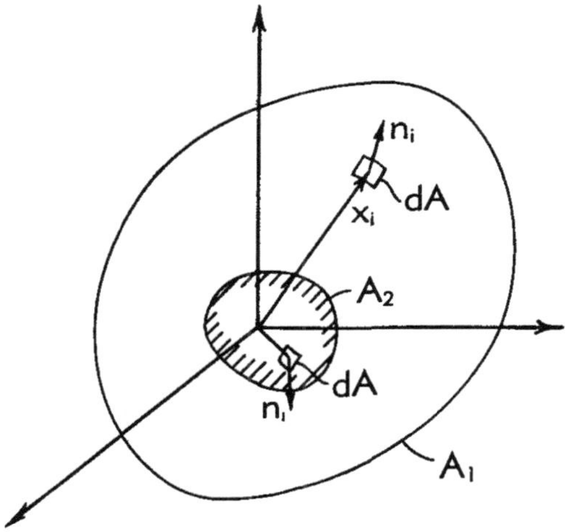

Fig. A,4b.

is the rate of change of potential energy, the fourth term is the rate of heat loss through conduction, and finally, the fifth term is the rate of work done by viscous stresses on the surface A of the region.

The integral forms of the basic equations can thus be very simply interpreted. In fact, if preferred, they could serve as the starting point of the discussion and the differential forms of the equations could be derived from them.

Flow over a body. As an example of application of the integral equations, take the problem of *steady* flow of fluid over a solid body (Fig. A,4b). For the fluid region, one chooses the space between a closed surface A_1, containing the solid body, and the body itself. The surface A of the region then consists of two parts: A_1 and the surface A_2 of the solid body. It is further assumed that there is no body force, and that $Q = 0$. In the momentum equation (4-3), the first terms on both the left and the right sides then vanish. Since the fluid velocity is zero on the solid surface, the second integral on the left of Eq. 4-3, the momentum flux integral, vanishes on A_2. Furthermore, from the principle of equality of action and reaction, the second integral on the right side of Eq. 4-3 over A_2 is the negative of resultant force F_i acting on the body. Thus

$$\int_{A_1} \rho u_i(u_j n_j)dA = \int_{A_1} (\pi_{ji} n_j)dA - F_i \tag{4-7}$$

This shows that the force F_i on the solid body can be computed as the integral of quantities on a surface A_1 enclosing the body. A favored technique is to choose the surface A_1 so far away from the body that the calculation of the integrals is greatly simplified.

By applying the same method to the equations of angular momentum (Eq. 4-4), one has equations for the moment M_i acting on the solid body, referred to the coordinate axes,

$$\int_{A_1} \rho \delta_{ijk} x_j u_k(u_l n_l)dA = \int_{A_1} \delta_{ijk} x_j(\pi_{kl} n_l)dA - M_i \tag{4-8}$$

By using Eq. 4-6 and denoting by H the amount of heat absorbed by the solid body, one has the energy equation

$$\int_{A_1} \rho(h + \tfrac{1}{2}u_j u_j)u_i n_i dA = -H + \int_{A_1} (\tau_{ij} u_j n_i)dA - \int_{A_1} (q_i n_i)dA \tag{4-9}$$

If the viscous stress τ_{ij} is used, then Eq. 4-7, 4-8, and 4-9 can be written as

$$F_i = \int_{A_1} [\tau_{ji} n_j - p n_i - \rho u_i(u_j n_j)]dA \tag{4-10}$$

$$M_i = \int_{A_1} \delta_{ijk} x_j[\tau_{lk} n_l - p n_k - \rho u_k(u_l n_l)]dA \tag{4-11}$$

$$H = \int_{A_1} [\tau_{ij} u_j n_i - q_i n_i - \rho(h + \tfrac{1}{2}u_j u_j)u_i n_i]dA \tag{4-12}$$

These equations show that the force on the body can be calculated as the difference of the stress integral over A_1 and the momentum integral over A_1. Similarly, the moment on the body is the difference of the moment of stress and the moment of momentum, integrated over A_1. Finally, the heat absorbed by the body is the difference of work done by the viscous stresses on the surface A_1 and the outflow of heat, enthalpy, and kinetic energy from the surface A_1.

A,5. Similarity and Flow Parameters.[1] For the sake of clarity, consider the definite problem of a solid body of typical linear dimension L, say its length, in a field of infinite extent with undisturbed velocity U

[1] For an alternate method of discussing similarity see VIII,D.

under the influence of gravity. The body is carrying out an oscillation of frequency f. The free stream quantities are denoted by the subscript ∞, and the nondimensional quantities by an asterisk. Then

$$\begin{aligned} t &= \frac{1}{f}t^* & T &= T_\infty T^* & X_i &= gX_i^* \\ x_i &= Lx_i^* & u_i &= Uu_i^* & \tau_{ij} &= \frac{\mu_\infty U}{L}\tau_{ij}^* \\ \rho &= \rho_\infty \rho^* & e &= e_\infty e^* & q_i &= \frac{k_\infty T_\infty}{L} q_i^* \\ p &= p_\infty p^* & Q &= \frac{Ue_\infty}{L} Q^* \end{aligned} \tag{5-1}$$

where

$$q_i^* = -\frac{k}{k_\infty}\frac{\partial T^*}{\partial x_i^*} \tag{5-2}$$

$$\tau_{ij}^* = \frac{\mu}{\mu_\infty}\left(\frac{\partial u_i^*}{\partial x_j^*} + \frac{\partial u_j^*}{\partial x_i^*}\right) + \left(\frac{\mu'}{\mu_\infty} - \frac{2}{3}\frac{\mu}{\mu_\infty}\right)\delta_{ij}\frac{\partial u_k^*}{\partial x_k^*} \tag{5-3}$$

and g is the gravitational constant. By using these nondimensional variables, the continuity equation (2-5) can then be written as

$$\left(\frac{Lf}{U}\right)\frac{\partial \rho^*}{\partial t^*} + \frac{\partial \rho^* u_i^*}{\partial x_i^*} = 0 \tag{5-4}$$

The dynamic equations (2-9) become

$$\left(\frac{Lf}{U}\right)\frac{\partial u_i^*}{\partial t^*} + u_j^*\frac{\partial u_i^*}{\partial x_j^*} = -\left(\frac{p_\infty}{\rho_\infty U^2}\right)\frac{1}{\rho^*}\frac{\partial p^*}{\partial x_i^*} + \left(\frac{gL}{U^2}\right)X_i^* + \left(\frac{\mu_\infty}{\rho_\infty UL}\right)\frac{1}{\rho^*}\frac{\partial \tau_{ij}^*}{\partial x_j^*} \tag{5-5}$$

Denote the specific heat at constant pressure by c_p, the value in the free stream by $(c_p)_\infty$. Then the energy equation (Eq. 2-12) is

$$\begin{aligned} \left(\frac{Lf}{U}\right)\frac{\partial e^*}{\partial t^*} + u_i^*\frac{\partial e^*}{\partial x_i^*} = Q^* &- \left(\frac{k_\infty}{c_{p_\infty}\mu_\infty}\right)\left(\frac{\mu_\infty}{\rho_\infty UL}\right)\left(\frac{c_{p_\infty}T_\infty}{e_\infty}\right)\frac{1}{\rho^*}\frac{\partial q_i^*}{\partial x_i^*} \\ &- \left(\frac{p_\infty}{\rho_\infty U^2}\right)\left(\frac{U^2}{e_\infty}\right)\frac{p^*}{\rho^*}\frac{\partial u_i^*}{\partial x_i^*} + \left(\frac{\mu_\infty}{\rho_\infty UL}\right)\left(\frac{U^2}{e_\infty}\right)\frac{\tau_{ij}^*}{\rho^*}\frac{\partial u_j^*}{\partial x_i^*} \end{aligned} \tag{5-6}$$

The factors of the above equations in parentheses are all nondimensional. They can be conveniently taken as parameters of the problem. In particular, Lf/U is generally called the "reduced frequency," U^2/gL the Froude number, and $\rho_\infty UL/\mu_\infty$ the Reynolds number. The ratio p_∞/ρ_∞ has the dimension of velocity squared. In fact $\sqrt{p_\infty/\rho_\infty}$ is of the order of the average speed of random molecular motion and is also of the order of the velocity of propagation of disturbances in the free stream.

$U/\sqrt{p_\infty/\rho_\infty}$ is then of the order of the ratio of free stream velocity to the speed of propagation of disturbances, the sound velocity. The ratio of flow velocity to the speed of sound is, of course, the Mach number. Therefore the quantity $p_\infty/\rho_\infty U^2$ is the inverse of the square of Mach number. e_∞ has the magnitude of the order of the kinetic energy of molecular agitation. Therefore e_∞ is again of the order of the square of average molecular speed or sound velocity. U^2/e_∞ is then again the square of Mach number. The factor $\mu_\infty c_{p_\infty}/k_\infty$ is a measure of the relative importance of the effects of heat conduction and the effects of viscosity. It is called the Prandtl number. The factor $c_{p_\infty}T_\infty/e_\infty$ is approximately the ratio of enthalpy and the internal energy and is a number of the order of 1.

A specific problem of great importance in fluid mechanics, however, is the problem of similarity. In particular, one is concerned with bodies of geometrical similarity, i.e. bodies which can be made to coincide by a translation and a uniform expansion or contraction. Consider again the previous problem of a body oscillating in an infinite region. The question is: How should one specify the free stream quantities U, p_∞, ρ_∞, T_∞, μ_∞, k_∞, e_∞ and the quantities L and f for geometrically similar bodies such that two sets of these parameters would lead to the same nondimensional differential equations (Eq. 5-4, 5-5, and 5-6)? If this question can be answered, the two individual flow fields corresponding to these two sets of parameters can be made to be "similar." Namely, for both flows, if all variables are reduced to the nondimensional form, the functional relationships between them are the same. This would give a tremendous simplification of the problem of gas dynamics and effect a great saving in time and effort. Such a gain cannot naturally be obtained without some sacrifice of generality. In other words, without some restriction on the fluid properties, such a powerful similarity law is not possible. Fortunately, for gases under ordinary temperature and pressure, it can be sufficiently approximated by the "perfect gas." A perfect gas is a gas having the following equation as the equation of state

$$p = \mathcal{R}\rho T \tag{5-7}$$

where $\mathcal{R}$ is the specific gas constant. The specific heat at constant volume c_v and the specific heat at constant pressure c_p can be functions of temperature, but they are related by

$$c_p = c_v + \mathcal{R} \tag{5-8}$$

If γ is the ratio of specific heats,

$$\gamma = \frac{c_p}{c_v} \tag{5-9}$$

according to Eq. 1-4 the velocity of sound a is given by

$$a^2 = \gamma\frac{p}{\rho} = \gamma\mathcal{R}T \tag{5-10}$$

The internal energy e is then

$$e = \int_0^T c_v dT \tag{5-11}$$

Then

$$\frac{e_\infty}{c_{p_\infty} T_\infty} = \int_0^1 \frac{c_p}{c_{p_\infty}} dT^* - \left(1 - \frac{1}{\gamma_\infty}\right) \tag{5-12}$$

and

$$e^* = \frac{\left[\int_0^{T^*} \frac{c_p}{c_{p_\infty}} dT^* - \left(1 - \frac{1}{\gamma_\infty}\right) T^*\right]}{\left[\int_0^1 \frac{c_p}{c_{p_\infty}} dT^* - \left(1 - \frac{1}{\gamma_\infty}\right)\right]} \tag{5-13}$$

If M is the Mach number, then

$$\frac{\rho_\infty U^2}{p_\infty} = \gamma_\infty M_\infty^2$$

and

$$\frac{U^2}{e_\infty} = (\gamma_\infty - 1) M_\infty^2 \frac{c_{p_\infty} T_\infty}{e_\infty}$$

Let the Reynolds number be denoted by Re and the Prandtl number by Pr, Froude number by Fr, reduced frequency by κ, i.e.

$$\frac{\rho_\infty U L}{\mu_\infty} = Re_\infty \tag{5-14}$$

$$\frac{\mu_\infty c_{p_\infty}}{k_\infty} = Pr_\infty \tag{5-15}$$

$$\frac{U^2}{gL} = Fr \tag{5-16}$$

$$\frac{Lf}{U} = \kappa \tag{5-17}$$

Then the continuity equation, the dynamic equations, and the energy equation become

$$\kappa \frac{\partial \rho^*}{\partial t^*} + \frac{\partial \rho^* u_i^*}{\partial x_i^*} = 0 \tag{5-18}$$

$$\kappa \frac{\partial u_i^*}{\partial t^*} + u_j^* \frac{\partial u_i^*}{\partial x_j^*} = -\frac{1}{\gamma_\infty M_\infty^2} \frac{1}{\rho^*} \frac{\partial p^*}{\partial x_i^*} + \frac{1}{Fr} X_i^* + \frac{1}{Re_\infty \rho^*} \frac{\partial \tau_{ij}^*}{\partial x_j^*} \tag{5-19}$$

$$\begin{aligned} \kappa \frac{\partial e^*}{\partial t^*} + u_i^* \frac{\partial e^*}{\partial x_i^*} = Q^* &- \frac{1}{Pr_\infty Re_\infty} \left(\frac{c_{p_\infty} T_\infty}{e_\infty}\right) \frac{1}{\rho^*} \frac{\partial q_i^*}{\partial x_i^*} \\ &- \frac{\gamma_\infty - 1}{\gamma_\infty} \left(\frac{c_{p_\infty} T_\infty}{e_\infty}\right) \frac{p^*}{\rho^*} \frac{\partial u_i^*}{\partial x_i^*} \\ &+ \frac{(\gamma_\infty - 1) M_\infty^2}{Re_\infty} \left(\frac{c_{p_\infty} T_\infty}{e_\infty}\right) \frac{\tau_{ij}^*}{\rho^*} \frac{\partial u_j^*}{\partial x_i^*} \end{aligned} \tag{5-20}$$

The equation of state (Eq. 5-7) can be written in nondimensional form as

$$p^* = \rho^* T^* \tag{5-21}$$

By examining Eq. 5-18 to 5-20 together with Eq. 5-12 and 5-13, it is seen that in order for two flows to be similar, the values of parameters must be such that the reduced frequency κ, the Mach number M_∞, the Reynolds number Re_∞, and the Froude number Fr are the same. In addition, the fluid properties must be such that the Prandtl number Pr_∞ and the ratio of specific heats γ_∞ are the same for both flows. Furthermore, the variation with temperature of the specific heats, the heat condition coefficient k, and the viscosity coefficients μ and μ' must be such that the ratios $c_p/c_{p\infty}$, k/k_∞, μ/μ_∞, and μ'/μ_∞ are unique functions of T^*. The last condition is satisfied if c_p, k, μ, μ' vary as powers of the temperature T. Needless to say, according to Eq. 5-1 it is implicitly assumed that the heat addition per unit mass Q is proportional to Ue_∞/L or UT_∞/L. When all these conditions are satisfied, the two flows will have the same nondimensional differential equations. Then the solutions in terms of nondimensional variables will be the same provided the nondimensionalized boundary conditions are also the same.

The important boundary conditions are the conditions on the surface of the solid body. These are the condition of vanishing relative velocity between the fluid and the solid surface, and the condition on the temperature of the surface. The condition on the fluid velocity, being a homogeneous condition, will not introduce a new similarity parameter. But the temperature of the surface of the body specifies the heat flux between the body and the fluid, and will introduce an additional condition for similarity. Let the heat flux vector at the solid surface be $(q_i)_w$. Then the nondimensional form of this boundary condition is

$$(q_i^*)_w = \frac{(q_i)_w L}{k_\infty T_\infty} = F((x_i^*)_w, t^*) \tag{5-22}$$

where $(x_i^*)_w$ are the nondimensional coordinates of the surface points. For similarity, the function F has to be the same. With this additional condition, the nondimensional wall temperature $T_w^* = T_w/T_\infty$ is fixed. The condition (Eq. 5-22) is then equivalent to

$$Nu = \frac{(q_i^*)_w}{T_w^* - 1} = \frac{(q_i)_w L}{k_\infty(T_w - T_\infty)} = Nu((x_i^*)_w, t^*) \tag{5-23}$$

Nu is called the Nusselt heat transfer number.

Summarizing, for a nonsteady, viscous, heat-conducting, perfect gas with heat addition of Q per unit mass per unit time and influence of gravity, the similarity condition is for the following parameters to be the same: Mach number M_∞, Reynolds number Re_∞, reduced frequency κ, Froude number Fr, Prandtl number Pr_∞, ratio of specific heats γ_∞, the ratios c_p/c_{p_∞}, k/k_∞, μ/μ_∞, μ'/μ_∞ as functions of T^*, QL/UT_∞ as functions of x_i^* and t^*, and finally the Nusselt number as functions of $(x_i^*)_w$ and t^*. Needless to say, when the motion of the body is not purely oscillatory, the analysis still applies if one replaces the frequency f by $1/\tau$, where τ is the characteristic time of the problem. Then the reduced frequency becomes $L/U\tau$. If the effects of viscosity and heat conduction can be neglected, the Reynolds number, the Prandtl number, the ratios k/k_∞, μ/μ_∞, μ'/μ_∞, and the Nusselt number can be dropped from the above list. If there is no external force field, the Froude number need not be considered. If the motion is steady, the reduced frequency will be identically zero. If there is no heat addition, Q will be identically zero. If the specific heats are constant, the ratio c_p/c_{p_∞} could be dropped from the list. Therefore, if one is concerned only with the steady motion of viscous, heat-conducting perfect gas with constant specific heats, a constant coefficient of heat conduction and viscosity, and without heat addition and force field, the similarity of flow is characterized by the Mach number, the Reynolds number, the Prandtl number, the Nusselt number, and the ratio of specific heats.

A,6. Ideal Gas. For any real gas or mixture of gases, the relative magnitudes of k and μ are such that the Prandtl number is of the order of 1. In fact, according to the kinetic theory of simple monatomic molecules, the Prandtl number should be exactly equal to 1. Therefore the effects of viscosity and heat conduction are of the same order of importance: When one is taken into account, the other must also be included in the analysis in order to have a correct evaluation. The coefficients of viscosity μ and μ' and the heat conduction coefficient k are very small in magnitude. For body dimensions of the order of feet, the velocity U of the order of tens or hundreds of feet per second, and pressure p_∞ of the order of 1 atmosphere, the Reynolds number is very large, approximately one million. Eq. 5-19 and 5-20 then show that in general the effects of viscosity and the effects of heat conduction are negligible. Only in regions where the temperature and the velocity gradients are large, will viscosity and heat conduction play an appreciable role. Where would the gradients of temperature and velocity be large? For subsonic flows, only the boundary layer gives these large gradients; for supersonic flows, shock waves also contain large temperature and velocity gradients. Therefore only in a boundary layer or in a shock wave, need one consider the effects of viscosity and heat conduction. Outside of these regions, the fluid can be considered as nonviscous and nonheat-conducting. Such a fluid is called an ideal compressible fluid. If the fluid is also a perfect gas, then it is called an ideal perfect gas. For ideal gas, the equations of motion are greatly simplified for the absence of the viscous stress and heat flux terms.

In the boundary layer and the shock wave, the viscous and heat conduction terms are of great importance. The treatment of these problems is thus rather complicated (see Sec. D and Vol. IV). If the solution of the viscous equations in the regions of boundary layer and shock wave has to be joined to the solution of the nonviscous equation outside the regions of boundary layer and shock wave so that the whole problem is solved simultaneously, then the problem of gas dynamics will not be appreciably simplified even at very large Reynolds number. Fortunately, in a great majority of problems this is not the case. The shock wave is generally so thin that its thickness can be considered as infinitesimal, and gas on either side of the shock, being outside of the shock, is an ideal gas. The shock is then simply a mathematical discontinuity. The magnitude of discontinuity can be computed rather easily without having to deal with the viscous differential equations, once the conditions before the shock and the configuration of the shock are specified. This means that one important element of the solution is readily available. The question of boundary layer is more difficult, since its thickness is small but not negligible. However, just because of this small thickness, the boundary layer generally introduces negligible disturbance to the flow outside of the boundary layer, the so-called potential flow. A logical way to treat the problems of gas dynamics is then to compute the flow, neglecting first viscosity and heat conduction, but including discontinuities of the shock waves if necessary. The pressure, velocity, and temperature distribution over the surface of the solid body of this first approximation is then used to determine the boundary layer over the solid body. When the boundary layer is calculated, the shear stress and the heat flux at the surface of the body can be computed. This iteration procedure simplifies the problem greatly, and is the basis of the classical Prandtl boundary layer theory.

The method of analysis stated in the preceding paragraph works satisfactorily for subsonic flows. For supersonic flows, the strong pressure rise in crossing a shock wave may seriously modify the boundary layer development and the modified boundary layer may appreciably change the shock configuration. This mutual influence is termed boundary layer and shock interaction, and is a subject of recent intensive research. For hypersonic flows, or flows of very large Mach number, the nose shock is so flattened toward the body surface and the shock and the boundary layer are so close to each other that even the small thickness of the boundary layer influences the configuration of the shock. This may be called hypersonic boundary layer-shock interaction. When such interactions occur, the classic boundary layer theory breaks down.

Another simplification is generally possible: The Froude number Fr is U^2/gL where L is the typical dimension of the body and U the typical velocity, and g is the gravitational constant if the external force field is that of gravity. Therefore, for gas dynamic problems of aircraft the

Froude number is generally so large as to make the external force term negligibly small in the dynamic equations (Eq. 5-19). Then the external force can be neglected.

From the discussions in the previous paragraphs, it is then apparent that the important problem in gas dynamics is the problem of flow of ideal gas, without external force field. Great simplification is then possible for the absence of viscous stresses and external forces in the dynamic equations, and the absence of heat flux terms and viscous dissipation in the energy equation. Concurrent with the simplification of the differential equation there is a relaxation in the boundary conditions at the surface of the solid body. The velocity of fluid relative to the surface is now required only to be tangential to the surface, but need not be zero. The condition on the temperature of the fluid at the surface has to be dropped. The temperature of the fluid is now uniquely determined by other variables and its value at the surface cannot be arbitrarily specified.

A,7. Diabatic Flow of an Ideal Gas. Circulation and Vorticity. For an ideal gas without external force, the appropriate differential equations are:

$$\frac{\partial \rho}{\partial t} + \frac{\partial(\rho u_i)}{\partial x_i} = 0 \tag{7-1}$$

$$\frac{Du_i}{Dt} = -\frac{1}{\rho}\frac{\partial p}{\partial x_i} \tag{7-2}$$

$$\frac{D}{Dt}(h + \tfrac{1}{2}u_i u_i) = Q + \frac{1}{\rho}\frac{\partial p}{\partial t} \tag{7-3}$$

This system of equations has been the basis of a study by Hicks [*8*] of the so-called diabatic flows, in contrast to adiabatic flows where there is no heat addition, i.e. $Q = 0$. Hicks seeks to find the effect of heat addition as a result of combustion and thus uses the diabatic flow as a model of reacting gas in a field of combustion. When the motion is *steady*, i.e. when the field variables are not functions of t, and thus $\partial p/\partial t = 0$, Eq. 7-3 shows that for any fluid element, the rate of change of the sum of enthalpy and the kinetic energy is exactly equal to the heat added Q. In this sense then, the sum of enthalpy and the kinetic energy can be considered as the "total energy" h^0 of the fluid. This is a very useful concept. For the special case of adiabatic flow, where $Q = 0$, the total energy along the path of any single fluid element, or along any streamline, is a constant. It is important, however, to observe the limitations of the theorem stated above: It is true only for nonviscous and nonheat-conducting gas in steady motion.

Another important field variable which has not yet been discussed is the vorticity vector $\mathbf{\Omega}$. The vorticity vector gives the intrinsic rotation of the fluid element and is defined as

$$\mathbf{\Omega} = \nabla \times \mathbf{u} \tag{7-4}$$

Here the conventional vector notation is used, $\mathbf{u}$ is the velocity vector whose magnitude is u,

$$\mathbf{u} = \mathbf{i}u_1 + \mathbf{j}u_2 + \mathbf{k}u_3 \tag{7-5}$$

and $\nabla \times$ is the curl operator such that

$$\mathbf{\Omega} = \nabla \times \mathbf{u} = \mathbf{i}\left(\frac{\partial u_3}{\partial x_2} - \frac{\partial u_2}{\partial x_3}\right) + \mathbf{j}\left(\frac{\partial u_1}{\partial x_3} - \frac{\partial u_3}{\partial x_1}\right) + \mathbf{k}\left(\frac{\partial u_2}{\partial x_1} - \frac{\partial u_1}{\partial x_2}\right) \tag{7-6}$$

In these equations, the vectors, $\mathbf{i}$, $\mathbf{j}$, and $\mathbf{k}$ are unit vectors in the x_1, x_2, and x_3 directions respectively. By comparing the quantities in Eq. 7-6 with those in Eq. 2-11, it is seen that the components of vorticity vector are equal to twice the three quantities in the antisymmetric part of deformation tensor. The vorticity vector is then equal to *twice* the vector of angular velocity of the fluid element.

With the vector notation, the continuity equation is

$$\frac{\partial \rho}{\partial t} + \nabla \cdot (\rho \mathbf{u}) = 0 \tag{7-7}$$

The operator $\nabla \cdot$ is the divergence operator such that

$$\nabla \cdot (\rho \mathbf{u}) = \frac{\partial(\rho u_i)}{\partial x_i} \tag{7-8}$$

The dynamic equation is then

$$\frac{D\mathbf{u}}{Dt} = \frac{\partial \mathbf{u}}{\partial t} + (\mathbf{u} \cdot \nabla)\mathbf{u} = -\frac{1}{\rho}\nabla p \tag{7-9}$$

where ∇p is the gradient vector of the pressure p. The energy equation can be written as

$$\frac{Dh^0}{Dt} = Q + \frac{1}{\rho}\frac{\partial p}{\partial t} \tag{7-10}$$

where h^0 is the total energy, or

$$h^0 = h + \tfrac{1}{2}u_i u_i \tag{7-11}$$

Now a few transformations of the basic equations are possible: According to vector analysis

$$\nabla(\tfrac{1}{2}u_i u_i) = \nabla(\tfrac{1}{2}\mathbf{u} \cdot \mathbf{u}) = (\mathbf{u} \cdot \nabla)\mathbf{u} + \mathbf{u} \times (\nabla \times \mathbf{u}) \tag{7-12}$$

Therefore Eq. 7-9 can be written as

$$\frac{\partial \mathbf{u}}{\partial t} + \nabla(\tfrac{1}{2}u_i u_i) - \mathbf{u} \times \mathbf{\Omega} = -\frac{1}{\rho}\nabla p \tag{7-13}$$

If s is the specific entropy of the gas, the first law of thermodynamics requires that

$$Tds = dh - \frac{1}{\rho} dp \tag{7-14}$$

and

$$T \frac{Ds}{Dt} = Q \tag{7-15}$$

For a perfect gas of constant specific heats, the equation of state can be written in terms of s, p, ρ as follows:

$$p = \text{const } \rho^{\gamma} e^{s/c_v} \tag{7-16}$$

By eliminating the gradient of pressure with Eq. 7-14, the dynamic equation can be written further as

$$\frac{D\mathbf{u}}{Dt} = T\nabla s - \nabla h \tag{7-17}$$

or

$$\frac{\partial \mathbf{u}}{\partial t} - \mathbf{u} \times \mathbf{\Omega} = T\nabla s - \nabla h^0 \tag{7-18}$$

This relationship is generally referred to as Crocco's theorem (see [9]). By replacing Q according to Eq. 7-15, the energy equation is

$$\frac{Dh^0}{Dt} = T \frac{Ds}{Dt} + \frac{1}{\rho} \frac{\partial p}{\partial t} \tag{7-19}$$

With these transformed equations, the production of vorticity in the flow of an ideal gas can be studied. By taking the curl of both sides of Eq. 7-13 and noting $\nabla \cdot \nabla \times \mathbf{u} = 0$,

$$\frac{\partial \mathbf{\Omega}}{\partial t} + (\mathbf{u} \cdot \nabla)\mathbf{\Omega} - (\mathbf{\Omega} \cdot \nabla)\mathbf{u} + \mathbf{\Omega}(\nabla \cdot \mathbf{u}) = -\nabla \times \left(\frac{1}{\rho} \nabla p\right)$$

or

$$\frac{D\mathbf{\Omega}}{Dt} = (\mathbf{\Omega} \cdot \nabla)\mathbf{u} - \mathbf{\Omega}(\nabla \cdot \mathbf{u}) - \nabla \times \left(\frac{1}{\rho} \nabla p\right) \tag{7-20}$$

Another form of this equation can be obtained by using Eq. 7-14,

$$\frac{D\mathbf{\Omega}}{Dt} = (\mathbf{\Omega} \cdot \nabla)\mathbf{u} - \mathbf{\Omega}(\nabla \cdot \mathbf{u}) + \nabla T \times \nabla s \tag{7-21}$$

If p is a function of ρ only, then the last term to the right of Eq. 7-20 can be written as the curl of the gradient of a scalar function and is thus zero. If the specific entropy is a constant, then the last term of Eq. 7-21 is zero because of $\nabla s = 0$. Both conditions are satisfied if the gas is isentropic, i.e. the gas has the same entropy everywhere. Then for the motion of an ideal gas, the rate of change of the vorticity of a fluid element is composed of the first two terms to the right of Eq. 7-20 or 7-21. If $\mathbf{\Omega}$ is zero for one instant at every point of the flow field, then $D\mathbf{\Omega}/Dt = 0$. Therefore the vorticity at every point of the field will remain zero. Such flows are called irrotational flows. On the other hand, if the flow is not isentropic, i.e. the fluid has different entropy at different points of the field, then even if the vorticity $\mathbf{\Omega}$ is zero everywhere at one time instant, the last term to the right of Eq. 7-21 will cause the vorticity to be different from zero in the next instant. Therefore nonisentropic flows cannot be irrotational. Hence irrotationality implies isentropy, but, of course, isentropy does not imply irrotationality.

Consider now an adiabatic continuous flow generated out of a uniform state of rest, i.e. $\mathbf{u} = 0$ at $t = 0$. According to Eq. 7-15, if the motion is adiabatic and continuous, the entropy for any fluid element is a constant. But the entropy is the same everywhere at $t = 0$, the uniform initial state. Therefore the motion will always be isentropic. Furthermore at $t = 0$, $\mathbf{\Omega} = 0$ because the velocity is zero everywhere. Hence under the assumption of adiabatic continuous motion of an ideal gas, any motion generated out of a uniform state of rest, is irrotational. This would include practically all potential flow problems, were it not for the occurrence of shock in supersonic flow. Shock in the ideal gas is a discontinuity and it could generate vorticity (see Sec. D and E). But even with shock, the vorticity generated is usually small; therefore irrotational flows are flows of great practical importance in spite of their very special nature.

The first term $(\mathbf{\Omega} \cdot \nabla)\mathbf{u}$ to the right of either Eq. 7-20 or 7-21 can be interpreted as the effect of the bending of the lines which follow the direction of the vorticity vector $\mathbf{\Omega}$, or vortex lines. To see this, let there be only one component Ω_1 of the vorticity vector in the direction of the x_1 axis at the time instant t, and there is no u_1 at the time instant t. Then the quantity $(\mathbf{\Omega} \cdot \nabla)\mathbf{u}$ has two components:

$$\Omega_1 \frac{\partial u_2}{\partial x_1} \quad \text{and} \quad \Omega_1 \frac{\partial u_3}{\partial x_1}$$

in the x_2 direction and x_3 direction. Now consider a line element (Fig. A,7) which at time t coincides with Ω_1 and has the length dx_1. After the time interval dt, the left end has moved from the point x_1, x_2, x_3 to x_1, $x_2 + u_2dt$, $x_3 + u_3dt$. The right end has moved from $x_1 + dx_1$, x_2, x_3 to

$$x_1 + dx_1, \quad x_2 + \left(u_2 + \frac{\partial u_2}{\partial x_1} dx_1\right) dt, \quad x_3 + \left(u_3 + \frac{\partial u_3}{\partial x_1} dx_1\right) dt$$

Therefore the original line element is now tilted with respect to the x_1 axis, and it has now a direction cosine with the x_2 axis equal to $(\partial u_2/\partial x_1)dt$ and a direction cosine with the x_3 axis equal to $(\partial u_3/\partial x_1)dt$. Now assume that the vorticity Ω_1 originally coincides with the x_1 axis, moves with the fluid, and maintains its strength. Then after the time interval dt, there is a component along the x_2 axis equal to $\Omega_1(\partial u_2/\partial x_1)dt$ and a component along the x_3 axis equal to $\Omega_1(\partial u_3/\partial x_1)dt$. The time rate of increase is then $\Omega_1(\partial u_2/\partial x_1)$ along the x_2 axis and $\Omega_1(\partial u_3/\partial x_1)$ along the x_3 axis. This is exactly the quantity $(\mathbf{\Omega} \cdot \nabla)\mathbf{u}$. Therefore this term expresses the rate of increase of vorticity due to the bending of the vortex line. Similarly the second term $-\mathbf{\Omega}(\nabla \cdot \mathbf{u})$ which is associated with divergence $\nabla \cdot \mathbf{u}$ or volume change can be interpreted as the effect of the stretching of a fluid

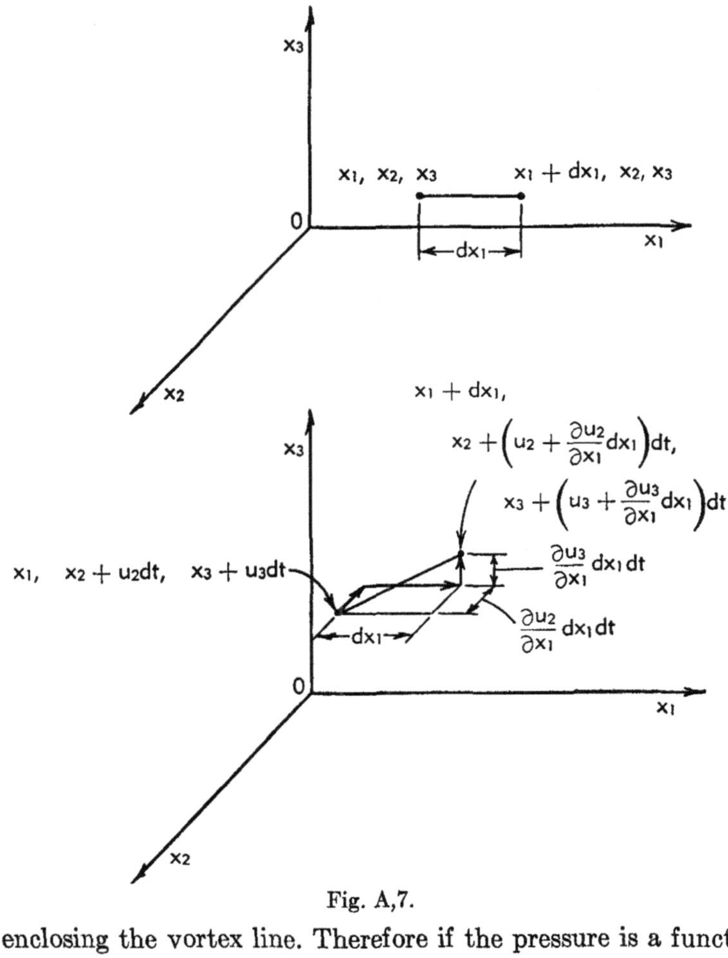

Fig. A,7.

tube enclosing the vortex line. Therefore if the pressure is a function of density only or if the entropy is a constant such that these two terms alone contribute to the rate of change of $\mathbf{\Omega}$ for a fluid element, then the result can be taken to indicate that the product of the cross-sectional area and of the vorticity $\mathbf{\Omega}$ of a vortex tube is constant both in space and time. This theorem was originally shown by Helmholtz for the incompressible fluid. Such a property of invariance indicates the usefulness of the concept of vorticity.

The content of Helmholtz' theorem can be demonstrated in another way, perhaps more directly: Let Γ be the circulation around any closed contour C in the space, defined as

$$\Gamma = \oint_C \mathbf{u} \cdot d\mathbf{l} \tag{7-22}$$

where $d\mathbf{l}$ is the line element of the contour C. By using the Stokes theorem of contour integral,

$$\Gamma = \int_A (\nabla \times \mathbf{u}) \cdot d\mathbf{A} = \int_A \mathbf{\Omega} \cdot d\mathbf{A} \tag{7-23}$$

where A is the simply connected surface whose boundary is C. Eq. 7-23 shows that the product of the cross-sectional area and of the vorticity $\mathbf{\Omega}$ of a vortex tube is the circulation Γ around the cross section of the vortex tube. The implication of Helmholtz' theorem is that this should be a constant by following the fluid. Thus one should compute $D\Gamma/Dt$. But

$$\frac{D\Gamma}{Dt} = \frac{D}{Dt}\oint_C \mathbf{u} \cdot d\mathbf{l} = \oint_C \frac{D\mathbf{u}}{Dt} \cdot d\mathbf{l} + \oint_C \mathbf{u} \cdot d\mathbf{u}$$

where $d\mathbf{u}$ is the differential velocity vector along the contour. However, $\mathbf{u} \cdot d\mathbf{u} = d(\frac{1}{2}u_i u_i)$; therefore if the velocity field is continuous, the second integral to the right vanishes. Then by using either Eq. 7-9 or 7-16 and then applying Stokes' theorem,

$$\frac{D\Gamma}{Dt} = -\int_A \left[\nabla + \left(\frac{1}{\rho}\nabla p\right)\right] \cdot d\mathbf{A} = \int_A (\nabla T \times \nabla s) \cdot d\mathbf{A} \tag{7-24}$$

Hence if the pressure is a function of density only or if the entropy is a constant, the circulation around any closed contour following the fluid is a constant. This is the Kelvin theorem.

For the general case, however, the circulation Γ is not a constant even for an ideal gas. This is quite different from the case of incompressible fluid which has constant density. In fact, Eq. 7-24 can be written as

$$\frac{D\Gamma}{Dt} = \int_A \left[\nabla p \times \nabla \frac{1}{\rho}\right] \cdot d\mathbf{A} \tag{7-25}$$

Eq. 7-25 immediately shows that if ρ is a constant, such as for incompressible fluid, Γ is a constant in general. For compressible flow, Eq. 7-25 can be interpreted as follows: Draw equidistant members of the families of surfaces p = const and $1/\rho$ = const, and so obtain a series of tubes bounded by these surfaces. Then the rate of change of circulation per unit time along a contour C following the fluid is proportional to the number of tubes surrounded by C. This is a theorem due to Bjerknes [*10*].

Stagnation quantities. During the discussion presented in this article, the quantity h^0 is introduced as the sum of enthalpy and kinetic energy, or total energy per unit mass. This quantity is often called the specific "stagnation enthalpy" of the fluid. The concept is as follows: The fluid at any point of the field is extracted by some means with its pressure, density, temperature, and velocity at the local value; then this parcel of fluid is compressed *isentropically* by decreasing its velocity until its velocity vanishes, i.e. until it reaches stagnant condition. The final state of the fluid at the stagnant condition is then called the stagnation condition of the fluid at the original point of extraction. Thus one can speak of stagnation enthalpy, stagnation temperature, stagnation pressure, and stagnation density. There is, of course, no such thing as stagnation velocity; that is by definition zero. In order to avoid possible confusion, the original conditions of the fluid before compression, i.e. the local values at the fluid field, are called static conditions. The local enthalpy is thus the static enthalpy; the pressure, static pressure, etc. One should note that for this conceptual experiment, the compression is supposed to be carried out isentropically. Thus the stagnation quantities can never be measured directly by instruments. At subsonic speed, the pressure measured by a Pitot tube only approximates the stagnation pressure. At supersonic speed, due to the presence of shock in front of the Pitot tube, the measured pressure is much lower than the stagnation pressure.

For a perfect gas with constant specific heats, the relation between the stagnation quantities and the static quantities is particularly simple. Thus if M is the local Mach number, i.e. the ratio of local speed to the velocity of sound at local conditions, then

$$M^2 = \frac{|\mathbf{u}|^2}{\gamma(p/\rho)}$$

and according to Eq. 1-5, 5-7, 5-8, 5-9, and 7-11 if we assume c_p to be a constant,

$$\frac{T^0}{T} = 1 + \frac{\gamma - 1}{2} M^2 \tag{7-26a}$$

Then by using the well-known relation between temperature and pressure in isentropic processes we have

$$\begin{aligned} \frac{p^0}{p} &= \left(1 + \frac{\gamma - 1}{2} M^2\right)^{\frac{\gamma}{\gamma-1}} \\ \frac{\rho^0}{\rho} &= \left(1 + \frac{\gamma - 1}{2} M^2\right)^{\frac{1}{\gamma-1}} \end{aligned} \tag{7-26b}$$

For a perfect gas, the ratio of stagnation enthalpy to static enthalpy is the same as the ratio of stagnation temperature to static temperature as the enthalpy is proportional to the temperature.

For small local Mach numbers, Eq. 7-26 can be expanded into a power series in M^2. As an example,

$$\frac{p^0}{p} = 1 + \frac{\gamma}{2} M^2 + \frac{\gamma}{8} M^4 + \frac{\gamma(2-\gamma)}{48} M^6 + \cdots$$

Therefore

$$p^0 - p = \tfrac{1}{2}\rho|\mathbf{u}|^2\left(1 + \tfrac{1}{4}M^2 + \frac{2-\gamma}{24} M^4 + \cdots\right) \tag{7-27}$$

The difference $p^0 - p$ between the stagnation pressure and the static pressure is the dynamic pressure rise. When M is very small, the dynamic pressure rise is approximately $\frac{1}{2}\rho|\mathbf{u}|^2$. The latter quantity is often referred to as the dynamic pressure.

A,8. Adiabatic Flow of an Ideal Gas. Bernoulli Equation. As discussed in the previous section, aside from the boundary layer, the effects of viscosity and heat conduction can be neglected for the majority of gas dynamic problems in aeronautics. Furthermore, the heat addition, except in problems involving combustion, is either zero or very small. Then under ordinary conditions, the gas behaves very much like an ideal gas. Therefore one of the fundamental problems of gas dynamics is to study the adiabatic flow of an ideal gas. Most of the following discussions in this section are devoted to this problem.

Let the flow be irrotational, such that there is a velocity potential $\phi(x_i, t)$ with

$$\mathbf{u} = \nabla\phi \tag{8-1}$$

Then $\mathbf{\Omega} = \nabla \times \mathbf{u} = \nabla \times (\nabla\phi)$ is automatically zero. The dynamic equation (7-13) becomes

$$\nabla\left(\frac{\partial\phi}{\partial t} + \tfrac{1}{2}|\mathbf{u}|^2 + \int \frac{dp}{\rho}\right) = 0$$

This means that the quantity within the parentheses is not a space function, but it could be a time function; and this time function can be absorbed into the definition of ϕ without influencing the relation as specified by Eq. 8-1. Therefore

$$\frac{\partial\phi}{\partial t} + \tfrac{1}{2}|\mathbf{u}|^2 + \int \frac{dp}{\rho} = \text{const} \tag{8-2}$$

This is the Bernoulli equation for nonsteady irrotational flow. It can be considered as the first integral of the dynamic equation, and gives a simple relation between the pressure or density and the velocity potential. The integral must be computed with isentropic pressure-density relation because, as discussed previously, irrotational motion can generally be maintained only by constant specific entropy throughout the field. For a perfect gas, Eq. 8-2 becomes

$$\frac{\partial\phi}{\partial t} + \tfrac{1}{2}|\mathbf{u}|^2 + \frac{\gamma}{\gamma - 1}\frac{p}{\rho} = \frac{\partial\phi}{\partial t} + h^0 = \text{const} \tag{8-3}$$

Therefore irrotational flows are not necessarily isoenergetic in the sense that the total energy h^0 is a constant throughout the field. Irrotational flows are isoenergetic only if they are steady, i.e. $\partial\phi/\partial t = 0$.

When the motion is rotational, i.e. when the vorticity is not zero, a general first integral is not possible. A first integral exists only when the motion is steady. Then for adiabatic flow, as shown previously, the total energy h^0 is a constant along the path of the fluid element or along any streamline but may be different for different streamlines.

$$h^0 = h + \tfrac{1}{2}|\mathbf{u}|^2 = \text{const on a streamline} \tag{8-4}$$

This is really an energy equation but it has the same form as the Bernoulli equation for steady irrotational flow, because then the flow is isoenergetic. The difference is of course that for steady isoenergetic flow, the motion is not necessarily isentropic; while for steady irrotational flow, the motion must be isentropic. An example for the isoenergetic but nonisentropic flow is the problem of steady supersonic flow over a body with curved detached shock.

The preceding discussions then indicate that there are four characteristics of the flow, namely: steadiness, rotationality, uniformity of entropy, and uniformity of energy. On this basis of classification there are eight possible types of adiabatic flow of an ideal gas as listed below, where + sign denotes positive for the property listed and − sign denotes negative for the property listed. For instance, + under steadiness means steady motion, − under rotationality means irrotational motion. The

Type	*Steadiness*	*Rotationality*	*Uniformity of entropy*	*Uniformity of energy*
1	+	−	+	+
2	−	−	+	−
3	+	+	−	+
4	−	+	−	−
5	+	+	+	−
6	+	+	−	−
7	−	+	+	−
8	−	+	−	−

subsonic flow over a wing if steady is of type 1; if nonsteady of type 2. The transonic flow over a wing if steady is of type 3; if nonsteady of type 4. The steady flow over a body with nonuniform free stream, i.e. shear flow, is of type 5, if the temperature and pressure of the free stream are uniform, but the velocity nonuniform; if otherwise, of type 6. If the motion is nonsteady, then it is of type 7 or type 8.

The mathematical complexity of the problem of analyzing the flow field is considerably reduced if a first integral of the system of equations of motion exists. As shown above, this is the case when the motion is either irrotational or steady. The simplest problem is of course that of steady irrotational and thus isentropic flow. These cases are treated in some detail in the following articles.

A,9. Irrotational Flows. Velocity Potential. When the flow is irrotational, there is a velocity potential $\phi(x_i, t)$ and the dynamic equations integrate to the Bernoulli equation (8-2). Since the motion must also be isentropic, the energy equation and the equation of state give the isentropic relations between any *two* of the three variables p, ρ, T. By combining the isentropic relations with the Bernoulli equation, one can obtain an equation involving the velocity potential and any *one* of the thermodynamic variables p, ρ, and T. For instance, in the case of a perfect gas of constant specific heats, if the flow is uniform and steady far from the solid body and if the uniform stagnation conditions far from the body are denoted by the superscript ^{0},

$$\frac{\partial\phi}{\partial t} + \tfrac{1}{2}|\mathbf{u}|^2 + \frac{\gamma}{\gamma - 1}\frac{p^0}{\rho^0}\left(\frac{p}{p^0}\right)^{\frac{\gamma-1}{\gamma}} = \frac{\gamma}{\gamma - 1}\frac{p^0}{\rho^0} = \frac{1}{\gamma - 1}(a^0)^2 \quad (9\text{-}1)$$

$$\frac{\partial\phi}{\partial t} + \tfrac{1}{2}|\mathbf{u}|^2 + \frac{\gamma}{\gamma - 1}\frac{p^0}{\rho^0}\left(\frac{\rho}{\rho^0}\right)^{\gamma-1} = \frac{\gamma}{\gamma - 1}\frac{p^0}{\rho^0} = \frac{1}{\gamma - 1}(a^0)^2 \quad (9\text{-}2)$$

$$\frac{\partial\phi}{\partial t} + \tfrac{1}{2}|\mathbf{u}|^2 + \frac{\gamma}{\gamma - 1}\frac{p^0}{\rho^0}\left(\frac{T}{T^0}\right) = \frac{\gamma}{\gamma - 1}\frac{p^0}{\rho^0} = \frac{1}{\gamma - 1}(a^0)^2 \quad (9\text{-}3)$$

$$\frac{\partial\phi}{\partial t} + \tfrac{1}{2}|\mathbf{u}|^2 + \frac{1}{\gamma - 1}a^2 = \frac{1}{\gamma - 1}(a^0)^2 \quad (9\text{-}4)$$

where a is the velocity of sound, as defined by Eq. 1-4. By using these equations, the pressure, density, and temperature can be computed when the velocity potential and hence $|\mathbf{u}|$ are determined.

The problem is then reduced to the calculation of the velocity potential. Since the Bernoulli equation is the result of integrating the dynamic equation together with the energy equation, the only equation of motion left unused is the continuity equation (7-7) which can be written as

$$\frac{1}{\rho}\frac{\partial\rho}{\partial t} + \nabla \cdot \mathbf{u} + \frac{1}{\rho}\mathbf{u} \cdot \nabla\rho = 0$$

But the motion is also isentropic, and therefore the pressure is a function of ρ alone. The previous equation can then be written as

$$\frac{1}{dp/d\rho}\left(\frac{1}{\rho}\frac{\partial p}{\partial t} + \mathbf{u} \cdot \frac{1}{\rho}\nabla p\right) + \nabla \cdot \mathbf{u} = 0 \quad (9\text{-}5)$$

However, in general, the derivative $dp/d\rho$ is simply related to the velocity of sound a, i.e. by Eq. 1-3,

$$\frac{dp}{d\rho} = a^2$$

By differentiating the Bernoulli equation with respect to time t,

$$\frac{\partial^2\phi}{\partial t^2} + \mathbf{u} \cdot \frac{\partial \mathbf{u}}{\partial t} + \frac{1}{\rho}\frac{\partial p}{\partial t} = 0 \qquad (9\text{-}6)$$

This equation can be solved for $(1/\rho)(\partial p/\partial t)$. By substituting this result and by eliminating $(1/\rho)\nabla p$ with the dynamic equation, Eq. 9-5 becomes finally

$$\nabla \cdot \mathbf{u} = \frac{1}{a^2}\frac{\partial^2\phi}{\partial t^2} + 2\frac{\mathbf{u}}{a^2} \cdot \frac{\partial \mathbf{u}}{\partial t} + \frac{1}{a^2}\mathbf{u} \cdot [(\mathbf{u} \cdot \nabla)\mathbf{u}] \qquad (9\text{-}7)$$

or

$$\nabla^2\phi = \frac{1}{a^2}\frac{\partial^2\phi}{\partial t^2} + \frac{2}{a^2}u_i\frac{\partial^2\phi}{\partial t\partial x_i} + \frac{1}{a^2}u_iu_j\frac{\partial^2\phi}{\partial x_j\partial x_i} \qquad (9\text{-}8)$$

The quantity a^2 can be computed from the Bernoulli equation in terms of ϕ. For a perfect gas, it is given by Eq. 9-4. Therefore Eq. 9-8 can be considered as a single equation for the single unknown ϕ. This is a great simplification from the original system of equations of motion.

When the velocity of sound is very large in comparison with the fluid velocity, the right side of Eq. 9-8 is negligibly small and can be taken to be zero. Then

$$\nabla^2\phi = 0, \qquad a \to \infty \qquad (9\text{-}9)$$

This is the differential equation for the velocity potential of an incompressible flow. Mathematically, it is fundamentally different from the general equation (9-8) in that while Eq. 9-9 is a linear equation, Eq. 9-8 is nonlinear in the unknown ϕ. The mathematical problem of solving Eq. 9-8 is thus far more difficult than the corresponding problem of incompressible flow. Physically, Eq. 9-9 shows that the problem of velocity distribution of an irrotational incompressible flow can be solved without reference to the Bernoulli equation. The Bernoulli equation is used to determine the pressure p after the velocity is determined. In fact, Eq. 9-9 can be obtained from the equation of continuity and irrotationality. Therefore for irrotational incompressible flows, the velocity field is determined by kinematical conditions alone, quite independent of the dynamics of the problem. Not so is the problem of compressible flow. Here the kinematics and the dynamics of the problem are intimately coupled since the equation for the velocity potential ϕ, Eq. 9-8, has to be solved simultaneously with the Bernoulli equation.

Another important difference between the compressible flow and the incompressible flow is the question of so-called "apparent mass" in the accelerated motion of a body through a fluid at rest far from the body. In the case of incompressible flow, the flow field, at every instant being determined by kinematic considerations alone, is identical to that of a steady motion of the body at a velocity equal to the instantaneous velocity of the accelerated motion. Only when the pressure on the surface of the body is calculated does the effect of nonsteady motion appear through the $\partial\phi/\partial t$ in the Bernoulli equation. But ϕ is proportional to the instantaneous velocity; therefore the additional pressure effect of acceleration is proportional to the time rate of change of the velocity or the acceleration of the body. By integrating the pressure over the surface of the body the resultant force on the body is obtained. The additional force due to acceleration is thus proportional to the instantaneous acceleration of the body. Consequently, the effects of accelerating the fluid around the body can be considered as an added mass moving with the body. This added mass is often called the apparent mass, a constant independent of the acceleration and determined by the shape of the body only. This greatly simplifies the computation of the accelerated motion of the body through the fluid. Physically this simplification is really the result of the instantaneous propagation of disturbance signals in the incompressible fluid. Hence the effect of acceleration is determined completely by the acceleration of the body at that instant. In compressible flow, the signal speed of disturbances is finite; past history of the motion is not forgotten immediately. Therefore the acceleration effects at one instant are determined not only by instantaneous situations but by the whole sequence of events prior to that instant. Then the accelerating effects bear little relation to the instantaneous acceleration of the body, and the concept of apparent mass is not helpful.

If the motion is steady, all time derivatives vanish. The equation for the velocity potential (Eq. 9-8) is then reduced to

$$\nabla^2\phi = \frac{1}{a^2} u_i u_j \frac{\partial^2\phi}{\partial x_j \partial x_i} \tag{9-10}$$

The velocity of sound a can again be computed from the Bernoulli equation, omitting now the $\partial\phi/\partial t$ term in that equation. For two-dimensional flows, one can write $x = x_1$, $y = x_2$; $u = u_1$, $v = u_2$, and Eq. 8-10 becomes

$$\left(1 - \frac{u^2}{a^2}\right)\frac{\partial^2\phi}{\partial x^2} - 2\,\frac{uv}{a^2}\,\frac{\partial^2\phi}{\partial x \partial y} + \left(1 - \frac{v^2}{a^2}\right)\frac{\partial^2\phi}{\partial y^2} = 0 \tag{9-11}$$

A,10. Variational Method for Irrotational Flows. Instead of solving the differential equation (9-8) for the velocity potential ϕ, it is possible to formulate the problem of irrotational flow as a variational problem where a certain integral for ϕ is to have a stationary value against small virtual variations of ϕ. This formulation was first presented by Bateman [*11*] and later extended by Wang [*12*].

To fix ideas, consider the problem of finding a nonsteady, irrotational, isentropic flow within a fixed, solid space boundary. At $t = t_1$, the initial instant, the velocity distribution within this space boundary is specified. At $t = t_2$, the final instant, the velocity distribution within the boundary is also specified. The pressure or density distribution is not specified at either t_1 or t_2 but must be found. But if the velocity distribution is specified at t_1 and t_2, the gradients of ϕ at t_1 and t_2 are specified and, at t_1 and t_2, ϕ itself is specified at every space point within the boundary except to a constant. This constant may be taken to be zero without altering the problem. In other words, the problem is transformed to one where ϕ is specified at t_1 and t_2, and in addition $(\partial\phi/\partial x_i)n_i$, where n_i is the unit vector normal to the boundary, must vanish on the boundary at all times. Now consider the integral

$$I = \iiiint p\,dx_1dx_2dx_3dt \tag{10-1}$$

extended for the space within the solid boundary and from $t = t_1$, to $t = t_2$. p is the pressure computed from the velocity potential ϕ by the Bernoulli equation (8-2). Bateman's statement is that the flow problem is equivalent to that of finding the stationary value of the integral I, i.e.

$$\delta I = 0 \tag{10-2}$$

Since the Bernoulli equation is the result of the dynamic equations and by using the velocity potential ϕ, the irrotationality conditions are satisfied, the only equation not considered is the continuity equation (2-5). The problem of proving Bateman's statement is to show the equivalence of Eq. 10-2 and the continuity equation (2-5). By using the Bernoulli equation in the form of Eq. 9-1, Eq. 10-1 is

$$I = p^0 \iiiint \left\{1 - \left(\frac{\gamma-1}{\gamma}\right)\frac{\rho^0}{p^0}\left[\frac{\partial\phi}{\partial t} + \frac{1}{2}\left(\frac{\partial\phi}{\partial x_i}\right)\left(\frac{\partial\phi}{\partial x_i}\right)\right]\right\}^{\frac{\gamma}{\gamma-1}} dx_1dx_2dx_3dt \tag{10-3}$$

If ϕ is given a virtual variation $\delta\phi$, then

$$\delta I = -\rho^0 \iiiint \left\{1 - \left(\frac{\gamma-1}{\gamma}\right)\frac{\rho^0}{p^0}\left[\frac{\partial\phi}{\partial t} + \frac{1}{2}\left(\frac{\partial\phi}{\partial x_i}\right)\left(\frac{\partial\phi}{\partial x_i}\right)\right]\right\}^{\frac{1}{\gamma-1}} \left(\frac{\partial\delta\phi}{\partial t} + \frac{\partial\phi}{\partial x_i}\frac{\partial\delta\phi}{\partial x_i}\right) dx_1dx_2dx_3dt$$

By using the Bernoulli equation in the form of Eq. 9-2, the first factor of the integrand in the above integral is seen to be equal to the density ρ. Therefore

$$\delta I = -\iiiint \rho\left(\frac{\partial\delta\phi}{\partial t} + \frac{\partial\phi}{\partial x_i}\frac{\partial\delta\phi}{\partial x_i}\right) dx_1dx_2dx_3dt \qquad (10\text{-}4)$$

But

$$\rho\frac{\partial\phi}{\partial x_i}\frac{\partial\delta\phi}{\partial x_i} = \frac{\partial}{\partial x_i}\left(\rho\delta\phi\frac{\partial\phi}{\partial x_i}\right) - \delta\phi\frac{\partial}{\partial x_i}\left(\rho\frac{\partial\phi}{\partial x_i}\right)$$

Then Eq. 10-4 can be transformed by using a partial integration with respect to t for the first term and the Stokes theorem of Eq. 3-1 on the second term. The result is

$$\delta I = -\iiint \left(\delta\phi\rho\right)_{t=t_1}^{t=t_2} dx_1dx_2dx_3 - \int dt \int dA(\delta\phi\, u_in_i)$$

$$+\iiiint \delta\phi\left(\frac{\partial\rho}{\partial t} + \frac{\partial\rho u_i}{\partial x_i}\right) dx_1dx_2dx_3dt \quad (10\text{-}5)$$

where the surface integral is to be extended over the surface of the solid boundary. The first integral vanishes because ϕ is specified at t_1 and t_2; therefore $\delta\phi$ has to be zero at all space points at t_1 and t_2. The second integral vanishes because $u_in_i = 0$ at the surface of the boundary. Therefore the meaning of Eq. 10-2 is simply that the third integral of Eq. 10-5 has to vanish. But $\delta\phi$ is arbitrary except for t_1 and t_2; Eq. 10-2 is possible only if the continuity equation is satisfied, i.e.,

$$\frac{\partial\rho}{\partial t} + \frac{\partial\rho u_i}{\partial x_i} = 0$$

Therefore Bateman's statement is proven.

For steady flow problems, no specification of ϕ at t_1 and t_2 needs to be made, and the integral I is simplified to

$$I = \iiint p\,dx_1dx_2dx_3 \qquad (10\text{-}6)$$

where p is to be computed from the Bernoulli equation, omitting of course the term $\partial\phi/\partial t$.

An application of the variational integral (Eq. 10-2) is the computation of flow by the direct variational method of Rayleigh and Ritz. The essence is to approximate the true velocity potential ϕ by a function chosen such that it satisfies the boundary conditions and conditions at $t = t_1$ and $t = t_2$, and furthermore contains several arbitrary constants. The procedure involves the actual substitution of the approximating function into either Eq. 10-1 or 10-6 as the case may be, and carrying out the integration. The integral I is then a function of the undetermined constants in the approximating function. Eq. 10-2 then finally fixes the constants. The function with the constant so determined is then the approximation to the true velocity potential.

Applying this method to steady flow of infinite extent, Wang found two difficulties to be overcome. The first difficulty is that the integral (Eq. 10-6) is now infinite and the variational problem is thus not properly put. The secondary difficulty is that the surface integral in Eq. 10-5 is now not zero because the surface itself is infinite. The first difficulty is overcome for subsonic flows by using instead of Eq. 10-1 the integral of the difference of p and the pressure p_{inc} of the corresponding incompressible flow, i.e. use

$$I = \iiint (p - p_{inc}) dx_1 dx_2 dx_3 \tag{10-7}$$

If the incompressible flow is known, the integral over p_{inc} is fixed and thus does not disturb the argument made above about the variation δI. But this artifice will make the integral to be varied finite. The second difficulty is overcome by actually computing the surface integral in terms of the approximating function for the velocity potential and using it to modify the integral of Eq. 10-7, so that the variations of the modified integral cancel out the troublesome surface integral in Eq. 10-5. (For the exact details, see [*12*].)

A,11. Adiabatic Steady Flows. As shown in the previous section, the total energy or the stagnation enthalpy h^0 is a constant along the streamlines of a steady adiabatic flow of an ideal gas. Then ∇h^0 must be a vector normal to the flow direction. But according to Eq. 7-18, for steady flows:

$$\mathbf{u} \times \mathbf{\Omega} = \nabla h^0 - T\nabla s \tag{11-1}$$

Therefore the vector $(\nabla h^0 - T\nabla s)$ is normal to the flow direction. However, ∇h^0 is already perpendicular to the flow, so ∇s must be also perpendicular to the flow. Consequently the entropy is also constant along the streamline. The exception is, of course, a discontinuity of the shock type. When that occurs, the entropy also changes discontinuously. Hence, between shocks, the steady adiabatic flow along a streamline is isentropic, although the entropy might change from streamline to streamline. The streamlines are then characterized by the stagnation enthalpy h^0 and the entropy s. Instead of the stagnation enthalpy, the "maximum velocity" u_{max} can be used. u_{max} is defined as

$$h^0 = \tfrac{1}{2} u_{max}^2 \tag{11-2}$$

Therefore u_{max} is the velocity of the gas when the enthalpy or temperature of the gas is zero, i.e. complete transformation of energy into kinetic energy. u_{max} is, of course, constant along any streamline.

Now let the velocity vector $\mathbf{u}$ be divided by the maximum velocity

$u_{\max}$ and the resultant nondimensional vector be denoted by $\mathbf{W}$; $\mathbf{W}$ may be called the reduced velocity vector,

$$\mathbf{W} = \frac{\mathbf{u}}{u_{\max}} \tag{11-3}$$

By using the energy equation (Eq. 8-4),

$$\frac{h}{h^0} = 1 - W^2 \tag{11-4}$$

where $W^2 = \mathbf{W} \cdot \mathbf{W}$, or the square of the magnitude of $\mathbf{W}$. If ρ^0 is stagnation density, again constant along a streamline, then the continuity equation is

$$0 = \nabla \cdot (\rho \mathbf{u}) = \nabla \cdot \left[(\rho^0 \sqrt{2h^0}) \left(\frac{\rho}{\rho^0} \right) \mathbf{W} \right]$$

$$= \rho^0 \sqrt{2h^0}\, \nabla \cdot \left[\left(\frac{\rho}{\rho^0} \right) \mathbf{W} \right] + \left(\frac{\rho}{\rho^0} \right) \mathbf{W} \cdot \nabla (\rho^0 \sqrt{2h^0})$$

But the quantity $\rho^0 \sqrt{2h^0}$ is a constant along a streamline; therefore its gradient must be perpendicular to the velocity vector or $\mathbf{W}$. Hence the second term to the right of the above equation is zero. Thus

$$\nabla \cdot \left[\left(\frac{\rho}{\rho^0} \right) \mathbf{W} \right] = 0 \tag{11-5}$$

For a perfect gas, Eq. 11-5 can be written, using Eq. 11-4, as

$$\nabla \cdot [(1 - W^2)^{1/(\gamma-1)} \mathbf{W}] = 0 \tag{11-6}$$

The dynamic equation now becomes

$$(\mathbf{W} \cdot \nabla)\mathbf{W} + \frac{\gamma - 1}{2\gamma} (1 - W^2) \nabla \ln p = 0 \tag{11-7}$$

If instead of the $\mathbf{W}$ vector, one divides the velocity vector $\mathbf{u}$ by the local velocity of sound or for the perfect gas by $\sqrt{\gamma p/\rho}$, another nondimensional vector, the Mach number vector $\mathbf{M}$, is obtained. In terms of this new vector, the continuity equation and the dynamic equation become, for a perfect gas,

$$\nabla \cdot \left[\left(1 + \frac{\gamma - 1}{2} M^2 \right)^{-\frac{\gamma+1}{2(\gamma-1)}} \mathbf{M} \right] = 0 \tag{11-8}$$

$$(\mathbf{M} \cdot \nabla)\mathbf{M} - \frac{\gamma - 1}{\gamma + 1} \mathbf{M}(\nabla \cdot \mathbf{M}) + \frac{1}{\gamma} \nabla \ln p = 0 \tag{11-9}$$

The use of vectors $\mathbf{W}$ and $\mathbf{M}$ was introduced by Munk and Prim [*13*] and Hicks, Guenther, and Wasserman [*14*]. In each case, the continuity equation and the dynamic equation are a system of two equations for the two unknowns $\mathbf{W}$ and p, or $\mathbf{M}$ and p. However, it is important to note that in Eq. 11-6 and 11-7, or 11-8 and 11-9, the stagnation enthalpy does not explicitly appear. A single solution of either Eq. 11-6 and 11-7 or 11-8 and 11-9 really consists of a family of solutions corresponding to different assignments of the stagnation enthalpy to streamlines. But for each family of solution, the pressure p is the same, independent of the assignment of stagnation enthalpy. Therefore, for instance, if a steady flow field is determined for specified body and stagnation enthalpy distribution among the streamlines, then a change in the stagnation enthalpy distribution requires only a change of the velocity field by a factor equal to the square root of the ratio stagnation enthalpy. The same pressure field will then keep the flow in equilibrium.

In terms of the vector $\mathbf{W}$, Eq. 11-1 becomes, for a perfect gas,

$$\frac{\mathbf{W} \times (\nabla \times \mathbf{W})}{1 - W^2} = \frac{\gamma - 1}{2\gamma} \nabla \ln p^0 \tag{11-10}$$

where p^0 is the stagnation pressure, again constant along a streamline. By taking the curl of Eq. 11-10, one has Crocco's equation, first derived by him in [*9*] for isoenergetic flows:

$$\nabla \times \left[\frac{\mathbf{W} \times (\nabla \times \mathbf{W})}{1 - W^2}\right] = 0 \tag{11-11}$$

Force on a body in uniform motion. As an application of the integral theorem derived in Art. 4, the forces, moments, and heat transfer for a body in uniform motion are computed by integrals over a control surface A_1 enclosing the body. These equations are Eq. 4-10, 4-11, and 4-12. If the control surface A_1 is chosen as a large surface far away from the body so that the pressure, density, temperature, and velocities settle down and vary very slowly with the space coordinates, the gradient of these quantities will be so small that the viscous stresses and heat flux will be negligible on A_1. This means that, on the surface A_1, the effects of boundary layer and shock waves are so diffused as to appear only as a broad wake behind the body. Then Eq. 4-10, 4-11, and 4-12 become

$$F_i = -\int_{A_1} [pn_i + \rho u_i(u_j n_j)]dA \tag{11-12}$$

$$M_i = -\int_{A_1} \delta_{ijk}[px_j n_k + \rho x_j u_k(u_l n_l)]dA \tag{11-13}$$

$$H = -\int_{A_1} \rho(h + \tfrac{1}{2}u_i u_i)u_j n_j dA \tag{11-14}$$

where F_i, M_i, and H are the force components, the moments about the coordinate axes acting on the body, and heat transfer to the body respectively. In addition there is the continuity equation from Eq. 4-2.

$$\int \rho u_i n_i dA = 0 \tag{11-15}$$

Now let p_∞, ρ_∞, T_∞, h_∞ be the uniform pressure, density, temperature, and specific enthalpy far ahead of the body, i.e. the undisturbed free stream quantities. Let U be the undisturbed free stream velocity in the x_1 direction and denote the small deviations from these uniform conditions by δp, δu_1, δh, etc. Then the resistance of the body or F_1 is

$$F_1 = -\int_{A_1} (p_\infty + \delta p) n_1 dA - \int_{A_1} (U + \delta u_1) \rho u_i n_i dA$$

The integral of $p_\infty n_1$ over the closed surface A_1 vanishes. Then because of the continuity equation (11-15) and the fact that U is a constant, the integral over A_1 of $U(\rho u_i n_i)$ vanishes. Furthermore on A_1 the deviations from uniform conditions are very small, and only first order quantities need be considered. The resistance F_1 is thus

$$F_1 = -\int_{A_1} (\delta p + \rho_\infty U \delta u_1) n_1 dA \tag{11-16}$$

A similar transformation of Eq. 11-14 gives

$$H = -\int_{A_1} \rho_\infty U (\delta h + U \delta u_1) n_1 dA \tag{11-17}$$

By eliminating δu_1 from Eq. 11-16 and 11-17, one has

$$UF_1 - H = \int_{A_1} \rho_\infty U \left(\delta h - \frac{\delta p}{\rho_\infty} \right) n_1 dA \tag{11-18}$$

However, if s_∞ is the uniform specific entropy of the free stream far ahead of the body, and δs is the deviation of s from s_∞ at the surface A_1, then under the present approximation of small deviations

$$T_\infty \delta s = \delta h - \frac{\delta p}{\rho_\infty} \tag{11-19}$$

By substituting Eq. 11-19 into Eq. 11-18,

$$UF_1 - H = T_\infty \int_{A_1} \rho_\infty U \delta s n_1 dA$$

On the other hand, it is easy to show by using the continuity equation that

$$\int_{A_1} \rho s (u_i n_i) dA = \int_{A_1} \rho_\infty U \delta s n_1 dA$$

Therefore finally,

$$UF_1 - H = T_\infty \int_{A_1} \rho s(u_i n_i) dA \tag{11-20}$$

Equation 11-20 is a very useful theorem first obtained by Oswatitsch [*15*]: The product of resistance and velocity of the body minus the heat absorbed by the body is equal to the entropy flux out of a large control surface enclosing the body multiplied by the free stream temperature. However, for the truth of the theorem, the control surface should be large enough to include all energy dissipative processes such as the boundary layer wake and the shocks. In fact, the theorem which is derived for infinitesimal deviations from the uniform state of undisturbed flow does not hold for a concentrated wake involving finite entropy jumps. In applying the theorem, this limitation must be kept in mind.

For an ideal compressible flow without viscosity, heat conduction, and shock wave, the heat H absorbed by the body is zero. Furthermore the flow must also be isentropic; then the entropy flux vanishes. Therefore for such cases, and up to the approximation of Eq. 11-19, the resistance also vanishes. This is the so-called *D'Alembert paradox*, first discussed for a compressible fluid by Theodorsen [*16*]. However, shocks are always present for transonic and supersonic flows, and the D'Alembert paradox is thus a much weaker proposition for compressible flow than for incompressible flow. Then, of course, there is the case of flow with a vortex wake such as the three-dimensional flow over a wing. For this case, the approximation of Eq. 11-19 is no longer sufficient and the resistance of a three-dimensional wing is not zero, even without shock.

A,12. Two-Dimensional Steady Isoenergetic Flows. Stream Functions. Steady isoenergetic, but not isentropic flow occurs when there is a curved shock in the flow field, e.g. the flow around a body in transonic flight with a detached shock. In such cases, the stagnation enthalpy is a constant throughout the field and is equal to the uniform stagnation enthalpy far away from the body. The entropy in the field, however, is different from streamline to streamline. By using Eq. 11-16 and 11-11, one has

$$\begin{aligned}
0 &= \nabla \times \left\{ [(1 - W^2)^{1/(\gamma-1)}\mathbf{W}] \times \left[\frac{\nabla \times \mathbf{W}}{(1 - W^2)^{\gamma/(\gamma-1)}} \right] \right\} \\
&= \left[\frac{\nabla \times \mathbf{W}}{(1 - W^2)^{\gamma/(\gamma-1)}} \cdot \nabla \right] (1 - W^2)^{1/(\gamma-1)}\mathbf{W} \\
&\qquad - [(1 - W^2)^{1/(\gamma-1)}\mathbf{W} \cdot \nabla] \frac{\nabla \times \mathbf{W}}{(1 - W^2)^{\gamma/(\gamma-1)}} \\
&\qquad + (1 - W^2)^{1/(\gamma-1)}\mathbf{W} \left[\nabla \cdot \frac{\nabla \times \mathbf{W}}{(1 - W^2)^{\gamma/(\gamma-1)}} \right]
\end{aligned} \tag{12-1}$$

For two-dimensional flows, there is only one component of $\nabla \times \mathbf{W}$, and it is perpendicular to the plane of flow. Therefore only the second term to the right in the above equation is not identically zero. In other words, if u and v are the components of $\mathbf{W}$ in the x and y directions, Eq. 12-1 reduces to

$$\left(u \frac{\partial}{\partial x} + v \frac{\partial}{\partial y}\right) \left[\frac{\dfrac{\partial v}{\partial x} - \dfrac{\partial u}{\partial y}}{(1 - W^2)^{\gamma/(\gamma-1)}} \right] = 0 \tag{12-2}$$

Now for isoenergetic flows, the reduced velocity vector $\mathbf{W}$ differs from the physical velocity vector $\mathbf{u}$ by a constant factor. Therefore Eq. 12-2 can be interpreted as that constancy of the strength of vorticity divided by $(1 - W^2)^{\gamma/(\gamma-1)}$ along the streamlines. But $(1 - W^2)^{\gamma/(\gamma-1)}$ is proportional to the local pressure if there is no discontinuity such as a shock. Therefore for steady two-dimensional isoenergetic flows, the strength of vorticity along any streamline is directly proportional to the local pressure. This theorem is due to Crocco [*9*].

Now following Crocco, define a stream function by

$$\frac{\partial \psi}{\partial y} = u(1 - W^2)^{1/(\gamma-1)} \tag{12-3}$$

and

$$\frac{\partial \psi}{\partial x} = -v(1 - W^2)^{1/(\gamma-1)}$$

Then the continuity equation (11-6) is automatically satisfied. Lines of constant values of the stream function are streamlines. If dn is the distance between two neighboring streamlines ψ and $\psi + d\psi$, then Eq. 12-3 gives

$$d\psi = (1 - W^2)^{1/(\gamma-1)} W dn = \frac{\rho}{\rho^0} W dn$$

Hence $\rho W dn$ or $\rho^0 d\psi$ is proportional to the mass rate of flow between the streamlines. But the entropy s is different from streamline to streamline and ρ^0 is not a constant for different streamlines. Therefore, $d\psi$ itself is not a measure of the mass rate of flow between streamlines ψ and $\psi + d\psi$. This results in a discontinuity in ψ when the flow crosses a shock, caused by the change in entropy and stagnation density ρ^0. Such a jump in the value of the stream function defined by Eq. 12-3 may be an inconvenience. Tollmien [*17*] tries to remedy this by using a different stream function ψ' related to Crocco's ψ by

$$d\psi' = \rho^0 d\psi.$$

The value of ψ' is now continuous across a shock, but the differential equation for ψ' is very complicated.

As indicated in the previous paragraph, Eq. 12-2 means that

$$\frac{\dfrac{\partial v}{\partial x} - \dfrac{\partial u}{\partial y}}{(1 - W^2)^{\gamma/(\gamma-1)}} = f(\psi) \tag{12-4}$$

The physical meaning of the function $f(\psi)$ can be obtained by analyzing Eq. 11-1 for the present problem. Let n be the direction normal to the streamlines, positive if n and the velocity vector are similarly situated as y and x. Then Eq. 11-1 gives

$$W\left(\frac{\partial v}{\partial x} - \frac{\partial u}{\partial y}\right) = \frac{1}{2c_p}(1 - W^2)\frac{ds}{dn}$$

But by the definition of the stream function (Eq. 12-3),

$$W(1 - W^2)^{1/(\gamma-1)} = \frac{d\psi}{dn}$$

Therefore

$$\frac{\dfrac{\partial v}{\partial x} - \dfrac{\partial u}{\partial y}}{(1 - W^2)^{\gamma/(\gamma-1)}} = f(\psi) = \frac{1}{2c_p}\frac{ds}{d\psi} \tag{12-5}$$

$f(\psi)$ is thus directly proportional to the rate of change of specific entropy from streamline to streamline. Here again, if the whole field is isentropic, $ds/d\psi$ vanishes, and the vorticity must be zero and the flow irrotational.

By eliminating the derivatives of velocity components in Eq. 12-5 with the stream function, the following equation for ψ is obtained (for details, cf. [9]):

$$\left(1 - \frac{u^2}{c^2}\right)\frac{\partial^2\psi}{\partial x^2} - 2\frac{uv}{c^2}\frac{\partial^2\psi}{\partial x\partial y} + \left(1 - \frac{v^2}{c^2}\right)\frac{\partial^2\psi}{\partial y^2}$$
$$= (1 - W^2)^{(\gamma+1)/(\gamma-1)}\left(\frac{W^2}{c^2} - 1\right)f(\psi) \tag{12-6}$$

where $f(\psi)$ is given by Eq. 12-5 and c is the ratio of local velocity of sound a to the maximum velocity u_{max} or

$$c^2 = \frac{a^2}{u_{max}^2} = \frac{\gamma - 1}{2}(1 - W^2) \tag{12-7}$$

And from Eq. 12-3

$$W^2(1 - W^2)^{2/(\gamma-1)} = \left(\frac{\partial\psi}{\partial x}\right)^2 + \left(\frac{\partial\psi}{\partial y}\right)^2 \tag{12-8}$$

When the entropy gradient $ds/d\psi$ is specified by, say, the curved shock wave, Eq. 12-6 is thus a single nonlinear equation for the stream function.

Variational method. Similar to the problem of irrotational flow given in Art. 10, two-dimensional rotational flows discussed above can also be solved by the variational method. Consider the following integral

$$I = \iint_A e^{-s(\psi)/\mathcal{R}}(1 - W^2)^{1/(\gamma-1)}\left(1 + \frac{\gamma+1}{\gamma-1}W^2\right)dxdy \qquad (12\text{-}9)$$

where W^2 is given by Eq. 12-8 and the region of integration A is bounded by streamlines and lines orthogonal to the streamlines. The basis of the variational method is to make

$$\delta I = 0 \qquad (12\text{-}10)$$

for all the virtual variations of the stream function ψ. This is the same as making ψ satisfy the differential equation (Eq. 12-6) or its original form (Eq. 12-5). The truth of the statement can be shown as follows:

$$\delta I = \iint_A e^{-s(\psi)/\mathcal{R}}\left[\frac{\gamma}{\gamma-1}\,\frac{(1-W^2)^{1/(\gamma-1)}}{1-W^2}\left(1 - \frac{\gamma+1}{\gamma-1}W^2\right)\delta W^2\right.$$
$$\left. - (1 - W^2)^{1/(\gamma-1)}\left(1 + \frac{\gamma+1}{\gamma-1}W^2\right)\frac{1}{\mathcal{R}}\frac{ds}{d\psi}\,\delta\psi\right]dxdy \qquad (12\text{-}11)$$

However, the definition of stream function (Eq. 12-3) gives the following relations:

$$\frac{\partial(\delta\psi)}{\partial y} = \delta u(1 - W^2)^{1/(\gamma-1)} - \frac{u}{\gamma-1}(1 - W^2)^{1/(\gamma-1)-1}\delta W^2$$

$$-\frac{\partial(\delta\psi)}{\partial x} = \delta v(1 - W^2)^{1/(\gamma-1)} - \frac{v}{\gamma-1}(1 - W^2)^{1/(\gamma-1)-1}\delta W^2$$

By multiplying the first equation by u and the second equation by v and then adding the results, one has

$$\delta W^2\frac{(1-W^2)^{1/(\gamma-1)}}{1-W^2}\left(1 - \frac{\gamma+1}{\gamma-1}W^2\right) = 2\left(u\frac{\partial\delta\psi}{\partial y} - v\frac{\partial\delta\psi}{\partial x}\right)$$

Therefore Eq. 12-11 can be written as

$$\delta I = \iint_A e^{-s(\psi)/\mathcal{R}}\left[\frac{2\gamma}{\gamma-1}\left(u\frac{\partial\delta\psi}{\partial y} - v\frac{\partial\delta\psi}{\partial x}\right)\right.$$
$$\left. - (1 - W^2)^{1/(\gamma-1)}\left(1 + \frac{\gamma+1}{\gamma-1}W^2\right)\frac{1}{\mathcal{R}}\frac{ds}{d\psi}\,\delta\psi\right]dxdy$$

This equation can be transformed into the following form by partial integration

$$\delta I = \iint_A e^{-s(\psi)/\mathcal{R}} \left[\frac{2\gamma}{\gamma - 1} \left(\frac{\partial v}{\partial x} - \frac{\partial u}{\partial y} \right) + \frac{2\gamma}{\gamma - 1} \frac{1}{\mathcal{R}} \frac{ds}{d\psi} \left(u \frac{\partial \psi}{\partial y} - v \frac{\partial \psi}{\partial x} \right) \right.$$

$$\left. - \frac{1}{\mathcal{R}} \frac{ds}{d\psi} (1 - W^2)^{1/(\gamma-1)} \left(1 + \frac{\gamma + 1}{\gamma - 1} W^2 \right) \right] \delta\psi dx dy$$

$$- \oint_C e^{-s(\psi)/\mathcal{R}} \left(\frac{2\gamma}{\gamma - 1} \right) \delta\psi \mathbf{W} \cdot d\mathbf{l}$$

The line integral is to be taken over the boundary of the region A. But the boundary is supposed to consist of streamlines and lines orthogonal to streamlines. On streamlines, $\delta\psi$ has to vanish. On lines orthogonal to streamlines $\mathbf{W} \cdot d\mathbf{l}$ vanishes. Therefore the contour integral vanishes. The derivatives $\partial\psi/\partial x$, $\partial\psi/\partial y$ in the area integral can be written in terms of velocities by Eq. 12-3. Then

$$\delta I = \frac{2\gamma}{\gamma - 1} \iint_A e^{-s(\psi)/\mathcal{R}} \left[\frac{\partial v}{\partial x} - \frac{\partial u}{\partial y} - \frac{1}{2c_p} (1 - W^2)^{\gamma/(\gamma-1)} \frac{ds}{d\psi} \right] \delta\psi dx dy \tag{12-12}$$

Since $\delta\psi$ is arbitrary in A_1 and the exponential factor cannot be zero, the condition (Eq. 12-10) requires the quantity within the square bracket of Eq. 12-12 to be zero in A. Therefore Eq. 12-10 is equivalent to Eq. 12-5 and 12-6.

Recently, Crocco [*18*] introduced a new potential function φ for steady nonisentropic but isoenergetic flows in two dimensions. It is defined as

$$p + \rho u^2 = -\frac{\partial^2 \varphi}{\partial y^2}, \quad \rho uv = \frac{\partial^2 \varphi}{\partial x \partial y}, \quad p + \rho v^2 = -\frac{\partial^2 \varphi}{\partial x^2} \tag{12-13}$$

When these values are introduced into the dynamic equations, the dynamic equations are identically satisfied. φ has the advantage of being continuous across the shock. The differential equation for φ obtained from the continuity equation is, however, a nonlinear equation of third order.

Axially symmetric flows. For axially symmetric flows, the most convenient coordinate system is the cylindrical coordinate system x and r (x along the axis of symmetry and r normal to the axis). Let the components of the reduced velocity $\mathbf{W}$ again be u and v in x and r directions, the stream function is then defined as (see Eq. 12-3)

$$ru(1 - W^2)^{1/(\gamma-1)} = \frac{\partial \psi}{\partial r}, \qquad -rv(1 - W^2)^{1/(\gamma-1)} = \frac{\partial \psi}{\partial x} \tag{12-14}$$

Crocco found that between discontinuities the strength of the vorticity along a streamline is now proportional to the product of local pressure p and r. The vorticity equation is now

$$\frac{\dfrac{\partial v}{\partial x} - \dfrac{\partial u}{\partial y}}{r(1 - W^2)^{\gamma/(\gamma-1)}} = f(\psi) \tag{12-15}$$

The equation for the stream function is

$$\left(1 - \frac{u^2}{c^2}\right)\frac{\partial^2\psi}{\partial x^2} - 2\frac{uv}{c^2}\frac{\partial^2\psi}{\partial x\partial r} + \left(1 - \frac{v^2}{c^2}\right)\frac{\partial^2\psi}{\partial r^2} - \frac{1}{r}\frac{\partial\psi}{\partial r}$$
$$= r^2(1 - W^2)^{(\gamma+1)/(\gamma-1)}\left(\frac{W^2}{c^2} - 1\right)f(\psi) \tag{12-16}$$

where the function $f(\psi)$ is related to the gradient $ds/d\psi$ as given by Eq. 12-5, and c^2 is given by Eq. 12-7. Corresponding to Eq. 12-8, the relation between the reduced velocity and the stream function is

$$r^2W^2(1 - W^2)^{2/(\gamma-1)} = \left(\frac{\partial\psi}{\partial x}\right)^2 + \left(\frac{\partial\psi}{\partial r}\right)^2 \tag{12-17}$$

The problem can also be formulated as a variational problem by using the integral

$$I = \iint e^{-s(\psi)/\mathfrak{R}}(1 - W^2)^{1/(\gamma-1)}\left(1 + \frac{\gamma+1}{\gamma-1}W^2\right)r\,dx\,dr \tag{12-18}$$

where W is related to ψ by Eq. 12-17 and the integration is to be extended over a region A bounded by the streamlines and lines orthogonal to streamlines. Then Eq. 12-15 or 12-16 is equivalent to making the integral I have a stationary value under virtual variation of the stream function ψ.

Nonisoenergetic flows. The previous treatment of isoenergetic flows through the use of the stream function ψ can easily be extended to nonisoenergetic flows. In fact, if the reduced velocity $\mathbf{W}$ and its components u and v are used, and the definition of the stream function given by Eq. 12-3, the equation for ψ (12-6) and the auxiliary equations (12-1) and (12-8) remain valid. The function $f(\psi)$, however, is not given by Eq. 12-5, but instead, due to Eq. 11-10, it becomes

$$f(\psi) = -\frac{\gamma-1}{2\gamma}\frac{d\ln p^0}{d\psi} \tag{12-19}$$

The same is true for the axially symmetric case. Of course, for nonisoenergetic flows, the maximum velocity $u_{\max}$ is different for different streamlines, the reduced velocity $\mathbf{W}$ and the components u and v differ from the physical velocity $\mathbf{u}$ and its component not by a constant factor. Therefore $(\partial v/\partial x) - (\partial u/\partial y)$, for instance, cannot be interpreted as a quantity proportional to the vorticity of the fluid.

A,13. Two-Dimensional Steady Irrotational Flows. Hodograph Transformation. If the entropy gradient $ds/d\psi$ vanishes, the flow is irrotational and isentropic. Then the right side of Eq. 12-6 is zero. Then due to Eq. 11-4

$$(1 - W^2)^{1/(\gamma-1)} = \frac{\rho}{\rho^0} \tag{13-1}$$

where ρ^0 is the stagnation density. ρ^0 is now not only constant along a streamline but constant throughout the field. With Eq. 13-1, the definition of stream function becomes

$$\frac{\partial\psi}{\partial y} = \frac{\rho}{\rho^0}u, \qquad \frac{\partial\psi}{\partial x} = -\frac{\rho}{\rho^0}v \tag{13-2}$$

Therefore the difference in the value of ψ for two streamlines is now a true measure of the mass rate of flow between these streamlines. In Eq. 12-2, the reduced velocity components u and v can be considered as the physical velocity components since only a constant factor is involved. Then the corresponding c^2 is a^2. Thus for two-dimensional steady irrotational flows, the equation for the stream function ψ defined by Eq. 13-2 is

$$\left(1 - \frac{u^2}{a^2}\right)\frac{\partial^2\psi}{\partial x^2} - 2\frac{uv}{a^2}\frac{\partial^2\psi}{\partial x\partial y} + \left(1 - \frac{v^2}{a^2}\right)\frac{\partial^2\psi}{\partial y^2} = 0 \tag{13-3}$$

By comparing Eq. 13-3 with 9-11, it is seen that the potential ϕ and the stream function ψ satisfy the same form of equation. The actual content of the equation for ϕ and for ψ is, of course, not the same; the relation between u, v, and ϕ and that between u, v, and ψ are quite different. For instance, Eq. 13-1 can be written as

$$\left(1 - \frac{\gamma-1}{2}\frac{|\mathbf{u}|^2}{(a^0)^2}\right)^{1/(\gamma-1)} = \frac{\rho}{\rho^0} \tag{13-4}$$

where a^0 is the sound speed corresponding to stagnation conditions, and Eq. 12-8 is

$$\left(\frac{\rho}{\rho^0}\right)^2 q^2 = \left(\frac{\partial\psi}{\partial x}\right)^2 + \left(\frac{\partial\psi}{\partial y}\right)^2 \tag{13-5}$$

Corresponding to Eq. 12-7, one has from Eq. 9-4

$$a^2 = (a^0)^2 - \frac{\gamma-1}{2}|\mathbf{u}|^2 \tag{13-6}$$

Both Eq. 9-11 and 13-3 are nonlinear differential equations and are very difficult to solve. The most effective way to attack the problem is first to "linearize" the equations and then solve the resulting linear equations. There are three methods used to linearize the equations. The first method is due to Rayleigh and Janzen and consists of expanding the solution in terms of power series in M_∞^2, where M_∞ is the Mach number of the uniform flow far from the body. The zeroeth order solution is thus the solution for incompressible flow and the method is appropriate when M_∞ is small.

The second method is to expand the solution in an ascending series of the "thickness ratio," and the angle of inclination of the body axis to the uniform flow far from the body. The thickness ratio is the ratio of the thickness of the body across the main flow to the maximum dimension of the body along the main flow. This method is thus appropriate for slender bodies and thin bodies, at subsonic or supersonic speeds. The first approximation was developed by Glauert, Prandtl, and Ackeret. (For details see Vol. VI and VII.)

The third method of linearization is the transformation of the equations by using the velocity components u and v, or their equivalent

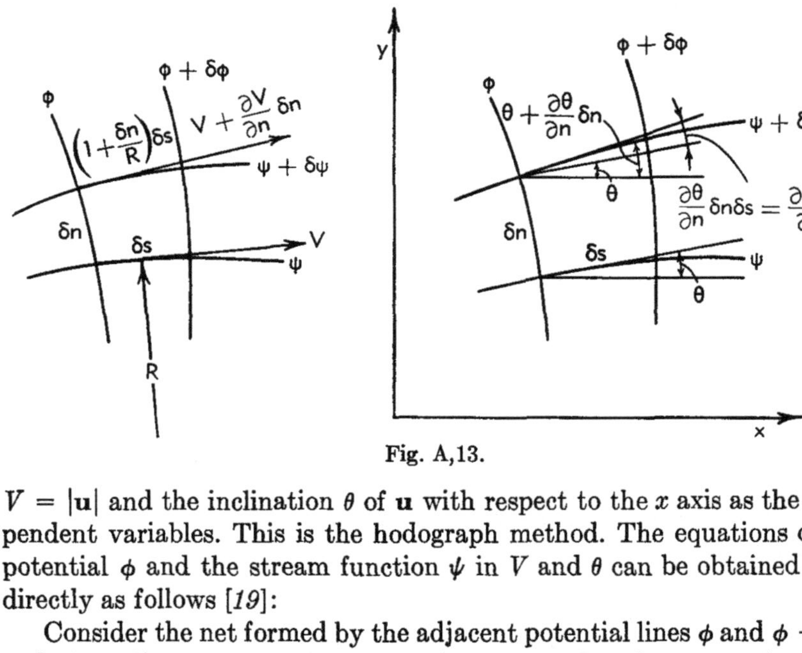

Fig. A,13.

$V = |\mathbf{u}|$ and the inclination θ of $\mathbf{u}$ with respect to the x axis as the independent variables. This is the hodograph method. The equations of the potential ϕ and the stream function ψ in V and θ can be obtained most directly as follows [*19*]:

Consider the net formed by the adjacent potential lines ϕ and $\phi + \delta\phi$, and the adjacent streamlines ψ and $\psi + \delta\psi$. The distances along the potential lines are denoted by n and along the streamlines by s. If R is the radius of curvature of the streamlines (Fig. A,13), then the condition of irrotationality is

$$-\frac{\partial V}{\partial n} - \frac{V}{R} = 0$$

But $1/R = -\partial\theta/\partial s$; therefore

$$\frac{1}{V}\frac{\partial V}{\partial n} - \frac{\partial \theta}{\partial s} = 0 \tag{13-7}$$

The continuity condition requires that for an elementary channel between ψ and $\psi + \delta\psi$, $\rho V \delta n = \text{const}$, i.e.

$$\frac{1}{\rho}\frac{\partial\rho}{\partial s} + \frac{1}{V}\frac{\partial V}{\partial s} + \frac{1}{\delta n}\frac{\partial \delta n}{\partial s} = 0 \tag{13-8}$$

However, the dynamic equation requires

$$\rho V \frac{\partial V}{\partial s} = -\frac{\partial p}{\partial s} = -a^2 \frac{\partial \rho}{\partial s}$$

Therefore

$$\frac{1}{\rho}\frac{\partial\rho}{\partial s} + \frac{1}{q}\frac{\partial V}{\partial s} = (1 - M^2)\frac{1}{V}\frac{\partial V}{\partial s}$$

where M is the local Mach number V/a, and a the local sound speed. But Fig. A,13 also shows that

$$\frac{1}{\delta n}\frac{\partial \delta n}{\partial s} = \frac{\partial \theta}{\partial n}$$

Therefore Eq. 13-8 becomes

$$(1 - M^2)\frac{1}{V}\frac{\partial V}{\partial s} + \frac{\partial \theta}{\partial n} = 0 \tag{13-9}$$

Now $d\phi = V ds$, and $d\psi = (\rho/\rho^0) V dn$; therefore Eq. 13-7 and 13-9 can be written as

$$\frac{\rho}{\rho^0}\frac{1}{V}\frac{\partial V}{\partial \psi} - \frac{\partial \theta}{\partial \phi} = 0 \tag{13-10}$$

and

$$(1 - M^2)\frac{1}{V}\frac{\partial V}{\partial \phi} + \frac{\rho}{\rho^0}\frac{\partial \theta}{\partial \psi} = 0 \tag{13-11}$$

V and θ can be chosen as the independent variables instead of ϕ and ψ. Then the following relations are obtained:

$$\begin{aligned} \frac{\partial V}{\partial \psi} &= \frac{1}{J}\frac{\partial \phi}{\partial \theta}, & \frac{\partial \theta}{\partial \psi} &= -\frac{1}{J}\frac{\partial \phi}{\partial V} \\ \frac{\partial V}{\partial \phi} &= -\frac{1}{J}\frac{\partial \psi}{\partial \theta}, & \frac{\partial \theta}{\partial \phi} &= \frac{1}{J}\frac{\partial \psi}{\partial V} \end{aligned} \tag{13-12}$$

where J is the Jacobian,

$$J = \frac{\partial(\phi, \psi)}{\partial(V, \theta)} = \frac{\partial \phi}{\partial V}\frac{\partial \psi}{\partial \theta} - \frac{\partial \phi}{\partial \theta}\frac{\partial \psi}{\partial V}$$

By substituting Eq. 13-12 into Eq. 13-10 and 13-11, the following basic hodograph equations are obtained:

$$\frac{\rho}{\rho^0}\frac{1}{V}\frac{\partial\phi}{\partial\theta} = \frac{\partial\psi}{\partial V}$$

$$\frac{\partial\phi}{\partial V} = -(1 - M^2)\frac{\rho^0}{\rho}\frac{1}{V}\frac{\partial\psi}{\partial\theta} \tag{13-13}$$

These equations are now linear, as the density ρ and the Mach number M are functions of V only. By eliminating one of the dependent variables from this pair of equations, one has a linear equation for either ϕ or ψ. The details of this development are treated in VI,F.

Axially symmetric flows. Equations for the stream function ψ for irrotational, axially symmetric flows can be deduced from Eq. 12-13, 12-16, and 12-17. Thus

$$\frac{\partial\psi}{\partial r} = \frac{\rho}{\rho^0} ru, \qquad \frac{\partial\psi}{\partial x} = -\frac{\rho}{\rho^0} rv \tag{13-14}$$

$$\left(\frac{\rho}{\rho^0}\right)^2 r^2V|\mathbf{u}|^2 = \left(\frac{\partial\psi}{\partial x}\right)^2 + \left(\frac{\partial\psi}{\partial r}\right)^2 \tag{13-15}$$

and

$$\left(1 - \frac{u^2}{a^2}\right)\frac{\partial^2\psi}{\partial x^2} - 2\frac{uv}{a^2}\frac{\partial^2 x}{\partial x\partial r} + \left(1 - \frac{v^2}{a^2}\right)\frac{\partial^2\psi}{\partial r^2} - \frac{1}{r}\frac{\partial\psi}{\partial r} = 0 \tag{13-16}$$

The equation for the velocity potential ϕ is

$$\left(1 - \frac{u^2}{a^2}\right)\frac{\partial^2\phi}{\partial x^2} - 2\frac{uv}{a^2}\frac{\partial^2\phi}{\partial x\partial r} + \left(1 - \frac{v^2}{a^2}\right)\frac{\partial^2\phi}{\partial r^2} + \frac{1}{r}\frac{\partial\phi}{\partial r} = 0 \tag{13-17}$$

Thus Eq. 13-16 and 13-17 are again nonlinear. The methods of linearization of Rayleigh and Janzen and the expanding of the solution in the thickness parameter still apply. However, unfortunately, the hodograph transformation now no longer linearizes the equations for ϕ and ψ.

A,14. Deviations from the Perfect Gas Law. During the discussion in the preceding sections, it is frequently assumed that the gas is a perfect gas, although many of the equations are written in forms that are true in general. As a matter of fact, if the equation does not involve c_v, c_p, or γ, then it is a general equation. If it does, then it is an equation for a perfect gas. It should be emphasized here that the fluid can be considered as an ideal gas, i.e. a gas which is both nonviscous and nonheat-conducting, but yet it is not a perfect gas, which must satisfy the equation of state of Eq. 5-7. For instance, outside the shock and the boundary layer, the effects of viscosity and heat conduction are negligible, but if the pressure is very high, if the temperature is high, or if the temperature is extremely low, the gas behavior deviates appreciably from that of a perfect gas. One example of such flows is the flow in a nozzle of a hypersonic wind tunnel where, due to the very large expansion ratio, extreme conditions are encountered. Such deviations are quite apart from the effects resulting from rate processes, that is, the processes of adjusting to the thermodynamic equilibrium conditions. Viscosities and heat conduction, on the other hand, are results of the finite rate of adjusting to the thermodynamic equilibrium and could occur even for a perfect gas.

There are two kinds of deviation from the perfect gas law: The gas may obey quite accurately the equation of state for a perfect gas and yet the specific heats may not be constants, in which case, the gas is called thermally perfect but calorically imperfect. Or, the gas may have constant specific heats but may not satisfy the equation of state of a perfect gas; in such cases, the gas is calorically perfect, but thermally imperfect. There is good reason to differentiate these two types of imperfections. If the density of the gas is such that the molecules on the average are relatively far apart, and at this average distance, there is only negligible interaction between them, then the equation of state has to be that of a perfect gas. On the other hand, the specific heat at constant volume, being determined by statistical distribution of the internal energy among the various quantum states of the molecule, could deviate appreciably from constant value both above and below the room temperature. At very low temperatures, the specific heat decreases because the rotational motion of the molecules is not excited; at very high temperatures it increases because the molecular vibrations are excited. Thus the calorical imperfection of a gas is the result of the particular internal structure of gas molecules. For temperatures near the room temperature, such effects do not occur; but if the gas density is high, so that the average distance of the molecules is reduced to small values, mutual influence between the molecules at their average distances takes place. Then the equation of state cannot be that of a perfect gas and the thermal imperfection of the gas is a result of the mutual interaction of the molecules.

The calorical imperfection and the thermal imperfection can be separated by considering the gas at zero pressure. Let the internal energy at zero pressure be denoted by e_0; then e_0 is only a function of temperature T. According to the second law of thermodynamics

$$e(T, v_{sp}) = \int_{\infty}^{v_{sp}} \left[T\left(\frac{\partial p}{\partial T}\right)_{v_{sp}} - p \right] dv_{sp} + e_0(T) \tag{14-1}$$

where v_{sp} is the specific volume of the gas, or

$$v_{sp} = \frac{1}{\rho} \tag{14-2}$$

The subscripts on the parentheses indicate the parameter held constant. Eq. 14-1 allows the computation of internal energy at any pressure (specific volume) and temperature, once the internal energy at zero pressure (infinite specific volume) $e_0(T)$ and the equation of state are specified. The enthalpy h is simply

$$h = e + pv_{sp} \tag{14-3}$$

For a perfect gas, the integral in Eq. 14-1 vanishes and the internal energy and the enthalpy are then independent of pressure.

For the isentropic process, there is no heat addition. Therefore

$$0 = de + pdv_{sp}$$

Or by using Eq. 14-1, the differential equation for the isentropic process is

$$\left(\frac{dT}{dv_{sp}}\right)_s = -\frac{T\left(\dfrac{\partial p}{\partial T}\right)_{v_{sp}}}{\dfrac{de_0}{dT}} \tag{14-4}$$

The subscript s denotes that the quantity is evaluated for isentropic process. When the isentropic process is determined, the integral in the Bernoulli equation (8-2) can be computed.

The velocity of sound a can be calculated as follows:

$$a^2 = \left(\frac{dp}{d\rho}\right)_s = -v_{sp}^2\left(\frac{dp}{dv_{sp}}\right)_s = -v_{sp}^2\left[\left(\frac{\partial p}{\partial T}\right)_{v_{sp}}\left(\frac{dT}{dv_{sp}}\right)_s + \left(\frac{\partial p}{\partial v_{sp}}\right)_T\right]$$

By introducing the derivative $(dT/dv_{sp})_s$ from Eq. 14-4, the final expression for the square of the sound velocity is

$$a^2 = v_{sp}\left[\frac{T\left(\dfrac{\partial p}{\partial T}\right)_{v_{sp}}^2}{\dfrac{de_0}{dT}} - \left(\frac{\partial p}{\partial v_{sp}}\right)_T\right] \tag{14-5}$$

In the case of a perfect gas, Eq. 14-5 reduces to the simple form of Eq. 1-4 and the sonic velocity is only a function of temperature. Eq. 14-5 shows, however, the general dependence of sonic velocity on both the temperature and the density of the gas.

Eq. 14-1 and 14-5 supply information for computing the quantities which are important in the analysis of the flow field so that the effects of deviations from the perfect gas law can then, in principle, be determined. The actual process of computing is, of course, tedious, and numerical and graphical methods are indicated. For instance, Busemann [*20*, p. 421] has shown how to construct graphically the characteristics diagrams needed to compute the two-dimensional supersonic flows. Since the deviations from the perfect gas law are usually not large, a more logical method may be that of perturbation theory, taking the solution for a perfect gas as the unperturbed solution. This procedure has been adopted by Tsien in his investigation of the one-dimensional flows of a van der Waals gas [*21*], and by Eggers [*22*] for the same problem, using Berthelot's equation of state.

A,15. Expression of the Equations of Motion in General Orthogonal Coordinates.

General orthogonal coordinates. Let the general orthogonal coordinates be α, β, γ and the elements of length at α, β, γ in the directions of increasing α, β, γ be $h_1 d\alpha$, $h_2 d\beta$, $h_3 d\gamma$ respectively. The arc length ds is then

$$(ds)^2 = h_1^2(d\alpha)^2 + h_2^2(d\beta)^2 + h_3^2(d\gamma)^2$$

If v_1, v_2, v_3 are the components of the vector $\mathbf{v}$ in the directions of increasing α, β, γ respectively, and if the components of $\nabla \times \mathbf{v}$ are ξ, η, ζ, then

$$\nabla \cdot \mathbf{v} = \frac{1}{h_1 h_2 h_3}\left[\frac{\partial}{\partial\alpha}(h_2 h_3 v_1) + \frac{\partial}{\partial\beta}(h_3 h_1 v_2) + \frac{\partial}{\partial\gamma}(h_1 h_2 v_3)\right]$$

and

$$\begin{aligned}
\xi &= \frac{1}{h_2 h_3}\left[\frac{\partial}{\partial\beta}(h_3 v_3) - \frac{\partial}{\partial\gamma}(h_2 v_2)\right] \\
\eta &= \frac{1}{h_3 h_1}\left[\frac{\partial}{\partial\gamma}(h_1 v_1) - \frac{\partial}{\partial\alpha}(h_3 v_3)\right] \\
\zeta &= \frac{1}{h_1 h_2}\left[\frac{\partial}{\partial\alpha}(h_2 v_2) - \frac{\partial}{\partial\beta}(h_1 v_1)\right]
\end{aligned}$$

The components of a gradient are

$$\frac{1}{h_1}\frac{\partial}{\partial\alpha}, \quad \frac{1}{h_2}\frac{\partial}{\partial\beta}, \quad \frac{1}{h_3}\frac{\partial}{\partial\gamma}$$

The components of the stress tensor can be written as follows:

$$\begin{aligned}
\pi_{\alpha\alpha} &= -p + \mu e_{\alpha\alpha} + (\mu' - \tfrac{1}{3}\mu)(e_{\alpha\alpha} + e_{\beta\beta} + e_{\gamma\gamma}) \\
\pi_{\beta\beta} &= -p + \mu e_{\beta\beta} + (\mu' - \tfrac{1}{3}\mu)(e_{\alpha\alpha} + e_{\beta\beta} + e_{\gamma\gamma}) \\
\pi_{\gamma\gamma} &= -p + \mu e_{\gamma\gamma} + (\mu' - \tfrac{1}{3}\mu)(e_{\alpha\alpha} + e_{\beta\beta} + e_{\gamma\gamma}) \\
\pi_{\beta\gamma} &= \pi_{\gamma\beta} = \mu e_{\beta\gamma}, \quad \pi_{\gamma\alpha} = \pi_{\alpha\gamma} = \mu e_{\gamma\alpha}, \quad \pi_{\alpha\beta} = \pi_{\beta\alpha} = \mu e_{\alpha\beta}
\end{aligned}$$

where the e's are expressed in terms of the velocity components u, v, w in directions of increasing α, β, γ.

$$\tfrac{1}{2}e_{\alpha\alpha} = \frac{1}{h_1}\frac{\partial u}{\partial \alpha} + \frac{v}{h_1h_2}\frac{\partial h_1}{\partial \beta} + \frac{w}{h_3h_1}\frac{\partial h_1}{\partial \gamma}$$

$$\tfrac{1}{2}e_{\beta\beta} = \frac{1}{h_2}\frac{\partial v}{\partial \beta} + \frac{w}{h_2h_3}\frac{\partial h_2}{\partial \gamma} + \frac{u}{h_1h_2}\frac{\partial h_2}{\partial \alpha}$$

$$\tfrac{1}{2}e_{\gamma\gamma} = \frac{1}{h_3}\frac{\partial w}{\partial \gamma} + \frac{u}{h_3h_1}\frac{\partial h_3}{\partial \alpha} + \frac{v}{h_2h_3}\frac{\partial h_3}{\partial \beta}$$

$$e_{\beta\gamma} = \frac{h_3}{h_2}\frac{\partial}{\partial \beta}\left(\frac{w}{h_3}\right) + \frac{h_2}{h_3}\frac{\partial}{\partial \gamma}\left(\frac{v}{h_2}\right)$$

$$e_{\gamma\alpha} = \frac{h_1}{h_3}\frac{\partial}{\partial \gamma}\left(\frac{u}{h_1}\right) + \frac{h_3}{h_1}\frac{\partial}{\partial \alpha}\left(\frac{w}{h_3}\right)$$

$$e_{\alpha\beta} = \frac{h_2}{h_1}\frac{\partial}{\partial \alpha}\left(\frac{v}{h_2}\right) + \frac{h_1}{h_2}\frac{\partial}{\partial \beta}\left(\frac{u}{h_1}\right)$$

The dynamic equations of Eq. 2-7 can be written as

$$\frac{\partial \mathbf{v}}{\partial t} + (\mathbf{v}\cdot\nabla)\mathbf{v} = \frac{\partial \mathbf{v}}{\partial t} + \nabla(\tfrac{1}{2}|\mathbf{v}|^2) - \mathbf{v}\times(\nabla\times\mathbf{v}) = X + \frac{1}{\rho}\nabla\cdot\boldsymbol{\pi}$$

where X is the body force vector and $\boldsymbol{\pi}$ is the stress tensor. The components of $\nabla\cdot\boldsymbol{\pi}$ are

α:

$$\frac{1}{h_1h_2h_3}\left[\frac{\partial}{\partial \alpha}(h_2h_3\pi_{\alpha\alpha}) + \frac{\partial}{\partial \beta}(h_3h_1\pi_{\alpha\beta}) + \frac{\partial}{\partial \gamma}(h_1h_2\pi_{\gamma\alpha})\right] + \pi_{\alpha\beta}\frac{1}{h_1h_2}\frac{\partial h_1}{\partial \beta}$$
$$+ \pi_{\gamma\alpha}\frac{1}{h_1h_3}\frac{\partial h_1}{\partial \gamma} - \pi_{\beta\beta}\frac{1}{h_1h_2}\frac{\partial h_2}{\partial \alpha} - \pi_{\gamma\gamma}\frac{1}{h_1h_3}\frac{\partial h_3}{\partial \alpha}$$

β:

$$\frac{1}{h_1h_2h_3}\left[\frac{\partial}{\partial \alpha}(h_2h_3\pi_{\alpha\beta}) + \frac{\partial}{\partial \beta}(h_3h_1\pi_{\beta\beta}) + \frac{\partial}{\partial \gamma}(h_1h_2\pi_{\beta\gamma})\right] + \pi_{\beta\gamma}\frac{1}{h_2h_3}\frac{\partial h_2}{\partial \gamma}$$
$$+ \pi_{\alpha\beta}\frac{1}{h_2h_1}\frac{\partial h_2}{\partial \alpha} - \pi_{\gamma\gamma}\frac{1}{h_2h_3}\frac{\partial h_3}{\partial \beta} - \pi_{\alpha\alpha}\frac{1}{h_2h_1}\frac{\partial h_1}{\partial \beta}$$

γ:

$$\frac{1}{h_1h_2h_3}\left[\frac{\partial}{\partial \alpha}(h_2h_3\pi_{\gamma\alpha}) + \frac{\partial}{\partial \beta}(h_3h_1\pi_{\beta\gamma}) + \frac{\partial}{\partial \gamma}(h_1h_2\pi_{\gamma\gamma})\right] + \pi_{\gamma\alpha}\frac{1}{h_3h_1}\frac{\partial h_3}{\partial \alpha}$$
$$+ \pi_{\beta\gamma}\frac{1}{h_3h_2}\frac{\partial h_3}{\partial \beta} - \pi_{\alpha\alpha}\frac{1}{h_3h_1}\frac{\partial h_1}{\partial \gamma} - \pi_{\beta\beta}\frac{1}{h_3h_2}\frac{\partial h_2}{\partial \gamma}$$

Cylindrical polar coordinates. Use coordinates r, ω, x such that

$$x_1 = r\cos\omega, \quad x_2 = r\sin\omega, \quad x_3 = x$$

If r, ω, x are taken as α, β, γ respectively, then $h_1 = 1$, $h_2 = r$, and $h_3 = 1$. Hence

$$\nabla \cdot \mathbf{v} = \frac{1}{r}\frac{\partial}{\partial r}(rv_1) + \frac{1}{r}\frac{\partial v_2}{\partial \omega} + \frac{\partial v_3}{\partial x}$$

$$\xi = \frac{1}{r}\frac{\partial v_3}{\partial \omega} - \frac{\partial v_2}{\partial x}, \quad \eta = \frac{\partial v_1}{\partial x} - \frac{\partial v_3}{\partial r}, \quad \zeta = \frac{1}{r}\frac{\partial}{\partial r}(rv_2) - \frac{1}{r}\frac{\partial v_1}{\partial \omega}$$

The components of $(\mathbf{v} \cdot \nabla)\mathbf{v}$ are

$$u\frac{\partial u}{\partial r} + \frac{v}{r}\frac{\partial u}{\partial \omega} + w\frac{\partial u}{\partial x} - \frac{v^2}{r} \qquad \text{in } r$$

$$u\frac{\partial v}{\partial r} + \frac{v}{r}\frac{\partial v}{\partial \omega} + w\frac{\partial v}{\partial x} + \frac{uv}{r} \qquad \text{in } \omega$$

$$u\frac{\partial w}{\partial r} + \frac{v}{r}\frac{\partial w}{\partial \omega} + w\frac{\partial w}{\partial x} \qquad \text{in } x$$

The Laplacian ∇^2 is

$$\nabla^2 = \frac{\partial^2}{\partial r^2} + \frac{1}{r}\frac{\partial}{\partial r} + \frac{1}{r^2}\frac{\partial^2}{\partial \omega^2} + \frac{\partial^2}{\partial x^2}$$

The equation of the velocity potential ϕ defined as

$$u = \frac{\partial \phi}{\partial r}, \quad v = \frac{1}{r}\frac{\partial \phi}{\partial \omega}, \quad w = \frac{\partial \phi}{\partial x}$$

can be written down by using the form given by Eq. 9-7.

The components of the rate of strain are

$$\tfrac{1}{2}e_{rr} = \frac{\partial u}{\partial r}, \quad \tfrac{1}{2}e_{\omega\omega} = \frac{1}{r}\frac{\partial v}{\partial \omega} + \frac{u}{r}, \quad \tfrac{1}{2}e_{xx} = \frac{\partial w}{\partial x}$$

$$e_{\omega x} = \frac{1}{r}\frac{\partial w}{\partial \omega} + \frac{\partial v}{\partial x}, \quad e_{xr} = \frac{\partial u}{\partial x} + \frac{\partial w}{\partial r}, \quad e_{r\omega} = r\frac{\partial}{\partial r}\left(\frac{v}{r}\right) + \frac{1}{r}\frac{\partial u}{\partial \omega}$$

The components of $\nabla \cdot \boldsymbol{\pi}$ are

$$\frac{\partial \pi_{rr}}{\partial r} + \frac{1}{r}\frac{\partial \pi_{r\omega}}{\partial \omega} + \frac{\partial \pi_{xr}}{\partial x} + \frac{\pi_{rr} - \pi_{\omega\omega}}{r} \qquad \text{in } r$$

$$\frac{\partial \pi_{r\omega}}{\partial r} + \frac{1}{r}\frac{\partial \pi_{\omega\omega}}{\partial \omega} + \frac{\partial \pi_{\omega x}}{\partial x} + 2\frac{\pi_{r\omega}}{r} \qquad \text{in } \omega$$

$$\frac{\partial \pi_{xr}}{\partial r} + \frac{1}{r}\frac{\partial \pi_{\omega x}}{\partial \omega} + \frac{\partial \pi_{xx}}{\partial x} + \frac{\pi_{xr}}{r} \qquad \text{in } x$$

Spherical polar coordinates. Use coordinates r, θ, ω such that

$$x_1 = r \sin\theta \cos\omega, \quad x_2 = r \sin\theta \sin\omega, \quad x_3 = r\cos\theta$$

if r, θ, ω are taken as α, β, γ respectively, then

$$h_1 = 1, \quad h_2 = r, \quad h_3 = r \sin \theta$$

Hence

$$\nabla \cdot \mathbf{v} = \frac{1}{r^2} \frac{\partial}{\partial r} (r^2 v_1) + \frac{1}{r \sin \theta} \frac{\partial}{\partial \theta} (v_2 \sin \theta) + \frac{1}{r \sin \theta} \frac{\partial v_3}{\partial \omega}$$

$$\xi = \frac{1}{r \sin \theta} \left[\frac{\partial}{\partial \theta} (v_3 \sin \theta) - \frac{\partial v_2}{\partial \omega} \right]$$

$$\eta = \frac{1}{r \sin \theta} \frac{\partial v_1}{\partial \omega} - \frac{1}{r} \frac{\partial}{\partial r} (r v_3)$$

$$\zeta = \frac{1}{r} \frac{\partial}{\partial r} (r v_2) - \frac{1}{r} \frac{\partial v_1}{\partial \theta}$$

The components of $(\mathbf{v} \cdot \nabla)\mathbf{v}$ are

$$u \frac{\partial u}{\partial r} + \frac{v}{r} \frac{\partial u}{\partial \theta} + \frac{w}{r \sin \theta} \frac{\partial u}{\partial \omega} - \frac{v^2 + w^2}{r} \qquad \text{in } r$$

$$u \frac{\partial v}{\partial r} + \frac{v}{r} \frac{\partial v}{\partial \theta} + \frac{w}{r \sin \theta} \frac{\partial v}{\partial \omega} + \frac{uv}{r} - \frac{w^2 \cot \theta}{r} \qquad \text{in } \theta$$

$$u \frac{\partial w}{\partial r} + \frac{v}{r} \frac{\partial w}{\partial \theta} + \frac{w}{r \sin \theta} \frac{\partial w}{\partial \omega} + \frac{wu}{r} + \frac{vw \cot \theta}{r} \qquad \text{in } \omega$$

The Laplacian ∇^2 is

$$\nabla^2 \equiv \frac{1}{r^2} \frac{\partial}{\partial r} \left(r^2 \frac{\partial}{\partial r} \right) + \frac{1}{r^2 \sin \theta} \frac{\partial}{\partial \theta} \left(\sin \theta \frac{\partial}{\partial \theta} \right) + \frac{1}{r^2 \sin^2 \theta} \frac{\partial^2}{\partial \omega^2}$$

The equation for the velocity potential ϕ defined as

$$u = \frac{\partial \phi}{\partial r}, \quad v = \frac{1}{r} \frac{\partial \phi}{\partial \theta}, \quad w = \frac{1}{r \sin \theta} \frac{\partial \phi}{\partial \omega}$$

can be written down by using Eq. 9-7.

The components of the rate of strain are

$$\tfrac{1}{2} e_{rr} = \frac{\partial u}{\partial r}, \quad \tfrac{1}{2} e_{\theta\theta} = \frac{1}{r} \frac{\partial v}{\partial \theta} + \frac{u}{r}, \quad \tfrac{1}{2} e_{\omega\omega} = \frac{1}{r \sin \theta} \frac{\partial w}{\partial \omega} + \frac{u}{r} + \frac{v \cot \theta}{r}$$

$$e_{\theta\omega} = \frac{\sin \theta}{r} \frac{\partial}{\partial \theta} \left(\frac{w}{\sin \theta} \right) + \frac{1}{r \sin \theta} \frac{\partial v}{\partial \omega}$$

$$e_{\omega r} = \frac{1}{r \sin \theta} \frac{\partial u}{\partial \omega} + r \frac{\partial}{\partial r} \left(\frac{w}{r} \right)$$

$$e_{r\theta} = r \frac{\partial}{\partial r} \left(\frac{v}{r} \right) + \frac{1}{r} \frac{\partial u}{\partial \theta}$$

The components of $\nabla \cdot \pi$ are

$$\frac{\partial \pi_{rr}}{\partial r} + \frac{1}{r}\frac{\partial \pi_{r\theta}}{\partial \theta} + \frac{1}{r \sin \theta}\frac{\partial \pi_{r\omega}}{\partial \omega} + \frac{1}{r}(2\pi_{rr} - \pi_{\omega\omega} - \pi_{\theta\theta} + \pi_{r\theta} \cot \theta) \qquad \text{in } r$$

$$\frac{\partial \pi_{r\theta}}{\partial r} + \frac{1}{r}\frac{\partial \pi_{\theta\theta}}{\partial \theta} + \frac{1}{r \sin \theta}\frac{\partial \pi_{\theta\omega}}{\partial \omega} + \frac{1}{r}[(\pi_{\theta\theta} - \pi_{\omega\omega}) \cot \theta + 3\pi_{r\theta}] \qquad \text{in } \theta$$

$$\frac{\partial \pi_{r\omega}}{\partial r} + \frac{1}{r}\frac{\partial \pi_{\theta\omega}}{\partial \theta} + \frac{1}{r \sin \theta}\frac{\partial \pi_{\omega\omega}}{\partial \omega} + \frac{1}{r}(3\pi_{r\omega} + 2\pi_{\theta\omega} \cot \theta) \qquad \text{in } \omega$$

Intrinsic coordinates for steady axially symmetric flow. Let s be the length along the meridional streamlines and n the length normal to it in the meridian plane; r is the distance from the axis of symmetry. The velocity vector $\mathbf{v}_m$ in the meridian plane is specified by two quantities, v_m the component of velocity in the meridian plane and θ, the inclination of $\mathbf{v}_m$ with respect to the axis of symmetry. The velocity component perpendicular to the meridian plane is v_n. Consider s, n, and circumferential direction ω as α, β, γ coordinate directions, and note that due to symmetry, there are only variations with respect to s and n. Then

$$\nabla \cdot \rho \mathbf{v} = \frac{\partial v_m \rho}{\partial s} + \frac{\rho v_m}{r} \sin \theta + \rho v_m \frac{\partial \theta}{\partial n}$$

$$\xi = \frac{1}{r}\frac{\partial (r v_n)}{\partial n}, \qquad \eta = -\frac{1}{r}\frac{\partial (r v_n)}{\partial s}, \qquad \zeta = -\frac{\partial v_m}{\partial n} + v_m \frac{\partial \theta}{\partial s}$$

The components of the rate of strain are:

$$\tfrac{1}{2}e_{ss} = \frac{\partial v_m}{\partial s}, \qquad \tfrac{1}{2}e_{nn} = v_m \frac{\partial \theta}{\partial n}, \qquad \tfrac{1}{2}e_{\omega\omega} = \frac{v_m}{r} \sin \theta$$

$$e_{n\omega} = r \frac{\partial}{\partial n}\left(\frac{v_n}{r}\right), \qquad e_{\omega s} = r \frac{\partial}{\partial s}\left(\frac{v_n}{r}\right), \qquad e_{ns} = \frac{\partial v_m}{\partial n} + v_m \frac{\partial \theta}{\partial s}$$

The dynamic equations are

$$\rho v_m \frac{\partial v_m}{\partial s} - \rho \frac{v_n^2}{r} \sin \theta = \frac{\partial \pi_{ss}}{\partial s} + \frac{\partial \pi_{sn}}{\partial n} + \pi_{ss}\left(\frac{1}{r} \sin \theta + \frac{\partial \theta}{\partial n}\right)$$
$$+ \pi_{sn}\left(\frac{1}{r} \cos \theta - 2\frac{\partial \theta}{\partial s}\right) - \pi_{nn}\frac{\partial \theta}{\partial n} - \pi_{\omega\omega}\frac{1}{r} \sin \theta + \rho X_s$$

$$\rho v_m^2 \frac{\partial \theta}{\partial s} - \rho \frac{v_n^2}{r} \cos \theta = \frac{\partial \pi_{sn}}{\partial s} + \frac{\partial \pi_{nn}}{\partial n} + \pi_{sn}\left(\frac{1}{r} \sin \theta + 2\frac{\partial \theta}{\partial n}\right)$$
$$+ \pi_{nn}\frac{1}{r} \cos \theta - \pi_{\omega\omega}\frac{1}{r} \cos \theta + \rho X_n$$

$$\frac{\rho v_m}{r}\frac{\partial (r v_n)}{\partial s} = \frac{\partial \pi_{\omega s}}{\partial s} + \frac{\partial \pi_{n\omega}}{\partial n} + \pi_{\omega s}\left(2\frac{1}{r} \sin \theta + \frac{\partial \theta}{\partial n}\right)$$
$$+ \pi_{n\omega}\left(2\frac{1}{r} \cos \theta - \frac{\partial \theta}{\partial s}\right) + \rho X_\omega$$

Intrinsic coordinates for steady two-dimensional flow. Let s be the length along the streamlines, and n the length normal to it. The velocity vector $\mathbf{v}$ is specified by its magnitude v and the inclination θ of $\mathbf{v}$ with respect to the x axis. Then

$$\nabla \cdot \rho \mathbf{v} = \frac{\partial \rho v}{\partial s} + \rho v \frac{\partial \theta}{\partial n}$$

$$\zeta = -\frac{\partial v}{\partial n} + v \frac{\partial \theta}{\partial s}$$

The components of the rate of strain are

$$\tfrac{1}{2} e_{ss} = \frac{\partial v}{\partial s}, \quad \tfrac{1}{2} e_{nn} = v \frac{\partial \theta}{\partial n}$$

$$e_{ns} = \frac{\partial v}{\partial n} + v \frac{\partial \theta}{\partial s}$$

The dynamic equations are

$$\rho v \frac{\partial v}{\partial s} = \frac{\partial \pi_{ss}}{\partial s} + \frac{\partial \pi_{sn}}{\partial n} + \pi_{ss} \frac{\partial \theta}{\partial n} - 2\pi_{sn} \frac{\partial \theta}{\partial s} - \pi_{nn} \frac{\partial \theta}{\partial n} + \rho X_s$$

$$\rho v^2 \frac{\partial \theta}{\partial n} = \frac{\partial \pi_{sn}}{\partial s} + \frac{\partial \pi_{nn}}{\partial n} + 2\pi_{sn} \frac{\partial \theta}{\partial n} + \rho X_n$$

Cartesian coordinates. Let the three Cartesian coordinates be x, y, z and the corresponding velocity components, u, v, w. Then the equation of continuity is

$$\frac{\partial \rho}{\partial t} + \frac{\partial (\rho u)}{\partial x} + \frac{\partial (\rho v)}{\partial y} + \frac{\partial (\rho w)}{\partial z} = 0$$

The components of vorticity are

$$\xi = \frac{\partial w}{\partial y} - \frac{\partial v}{\partial z}, \quad \eta = \frac{\partial u}{\partial z} - \frac{\partial w}{\partial x}, \quad \zeta = \frac{\partial v}{\partial x} - \frac{\partial u}{\partial y}$$

The viscous stresses are

$$\tau_{xx} = 2\mu \frac{\partial u}{\partial x} + (\mu' - \tfrac{2}{3}\mu)\left(\frac{\partial u}{\partial x} + \frac{\partial v}{\partial y} + \frac{\partial w}{\partial z}\right)$$

$$\tau_{yy} = 2\mu \frac{\partial v}{\partial y} + (\mu' - \tfrac{2}{3}\mu)\left(\frac{\partial u}{\partial x} + \frac{\partial v}{\partial y} + \frac{\partial w}{\partial z}\right)$$

$$\tau_{zz} = 2\mu \frac{\partial w}{\partial z} + (\mu' - \tfrac{2}{3}\mu)\left(\frac{\partial \mu}{\partial x} + \frac{\partial v}{\partial y} + \frac{\partial w}{\partial z}\right)$$

$$\tau_{xy} = \tau_{yx} = \mu\left(\frac{\partial v}{\partial x} + \frac{\partial u}{\partial y}\right)$$

$$\tau_{yz} = \tau_{zy} = \mu\left(\frac{\partial w}{\partial y} + \frac{\partial v}{\partial z}\right)$$

$$\tau_{zx} = \tau_{xz} = \mu\left(\frac{\partial u}{\partial z} + \frac{\partial w}{\partial x}\right)$$

If X, Y, Z are the components of the body force, the dynamic equations are

$$\rho \frac{\partial u}{\partial t} + \rho u \frac{\partial u}{\partial x} + \rho v \frac{\partial u}{\partial y} + \rho w \frac{\partial u}{\partial z} = -\frac{\partial p}{\partial x} + \rho X + \frac{\partial \tau_{xx}}{\partial x} + \frac{\partial \tau_{xy}}{\partial y} + \frac{\partial \tau_{xz}}{\partial z}$$

$$\rho \frac{\partial v}{\partial t} + \rho u \frac{\partial v}{\partial x} + \rho v \frac{\partial v}{\partial y} + \rho w \frac{\partial v}{\partial z} = -\frac{\partial p}{\partial y} + \rho Y + \frac{\partial \tau_{yx}}{\partial x} + \frac{\partial \tau_{yy}}{\partial y} + \frac{\partial \tau_{yz}}{\partial z}$$

$$\rho \frac{\partial w}{\partial t} + \rho u \frac{\partial w}{\partial x} + \rho v \frac{\partial w}{\partial y} + \rho w \frac{\partial w}{\partial z} = -\frac{\partial p}{\partial z} + \rho Z + \frac{\partial \tau_{zx}}{\partial x} + \frac{\partial \tau_{zy}}{\partial y} + \frac{\partial \tau_{zz}}{\partial z}$$

The dissipation function Φ is now

$$\Phi = \mu \left\{ 2\left[\left(\frac{\partial u}{\partial x}\right)^2 + \left(\frac{\partial v}{\partial y}\right)^2 + \left(\frac{\partial w}{\partial z}\right)^2\right] + \left(\frac{\partial w}{\partial y} + \frac{\partial v}{\partial z}\right)^2 + \left(\frac{\partial u}{\partial z} + \frac{\partial w}{\partial x}\right)^2 + \left(\frac{\partial v}{\partial x} + \frac{\partial u}{\partial y}\right)^2 \right\} + (\mu' - \tfrac{2}{3}\mu)\left(\frac{\partial u}{\partial x} + \frac{\partial v}{\partial y} + \frac{\partial w}{\partial z}\right)^2$$

The energy equation is

$$\frac{D}{Dt}(h) = \left(\frac{\partial}{\partial t} + u\frac{\partial}{\partial x} + v\frac{\partial}{\partial y} + w\frac{\partial}{\partial z}\right)h$$
$$= Q + \frac{1}{\rho}\left[\frac{\partial}{\partial x}\left(k\frac{\partial T}{\partial x}\right) + \frac{\partial}{\partial y}\left(k\frac{\partial T}{\partial y}\right) + \frac{\partial}{\partial z}\left(k\frac{\partial T}{\partial z}\right)\right] + \frac{1}{\rho}\left[\frac{\partial p}{\partial t} + u\frac{\partial p}{\partial x} + v\frac{\partial p}{\partial y} + w\frac{\partial p}{\partial z}\right] + \frac{\Phi}{\rho}$$

A,16. Cited References.

1. Stewart, H. J. *Proc. Natl. Acad. Sci. 28*, 161 (1942).
2. Irving, J. H., and Kirkwood, J. G. *J. Chem. Phys. 18*, 817 (1950).
3. Chapman, S., and Cowling, T. G. *The Mathematical Theory of Non-Uniform Gases*. Cambridge Univ. Press, 1939.
4. Grad, H. *Commun. on Pure and Appl. Math. 2*, 331 (1949).
5. Truesdell, C. *J. math. pures et appl. 29*, 215–244 (1950); *30*,111–158(1951).
6. Kantrowitz, A. *NACA Wartime Rept. L457*, 1944; *J. Chem. Phys. 14*, 150 (1946).
7. Goldstein, S. *Modern Developments in Fluid Mechanics*. Oxford Univ. Press, 1938.
8. Hicks, B. L. *Quart. Appl. Math. 6*, 221, 405 (1948).
9. Crocco, L. *Z. angew. Math. u. Mech. 17*, 1 (1937).
10. Bjerknes, V. *Z. Meteorol. 17*, 97 (1900).
11. Bateman, H. *Proc. Roy. Soc. London A125*, 598 (1939).
12. Wang, C. T. *J. Aeronaut. Sci. 15*, 675 (1948); *17*, 343 (1950).
13. Munk, M., and Prim, R. C. *Proc. Natl. Acad. Sci. 33*, 137 (1947).
14. Hicks, B. L., Guenther, P. E., and Wasserman, R. H. *Quart. Appl. Math. 5*, 357 (1947).
15. Oswatitsch, K. *Nachr. Akad. Wiss. Göttingen Math. -physik. Kl.*, 88 (1945).
16. Theodorsen, T. *J. Aeronaut. Sci. 4*, 239 (1937).
17. Tollmien, W. *Luftfahrtforschung 19*, 145 (1942).
18. Crocco, L. *Aerotecnica 29*, 347 (1949).
19. von Kármán, Th. *J. Aeronaut. Sci. 8*, 345 (1941).
20. Busemann, A. *Handbuch der Experimentalphysik*, Vol. IV, Part 1. Teil, Akad. Verlagsgesellschaft, Leipzig, 1931.
21. Tsien, H. S. *J. Math. and Phys. 25*, 301 (1946); *26*, 76 (1947).
22. Eggers, A. J. *NACA Rept. 959*, 1950.

SECTION B

ONE-DIMENSIONAL TREATMENT OF STEADY GAS DYNAMICS

L. CROCCO

B,1. Meaning of the One-Dimensional Approach. The one-dimensional approximation represents the oldest and simplest approximate solution of the equations of gas dynamics. It can be applied to the study of the flow in ducts of small divergence and curvature. The consequence of the small divergence is that a flow configuration can exist where the velocities at every point of each section are nearly parallel. In this case if one defines an axis of the section, the velocity components normal to this axis are small compared to the axial components, and so are the transversal accelerations. If in addition the width of the duct is small compared with the radius of curvature of the axial line, the transversal gradients of pressure can be neglected, so that the pressure can be assumed to be a constant on each section and therefore a function only of the axial coordinate. Also, the velocities need not be distinguished from their axial components.

The assumption of small obliquity of the velocities is fundamental in the one-dimensional treatment. It is common to another approximate method, the boundary layer approximation (Vol. IV), which considers the case of flows only partially bounded by solid walls. The fact that in ducts the flow is completely delimited by solid boundaries, so that the amount of fluid participating in the processes is well defined, represents the great advantage of the one-dimensional treatment. In fact, it makes it possible to take into account effects due to viscosity, heat conductivity, diffusion, and so on, in a much simpler fashion than in any other type of treatment, through the additional assumption of uniformity. More specifically one assumes that not only the relative obliquity of the velocities is small, but also that the magnitude is constant on each cross section. Similarly, for small curvatures, not only the pressure at each point of the section is the same, but the temperature, the concentration of different chemical species, and any other parameter related to the physical or chemical conditions are also distributed uniformly. The uniformity assumption allows the mechanisms of momentum, energy, and mass transfer to be ignored, because all frictional effects on the wall or exchanges of heat and matter through the wall are supposed to be immediately distributed evenly throughout the limited amount of mass flowing in the duct.

One realizes immediately that while the obliquity assumption can be very closely satisfied in practical cases, although exceptions do exist, the uniformity assumption never represents the actual physical conditions, even for small or zero curvature. The velocity is never uniform in the duct because it must always vanish at the boundaries of the duct. Similarly, heat or mass exchanges can only be present if a transversal gradient of temperature or concentration exists at the walls. However, it is intuitively felt that if one defines proper mean values of the velocity, temperature, and other physical parameters, one can approximately replace the actual nonuniform flow by an equivalent uniform flow characterized by values of the velocity, temperature, and other physical parameters equal to these mean values. The validity and the limitations of this substitution are discussed in Art. 16, where it is shown that nonuniformities of pressure can also be taken into account in the same fashion. The important consequence of the uniformity assumption is that changes of the physical and chemical quantities take place only in the axial direction, so that each quantity is a function of a single spatial coordinate defined along the axial line. This is the reason for the name "one-dimensional" used to characterize this type of flow, in analogy to the denomination of two-dimensional and three-dimensional given to plane or spatial flows.

The uniformity assumption can be widened so as to include in the one-dimensional treatment more general types of flow, by prescribing only a piecewise uniformity. If the flow on each section can be divided into two or more portions and each portion can be replaced by a uniform flow region, the different quantities characterizing each portion are functions only of the axial coordinate and therefore the one-dimensional treatment applies. Examples of this type of flow are given in Art. 15. It is also possible to obtain interesting general information on the flow in ducts without the uniformity assumption (Art. 17).

As already mentioned, the obliquity assumption can correspond fairly well to the real conditions. However, there are types of flow, interesting for their practical importance, in which large obliquities of the velocities and large transversal accelerations are certainly present. Ducts with sharp or sudden variations of the section, or with strong curvatures, are characteristic examples. Another important case is represented by the mixing of a supersonic jet with a secondary flow when the supersonic nozzle is not correctly expanded, and large transversal velocities and nonuniformities of pressure are created. In addition to these cases where the failure of the obliquity assumption is an evident result of the geometry of the duct, there are cases in which the occurrence of large transversal velocities is not so evident. Such is the case when in a "well-behaved" duct, for more subtle fluid dynamical reasons, "separation" occurs or a highly turbulent condition is generated. In both cases the local direction of the velocities can deviate considerably from the direction of the walls.

Where the obliquity assumption fails, the application of one-dimensional methods is subject to caution. However, it is still possible in many instances to obtain interesting information, especially if one concentrates on the over-all effects between two sections where the obliquity assumption is more or less satisfied, without asking details of the intermediate stages where the obliquities are large.

An important advantage of the simplifications inherent in the one-dimensional treatment is that the shape of the section of the duct and its variations can be quite arbitrary. Simply connected and multiconnected sections can be considered. Among the last ones particularly interesting for jet propulsion is the case of toroidal ducts with annular sections. The interpretation of the concepts of curvature, divergence, section, and axis of the section for such ducts is evident and does not need detailed explanation. In these axially symmetric ducts the velocities can either be contained in the meridional plane, that is, practically perpendicular to the section, or an important tangential component can exist, in which case the obliquity assumption needs to be satisfied for the meridional component alone. The tangential component may only be produced at the entrance section of the duct and thereupon follow a natural development; or it can be determined by the presence of blades or vanes subdividing the toroidal duct into a number of elementary ducts symmetrically distributed around the axis. Finally a bladed duct can be set in rotation around the axis of symmetry. Ducts of these types are of fundamental importance in compressors and turbines, and the one-dimensional method can be extended to cover these cases. However, the corresponding treatment is not developed in this section, where we consider only stationary ducts without tangential flow. Applications of one-dimensional methods (sometimes indicated as two-dimensional because of the presence of the tangential flow) to the theory of compressors and turbines can be found in Vol. X. In this section we only consider the case in which the flow conditions at each section are the same at each instant; that is the flow is steady, and the various parameters are only functions of the spatial variable. The nonsteady case in which the parameters depend on the time is treated in Sec. C. Observe, however, that the steadiness or nonsteadiness applies to the conditions averaged on each section and does not include the possible effects of turbulence.

B,2. Thermodynamics of the Working Substance.

Equation of state. The most common substances considered in flow problems are homogeneous fluids; in particular, in the case of gas dynamics, gases. However, in some practical cases the heterogeneous flow of two-phase mixtures of a gas and a liquid, or a gas and a solid, have to be considered (Art. 13).

The gases are in general supposed to obey the perfect gas law

$$p = \rho \mathcal{R} T \tag{2-1}$$

Here p, ρ, T are the pressure, density, and absolute temperature, and $\mathcal{R}$ is the specific gas constant, obtained from the universal gas constant R (I,A,3) by dividing by the mean molecular weight of the gas.

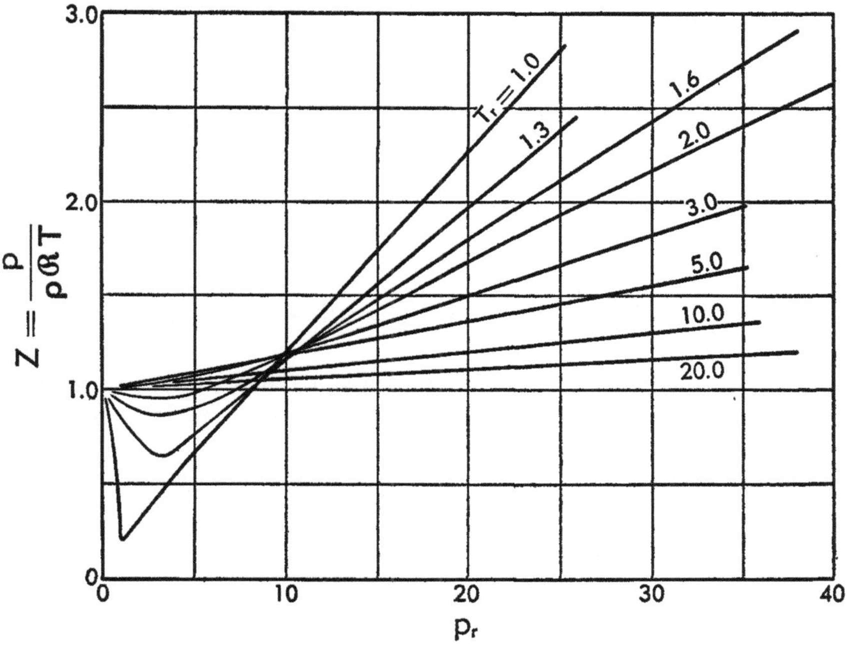

Fig. B,2a. Compressibility factor versus reduced pressures $p_r = p/(p_{cr} + 8 \text{ atm})$ for a series of reduced temperatures $T_r = T/(T_{cr} + 8°\text{C})$.

Actually for any real gas this equation is only an approximation to the correct equation of state, expressed by a relation of the type:

$$p = p(\rho, T); \qquad \rho = \rho(p, T) \tag{2-2}$$

the deviations becoming especially important at very high pressures and at low temperatures. A check on the accuracy of Eq. 2-1 for the given nature and state of the gas can be obtained from an examination of the departure from unity of the compressibility factor $Z = p/\rho \mathcal{R} T$. General plots of Z which can be applied approximately to most of the practical gases of interest in flow systems are currently used by chemical engineers (see I,C,2, Fig. C,2c and C,2d; also [*1*, pp. 159–164]). In these plots Z is given as a function of a reduced pressure p_r and a reduced temperature T_r, the reduced values being obtained by dividing the pressure and the temperature by their critical values (possibly empirically corrected for best coincidence; see [*1*, pp. 159–164]). The accuracy of these plots is claimed to be good for most of the gases of practical interest. Fig. B,2a reproduces in small scale and without details the plots of [*1*, pp. 159–164]. It shows that if p_r and T_r are around unity, very large deviations from the perfect gas law can be present, Z becoming smaller than unity, substantially because of the effect of the cohesive forces between molecules. When T_r is around unity and $p_r \gg 1$, important deviations in the other direction can appear, due essentially to the effect of the covolume of the molecules. For gases in these conditions it would be clearly wrong to use Eq. 2-1; many special forms of Eq. 2-2 have been suggested, one of the most completely studied being the Beattie-Bridgeman equation of state, which can be written in the form [*2*]

$$p = \rho \mathcal{R} T(1 + e_1\rho + e_2\rho^2 + e_3\rho^3) \tag{2-3}$$

where the coefficients in parentheses are the following functions of the temperature

$$e_1 = B_0 - \frac{A_0}{\mathcal{R}T} - \frac{c}{T^3}; \quad e_2 = \frac{aA_0}{\mathcal{R}T} - bB_0 - \frac{cB_0}{T^3}; \quad e_3 = \frac{bcB_0}{T^3}$$

and A_0, B_0, a, b, c are constants for each gas. These constants coincide with those of I,C,6 if ρ is defined as the inverse of the molal volume (instead of the specific volume) and $\mathcal{R}$ is replaced by the universal molal constant R. If, on the contrary, the density is defined in the ordinary way the values of a, b, c, and B_0 are obtained from those of I,C,6, dividing by the molecular weight and that of A_0 by its square.

Fortunately, in most of the flow systems of interest in aerodynamics and jet propulsion, high pressures are generally associated with very high temperatures, and low temperatures with very low pressures. In both cases Z tends to stay close to unity and the perfect gas law can be used without significant errors. Only in the exceptional cases when the above conditions are not satisfied must the correct equation of state be considered (see Art. 13).

Caution must be used, however, when the temperature is raised above the level where chemical dissociations become appreciable. In this case the composition of the gas becomes variable, and can be calculated as a function of pressure and temperature if one assumes that thermochemical equilibrium is achieved in every condition (II,A). As a result of these calculations it is possible to find numerically the relation between the state quantities p, ρ, and T; but this correlation cannot be approximated by the perfect gas equation (Eq. 2-1), despite the fact that each individual component of the mixture obeys the perfect gas law. It is interesting to notice that for this case the temperature at which the perfect gas law ceases being applicable increases with the pressure, because of the adverse effects of pressure on dissociations.

With the exclusion of the exceptions that have been mentioned, the perfect gas law can be accepted as sufficiently correct for practical purposes. It has the great advantage of simplifying radically the formulation of flow problems, and because of this simplification it is sometimes applied even beyond its range of applicability.

Thermodynamic relations for perfect gases. We shall consider here the specific values of the quantities, that is, we refer to the unit weight of substance, and, except when explicitly mentioned, we shall consider only perfect gases. The specific internal energy e and the specific enthalpy h are then, for fixed composition, functions of the temperature alone (see I,A,6 and 7), related by the relation

$$h(T) = e(T) + \frac{p}{\rho} = e(T) + \mathcal{R}T$$

and so are their derivatives, the specific heats at constant pressure and volume, connected by the relation

$$c_p(T) = c_v(T) + \mathcal{R} \tag{2-4}$$

The ratio

$$\frac{c_p}{c_v} = \gamma(T) \tag{2-5}$$

is called the adiabatic index. From Eq. 2-4 and 2-5 we obtain

$$c_p = \frac{\gamma\mathcal{R}}{\gamma - 1}; \qquad c_v = \frac{\mathcal{R}}{\gamma - 1} \tag{2-6}$$

The internal energy and the enthalpy play an important role in thermodynamics. They appear in the first law of thermodynamics (see I,A,6) which for the unit weight of substance can be written in a differential form as

$$\delta q = de + pd\left(\frac{1}{\rho}\right) = dh - \frac{1}{\rho}\,dp \tag{2-7}$$

where δq represents the element of heat supplied from outside.

Observe that the first law of thermodynamics expresses the conservation of energy for a substance at rest, or, more generally, in a frame of reference which moves with the substance, with only pressure work present. A more general expression of the law of conservation of energy for a flowing fluid is given in Art. 3. An important difference between the two expressions resides in the presence of the kinetic energy of the fluid. If the local velocity of the fluid is u, its kinetic energy per unit mass is $u^2/2$. A useful concept is obtained when one assimilates this kinetic energy to an increase of internal energy introducing the total internal energy, defined as

$$e_t = e + \frac{u^2}{2}$$

and the stagnation enthalpy, defined as

$$h^0 = h + \frac{u^2}{2} \tag{2-8}$$

The temperature T^0 corresponding to $h^0 = h(T^0)$ (different from that corresponding to e_t) is called the stagnation temperature. By opposition e, h, and T are generally denominated "static."

If a reference temperature T_r is chosen one can write

$$e = e_r + \int_{T_r}^{T} c_v dT; \quad h = h_r + \int_{T_r}^{T} c_p dT \tag{2-9}$$

where e_r and h_r represent the values of e and h at T_r, related by $h_r = e_r + \mathcal{R}T_r$. Although in principle the absolute values of e_r and h_r can be experimentally obtained (I,A,5), in practice this cannot be done with the necessary accuracy. Moreover the knowledge of the absolute values is not essential in thermodynamics, where only the variations of e and h for changing conditions are required. If the change of conditions is purely physical, the value of e_r, or h_r, does not change during the transformation, and may as well be taken zero. In this case e and h coincide with the so-called sensible internal energy and enthalpy.

If, however, the transformation involves a change in the chemical nature of the gas (for instance, combustion), it is necessary to determine the values of e_r and h_r with respect to a reference chemical system. In general this chemical system is represented by the chemical elements in a fixed reference state. In this case e_r and h_r include the potential energy due to chemical affinity, and represent the heat of formation from the elements at constant volume or pressure. It is general use, however, to reserve the denomination "heat of formation" to h_r without specifying that it applies to the constant pressure case. By definition the heat of formation of the elements in the reference state is zero.

The entropy of the unit weight of substance satisfies the relation (see I,A,13)

$$Tds = dh - \frac{1}{\rho} dp \tag{2-10}$$

It is important to observe that this equation applies only to substances with fixed composition (in frozen conditions) or in thermochemical equilibrium. For instance it applies equally well to a combustible mixture, which, though being in metastable equilibrium, has a fixed composition, and to the burned gases obtained after combustion, if the transformations are sufficiently slow so that the gases are practically in equilibrium conditions, or sufficiently fast so that the composition does not change appreciably and can be considered frozen. However, Eq. 2-10 cannot be applied to the substances developed during the intermediate stages of combustion, where the processes are controlled by chemical kinetics, and the time enters as an essential parameter, independently of the state parameters.

For perfect gases, Eq. 2-10, integrated with the help of Eq. 2-1, gives

$$s = s_r + \int_{T_r}^{T} c_p d(ln\ T) - \mathcal{R} \ln p \tag{2-11}$$

Here s_r is the value of the entropy at the reference conditions $T = T_r$ and $p = p_r = 1$. Again we are not concerned with the absolute values, but with the variations of entropy, so that if s_r is constant (purely physical transformation) it may be taken zero. When chemical transformations are present, however, the value of s_r varies, and therefore it must be referred to a fixed chemical system. Again this reference chemical system may be provided by the chemical elements in the reference state, in which case s_r represents the entropy variation when the working substance is produced reversibly, starting from the elements in the reference conditions.

Introducing the quantities

$$\sigma(T) = \int_{T_r}^{T} c_p d(\ln T); \quad \varphi(T) = e^{\sigma/\mathcal{R}} \tag{2-12}$$

Eq. 2-11 can be rewritten as

$$s - s_r = \sigma(T) - \mathcal{R} \ln p = \mathcal{R} \ln \frac{\varphi(T)}{p} \tag{2-13}$$

For a given perfect gas, and for a prescribed T_r, σ and φ are functions only of T. Observe that a change in T_r merely changes s_r and σ by a constant amount, and φ by a constant factor. The relations between p and T in the isentropic transformation of a perfect gas of fixed composition can be obtained from Eq. 2-13 with s = const, and is expressed by $p = \varphi(T) \times$ const. If the initial and final temperatures T_i and T_f are prescribed, one has, with the aid of Eq. 2-1,

$$\frac{p_f}{p_i} = \frac{\varphi(T_f)}{\varphi(T_i)}; \quad \frac{\rho_f}{\rho_i} = \frac{\varphi(T_f)}{T_f} \frac{T_i}{\varphi(T_i)} \tag{2-14}$$

The corresponding change in enthalpy is given, from Eq. 2-9, by

$$h_f - h_i = \int_{T_i}^{T_f} c_p dT \tag{2-15}$$

An important quantity in the flow of gases is the velocity of sound, defined as (Sec. A)

$$a^2 = \left(\frac{dp}{d\rho}\right)_{s=\text{const}} \tag{2-16}$$

The explicit expression for perfect gases is obtained by writing Eq. 2-10, with the help of Eq. 2-1, in the form

$$ds = c_v d(\ln p) - c_p d(\ln \rho)$$

from which we obtain for $ds = 0$

$$a^2 = \gamma \frac{p}{\rho} = \gamma \mathcal{R} T \tag{2-17}$$

Here γ must be evaluated at the temperature T.

Thermodynamic tables for many of the above functions can be found in several publications. See, for instance, [*3*].

Combustion. When a chemical transformation, such as combustion, takes place, h_r changes. If we indicate with the subscripts u and b the substance before (unburned) and after the transformation (burned), and with the subscripts i and f the states before and after the transformation, we can write

$$(h_b)_f - (h_u)_i = (h_r)_b - (h_r)_u + \int_{T_r}^{T_f} c_{p_b} dT - \int_{T_r}^{T_i} c_{p_u} dT \tag{2-18}$$

In a constant pressure, adiabatic transformation ($\delta q = 0$), h stays constant because of the first law of thermodynamics (Eq. 2-7), and the two members of Eq. 2-18 vanish. The quantity

$$q_{th} = \int_{T_i}^{T_f} c_{p_b} dT = (h_r)_u - (h_r)_b + \int_{T_r}^{T_i} (c_{p_u} - c_{p_b}) dT \tag{2-19}$$

is the thermal effect of the combustion (at constant pressure). In general $c_{p_u} \neq c_{p_b}$ so that q_{th} varies with T_i. However, if T_i does not change too much, the variations of q_{th} are not too large and are often neglected. In this case one can take for q_{th} its value at the reference temperature, equal to the difference between the heats of formation of the unburned and of the burned substance.

In the absence of appreciable dissociations in the products, the thermal effect coincides with the amount of heat that has to be extracted from a constant pressure calorimeter in order to bring the burned substance to the initial temperature T_i. If, as customary in the case of the combustion of a given fuel with oxygen or air in excess, we refer the heat developed to the unit weight of fuel, and the products of combustion are all considered to be in gaseous form, the result represents the upper heating value q_H of the fuel or, simply, the heating value. The minimum amount of oxygen or air required for complete combustion of the unit weight of fuel is called the stoichiometric ratio μ_{st}. When the air-fuel ratio μ is larger than μ_{st}, that is the oxygen is in sufficient excess, the fuel is able to develop its heating value almost completely. Therefore the thermal effect is

$$q_{th} = \frac{q_H}{1 + \mu} \qquad (\mu > \mu_{st}) \tag{2-20}$$

If, on the contrary, the oxygen is in defect one can approximately say that only a fraction μ/μ_{st} of the fuel present can develop its full heating value. Thus the thermal effect is

$$q_{th} = \frac{\mu}{\mu_{st}} \frac{q_H}{1 + \mu} \qquad (\mu < \mu_{st}) \tag{2-21}$$

For $\mu = \mu_{st}$, both equations give $q_{th} = q_H/(1 + \mu_{st})$. However, the latter relation is likely to be in larger error than Eq. 2-20 and 2-21, because the presence of dissociation in the burned gases may appreciably reduce the thermal effect of the combustion, the corresponding reduction being found as an increase of the term $(h_r)_b$ of Eq. 2-19. This is especially true if the combustion takes place with pure oxygen, and for high initial temperatures T_i. For off-stoichiometric mixtures the dissociations are reduced, and for μ sufficiently different from μ_{st}, and T_i not too high, Eq. 2-20 and 2-21 may become sufficiently accurate for most of the practical needs.

In all other cases, when appreciable dissociations are present, q_{th} and T_f can be obtained only by involved calculations of the kind indicated in II,C. Analogous results can be derived for other oxidizers than air or oxygen. When q_{th} is known, and c_{p_b} is a known function of the temperature, T_f can be obtained from Eq. 2-19, which can also be written as

$$T_f - T_i = \frac{q_{th}}{(c_{p_b})_m} \tag{2-22}$$

$(c_{p_b})_m$ representing the proper mean value of c_{p_b} for the range T_i to T_f. In the general case where c_{p_u} and c_{p_b} are different and vary with the temperature, q_{th} is a function of T_i and $(c_{p_b})_m$ of T_i and T_f. Thus the simple form of Eq. 2-22 is illusory, because the temperature increase cannot be explicitly evaluated from it. However, if the difference between c_{p_u} and c_{p_b} is neglected and c_{p_b} is supposed constant, and hence equal to $(c_{p_b})_m$, Eq. 2-22 provides an explicit expression for $T_f - T_i$. Because of its simplicity this expression is often used in flow problems. Alternately one can use the relation

$$\frac{T_f}{T_i} = 1 + \frac{q_{th}}{(c_{p_b})_m T_i} \tag{2-23}$$

for the temperature ratio.

Even in the aforementioned assumption that q_{th} and $(c_{p_b})_m$ are known constants, the temperature ratio is a decreasing function of T_i, and therefore cannot be assigned a fixed value for a given chemical transformation, but depends on the initial conditions. For instance, for hydrocarbon fuels in stoichiometric ratio with air, the temperature ratio calculated from Eq. 2-23 with $\mu = \mu_{st}$ varies roughly between 10 and 3.5 when T_i goes from 220 to 800°K.

The temperature ratio enters in many equations of one-dimensional gas dynamics. Its effects on the corresponding flow problems can be studied independently of its derivation from Eq. 2-19, which expresses the law of conservation of energy by prescribing its values or the values of the related quantity

$$\lambda = \frac{(c_{p_b})_f T_f}{(c_{p_u})_i T_i} \tag{2-24}$$

In other words instead of deriving the value of λ from the energy equation

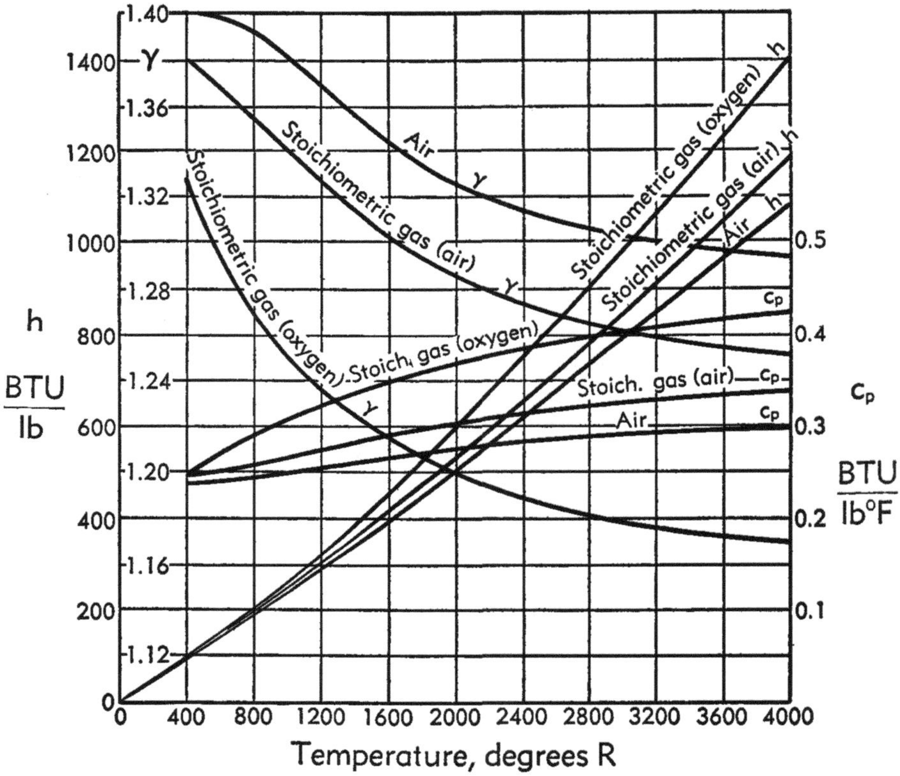

Fig. B,2b. Specific heat, adiabatic index, and sensible enthalpy for air and for stoichiometric gas of combustion of the representative fuel $(C_3H_5)_x$ with air or oxygen. The sensible enthalpy is referred to $T_r = 0°K$.

for any given system and set of conditions, one can replace the energy equation by assigning the value of λ, with the implicit assumption that the energy equation is satisfied for the given value of λ.

Specific heats and polytropic gases. Particularly simple expressions for the transformations of a gas are found when c_p is approximately assumed to be constant, in which case the gas is called polytropic. Disregarding the effects of the gas imperfections, this assumption is exactly satisfied only for monatomic gases, for which $c_p = \frac{5}{2}\mathcal{R}$ at all temperatures (with the exception of extremely high temperatures, where electronic levels may be excited). For other gases it represents only a useful approximation when the range of variation of temperature is small, and its accuracy varies with the particular temperature around which it is applied. For instance, for most of the diatomic gases the approximation of constant c_p is quite good below the ambient temperature in a range that extends down to very low temperatures (~20°K). Hydrogen and hydrides represent an exception to this property. Above ambient temperature, however, the approximation is worse, as shown by the curve of Fig. B,2b, relative to air. On the same figure another curve corresponds to the gases obtained from the combustion with air of a representative hydrocarbon fuel of gross composition $(C_3H_5)_x$ in the stoichiometric ratio $\mu_{st} = 14.722$. These combustion gases contain the triatomic gases H_2O and CO_2 for which the variation of c_p with temperature is larger than for diatomic gases; however, they still contain a majority of diatomic gases (around 72 per cent of N_2 in weight), so that the variations of c_p are contained in moderate limits. In the absence of diatomic gases the variations are much more important, as the curve of Fig. B,2b shows for the gases formed by the stoichiometric combustion of the representative fuel with pure oxygen. We see from the figures that the maximum slope of c_p takes place around 500°C. Thus one can expect that the approximation of constant c_p introduces the largest errors when the transformations are centered around this temperature, as will be verified in the following. For temperatures higher or lower, the error must decrease. However, beyond 2000°C the appearance of dissociations provides another cause for errors, which may become more important than the variation of c_p.

On Fig. B,2b are also drawn the curves of h, which should be straight lines for polytropic gases, and those of γ which, instead of remaining constant, decreases appreciably with increasing temperature.

The simplifications resulting from the approximation of constant c_p are so radical that this assumption is nearly always used in the treatment of flow problems. For purely physical transformations one obtains, instead of Eq. 2-15, 2-12, and 2-13

$$h_f - h_i = c_p(T_f - T_i) \tag{2-25}$$

$$\frac{\varphi_f}{\varphi_i} = \left(\frac{T_f}{T_i}\right)^{\frac{c_p}{\mathcal{R}}} = \left(\frac{T_f}{T_i}\right)^{\frac{\gamma}{\gamma-1}} \tag{2-26}$$

$$s_f - s_i = c_p \ln\frac{T_f}{T_i} - \mathcal{R}\ln\frac{p_f}{p_i} = c_v \ln\frac{p_f}{p_i} - c_p \ln\frac{\rho_f}{\rho_i} \tag{2-27}$$

where γ is now a constant.

One can eliminate T_f from Eq. 2-26 and 2-27 and obtain the following relation directly connecting h to σ or φ

$$\frac{h_f - h_i}{c_p T_i} = e^{\frac{\sigma_f - \sigma_i}{c_p}} - 1 = \left(\frac{\varphi_f}{\varphi_i}\right)^{\frac{\gamma-1}{\gamma}} - 1 \qquad (2\text{-}28)$$

If the transformation is isentropic the particular form of Eq. 2-14 is therefore

$$\frac{p_f}{p_i} = \left(\frac{T_f}{T_i}\right)^{\frac{\gamma}{\gamma-1}} = \left(\frac{\rho_f}{\rho_i}\right)^{\gamma} \qquad (2\text{-}29)$$

and Eq. 2-28 becomes

$$\frac{h_f - h_i}{c_p T_i} = \left(\frac{p_f}{p_i}\right)^{\frac{\gamma-1}{\gamma}} - 1 \qquad (2\text{-}30)$$

Approximate relations for variable specific heats. In general the variability of c_p cannot be taken into account in a simple analytical form, even when a relatively simple analytical expression is available for c_p [*2,4*]. Therefore the evaluation of transformations with variable c_p is generally made numerically or graphically, determining the values of e, h, σ and φ (Eq. 2-9 and 2-12) from tables or graphs, which are readily available for all simple gases. Easy methods have also been devised for the determination of the same quantities for gas mixtures of arbitrary composition [*5,6*].

For more accurate evaluations, use can be made of correction factors to be applied to the result of the calculations made as if the specific heats were constant. The advantage is that these correction factors are generally close to unity and can be plotted in a wide scale, thus allowing high-accuracy reading. Such a method will be discussed later for the case of real gases.

For expansions of air or conventional gases of combustion in the range which may be of interest in jet propulsion, one can make use of an approximate procedure [*7*] based on the fact that, if properly chosen, the exact correction factors may be approximately replaced by a unique function, which can also be expressed analytically. Assuming a transformation from a fixed initial temperature T_i to an arbitrary final temperature $T_f < T_i$, by analogy to Eq. 2-26 we can define a correction factor $\mathcal{A}$ such that

$$\frac{T_i}{T_f} = \mathcal{A}\left(T_i, \frac{\varphi_f}{\varphi_i}\right)\left(\frac{\varphi_i}{\varphi_f}\right)^{\frac{\gamma_i-1}{\gamma_i}} \qquad (2\text{-}31)$$

where γ_i represents $\gamma(T_i)$ and φ_i, φ_f stand for $\varphi(T_i)$, $\varphi(T_f)$. The correction factor $\mathcal{A}$ (1 for γ = const) is in general different from unity, and can be computed for each given gas and for various initial temperatures as a function of φ_f/φ_i. The values of $\mathfrak{A} - 1$ are shown in Fig. B,2c and B,2d for air, and for the gases obtained by combustion of stoichiometric mixtures of the representative fuel $(C_3H_5)_x$ with air. Examination of these figures shows that for the air, and for the gas of stoichiometric combustion with air, a single mean curve can approximately replace the curves obtained for various T_i.

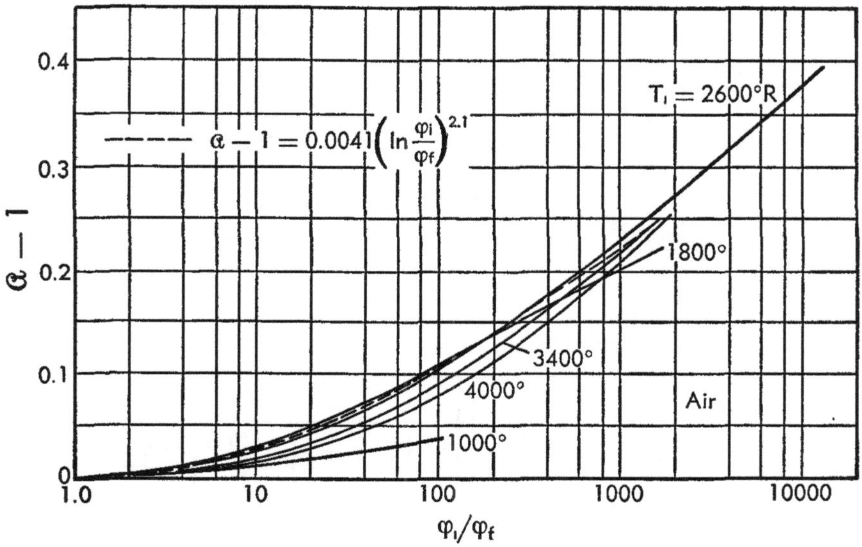

Fig. B,2c. Correction factor of Eq. 2-31 for expansions of air.

For instance, a good mean value is obtained, as the figures show, from the expression

$$\mathfrak{A}\left(\frac{\varphi_f}{\varphi_i}\right) = 1 + 0.0041\left(\ln \frac{\varphi_i}{\varphi_f}\right)^{2.1} \tag{2-32}$$

This approximate property is a consequence of the fact that for a given abscissa $\mathfrak{A}$ has a maximum around which its behavior is quite flat. This maximum, for not too large values of the abscissa, takes place around 1400–1800°R, that is around the temperature for which one obtains the maximum slope of the $c_p(T)$ curves.

The error committed using Eq. 2-32 is below a few per cent in all the range examined for the stoichiometric combustion gas, and stays in the same limits with air, provided the range of temperature variation is limited below a certain value when the initial temperature is low. The importance of these limitations can be estimated by observing that, from Eq. 2-14, φ_i/φ_f coincides with the pressure ratio of an isentropic expansion. It is clear that in the range of practical expansion ratios the approximate relation (Eq. 2-32) can be used with confidence. The fact that the approximation (Eq. 2-32) can be used both for air and for the gases of stoichio-

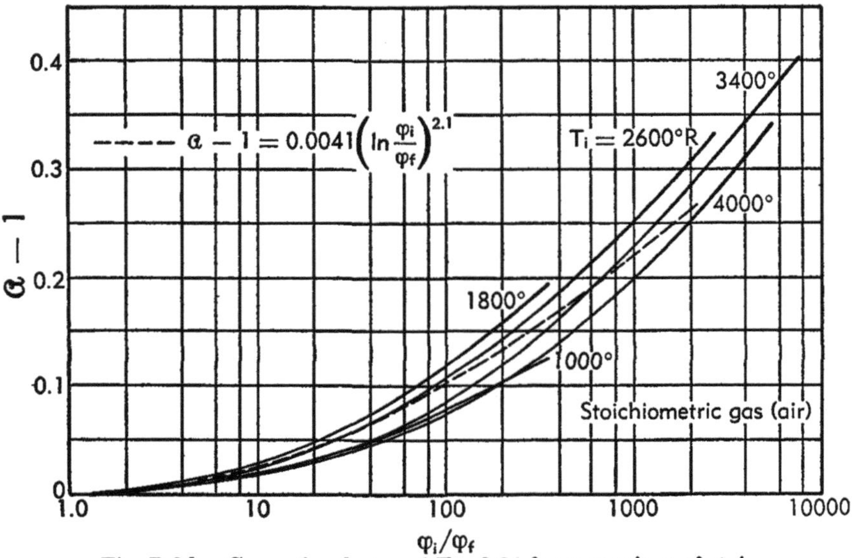

Fig. B,2d. Correction factor of Eq. 2-31 for expansions of stoichiometric gases of combustion with air of $(C_3H_5)_x$.

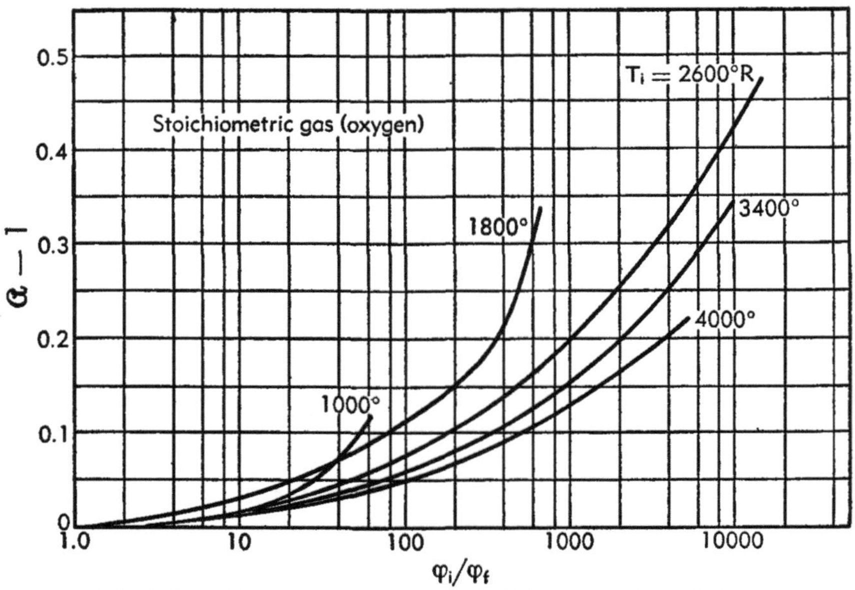

Fig. B,2e. Correction factor of Eq. 2-31 for expansions of stoichiometric gas of combustion with oxygen of $(C_3H_5)_x$.

metric combustion with air indicates that it is also a good approximation for arbitrary mixtures of the two, that is for gases of combustion with an arbitrary air excess. It shows further that the effect of the presence of triatomic combustion products does not fundamentally alter the correction factor, which is determined substantially by the large proportion of

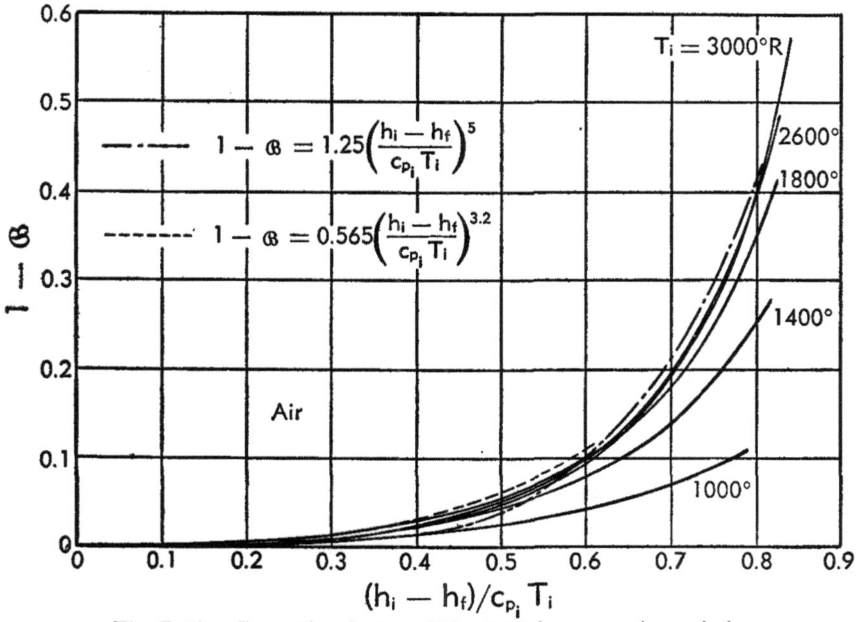

Fig. B,2f. Correction factor of Eq. 2-33 for expansions of air.

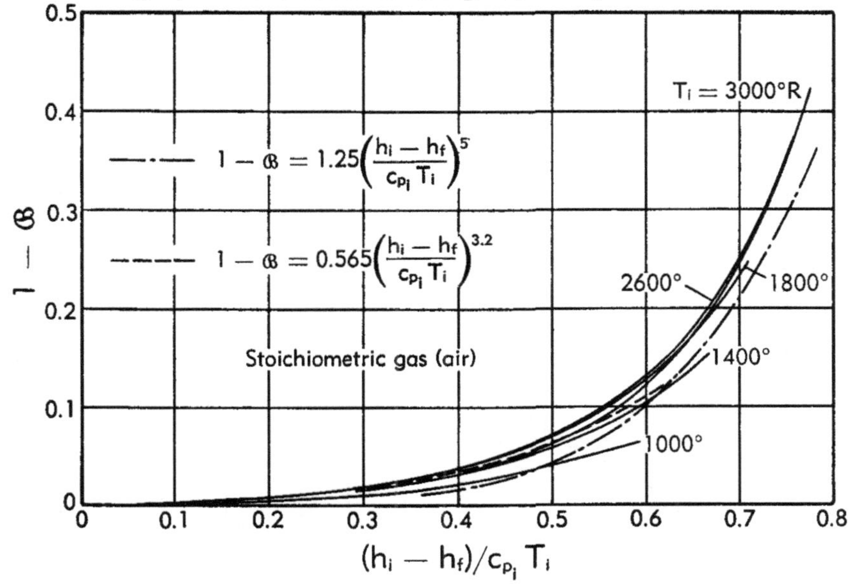

Fig. B,2g. Correction factor of Eq. 2-33 for expansions of stoichiometric gas of combustion with air of $(C_3H_5)_x$.

diatomic species present in the gases. Thus it may be expected that the same correction factor can be applied to the products of combustion of other conventional fuels.

For the gases of stoichiometric combustion of the representative fuel with pure oxygen, the effect of T_i on the correction factor is larger than

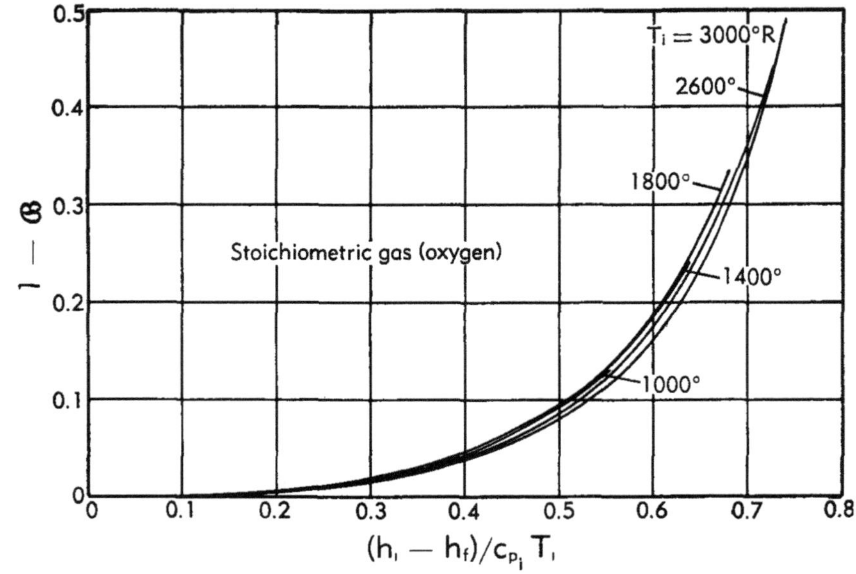

Fig. B,2h. Correction factor of Eq. 2-33 for expansions of stoichiometric gas of combustion with oxygen of $(C_3H_5)_x$.

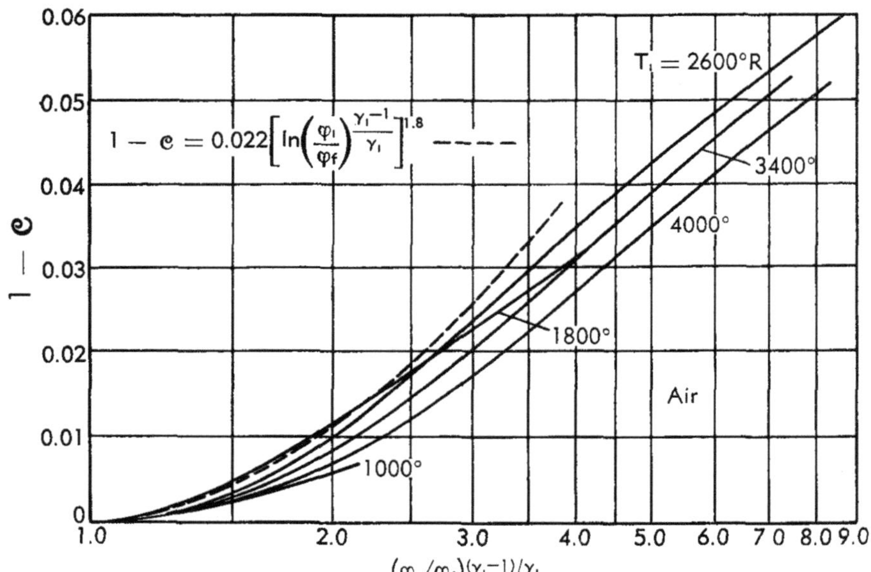

Fig. B,2i. Correction factor of Eq. 2-35 for expansions of air.

in the previous cases, as Fig. B,2e shows clearly. The deviations from the relation (Eq. 2-32) are still small for low T_i, but become large at higher initial temperatures. A good approximate expression for this case could be obtained by using in Eq. 2-32, instead of the constant coefficient 0.0041, a function of T_i.

Another correction factor allows one to relate the temperature and the enthalpy through the relation

$$\frac{T_f}{T_i} = \mathcal{B}\left(T_i, \frac{h_f - h_i}{c_{p_i}T_i}\right)\left(1 + \frac{h_f - h_i}{c_{p_i}T_i}\right) \tag{2-33}$$

which is immediately verified to be true, with $\mathcal{B} = 1$, for polytropic gases. The values of $1 - \mathcal{B}$ are shown in Fig. B,2f, B,2g, and B,2h for the same representative cases examined previously. Similar observations can be

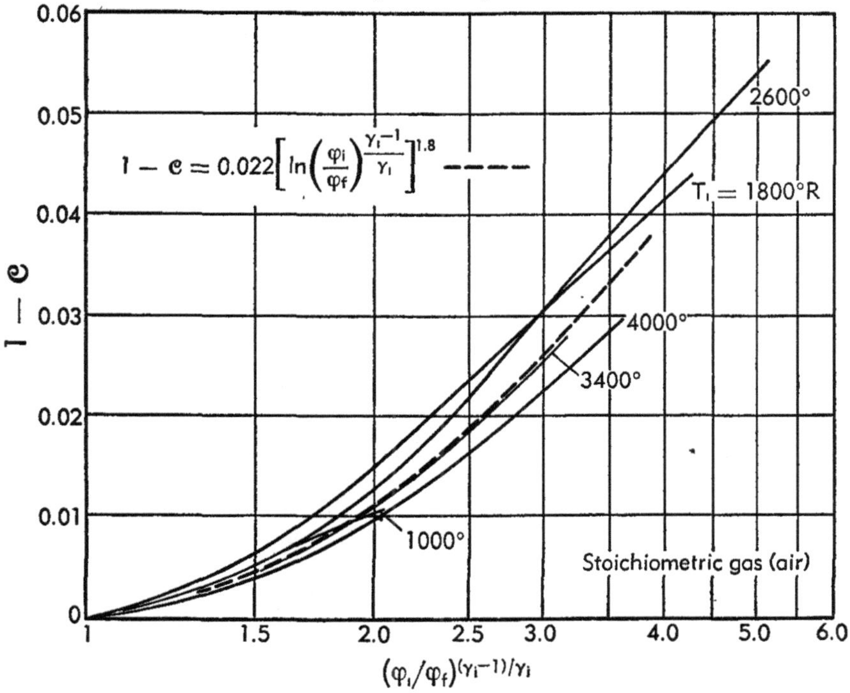

Fig. B,2j. Correction factor of Eq. 2-35 for expansions of stoichiometric gas of combustion with air of $(C_8H_5)_x$.

made here. Again, for air and for the gases of stoichiometric combustion with air, one of the mean relations

$$\mathcal{B}\left(\frac{h_f - h_i}{c_{p_i}T_i}\right) = 1 - 1.25\left(\frac{h_i - h_f}{c_{p_i}T_i}\right)^5;$$

$$\mathcal{B}\left(\frac{h_f - h_i}{c_{p_i}T_i}\right) = 1 - 0.565\left(\frac{h_i - h_f}{c_{p_i}T_i}\right)^{3.2} \tag{2-34}$$

depending on the range, can be approximately used and applied also to other gases of combustion with air.

Another useful correction factor can be applied to Eq. 2-28, which for a nonpolytropic gas can be written as

$$\frac{h_f - h_i}{c_{p_i}T_i} = \mathcal{C}\left(T_i, \frac{\sigma_f - \sigma_i}{c_{p_i}}\right)\left(e^{\frac{\sigma_f - \sigma_i}{c_{p_i}}} - 1\right) = \mathcal{C}\left[\left(\frac{\varphi_f}{\varphi_i}\right)^{\frac{\gamma_i - 1}{\gamma_i}} - 1\right] \tag{2-35}$$

Fig. B,2i, B,2j, and B,2k show the values of $1 - \mathfrak{C}$ calculated for the same cases already discussed. One can replace in each figure a single mean curve for $\mathfrak{C}$, from which the deviations are not large if the expansion ratios are contained in certain limits well above the practical range. Actually,

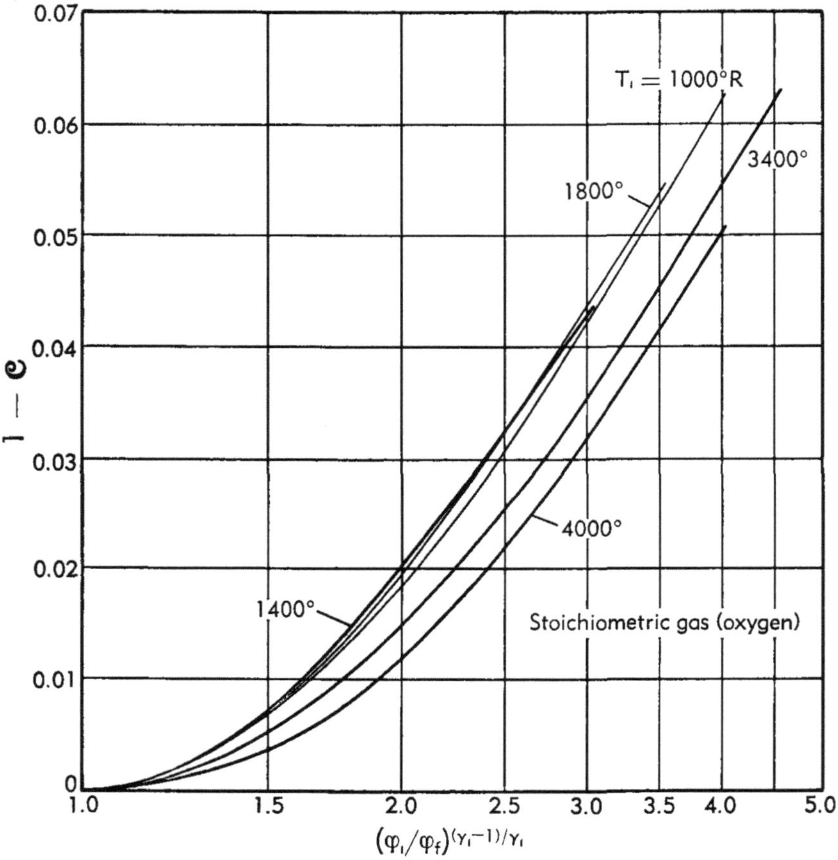

Fig. B,2k. Correction factor of Eq. 2-35 for expansions of stoichiometric gas of combustion with oxygen of $(C_3H_5)_x$.

without too much error, one can select the same curve for Fig. B,2i and B,2j, for instance the curve corresponding to the equation

$$\mathfrak{C}\left(\frac{\sigma_f - \sigma_i}{c_{p_i}}\right) = 1 - 0.022\left(\frac{\gamma_i - 1}{\gamma_i}\ln\frac{\varphi_i}{\varphi_f}\right)^{1.8} \tag{2-36}$$

represented in the two figures. Again, with the exception of extreme cases, the errors are below 1 per cent. Since Eq. 2-36 can be used for both air and stoichiometric combustion gases, it can be concluded as before that it is also a good approximation for combustion gases with arbitrary air excess and for other conventional fuels. For gases of combustion with oxygen, Eq. 2-36 gives values that are definitely too low, but a similar equation with the numerical coefficient equal to 0.033 would give excellent results. However, the corresponding expression for $\mathfrak{C}$ does not exhibit the same character of generality as Eq. 2-36, and can be expected to change with mixture ratio and constitution of fuel.

The reciprocal relation of Eq. 2-35 can be treated in a similar fashion, introducing a new correction factor $\mathfrak{D}$:

$$\frac{\sigma_f - \sigma_i}{c_{p_i}} = \frac{\gamma_i - 1}{\gamma_i} \ln \frac{\varphi_f}{\varphi_i} = \mathfrak{D}\left(T_i, \frac{h_f - h_i}{c_{p_i} T_i}\right) \ln\left(1 + \frac{h_f - h_i}{c_{p_i} T_i}\right) \qquad (2\text{-}37)$$

The correction factor $\mathfrak{D}$ is shown in Fig. B,2l, B,2m, and B,2n for the same

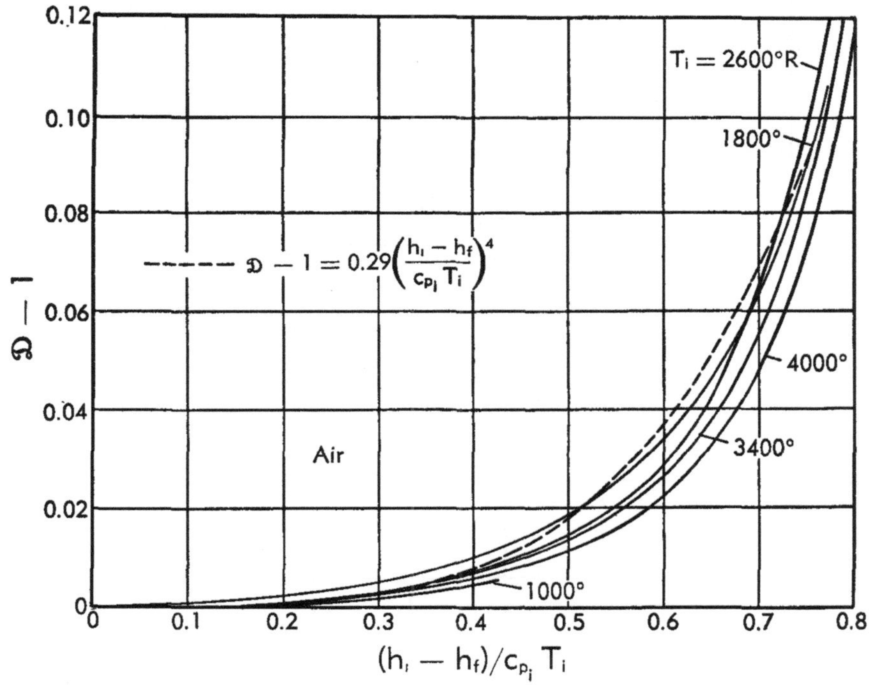

Fig. B,2l. Correction factor of Eq. 2-37 for expansions of air.

representative cases as before. Again a single curve, corresponding to the equation

$$\mathfrak{D}\left(\frac{h_i - h_f}{c_{p_i} T_i}\right) = 1 + 0.29\left(\frac{h_i - h_f}{c_{p_i} T_i}\right)^4 \qquad (2\text{-}38)$$

represents well the results for air and gases of combustion with air. For Fig. B,2n the numerical coefficient should be higher.

An interesting observation can be made about the relative importance of the correction factors $\mathfrak{A}$ and $\mathfrak{C}$ for an identical transformation. In the following table examples have been calculated from Eq. 2-32 and 2-36 for two widely different ranges of temperature variation, the first corresponding to an isentropic expansion ratio of about 7.4 and the second to about

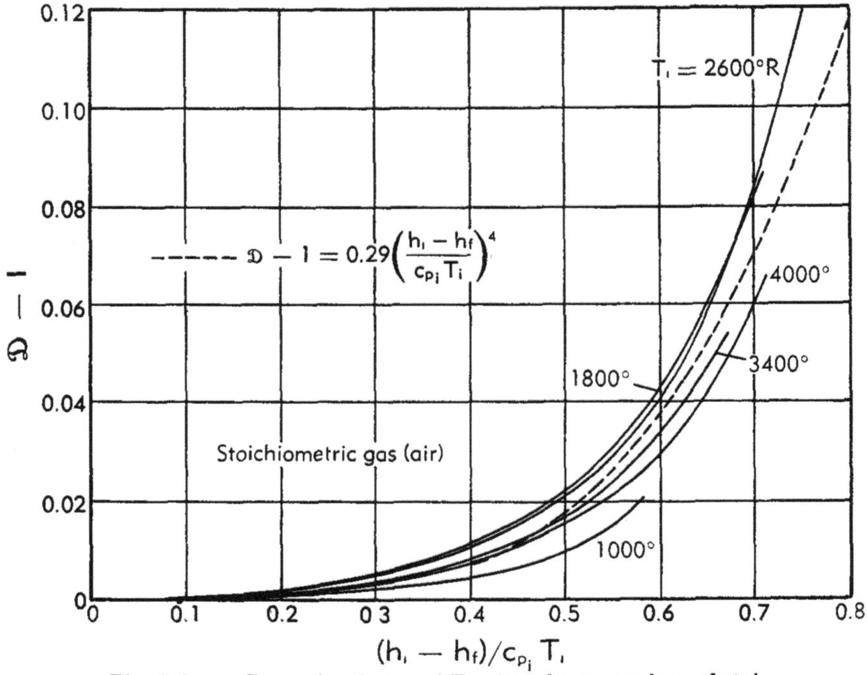

Fig. B,2m. Correction factor of Eq. 2-37 for expansions of stoichiometric gas of combustion with air of $(C_3H_5)_x$.

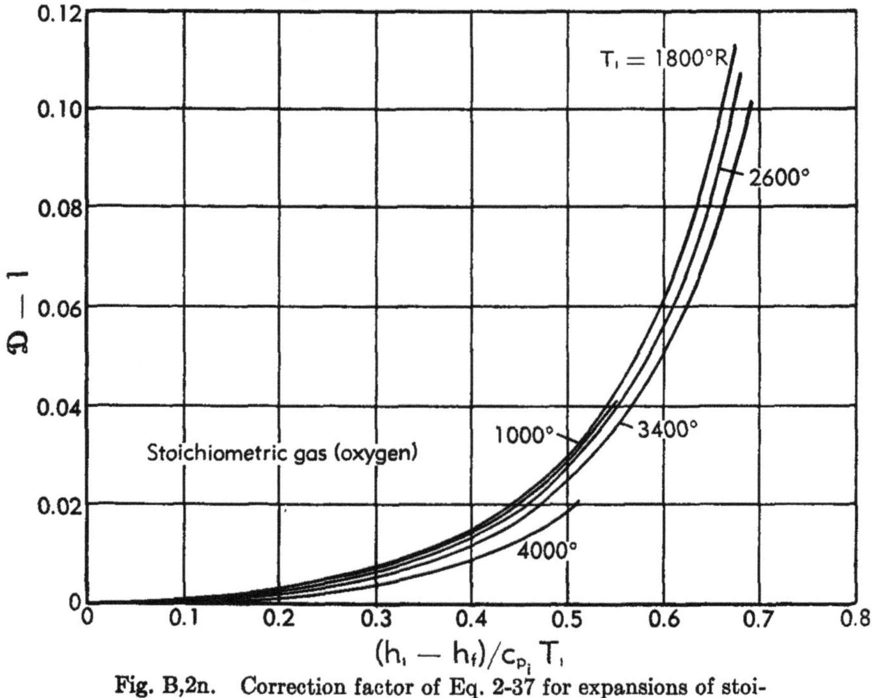

Fig. B,2n. Correction factor of Eq. 2-37 for expansions of stoichiometric gas of combustion with oxygen of $(C_3H_5)_x$.

150. For the calculation of $\mathcal{C}$ the two values $\gamma_i = 1.25$ and 1.333 have been used in order to show the influence of γ_i.

Table B,2

$\ln \frac{\varphi_i}{\varphi_f}$	$\left(\frac{p_i}{p_f}\right)_{is}$	$100(\mathcal{A} - 1)$	$100(\mathcal{C} - 1)$	
			$\gamma_i = 1.250$	$\gamma_i = 1.333$
2.5	7.4	2.9	0.73	0.95
5	150.	12.6	2.2	3.3

It is clear that the values of $\mathcal{A} - 1$ are in all cases several times larger than $1 - \mathcal{C}$, with the consequence that Eq. 2-35 and also, therefore, Eq. 2-37 are always closer to the relation obtained for polytropic gases than are Eq. 2-31 and 2-33. For instance, for the lower isentropic expansion ratio considered, the error deriving from the assumption that the

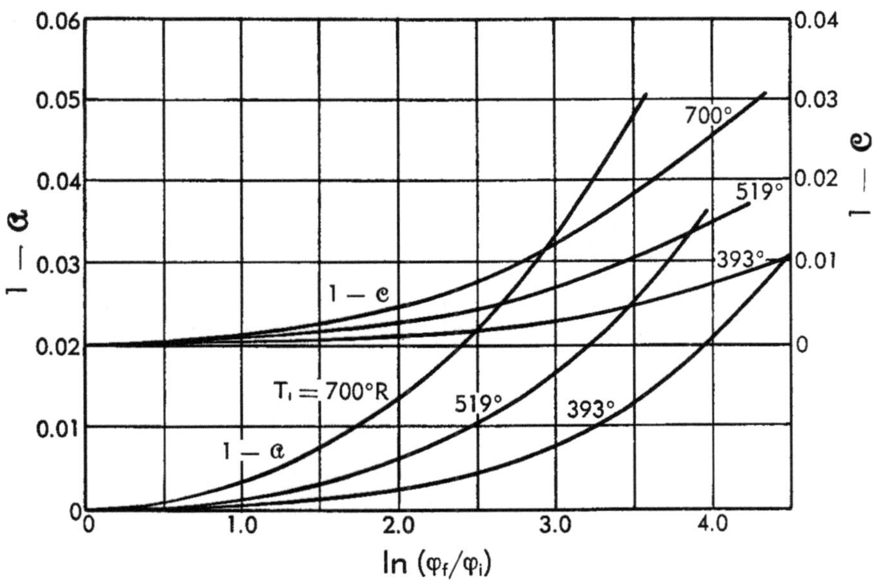

Fig. B,2o. Correction factors of Eq. 2-31 and 2-35 for compressions of air.

specific heat stays constant and equal to the initial value would be below 1 per cent for Eq. 2-35, and about 3 per cent for Eq. 2-31. In other words the assumption of polytropic gas can be accepted in a wider range, when computing the relation between the enthalpy and the temperature-dependent part of the entropy, than when calculating the relation between the latter and the temperature. This interesting property derives from the fact that the variation of specific heats affects only the right-hand side of

Eq. 2-31, while it affects both sides of Eq. 2-35 in the same direction, thus introducing a kind of compensation.

Of course the same correction factors used for expansion processes can also be introduced for compressions. In this case, however, the effect of T_1 on the correction factors cannot be eliminated even approximately, as Fig. B,2o and B,2p clearly show for the quantities $\mathcal{A}$ and $\mathcal{C}$, $\mathcal{B}$ and $\mathcal{D}$ relative to air compressions. Notice, however, that both $1 - \mathcal{C}$ and $\mathcal{D} - 1$ remain quite small in a wide range of isentropic compression ratios, and become of the order of 1 per cent or more only for extremely high compression ratios. The quantities $\mathcal{A} - 1$ and $1 - \mathcal{B}$, though also small, are appreciably larger than $1 - \mathcal{C}$ and $\mathcal{D} - 1$.

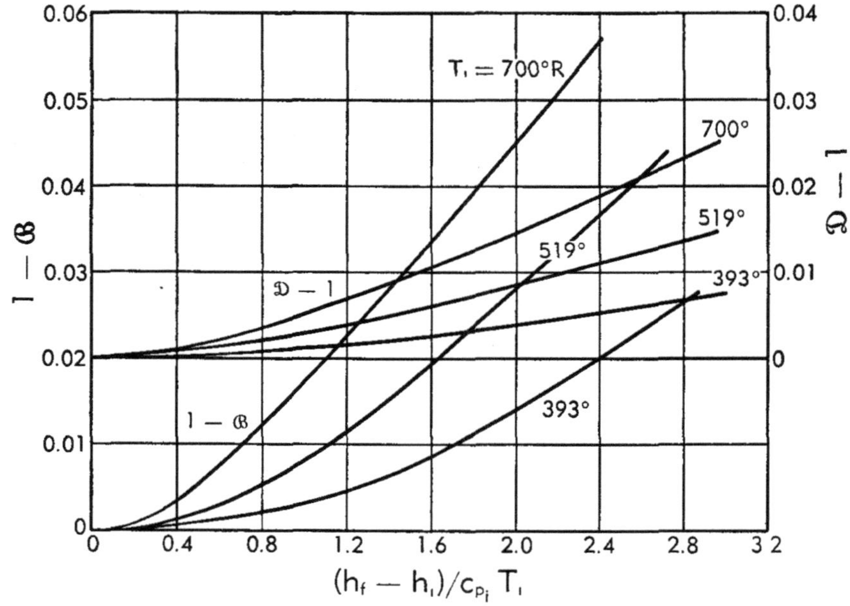

Fig. B,2p. Correction factors of Eq. 2-33 and 2-37 for compressions of air.

Real gases. As already observed at the beginning of this article, deviations from the perfect gas law (Eq. 2-1) take place when either the temperature is very low or the pressure very high, and also in the high temperature range when chemical dissociations appear, even though each chemical constituent follows the perfect gas law. In these cases the specific heats, internal energy, and enthalpy cease to be merely functions of the temperature and depend also on the pressure. As a consequence, the entropy also depends on temperature and pressure in a more intricate way than for perfect gases (I,C).

The most classic procedure for the presentation of the properties of a real gas is through an h, s diagram (Mollier diagram) containing isothermal, isobaric, and isochoric lines. This procedure has been in use for a long time in the calculations of steam turbines, and Mollier diagrams for water vapor, above and below saturation, are easily available. Analogous diagrams have been calculated for combustion gases [*8,9,10,11*] to take into account, essentially, the effect of chemical dissociations.

In addition to problems of flow involving steam or combustion gases, most common in engineering practice, problems concerned with the flow of air at very large velocities (hypersonic) in which the air may undergo extreme changes of pressure and temperature have to be considered. These problems also may be solved with the use of the Mollier diagram for air, which may be extended to the range of very high temperatures where dissociations of the oxygen and nitrogen molecules become important and of the even higher temperatures where ionization is produced, by assuming that in each condition equilibrium is reached and, therefore, by neglecting relaxation effects.

The inconvenience of the use of Mollier diagrams resides in the difficulty of obtaining highly accurate results because of the limited size of the diagram. A more accurate method, based on the use of correction factors, has been presented by Crown [*2*]. The correction factors are defined by a comparison of the desired thermodynamic functions for the real air with the corresponding value for a perfect, polytropic reference gas with specific heats ${}_0c_p$ and ${}_0c_v$, specified by the following considerations. The actual values of $c_p(p, T)$ and $c_v(p, T)$ for the real gas can be obtained from the values of $c_p(0, T)$ and $c_v(0, T)$ once the equation of state is known (I,C,17). On the other hand, excluding the case of extremely high temperatures, $c_v(0, T)$ consists of three parts

$$c_v(0, T) = (c_v)_{\text{trans}} + (c_v)_{\text{rot}} + (c_v)_{\text{vib}}$$

of which the first two can be considered constant down to 20°K for diatomic gases (with the exception of hydrogen and hydrides), their sum being equal to $\frac{5}{2}\mathfrak{R}$. Therefore Crown chooses, for the perfect polytropic reference gas,

$$ {}_0c_v = \tfrac{5}{2}\mathfrak{R}; \quad {}_0c_p = {}_0c_v + \mathfrak{R} = \tfrac{7}{2}\mathfrak{R}; \quad {}_0\gamma = \frac{{}_0c_p}{{}_0c_v} = 1.4 \tag{2-39}$$

It is to be observed that, for the constituents of air, $(c_v)_{\text{vib}}$, which can be calculated from quantum mechanics (I,A,15), is only appreciable above room temperature, so that the values of Eq. 2-39 are practically those of air at room temperature.

For the reference gas one has, after a suitable choice of the additive constant of the enthalpy,

$$\left.\begin{aligned} {}_0\rho &= \frac{p}{\mathfrak{R}T}; \quad {}_0h = {}_0c_pT; \quad a_0^2 = {}_0\gamma\mathfrak{R}T \\ {}_0s &= s_r + {}_0c_p \ln\frac{T}{T_r} - \mathfrak{R}\ln p \end{aligned}\right\} \tag{2-40}$$

We define the following correction factors, all functions of p and T,

$$\mathcal{E} = \frac{\rho}{{}_0\rho}; \quad \mathcal{F} = \frac{h}{{}_0h}; \quad \mathcal{G} = \frac{a^2}{{}_0a^2}; \quad \mathcal{I} = \frac{{}_0\rho}{\rho}\frac{{}_0h}{h} = \frac{1}{\mathcal{E}\mathcal{F}} \tag{2-41}$$

$$\mathcal{J} = \exp \frac{s(p, T) - {}_0s(p, T)}{\mathcal{R}}; \quad \mathcal{K} = \mathcal{E}\mathcal{J} \tag{2-42}$$

In the first (Eq. 2-42) the additive constant of the entropy is chosen so as to make s coincide with ${}_0s$ when the deviations of the real gas from the reference gas vanish.

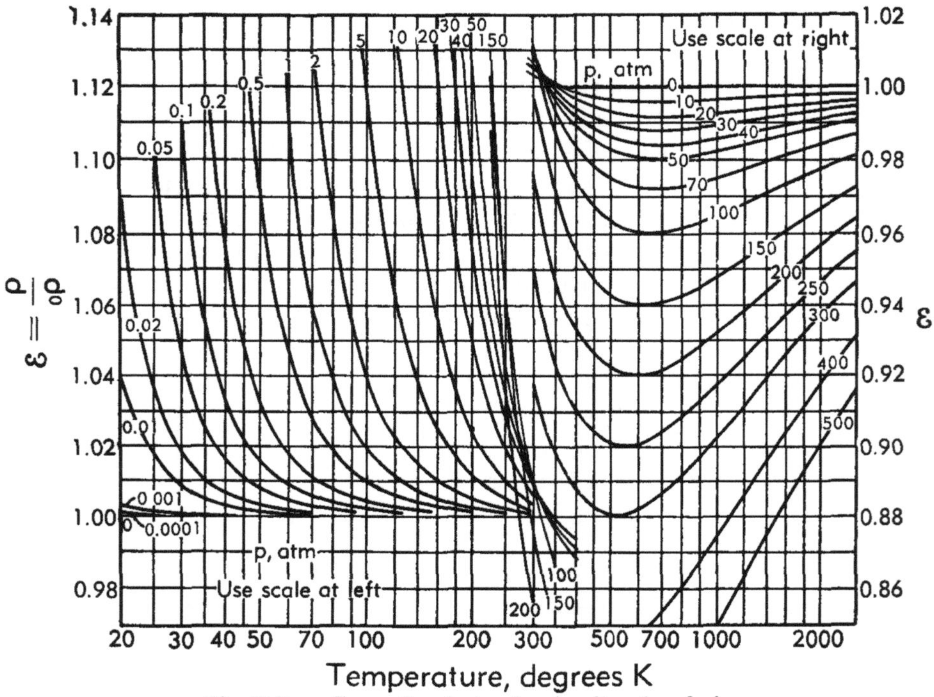

Fig. B,2q. Correction factor for density of real air.

The values of these correction factors computed for air by Crown, using the Beattie-Bridgeman equation of state (Eq. 2-3), are shown in Fig. B,2q, B,2r, B,2s, B,2t, B,2u, and B,2v. Using these correction factors, it is easy to perform the calculation of flow problems, as shown in Art. 13.

The calculation of the correction factors (Eq. 2-41 and 2-42) can in principle be extended into the higher temperature range, where the dissociation of molecular oxygen and nitrogen into atoms cannot be neglected, and even further, in the range of temperature where ionization becomes important. In both cases, however, one has to assume quasi-steady conditions, that is, at each pressure and temperature both dis-

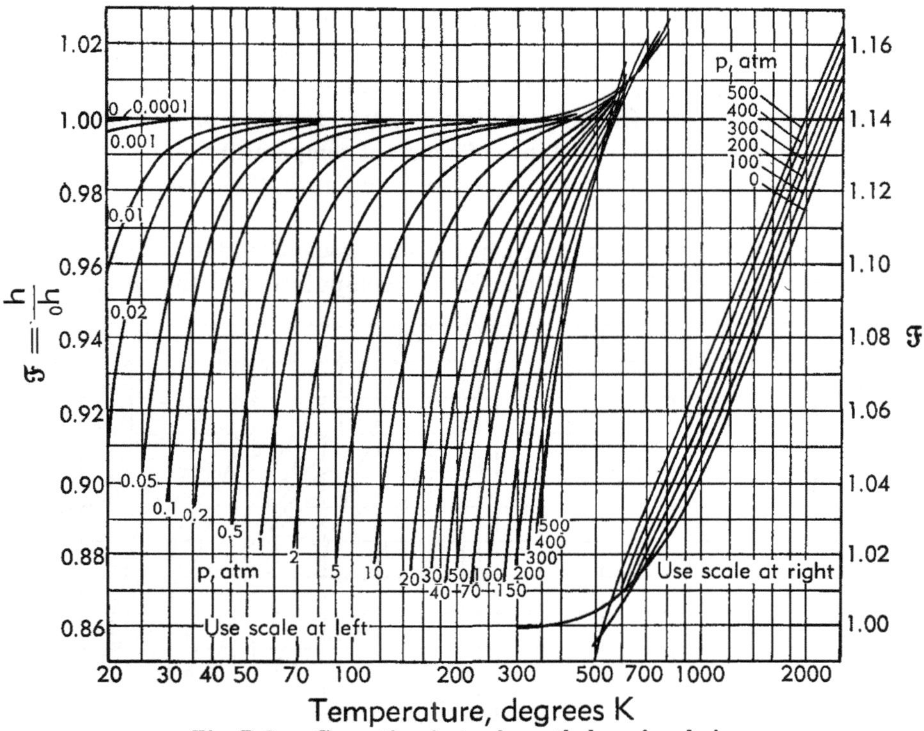

Fig. B,2r. Correction factor for enthalpy of real air.

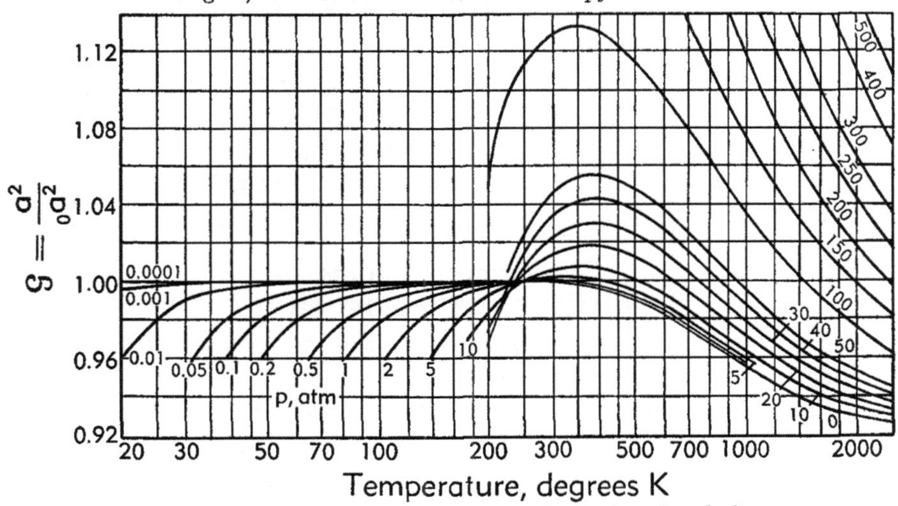

Fig. B,2s. Correction factor for sound velocity of real air.

sociation and ionization take their equilibrium values. The consideration of relaxation effects is not possible without the introduction of substantial complications.

Here, instead of giving the correction factors, we give in Fig. B,2w the Mollier diagram of air at extremely high temperatures, where the dissociations are important, but the ionization still negligible, as obtained from the properties calculated in [12]. Here, however, the different chemical species (O_2, N_2, O, N, NO, and A) are assumed to behave individually as perfect gases, and the only gas imperfection results from the shifting composition of the mixture. Lines of constant temperature, pressure, and specific volume ($=(g\rho)^{-1}$) are traced on the figure. The unit of weight used here is the gram.

For gases other than air, one can use analogous procedures. The paper by Crown provides the equations for the computation of the correction

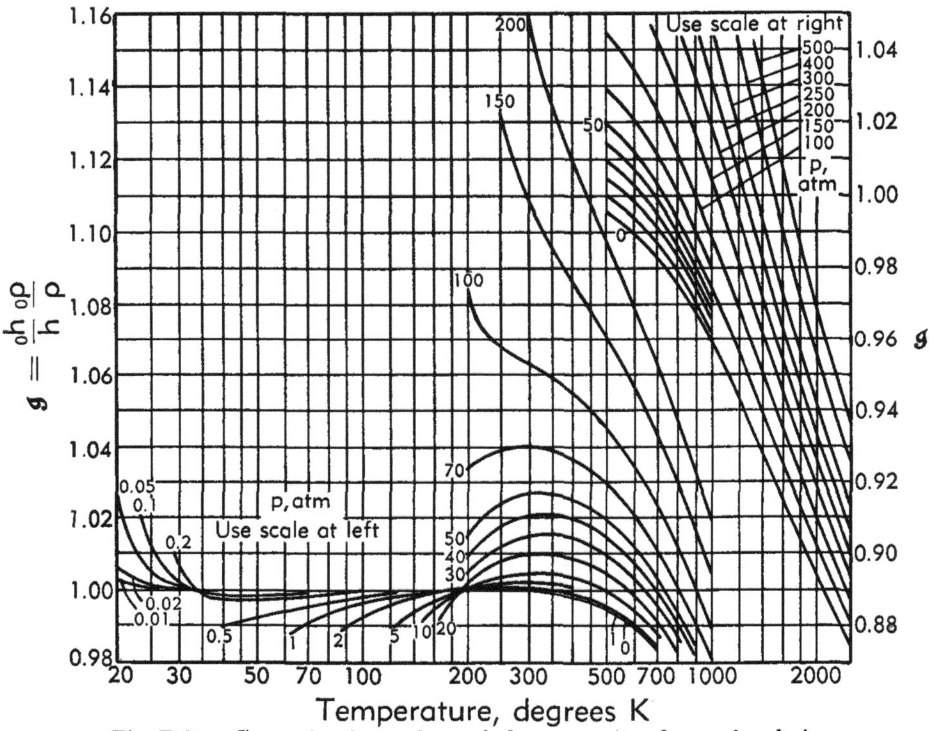

Fig. B,2t. Correction factor for enthalpy per unit volume of real air.

factors for a gas of fixed composition when the coefficients of the Beattie-Bridgeman equation are assigned.

Two-phase mixtures. No simple thermodynamic relations can be written in general for a substance obtained from the nonhomogeneous mixture of two homogeneous phases. Simple relations can indeed be expressed for each phase, but in general the rate of mass and heat exchanges between the two phases depends on the rate at which the particular transformation takes place in a way which cannot be determined on purely thermodynamic grounds.

A purely thermodynamic treatment, however, is possible in the two extreme cases when the rate of the transformation is very low or very large. In the first case the substance can at each moment be considered to be in a state of thermodynamic equilibrium, all exchanges taking place in a reversible fashion, and the mechanism which produces the exchanges can be ignored. The exchanges play a secondary role and can be disregarded. It must be observed, however, that the magnitudes of the transformation rates, at which one can accept these simplifications, depend strongly on the extent of the interface between the two phases, and therefore on the dispersion of the mixture. The same rate may be considered high for small dispersions and low for high dispersions. As a consequence a check on the plausibility of the above simplifications is only possible

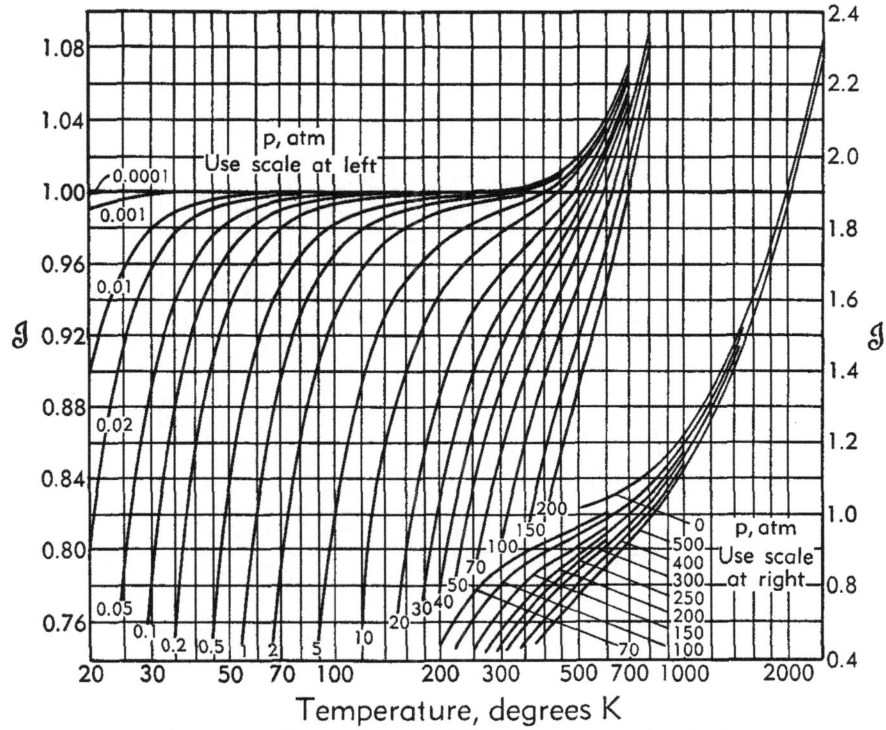

Fig. B,2u. First correction factor for entropy of real air.

if one knows the actual conditions of the mixture and the laws governing the exchanges. Nevertheless these simplifications are often used without any kind of check, or with only very rudimentary ones, because of the considerable advantages deriving from their introduction.

Of particular interest in practical applications is the case in which no mass transfer takes place between the two phases, so that the composition of the mixture is fixed. In general, the mixtures of a gas and solid or liquid particles can be considered as such when, at the temperatures under consideration, the saturation pressure of the solid or liquid is much lower than the pressure of the gas. This case is considered in some detail in II,B,5.

Here we shall only consider another type of mixture [*13*,*14*] which has some interesting practical applications in jet propulsion, i.e. a mixture of a liquid with gas bubbles. We assume that the vapor pressure of the liquid phase is negligible, so that the mass exchanges can be neglected and the composition is "frozen." On the other hand, if the dispersion of the bubbles is large and the rates of transformation not exceedingly high, the small gas bubbles can be considered in thermal equilibrium with the liquid, and heat exchanges, taking place reversibly, do not need to be considered explicitly.

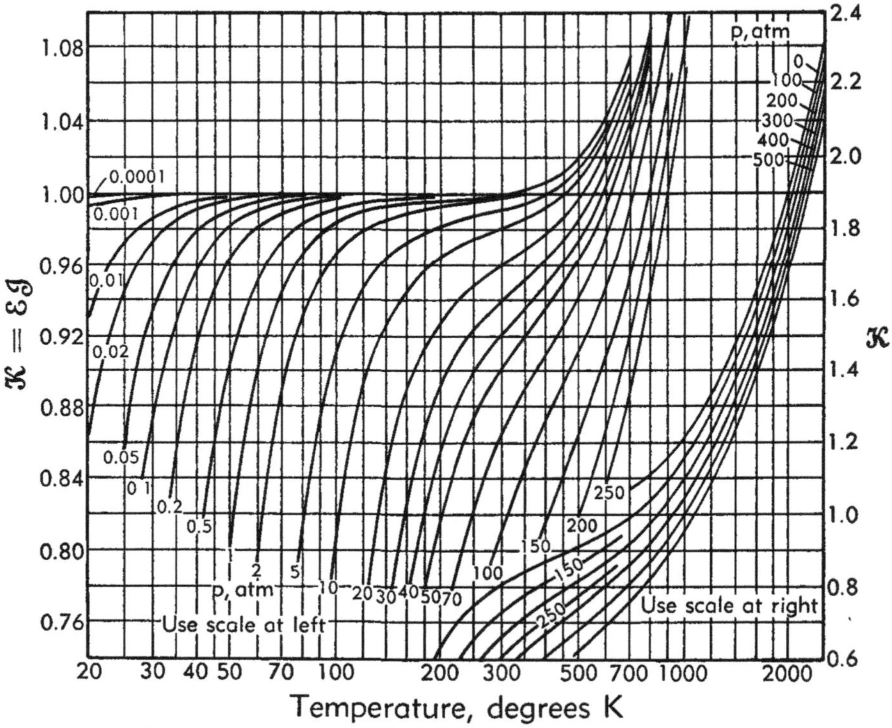

Fig. B,2v. Second correction factor for entropy of real air.

If the unit volume contains a mass M_{liq} of liquid of density ρ_{liq}, and M_{g} of gas of density ρ_{g}, and we define the composition of the mixture through

$$\mu = \frac{M_{\text{g}}}{M_{\text{liq}}} = \text{const} \tag{2-43}$$

the mean density ρ can be obtained from the relation

$$\frac{M_{\text{g}}}{\rho_{\text{g}}} + \frac{M_{\text{liq}}}{\rho_{\text{liq}}} = \frac{M_{\text{g}} + M_{\text{liq}}}{\rho}$$

as

$$\frac{1}{\rho} = \frac{1}{1+\mu}\left(\frac{\mu}{\rho_{\text{g}}} + \frac{1}{\rho_{\text{liq}}}\right) \tag{2-44}$$

We assume, for simplicity, that the liquid is incompressible, and the gas is perfect, with the gas constant $\mathcal{R}_g$. Then

$$\rho_{liq} = \text{const}; \qquad \rho_g = \frac{p}{\mathcal{R}_g T} \tag{2-45}$$

Replacing in Eq. 2-44 we obtain the equation of state of the mixture under the form

$$p\left[\frac{1}{\rho} - \frac{1}{(1+\mu)\rho_{liq}}\right] = \frac{\mu \mathcal{R}_g}{1+\mu} T = \mathcal{R} T \tag{2-46}$$

an equation which differs from Eq. 2-1 only for the presence of the

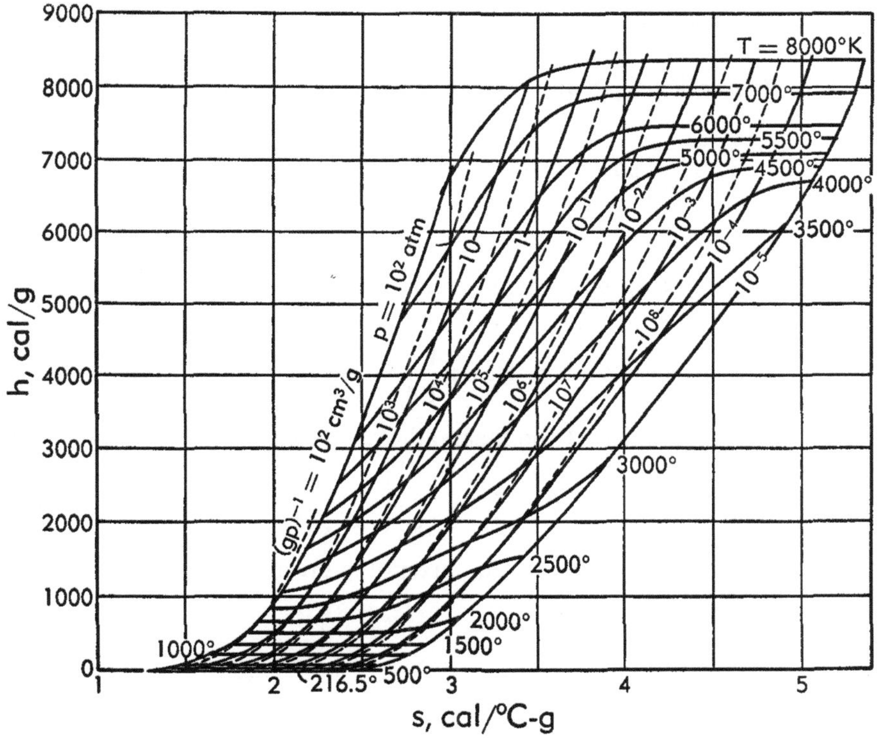

Fig. B,2w. Mollier diagram for equilibrium air at very high temperatures, taking into account the dissociations and combinations of O_2 and N_2.

"covolume" $1/(1+\mu)\rho_{liq}$, which represents the limiting specific volume $1/\rho$ of the mixture for infinite pressure. For very low pressures $1/\rho \gg 1/(1+\mu)\rho_{liq}$ and the behavior of the mixture tends to that of a perfect gas with gas constant $\mathcal{R}$.

If, again for simplicity, we assume that the specific heat of the liquid, c_{liq}, and the specific heats of the gas, c_{v_g} and $c_{p_g} = c_{v_g} + \mathcal{R}_g$, are constant, the specific internal energy of the mixture is given by

$$e = \frac{c_{liq} + \mu c_{v_g}}{1+\mu} T = c_v T \tag{2-47}$$

where we have introduced the specific heat c_v of the mixture, and we have chosen the arbitrary constant e_r of the internal energy so as to make $e \sim T$.

The specific enthalpy can be calculated from Eq. 2-46 and 2-47 as

$$h = e + \frac{p}{\rho} = \frac{c_{\text{liq}} + \mu c_{p_g}}{1 + \mu} T + \frac{p}{(1 + \mu)\rho_{\text{liq}}} = c_p T + \frac{p}{(1 + \mu)\rho_{\text{liq}}} \tag{2-48}$$

where the mean specific heat c_p is related to c_v by the relation

$$c_p = c_v + \mathcal{R} \tag{2-49}$$

as for a perfect gas. Observe that Eq. 2-47 is identical to that of a polytropic perfect gas, but Eq. 2-48 contains a pressure-dependent term which is absent for a perfect gas.

Introducing Eq. 2-48 and 2-46 in Eq. 2-10 we obtain

$$ds = c_p \frac{dT}{T} - \mathcal{R} \frac{dp}{p} = c_v \frac{dp}{p} + c_p \frac{d\left(\frac{1}{\rho}\right)}{\frac{1}{\rho} - \frac{1}{(1 + \mu)\rho_{\text{liq}}}} \tag{2-50}$$

and, therefore, the specific entropy can be expressed as

$$s = \mathcal{R} \ln \frac{T^{\frac{\Gamma}{\Gamma - 1}}}{p} + \text{const} = c_v \ln p \left[\frac{1}{\rho} - \frac{1}{(1 + \mu)\rho_{\text{liq}}}\right]^{\Gamma} + \text{const} \tag{2-51}$$

where we have introduced the mean adiabatic index for the mixture

$$\Gamma = \frac{c_p}{c_v} = \frac{c_{\text{liq}} + \mu c_{p_g}}{c_{\text{liq}} + \mu c_{v_g}} \tag{2-52}$$

Finally, the sound velocity in the mixture is given by Eq. 2-16, that is, using Eq. 2-50, by

$$\left.\begin{aligned} a^2 = \left(\frac{dp}{d\rho}\right)_{s=\text{const}} &= \frac{\Gamma \frac{p}{\rho}}{1 - \frac{\rho}{(1 + \mu)\rho_{\text{liq}}}} = \frac{\Gamma \mathcal{R} T}{\left[1 - \frac{\rho}{(1 + \mu)\rho_{\text{liq}}}\right]^2} \\ &= \Gamma \mathcal{R} T \left[1 - \frac{1}{1 + (1 + \mu)\rho_{\text{liq}} \frac{\mathcal{R} T}{p}}\right]^{-2} \end{aligned}\right\} \tag{2-53}$$

which is reduced to Eq. 2-17 for very low pressures, and tends to infinite for very high pressures, when the volume of the gas becomes negligible and the mixture becomes practically incompressible.

B,3. The Laws of Conservation. For the most general type of flow, if one chooses an arbitrary closed control surface, the balance of mass, momentum, and energy can be expressed by the following conservation laws (A,4):

1. The rate of increase of the mass enclosed within the control surface is equal to the net inflow of mass through the surface itself.
2. The rate of increase of the momentum contained within the control surface is equal to the net inflow per unit time of momentum through the surface plus the resultant of all the forces acting on the surface, the mass forces acting on the fluid within the surface, and the forces exerted on the fluid by bodies present within the surface.
3. The rate of increase of the kinetic plus the internal energy contained within the control surface is equal to the net inflow per unit time of stagnation enthalpy through the surface plus (a) the rate of heat supply to the fluid within the surface, (b) the work performed by mass forces on the fluid within the surface per unit time, (c) any amount of heat or work supplied to the fluid from bodies present within the surface, and (d) the work performed in the unit time by the viscous stresses acting on the control surface.[1] The contribution to the last item vanishes on solid walls, where the fluid velocity is zero.

These laws can be applied to the flow in a duct by choosing a control surface made of two arbitrary sections plus the portion of duct between the two sections. Let us consider the case of flows that are rigorously steady. In this case the rates of increase of mass, momentum, or energy enclosed are identically zero.

Mass balance. The mass flux m through a given section A of the duct is

$$m = \int_m d\mu = \int_A \rho u d\eta \tag{3-1}$$

In this expression the elementary flux of mass $d\mu$, through the elementary surface $d\eta$, is computed using for u the component of the velocity normal to the surface. Due to the assumption of small obliquity, the normal component u can be identified with the velocity q itself, so that in general u will also be used to represent the total magnitude of the local velocity. Only in exceptional cases, when the obliquity is not so small, shall we distinguish between q and u. An alternate expression, complementary to Eq. 3-1, is

$$A = \int_A d\eta = \int_m \frac{d\mu}{\rho u} \tag{3-2}$$

In terms of m, the mass balance, expressing the fact that the net inflow of mass through the control surface is zero, can be written as

$$m_1 - m_2 + m_i = 0 \tag{3-3}$$

[1] One assumes here that no other forms of energy are being exchanged.

Here m_1 represents the mass inflow through the first section (subscript 1), m_2 the mass outflow through the second section (subscript 2), and m_i the rate of mass injection (positive) or extraction (negative) through the walls of the duct between the two sections.

Momentum balance. Law 2 expresses an equality between vectors. However, it is evident that if the duct presents a certain character of symmetry around a straight axis, around which the mass forces and the enclosed bodies are also symmetrical, all the vectors will be directed along the axis, and law 2 can be expressed by a single scalar equation. The momentum flux through each section is given by

$$I = \int_m u d\mu = \int_A \rho u^2 d\eta \tag{3-4}$$

The force acting on each section is essentially due to the static pressure, and since this is assumed constant on the section, it is given by pA. Observe that if there is an axial gradient of velocity an axial viscous stress should also be considered. However, if the axial gradients are small (consistent with the small divergence of the duct) these stresses are negligible, a statement that can be proved quantitatively through an order-of-magnitude analysis analogous to that used in boundary layer theory (IV,B).

The forces developed between the walls of the duct and the fluid consist first of the static pressure directed along the normal to each element of the wall. Consider two sections of areas A and $A + dA$, separated by an axial distance dx. If the wall area between the two sections is dA_w, one generally defines a hydraulic diameter d_h such that

$$dA_w = 4A \frac{dx}{d_h}$$

For circular sections d_h coincides with the actual diameter d. The projection of dA_w on the plane of the section is obviously dA. The resultant in the axial direction of the uniform pressure acting on dA_w can be obtained by multiplying the pressure by each element of dA_w, projecting from the normal to the element to the axial direction, and summing up. For uniform pressure, the result is the same if the elements of dA_w are first projected on the plane of the section, summed, and finally multiplied by the pressure. Therefore the axial resultant of the pressures acting on dA_w, or that of the pressures exerted by dA_w on the fluid, is pdA; and the corresponding resultant force acting on the fluid between section 1 and section 2 is obtained by integration as $\int_{x_1}^{x_2} pdA = \int_{x_1}^{x_2} p(dA/dx)dx$, directed from x_1 to x_2.

In addition to the static pressure acting normal to the wall, there is also a tangential stress τ due to viscous forces, which, contrary to what has been told about the axial viscous stresses, cannot be neglected in general because the transversal gradients of velocity near the wall can be quite large. Assuming that τ is uniform on all elements dA_w of the wall, or that it represents a proper mean value, and identifying the tangential stress with its axial component in accordance with the assumption of small divergence, $\tau dA_w = 4\tau_w A dx/d_h$ represents the elementary axial force due to viscous stresses acting on the wall in the flow direction or on the fluid in the opposite direction. The total force acting on the fluid between sections 1 and 2 is therefore

$$X_w = 4\int_{x_1}^{x_2} \tau_w A \frac{dx}{d_h} \tag{3-5}$$

directed from x_2 to x_1 (positive for drag).

Finally, in evaluating the terms for the momentum balance, we shall indicate with X_m the axial mass forces acting on the fluid between sections 1 and 2 and with X_b the axial force exerted on the fluid by bodies immersed in the flow between the two sections. Both forces shall be considered positive when directed from section 2 to section 1. Observe that if a multiconnected control surface were chosen, obtained by adding the surface of the immersed bodies to the previously defined control surface, the force X_b would not need to be taken into account separately but would be obtained from the integration of the pressures and viscous stresses on the additional surface. Observe also that in general X_b represents a drag and is, therefore, positive.

We can now give to the momentum balance the following expression

$$I_1 - I_2 + p_1A_1 - p_2A_2 + \int_{x_1}^{x_2} pdA - X_w - X_m - X_b + I_i = 0 \tag{3-6}$$

The quantity I_i represents here the rate of injection (or extraction) of axial momentum through the walls of the duct between sections 1 and 2, that is, the axial momentum of the mass m_i.

Eq. 3-6 has been derived under the assumptions that the axis of the duct is straight and that certain symmetry conditions are verified. If these assumptions are not satisfied, the vectors considered by law 2 are not necessarily parallel to the axis, so that the law itself cannot be expressed by a single scalar relation. However, Eq. 3-6 can still be derived, assuming that the pressure on each section is nearly uniform, by integrating the axial component of the different vectors between the first and the second section. On the other hand, from the integration of the other components one can obtain other equations from which, for instance, one can evaluate such quantities as the total force exerted by the flow on a curved duct. In Eq. 3-6 the quantities X_m, X_b, and I_i now represent *only* the axial components of the corresponding vectors.

If the distribution of the pressure on the section is not uniform, Eq. 3-6 can still be written with some modifications. The axial force on each section is now $\int_A p d\eta$ instead of pA, so that the corresponding terms $p_1A_1 - p_2A_2$ can be written using in place of the uniform pressure the mean pressure

$$\bar{p} = \frac{1}{A}\int_A p d\eta \tag{3-7}$$

The pressure p_w acting on the wall is now different for different elements of dA_w. The corresponding axial force to be inserted in Eq. 3-6 is now

$$\int_{A_1}^{A_2} \bar{p}_w dA = \int_{A_1}^{A_2} \int_{dA} p_w d(dA) \tag{3-8}$$

where $\bar{p}_w$ represents a mean value of the wall pressure, in general different from $\bar{p}$. The value of $\bar{p}_w$ depends on the arbitrary choice of the axis of the duct. If the direction of the axis is changed by a small angle, dA remains the same but the distribution of the elements $d(dA)$ around the periphery changes, and therefore $\bar{p}_w$ also changes. In certain cases, as for instance for ducts with strong curvature, it is possible to define the axis of the duct in such a way as to make $\bar{p}_w = \bar{p}$. In this case, and for this choice of the axis, the only change in Eq. 3-6 is the replacement of p by $\bar{p}$.

The energy balance. The flux of enthalpy through a section is given by

$$H = \int_m h d\mu = \int_A h\rho u d\eta \tag{3-9}$$

and the flux of kinetic energy by

$$E_k = \tfrac{1}{2}\int_m u^2 d\mu = \tfrac{1}{2}\int_A \rho u^3 d\eta \tag{3-10}$$

The flux of stagnation enthalpy is the sum of the two (Eq. 2-8):

$$H^0 = \int_m h^0 d\mu = H + E_k \tag{3-11}$$

In order to formulate the energy balance, the amount of heat supplied to the fluid must be known. Using for the enthalpy the definition given in Art. 2, the heat released by chemical transformations or physical changes of state is already accounted for in the evaluation of the enthalpy flux (Eq. 3-9). On the other hand the heat exchanged by conductivity through sections 1 and 2 is negligible if the axial gradients are, as assumed, small. Thus we have to consider only the rate at which heat is exchanged with the surroundings through the wall of the duct, given by

$$Q_w = \int_{x_1}^{x_2} q_w dA_w = 4\int_{x_1}^{x_2} q_w A \frac{dx}{d_h} \tag{3-12}$$

where q_w represents the heat supplied through the walls to the fluid, per unit time and unit area. The work performed on the fluid by the mass forces can be obtained from the net inflow E_m of potential energy through the control surface, a quantity which can be computed from the known distributions of flow rate and potential energy. The work performed by the viscous stresses acting on the control surface vanishes, because it is zero on the wall and because the axial viscous stresses on the two sections are negligible.

The energy balance for the steady case can now be written as

$$H_1^0 - H_2^0 + E_m + Q_w + Q_b + W_b + H_i^0 = 0 \tag{3-13}$$

Here Q_b and W_b represent the heat and the work supplied to the fluid by immersed bodies and H_i^0 represents the stagnation enthalpy injected (or extracted) through the walls with the mass m_i, i.e. the sum of the corresponding kinetic energy and enthalpy, the last one taking into account, according to Art. 2, the latent heat and the heat of formation of the injected (or extracted) substances. Observe that W_b may, in general, represent mechanical energy supplied to the flowing fluid in any form whatever.

The derivation of Eq. 3-3, 3-6, and 3-13 is based on the assumption of complete steadiness of the flow.[2] There is, however, the very important case of turbulent flow characterized by the fact that even when, macroscopically, the flow looks steady, rapid local fluctuations of velocity, pressure, temperature, and composition take place around their steady mean values. Without entering here into details, which can be found in Vol. IV, it will suffice to say that the volume and surface integrals appearing for this nonsteady case in the expression of the laws of conservation of mass, momentum, and energy, can be split into a steady part, computed through the mean values, and a nonsteady part, containing nonlinear terms due to the fluctuations. These additional terms are in average independent of the time, so that the rates of increase of the mass, momentum, and energy enclosed are still zero. However, the additional terms originating in the calculation of the flux of mass, momentum, or stagnation enthalpy are generally different from zero. Therefore, if Eq. 3-3, 3-6, and 3-13 are written in terms of the quantities m, I, and H^0 calculated from Eq. 3-1, 3-4, and 3-11 with the mean values of ρ, u, and h^0, as if the flow were really steady, the additional terms must appear separately in the corresponding equations as if they represented an additional mass introduced, or stress applied, or energy supplied to the fluid within the control surface. The most important terms are likely to be: (1) a

[2] Strictly speaking, this is not compatible with the production of the work from immersed bodies, because in practice this work is obtained through the motion of a discrete number of parts and thus determines a certain nonsteadiness. Steady conditions would only be possible in the limit when the number of moving parts tends to infinity.

turbulent stress X_t in the axial direction on the two sections, to be introduced in Eq. 3-6, (2) the work due to this X_t to be introduced in Eq. 3-13, and (3) the kinetic energy E_t of the fluctuations, to be introduced in Eq. 3-13. Too little is known in general form about these quantities to allow their introduction in the one-dimensional treatment. Thus these turbulent terms are generally neglected, and the equations are written in the form derived for truly steady flow in terms of the mean quantities without any additional terms. However, it is good to bear in mind that they exist, and that in certain cases they may produce appreciable deviations from the results of the one-dimensional theory.

B,4. Uniform Flow. Particular Cases. The equations of the preceding article become simpler when the assumption is made of uniform velocity, temperature, and composition distributions on each section. In this case, instead of Eq. 3-1, 3-4, and 3-11 one can write

$$m = A\rho u, \quad I = mu = A\rho u^2, \quad H^0 = mh^0 \tag{4-1}$$

in terms of the uniform density ρ, velocity u, and stagnation enthalpy h^0 on the section considered.

Assuming that the mass forces X_m and the variations of potential energy E_m are negligible and that no work W_b is being supplied (true in all practical applications with stationary ducts), one can rewrite Eq. 3-3, 3-6, and 3-13 in the form

$$m_2 - m_1 = m_i \tag{4-2}$$

$$I_2 - I_1 = -\int_{x_1}^{x_2} A dp - X + I_i \tag{4-3}$$

$$H_2^0 - H_1^0 = Q + H_i^0 \tag{4-4}$$

where we have written for brevity

$$X = X_w + X_b; \quad Q = Q_w + Q_b$$

Eq. 4-3 can be also written in the alternate form

$$J_2 - J_1 = \int_{x_1}^{x_2} p dA - X + I_i \tag{4-5}$$

where

$$J = I + pA \tag{4-6}$$

is an important quantity, sometimes called the "thrust function" or the "impulse function."

These equations can also be written in differential form as

$$dm = dm_i \tag{4-2a}$$

$$dI = -A dp - dX + u_i dm_i \tag{4-3a}$$

$$dH^0 = dQ + h_i^0 dm_i \tag{4-4a}$$

$$dJ = p dA - dX + u_i dm_i \tag{4-5a}$$

Here we have introduced the axial component u_i of the velocity of injection and the stagnation enthalpy h_i^0 of the injected gas. A few observations are necessary about these quantities. We assume here that the injection is performed continuously without altering the geometry of the duct, that is through perforations, slots, or pores present in the walls;[3] and that after injection each element of the injected gas is instantly mixed with the rest of the flow so that the uniformity of velocity, temperature, and composition is not disturbed. Naturally both u_i and h_i^0 can be arbitrary functions of x. If they are constant, one can write $I_i = m_i u_i$ and $H_i^0 = m_i h_i^0$. The case in which dm_i is negative (mass extraction) must be considered with particular attention, because of ambiguity in the choice of u_i and h_i^0. These might in fact be taken equal to the uniform of u and h^0 of the flow. However, the actual conditions of the fluid reaching the wall to be withdrawn do not coincide with the mean conditions of the flow. For instance, for pores or holes with an orientation normal to the axis, u_i is zero. On the other hand, the stagnation temperature corresponding to h_i^0 is determined by the temperature of the wall, which may take arbitrary values, independently of the stagnation temperature T^0 of the flow. The question is discussed in some detail in Art. 14.

Application of the second law. In writing the equations of conservation, we have not made use of the second law of thermodynamics. This law does provide an inequality which cannot be violated, and therefore introduces a limitation to the possible solutions of a flow system. For such systems as considered in this article, this inequality is expressed by the fact that the total outflow of entropy ΔS through the control surface cannot be negative. The inflow of entropy through section 1 is $m_1 s_1$ and the outflow through section 2 is $m_2 s_2$. The entropy flowing in through the walls with the injected material is $\int_{x_1}^{x_2} s_i dm_i$ (s_i representing the corresponding specific entropy). Finally if the local temperature of the wall is T_w the inflow of entropy relative to the heat Q_w is $\int_{x_1}^{x_2} dQ_w/T_w$, and that relative to Q_b is Q_b/T_b, T_b representing the temperature of the corresponding body, assumed uniform. Thus we can write the second law in the form

$$\Delta S = m_2 s_2 - m_1 s_1 - \int_{x_1}^{x_2} s_i dm_i - \int_{x_1}^{x_2} \frac{dQ_w}{T_w} - \frac{Q_b}{T_b} \geqq 0 \qquad (4\text{-}7)$$

In differential form we have

$$dS = d(ms) - s_i dm_i - \frac{dQ_w}{T_w} - \frac{dQ_b}{T_b} \geqq 0 \qquad (4\text{-}7a)$$

Special cases with constant flow rate. The most common cases of flow are without injection. We list below a number of cases of special interest.

[3] If the gas is injected or withdrawn through concentrated ports protruding into the duct, such as scoops, injection nozzles, and sampling probes, one must apply the kind of treatment used for the study of mixing processes (Art. 15).

For $m_i = dm_i = 0$ the above equations are reduced to the following form:

$$m_2 = m_1 = m = A\rho u = \text{const}; \quad dm = 0 \tag{4-2b}$$

$$I_2 - I_1 = m(u_2 - u_1) = -\int_{x_1}^{x_2} A dp - X; \qquad dI = m du = -A dp - dX \tag{4-3b}$$

$$H_2^0 - H_1^0 = m(h_2^0 - h_1^0) = Q; \qquad dH^0 = m dh^0 = dQ \tag{4-4b}$$

$$J_2 - J_1 = \int_{x_1}^{x_2} p dA - X; \qquad dJ = p dA - dX \tag{4-5b}$$

$$\Delta S = m(s_2 - s_1) - \int_{x_1}^{x_2} \frac{dQ_w}{T_w} - \frac{Q_b}{T_b} \geqq 0; \qquad dS = m ds - \frac{dQ_w}{T_w} - \frac{dQ_b}{T_b} \geqq 0 \tag{4-7b}$$

From Eq. 4-5b we observe that the change in the impulse function equals the total force exerted by the walls on the fluid (including the body forces). Similarly, Eq. 4-4b shows that the change of stagnation enthalpy flux is equal to the total heat introduced in the fluid through the walls (including those of the bodies). This observation justifies the importance of the impulse function and of the stagnation enthalpy.

No injection, no friction. Now $X = dX = 0$ and, therefore, Eq. 4-3b and 4-5b become

$$I_2 - I_1 = m(u_2 - u_1) = -\int_{x_1}^{x_2} A dp; \quad dI = m du = -A dp \tag{4-3c}$$

$$J_2 - J_1 = \int_{x_1}^{x_2} p dA; \qquad dJ = p dA \tag{4-5c}$$

Replacing m from Eq. 4-2b, the second Eq. 4-3c becomes

$$dp = -\rho u du \tag{4-8}$$

which is the special Bernoulli form of the energy equation for steady flow (A,8).

Adiabatic flow. Here Q, dQ, and their components Q_w, Q_b, dQ_w, and dQ_b vanish, and Eq. 4-4b reduces to

$$H_2^0 = H_1^0, \qquad dH^0 = 0 \tag{4-4c}$$

or, remembering Eq. 2-8, to

$$h_2^0 = h_1^0 = h_0 = h + \frac{u^2}{2} = \text{const}, \quad dh^0 = dh + u du = 0 \tag{4-9}$$

By definition of the stagnation temperature T^0, h^0 is related to T^0 by the same equation (Eq. 2-9) which relates the static values. In the general case when chemical changes are present one can write, instead of Eq. 2-18,

$$h_{b2}^0 - h_{u1}^0 = (h_r)_b - (h_r)_u + \int_{T_r}^{T_2^0} c_{p_b} dt - \int_{T_r}^{T_1^0} c_{p_u} dt \tag{4-10}$$

For adiabatic flows this quantity vanishes because of Eq. 4-9 and we obtain the equation similar to Eq. 2-19

$$\int_{T_1^0}^{T_2^0} c_{p_b} dT = q_{th} = (h_r)_u - (h_r)_b + \int_{T_r}^{T_1^0} (c_{p_u} - c_{p_b}) dT \qquad (4\text{-}11)$$

where q_{th} is the thermal effect of the transformation at the temperature T_1^0. One can make about Eq. 4-11 the same observations made on Eq. 2-19, and develop the same approximate relations with the only difference that the static temperatures must be replaced by their stagnation values. One can thus define a quantity

$$\lambda = \frac{(c_{p_b})_2 T_2^0}{(c_{p_u})_1 T_1^0} \qquad (4\text{-}12)$$

and replace the solution of the energy equation (Eq. 4-11) by assigning the value of λ, with the implicit assumption that the energy equation is satisfied for this value of λ. Here $(c_{p_b})_2$ and $(c_{p_u})_1$ are the values relative to the corresponding stagnation temperatures.

For purely physical transformations without changes of state an adiabatic flow is also called isoenergetic. In this case $q_{th} = 0$ and, therefore, instead of the energy equation (Eq. 4-9) one can write

$$T_2^0 = T_1^0 = T^0 = \text{const}; \qquad dT^0 = 0 \qquad (4\text{-}13)$$

If at section 2 the velocity u_2 vanishes, h_2^0 coincides with h_2, and T_2^0 with T_2. Therefore the stagnation temperature in section 1 can be evaluated by measuring the static temperature of the gas after it has been brought to rest adiabatically. The instrument based on this property is called the "stagnation probe" (see IX,D,2). For polytropic gases one can write explicitly

$$T^0 = T + \frac{u^2}{2c_p} \qquad (4\text{-}14)$$

For adiabatic flows the second law is simply expressed by the inequalities

$$s_2 \geqq s_1; \qquad ds \geqq 0 \qquad (4\text{-}7c)$$

Frictionless adiabatic flow. In this case the Bernoulli equation (Eq. 4-8) applies. Combining with Eq. 4-9 we obtain

$$dh - \frac{1}{\rho} dp = 0$$

and, therefore, for frozen or equilibrium systems we find from Eq. 2-12,

$$ds = 0$$

Thus in this case the flow is isentropic, and the constancy of entropy can be used as an equation instead of one of the other equations of conservation. Obviously this property does not apply when an irreversible transformation, such as combustion, is present.

Stagnation pressure. The abstract concept of entropy can be supplemented by the more mechanical concept of stagnation pressure. The stagnation temperature is the temperature obtained when the gas is brought to rest adiabatically. The resulting temperature is the same, independent of the kind of process used to reduce the velocity to zero. However, the final pressure depends on the amount of energy dissipated during the process, for instance on the value of the friction force X, and is always smaller than the pressure obtained when no dissipative forces are present. In the latter case the process is isentropic, and the corresponding pressure p^0, i.e. the pressure obtained when the gas is brought to rest isentropically is called the stagnation pressure. Contrary to the stagnation temperature, an instrument directly measuring the stagnation pressure is in principle possible only in subsonic flow. For supersonic flow the stagnation pressure can be obtained only indirectly from simultaneous measurements of the so-called Pitot or impact pressure and of some other quantity, such as the static pressure.[4]

By definition the entropy of the gas is the same, whether expressed in terms of static or stagnation temperature and pressure, and from Eq. 2-13 is given by

$$s - s_r = \sigma(T) - \mathcal{R} \ln p = \sigma(T^0) - \mathcal{R} \ln p^0 = \mathcal{R} \ln \frac{\varphi(T)}{p} = \mathcal{R} \ln \frac{\varphi(T^0)}{p^0} \tag{4-15}$$

Therefore, we obtain, as a particular case of Eq. 2-14,

$$\frac{p^0}{p} = \frac{\varphi(T^0)}{\varphi(T)} \tag{4-16}$$

For polytropic gases this relation takes the form of Eq. 2-29. Recalling Eq. 4-14 we can write

$$\frac{p^0}{p} = \left(\frac{T^0}{T}\right)^{\frac{\gamma}{\gamma-1}} = \left(1 + \frac{u^2}{2c_pT}\right)^{\frac{\gamma}{\gamma-1}} \tag{4-16a}$$

The corresponding expression for the entropy is given by

$$s - s_r = c_p \ln T - \mathcal{R} \ln p = c_p \ln T^0 - \mathcal{R} \ln p^0 \tag{4-15a}$$

Since T^0 remains constant for an adiabatic flow without chemical changes (isoenergetic flow), the changes in entropy are only produced by changes of stagnation pressure, the latter decreasing when the entropy increases.

Isoenergetic, isentropic flows are distinguished by the constancy of the stagnation temperature and pressure.

[4] See IX,B,1, where, however, the stagnation pressure is identified with the impact pressure, which is not the conventional definition. See also A,7.

Adiabatic constant area flow. In this particular case one has

$$h^0 = \text{const}, \quad A = \text{const}$$

For frozen or equilibrium systems this type of flow is related to the so-called Fanno line, discussed in Art. 13.

Frictionless constant pressure flow. Eq. 4-3c shows that in this case

$$I_2 = I_1 = I = mu = \text{const}, \quad p = \text{const}$$

and therefore

$$u = \text{const} \tag{4-17}$$

The momentum and the velocity cannot vary because friction and pressure forces vanish by assumption.

Frictionless constant area flow. From Eq. 4-5c we obtain in this case

$$J_2 = J_1 = J = \text{const}, \quad A = \text{const} \tag{4-18}$$

Therefore this category of flows is characterized by the constancy of the impulse function, and, for frozen or equilibrium systems, is related to the so-called Rayleigh line, discussed in Art. 13.

Ducts of the pressure-area power family. The two latter types of flow considered are distinguished by the constancy of a certain quantity throughout the flow. Both types can be included as particular cases in a more general family of flows obtained by prescribing a power relation between pressure and area. We assume that the duct is such that at every point the following relation is satisfied

$$pA^{\frac{\varepsilon}{\varepsilon-1}} = p_1A_1^{\frac{\varepsilon}{\varepsilon-1}} = p_2A_2^{\frac{\varepsilon}{\varepsilon-1}} = \text{const} \tag{4-19}$$

with a prescribed value of ε. Then the integral in Eq. 4-3 can be evaluated giving

$$\int_{x_1}^{x_2} A\,dp = \varepsilon(p_2A_2 - p_1A_1) \tag{4-19a}$$

Introducing the quantity

$$J^{(\varepsilon)} = I + \varepsilon pA \tag{4-20}$$

Eq. 4-3 becomes

$$J_2^{(\varepsilon)} - J_1^{(\varepsilon)} = -X + I_i \tag{4-21}$$

Hence, for constant flow rate and no friction, the flow in ducts of the p-A power family is characterized by

$$J_2^{(\varepsilon)} = J_1^{(\varepsilon)} = J^{(\varepsilon)} = \text{const}, \quad (p^{\varepsilon-1}A^{\varepsilon} = \text{const}) \tag{4-22}$$

We shall call this quantity the "generalized impulse function."

Obviously the constant pressure and the constant area cases correspond to the particular values $\varepsilon = 0$ and $\varepsilon = 1$, and in these cases we have $J^{(0)} = I$ and $J^{(1)} = J$. It is clear that Eq. 4-19 does not define a particular geometry of duct, the relation between A and p being determined a posteriori by the transformation which takes place in the duct, a feature which will be better seen in the discussion of particular examples. Only for $\varepsilon = 1$ does the duct become a pipe of constant area, therefore having a shape independent of the transformation. Despite this lack of geometrical definition the consideration of ducts of the power family leads to interesting results.

Reduced velocities and Mach number. Nondimensional expressions for the velocity are very useful. The most commonly used quantity is the Mach number

$$M = \frac{u}{a} \tag{4-23}$$

obtained by dividing by the local sound velocity (Eq. 2-16). However, for the purpose of practical calculations it is sometimes better to use a more convenient quantity as the reference velocity. Such a quantity is suggested by the energy equation (Eq. 4-8), which shows that there exists an upper limit to the velocity, obtained when h takes its minimum possible value. If, in Eq. 2-9, h_r is chosen so as to make $h = 0$ at $T = 0$, this is the minimum value of h and thus

$$u_{max} = \sqrt{2h^0} \tag{4-24}$$

represents the maximum velocity, attainable through an adiabatic (isoenergetic) expansion down to the absolute zero. For polytropic gases we have

$$u_{max} = \sqrt{2c_pT^0} \tag{4-24a}$$

Based on the last expression for u_{max} we define a reduced velocity

$$w = \frac{u}{\sqrt{2c_pT^0}} \tag{4-25}$$

In agreement with the considerations developed at the end of this article, we shall also employ this definition, instead of the exact definition u/u_{max}, for nonpolytropic gases, c_p being the value corresponding to the stagnation temperature.

Two other reference velocities are also in use. One is the sound velocity at the stagnation temperature

$$a^0 = \sqrt{\gamma(T^0)\mathcal{R}T^0}$$

The other is the so-called critical velocity a^*, i.e. the value of u when $u = a$, or $M = 1$. The corresponding reduced velocities

$$M^0 = \frac{u}{a^0}, \qquad M^* = \frac{u}{a^*} \tag{4-26}$$

are both used in practice.

For polytropic gases one can write Eq. 4-9 in the corresponding equivalent forms

$$c_pT + \frac{u^2}{2} = \frac{a^2}{\gamma - 1} + \frac{u^2}{2} = c_pT^0 = \frac{a^{02}}{\gamma - 1} = \frac{u^2_{\max}}{2} = \frac{\gamma + 1}{\gamma - 1}\frac{a^{*2}}{2} \quad (4\text{-}27)$$

from which one easily obtains the ratio between stagnation and static temperature (and from Eq. 4-16a, between the corresponding pressures) as

$$\frac{T^0}{T} = \left(\frac{p^0}{p}\right)^{\frac{\gamma-1}{\gamma}} = 1 + \frac{\gamma - 1}{2}M^2 = \frac{1}{1 - w^2} = \frac{1}{1 - \frac{\gamma - 1}{2}M^{02}} = \frac{1}{1 - \frac{\gamma - 1}{\gamma + 1}M^{*2}} \quad (4\text{-}28)$$

The following relations between the various reduced velocities are obtained from Eq. 4-28:

$$M^2 = \frac{2}{\gamma - 1}\frac{w^2}{1 - w^2}, \quad M^{02} = \frac{2}{\gamma - 1}w^2, \quad M^{*2} = \frac{\gamma + 1}{\gamma - 1}w^2 \quad (4\text{-}29)$$

The choice of the proper reduced velocity is a matter of formal convenience or personal preference; and actually, with the exclusion of M, the reduced velocities are simply proportional to each other. The following table compares the subsonic and the supersonic ranges in terms of the various reduced velocities. In the rest of this section we shall make use only of w and M.

Table B,4

Range	M	w	M^0	M^*
Subsonic	0	0	0	0
Subsonic / Supersonic	1	$\sqrt{\frac{\gamma - 1}{\gamma + 1}}$	$\sqrt{\frac{2}{\gamma + 1}}$	1
Supersonic	∞	1	$\sqrt{\frac{2}{\gamma - 1}}$	$\sqrt{\frac{\gamma + 1}{\gamma - 1}}$

Useful relations for polytropic gases. The various quantities appearing in the conservation laws can be expressed in different forms, useful in developments. The velocity is given by

$$u = M\sqrt{\gamma \mathcal{R} T} = \sqrt{\mathcal{R}T^0}\left(\sqrt{\gamma}\, M\sqrt{\frac{T}{T^0}}\right) \quad (4\text{-}30)$$

For polytropic gases the quantity in parentheses can be expressed explicitly as a function of only γ and M through Eq. 4-28. The flow rate

(Eq. 4-1) can be obtained, using the equation of state and Eq. 4-30, as

$$m = \frac{pA}{\sqrt{\mathcal{R}T^0}}\left[\sqrt{\gamma}\,M\sqrt{\frac{T^0}{T}}\right] = \frac{p^0A}{\sqrt{\mathcal{R}T^0}}\left[\sqrt{\gamma}\,M\left(\frac{T}{T^0}\right)^{\frac{1}{2}\frac{\gamma+1}{\gamma-1}}\right] \quad (4\text{-}31)$$

Here use has been made of Eq. 4-16a. The momentum flow rate (Eq. 4-1) is given by

$$I = m\sqrt{\mathcal{R}T^0}\left[\sqrt{\gamma}\,M\sqrt{\frac{T}{T^0}}\right] = pA[\gamma M^2] = p^0A\left[\gamma M^2\left(\frac{T}{T^0}\right)^{\frac{\gamma}{\gamma-1}}\right] \quad (4\text{-}32)$$

the last expression being obtained with the help of Eq. 4-16a. Similarly the impulse function (Eq. 4-6) and the generalized impulse function (Eq. 4-20) are expressed by

$$J = m\sqrt{\mathcal{R}T^0}\left[\frac{1+\gamma M^2}{\sqrt{\gamma}\,M}\sqrt{\frac{T}{T^0}}\right] = pA[1+\gamma M^2]$$

$$= p^0A\left[(1+\gamma M^2)\left(\frac{T}{T^0}\right)^{\frac{\gamma}{\gamma-1}}\right] \quad (4\text{-}33)$$

$$J^{(\varepsilon)} = m\sqrt{\mathcal{R}T^0}\left[\frac{\varepsilon+\gamma M^2}{\sqrt{\gamma}\,M}\sqrt{\frac{T}{T^0}}\right] = pA[\varepsilon+\gamma M^2]$$

$$= p^0A\left[(\varepsilon+\gamma M^2)\left(\frac{T}{T^0}\right)^{\frac{\gamma}{\gamma-1}}\right] \quad (4\text{-}34)$$

Again all quantities in brackets contain the whole dependence on γ and M. Many of these quantities in brackets, or quantities closely related to them, have been tabulated [*3,15,16,17*].

The corresponding expressions in terms of w are

$$u = \sqrt{\mathcal{R}T^0}\left[\sqrt{\frac{2\gamma}{\gamma-1}}\,w\right] \quad (4\text{-}30a)$$

$$m = \frac{pA}{\sqrt{\mathcal{R}T^0}}\left[\sqrt{\frac{2\gamma}{\gamma-1}}\,\frac{w}{1-w^2}\right] = \frac{p^0A}{\sqrt{\mathcal{R}T^0}}\left[\sqrt{\frac{2\gamma}{\gamma-1}}\,w(1-w^2)^{\frac{1}{\gamma-1}}\right] \quad (4\text{-}31a)$$

$$I = m\sqrt{\mathcal{R}T^0}\left[\sqrt{\frac{2\gamma}{\gamma-1}}\,w\right] = pA\left[\frac{2\gamma}{\gamma-1}\,\frac{w^2}{1-w^2}\right]$$

$$= p^0A\left[\frac{2\gamma}{\gamma-1}\,w^2(1-w^2)^{\frac{1}{\gamma-1}}\right] \quad (4\text{-}32a)$$

$$J = m\sqrt{\mathcal{R}T^0}\left[\sqrt{\frac{2\gamma}{\gamma-1}}\left(\frac{\gamma+1}{2\gamma}\,w + \frac{\gamma-1}{2\gamma}\,\frac{1}{w}\right)\right] = pA$$

$$\left[1+\frac{2\gamma}{\gamma-1}\,\frac{w^2}{1-w^2}\right] = p^0A\left[\left(1+\frac{\gamma+1}{\gamma-1}\,w^2\right)(1-w^2)^{\frac{1}{\gamma-1}}\right] \quad (4\text{-}33a)$$

$$J^{(\varepsilon)} = m\sqrt{\mathfrak{R}T^0}\left[\sqrt{\frac{2\gamma}{\gamma-1}}\left(w + \varepsilon\frac{\gamma-1}{2\gamma}\frac{1-w^2}{w}\right)\right]$$

$$= pA\left[\varepsilon + \frac{2\gamma}{\gamma-1}\frac{w^2}{1-w^2}\right] = p^0A\left[\left(\varepsilon + \left(\frac{2\gamma}{\gamma-1} - \varepsilon\right)w^2\right)(1-w^2)^{\frac{1}{\gamma-1}}\right] \quad (4\text{-}34a)$$

Observe that only the expressions where p^0 appears contain an irrational quantity, all other expressions being rational in w. This is not true when M is used instead of w, and may constitute an advantage in practical manipulations. In the following articles we shall show some applications of the equations developed here.

Approximate treatment for nonpolytropic gases. In Art. 13 we shall discuss exact and approximate methods of accounting for the variability of the specific heats. All these treatments are substantially more complicated than those that are obtained for polytropic gases, based on the simple relations developed in this article. This is the reason why, quite generally, the treatment of flow problems for actual gases is made using these relations, despite the fact that with the exception of monatomic gases, or of particular temperature ranges, no real gas is polytropic. This approximation can be justified to a certain extent by the following considerations.

During a purely physical process the temperature of the gas may vary in a wide range, with the consequence that the specific heat changes substantially, and taking it invariable during the whole process would represent a poor approximation. If, however, at every point of the duct the Mach number is sufficiently low so that the ratio between stagnation and static temperature does not depart considerably from unity, then one can neglect the variation of specific heats between the temperatures T^0 and T proper of each section, taking for instance the value of the specific heat proper of T^0, and apply to each section the relations of Eq. 4-28. The result is that Eq. 4-30, 4-31, 4-32, 4-33, and 4-34 can again be expressed explicitly in terms of the local values of the various quantities, where, however, γ would be variable from section to section. Similarly one can apply Eq. 4-30a, 4-31a, 4-32a, 4-33a, and 4-34a with w defined by Eq. 4-25 with $c_p = c_p(T^0)$ variable along the process.

When chemical transformations are present the specific heat varies, not only as a consequence of the temperature variation but also because of the change in composition, which involves a change of c_p and of $\mathfrak{R}$ as well (the latter being invariable in the previous case) and thus of γ and $\mathfrak{R}$. Obviously, however, $\mathfrak{R}$ is constant on each section between the static and the stagnation conditions; and if the same approximate assumption is made for c_p or γ, again we can use the relations derived for polytropic gases with $\mathcal{R}$, c_p, and γ variable from section to section.

It is clear that this useful approximation can be used confidently for subsonic velocities. In the range of supersonic velocities, however, it becomes worse with increasingly high Mach number, and for very high M it may lead to considerable errors.

B,5. Shock Waves and Pseudo-Shocks in Ducts. The simplest and most common application of the one-dimensional theory is represented by the treatment of isentropic flows in ducts, in which the variations of the properties are determined by variations of the duct cross-sectional area. However, for the understanding of important effects that may be present in a duct when the velocity is supersonic, it is convenient to discuss first a phenomenon of fundamental importance.

For simplicity let us assume that the duct has constant area, and let us neglect the frictional effects. We take a control surface constituted of two sections, 1 and 2, and the cylindrical wall between the two. In the absence of mass or heat exchanges, the three conservation equations can be written in the form of Eq. 4-2b, 4-13, and 4-18, that is

$$m_2 = m_1, \quad T_2^0 = T_1^0, \quad J_2 = J_1 \tag{5-1}$$

Using Eq. 4-33a and the fact that $A_2 = A_1$, the following two equations are obtained from Eq. 5-1:

$$\frac{\gamma - 1}{2\gamma}\frac{1}{w_2} + \frac{\gamma + 1}{2\gamma} w_2 = \frac{\gamma - 1}{2\gamma}\frac{1}{w_1} + \frac{\gamma + 1}{2\gamma} w_1;$$

$$p_2\left(1 + \frac{2\gamma}{\gamma - 1}\frac{w_2^2}{1 - w_2^2}\right) = p_1\left(1 + \frac{2\gamma}{\gamma - 1}\frac{w_1^2}{1 - w_1^2}\right) \tag{5-2}$$

Also, from Eq. 4-28 and 4-1, observing that u_{max}, defined by Eq. 4-24a, is constant because of Eq. 5-1, one has

$$\frac{T_2}{1 - w_2^2} = \frac{T_1}{1 - w_1^2}, \qquad \rho_2 w_2 = \rho_1 w_1 \tag{5-3}$$

Obviously a solution of Eq. 5-2 and 5-3 is $w_2 = w_1$, $p_2 = p_1$, $T_2 = T_1$, and $\rho_2 = \rho_1$. However, independently of this trivial solution, the existence of which does not need to be established through equations, a second solution of the first of Eq. 5-2 appears to be possible, expressed by the relation

$$w_2 = \frac{\gamma - 1}{\gamma + 1}\frac{1}{w_1} = \frac{w^{*2}}{w_1} \quad \left(w^* = \frac{a^*}{u_{max}} = \sqrt{\frac{\gamma - 1}{\gamma + 1}}\right) \tag{5-4}$$

known as the Prandtl relation. The reduced sound velocity, w^*, plays an important role in compressible flow treatment. From the other equations (Eq. 5-2 and 5-3) one then obtains

$$\frac{p_2}{p_1} = \frac{\dfrac{w_1^2}{w^{*2}} - w^{*2}}{1 - w_1^2}, \quad \frac{\rho_2}{\rho_1} = \frac{w_1^2}{w^{*2}}, \quad \frac{T_2}{T_1} = \frac{1 - \dfrac{w^{*4}}{w_1^2}}{1 - w_1^2} \tag{5-5}$$

Through Eq. 4-29 one can write Eq. 5-4 and 5-5 in terms of M instead of w, obtaining, after replacing w^* by its value in terms of γ (Eq. 5-4)

$$\left(M_1^2 - \frac{\gamma - 1}{2\gamma}\right)\left(M_2^2 - \frac{\gamma - 1}{2\gamma}\right) = \left(\frac{\gamma + 1}{2\gamma}\right)^2 \tag{5-4a}$$

$$\frac{p_2}{p_1} = \frac{2\gamma}{\gamma + 1} M_1^2 - \frac{\gamma - 1}{\gamma + 1}, \quad \frac{\rho_2}{\rho_1} = \frac{\gamma - 1}{\gamma + 1} + \frac{2}{\gamma + 1}\frac{1}{M_1^2} \tag{5-5a}$$

and T_2/T_1 as the product of the last two expressions. Thus the conservation equations (Eq. 5-1) can be satisfied by a particular set of conditions of the gas leaving the control surface, different from the entrance conditions. Without bothering for the moment with the physical interpretation of this solution, let us check if it is consistent with the limitations imposed by the second law of thermodynamics expressed in this case by Eq. 4-7c, that is by the condition that the entropy must not decrease. For this purpose let us eliminate w_1 from the first two equations (Eq. 5-5) and solve for the density ratio. We obtain the so-called Rankine-Hugoniot relation

$$\frac{\rho_2}{\rho_1} = \frac{\dfrac{p_2}{p_1} + \dfrac{\gamma - 1}{\gamma + 1}}{1 + \dfrac{\gamma - 1}{\gamma + 1}\dfrac{p_2}{p_1}} \tag{5-6}$$

Since this relation is different from the isentropic relation (Eq. 2-29), we see that the transition from conditions 1 to conditions 2, if possible, is always associated with an entropy change. In Fig. B,5a we have represented, for $\gamma = 1.4$, the relation of Eq. 5-6, the isentropic relation, and the entropy change, related to the pressure and density ratios and to the stagnation pressure ratio by the relations

$$\frac{s_2 - s_1}{\mathcal{R}} = \frac{1}{\gamma - 1}\left(\ln\frac{p_2}{p_1} - \gamma \ln\frac{\rho_2}{\rho_1}\right) = \ln\frac{p_1^0}{p_2^0} \tag{5-7}$$

derived from Eq. 2-27 and 4-15a. We see that $s_2 - s_1$ is zero at $M_1 = 1$ and increases with M_1 as shown by the corresponding values on the curves. For other values of γ the qualitative behavior is the same.

We immediately see that if $M_1 < 1$, that is ρ_2/ρ_1 and p_2/p_1 are smaller than one, the transition to conditions 2 always corresponds to an entropy decrease, and hence is impossible. However, for $M_1 > 1$ we have a positive entropy change, allowed by the second law. In this case p_2/p_1 and

ρ_2/ρ_1 are larger than one; also since $w_1 > w^*$, Eq. 5-4 shows that $w_2 < w^*$. Observe that at $M_1 = 1$ the curve corresponding to Eq. 5-6 appears to have a contact higher than simple tangency with that corresponding to the isentropic transformation, so that the entropy change stays quite small for a considerable range of M_1. This property is shown analytically by the expansion of Eq. 5-7, after introduction of Eq. 5-6, in powers of

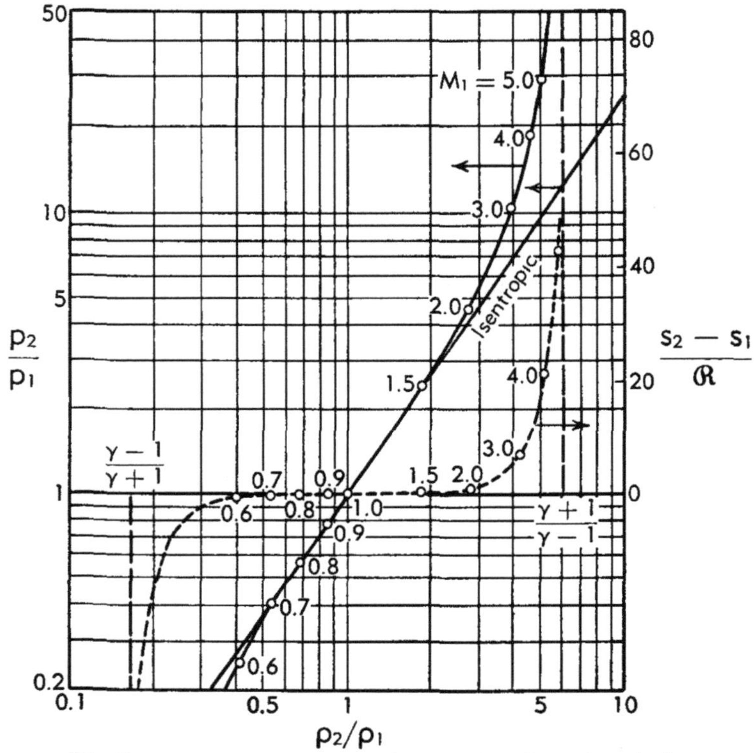

Fig. B,5a. Representation of the Rankine-Hugoniot relation and associated entropy variation.

$(p_2/p_1) - 1$ or, through Eq. 5-5a, of $M_1^2 - 1$; or, finally, by the expansion of Eq. 5-7 in powers of $(w_1^2/w^{*2}) - 1$:

$$\left.\begin{aligned}\frac{s_2 - s_1}{\mathcal{R}} &= \frac{\gamma + 1}{12\gamma^2}\left(\frac{p_2}{p_1} - 1\right)^3 - \frac{\gamma + 1}{8\gamma^2}\left(\frac{p_2}{p_1} - 1\right)^4 + \cdots \\ &= \frac{2\gamma}{3(\gamma + 1)^2}(M_1^2 - 1)^3 - \frac{2\gamma^2}{(\gamma + 1)^3}(M_1^2 - 1)^4 + \cdots\end{aligned}\right\} \tag{5-8}$$

$$\frac{s_2 - s_1}{c_p} = \frac{w^{*2}}{3(1 - w^{*2})^2}\left(\frac{w_1^2}{w^{*2}} - 1\right)^3 - \frac{w^{*2}}{2(1 - w^{*2})^2}\left(\frac{w_1^2}{w^{*2}} - 1\right)^4 + \cdots \tag{5-8a}$$

The observed behavior near $M_1 = 1$ comes from the fact that the first two terms of these expansions are identically zero.

It results from the above discussion that on the basis of the conservation equations, and of the second law of thermodynamics, the only possible flow in a frictionless, adiabatic, constant area duct is the "trivial" solution (constant conditions) for subsonic velocities; however, if the flow is initially supersonic we cannot exclude the possibility of the above-analyzed transition to subsonic velocity with the corresponding increase of pressure, density, and temperature. Numerical values can be obtained from published tables (see, for instance, [*3,15,17*]).

Observe that if the assumptions on which the derivation of this solution is based were satisfied for each section between 1 and 2, at each intermediate section the velocity should be either w_1 or w_2. Thus a transition from conditions 1 to conditions 2 is possible only if a deviation from such assumptions is present. Moreover, this deviation must involve a dissipative process, necessary to produce the required entropy increase. In the rest of this article we shall discuss in some detail two mechanisms through which the transition can take place.

The shock wave. One of these mechanisms is the shock wave. We shall make here only a few remarks on this phenomenon, which is thoroughly discussed and analyzed in Sec. D. In the derivation of the conservation equations it has been assumed that the axial gradients are small so that the axial viscous stress and the axial heat exchange by conduction can be neglected. Of course, this simplification is not justified if the limitation about the magnitude of the axial gradients is relieved; and for very large gradients the presence of substantial local axial stresses and heat exchanges must be taken into account, thus altering the momentum and energy equations.

The treatment of these modified equations (Sec. D) confirms the possibility of gradual transition only in the direction from a supersonic to a subsonic flow. The entropy increases as a result of both the dissipation due to the viscous stressses and the heat transferred with finite temperature gradients. The region where practically the whole transition from the initial to the final conditions takes place has a thickness which, though finite, becomes exceedingly small as soon as the initial velocity is even slightly supersonic. For all practical purposes this thickness can be neglected, and the transition assimilated to a discontinuity, called shock wave. Because of the one-dimensionality of the flow the discontinuity makes a right angle with the flow direction. This type of shock wave is therefore called a normal shock wave to distinguish it from the more general type, the oblique shock wave.

The properties of oblique shock waves are discussed in detail in Sec. E. Here we shall only mention some of its fundamental features which are necessary for the discussions of this section. Let us consider an indefinite normal shock wave, with velocity u_{n1} before and u_{n2} after the shock. The Mach numbers before and after the shock are $M_{n1} = u_{n1}/a_1$ and $M_{n2} =$

u_{n2}/a_2 and are related by Eq. 5-4a written in terms of M_{n1}, M_{n2} instead of M_1, M_2. The pressure, density, and temperature ratios can be calculated from Eq. 5-5a, expressed in terms of M_{n1} instead of M_1. Suppose now (Fig. B,5b) we apply to the one-dimensional flow field constituted by this normal shock wave a uniform velocity u_t parallel to the shock wave. Obviously this uniform translation does not affect the pressure, density, and temperature ratios. However, the resulting initial and final velocities, u_1 and u_2, are no more at right angle to the shock wave and, in general, are not aligned. If θ_1 and θ_2 represent the angles between u_1, u_2 and the shock wave, and M_1 and M_2 are the corresponding Mach numbers, then

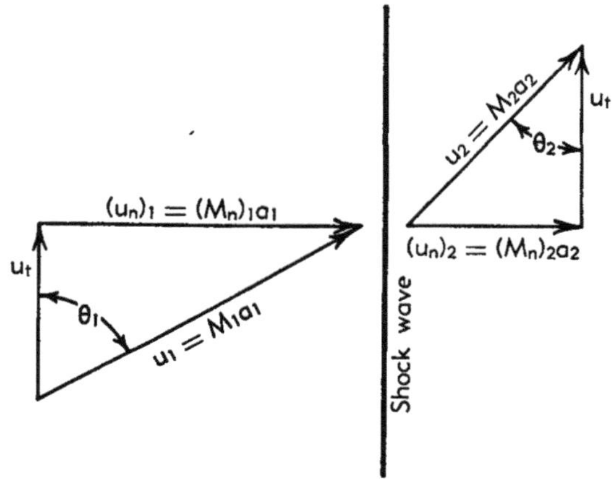

Fig. B,5b. Triangles of velocity before and after an oblique shock.

$M_{n1} = M_1 \sin \theta_1$ and $M_{n2} = M_2 \sin \theta_2$ now represent the normal components of the Mach numbers. Eq. 5-4a and 5-5a, written in terms of these normal components, plus the relation

$$\frac{\tan \theta_2}{\tan \theta_1} = \frac{u_{n2}}{u_{n1}} = \frac{\rho_1}{\rho_2}$$

immediately derived from geometrical considerations and from Eq. 5-3 (written with u_{n1} and u_{n2} in place of w_1 and w_2), are sufficient to determine completely the conditions after the shock in terms of the conditions before and of the obliquity θ_1 of the shock. The angle $\theta_2 - \theta_1$ represents the deviation of the velocity through the shock.

If we change θ_1 for fixed initial conditions the final conditions vary. The minimum θ_1 is determined by the condition $M_{n1} \geqq 1$, and is therefore given by $\sin \theta_1 = 1/M_1$. This value of θ_1 is called the "Mach angle." At the Mach angle we have $M_{n1} = 1$, and also therefore $M_{n2} = 1$ and $p_2/p_1 = \rho_2/\rho_1 = T_2/T_1 = 1$, that is $M_2 = M_1$ and the shock wave vanishes. The maximum M_{n1} and, therefore, the maximum pressure, density, and temperature ratios and the minimum M_2 are obtained when the shock is normal ($\theta_1 = 90°$). Thus for fixed initial conditions a whole range of shock waves is possible, ranging from a compression wave of vanishing strength (Mach wave; $\theta_1 = \theta_2 =$ Mach angle), to the maximum strength of a normal shock ($\theta_1 = \theta_2 = 90°$). The deviation is zero only in these two extreme cases, and reaches a maximum for some intermediate strength. In the range near the Mach wave the final conditions are supersonic, p_2/p_1 is close to 1 and therefore, as already observed, the entropy increase is small. Shock waves in this range are called weak shock waves. They are, on the contrary, called strong shock waves in the range near the normal shock wave, where the entropy increase is large (if M_1 is not close to one) and the final velocity is subsonic.

Shock waves are present very frequently in supersonic flows, around bodies or in ducts. However, a simple normal shock wave is not, in most cases, the proper mechanism of transition from the supersonic to the subsonic conditions as observed in ducts. The actual mechanism, discussed below, is in general more complicated and leads to a more gradual transition which is substantially different from the simple discontinuity of a shock wave.

The pseudo-shock in ducts. The reason for the discrepancy resides in the presence of the boundary layers adherent to the walls of the duct, in which the flow is retarded substantially by viscous actions. On the walls themselves, in particular, the velocity must vanish. The longer the fluid is subject to the retarding effects of the viscous forces, the more the boundary layer develops. If at the entrance of a pipe the velocity is uniformly distributed, boundary layers start growing at once. Very thin at the beginning, they grow thicker and thicker until at a certain distance from the entrance section the boundary layers from opposite walls merge together, and fill the whole section. In the same time the nondissipative core, in which the velocity is still uniform, is reduced to a smaller and smaller portion of the section until it disappears completely. The region following the entrance where the development of the boundary layer takes place is called the "entrance stretch." After the entrance stretch the velocity distribution reaches a kind of "regime" condition, characteristic of the so-called fully developed pipe flow. Analogous effects are present in variable area ducts, but often the limitation of the length does not allow the establishment of fully developed conditions.

The one-dimensional treatment disregards, as already stated in Art. 1, the nonuniformities deriving from the presence of the boundary layers, or better it replaces approximately the nonuniform flow by a fictitious uniform mean flow. However, a shock wave is a local phenomenon, and the relations developed above must be exactly satisfied at each point of the flow. Therefore whereas a discontinuity, separating two regions with uniform values of pressure, is possible in the central uniform core, such discontinuity is incompatible with the boundary layer regions where the decreased Mach number makes impossible the realization, through a shock, of the same pressure ratio as in the core. Moreover, near the walls the velocities are certainly subsonic and shocks are impossible. Thus the simple mechanism of the normal shock is incompatible with the presence of boundary layers.

The experimental observations confirm the above results. Plate B,5a (from [*18*]) shows a case where, the boundary layers having been removed, a shock can be established throughout the duct. This shock is curved, but in a case like this it is possible to obtain, by proper adjustment, a practically normal shock. Plate B,5b from [*19*] shows that a nearly normal shock apparently filling the whole section can be established when the Mach number is only slightly supersonic and the boundary layers are sufficiently thin, i.e. the Reynolds number is sufficiently high[5] (the thickness of the boundary layer, for a given duct geometry, decreases with

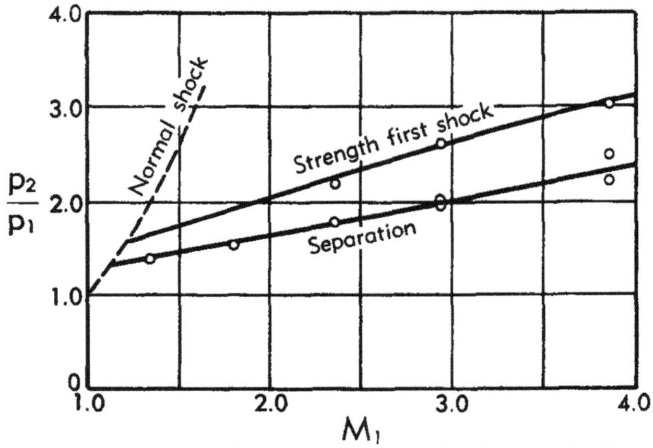

Fig. B,5c. Pressure ratio through an oblique shock emerging from a turbulent boundary layer. Pressure ratio at the separation point.

increasing Reynolds number). In this case it is still true that the normal shock cannot be established right to the walls; however, the pressure ratio being quite close to 1, the boundary layer can adjust itself to this pressure increase in a sufficiently short length without undergoing separation, so that the corresponding gradual process, made possible by viscous forces, is not clearly detectable by optical means.

There is a limit, however, to the possibility of adjustment of a boundary layer without separation, and above a certain pressure ratio (around 2 for turbulent boundary layers, see Fig. B,5c [*20*]), separation is produced. This involves a profound distortion of the flow downstream of the separation point, with the consequence that the foot of the normal shock, still present in the central part of the duct, becomes bifurcated near the wall and is replaced by two more or less straight oblique shocks (λ shock), as shown in Plate B,5c from [*19*], relative to a duct with rectangular section.

[5] See Eq. 10-2 for a definition of the Reynolds number for duct flows.

It must be noticed that under these conditions the configuration of the shock pattern becomes strongly dependent on the shape of the section. In the following discussion we assume a duct with a two-dimensional behavior, such as that which can be obtained for an elongated rectangular section in which the flow between the central parts of the wider walls is not very much influenced by the presence of the two shorter walls.

The strength of the first oblique shock of the λ shock is, for a turbulent boundary layer, a function only of the geometry of the system and of the Mach number [*20,21*] and is represented in Fig. B,5c for the case in which the wall is a flat plate. For comparison the strength of a normal shock for the same Mach number is also shown on the figure. We see that the difference between the two strengths undergoes a fast increase with M. The residual pressure increase, after the first oblique shock, is produced in a complicated way in the flow region included between the two branches of the λ shock.

If, however, this residual pressure increase is too large, only part of it can be realized in this way. In this case the flow after the normal shock must adjust itself to the lower pressure existing on the walls through an expansion which may again create supersonic conditions, and allow the whole pattern of λ shocks to be repeated more than once (Plate B,5d from [*15,19*]).

For a given Mach number the longitudinal and transversal dimensions of the region within the bifurcation of the λ shock increase with the boundary layer thickness, and therefore for a given height of the duct, the extent of the normal shock in the central part of the duct is reduced, until, above a certain thickness of the boundary layer, it completely disappears. The resulting pattern, shown in Plate B,5e, from [*22*], contains only a series of X shocks, and no more normal shocks. For other shapes of section this pattern is complicated by the interaction between the various walls. In particular, for circular section the plane oblique shocks must be replaced by conical shocks, and their intersection on the axis can only be produced through normal shocks. However, it is probably still true that the possible presence of portions of normal shocks is quite unimportant, and that for the sake of the discussion one can completely neglect them and assume that the shock pattern contains only oblique shocks of moderate strength.

The previous discussion can be summarized as follows. Only for low supersonic velocities and thin boundary layers is a quasi-normal shock possible in a duct. Otherwise, a more complicated, non-one-dimensional pattern is produced, which, for sufficiently high Mach numbers, becomes multiple. Above a certain boundary layer thickness this multiple pattern contains only oblique shocks crossing in the center of the duct and reflected back and forth in the central supersonic region of flow, while the region adjacent to the wall is adjusted gradually to the corresponding pressure increase. This adjustment is produced by strong exchanges of momentum due to the high turbulence generated in these regions. As a result of the turbulence the subsonic regions spread more and more into the supersonic region, until finally the latter disappears completely and the shock pattern terminates. The subsequent process is only one of adjustment of the subsonic velocity distribution, and does not produce further appreciable increases in pressure. We shall call this process "pseudo-shock."

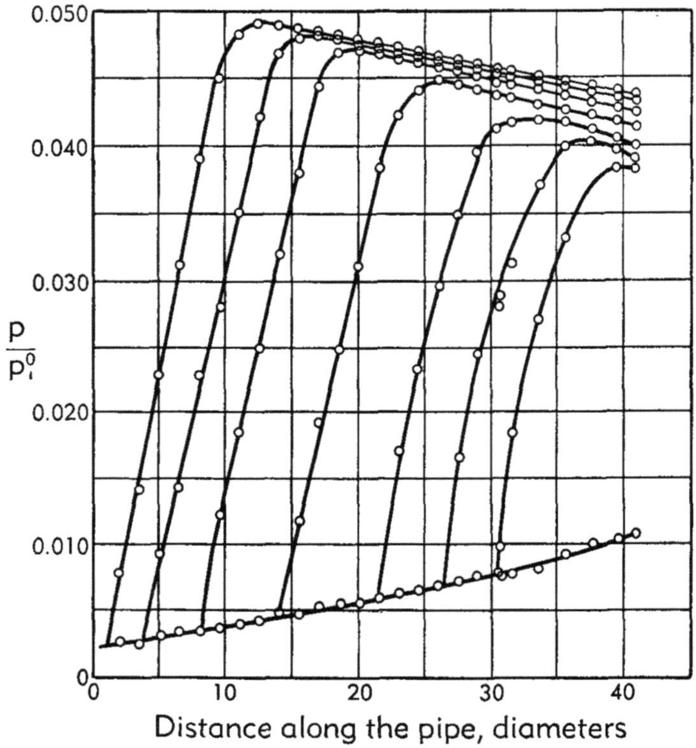

Fig. B,5d. Static pressure distributions in pseudo-shocks obtained in a pipe with fixed supersonic initial conditions and varying back pressure (p_1^0 = pressure in the supply tank).

Since throughout the whole process the velocity of the regions adjacent to the walls is subsonic, the corresponding frictional effects are small, and therefore the relations between the initial supersonic and the final subsonic conditions are very close to those derived in this article, on assumptions that are exactly true only for a normal shock. Thus the total entropy increase must also be nearly the same as in a normal shock. However, in the pseudo-shock only a part of this entropy increase is produced by the shock pattern, and the rest is due to the dissipative processes present in the strongly turbulent regions adjacent to the walls. It can easily be checked that it may very well be that the total entropy increase due to the series of oblique shocks of small strength constituting the shock pattern of Plate B,5e is only a small fraction of the total entropy rise required, in which case the predominant cause for the entropy increase resides in the dissipative turbulent regions.

The length of the region occupied by the pseudo-shock is a function of the flow characteristics. Fig. B,5d [*23*] shows the distributions of the static pressures measured on the wall of a pipe with fixed supersonic initial conditions and varying back pressure. The frictional effects and

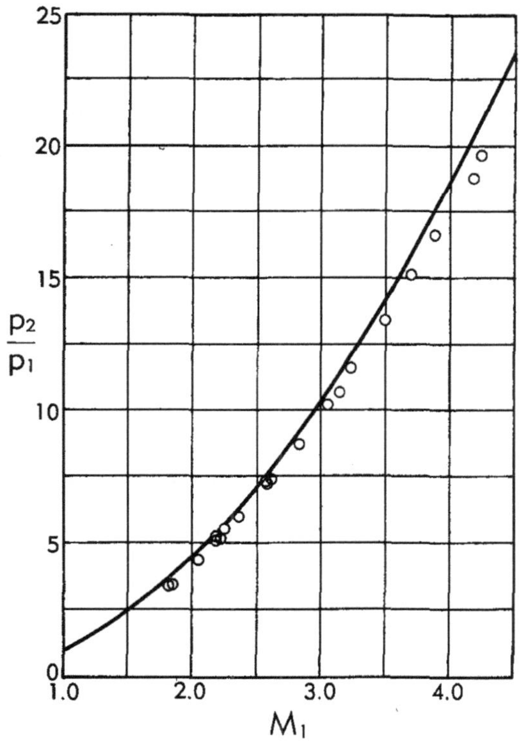

Fig. B,5e. Experimental pressure ratios of pseudo-shocks (circles) and theoretical pressure ratio of normal shocks (curve) at various Mach numbers.

the boundary layer development in the supersonic region are responsible for the gradual pressure increase (and velocity decrease) in the initial part of the pipe. These effects are discussed in detail in Art. 10. At a certain position along the pipe (determined by the value of the back pressure) a sudden increase of the pressure gradient is observed, and the pseudo-shock starts. When the starting point is at a sufficient distance from the entrance section, so that the boundary layers are sufficiently thick, as happens in the cases of the figure, the pressure reaches a maximum after several pipe diameters. Here the pseudo-shock is practically complete, and the ordinary effects of friction in a subsonic flow take over, producing a slow pressure decrease and velocity increase (Art. 10).

The values of the ratio p_2/p_1 between the pressure maximum p_2 and the pressure p_1 in the starting section have been plotted in [*23*] against the Mach number M_1 in the starting section itself (which varies from section to section as a result of frictional effects), and is reproduced in Fig. B,5e, compared to the curve relative to normal shocks. The small divergence is essentially due to the neglected frictional effects in the pseudo-shock region.

Finally, Fig. B,5f, again from [*23*], shows the length of the pseudo-shock as a function of M_1. These values apply to the particular values of Reynolds number of these tests and to circular sections.

Shockless model of pseudo-shock at constant area. The phenomena involved in the pseudo-shock are too complicated and incompletely understood to be amenable to exact analytical treatment. However, an

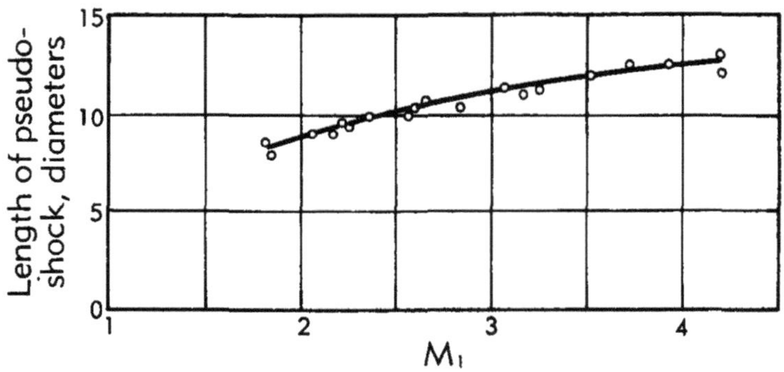

Fig. B,5f. Length of pseudo-shocks at various Mach numbers.

approximate treatment through simple one-dimensional methods is possible in the case of the fully developed pseudo-shock, based on the previous observation that the total entropy increase due to a number of oblique shocks can be substantially smaller than that connected with a single normal shock. If that is true, the essential dissipative phenomenon for the pseudo-shock does not reside in the shocks but in the turbulence generated in the dissipative region. In the limit we may disregard entirely the presence of the shocks, and make the simplifying assumption that the flow in the central core is uniform and isentropic. In addition, we may replace the nonuniform dissipative region with an approximately equivalent uniform region, the properties of which have to be considered as proper average values of the actual properties (Art. 16). With these simplifications, we arrive at the model of Fig. B,5g, where the respective quantities are clearly defined, the primed and double-primed quantities concerning the isentropic and the dissipative regions, respectively. Starting from the initial section where the flow is still undisturbed, the dissipative region spreads through the isentropic region, until eventually it occupies the whole final section. The pressure in the two regions is the same and will be simply designated by p. The duct is assumed to have constant area.

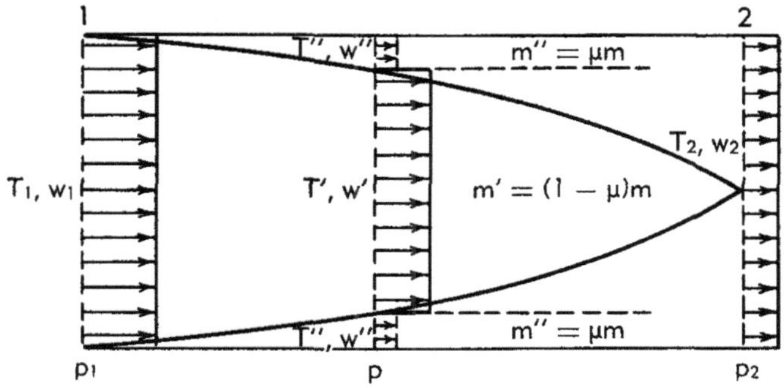

Fig. B,5g. Shockless model of pseudo-shock.

The conservation equations, for a control surface delimited by the initial section (subscript 1) and a generic section (no subscript) are immediately written in the following form, the friction force on the wall being assumed negligibly small.

Continuity:

$$m = m' + m'' = m_1 \tag{5-9}$$

Momentum (from Eq. 4-5 with $dA = X = I_i = 0$):

$$J = J' + J'' = J_1 \tag{5-10}$$

Energy (from Eq. 4-4 with $Q = H_i^0 = 0$):

$$H^0 = H^{0\prime} + H^{0\prime\prime} = H_1^0 \tag{5-11}$$

Using Eq. 4-1, Eq. 5-11 can be written as

$$m'h^{0\prime} + m''h^{0\prime\prime} = m_1h_1^0$$

In the isentropic region the flow is also isoenergetic, so that $h^{0\prime} = h_1^0$. Thus, using Eq. 5-9, we obtain $h^{0\prime\prime} = h_1^0$. The stagnation enthalpy and the stagnation temperature are therefore the same throughout the flow. We shall indicate the latter simply by T^0. Thus from the energy equation we have

$$T^{0\prime} = T^{0\prime\prime} = T_1^0 = T^0 \tag{5-11a}$$

Eq. 5-9 can also be written as

$$\frac{m'}{m_1} = 1 - \mu; \qquad \frac{m''}{m_1} = \mu \tag{5-9a}$$

Here we have introduced the fractional amount of flow μ, flowing in the dissipative region; μ increases from 0 at the initial section to 1 at the final section. The flow rate in each region can be related to the relative area through Eq. 4-31a. We obtain

$$\left.\begin{aligned} A' &= \frac{m'\sqrt{\mathcal{R}T^0}}{p}\sqrt{\frac{\gamma-1}{2\gamma}}\frac{1-w'^2}{w'} \\ A'' &= \frac{m''\sqrt{\mathcal{R}T^0}}{p}\sqrt{\frac{\gamma-1}{2\gamma}}\frac{1-w''^2}{w''} \end{aligned}\right\} \tag{5-12}$$

and

$$A = A' + A'' = A_1 = \frac{m_1\sqrt{\mathcal{R}T^0}}{p_1}\sqrt{\frac{\gamma-1}{2\gamma}}\frac{1-w_1^2}{w_1} \tag{5-13}$$

Replacing from Eq. 5-9a and 5-12 and eliminating the common factors, Eq. 5-13 becomes

$$(1-\mu)\frac{1-w'^2}{w'} + \mu\frac{1-w''^2}{w''} = \frac{p}{p_1}\frac{1-w_1^2}{w_1} \tag{5-13a}$$

The values of J', J'', and J_1 can be expressed through Eq. 4-33a. After substitution and elimination of the common factors, Eq. 5-10 becomes

$$(1-\mu)\left(\frac{\gamma-1}{2\gamma}\frac{1}{w'} + \frac{\gamma+1}{2\gamma}w'\right) + \mu\left(\frac{\gamma-1}{2\gamma}\frac{1}{w''} + \frac{\gamma+1}{2\gamma}w''\right) = \frac{\gamma-1}{2\gamma}\frac{1}{w_1} + \frac{\gamma+1}{2\gamma}w_1 \tag{5-10a}$$

An additional relation is obtained from the condition of constant entropy (or stagnation pressure) in the isentropic region. From Eq. 4-28 one obtains

$$\frac{p}{p_1} = \left(\frac{1-w'^2}{1-w_1^2}\right)^{\frac{\gamma}{\gamma-1}} \tag{5-14}$$

With the help of Eq. 5-14, the relations (Eq. 5-13a and 5-10a) constitute, for given w_1, two equations between w', w'', and μ, sufficient to determine two of these quantities in terms of the third. Obviously for $\mu = 0$ the system is satisfied by $p = p_1$, and $w' = w_1$. For $\mu = 1$, Eq. 5-10a becomes identical with the first Eq. 5-2 and with its help Eq. 5-13a can be reduced to the second Eq. 5-2. Hence for $\mu = 1$, p and w'' coincide with the values p_2 and w_2 of the normal shock. Between these two conditions one obtains an explicit solution by eliminating μ from Eq. 5-10a and 5-13a, thus obtaining an equation between w' and w'' which is of the second degree in w''. The larger root of this equation is $w'' = w'$, and can be easily checked to correspond to the trivial solution $w' = w_1$ and $p = p_1$, with μ arbitrary. The smaller root is given by

$$w'' = \frac{w^{*2}F(w_1, w') - 1}{w'[F(w_1, w') + 1]} \tag{5-15}$$

with

$$F(w_1, w') = \frac{\dfrac{p}{p_1}\dfrac{1 - w_1^2}{w_1} - \dfrac{1 - w'^2}{w'}}{(w_1 - w')\left(1 - \dfrac{w^{*2}}{w_1 w'}\right)} \tag{5-16}$$

When $w' \to w_1$, $(\mu \to 0)$ it can be shown from Eq. 5-16 that $F(w_1, w') \to 1/w^{*2}$, and therefore from Eq. 5-15, $w'' \to 0$. For some value $w_2' < w_1$ of w', Eq. 5-15 gives $w'' = w_2$, and therefore $\mu = 1$. Thus between $\mu = 0$

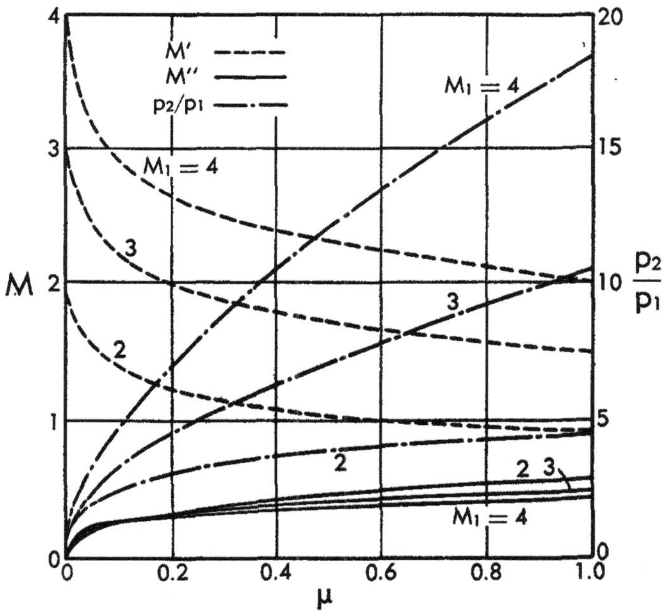

Fig. B,5h. Shockless model of pseudo-shock. Mach numbers M' in the isentropic core and M'' in the dissipative region, and static pressure, against the fractional amount of dissipative flow for three initial Mach numbers M_1.

and $\mu = 1$, w' decreases from w_1 to w_2' and w'' increases from 0 to w_2. Since for $\mu = 1$ the pressure ratio is given by Eq. 5-5, the value of w_2' can be calculated in terms of w_1 from Eq. 5-14, with $p = p_2$. Fig. B,5h shows the results of the calculations for $M_1 = 2$, 3, and 4. The values of the Mach numbers M', M'' (instead of those of the corresponding reduced velocities), and those of p/p_1 have been plotted in function of μ. The velocity in the isentropic core is considerably reduced by the compression. It is still supersonic at the end of the pseudo-shock for $M_1 = 3$ and 4, but it becomes subsonic for $M_1 = 2$. The velocity in the dissipative region is, in average, always smaller than the final subsonic velocity. Since the dissipative region is generated at the wall and at all times stays adjacent to it, the friction forces are substantially smaller than those present in the supersonic flow preceding the pseudo-shock. This fact provides a good justification for the already observed good agreement between the experimental pressure ratios and those calculated neglecting friction.

The results in function of μ could be related to the axial distance along the duct if we knew the rate dm''/dx at which the dissipative region spreads into the isentropic core. Conversely, from the knowledge of the length L of the pseudo-shock we gain some information on the spreading rate. Since, from Fig. B,5f, L/D is around 10, we are forced to conclude that, for the experiments considered, the rate of spreading of the dissipative region is of the same order as the corresponding rate at which the ordinary turbulent boundary layer spreads, because the practical entrance stretch is around 10 diameters long (Art. 10).

Pseudo-shock with variable area. The variability of the cross-sectional area would have no effect on a normal shock, because its extremely small thickness makes it a purely local phenomenon, unaffected by the gradients of the various quantities. On the contrary, the pseudo-shock, due to its finite axial extension, is definitely affected by area changes. For arbitrary area variation a precise analytical treatment is again only possible through the shockless model. For this purpose, one can use Eq. 4-2b, 4-4b, and 4-3b, or 4-5b with $X = dX = 0$, assuming that $A(x)$ and the fractional flow rate in the dissipative region, $\mu(x)$, are known. The resulting system of equations cannot in general be integrated in closed form, and one can get a result only through numerical integration.

A more elegant treatment can be obtained for ducts of the p-A power family introduced in Art. 4. Independent of the actual phenomena within the pseudo-shock region, one can correlate the condition before and after the transition process in a way very similar to that developed at the beginning of this article for the constant area transition. The first two relations (Eq. 5-1) are unaffected by the duct shape. The third is replaced by Eq. 4-22. Thus the equations are

$$m_2 = m_1, \quad T_2^0 = T_1^0, \quad J_2^{(\varepsilon)} = J_1^{(\varepsilon)} \tag{5-17}$$

From Eq. 4-34a and 4-19 one obtains the following equations, replacing Eq. 5-2,

$$\left.\begin{aligned} \varepsilon\frac{\gamma-1}{2\gamma}\frac{1}{w_2} + \left(1-\varepsilon\frac{\gamma-1}{2\gamma}\right)w_2 &= \varepsilon\frac{\gamma-1}{2\gamma}\frac{1}{w_1} \\ &\quad + \left(1-\varepsilon\frac{\gamma-1}{2\gamma}\right)w_1 \\ p_2^{1/\varepsilon}\left(\varepsilon + \frac{2\gamma}{\gamma-1}\frac{w_2^2}{1-w_2^2}\right) &= p_1^{1/\varepsilon}\left(\varepsilon + \frac{2\gamma}{\gamma-1}\frac{w_1^2}{1-w_1^2}\right) \end{aligned}\right\} \tag{5-18}$$

The first Eq. 5-3 is unchanged, and the second becomes $\rho_2 w_2 A_2 = \rho_1 w_1 A_1$, that is, using Eq. 4-19,

$$\rho_2 w_2 p_2^{(1-\varepsilon)/\varepsilon} = \rho_1 w_1 p_1^{(1-\varepsilon)/\varepsilon} \tag{5-19}$$

The equations have again a trivial solution, $w_2 = w_1$, $p_2 = p_1$, $T_2 = T_1$,

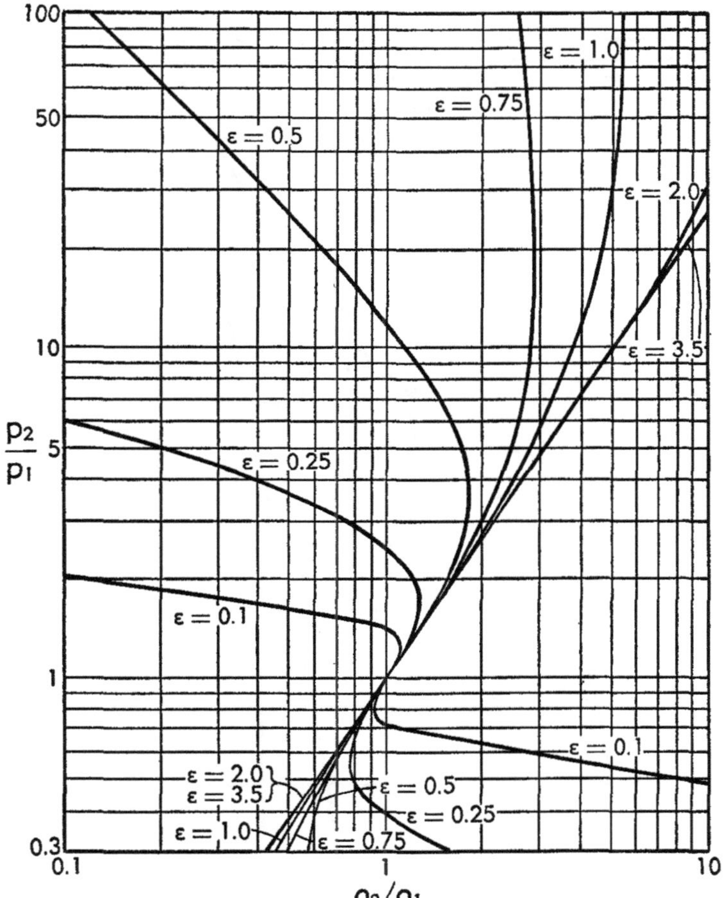

Fig. B,5i. Representation of the generalized Rankine-Hugoniot relation concerning the pseudo-shocks in ducts of the p-A power family. $\gamma = 1.4$.

$\rho_2 = \rho_1$, and $A_2 = A_1$, and a nontrivial solution, expressed by

$$w_2 = \frac{w_{cr}^2}{w_1}, \quad \left(\frac{1}{w_{cr}^2} = \frac{2\gamma}{\varepsilon(\gamma - 1)} - 1\right) \tag{5-20}$$

$$\left(\frac{A_2}{A_1}\right)^{\frac{1}{1-\varepsilon}} = \left(\frac{p_2}{p_1}\right)^{\frac{1}{\varepsilon}} = \frac{\dfrac{w_1^2}{w_{cr}^2} - w_{cr}^2}{1 - w_1^2}, \quad \frac{\rho_2}{\rho_1} = \frac{w_1^2}{w_{cr}^2}\left(\frac{p_2}{p_1}\right)^{\frac{\varepsilon-1}{\varepsilon}}, \quad \frac{T_2}{T_1} = \frac{1 - \dfrac{w_{cr}^4}{w_1^2}}{1 - w_1^2} \tag{5-21}$$

Observe that this solution is possible only in a certain range of ε, because, since both w_1 and w_2 are positive and smaller than unity, so must be the quantity w_{cr}^2 defined by Eq. 5-20. Thus the solution may have a physical meaning only for

$$0 \leqq \varepsilon \leqq \frac{\gamma}{\gamma - 1} \tag{5-22}$$

The elimination of w_1^2 from the first two Eq. 5-21 leads to the following

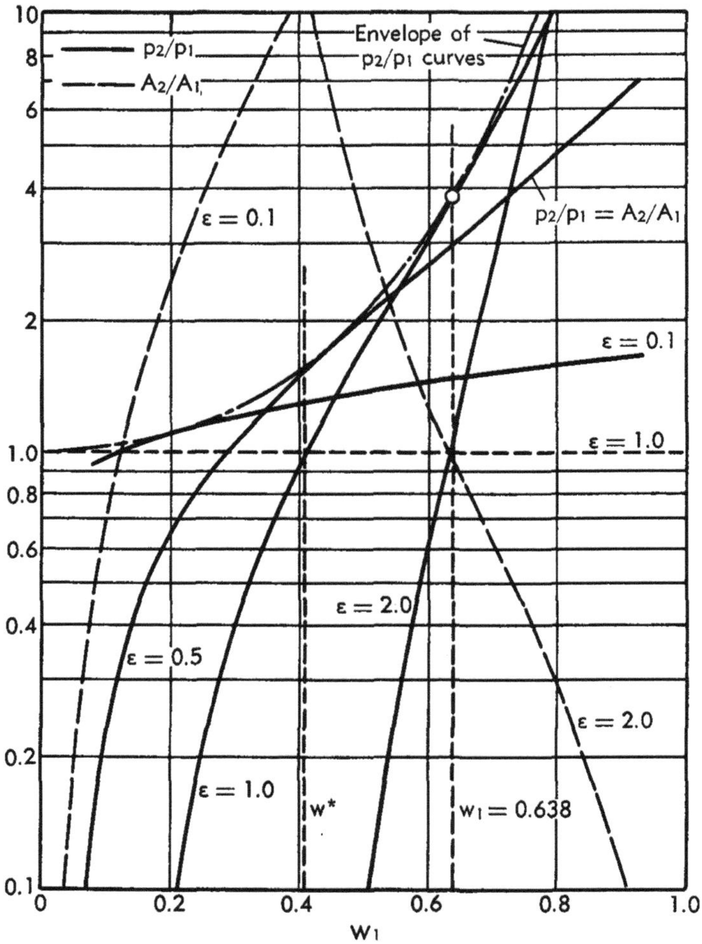

Fig. B,5j. Pressure and area ratios for pseudo-shocks in ducts of the p-A power family. $\gamma = 1.4$.

generalization of the Rankine-Hugoniot relation:

$$\frac{\rho_2}{\rho_1} = \frac{p_2}{p_1} \frac{1 + w_{cr}^2 \left(\frac{p_1}{p_2}\right)^{\frac{1}{\varepsilon}}}{1 + w_{cr}^2 \left(\frac{p_2}{p_1}\right)^{\frac{1}{\varepsilon}}} \tag{5-23}$$

Fig. B,5i shows this relation for $\gamma = 1.4$ and various values of ε in the range of Eq. 5-22. At the upper limit, $\varepsilon = \gamma/(\gamma - 1)$, Eq. 5-23 coincides with the isentropic relation, but for all other values of ε there is an entropy increase for $p_2/p_1 > 1$, and there would be a decrease of entropy in the range $p_2/p_1 < 1$. Thus, as a consequence of the second law of

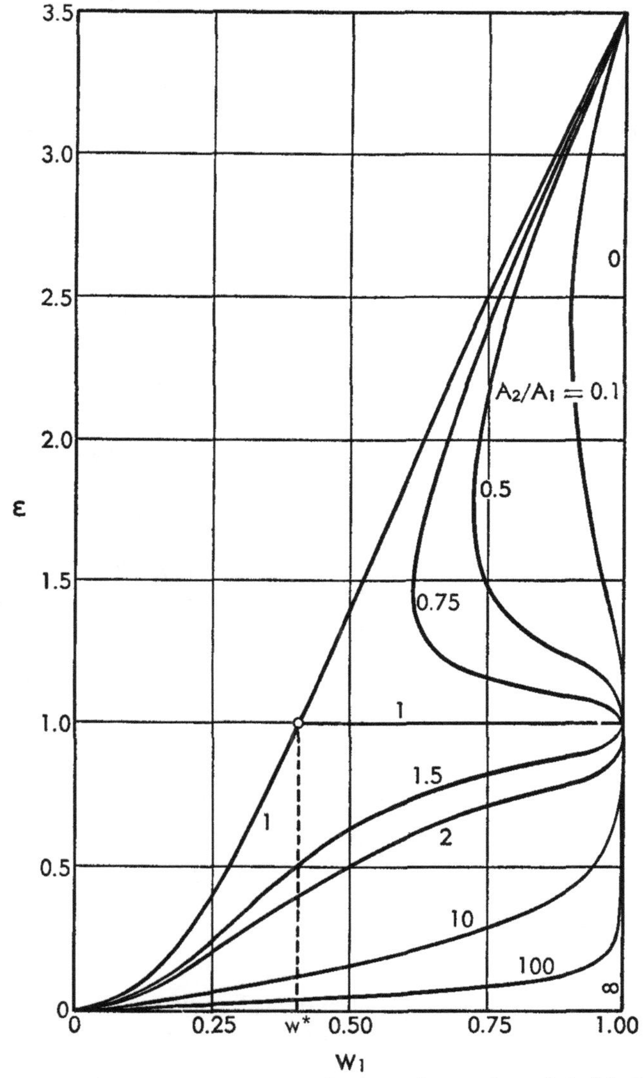

Fig. B,5k. Values of ε necessary to obtain a given value of A_2/A_1 through a pseudo-shock in ducts of the p-A power family. $\gamma = 1.4$.

thermodynamics, only the solutions corresponding to a compression can have a physical existence, and, since ε is positive, the first Eq. 5-21 shows that w_1 must be larger than w_{cr}. This result can also be obtained by introducing Eq. 5-21 in Eq. 5-7 and expanding in powers of $(w_1^2/w_{cr}^2) - 1$. The resulting series is the same as Eq. 5-8a, with w_{cr} instead of w^*. If we call "supercritical" the range $w > w_{cr}$, and "subcritical" the other, we conclude that the pseudo-shock is impossible in the subcritical range and is only allowed in the supercritical range. As already observed, Eq. 5-23 is reduced to the isentropic relation for $\varepsilon = \gamma/(\gamma - 1)$. However, in this case $w_{cr} = 1$ and the supercritical range vanishes. Hence the pseudo-shock is always accompanied by an entropy increase.

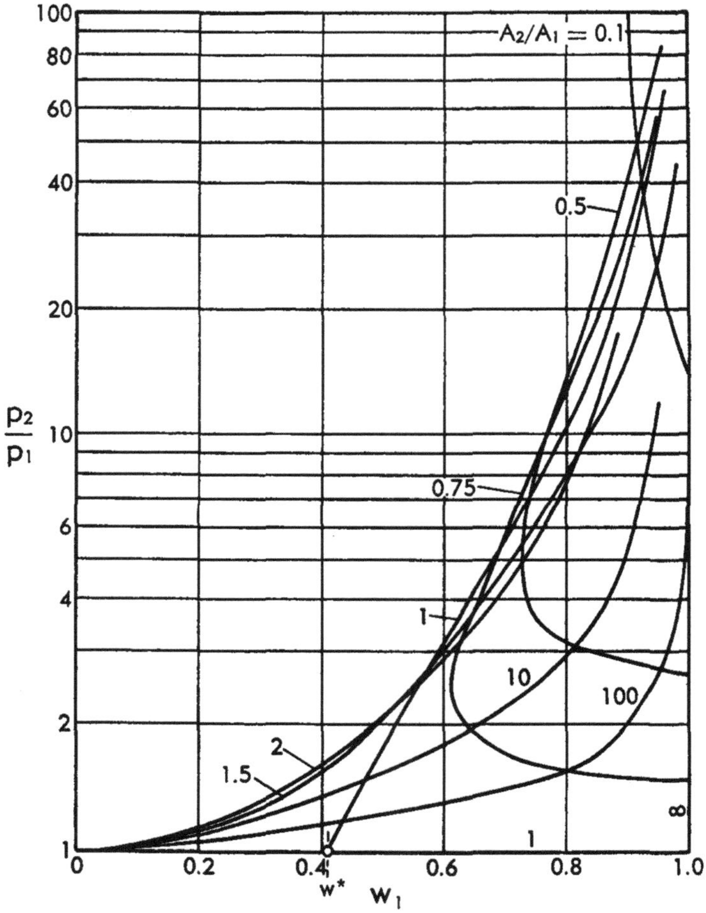

Fig. B,5l. Values of p_2/p_1 obtained through pseudo-shocks in ducts of the p-A power family for assigned values of A_2/A_1. $\gamma = 1.4$.

For $\varepsilon = 1$, evidently, the preceding relations reproduce those relative to the constant area case, and $w_{cr} = w^*$. For $\varepsilon \gtrless 1$, w_{cr} is $\gtrless w^*$. Thus for $\varepsilon < 1$ the supercritical range includes a range of subsonic velocities, in which the existence of pseudo-shocks may be allowed, a result that would never be possible if the pseudo-shock consisted of a normal shock.

Fig. B,5j illustrates the first Eq. 5-21, showing the values of the pressure and area ratios as functions of w_1 for various values of ε. One

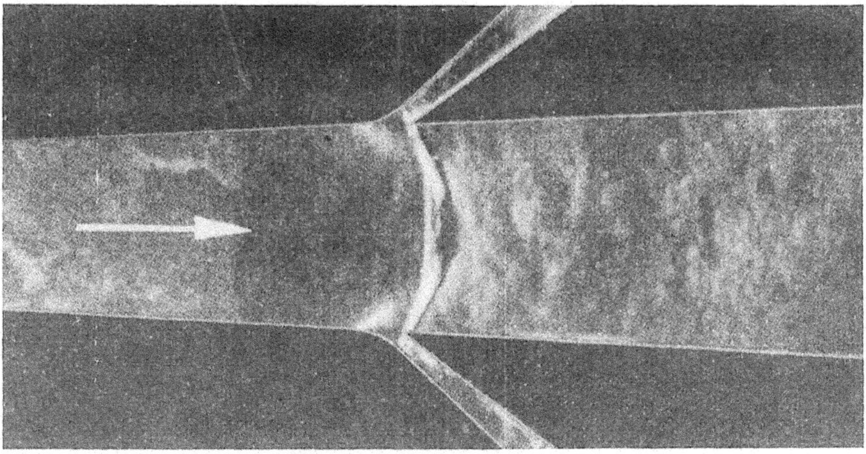

Plate B,5a. Quasi-normal shock in a duct after removal of boundary layers.

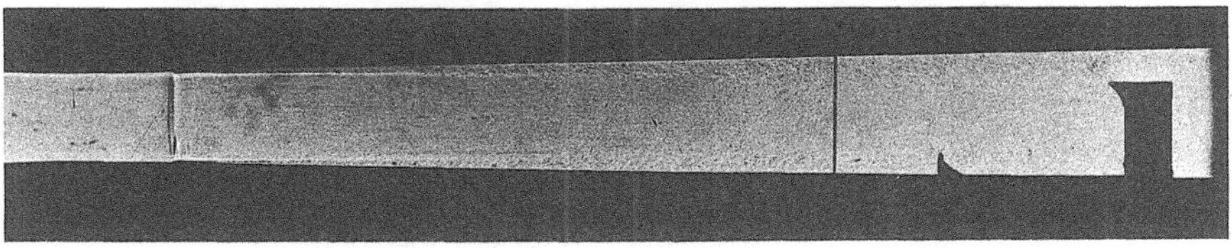

Plate B,5b. Quasi-normal shock in a duct with thin boundary layers at slight supersonic velocity. $M = 1.18$.

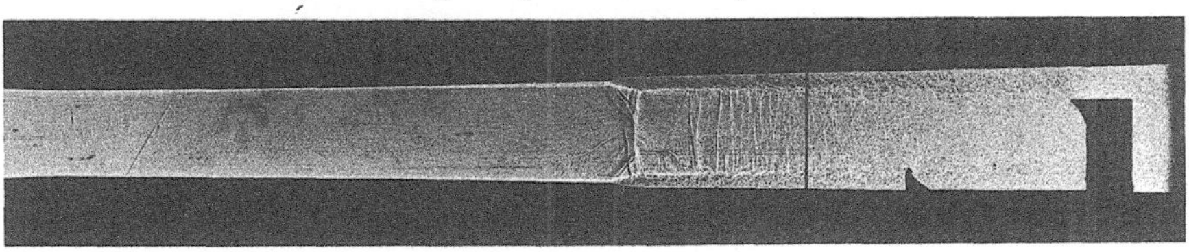

Plate B,5c. Bifurcated shock in the duct of Plate B,5b at higher Mach number. $M = 1.46$.

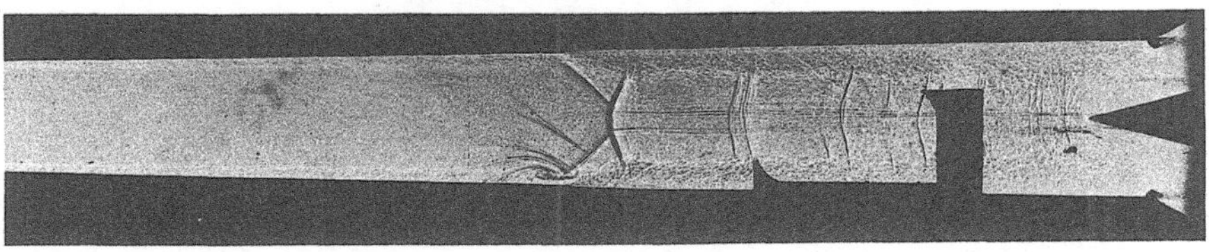

Plate B,5d. Multiple bifurcated shocks in the duct of Plate B,5b at still higher Mach number. $M = 1.70$.

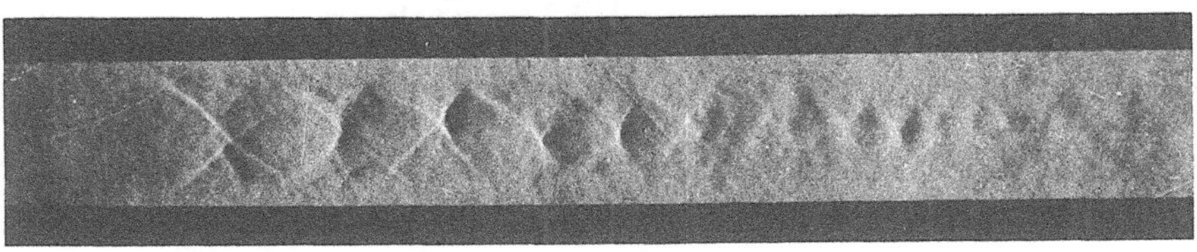

Plate B,5e. Multiple X shocks produced with well-developed boundary layers.

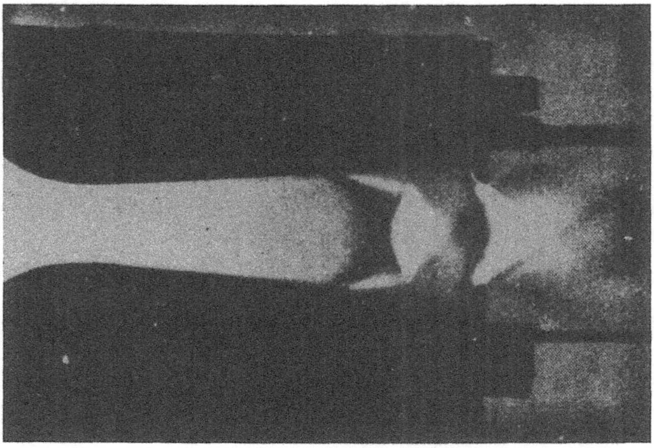

Plate B,8. Experimentally observed wave pattern
in an over-expanded nozzle.

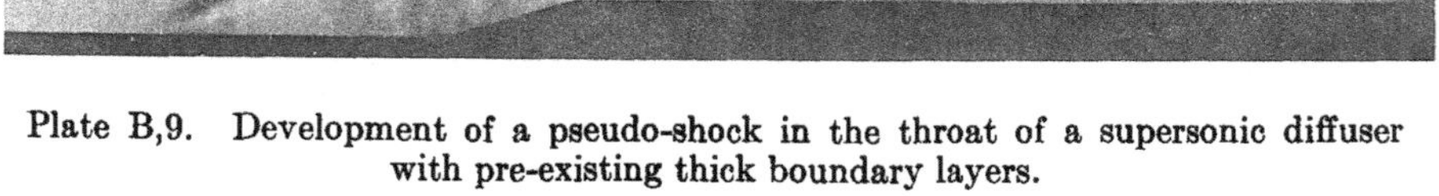

Plate B,9. Development of a pseudo-shock in the throat of a supersonic diffuser
with pre-existing thick boundary layers.

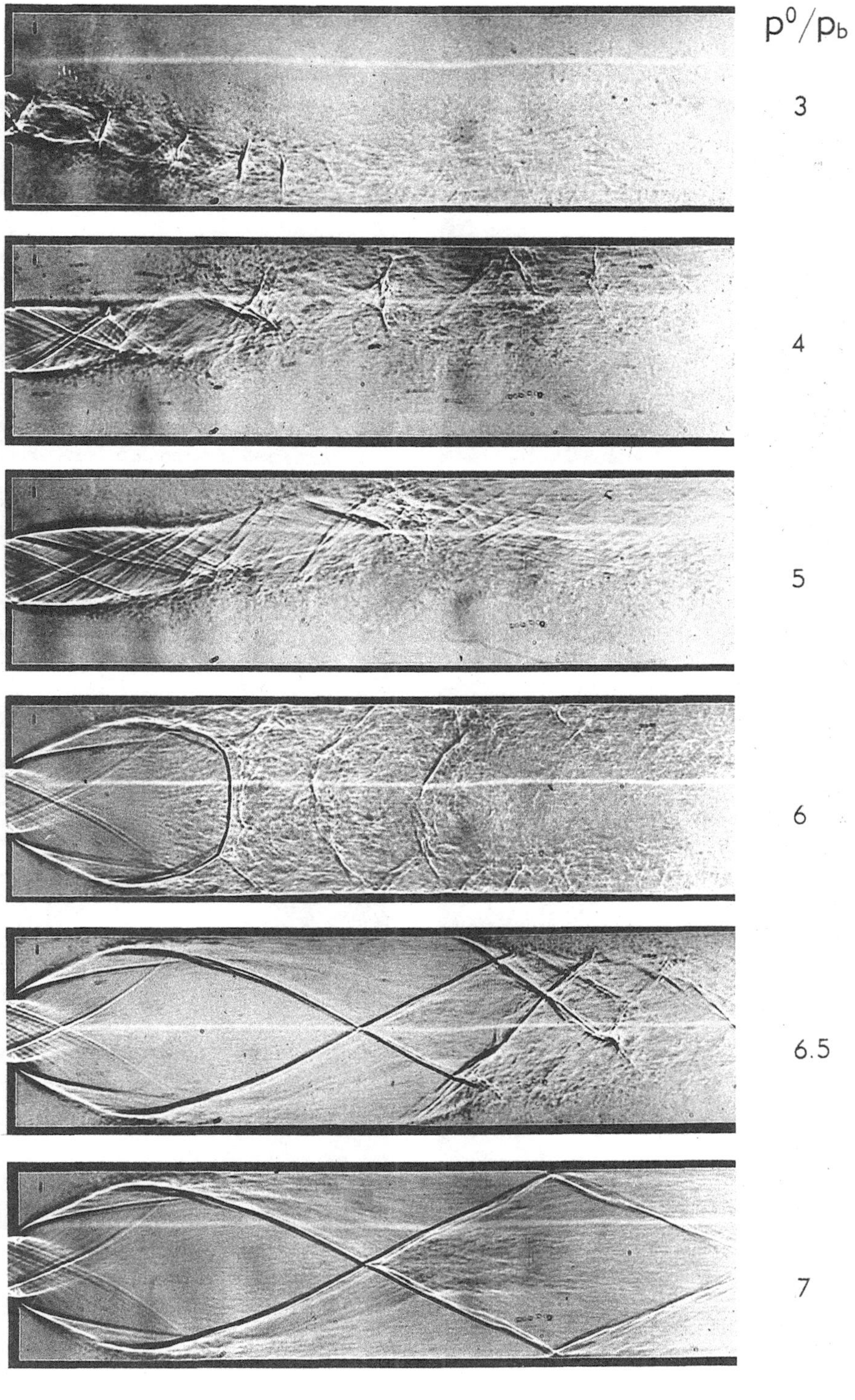

Plate B,15. Sequence of patterns in a duct of the kind shown in Fig. B,15f at different back pressures.

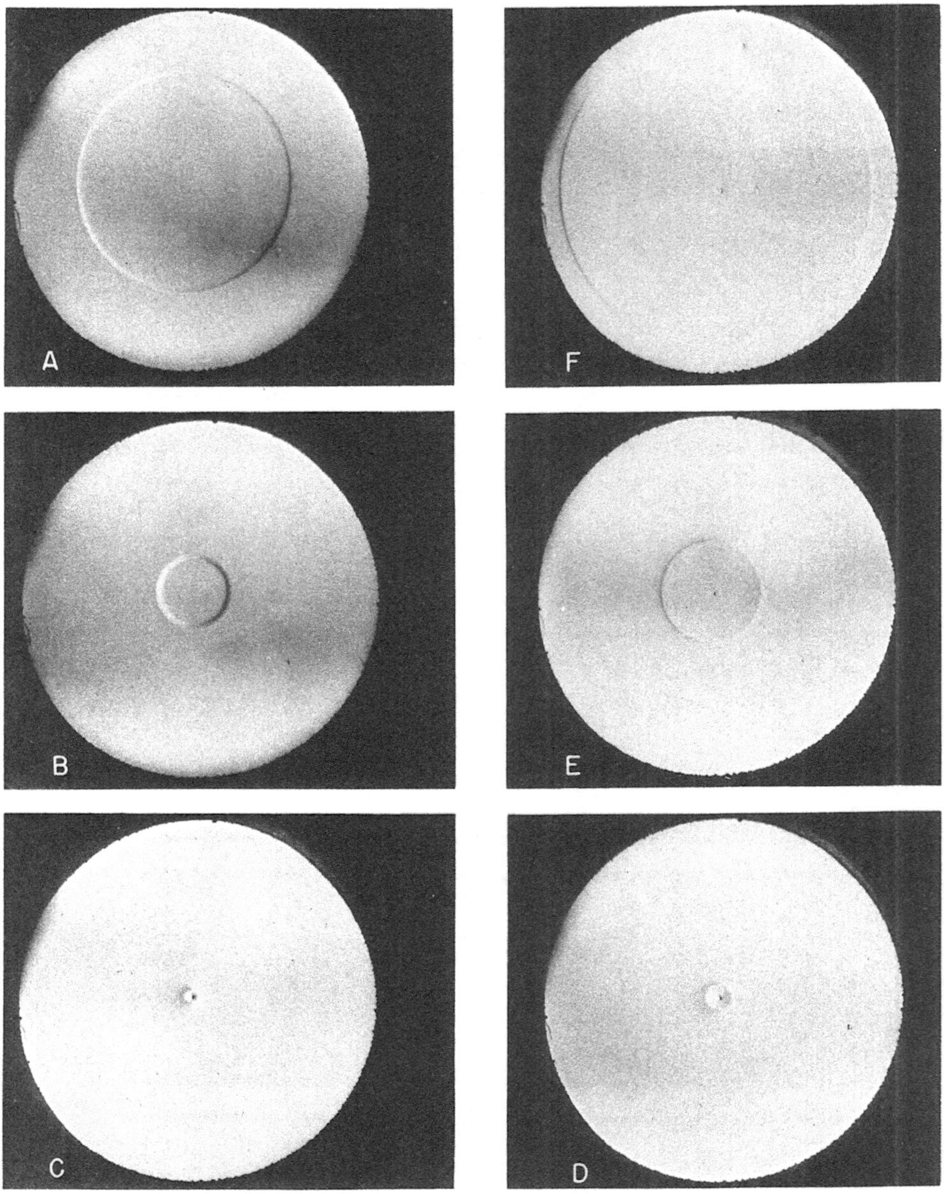

Plate C,8. Schlieren photographs of converging cylindrical shock waves in air (obtained from plane shocks with $M = 1.1$). *A*, *B*, and *C* indicate the incident waves and *D*, *E*, and *F* indicate the reflected waves. Each photograph is of a different shock wave. The glass window was $1\frac{1}{4}$ in. in diameter.

can see that in the subsonic range the attainable pressure ratios are very close to 1; that for each given value of w_1 the maximum p_2/p_1 is obtained for a well-determined value of ε, smaller than 1 below $w_1 = 0.638$ ($M_1 = 1.85$), and >1 above this w_1; and that $A_2/A_1 \gtrless 1$ for $\varepsilon \lessgtr 1$, so that the duct is divergent for $\varepsilon < 1$ and convergent for $\varepsilon > 1$. For fixed values of A_2/A_1, one obtains the curves of Fig. B,5k and B,5l, showing the necessary values of ε and the resulting compression ratios as functions of w_1.

The previous results apply to the pseudo-shock phenomenon in ducts of the p-A power family. For arbitrary shapes of the duct one can define a value of ε which makes Eq. 4-19 true at stations 1 and 2, but not at intermediate stations. Despite this, Eq. 4-19a can still provide an approximate value of the integral, with the consequence that all of the previous results can be used approximately for a duct of arbitrary geometry.

Shockless model of pseudo-shock in ducts of the p-A power family. The above treatment provides the relations between the conditions before and after the pseudo-shock independently of the details of the phenomena within the transition region. However, following the same schematization used for the constant area case, one can develop an analytical treatment of the transition region in ducts of the p-A power family by assuming that in the case of well-developed boundary layers one can replace the actual mechanism by a shockless model with an isentropic region and a uniform dissipative region gradually spreading on the entire section. Eq. 5-9 and 5-11 are unchanged, but Eq. 5-10 is replaced by

$$J^{(\varepsilon)} = J^{(\varepsilon)\prime} + J^{(\varepsilon)\prime\prime} = J_1^{(\varepsilon)} \tag{5-24}$$

Again Eq. 5-9 is identically satisfied by Eq. 5-9a; Eq. 5-13 is replaced by

$$pA = p(A' + A'') = \left(\frac{p}{p_1}\right)^{\frac{1}{\varepsilon}} p_1 A_1 \tag{5-25}$$

which follows from Eq. 4-19. Replacing in Eq. 5-25 the expressions (Eq. 5-12 and 5-13) for A', A'', and A_1 we obtain

$$(1 - \mu)\frac{1 - w'^2}{w'} + \mu\frac{1 - w''^2}{w''} = \left(\frac{p}{p_1}\right)^{\frac{1}{\varepsilon}} \frac{1 - w_1^2}{w_1} \tag{5-25a}$$

In this equation, the value of p/p_1 can be expressed through w' and w_1 by Eq. 5-14, which still holds in the isentropic region.

Using Eq. 4-34a for $J^{(\varepsilon)}$, Eq. 5-24 becomes

$$(1 - \mu)\left(w' + \frac{w_{cr}^2}{w'}\right) + \mu\left(w'' + \frac{w_{cr}^2}{w''}\right) = w_1 + \frac{w_{cr}^2}{w_1} \tag{5-24a}$$

with w_{cr}^2 defined by Eq. 5-20. One can eliminate μ from Eq. 5-24a and 5-25a, and obtain an equation of the second degree in w'' with a root $w'' = w'$, leading to the trivial solution (constant area duct with constant conditions) and a second root given by the same relations (Eq. 5-15 and 5-16), with w^* replaced by w_{cr} and p/p_1 by $(p/p_1)^{1/\varepsilon}$. Again, as in the constant area case, this root provides, for given w_1 and ε, w'' as a function of w'. The corresponding value of μ can then be computed. This solution provides the desired gradual transition between p_1 and p_2 when μ goes from 0 to 1, A from A_1 to A_2, and w'' from 0 to w_2. Again the fact that w'', which represents the mean velocity in the region adjacent to the walls, stays always at such low values justifies a posteriori the assumption that the friction forces on the walls can be neglected.

Observe that if we know the rate of spreading of the dissipative region, and therefore $\mu(x)$, the previous results can be related directly to x, and among other things the function $A(x)$ representing the geometry of the duct can be obtained. We have here an example of the fact, indicated after Eq. 4-22, that the choice of ε is not sufficient to define the geometry of the duct, and that this geometry can only be determined if one knows the processes that are taking place in the duct. In other words the shape of the duct is determined by the transformation and not the transformation by the shape, which is the case in the types of flow considered in the coming articles. For instance, in the present example, for no matter what value of ε, the resulting duct is a constant section pipe with constant conditions when no dissipative cause is present (trivial solution), so that the only type of flow compatible with both the p-A power law and the isentropic relation is a flow with constant conditions. In order to be able to associate a nontrivial flow with the power law, there must be present a process of which the development is controlled by some kind of mechanism primarily unrelated to the shape of the duct. In the example of the pseudo-shock such a process consists in the spreading of the dissipative region in the isentropic core, and is controlled by the turbulent condition in the dissipative region itself.

Other types of process are discussed in later articles.

B,6. Flow in Ideal Nozzles. One of the fundamental purposes of the one-dimensional treatment is the study of flows in ducts of variable cross-sectional area, in which the main cause for the change in flow conditions is precisely the variability of the area, and other causes, such as boundary layer development and skin friction, have only a secondary influence. Clearly in order that the skin friction be only a secondary agent in the determination of the flow, the total friction forces must be small compared to the pressure forces, and therefore the duct must be short.

There are two kinds of short ducts that are of primary interest: the effusers or nozzles, in which a pressure drop is used to produce a velocity increase, and the diffusers, in which a pressure increase is generated through a decrease of velocity. The condition that the development of the boundary layers must have little influence on the flow is easy to realize in nozzles, because no matter how much the velocities in the boundary layer have been reduced by friction forces, the retarded layers have no difficulty in following the favorable pressure gradients of the expansion. For diffusers, however, the boundary layers are more critical, because the adverse pressure gradients may affect considerably the flow of the retarded layers, and produce thickening of the boundary layer and possibly flow separation. For this reason, and because of the problem of "starting" supersonic diffusers (Art. 9), the two types of ducts are distinguished by different requirements and must be treated separately.

Since friction forces have but little influence on the flow in effusers, one can assume in first approximation that the friction is absent. In this case, as shown in Art. 4, the momentum equation (Eq. 4-3c) is equivalent to the condition of constant entropy. Thus the problem is reduced to the study of the isoenergetic, isentropic flow in a variable area duct.

Isentropic flow in ideal nozzles. The constancy of stagnation enthalpy and entropy can be simply expressed (Art. 4) by the relations

$$T^0 = \text{const} = T_i^0; \qquad p^0 = \text{const} = p_i^0 \tag{6-1}$$

T_i^0 and p_i^0 represent here the corresponding initial values. They may for instance coincide with the static conditions in the tank, if the nozzle is directly attached to a large volume reservoir of pressurized gas where the gas velocity w_i is practically zero; or they may indicate the stagnation value at the entrance section of the nozzle, when the nozzle is preceded by a pipe, where, however, w_i, though finite, has a value lower than the velocities attained in the nozzle.

The continuity equation (Eq. 4-2b), with m given by Eq. 4-31a, can be written as

$$\begin{aligned} Aw(1 - w^2)^{\frac{1}{\gamma-1}} &= \sqrt{\frac{\gamma - 1}{2\gamma}} \frac{m\sqrt{\mathcal{R}T_i^0}}{p_i^0} = \text{const} \\ &= A_{is}^* w^*(1 - w^{*2})^{\frac{1}{\gamma-1}} = \sqrt{\frac{\gamma-1}{\gamma+1}}\left(\frac{2}{\gamma+1}\right)^{\frac{1}{\gamma-1}} A_{is}^* \end{aligned} \tag{6-2}$$

where w^* is the reduced sound velocity and, for fixed flow conditions, A_{is}^* is a constant with the dimensions of an area which we shall call, for reasons apparent later, the "isentropic critical area." Eq. 6-2 produces the following fundamental relation between the "reduced area" A/A_{is}^* and the reduced velocity, or, through Eq. 4-29, the Mach number

$$\frac{A}{A_{is}^*} = \frac{w^*}{w}\left(\frac{1 - w^{*2}}{1 - w^2}\right)^{\frac{1}{\gamma-1}} = \frac{1}{M}\left[\frac{2}{\gamma+1}\left(1 + \frac{\gamma-1}{2}M^2\right)\right]^{\frac{1}{2}\frac{\gamma+1}{\gamma-1}} \tag{6-3}$$

For given A_{is}^* a value of A corresponds to each w or M and, through Eq. 6-1 and 4-28, of T and p

$$\frac{T}{T^0} = \frac{T}{T_i^0} = 1 - w^2 = \left(1 + \frac{\gamma - 1}{2} M^2\right)^{-1} \tag{6-4a}$$

$$\frac{p}{p^0} = \frac{p}{p_i^0} = (1 - w^2)^{\frac{\gamma}{\gamma-1}} = \left(1 + \frac{\gamma - 1}{2} M^2\right)^{-\frac{\gamma}{\gamma-1}} \tag{6-4b}$$

The value of ρ, divided by the initial stagnation density ρ_i^0 is immediately

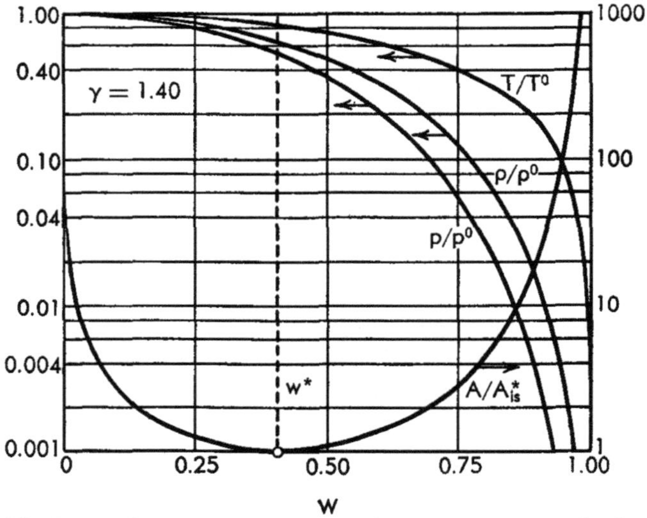

Fig. B,6a. Pressure, temperature, density, and area ratios for isentropic flow in nozzles. $\gamma = 1.4$.

obtained from the equation of state

$$\frac{\rho}{\rho^0} = \frac{\rho}{\rho_i^0} = (1 - w^2)^{\frac{1}{\gamma-1}} = \left(1 + \frac{\gamma - 1}{2} M^2\right)^{-\frac{1}{\gamma-1}} \tag{6-4c}$$

Numerical values computed from these relations can be found in various published tables (for instance, [*15,16,17*]). Their behavior is represented in Fig. B,6a as a function of w for $\gamma = 1.4$. The most interesting peculiarity of these curves concerns the reduced area. As the figure shows, A/A_{is}^* has a minimum, equal to 1, at the sound velocity ($w = w^*$). This is a consequence of the fact that $w(1 - w^2)^{1/(\gamma-1)}$ has there a maximum equal to the coefficient of A_{is}^* in Eq. 6-2. For fixed stagnation conditions, this quantity is proportional to the mass flow per unit area ρu, which is, therefore, for given stagnation conditions, maximum at the sonic velocity. The presence of this maximum plays an important role in the flow of gases. The corresponding values of w, T, p, and ρ are ordinarily called the critical values. They are given by

$$w^* = \sqrt{\frac{\gamma - 1}{\gamma + 1}}, \quad \frac{T^*}{T^0} = \frac{2}{\gamma + 1}, \quad \frac{p^*_{is}}{p^0} = \left(\frac{2}{\gamma + 1}\right)^{\frac{\gamma}{\gamma - 1}}, \quad \frac{\rho^*_{is}}{\rho^0} = \left(\frac{2}{\gamma + 1}\right)^{\frac{1}{\gamma - 1}} \tag{6-5}$$

and are represented as functions of γ in Fig. B,6b.

Observe that the first two Eq. 6-5 are based on the energy equation, and are correct for the most general isoenergetic flow, even if losses and entropy variations are present; but the last two Eq. 6-5 are only true for isentropic flow. For this reason, contrary to the usual practice, we use the subscript emphasizing the isentropic condition, in order to distinguish these isentropic critical values from those relative to flows with losses.

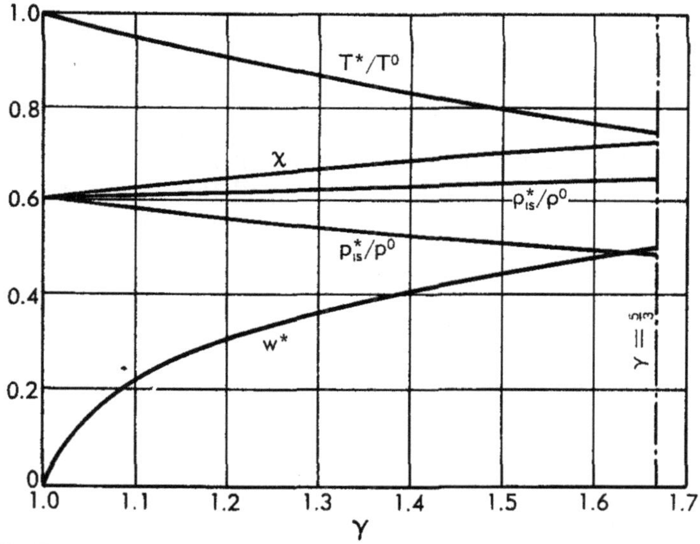

Fig. B,6b. Reduced sound velocity, critical pressure, temperature and density ratios, and mass flow coefficient χ in function of γ.

For the same reason the subscript is used in A^*_{is}. Observe also that the denomination "critical" is also used in this treatment in a broader sense (see Art. 5, 11, and 17) which does not necessarily refer to the sonic condition.

Isentropic mass flow rate. Convergent and Laval nozzles. Suppose now that a gas flows isentropically in a nozzle, from an initial stagnation pressure p_i^0 to a final pressure p_f. The reduced velocity in the final section can be calculated from Eq. 6-4b as

$$w_f = \left[1 - \left(\frac{p_f}{p_i^0}\right)^{\frac{\gamma-1}{\gamma}}\right]^{\frac{1}{2}} \tag{6-6}$$

Suppose first that $p_f > p^*_{is}$, then $w_f < w^*$. The mass flow rate is given by

Eq. 4-31a as

$$\left.\begin{aligned} m &= \frac{p_{\mathrm{i}}^0 A_{\mathrm{f}}}{\sqrt{\mathfrak{R}T_{\mathrm{i}}^0}} \sqrt{\frac{2\gamma}{\gamma - 1}}\, w_{\mathrm{f}}(1 - w_{\mathrm{f}}^2)^{\frac{1}{\gamma - 1}} \\ &= \frac{p_{\mathrm{i}}^0 A_{\mathrm{f}}}{\sqrt{\mathfrak{R}T_{\mathrm{i}}^0}} \sqrt{\frac{2\gamma}{\gamma - 1}} \left[\left(\frac{p_{\mathrm{f}}}{p_{\mathrm{i}}^0}\right)^{\frac{2}{\gamma}} - \left(\frac{p_{\mathrm{f}}}{p_{\mathrm{i}}^0}\right)^{\frac{\gamma+1}{\gamma}}\right]^{\frac{1}{2}} \\ &= \frac{p_{\mathrm{i}}^0 A_{\mathrm{is}}^*}{\sqrt{\mathfrak{R}T_{\mathrm{i}}^0}} \sqrt{\frac{2\gamma}{\gamma - 1}}\, w^*(1 - w^{*2})^{\frac{1}{\gamma - 1}} = \chi(\gamma) \frac{p_{\mathrm{i}}^0 A_{\mathrm{is}}^*}{\sqrt{\mathfrak{R}T_{\mathrm{i}}^0}} \end{aligned}\right\} \tag{6-7}$$

The coefficient χ is shown as a function of γ in Fig. B,6b. One can also calculate $A_{\mathrm{f}}/A_{\mathrm{is}}^*$ from Eq. 6-3. The corresponding value is >1. The velocity w_{f} is reached in the nozzle starting from a velocity $w_{\mathrm{i}} < w_{\mathrm{f}}$. For each $w < w_{\mathrm{f}}$ one finds from Eq. 6-3 a value of $A/A_{\mathrm{is}}^* > A_{\mathrm{f}}/A_{\mathrm{is}}^* > 1$, and from Eq. 6-4b a value of $p/p_{\mathrm{i}}^0 > p_{\mathrm{f}}/p_{\mathrm{i}}^0$. Thus to each section $A > A_{\mathrm{f}}$ of the nozzle one can associate the proper velocity and the corresponding pressure, the former increasing and the latter decreasing monotonically from the initial to the final values. A simply convergent nozzle is therefore sufficient for the realization of the prescribed isentropic expansion. For $p_{\mathrm{f}} = p_{\mathrm{i}}^0$ obviously both w_{f} and m vanish and the flow is stopped. For $p_{\mathrm{f}} = p_{\mathrm{is}}^*$, w_{f} is equal to the reduced sound velocity w^*, and m takes the maximum value compatible with the given A_{f}, which in this case coincides with A_{is}^*. The maximum is given by

$$m_{\mathrm{is}}^* = \chi(\gamma) \frac{p_{\mathrm{i}}^0 A_{\mathrm{f}}}{\sqrt{\mathfrak{R}T_{\mathrm{i}}^0}} \qquad (p_{\mathrm{f}} = p_{\mathrm{si}}^*) \tag{6-8}$$

In the case $p_{\mathrm{f}} < p_{\mathrm{is}}^*$, w_{f} is in the supersonic range and m is again below m_{is}^*. Again one can calculate $A_{\mathrm{f}}/A_{\mathrm{is}}^* > 1$, and the values of A/A_{is}^* corresponding to any $w < w_{\mathrm{f}}$. The difference from the case $p_{\mathrm{f}} > p_{\mathrm{is}}^*$ is that now A/A_{is}^* decreases when w decreases from w_{f} to w^*, reaches its minimum value 1 at w^*, and increases again for $w < w^*$. Thus, in order to attain the value w_{f} from the lower value w_{i}, the cross-sectional area must first decrease, go through the value $A_{\mathrm{is}}^* < A_{\mathrm{f}}$ at the sonic velocity, and increase again to A_{f}. We conclude that, for $p_{\mathrm{f}} < p_{\mathrm{is}}^*$, the nozzle cannot be the simple convergent duct of Fig. B,6d but must have the characteristic convergent-divergent shape of Fig. B,6e, with a throat of area $A_{\mathrm{t}} = A_{\mathrm{is}}^*$. We see now the physical meaning of the isentropic critical area introduced with Eq. 6-2. A duct of this shape is called a Laval nozzle.

Fig. B,6c reproduces in graphical form the results of the above discussion. When $p_{\mathrm{f}}/p_{\mathrm{i}}^0$ is varied, m/m_{is}^* changes according to the corresponding curve which represents the relation $m/m_{\mathrm{is}}^* = A_{\mathrm{is}}^*/A_{\mathrm{f}}$ obtained from Eq. 6-7 and 6-8. For a given value of A_{f}, and fixed initial conditions, m_{is}^* is fixed. Thus the same value of m, (and of A_{is}^*) is obtained for two different levels of p_{f}, the level I corresponding to a subsonic w_{f} and the level II to a supersonic w_f. The value of A^*_{is}/A_f is the same for the two levels, and one can draw the curve of $A/A_f = (A^*_{is}/A_f)(A/A^*_{is})$, the latter factor being calculated as a function of p/p_i^0 from Eq. 6-3 and 6-4b. The figure shows immediately that in the supersonic case the pressure level II can be reached from higher pressures only when A/A_f goes through values less than 1, that is when one uses a Laval nozzle with the appropriate throat area.

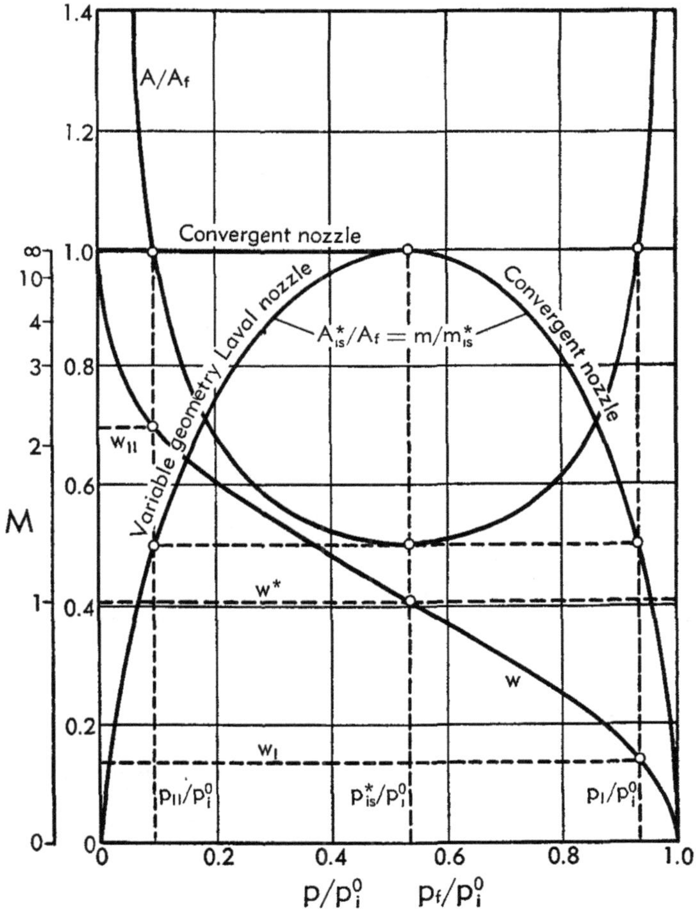

Fig. B,6c. Representation of isentropic transformation along a nozzle with given final area. $\gamma = 1.4$.

From the above discussion we conclude that it is impossible to reach supersonic velocities and values of $p_f < p^*_{is}$ in a purely convergent nozzle. We also conclude that in the range $p^*_{is} < p_f < p_i^0$ all flow rates between 0 and m^*_{is} can be realized with a convergent nozzle of fixed geometry. On the contrary, in the range $0 < p_f < p^*_{is}$ each m is realized with a Laval nozzle of different geometry, distinguished by a particular value of the throat area. The throat area decreases steadily for decreasing p_f, from the value $A^*_{is} = A_f$ relative to $m = m^*_{is}$, $p_f = p^*_{is}$, and $M_f = 1$, to $A^*_{is} = 0$ for $m = 0$, $p_f = 0$, and $M_f = \infty$.

Nozzle of fixed geometry with varying back pressure. We can now discuss what happens when a nozzle of given shape is subject to fixed upsteam conditions and varying downstream conditions. In the above discussion we have characterized the downstream conditions through the pressure in the final section. In practice we can act on this final pressure only indirectly, by modification of the ambient downstream pressure or "back pressure" p_b in the space where the gas is discharged. Let us examine the effect of changing p_b. If the nozzle is purely convergent, as in Fig. B,6d, and if the area upstream of the nozzle is large, the initial pressure will coincide practically with p_i^0 for all flow rates. On the right of the

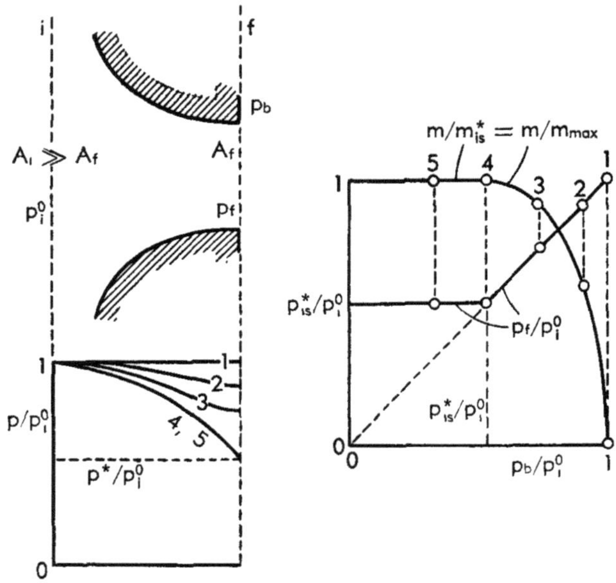

Fig. B,6d. Behavior of a convergent nozzle with varying back pressure.

figure the values of p_f/p_i^0 and m/m^*_{is} are plotted against p_b/p_i^0; on the left are represented schematically the pressure distributions in the nozzle for a certain number of representative cases. In case 1 ($p_b = p_i^0$), m is zero and $p = p_i^0$ everywhere. In case 2 and 3 ($p^*_{is} < p_b < p_i^0$) one has $0 < m < m^*_{is}$ and $p_f = p_b$. In case 4 ($p_b = p^*_{is}$), m reaches m^*_{is} and $p_f = p_b = p^*_{is}$; the final velocity is now sonic. Since this is the maximum velocity attainable in a convergent duct, a further decrease of p_b cannot be followed by p_f; therefore for $p_b < p^*_{is}$ (case 5) the conditions in A_f remain at $p_f = p^*_{is}$, $w_f = w^*$, and $m = m^*_{is}$; in other terms decreasing the back pressure below the critical value has no further effect on the flow within the nozzle and the flow rate stays constant at its maximum value m^*_{is}. Since the pressure in the final section remains higher than the ambient pressure, the corresponding adjustment must take place outside of the nozzle, where the finite transverse pressure differential will produce important transversal accelerations and pressure nonuniformities, which cannot be accounted for by the one-dimensional treatment.

A similar discussion for a Laval nozzle leads to the results of Fig. B,6e. Here not only the upstream conditions and A_f are given, but also the area A_t of the throat has a prescribed value. Again in case 1 ($p_b = p_i^0$), m is zero and the pressure is uniform. In cases 2 and 3 (p_b somewhat smaller than p_i^0), $p_f = p_b$ and a finite flow rate m is established. The maximum velocity in the nozzle is attained at the point of minimum section, and there one obtains a pressure $p_t < p_b$. The Laval nozzle acts in this case as a subsonic "venturi" in the divergent portion of which compression rather than expansion takes place. For a certain value $p_b = p_I$ (case 4)

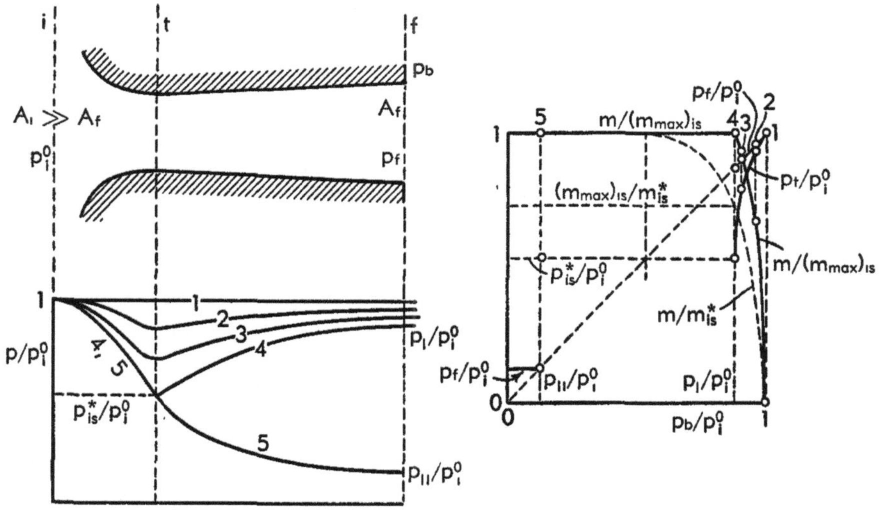

Fig. B,6e. Behavior of a Laval nozzle for varying back pressure.

the conditions at the throat become sonic, $p_t = p_{is}^*$ and $w_t = w^*$. This case can be represented in Fig. B,6c by the portion of the A/A_f curve which is to the right of the minimum. The representative point moves to the left until it reaches the minimum at the throat, and thereafter moves again to the right up to the point $A/A_f = 1$. Since the maximum flow rate per unit area is reached at the sonic velocity, the flow rate in case 4 represents the maximum possible flow rate for the given throat area, and it is given by Eq. 6-7 with $A_{is}^* = A_t$

$$(m_{max})_{is} = \chi(\gamma)\frac{p_i^0 A_t}{\sqrt{\mathcal{R}T_i^0}} = m_{is}^* \frac{A_t}{A_f} \tag{6-9}$$

Obviously $(m_{max})_{is} < m_{is}^*$, if the latter is referred to A_f.

Clearly in the range $p_I < p_b < p_i^0$ the presence of the throat does not affect the flow rate and the final pressure, and the relative curves coincide with a portion of the corresponding curves of Fig. B,6d. When $p_b < p_I$, however, the presence of the throat becomes determinant because the flow rate cannot increase beyond the value $(m_{max})_{is}$. For $p_b < p_I$, therefore, the condition at the throat remains sonic, and $m = (m_{max})_{is}$, with the consequence that the conditions upstream of the throat are not affected further by changes in p_b. Conversely, the flow conditions downstream of the throat are still determined by the value of p_b. When p_b takes the particular value p_{II} defined by Fig. B,6c (case 5), the flow downstream of the throat is again isentropic, but it is now all supersonic and in expansion, while it was all subsonic and in compression in case 4. This is the type of flow already considered in the discussion of Fig. B,6c, for which $p_f = p_b = p_{II}$. When this condition is realized we say that the supersonic nozzle is "correctly expanded" for the particular pressure ratio p_b/p_i^0. The correct area ratio A_f/A_t for a given p_b/p_i^0 is given by the corresponding value of the reduced area (Eq. 6-3 and 6-4b). When, on the contrary, $p_{II} \neq p_b$ we say that the nozzle is incorrectly expanded—more precisely, overexpanded for $p_{II} < p_b$ and underexpanded for $p_{II} > p_b$.

The phenomena in overexpanded nozzles are discussed in detail in Art. 8. For the case of underexpansion the conditions are simpler, and analogous to those discussed for the convergent nozzle with $p_b < p_{is}^*$. Because w_f is supersonic, downstream perturbations cannot be propagated upstream and therefore the conditions within the nozzle cannot be affected by a decrease of p_b. Thus p_f remains at its value p_{II} and the adjustment to p_b takes place outside of the nozzle in a way that involves non-one-dimensional effects.

The phenomenon of choking. We have pointed out in the foregoing discussion an important phenomenon present both in convergent and Laval nozzles. Once the sonic velocity is attained at a point of the nozzle, the flow rate cannot increase any further for fixed upstream conditions. This is the phenomenon known as "choking" and it can be physically visualized as the consequence of the simple fact that, since perturbations are propagated with the velocity of sound, upstream-moving disturbances remain stationary at a sonic section and cannot be transmitted upstream any further. Therefore, changing the conditions downstream of a sonic section cannot produce any change in the conditions upstream.

Choking can also be obtained for fixed back pressure by changing the upstream conditions p_i^0 or T_i^0. For a given nozzle, however, the pressure ratio p_i^0/p_b necessary for choking is only a function of γ. Once the flow is choked the flow rate follows Eq. 6-9 and the effect of p_i^0 becomes one of simple proportionality.

Clearly the cause for the phenomenon of choking, as described, is the presence of a minimum area, or throat. For given p_i^0 and T_i^0 the flow rate at which the flow becomes choked is determined by the throat area. As an illustration, let us consider a frictionless pipe of cross-sectional area A_i in which a gas is flowing with conditions represented by w_i, p_i^0, and T_i^0, with the corresponding flow rate m. If at the end of the pipe we apply a nozzle with a throat of area $A_t < A_i$, the same flow can be maintained in the pipe only if $m < (m_{max})_{is}$ (given by Eq. 6-9). The opposite case, $m > (m_{max})_{is}$, results in an impossible situation which can be corrected only by a change in the initial conditions in the pipe.

The mass flow rate is not the best quantity for the characterization of the condition for choking, because it depends on the values of p_i^0 and T_i^0. The reduced velocity w_i or the Mach number M_i represents a better criterion, independent of the state of the gas. If the value of A_i/A_t is assigned, the flow is possible if this ratio is smaller than, or equal to, the reduced area calculated from Eq. 6-3 for w_i or M_i, and is impossible otherwise. Conversely one can calculate from Eq. 6-3 a value of $(w_i)_{max}$ or $(M_i)_{max}$ corresponding to the reduced area A_i/A_t. The flow in the pipe is possible only with $w_i \leqq (w_i)_{max}$ or $M_i \leqq (M_i)_{max}$. The conditions for choking are therefore well characterized by the maximum value of the reduced velocity or of the Mach number.

We shall see in the following articles that many other causes can produce choking. The particular type of choking described here can be distinguished as choking by constriction.

Conditions at the sonic section. It has long been recognized [*24*; *25*, pp. 98–100] that certain analytical properties of the flow equations are connected with the phenomenon of choking. From Eq. 4-8 and 2-16 one obtains

$$\frac{d\rho}{\rho} = -\frac{u^2}{a^2}\frac{du}{u} = -M^2\frac{du}{u} \tag{6-10}$$

replacing this expression in the relation

$$\frac{dA}{A} + \frac{d\rho}{\rho} + \frac{du}{u} = 0 \tag{6-11}$$

derived by logarithmic differentiation from Eq. 4-2b, one has

$$\frac{dA}{A} = (1 - M^2)\frac{du}{u} \tag{6-12}$$

which again shows that $dA = 0$ at $M = 1$. In terms of reduced velocity Eq. 6-12, divided by dx, becomes

$$\frac{dw}{dx} = -\frac{w(1 - w^2)}{(1 - w^2/w^{*2})}\frac{1}{A}\frac{dA}{dx} \tag{6-12a}$$

If $A(x)$ is assigned, Eq. 6-12a is an equation between w and x, the integral of which is given by Eq. 6-2. Examination of Eq. 6-12a shows that at a throat, where $dA/dx = 0$, dw/dx is zero, except if $w = w^*$; vice versa at the sonic velocity, $w = w^*$, dw/dx is infinite, except if $dA/dx = 0$. Thus the integral curves of Eq. 6-12a in the plane w, x have a horizontal tangent at a throat and a vertical tangent at the sonic velocity. The same is true for the integral curves in the plane p, x, because from Eq. 6-4b we have

$$\frac{1}{p}\frac{dp}{dx} = -\frac{2\gamma}{\gamma - 1}\frac{w}{1 - w^2}\frac{dw}{dx} \tag{6-13}$$

If we have $w = w^*$ and $dA/dx = 0$ simultaneously, i.e. if the sonic velocity is attained at the throat, dw/dx and dp/dx become indeterminate. The corresponding values can be obtained using the l'Hospital rule, which gives

$$\left(\frac{dw}{dx}\right)^* = \frac{w^*(1 - w^{*2})}{A_t}\frac{\left(\dfrac{d^2A}{dx_2}\right)_t}{\dfrac{2}{w^*}\left(\dfrac{dw}{dx}\right)^*}$$

the subscript $_t$ indicating the values at the throat. Hence

$$\left(\frac{dw}{dx}\right)^* = \pm\sqrt{\frac{w^{*2}(1 - w^{*2})}{2A_t}\left(\frac{d^2A}{dx^2}\right)_t} \tag{6-14}$$

and from Eq. 6-13,

$$\frac{1}{p^*_{is}}\left(\frac{dp}{dx}\right)^* = \mp\gamma\sqrt{\frac{1}{(\gamma + 1)A_t}\left(\frac{d^2A}{dx^2}\right)_t} \tag{6-15}$$

In mathematical language we express these results by saying that the point $w = w^*$, $dA/dx = 0$ is a singular point of Eq. 6-12a; while a single integral curve goes through each point of the w, x or the p, x planes, two such curves pass through the singular point, with symmetrical slopes given by Eq. 6-14 and 6-15. This kind of singularity is called a saddle point. The behavior of the integral curves is shown in Fig. B,6f. In addition to the curves 1, 2, 3, and 5 which coincide with those already discussed on the p, x plane of Fig. B,6e, there are new curves which require explanations. Case 4 of the preceding figure consisted of the subsonic portions of the curves 4 and 5 intersecting at the saddle point of Fig. B,6f. The curve 4, however, also contains a supersonic part which has no meaning in the discussion of the nozzle behavior and actually represents a compression from an initially supersonic condition to a finally subsonic condition, going through the sonic velocity at the throat. The curve 4 would therefore represent the operation of an isentropic supersonic diffuser (see Art. 8). Starting from the same initial conditions, however, there is also the possibility of an isentropic flow along the supersonic portions of the curves 4 and 5. All the flows represented by combination of the curves 4 and 5 correspond to the same flow rate $m = (m_{max})_{is}$ and to $A^*_{is} = A_t$. Both quantities are related to the constant of integration of Eq. 6-12a, as clearly shown by Eq. 6-2.

The integral curves 6 and 7 contained between the supersonic portions of curves 4 and 5 represent entirely supersonic flows with $m < (m_{max})_{is}$

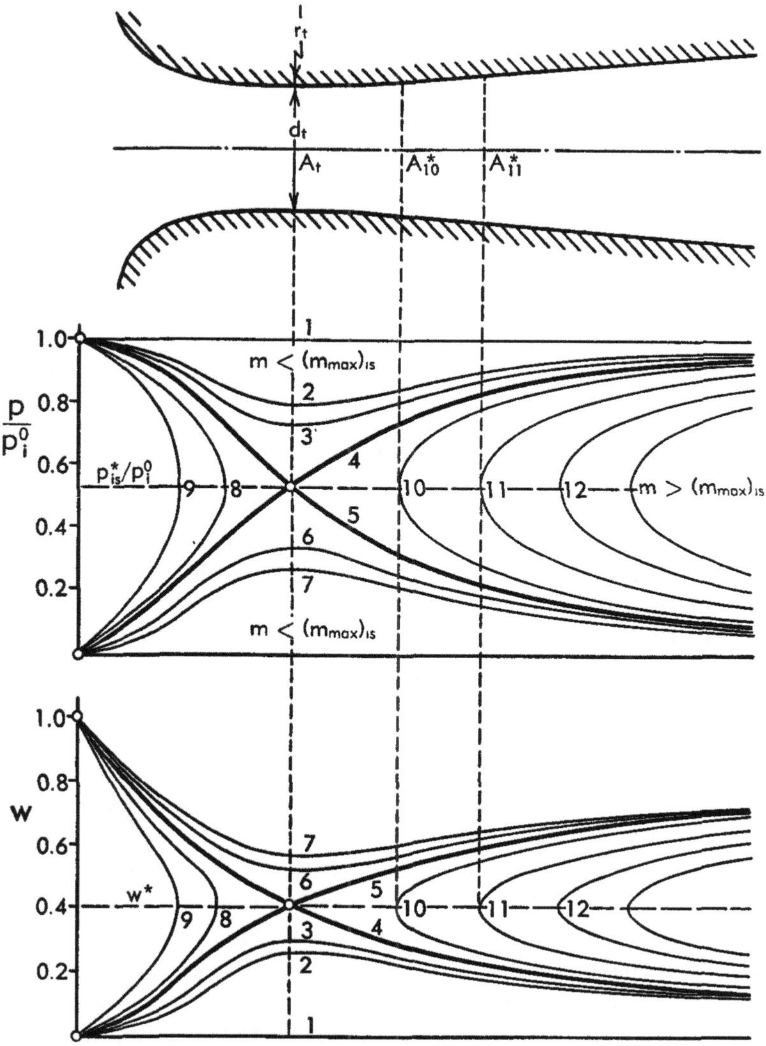

Fig. B,6f. Integral curves of Eq. 6-12a and 6-13.

and $A^*_{is} < A_t$, and correspond, in the supersonic range, to the unchoked subsonic flows represented by the curves 2 and 3.

Finally the integral curves 8, 9, 10, 11, and 12 correspond to the case $m > (m_{max})_{is}$, $A^*_{is} > A_t$ which, as already noticed, is physically impossible. It is clear that these curves cannot correspond to a real flow from the fact that to each value of x correspond two velocities and two pressures, or no solution at all. A flow starting subsonic on curves 8 and 9 from the initial section would only reach a certain maximum abscissa, where the velocity is sonic and $A = A_{is}^*$. From this point any change of area must be an increase, and hence the flow should reverse its direction toward increasing A values. Clearly this flow could be given a physical reality only by chopping off the part of the nozzle to the right of the maximum abscissa, or by boring a larger throat of area A_{is}^*. Similar considerations apply to curves 10 and 11.

Observe that the direction of flow on each integral curve is arbitrary, in agreement with the reversibility of isentropic transformations. For straight nozzles with circular section, if d_t represents the diameter of the throat section, and r_t the corresponding radius of curvature of the generatrix (see Fig. B,6f), Eq. 6-15 becomes

$$\frac{d_t}{p_{is}^*}\left(\frac{dp}{dx}\right)^* = \mp 2\gamma\sqrt{\frac{1}{\gamma+1}\frac{d_t}{r_t}}$$

so that this nondimensional pressure gradient is a function of the ratio between the diameter of the section and the radius of curvature of the generatrix at the throat. For $\gamma = 1.4$ and $d_t = r_t$ its value is ± 1.806.

B,7. Flow in Real Nozzles. The presence of friction is not the only cause for the different behavior of real nozzles with respect to the behavior calculated in the isentropic case. A second reason for the divergence consists in the important deviations from the assumptions of obliquity and uniformity that may be present when the section of the nozzle changes too rapidly or the curvature of the axis is too strong.

In general, however, even when the local characteristics of the flow deviate considerably from those obtained for the one-dimensional isentropic flow, these still provide a sufficiently good approximation from which the over-all characteristics of the real nozzle can be obtained by applying relatively small corrections.

The values of the corrections originated by the departure from the one-dimensional assumptions can be obtained theoretically through procedures other than the one-dimensional theory, and, in general, are out of the scope of this section (see VI,E). However, within the one-dimensional approximation one can estimate the corrections due to the presence of friction. We observe again that in the real flow the effects of friction are limited to the boundary layers adjacent to the walls of the nozzle, and that the flow outside of the boundary layers is still isentropic (except when shock waves or strong turbulent effects are present). Nevertheless the calculation of the frictional effects can be based on the assumptions of one-dimensional flow, provided that the parameters of the flow are properly chosen mean values (Art. 16).

Of course, calculations based on the one-dimensional approximation will not be sufficient, and more refined procedures taking into account the departure from the one-dimensional assumptions are required, when the actual details of the flow are desired. Such is the case, for instance, for the wind tunnel nozzles which must fulfill stringent conditions of uniformity in the region corresponding to the test section. In the other cases, however, where the details of the flow are not of primary importance, such as in propulsion or turbine nozzles, the calculations based on the one-dimensional approximation may be sufficient. Moreover, independently of the calculations, the corresponding corrections can be based on empirically determined coefficients, of which the most commonly used are discussed below.

Nozzle coefficients and efficiencies. The over-all quantities which are important for the evaluation of nozzles are the mass flow rate and the axial momentum or the axial component of the velocity at the exit. The corresponding values in a real nozzle are generally compared with those of the isentropic case for the same value of the expansion ratio p_i^0/p_f, and the following coefficients are defined.

The discharge coefficient

$$C_m = \frac{m}{m_{is}} \tag{7-1}$$

is generally a function of the nozzle shape and size, of the conditions of the walls (roughness), and of the nature and conditions of the gas. It is, moreover, a function of the expansion ratio, but becomes independent of it when the critical conditions are reached at the throat and the flow rate stays constant. In these conditions the discharge coefficient becomes independent of the flow downstream of the throat[6] and depends only on the phenomena taking place in the convergent portion of the nozzle.

One can distinguish between frictional effects within the one-dimensional approximation, deviations from the one-dimensional assumptions due to frictional effects, and deviations due to inviscid effects. C_m is affected by friction because of the cumulative effects of friction in building up a boundary layer in the convergent part of the nozzle. The larger the relative boundary layer thickness with respect to the radius at the throat, the lower will be C_m. Only part of these effects can be accounted for by the one-dimensional treatment. The blocking effect due to the increased thickness of the boundary layer with respect to that corresponding, if there were no friction, to the same flow rate which flows in the boundary layer, cannot be described correctly by the one-dimensional theory. However, from general principles of the boundary layer theory one obtains the fol-

[6] Rigorously, due to non-one-dimensional effects a short portion of the nozzle downstream of the throat affects the flow at the throat and therefore the discharge coefficient.

lowing simple qualitative indications about the frictional effects on C_m. For geometrically similar nozzles C_m increases with the size and the density of the gas and decreases with the viscosity; for a nozzle of given throat diameter C_m decreases if the length of the converging part of the nozzle is increased homologously. In particular, if the generatrix of an axially symmetric nozzle has a radius of curvature r_t at the throat, increasing r_t with fixed throat diameter tends to decrease C_m.

These frictional effects must be combined with the inviscid effects due to the fact that the velocity and density are not uniformly distributed at the throat. This has the effect of reducing the flow rate with respect to that corresponding to uniform distributions, and the reduction increases with increasing nonuniformity. For large r_t the deviations from uniformity are small, and they increase when r_t decreases because the transversal pressure gradients necessary to balance the centrifugal force become larger. Thus decreasing r_t with fixed throat diameter tends to reduce C_m.

The effect of r_t is therefore in opposite direction on the frictional effects and on the inviscid effects on C_m; too large values of r_t increase the former, and too small values of r_t increase the latter. It is to be expected that a maximum of C_m is obtained for a proper value of r_t.

For large nozzles or high densities the values of C_m tend to be very close to 1. Values of the order of 0.99 or more are easily obtained. For small nozzles, or very low densities, or nozzles with a curved axis, C_m can become considerably lower than unity.

The previous considerations apply not only to Laval nozzles, but also to the simple convergent nozzles that can be obtained from these by cutting the nozzle at the throat. However, for convergent nozzles with $dA/dx \neq 0$ at the final section another inviscid effect is present, known as the phenomenon of the "vena contracta." In this case the streamlines leave the final section with a definite obliquity toward the axis which cannot be canceled immediately; therefore the section of the issuing jet must decrease and reach a minimum area, smaller than the exit area A_f, at a certain distance from the final section. Thus A_f can never constitute the throat of the nozzle, and the actual throat of the flow is always reached outside of the nozzle with $A_t < A_f$. This phenomenon may produce a substantial decrease of C_m, and it may be particularly important for certain types of adjustable exhaust nozzles for jet engines.

Another parameter often used to characterize the maximum flow rate of rocket nozzles is the characteristic velocity c^* defined as

$$c^* = \frac{p_i^0 A_t}{m_{\max}} = \frac{p_i^0 A_t}{C_m (m_{\max})_{is}} = \frac{\sqrt{\mathcal{R} T_i^0}}{C_m \chi(\gamma)} \tag{7-2}$$

with $\chi(\gamma)$ given by Eq. 6-7.[7] It is interesting to notice that c^* is generally

[7] In practical units the last expression must be multiplied by $\sqrt{g}$.

used as an experimental quantity which shows substantially how high the temperature is before the nozzle, and therefore gives an indication of the completeness of the combustion in the rocket chamber. We observe also that low values of c^* compared with the theoretical are not only a result of low T_i^0, but also of the fact that when the combustion is far from being completed in the rocket chamber substantial amounts of heat are released during the expansion in the rocket itself. This is indeed true, but to a smaller extent, even when the combustion gases reach the nozzle in a state of thermochemical equilibrium corresponding to the maximum possible heat release in the chamber; the reason for this being that during the expansion in the nozzle the equilibrium shifts, with a recombination of dissociated species and heat development. Obviously when heat is evolved during the expansion the last expression of Eq. 7-2 is not correct, and a T_i^0 computed from it on the basis of the experimental c^* would be in error.

Another important coefficient is the "velocity coefficient" defined as

$$C_V = \frac{q_f}{(q_f)_{is}} \tag{7-3}$$

Again the comparison between the final velocity q_f in the real nozzle and the final velocity for isentropic flow is made for the same expansion ratio. Here we have used the total velocity q instead of its axial component because we want to include cases in which the obliquity θ of the velocity is such that the assumption $\cos\theta = 1$ is not compatible with the accuracy required from the calculations. Such may be the case, for instance, for a Laval nozzle, the divergent supersonic portion of which, following a common practice, is conical. Away from the throat one can approximately assume that the streamlines are straight lines diverging from the cone apex, and that the velocities q are normal to each spherical cap with center in the cone apex. Integrating with respect to the elementary area $2\pi R^2 \sin\theta d\theta$ of the outer spherical cap of radius R, and assuming that velocity and density on it are uniform and equal to q_f and ρ_f, one finds for the flow rate

$$m_{max} = \int_0^{\theta_c} \rho_f q_f \cdot 2\pi R^2 \sin\theta d\theta = \rho_f q_f 2\pi R^2(1 - \cos\theta_c)$$

and for the axial component of the momentum

$$I_f = \int_0^{\theta_c} q_f \cos\theta \cdot \rho_f q_f \cdot 2\pi R^2 \sin\theta d\theta = \rho_f q_f^2 \pi R^2(1 - \cos^2\theta_c)$$

Here θ_c represents the half-opening of the cone.

The mean value of the axial component of the velocity on the final section can be calculated as

$$u_f = \frac{I_f}{m_{max}} = q_f \frac{1 + \cos\theta_c}{2} \tag{7-4}$$

One can therefore define a conicity coefficient

$$C_c = \frac{1 + \cos\theta_c}{2} \cong 1 - \frac{1}{4}(\sin\theta_c)^2 \tag{7-5}$$

and, introducing Eq. 7-3, obtain

$$u_f = (q_f)_{is} C_V C_c \tag{7-6}$$

Similarly one finds, with the help of Eq. 7-1,

$$I_f = (q_f)_{is}(m_{max})_{is} C_m C_V C_c = (I_f)_{is} C_m C_V C_c \tag{7-7}$$

For practical Laval nozzles the angle θ_c is generally below 20°, which corresponds to $C_c = 0.97$. For smaller θ_c, C_c approaches unity rapidly. For instance at $\theta_c = 10°$, $C_c = 0.9924$. Hence, except for large θ_c or refined thrust calculations, one can in general take $C_c = 1$ and $u_f = q_f$, in agreement with the one-dimensional assumptions.

The value of C_V also can be quite close to unity. It can be above 0.99 for large straight nozzles. However, for small nozzles, and especially for curved nozzles, it can take values considerably smaller for the reasons discussed later. It is interesting to observe that while C_m, as stated above, depends only on the conditions upstream of the throat, for straight Laval nozzles C_V is practically determined by the conditions downstream of it. This is due to the fact, to be checked later, that the weight of the frictional effects increases with the velocity.

In addition to the above-defined flow coefficients, one can consider the efficiency of the nozzle. Following the ordinary concept, the adiabatic (or isentropic) efficiency can be defined as the ratio of the final kinetic energy in the real nozzle to the kinetic energy obtainable, for the same expansion ratio, in an isentropic nozzle:

$$\eta_{ad} = \frac{q_f^2}{(q_f)_{is}^2} = C_V^2 \tag{7-8}$$

and coincides with the square of the velocity coefficient. Using Eq. 4-8, with u replaced by q, we obtain

$$\eta_{ad} = \frac{h_i^0 - h_f}{h_i^0 - (h_f)_{is}} = \frac{T_i^0 - T_f}{T_i^0 - (T_f)_{is}} = \frac{1 - \dfrac{T_f}{T_i^0}}{1 - \left(\dfrac{p_f}{p_i^0}\right)^{\frac{\gamma-1}{\gamma}}} \tag{7-9}$$

the last two steps being true for polytropic gases.

Another interesting definition of the efficiency, which has certain advantages (to be discussed in Art. 13), is that of the "elemental" or "polytropic" efficiency:

$$\eta_{el} = \frac{\ln T_i^0 - \ln T_f}{\ln T_i^0 - \ln (T_f)_{is}} = \frac{\gamma}{\gamma - 1} \frac{\ln \dfrac{T_f}{T_i^0}}{\ln \dfrac{p_f}{p_i^0}} \tag{7-10}$$

From Eq. 7-9 and 7-10 the following relation is found between the two efficiencies

$$\eta_{ad} = \frac{1 - \left(\frac{p_f}{p_i^0}\right)^{\eta_{el}\frac{\gamma-1}{\gamma}}}{1 - \left(\frac{p_f}{p_i^0}\right)^{\frac{\gamma-1}{\gamma}}} \tag{7-11}$$

This relation is represented in Fig. B,7a. For efficiencies close to unity,

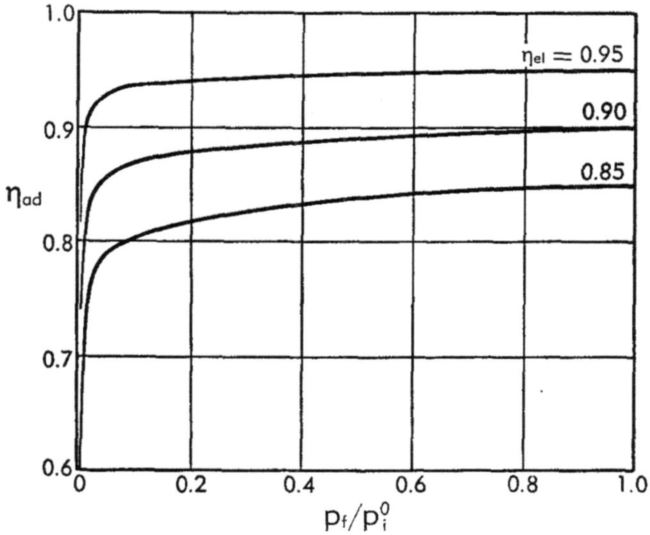

Fig. B,7a. Relation between adiabatic and elemental (polytropic) efficiency for a nozzle.

one has

$$\left(\frac{p_f}{p_i^0}\right)^{\eta_{el}\frac{\gamma-1}{\gamma}} \cong \left(\frac{p_f}{p_i^0}\right)^{\frac{\gamma-1}{\gamma}} \left[1 - (1 - \eta_{el}) \ln \left(\frac{p_f}{p_i^0}\right)^{\frac{\gamma-1}{\gamma}}\right]$$

and hence

$$\frac{1 - \eta_{ad}}{1 - \eta_{el}} \cong \frac{\ln \left(\frac{p_i^0}{p_f}\right)^{\frac{\gamma-1}{\gamma}}}{\left(\frac{p_i^0}{p_f}\right)^{\frac{\gamma-1}{\gamma}} - 1} = \frac{\ln \frac{T_i^0}{(T_f)_{is}}}{\frac{T_i^0}{(T_f)_{is}} - 1} \tag{7-12}$$

Only for isentropic temperature ratios close to unity do the values of the two efficiencies practically coincide. From Fig. B,7a and from Eq. 7-12 we see that for nozzles η_{ad} is always larger than η_{el}.

Flow with constant adiabatic efficiency. The same definitions of the efficiencies can be used not only for the final conditions at the nozzle exit, but also for any intermediate point during the expansion. One can then investigate the flow in a nozzle in which the losses are distributed in such a way that a constant value of η_{ad} is maintained, as defined by Eq. 7-9 with p_t and T_t replaced by p and T. The energy equation (Eq. 4-27) and the derived equation (Eq. 4-28) are independent of the presence of the losses; therefore one can replace T/T_i^0 by $1 - w^2$ and solve for p/p_i^0 which is given by

$$\frac{p}{p_i^0} = \left(1 - \frac{w^2}{\eta_{ad}}\right)^{\frac{\gamma}{\gamma-1}} \tag{7-13}$$

Replacing this value of p in the first equation (Eq. 4-31a) we obtain the relation

$$\frac{Aw}{1 - w^2}\left(1 - \frac{w^2}{\eta_{ad}}\right)^{\frac{\gamma}{\gamma-1}} = \sqrt{\frac{\gamma - 1}{\gamma}}\,\frac{m\sqrt{\mathcal{R}T_i^0}}{p_i^0} \tag{7-14}$$

which coincides with Eq. 6-2 for $\eta_{ad} = 1$. The second member of this expression is constant. Therefore, differentiating the first member logarithmically, one obtains

$$\frac{dA}{A} + \left(\frac{1 + w^2}{1 - w^2} - \frac{\gamma}{\gamma - 1}\,\frac{2w^2}{\eta_{ad} - w^2}\right)\frac{dw}{w} = 0$$

At the throat, $dA = 0$, the quantity in parenthesis must be zero. Thus one obtains the velocity at the throat w_t from a second degree equation in w_t^2, of which the larger root must be discarded because it is always larger than 1, and the other is

$$w_t^2 = \tfrac{1}{2}\{1 + (2 - \eta_{ad})w^{*2} - \sqrt{[1 + (2 - \eta_{ad})w^{*2}]^2 - 4\eta_{ad}w^{*2}}\} \tag{7-15}$$

One can show that $w_t < w^*$, so that the velocity is subsonic at the throat, and the sonic velocity is reached in the divergent portion of the nozzle.

The quantity which multiplies A on the left-hand side of Eq. 7-14 has a maximum for $w = w_t$. Hence for a fixed value of the throat area one can calculate from Eq. 7-14 the corresponding value of m_{max} and compare it with the value of Eq. 6-9 relative to the isentropic case, thus obtaining C_m.

Expanding Eq. 7-15 in powers of $1 - \eta_{ad}$ one obtains

$$w_t^2 = w^{*2}\{1 - \gamma(1 - \eta_{ad}) + O[(1 - \eta_{ad})^2]\}$$

One has also

$$\frac{p_t}{p_i^0} = \frac{1}{1 - w_t^2}\left(1 - \frac{w_t^2}{\eta_{ad}}\right)^{\frac{\gamma}{\gamma-1}} = (1 - w^{*2})^{\frac{1}{\gamma-1}}\{1 + O[(1 - \eta_{ad})^2]\}$$

For η_{ad} so close to unity that terms of order $(1 - \eta_{ad})^2$ can be neglected, the density at the throat coincides with the critical value relative to the isentropic flow and, as Eq. 7-14 shows, the discharge coefficient coincides with w_t/w^* and is given by

$$C_m = 1 - \frac{\gamma}{2}(1 - \eta_{ad})$$

Within the same approximation the velocity coefficient is

$$C_V = \sqrt{\eta_{ad}} = 1 - \tfrac{1}{2}(1 - \eta_{ad})$$

Flow with constant polytropic efficiency. The definition of polytropic efficiency (Eq. 7-10) can be applied to any intermediate p and T, and can be written, using Eq. 4-28, as

$$\frac{p}{p_i^0} = \left(\frac{T}{T_i^0}\right)^{\frac{1}{\eta_{el}}\frac{\gamma}{\gamma-1}} = (1 - w^2)^{\frac{1}{\eta_{el}}\frac{\gamma}{\gamma-1}} \tag{7-16}$$

This relation evidently is reduced to the ordinary isentropic relation for $\eta_{el} = 1$. For constant η_{el} the exponent of Eq. 7-16 is constant, and one can write

$$\frac{1}{\eta_{el}}\frac{\gamma}{\gamma - 1} = \frac{n}{n - 1}$$

n representing the exponent of a polytropic transformation $p \sim \rho^n$. This polytropic exponent

$$n = \frac{\gamma}{1 + (\gamma - 1)(1 - \eta_{el})} = \gamma\{1 - (\gamma - 1)(1 - \eta_{el}) + O[(1 - \eta_{el})^2]\} \tag{7-17}$$

is, for a nozzle, always smaller than the adiabatic exponent γ.

Using Eq. 4-28 again, one gets from Eq. 7-16

$$\frac{p^0}{p_i^0} = \left(\frac{T}{T_i^0}\right)^{\frac{\gamma}{\gamma-1}\cdot\left(\frac{1}{\eta_{el}}-1\right)} = (1 - w^2)^{\frac{1}{n-1}-\frac{1}{\gamma-1}} \tag{7-18}$$

Replacing Eq. 7-18 in the second equation (Eq. 4-31a) one has the following relation, similar to Eq. 6-2,

$$Aw(1 - w^2)^{\frac{1}{n-1}} = \sqrt{\frac{\gamma - 1}{2\gamma}}\,\frac{m\sqrt{\mathcal{R}T_i^0}}{p_i^0} \tag{7-19}$$

For a given Laval nozzle A is a minimum when the function of w in the left member is maximum. This happens for

$$w_t^2 = \frac{n - 1}{n + 1} = w^{*2}\left\{1 - \frac{2\gamma}{\gamma + 1}(1 - \eta_{el}) + O[(1 - \eta_{el})^2]\right\} \tag{7-19a}$$

Thus $w_t < w^*$ represents the velocity at the throat for which the flow rate takes its maximum value

$$m_{max} = \frac{p_i^0 A_t}{\sqrt{\mathcal{R}T_i^0}}\sqrt{\frac{2\gamma}{\gamma - 1}}\sqrt{\frac{n - 1}{n + 1}}\left(\frac{2}{n + 1}\right)^{\frac{1}{n-1}}$$

Comparing with Eq. 6-9 one has therefore

$$C_m = \sqrt{\frac{n-1}{n+1}}\left(\frac{2}{n+1}\right)^{\frac{1}{n-1}} \sqrt{\frac{\gamma+1}{\gamma}}\left(\frac{\gamma+1}{2}\right)^{\frac{1}{\gamma-1}}$$

From Eq. 7-11 one can obtain the velocity coefficient $C_V = \sqrt{\eta_{ad}}$. Again, neglecting terms of order $(1 - \eta_{el})^2$, one finds

$$C_m = 1 - \frac{\gamma}{\gamma - 1} \ln \frac{\gamma + 1}{2} (1 - \eta_{el})$$

$$C_V = 1 - \frac{1}{2} \frac{\ln \left(\frac{p_1^0}{p_f}\right)^{\frac{\gamma-1}{\gamma}}}{\left(\frac{p_1^0}{p_f}\right)^{\frac{\gamma-1}{\gamma}} - 1} (1 - \eta_{el})$$

General relations for flows with friction. Obviously the assumption that the losses are distributed in such a way as to make η_{ad} or η_{el} constant cannot represent exactly the real flow conditions, where the losses are determined by the actual magnitude and distribution of the skin friction τ_w, which according to Eq. 3-5 determines the value of the friction force X_w. One can define a skin friction coefficient c_f[8] connected to the skin friction by the relation

$$\tau_w = \tfrac{1}{2}\rho u^2 c_f = \frac{\gamma}{2} p M^2 c_f = \frac{\gamma}{\gamma - 1} \frac{p w^2}{1 - w^2} c_f \tag{7-20}$$

The skin friction coefficient has the advantage that its dependence on the gas physical conditions and on the flow velocity is much less marked than that of τ_w itself.

The values of c_f can be based on empirical determinations or on the theoretical results of boundary layer theory. For flows with pressure approximately constant on flat plates or in tubes, a large mass of data of the two kinds is available. They show that c_f is principally affected by the so-called Reynolds number (proportional to the density, the velocity, and a properly chosen length; and inversely proportional to the viscosity coefficient) by the Mach number, by the heat transfer condition, and by the roughness of the walls (Vol. IV). Here we shall only remark that for purely laminar or well-established turbulent boundary layer c_f decreases both with the Reynolds number and the Mach number. For flows with pressure gradients, such as those present in nozzles, the data are less

[8] This coefficient must be distinguished from the usual skin friction coefficient c_f of boundary layer theory. The distinction is due to the fact that the values of ρ and u used in Eq. 7-20 are the mean values on the section as considered by the one-dimensional treatment, while in boundary layer theory the values of the density and velocity at the outer edge of the boundary layer are used. Of course if the boundary layer development is known in detail the two coefficients can be easily correlated to each other.

abundant, particularly in the turbulent case. However, it can be safely assumed that c_f increases for favorable pressure gradients, and decreases in the opposite case.

The presence of the dissipative viscous forces produces an increase of entropy and, for adiabatic flows, a decrease of stagnation pressure. These effects can be easily derived for the one-dimensional theory from Eq. 4-3b which, with the help of Eq. 4-2b, 4-9, and 2-10 can be successively written as

$$\begin{aligned} -dX &= Adp + mdu = Adp + A\rho udu \\ &= -A\rho\left(dh - \frac{dp}{\rho}\right) = -Ap\frac{ds}{\mathcal{R}} \end{aligned} \tag{7-21}$$

For perfect gases and adiabatic flow this relation can be expressed in terms of the stagnation pressure through Eq. 4-15 as

$$\frac{dp^0}{p^0} = -\frac{ds}{\mathcal{R}} = -\frac{dX}{Ap} \tag{7-21a}$$

Eq. 7-21 has a simple physical interpretation. The energy dissipated by the friction forces per unit mass is udX/m. Dividing by T one must obtain the entropy increase

$$ds = \frac{udX}{Tm} \tag{7-21b}$$

It is immediately checked that this equation coincides with Eq. 7-21. Clearly, in the real flow, where the dissipative forces are concentrated in the boundary layer, the total entropy increase is still given by udX/T, but it affects only the boundary layer flow.[9] If this nonuniform flow is converted into a uniform one by mixing the boundary layer with the rest of the flow, additional dissipative forces have to be used and an additional entropy increase due to the mixing process follows. Only when thoroughly mixed does the flow correspond to a truly one-dimensional flow and can it be rigorously treated by the corresponding equations. Thus, if one wants to make use of Eq. 7-21 in the one-dimensional treatment of the flow with friction, one has to apply the proper increase in the value of ds, or on the corresponding value of dX. This can be done by using in the evaluation of the friction force an "apparent" skin friction coefficient $f > c_f$. We shall discuss this point further in Art. 10.

If $dX = dX_w (dX_b = 0)$, from Eq. 3-5 and 7-20 we obtain

$$\frac{dp^0}{p^0} = -2c_f\frac{dx}{d_h}\gamma M^2 = -2c_f\frac{dx}{d_h}\frac{2\gamma}{\gamma - 1}\frac{w^2}{1 - w^2} \tag{7-22}$$

[9] This is not rigorously true. To be exact, when dX is given by Eq. 3-5, one must replace $\tau_w u$ by the integral $\int \tau du$ extended through the boundary layer. This integral can be evaluated from boundary layer theory and is close to, but smaller than, $\tau_w u_e$, where u_e is the velocity in the isentropic core outside of the boundary layer. Finally the mean entropy increase for the whole flow is not far from Eq. 7-21b.

Sonic conditions with friction. From Eq. 4-31a we obtain by logarithmic differentiation

$$\frac{dp^0}{p^0} + \frac{dA}{A} + \frac{1 - \frac{w^2}{w^{*2}}}{1 - w^2}\frac{dw}{w} = 0 \tag{7-23}$$

Thus from Eq. 7-22 and 7-23 one gets the equation

$$\frac{dw}{dx} = \frac{w(1 - w^2)\left[\frac{2c_f}{d_h}\gamma M^2 - \frac{1}{A}\frac{dA}{dx}\right]}{1 - \frac{w^2}{w^{*2}}} \tag{7-24}$$

This equation can be discussed in the same way as in the case $c_f = 0$, in which case it coincides with Eq. 6-12a. We see that dw/dx is infinite again for $w = w^*$ and zero when the last factor of the numerator vanishes. It becomes indeterminate when both numerator and denominator vanish simultaneously, which implies $M = 1$ and

$$\left(\frac{d_h}{2A}\frac{dA}{dx}\right)^* = \gamma c_f = \left(\frac{dd}{dx}\right)^* \tag{7-25}$$

The last expression being obtained for circular section ducts, for which $A = \pi d^2/4$ and $d_h = d$. This singular point of Eq. 7-24 is again a saddle point, and the corresponding two values of dw/dx are obtained for the circular section from Eq. 7-24 using the l'Hospital rule,

$$\left(\frac{1}{w}\frac{dw}{dx}\right)^* = -\frac{\gamma c_f}{d^*} \pm \sqrt{\left(\frac{\gamma c_f}{d^*}\right)^2 + \frac{2}{\gamma + 1}\left(\frac{1}{d}\frac{d^2d}{dx^2}\right)^*} \tag{7-26}$$

Eq. 7-25 shows that the sonic velocity is not reached at the throat but at a well-defined value of the angle of divergence of the divergent part of the nozzle. The corresponding value of the diameter, d^*, is therefore different from d_t. The difference can be found using the first terms of the Taylor expansions

$$\frac{dd}{dx} = (x - x_t)\left(\frac{d^2d}{dx^2}\right)_t; \quad d = d_t + \left(\frac{d^2d}{dx^2}\right)_t\frac{(x - x_t)^2}{2} \tag{7-27}$$

which give the divergence and the diameter for a section at a distance $x - x_t$ from the throat. Within this approximation, observing that $(d^2d/dx^2)_t = 2/r_t$, where r_t again represents the radius of curvature of the generatrix at the throat, from Eq. 7-25 and 7-27 one gets at the sonic section ($x = x^*$)

$$x^* - x_t = \frac{\gamma c_f}{2}r_t; \quad \frac{d^*}{d_t} = 1 + \left(\frac{\gamma c_f}{2}\right)^2\frac{r_t}{d_t}$$

In practical cases c_f is smaller than 0.01, and r_t/d_t is of the order of unity. Thus the distance of the sonic section from the throat is a fraction of a hundredth of the throat diameter, and d^* practically coincides with d_t. For such small values of c_f, identifying the value of the second derivative

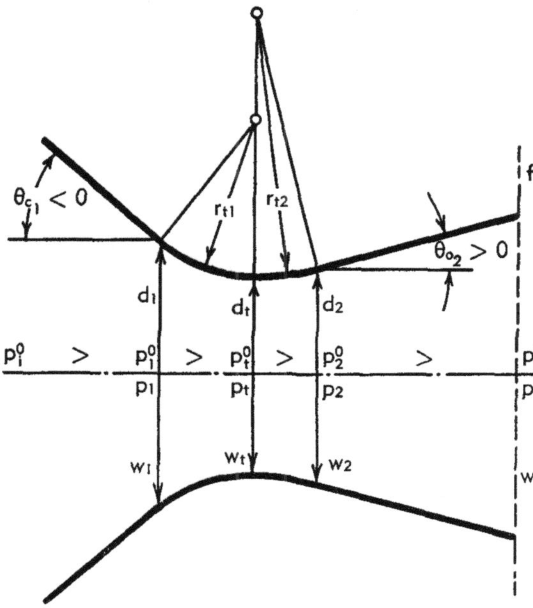

Fig. B,7b. Model of Laval nozzle used in the calculation of frictional effects.

of d at the sonic section and at the throat, Eq. 7-26 can be written

$$\left(\frac{d_t}{w}\frac{dw}{dx}\right)^* = \pm 2\sqrt{\frac{1}{\gamma+1}\frac{d_t}{r_t}} - \gamma c_f$$

and one obtains approximately, using the upper sign which corresponds to expansion,

$$\frac{w_t}{w^*} = 1 - (x^* - x_t)\left(\frac{1}{w}\frac{dw}{dx}\right)^* = 1 - \gamma c_f\sqrt{\frac{1}{\gamma+1}\frac{r_t}{d_t}}$$

For a more complete discussion of other singularities that may appear in the flow with friction the reader can consult [*26*].

Flow with constant friction coefficient. The integration of Eq. 7-24 when $A(x)$ and $c_f(x)$ are known is not feasible in a simple way. Here we shall obtain an approximate solution for c_f = const and for the particular shape of the Laval nozzle of Fig. B,7b, composed of two cones connected by two surfaces of revolution with circular generatrix. We can rewrite Eq. 7-22 in the form

$$\frac{dp^0}{p^0} = -\frac{4\gamma}{\gamma-1}c_f\frac{1}{d}\frac{dl}{dw}\frac{w^2dw}{1-w^2} \tag{7-28}$$

where we have identified the axial element dx with the actual element of length of the wall dl, which is allowed when the obliquity of the wall is quite small. We can write now

$$\frac{1}{d}\frac{dl}{dw} = \frac{dl}{dd}\frac{1}{d}\frac{dd}{dw} = \frac{dl}{dd}\frac{1}{2A}\frac{dA}{dw}$$

For a conical duct $dd = 2dl \sin \theta_c$, where θ_c is the half-opening of the cone; moreover we can approximately replace the second factor of the preceding expression by the value it has for isentropic flow, given by Eq. 7-23 with $dp^0 = 0$.[10] Finally Eq. 7-28 becomes

$$\frac{dp^0}{p^0} = -\frac{\gamma}{\gamma - 1}\frac{c_f}{\sin \theta_c}\frac{\left(\frac{w^2}{w^{*2}} - 1\right) w\,dw}{1 - w^2}$$

Integrating for constant θ_c and c_f one obtains

$$\left.\begin{aligned} &\frac{\sin \theta_c}{c_f} \ln p^0 \\ &= -\frac{\gamma}{2(\gamma - 1)}\left[\frac{\gamma + 1}{\gamma - 1}\ln (1 - w^2) + \frac{2}{\gamma - 1}\frac{w^2}{1 - w^2}\right] + \text{const} \\ &= F(w) + \text{const} \end{aligned}\right\} \quad (7\text{-}29)$$

For an infinite conical nozzle starting from $p^0 = p_i^0$, $w_i = 0$ we have

$$\frac{\sin \theta_c}{c_f} \ln \frac{p^0}{p_i^0} = F(w) \quad (7\text{-}30)$$

$F(w)$ is always positive, and for small w it is approximately equal to $\gamma w^2/2(\gamma - 1)$. Incidentally, one can compare Eq. 7-30 with the expression obtained by developing in powers of w^2 the logarithm of p^0/p_i^0 derived from Eq. 7-20. For small w one gets a constant value for the elemental efficiency given by

$$\frac{1}{\eta_{el}} - 1 = \frac{c_f}{2|\sin \theta_c|}$$

Eq. 7-30 can be used to calculate the losses in a convergent infinite cone ($\theta_{c_1} < 0$), up to a final velocity $w_1 < w^*$. This velocity is determined by the value of the final diameter d_1 because one obtains from Eq. 6-3 (applied here approximately)

$$\frac{d_1}{d_t} = \left(\frac{w^*}{w_1}\right)^{\frac{1}{2}} \left(\frac{1 - w^{*2}}{1 - w_1^2}\right)^{\frac{1}{2(\gamma - 1)}} \quad (7\text{-}31)$$

[10] For a conical duct and constant c_f it is actually possible to find an exact solution of the flow problem [*27*].

On the other hand one has from Fig. B,7b

$$\frac{d_1}{d_t} - 1 = 2\frac{r_{t1}}{d_t}(1 - \cos\theta_{c1}) \tag{7-32}$$

so that w_1 is a function only of the second member of this equation. Finally one can calculate

$$\frac{\sin\theta_{c1}}{c_f}\ln\frac{p_1^0}{p_i^0} = F\left\{w_1\left[2\frac{r_{t1}}{d_t}(1 - \cos\theta_{c1})\right]\right\} = G_0\left[2\frac{r_{t1}}{d_t}(1 - \cos\theta_{c1})\right] \tag{7-33}$$

Observe that most of the losses represented by Eq. 7-33 are concentrated near the end of the conical duct where the velocities are high, and that a change of the shape of the nozzle in the region where the velocities are small would leave the losses practically unchanged. Thus Eq. 7-33 may be used for other shapes of the convergent part of the nozzle provided that they are conical in a sufficient range near w_1.

In the rest of the convergent part of the nozzle, one can integrate Eq. 7-28, replacing dl/dw by its mean value

$$\left(\frac{dl}{dw}\right)_m = -\frac{r_{t1}\theta_{c1}}{w^* - w_1} \tag{7-34}$$

Observe that this value would be exact if dw/dl were constant. One can calculate the velocity for a duct of the shape considered, with isentropic flow. The fact that the resulting $w(l)$ is not far from linear, indicates that the approximation (Eq. 7-34) is not too bad. Replacing in Eq. 7-28 dl/dw by Eq. 7-34 and d by the value corresponding to Eq. 7-31 with w_1 substituted by w, and integrating between w_1 and $w_t \cong w^*$, one finds

$$\frac{\sin\theta_{c1}}{c_f}\ln\frac{p_t^0}{p_1^0} = \frac{4\gamma}{\gamma - 1}\left(\frac{\gamma - 1}{\gamma + 1}\right)^{\frac{1}{4}}\left(\frac{\gamma + 1}{2}\right)^{\frac{1}{2(\gamma-1)}}\frac{r_{t1}}{d_t}\theta_{c1}\sin\theta_{c1}\frac{\Phi(w_1)}{w^* - w_1} \tag{7-35}$$

where

$$\Phi(w_1) = \int_{w_1}^{w^*} w^{\frac{3}{2}}(1 - w^2)^{\frac{1}{2(\gamma-1)}}dw$$

For $\gamma = 1.25$ and $\gamma = 1.5$, $\Phi(w_1)$ can be easily evaluated in finite terms. The right-hand side of Eq. 7-35 is a function of $(r_{t1}/d_t)\theta_{c1}\sin\theta_{c1}$ and of w_1, which in turn is a function of

$$2\frac{r_{t1}}{d_t}(1 - \cos\theta_{c1}) = \frac{r_{t1}}{d_t}\left(\theta_{c1}\sin\theta_{c1} + \frac{\theta_{c1}^4}{12} + \cdots\right)$$

If, in this expansion, the terms beyond the first are neglected, which does not involve serious errors even for quite large θ_{c1}, we can write

$$\frac{\sin\theta_{c1}}{c_f}\ln\frac{p_t^0}{p_1^0} = G_1\left(\frac{r_{t1}}{d_t}\theta_{c1}\sin\theta_{c1}\right)$$

and, combining with Eq. 7-33,

$$\frac{\sin\theta_{c1}}{c_f}\ln\frac{p_{t1}^0}{p_1^0} = G_0 + G_1 \tag{7-36}$$

where both G_0 and G_1 can be considered to be only functions of $(r_{t1}/d_t)\theta_{c1}$ $\sin\theta_{c1}$.

For the supersonic side one can use the same procedure. The connection between d_2 and the velocity $w_2 > w^*$ reached at the entrance of the

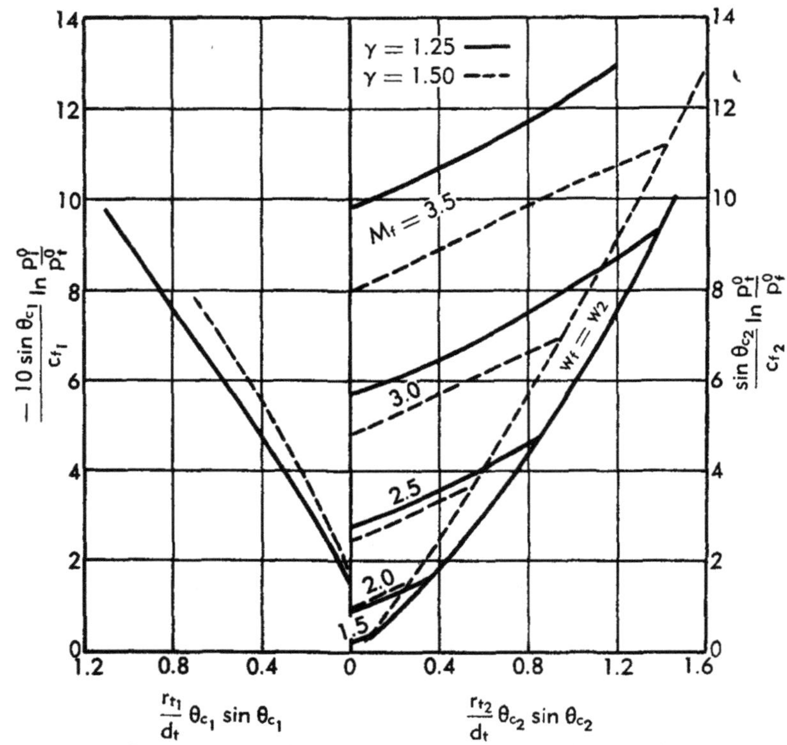

Fig. B,7c. Frictional effects for the nozzle of Fig. B,7b.

divergent conical duct is the same as Eq. 7-31. Also the equation corresponding to Eq. 7-32 can be written. Thus w_2 is only a function of

$$2\frac{r_{t2}}{d_t}(1 - \cos\theta_{c2}) \cong \frac{r_{t2}}{d_t}\theta_{c2}\sin\theta_{c2}$$

The same is true for

$$\frac{\sin\theta_{c2}}{c_f}\ln\frac{p_2^0}{p_t^0} = -G_2\left(\frac{r_{t2}}{d_t}\theta_{c2}\sin\theta_{c2}\right)$$

where G_2 is given by the second member of Eq. 7-35 with the subscript 1 replaced by 2.

In the divergent conical duct one can again apply Eq. 7-29, with the conditions w_2, p_2^0 at the entrance and w_f, p_f^0 at the final section. Thus one has, with evident definition of G_3,

$$\frac{\sin \theta_{c2}}{c_f} \ln \frac{p_f^0}{p_2^0} = F(w_f) - F(w_2) = F(w_f) - G_3\left(\frac{r_{t2}}{d_t}\theta_{c2} \sin \theta_{c2}\right)$$

and finally for the whole divergent duct one gets

$$\frac{\sin \theta_{c2}}{c_f} \ln \frac{p_f^0}{p_t^0} = F(w_f) - G_2 - G_3 \tag{7-37}$$

From Eq. 7-36 and 7-37 one can calculate the value of p_f^0/p_i^0 for given values of the geometrical parameters of the duct, of the final velocity w_f or the corresponding M_f, and of c_f. The calculation of the elemental efficiency of the nozzle follows from Eq. 7-20, and η_{ad} can also be easily derived.

As an example Fig. B,7c shows the calculated values of the second member of Eq. 7-36 as a function of $(r_{t1}/d_t)\theta_{c1} \sin \theta_{c1}$, and of Eq. 7-37 as a

Table B,7

M_f	$\frac{1}{c_f} \ln \frac{p_i^0}{p_t^0}$	$\frac{1}{c_f} \ln \frac{p_t^0}{p_f^0}$	$\ln p_i^0/p_f^0$; $c_{f1} = 0.006$; $c_{f2} = 0.004$	η_{el}	η_{ad}	C_V	C_m
1	0.72	—	0.0043	0.9925	0.993	0.9965	0.998
2	0.72	3.8	0.0196	0.991	0.992	0.996	0.998
2.5	0.72	11.1	0.0487	0.983	0.988	0.994	0.998
3	0.72	22.2	0.0930	0.975	0.983	0.9915	0.998

function of $(r_{t2}/d_t)\theta_{c2} \sin \theta_{c2}$ for various values of M_f. Two values of γ have been used: $\gamma = 1.25$, and $\gamma = 1.5$. As an example we take $r_{t1} = r_{t2} = d_t$ and $\theta_{c1} = -30°$, $\theta_{c2} = 15°$. Table B,7 gives some of the results for $\gamma = 1.25$, and for various M_f. Columns 2 and 3 show that for M_f substantially above unity the contribution of the convergent part to the losses is small. This may be somewhat compensated by the fact that c_f decreases, as already mentioned, with increasing Mach number. This fact can be approximately taken into account by taking in the convergent portion a value of c_{f1} larger than the c_{f2} in the divergent portion. In column 4 the logarithm of the total stagnation pressure ratio has been calculated for $c_{f1} = 0.006$ and $c_{f2} = 0.004$. The corresponding values of η_{el}, η_{ad}, C_V, and C_m are shown in the last four columns.

Inadequacy of the one-dimensional treatment for short ducts with friction. One difficulty in the application of the above developments comes from our incomplete knowledge about c_f in the flow at high Mach numbers. First, in several cases of practical importance, heat exchanges take place from the hot gases to the cooler nozzle, the distribution of which cannot be predicted accurately. Second, in certain Reynolds number ranges it cannot be decided a priori if the boundary layer is turbulent or laminar, especially with favorable pressure gradients and heat exchanges. Finally, the actual values of c_f in the presence of pressure gradients and heat exchanges are not well established, particularly for the turbulent case.

However, even in the assumption that reasonable values of c_f are used, the resulting values of the velocity and discharge coefficients are in excess with respect to those determined experimentally. The main reason for the deviation comes from the uniformity assumption. This assumption, in the case of short ducts where relatively thin boundary layers are rapidly building up, leads to errors of the same order as the corrections to the isentropic flow that are the purpose of the calculation. Only when the boundary layers reach some kind of regime condition (as in fully developed pipe flow) the error becomes small, with the result that the one-dimensional predictions can be used to evaluate the effects of friction in long pipes.

The inaccuracy of frictional corrections based on one-dimensional theory may be particularly important for curved nozzles or flow in cascades of turbine blades. This is due to two important effects, not considered by the one-dimensional treatment. The first is the creation of secondary flows with transverse velocities which absorb a part of the available kinetic energy and may affect the pattern of the boundary layers. The second is the possibility that, even when the mean pressure evolves in a favorable direction (expansion), the local pressure distribution on the walls may present some regions of adverse pressure gradients where the boundary layer may be considerably thickened and possibly separated. Thicker or separated boundary layers result in lower discharge coefficients and velocity coefficients than those calculated on the basis of the friction coefficient using the uniformity assumption.

Evidently it is possible to use an "apparent" friction coefficient, larger than the actual coefficient, to include these effects. However, the magnification of the coefficient depends on the nozzle shape and cannot be reasonably assigned a priori. One could base the value of this apparent coefficient on experimental data, a procedure substantially equivalent to the use of values of C_V and C_m based on experimental determinations.

B,8. Operation of Incorrectly Expanded Nozzles. Thrust.

Incorrectly expanded ideal Laval nozzle. A frictionless Laval nozzle is operating with known upstream conditions p_i^0, T_i^0, and back pressure p_b. Assume that $m = (m_{max})_{is}$ and that the pressure p_f in the final section coincides with the back pressure. From Eq. 6-9 and 4-31a one can write

$$\sqrt{\mathcal{R}T_i^0}\,(m_{max})_{is} = \chi(\gamma)p_i^0A_t = \sqrt{\frac{2\gamma}{\gamma-1}}\,p_fA_f\,\frac{w_f}{1-w_f^2}$$

$$= \sqrt{\frac{2\gamma}{\gamma-1}}\,p_f^0A_fw_f(1-w_f^2)^{\frac{1}{\gamma-1}}$$

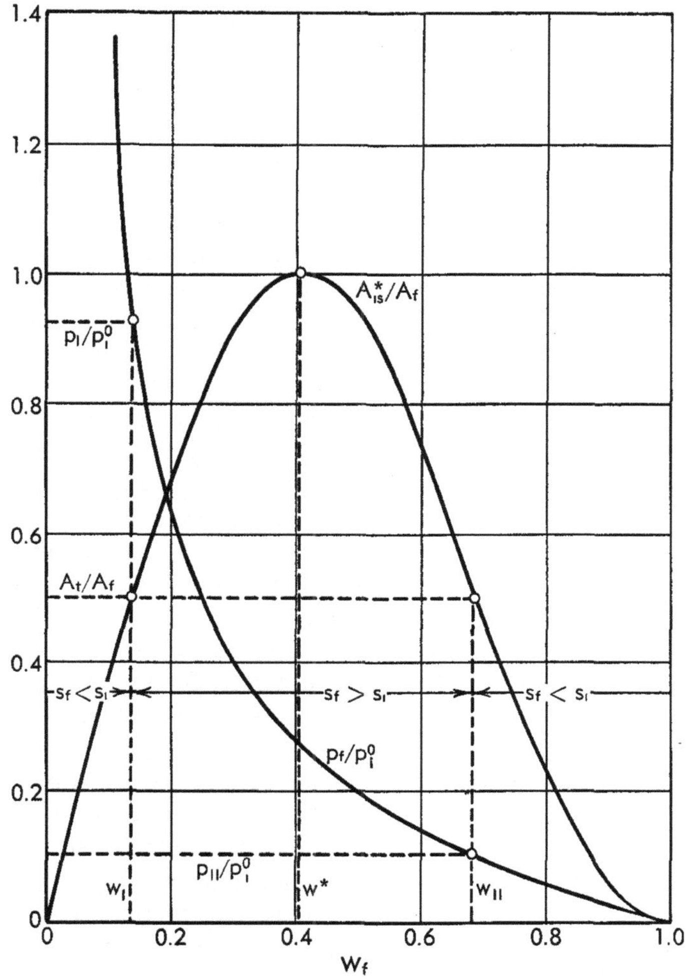

Fig. B,8a. Application of the second law to incorrectly expanded ideal Laval nozzles.

Therefore one has, inserting the value of $\chi(\gamma)$ (Eq. 6-7), and comparing with Eq. 6-3,

$$s_f - s_i = \mathcal{R} \ln \frac{p_i^0}{p_f^0} = \mathcal{R} \ln \left[\frac{A_f}{A_t} \sqrt{\frac{\gamma + 1}{\gamma - 1}} \left(\frac{\gamma + 1}{2} \right)^{\frac{1}{\gamma - 1}} w_f (1 - w_f^2)^{\frac{1}{\gamma - 1}} \right] \quad \text{(8-1)}$$

$$= \mathcal{R} \ln \left[\frac{A_f}{A_t} \frac{A^*_{is}(w_f)}{A_f} \right]$$

and

$$\frac{{}_{f}}{p_i^0} = \frac{A_t}{A_f} \sqrt{\frac{\gamma - 1}{\gamma + 1}} \left(\frac{2}{\gamma + 1} \right)^{\frac{1}{\gamma - 1}} \frac{1 - w_f^2}{w_f} \quad \text{(8-2)}$$

Fig. B,8a shows, as a function of w_f, the values of $A^*_{is}(w_f)/A_f$. For fixed A_t/A_f, p_f/p_i^0 is a monotonically decreasing function of w_f, and is shown in the figure for $A_t/A_f = 0.5$.

A horizontal line with ordinate equal to the assigned value of A_t/A_f (0.5 in Fig. B,8a) cuts the curve $A_{is}^*(w_f)/A_f$ in two points where, because of Eq. 8-1, $s_f - s_i = 0$ and the flow is isentropic. At the values w_I and w_{II} of w_f, relative to these two intersections, Eq. 8-2 gives the values p_I and p_{II} of p_f, values which evidently coincide with those already determined for isentropic flow in Art. 6. For $w_I < w_f < w_{II}$ and, therefore, $p_{II} < p_f < p_I$, one has $A_{is}^*(w_f)/A_f > A_t/A_f$ and so from Eq. 8-1, $s_f - s_i > 0$. Hence in the range of p_f considered, the assumptions made about the flow are not in contradiction with the second law of thermodynamics. However, outside of that range $s_f - s_i$ would be negative, a result which is not allowed by the second law. We can conclude that if p_b lies outside of the range between p_I and p_{II} it is impossible to obtain a flow satisfying both the conditions $m = (m_{max})_{is}$ and $p_f = p_b$. In fact we know already from the discussions of Art. 6 that for $p_b > p_I$ we have $p_f = p_b$, but $m < (m_{max})_{is}$ and the nozzle operates as a subsonic venturi; while for $p_b < p_{II}$ we have $m = (m_{max})_{is}$, but $p_f = p_{II} > p_b$ and the nozzle is underexpanded.

Let us examine now the operating conditions in the range $p_{II} < p_b < p_I$ and discuss if and how our initial assumptions can be fulfilled. In this range the flow, though frictionless, must include some nonisentropic process. A plausible process for ideal nozzles when the boundary layers are absent and the flow is uniform is immediately found in the possibility of stationary normal shock waves being present in the supersonic region. These shock waves, being downstream of the throat, do not affect the flow rate. They must produce the entropy increase required by Eq. 8-1. Now the entropy increase through a normal shock increases with the Mach number in front of the shock, and therefore is a function of the shock location and can take any value up to a maximum corresponding to the maximum, M_f, reached isentropically in the exit section. The result of the presence of a normal shock at different locations in the nozzle is shown in Fig. B,8b. From the shock relations one can immediately find the equation of the curve B, locus of the pressures $p_2(x)$, immediately after a shock located at station x and starting from the pressure $p_1(x)$ corresponding to the isentropic curve A. The equation is

$$\frac{p_2}{p_i^0} = \frac{4\gamma}{\gamma^2 - 1}\left(\frac{p_1}{p_i^0}\right)^{\frac{1}{\gamma}} - \frac{\gamma + 1}{\gamma - 1}\frac{p_1}{p_i^0}$$

The maximum of $p_2(x)$ takes place for

$$\frac{p_1}{p_i^0} = \left(\frac{2}{\gamma + 1}\right)^{\frac{2\gamma}{\gamma - 1}} = \left(\frac{p_{is}^*}{p_i^0}\right)^2; \quad M_1^2 = \frac{\gamma + 3}{2}$$

that is for a p_1/p_i^0 close to 0.3 and an M_1 somewhat higher than $\sqrt{2}$.

After the shock the flow follows an isentropic subsonic diffusion in the divergent duct, as shown in Fig. B,8b for a few cases. The continuation of the corresponding isentropic curves is also shown in the figure. They show a behavior similar to the lines of constant $m > (m_{max})_{is}$ of Fig. B,6f with a vertical tangent at the sonic velocity corresponding to an A_{min} and a p_2^* given in terms of p_1 by the relations

$$\frac{A_t}{A_{min}} = \frac{p_2^*}{p_{is}^*} = \frac{p_2^0}{p_i^0} = \frac{p_2}{p_i^0}\left[1 + \frac{w^{*4}}{1 - w^{*4} - \left(\frac{p_1}{p_i^0}\right)^{\frac{\gamma-1}{\gamma}}}\right]^{\frac{\gamma}{\gamma-1}}$$

obtained immediately for $w = w^*$ from Eq. 4-31a and from the shock relations. The curve C of Fig. B,8b connects the sonic points of the various

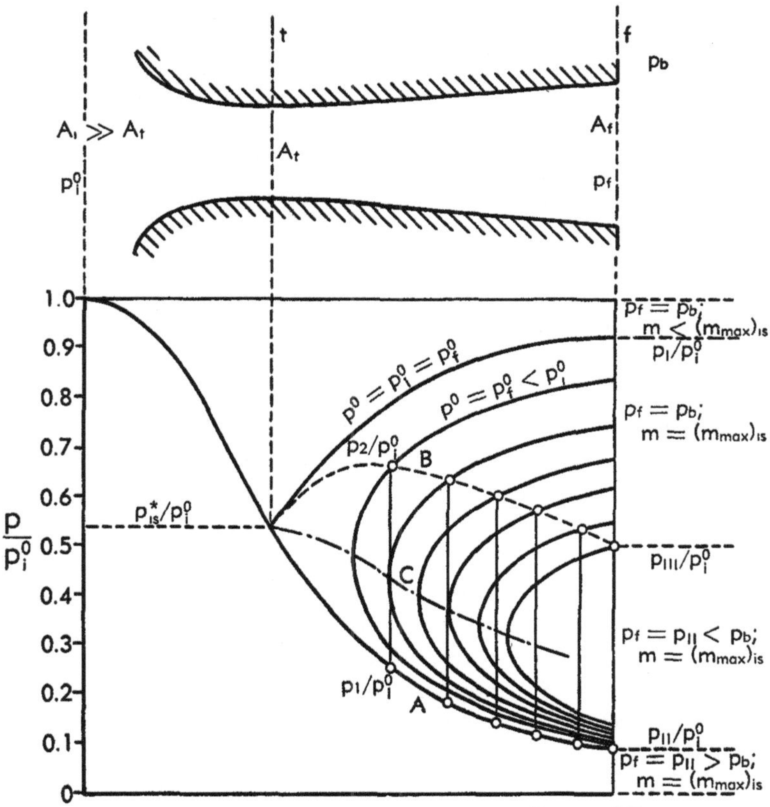

Fig. B,8b. Pressure distributions corresponding to different back pressures in an ideal Laval nozzle.

isentropic curves following the shocks. Similar to the curves $m > (m_{max})_{is}$ of Fig. B,6f, these complete isentropic curves cannot represent a physical flow; but it is now clear that a part of their subsonic portion represents a well-defined physical flow.

We have now a model which may explain the mechanism through which the flow adapts itself to variations of p_b in the range between p_I and p_{III}, the latter being the value of p_2 when the shock is at the exit section. Observe that when $p_b = p_{III}$ we have in the exit section two values of p_f: one, after the shock, equal to p_b, the other, before the shock, coinciding with p_{II}.

When p_b decreases below p_{III} the only possibility for the gases to go from the pressure p_{II} to p_b at the periphery of the exit section is through a shock wave which must be oblique, because the pressure increase must be smaller than for a normal shock. Thus an oblique shock wave must start from the periphery of the exit section directed downstream toward the axis. The way these oblique shock waves interact with each other may be quite involved, especially for complicated shapes of section, and thus escapes the limits of the one-dimensional treatment. We may just mention here that, in general, the resulting patterns of shock waves and expansion waves are reproduced almost periodically in the issuing jet and constitute the well-known "diamond" formation characteristic of incorrectly expanded nozzles. The important point for us is that these patterns do not interfere with the flow in the exit section where the conditions for $p_b < p_{III}$ coincide with the isentropic flow conditions, the adjustment to the external pressure taking place entirely outside of the nozzle.

This is, of course, true not only for the overexpanded nozzles corresponding to the range $p_{II} < p_b < p_{III}$ but also, as already established in Art. 6, for the underexpanded nozzles with $p_b < p_{III}$. When we go from the overexpanded to the underexpanded condition, the oblique shock waves issuing from the exit section periphery gradually become weaker, disappear at $p_b = p_{II}$, and are replaced by expansion wave fans for $p_b < p_{II}$. At the same time the diamond formation in the issuing jet becomes less marked, vanishes when the expansion is correct, and grows again in a complementary form for increasing underexpansion.

Thrust of ideal nozzles. The thrust produced by the reaction of the issuing jet can be evaluated from the law of conservation of momentum. Let us take a control surface around the nozzle such that the exit section is a part of it (Fig. B,8c). Suppose that the nozzle is supplied with gas in such a way that the axial component of the momentum of the supply gas is zero. In Fig. B,8c this is accomplished by bringing the gas to the nozzle without axial velocity. The same would be true for a rocket in which the gas would be generated in the rocket chamber, without preexisting axial momentum. Of course this is not the most general condition for a nozzle, since in practice the gas may contain an initial axial momentum. However, with this assumption one can separate the thrust due to the nozzle from its variations due to other particularities of the flow, which can be taken into account separately.

The momentum conservation law states that the flux of the axial component of momentum out of the control surface, that is the momentum I_f of the flow in the exit section, must be equal to the total force acting on the control surface in the axial direction, which is equal to the force F applied to the nozzle to balance the thrust, plus the axial component of the resultant of the pressures acting on the outside of the control surface. Since all of the control surface is subject to the ambient pressure p_b except the exit section A_f where the pressure is p_f, this resultant is $A_f(p_b - p_f)$ in the axial direction. One can therefore obtain the thrust F as

$$F = I_f + A_f(p_f - p_b) = J_f - A_f p_b \tag{8-3}$$

For fixed ambient conditions the thrust depends only on the impulse or thrust function J (Eq. 4-6) computed at the final section. This is actually the reason for one of the names given to this function.

For correctly expanded nozzles ($p_f = p_b$) the thrust is simply equal to the momentum in the final section. If the nozzle is operating in the range where $m < (m_{max})_{is}$, this condition is always satisfied, so that the thrust

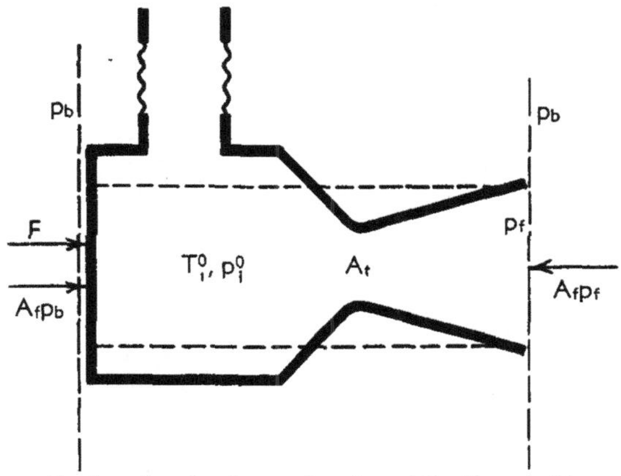

Fig. B,8c. Schematic chamber for the evaluation of the thrust of incorrectly expanded nozzles. The two vertical dashed lines represent a part of the control surface. The rest of this surface may be drawn arbitrarily.

is simply obtained from the last of Eq. 4-32a as the product of $p_i^0 A_f$ times a function of w_f, which is in turn only a function of p_i^0/p_f; therefore the, thrust, for a given area A_f, depends only on p_i^0 and p_f, and not on T_i^0.

In particular the assumption $p_f = p_b$ is true for simply convergent nozzles with $p_b > p_{is}^*$. However, this is not the most interesting case, because the nozzles of most of the jet devices in practice operate at the maximum flow rate, and have a fixed A_f/A_t which corresponds to a well-determined value of p_i^0/p_f; for supersonic final velocity ($p_b < p_{III}$) this can coincide with the ratio p_i^0/p_b only for a particular operating condition, so that for all other operating conditions the nozzle is incorrectly expanded. Observe that for subsonic $w_f(p_b > p_{III})$ the condition $p_f = p_b$ is always fulfilled, but this case has no practical importance.

For the more general case when the nozzle is incorrectly expanded the thrust is obtained from Eq. 8-3 with the help of Eq. 4-33a in terms of

$w = w_f$, $p^0 = p_i^0$, $A = A_f$, and p_b. Since from Eq. 6-3 one has

$$\frac{A_f}{A_t} = \sqrt{\frac{\gamma - 1}{\gamma + 1}} \left(\frac{2}{\gamma + 1}\right)^{\frac{1}{\gamma - 1}} \frac{1}{w_f(1 - w_f^2)^{1/(\gamma - 1)}} \tag{8-4}$$

the expression for F in terms of w_f, p_i^0, A_t, and p_b becomes

$$F = p_i^0 A_t \left[\left(\frac{2}{\gamma + 1}\right)^{\frac{1}{\gamma - 1}} \left(\frac{w_f}{w^*} + \frac{w^*}{w_f}\right) - \frac{p_b}{p_i^0}\frac{A_f}{A_t}\right] \tag{8-5}$$

with A_f/A_t given by Eq. 8-4. The so-called "thrust coefficient" for the ideal nozzle is given by

$$C_F = \frac{F}{p_i^0 A_t} = \left(\frac{2}{\gamma + 1}\right)^{\frac{1}{\gamma - 1}} \left(\frac{w_f}{w^*} + \frac{w^*}{w_f}\right) - \frac{p_b}{p_i^0}\frac{A_f}{A_t} \tag{8-6}$$

and can be considered a function of p_b/p_i^0 and w_f or of p_b/p_i^0 and A_f/A_t. For a fixed value of p_b/p_i^0 both terms of Eq. 8-6, and also C_F, are minimum at $w_f = w^*$, that is, for a simply convergent nozzle. This minimum $C_{F_{conv}}$ is given by

$$C_{F_{conv}} = 2\left(\frac{2}{\gamma + 1}\right)^{\frac{1}{\gamma - 1}} - \frac{p_b}{p_i^0} \tag{8-7}$$

The maximum value of C_F for given A_f/A_t is reached in the vacuum ($p_b = 0$); and the maximum thrust coefficient in the vacuum is obtained for $w_f = 1$ ($A_f = \infty$), as

$$C_{F_{max}} = \frac{2\gamma}{\sqrt{\gamma^2 - 1}} \left(\frac{2}{\gamma + 1}\right)^{\frac{1}{\gamma - 1}} \tag{8-8}$$

From Eq. 8-7 we see that for a convergent nozzle the maximum thrust coefficient is obtained for $p_b = 0$; conversely, the minimum is obtained for the maximum p_b compatible with the assumption that the flow is choked, $p_b = p_{is}^*$. The maximum and minimum of $C_{F_{conv}}$ are therefore given by

$$(C_{F_{conv}})_{max} = 2\left(\frac{2}{\gamma + 1}\right)^{\frac{1}{\gamma - 1}}, \quad (C_{F_{conv}})_{min} = \gamma\left(\frac{2}{\gamma + 1}\right)^{\frac{\gamma}{\gamma - 1}} \tag{8-9}$$

The values Eq. 8-8 and 8-9 are functions only of γ and are represented in Fig. B,8d. It is interesting to observe that the gain in thrust that can be obtained with the addition of a divergent duct increases sharply for decreasing γ.

In Fig. B,8e are represented the values of $C_F/C_{F_{conv}}$ as a function of p_b/p_i^0 and A_f/A_t for $\gamma = 1.2$ [*28*]. Each curve $p_b/p_i^0 = \text{const}$ represents the gain in thrust attainable through the addition of a divergent to a purely convergent nozzle for fixed operating conditions. The maximum gain is obtained at the points connected by the dash-dot line. The con-

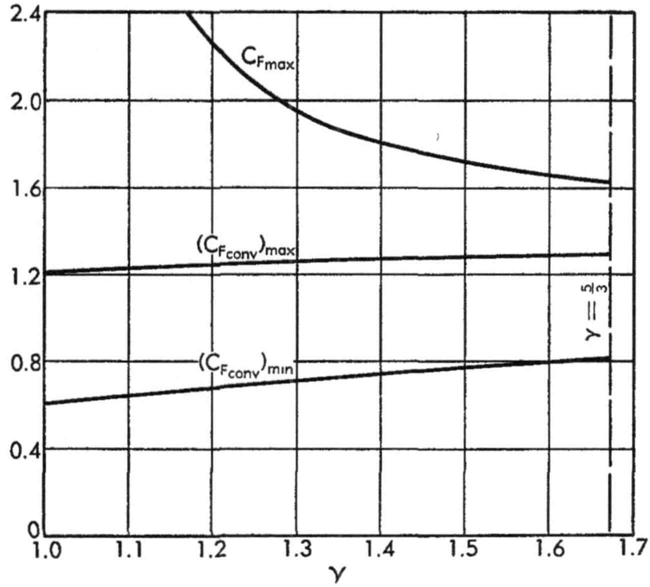

Fig. B,8d. Values of the maximum thrust coefficient of ideal Laval nozzles in the vacuum and of the maximum and minimum thrust coefficients for a choked ideal convergent nozzle in function of γ.

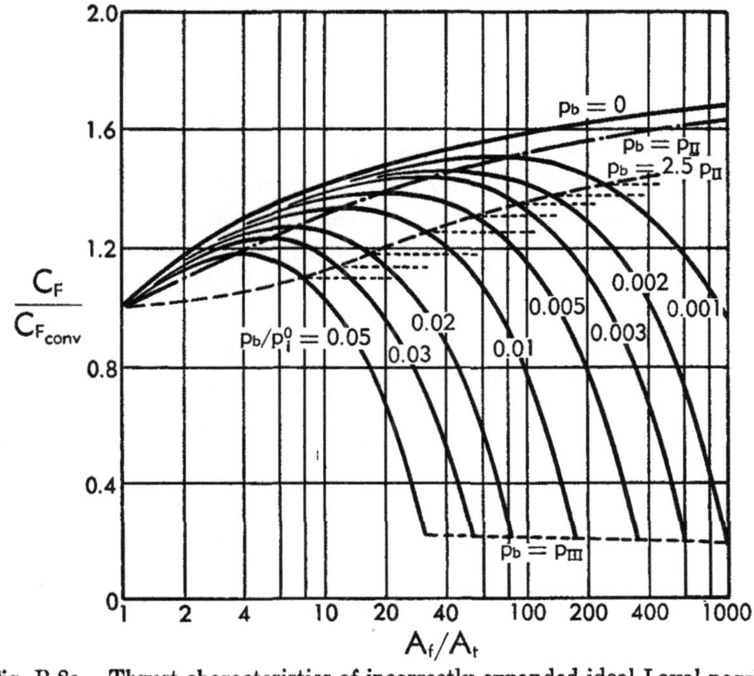

Fig. B,8e. Thrust characteristics of incorrectly expanded ideal Laval nozzles.

ditions at the maximum are easily derived differentiating Eq. 8-3 with respect to the axial distance along the nozzle for constant p_b:

$$\frac{dF}{dx} = \frac{dJ_f}{dx} - p_b \frac{dA_f}{dx} \tag{8-10}$$

Replacing dJ_f/dx from Eq. 4-5c this expression becomes equal to $(p_f - p_b)(dA_f/dx)$. Thus F is maximum for $p_f = p_b$, that is when the nozzle is correctly expanded.

The curves of Fig. B,8e are calculated from Eq. 8-6, and 8-7, and 8-4 in the range where the value of p_{III} (reached behind a shock located in the isentropic flow at A_f) stays larger than p_b. When $p_{III} < p_b$ the thrust must be calculated from different relations that could easily be obtained from our discussion of the operation of incorrectly expanded nozzles. This case, however, has scarce practical importance for thrust-producing devices.

Incorrectly expanded real Laval nozzles. While the effects of friction on the operation of correctly expanded nozzles are small, the presence of friction and boundary layers may considerably affect the operation of incorrectly expanded nozzles. This is due to the interaction between shock waves and boundary layers, already discussed in Art. 5. When the nozzle is underexpanded there are no shock waves in the nozzle, and therefore the operation is close to that of a frictionless nozzle (Fig. B,8f(1)). For overexpanded nozzles, however, the situation is different. Starting from the correctly expanded case, with a pressure $p_f = p_b = p'_{II}$ close to the value of p_{II} of the frictionless case, and increasing p_b, gradually the adjustment of the flow near the wall from p_f to p_b takes place through an oblique shock. The flow in the boundary layer must also overcome the same pressure differential, but due to its smaller momentum it can only do so up to a certain value r_{max} of the ratio of the two pressures. This maximum value has been determined experimentally and theoretically (see [*20,21, 29*]) for a turbulent boundary layer without pre-existent pressure gradient and is shown, as a function of w_f for $\gamma = 1.4$ and $\gamma = 1.2$ (theoretical) in Fig. B,8g. We assume here that the same curves can be applied to the nozzle flow despite the presence of pressure gradients.

Up to $p_b = p'_{II}r_{max}$ the divergence between the actual flow pattern and the frictionless flow is quite small (Fig. B,8f(2)). Above this value, however, the boundary layers in the final section are unable to reach the pressure p_b after having expanded down to p_f, and since they must in some way reach the prefixed back pressure, before reaching p_f they will cease to expand at a section preceding the exit section where the value of p and w are such as to provide the correct value of p_b/p corresponding to w. In other words the flow becomes separated from the wall before the exit section, at the section where the condition $p_b = pr_{max}(w)$ is fulfilled (Fig. B,8f(3)). Oblique shock waves will emerge from the separation

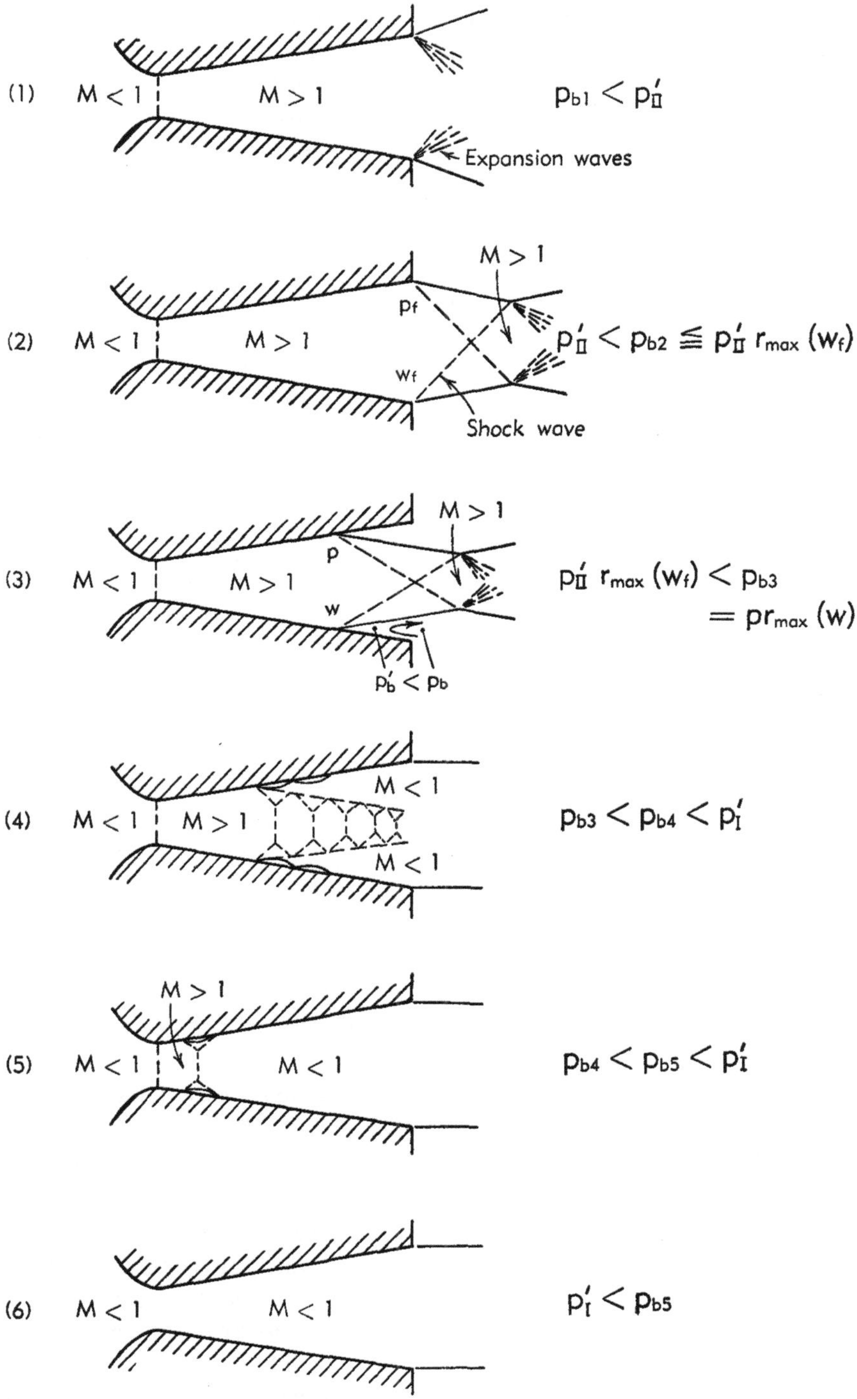

Fig. B,8f. Flow patterns in a real Laval nozzle for varying back pressure.

point and the remaining pattern of the jet will be similar to the one that would have been found if the nozzle had been cut off at this point. In reality this is an oversimplification, because some fluid is entrained in the space between the jet and the walls in the way sketched in the lower part of Fig. B,8f(3); as a result the pressure p_b' at the separation point is lower than p_b and the quantitative results should be slightly modified.

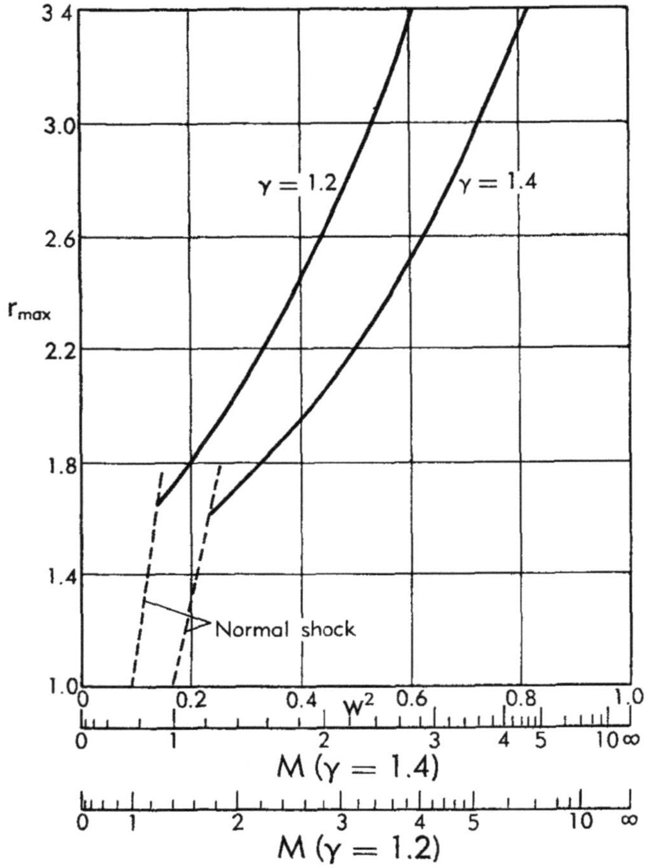

Fig. B,8g. Maximum theoretical pressure ratio of a shock wave emerging from a turbulent boundary layer.

When p_b is further increased, the separation point moves upstream in the nozzle. It may happen particularly for small angles of divergence and thin boundary layers (high Reynolds number) that, after a sufficient upstream displacement of the separation point, the conditions switch from those of a nearly free supersonic jet to those of a pseudo-shock (Art. 5). The separated regions in this case will not reach the exit section but have a limited extension, and the flow again fills the exit section with subsonic velocities (Fig. B,8f(4)). New increases of p_b make the pseudo-shock move farther upstream where the configuration of the pseudo-shock changes because of the decrease of its strength and the thinning of the boundary layers. Following the discussion of Art. 5, the pseudo-shock tends to become more concentrated until finally it may be reduced to a single shock with bifurcated ends (Fig. B,8f(5)). For the reasons discussed in Art. 5 the bifurcation may disappear when the shock approaches the throat. Finally, when p_b becomes larger than a value p'_I not very different from the isentropic p_I the pseudo-shock disappears, the velocity at the throat is reduced to a subsonic value with a corresponding flow rate decrease, and the divergent part of the nozzle operates as a subsonic diffuser (Fig. B,8f(6)). Observe that because of the larger losses relative to flows with adverse pressure gradients, the difference between p'_I and p_I is relatively larger than that between p'_{II} and p_{II}.

Concluding this discussion, we see that cases 4, 5, and 6 just considered are closely similar to the analogous cases for a frictionless nozzle, the difference residing mainly in the replacement of a shock without thickness by a more or less thick pseudo-shock. Similarly, the cases 1 and 2 are very closely similar to the corresponding frictionless cases. Therefore the only new type of flow which appears in real, and not in ideal, nozzles is the case 3 involving separation. Observe that in the figure the separation has been supposed to be symmetrical. However, it may also happen in a nonsymmetric way, particularly with large divergence angles, in which case the flow may stay attached to one side of the walls and the jet can deviate considerably from the axial direction.

The patterns schematically shown by Fig. B,8f are actually observed experimentally as shown in Plate B,8, taken from [*30,31*] and in Plates B,5b, B,5c and B,5d. Fig. B,8h [*29*] shows the axial pressure distributions measured for different back pressures. From the results of [*31*] the values of p_b/p at which separation occurs have been shown to be roughly equal to 2.5, in rough agreement with the values of r_{max} of Fig. B,8g in the Mach number range of the experiments.

Thrust of real nozzles. Eq. 8-3 still provides the expression of the thrust, so that the thrust coefficient is

$$C_F = \frac{F}{p_i^0 A_t} = \frac{I_f}{p_i^0 A_t} + \frac{A_f p_f}{A_t p_i^0} - \frac{A_f}{A_t}\frac{p_b}{p_i^0}$$

Using Eq. 7-7, 6-9, 4-32a, and 4-31a, C_F can be expressed as a function of p_f/p_i^0, p_b/p_i^0 and of the discharge, velocity, and conicity coefficients defined in Art. 7. One finds

$$C_F = C_m\chi(\gamma)\left\{C_c C_V\sqrt{\frac{2\gamma}{\gamma-1}}\,(w_f)_{is} + \sqrt{\frac{\gamma-1}{2\gamma}}\left[\frac{1}{C_V(w_f)_{is}} - C_V(w_f)_{is}\right]\right\} - \frac{A_f}{A_t}\frac{p_b}{p_i^0} \quad \text{(8-11)}$$

with

$$(w_{\mathrm{f}})_{\mathrm{is}}^2 = 1 - \left(\frac{p_{\mathrm{f}}}{p_{\mathrm{i}}^0}\right)^{\frac{\gamma-1}{\gamma}} \tag{8-12}$$

and

$$\frac{A_{\mathrm{f}}}{A_{\mathrm{t}}} = C_m\chi(\gamma)\sqrt{\frac{\gamma-1}{2\gamma}}\left[\frac{1}{C_V(w_{\mathrm{f}})_{\mathrm{is}}} - C_V(w_{\mathrm{f}})_{\mathrm{is}}\right]\frac{p_{\mathrm{i}}^0}{p_{\mathrm{f}}} \tag{8-13}$$

Obviously Eq. 8-11 is reduced to Eq. 8-6 for $C_m = C_V = C_c = 1$.

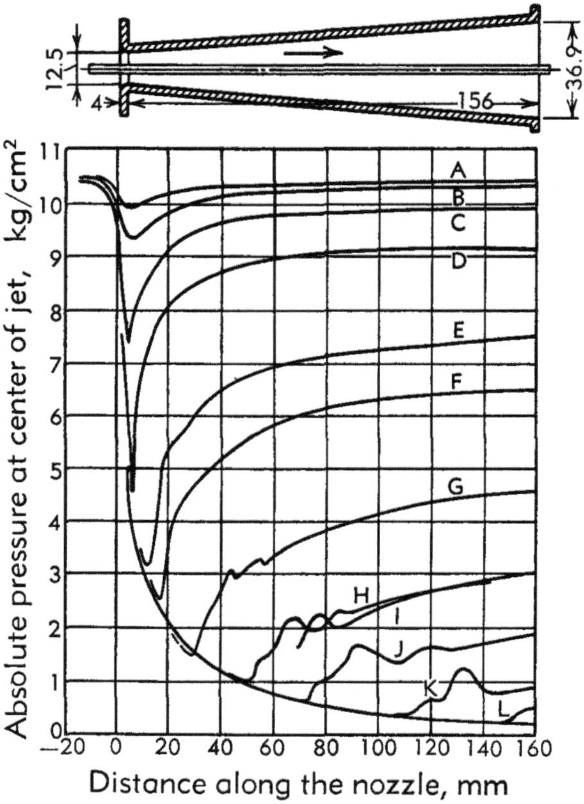

Fig. B,8h. Pressure distributions measured with an axial static pressure probe in a Laval nozzle for varying back pressure (steam).

In turn $p_{\mathrm{f}}/p_{\mathrm{i}}^0$ can be considered a function of $A_{\mathrm{f}}/A_{\mathrm{t}}$ through Eq. 8-12 and 8-13, so that C_F can be looked upon as a function of A_f/A_{t}, $p_{\mathrm{b}}/p_{\mathrm{i}}^0$ and for given C_m, C_V, C_c its representation will be similar to that of Fig. B,8e provided the flow does not suffer separation. However, large deviations will appear when separation is present, i.e. for $p_{\mathrm{b}}/p_{\mathrm{II}} > r_{\max}$. A schematic idea of the actual behavior of a real nozzle in this range can be obtained from Fig. B,8e where a line corresponding to $p_{\mathrm{b}} = r_{\max}p_{\mathrm{f}}$, with $r_{\max}$ equal to 2.5 in agreement with the above-mentioned results of [31], has been traced as the approximate condition at which separation occurs. As already indicated, if the value of A_f/A_t is increased in a certain range beyond that corresponding to separation, no important change in the separated jet takes place and the thrust stays approximately constant, as indicated by the dotted lines in the figure. Only for substantial increases of A_f/A_t would the flow become attached again, with a consequent sharp decrease in thrust due to the switch to subsonic jet velocities.

For unseparated nozzles with constant values, close to unity, of η_{ad} or η_{el}, C_m and C_V can be replaced by their values (Eq. 7-16 and 7-17) in terms of these quantities.

A consequence of the presence of the losses is that for given p_i^0/p_b the maximum thrust is not obtained for correct expansion, but for a slightly underexpanded nozzle. This can be shown easily by replacing in Eq. 8-10 the value of dI_f/dx given by Eq. 4-3b, with $dX = dX_w$. One obtains the maximum thrust for

$$\frac{dF}{dx} = (p_f - p_b)\frac{dA_f}{dx} - \frac{dX_w}{dx} = 0 \tag{8-14}$$

For small divergence angles θ_c we have

$$\frac{dA_f}{dx} = \pi d_f \theta_c$$

From Eq. 3-5 and 7-20 we have

$$\frac{dX_w}{dx} = \pi d_f \tau_w = \pi d_f \frac{\gamma}{2} p_f M_f^2 c_f$$

Thus the condition (Eq. 8-14) for maximum thrust becomes

$$\frac{2}{\gamma}\frac{p_f - p_b}{p_f M_f^2} = \frac{c_f}{\theta_c}$$

Observe that if c_f stands for the real friction coefficient of the boundary layer, M_f does not represent the mean Mach number of the flow (considered one-dimensional), but rather the true Mach number out of the boundary layer in the isentropic core, and therefore it is given by Eq. 4-28 with p_i^0/p_f in place of p^0/p. Finally, we obtain the maximum thrust condition under the form

$$\frac{\gamma - 1}{\gamma}\,\frac{1 - \dfrac{p_b}{p_f}}{\left(\dfrac{p_i^0}{p_b}\dfrac{p_b}{p_f}\right)^{\frac{\gamma-1}{\gamma}} - 1} = \frac{c_f}{\theta_c}$$

The optimum underexpansion, represented by the value of p_b/p_f, can be determined as a function of p_i^0/p_b and c_f/θ_c from this equation and is represented in Fig. B,8i.

B,9. Flow in Diffusers. The presence of the boundary layers has a much more important effect on the flow in diffusers than on the flow in nozzles. Due to the fact that the flow in retarded layers under an adverse pressure gradient undergoes a larger reduction of velocity than the rest of the flow, the boundary layers have a tendency to grow thicker and, possibly, to become separated from the wall. In these conditions the mean transversal velocity gradients in the boundary layers become smaller, and therefore the skin friction and the corresponding dissipative effects decrease. However, despite the decrease of the wall friction, the deviations from the isentropic one-dimensional flow grow larger because of the increase of the flow nonuniformities. Only with special precautions is it possible to obtain a flow that fills the section at the end of a diffuser, and even then the velocity distributions are far from approaching uniformity, having a pronounced maximum on the axis of the duct.

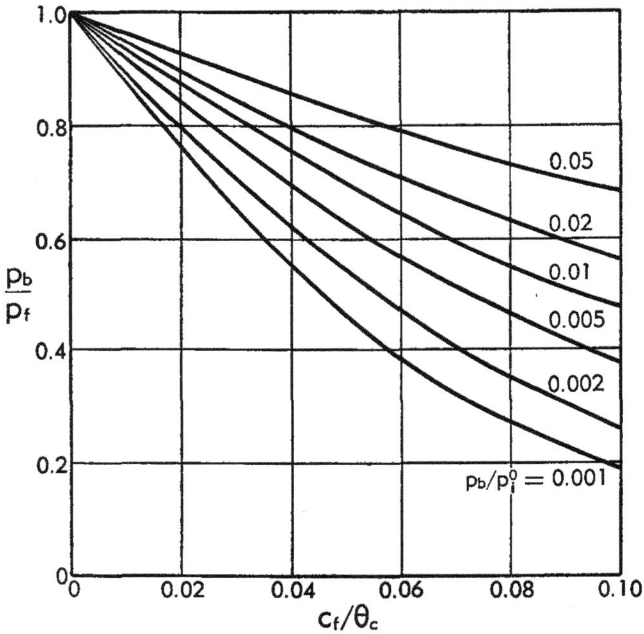

Fig. B,8i. Optimum underexpansion of real Laval nozzles.

It is, of course, possible to approach again the conditions of uniformity at the end of a diffuser by adding a sufficiently long, constant section duct. The approach to a uniform distribution, however, is accompanied by internal losses (mixing losses) which increase the entropy and result in considerable deviations of the final conditions from those of isentropic flow.

Another reason for increased dissipation resides in the kinetic energy diverted to the strong turbulent motion which results from nonuniformity of the flow and separation; this kinetic energy is finally dissipated by friction and is therefore ineffective in producing pressure increases.

The one-dimensional theory is inadequate to analyze quantitatively the details of the resulting flow, so that the one-dimensional treatment of diffusers must be based on empirically determined coefficients, such as the efficiencies defined below. Observe that the same conclusion was reached in Art. 7 for the deviations of the flow in real nozzles from the isentropic flow. However, the deviations are more serious in the case of diffusers and in general cannot be overlooked, even for rough calculations.

Diffuser efficiencies and related parameters. In order to evaluate the losses in diffusers it is customary to make use of the concept of efficiency. Due to the larger losses, this concept has to be discussed in more detail than for nozzles.

There are several definitions of the diffuser efficiency and related quantities. The adiabatic efficiency (sometimes called ram efficiency) is defined on an energy basis in a way similar to Eq. 7-8

$$\eta_{ad} = \frac{(u_i^2)_{is}}{u_i^2} = \frac{(h_f^0)_{is} - h_i}{h_f^0 - h_i} \tag{9-1}$$

where the initial velocity has been assumed to be practically axial ($q_i \equiv u_i$), $(u_i)_{is} < u_i$ represents the velocity that would be necessary to reach the same final pressure if there were no losses, and $(h_f)_{is}$ stands for the corresponding stagnation enthalpy. Here it is necessary to observe that η_{ad} is defined by Eq. 9-1 as the ratio of the two initial kinetic energies, rather than the ratio of the two variations of kinetic energy from the initial to the final section of the diffuser. The reason is that in general the kinetic energy at the end of a diffuser is quite small, and even when it cannot be completely disregarded, its presence is usually taken into account by considering the final stagnation pressure rather than the static pressure, and comparing the values of $(u_i)_{is}$ and u_i for the same final stagnation pressure p_f^0. When the diffusion is partial, however, so that only a fraction of the initial kinetic energy is converted into pressure, the definition (Eq. 9-1) should be modified so as to depend on the variations of the kinetic energy, rather than the initial values of them. The corresponding expression in terms of enthalpies is obtained by canceling the superscript ^{0} from the last expression (Eq. 9-1).

For polytropic gases we obtain from Eq. 9-1 the following expressions for η_{ad}:

$$\begin{aligned}\eta_{ad} &= \frac{(T_f^0)_{is} - T_i}{T_f^0 - T_i} = \frac{\left(\frac{p_f^0}{p_i}\right)^{\frac{\gamma-1}{\gamma}} - 1}{\frac{T_f^0}{T_i} - 1} \\ &= \frac{\left(\frac{p_f^0}{p_i}\right)^{\frac{\gamma-1}{\gamma}} - 1}{\frac{\gamma-1}{2} M_i^2} = \frac{\left(1 + \frac{\gamma-1}{2} M_i^2\right)\left(\frac{p_f^0}{p_i^0}\right)^{\frac{\gamma-1}{\gamma}} - 1}{\frac{\gamma-1}{2} M_i^2}\end{aligned} \tag{9-1a}$$

the third expression being a consequence of the fact that $T_f^0 = T_i^0$. Thus η_{ad} can be obtained experimentally from pressure and temperature determinations. Observe that η_{ad} can also be written as

$$\eta_{ad} = \frac{\left(\frac{p_f^0}{p_i}\right)^{\frac{\gamma-1}{\gamma}} - 1}{\left(\frac{(p_f^0)_{is}}{p_i}\right)^{\frac{\gamma-1}{\gamma}} - 1} \tag{9-1b}$$

where $(p_f^0)_{is} = p_i^0$ stands for the final stagnation pressure obtainable from an isentropic diffusion with the same initial velocity u_i.

A different, less orthodox, definition of efficiency on an energy basis has been used by Kantrowitz and Donaldson [32] and a few other authors. It represents the ratio of the kinetic energy $(u_{exp}^2)_{is}/2$ obtainable by isentropic expansion between the final condition p_f^0 and the initial pressure p_i, to the initial kinetic energy, and is given by

$$\eta_{KD} = \frac{(u_{exp}^2)_{is}}{u_i^2} = \frac{T_f^0 - (T_{exp})_{is}}{T_f^0 - T_i} = \frac{1 - \left(\frac{p_i}{p_f^0}\right)^{\frac{\gamma-1}{\gamma}}}{1 - \frac{T_i}{T_f^0}}$$

Another useful definition of efficiency, which has the advantage of indicating the quality of a diffuser better than η_{ad}, independent of the operating conditions (Art. 10), is the elemental or polytropic efficiency, defined for polytropic gases by the equations

$$\eta_{el} = \frac{\ln (T_f^0)_{is} - \ln T_i}{\ln T_f^0 - \ln T_i} = \frac{\gamma - 1}{\gamma} \frac{\ln \frac{p_f^0}{p_i}}{\ln \frac{T_f^0}{T_i}}$$

$$= \frac{\gamma - 1}{\gamma} \frac{\ln \frac{p_f^0}{p_i}}{\ln\left(1 + \frac{\gamma - 1}{2} M_i^2\right)} = 1 + \frac{\gamma - 1}{\gamma} \frac{\ln \frac{p_f^0}{p_i^0}}{\ln\left(1 + \frac{\gamma - 1}{2} M_i^2\right)} \tag{9-2}$$

Also, η_{el} can be obtained experimentally from determinations of pressure and temperature. The elemental efficiency is directly related to the entropy increase by the relation

$$\frac{s_f - s_i}{c_p} = \ln \frac{T_f^0}{T_i} - \frac{\gamma - 1}{\gamma} \ln \frac{p_f^0}{p_i} = (1 - \eta_{el}) \ln \frac{T_f^0}{T_i} \tag{9-3}$$

As already observed, if the diffusion is incomplete, one obtains the proper definition of η_{ad} by canceling the superscript ^{0} from the first two expressions of Eq. 9-1a. Similarly, one obtains the definition of η_{el} by eliminating the superscript in the first two expressions (Eq. 9-2 or 9-3). The relation between η_{ad} and η_{el} is

$$\eta_{ad} = \frac{\left(\frac{p_f^0}{p_i}\right)^{\frac{\gamma-1}{\gamma}} - 1}{\left(\frac{p_f^0}{p_i}\right)^{\frac{\gamma-1}{\gamma\eta_{el}}} - 1} = \frac{\left(1 + \frac{\gamma-1}{2} M_i^2\right)^{\eta_{el}} - 1}{\frac{\gamma-1}{2} M_i^2}$$

and depends on the operating conditions. The last relation is illustrated

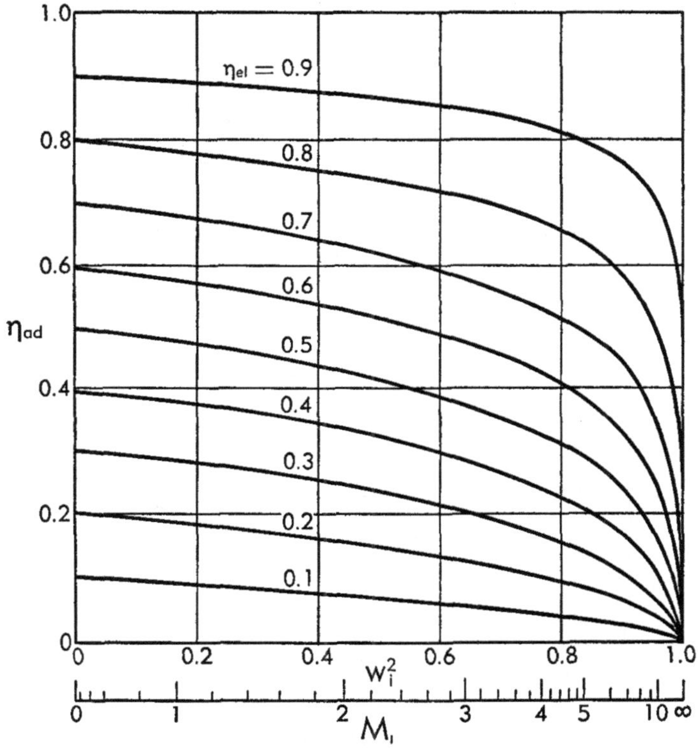

Fig. B,9a. Relation between adiabatic and elemental polytropic efficiency for a diffuser. $\gamma = 1.4$.

by Fig. B,9a. Similarly, the relation between η_{KD} and η_{el} is given by

$$\eta_{KD} = \frac{1 - \left(\frac{p_i}{p_f^0}\right)^{\frac{\gamma-1}{\gamma}}}{1 - \left(\frac{p_i}{p_f^0}\right)^{\frac{\gamma-1}{\gamma\eta_{el}}}} = \frac{1 - (1 - w_i^2)^{\eta_{el}}}{w_i^2}$$

with w_i related to M_i by Eq. 4-28. This relation is represented in Fig. B,9b.

For fixed p_f^0/p_i or M_i, the efficiencies η_{ad}, η_{KD}, and η_{el} vary in the same direction. Thus when comparing diffusers operating at the same Mach number, it does not matter which one of the efficiencies is taken to represent the quality of the diffuser. However, the correct comparison between two diffusers operating at different M_i depends on using the right definition of the efficiency. For instance we see from Fig. B,9a that at $M_i = 2$ a diffuser with $\eta_{el} = 0.9$ is better than one with $\eta_{ad} = 0.8$, but at $M_i = 6$ the inverse is true. As already mentioned the actual quality of a diffuser seems to be better represented by η_{el} (see Art. 10), so that, for a diffuser of constant quality, η_{ad} is a decreasing function, and η_{KD}

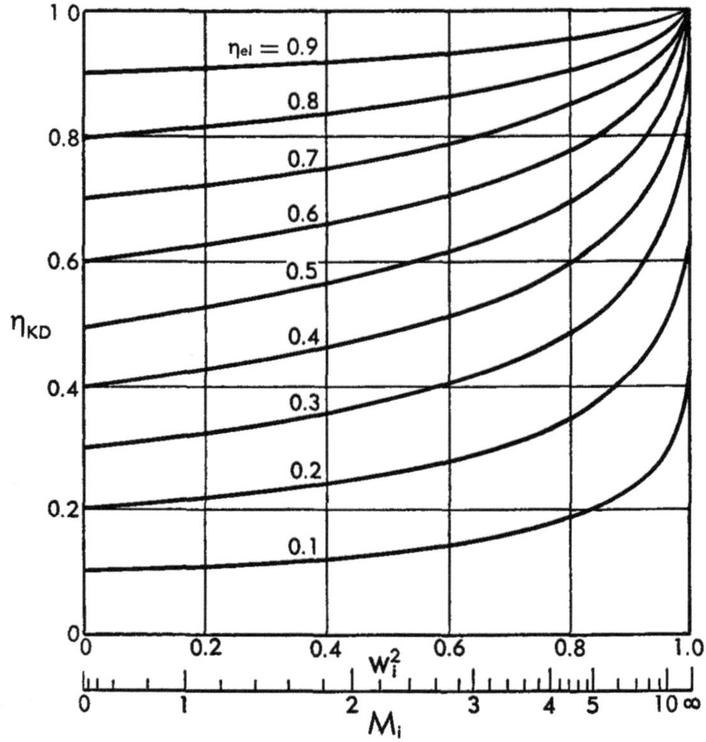

Fig. B,9b. Relation between the Kantrowitz-Donaldson efficiency and the elemental (polytropic) efficiency of a diffuser. $\gamma = 1.4$.

an increasing function, of M_i (or of p_f^0/p_i). For large M_i the difference between the three efficiencies becomes considerable; it becomes difficult to reach high values of η_{ad}, while the attainment of large η_{KD} does not necessarily mean high quality.

Another efficiency which is sometimes used is the pressure efficiency η_p defined by

$$\eta_p = \frac{p_f^0 - p_i}{p_i^0 - p_i} = \frac{\left(1 + \frac{\gamma - 1}{2} M_i^2\right)^{\frac{\gamma\eta_{el}}{\gamma - 1}} - 1}{\left(1 + \frac{\gamma - 1}{2} M_i^2\right)^{\frac{\gamma}{\gamma - 1}} - 1}$$

For small M_i, η_p practically coincides with η_{ad} and η_{el}; however, for increasing M_i its values depart from those of η_{el} even faster than those of η_{ad}. Fig. B,9c shows the relation between η_p and η_{el}.

In addition to the efficiencies, other quantities have been introduced to characterize the performance of a diffuser. The most commonly used is the "pressure recovery" p_f^0/p_i^0, which is related to η_{ad} and η_{el} by the last expressions of Eq. 9-1a and 9-2. Obviously for large values of p_f^0/p_i or M_i the pressure recovery practically coincides with η_p. When, however, p_f^0/p_i is close to unity (or M_i is small) it behaves very differently, and

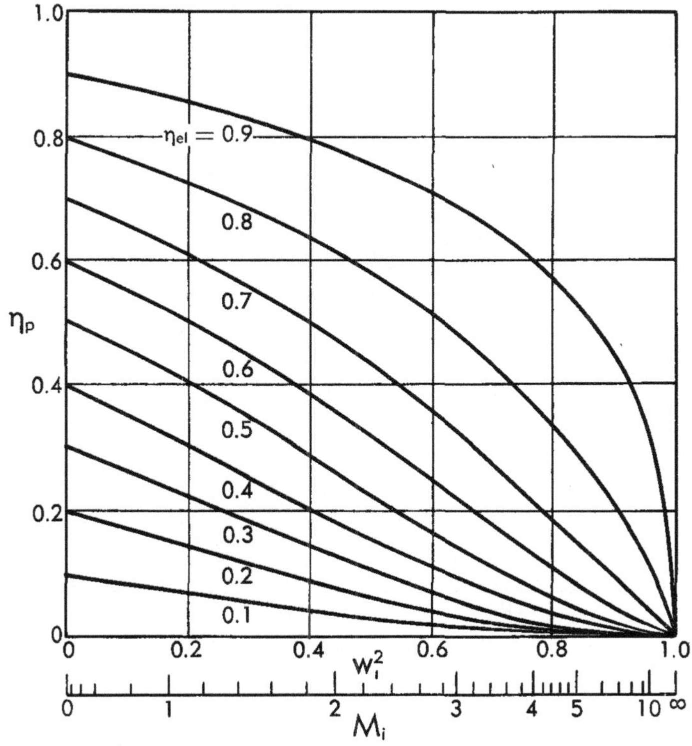

Fig. B,9c. Relation between the pressure efficiency and the elemental (polytropic) efficiency for a diffuser. $\gamma = 1.4$.

becomes inadequate to represent the performance of the diffuser because it tends to unity in all cases. This is clearly shown by Fig. B,9d. From this figure and from Fig. B,9c we also derive the interesting result that for large M_i one obtains very low values of p_f^0/p_i^0 and η_p, even if η_{el} is quite good. Since η_{el} is a better representation of the absolute quality of the diffuser than the other quantities, it follows that the task of substantially improving the values of p_f^0/p_i^0 or η_p is a very hard one, when initially η_{el} is already quite good; moreover the benefit to be derived in devices including diffusers from a substantial improvement of p_f^0/p_i^0 or

η_p only depends on the relative improvement of η_{el}, which may be confined to a few per cent and this may not justify the effort.

Other quantities, also used in connection with diffusers, may be reconducted to those already defined. For instance, if one compares a real diffuser with an isentropic one for the same values of the initial and final velocities, the ratio $p_f/(p_f)_{is}$ coincides with the pressure recovery; with the same assumption the so-called "pressure loss coefficient" $[(p_f)_{is} - p_f]/(p_f)_{is}$ [*33*, p. 188] is the complement to unity of the pressure recovery.

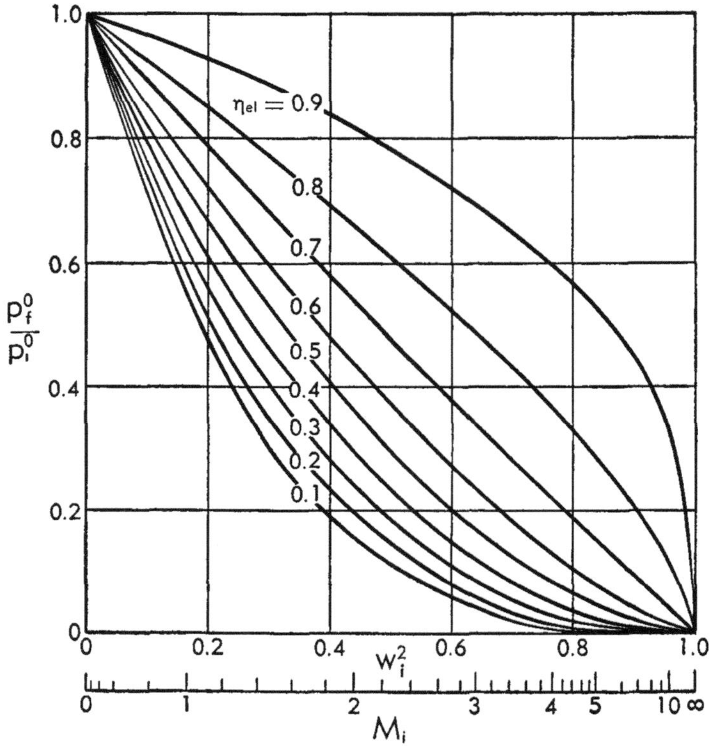

Fig. B,9d. Relation between the pressure recovery and the elemental (polytropic) efficiency of a diffuser. $\gamma = 1.4$.

Subsonic diffusers. A simple divergent duct constitutes a subsonic diffuser. However, the geometry of the duct may affect considerably the efficiency of the diffuser, a few indications of which will be given here.

The best results are obtained for straight ducts with circular section. Because of their simplicity, conical ducts are often used although they do not correspond to the maximum efficiencies. The value of the efficiency —it is immaterial which definition is chosen, because in the subsonic range they all practically coincide—depends on the initial conditions, the Reynolds number, and the angle. For Reynolds numbers based on the diameter in the range around 10^5, and with regular initial conditions, the optimum efficiency is around 90 per cent and is obtained for half-opening of the cone around 5–6°. For smaller angles the flow becomes more uniform because the pressure gradient is lower, but the friction losses increase because of the increased length of the diffuser. For larger angles the friction losses are reduced, but the deviations from uniformity increase. The optimum efficiency increases, and the optimum angle decreases with increasing Reynolds number. For very large diffusers (such as those used in large wind tunnels) and sufficiently smooth walls, the optimum efficiency can reach 95 per cent with half-openings around 3°.

The efficiency is adversely affected by anything which tends to increase the nonuniformity of the flow, such as initial nonuniformity or additional local pressure gradients present in curved diffusers. The efficiency is lower for noncircular sections and especially for polygonal sections, due partly to the increase of the wall perimeter per given cross-sectional area, and partly to the unfavorable conditions of the boundary layers in regions with small radius of curvature and particularly in corners. Changes in the shape of the section along the diffuser also have unfavorable effects because of the additional local pressure gradients introduced.

The efficiency of subsonic diffusers may be considerably improved by controlling the boundary layer, with the purpose of reducing the nonuniformity or preventing separation. The control can be performed in several ways, for instance, by suction of the retarded layers through perforated or porous walls, or by blowing high velocity air through properly directed slots, or by proper use of centrifugal forces. However, the boundary layer control generally involves a supplementary expense of energy, which should be taken into account in the evaluation of the efficiency, thus decreasing its effective value.

Supersonic diffusers. Constant section duct followed by divergence. The simplest type of supersonic diffuser is based on the use of the pseudo-shock in a constant section duct (Art. 5), which brings the velocity from the initial supersonic to a subsonic value, followed by a conventional subsonic diffuser, when necessary to reach the prescribed final value.

We can easily compute the elemental efficiency of this diffuser, if we assume a known efficiency η_{sub} of the subsonic diffuser and, according to Art. 5, we neglect the small friction losses in the pseudo-shock region. We can first calculate the elemental efficiency of the incomplete diffusion in the pseudo-shock itself from the relation

$$\eta_{sh} = \frac{\gamma - 1}{\gamma} \frac{\ln \dfrac{p_2}{p_i}}{\ln \dfrac{T_2}{T_i}} = \eta_{sh}(w_i) \tag{9-4}$$

where $p_i \equiv p_1$ and p_2, $T_i \equiv T_1$ and T_2 represent the condition before and after the pseudo-shock. With p_2/p_i and T_2/T_i given by Eq. 5-5 or 5-5a, η_{sh} is a function of w_i or M_i. Its values for $\gamma = 1.4$ are shown in Fig. B,9e. We see that the efficiency of a shock or pseudo-shock is quite high near the sonic velocity, but it decreases rapidly with M_i. This is due to the

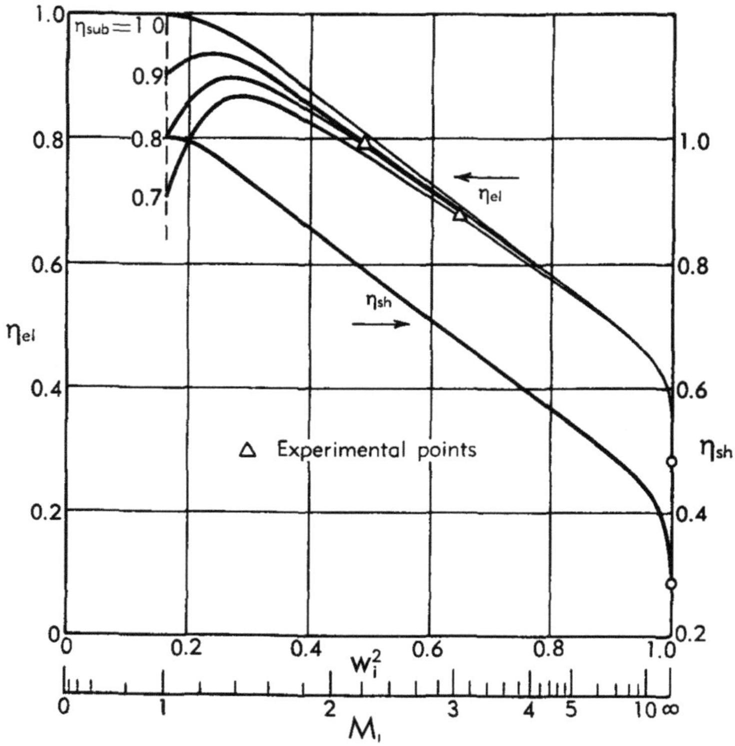

Fig. B,9e. Elemental efficiencies of the incomplete diffusion process through a shock or pseudo-shock, and of supersonic diffusers using a pseudo-shock followed by a subsonic diffuser of efficiency η_{sub}. $\gamma = 1.4$.

behavior (discussed in Art. 5) of the entropy increase and to Eq. 9-3, which for the present case of incomplete diffusion must be written

$$\frac{s_2 - s_i}{c_p} = (1 - \eta_{sh}) \ln \frac{T_2}{T_i} = (1 - \eta_{sh}) \ln \frac{1 - w_2^2}{1 - w_i^2} \tag{9-5}$$

The rest of the entropy increase takes place in the subsonic diffuser and is given by Eq. 9-3:

$$\frac{s_f - s_2}{c_p} = (1 - \eta_{sub}) \ln \frac{T_{f}^0}{T_2} = -(1 - \eta_{sub}) \ln (1 - w_2^2) \tag{9-6}$$

The total entropy increase is therefore

$$\frac{s_f - s_i}{c_p} = -(1 - \eta_{sh}) \ln (1 - w_i^2) - (\eta_{sh} - \eta_{sub}) \ln (1 - w_2^2)$$

Finally we can again apply Eq. 9-3 and obtain for the elemental efficiency of the complete diffuser the following relations:

$$\left.\begin{aligned} \frac{\eta_{sh} - \eta_{el}}{\eta_{sh} - \eta_{sub}} &= \frac{\ln(1 - w_2^2)}{\ln(1 - w_i^2)} \\ \eta_{el} = \frac{\gamma - 1}{\gamma}\left[1 - \frac{\ln\dfrac{w_i^2}{w^{*2}}}{\ln(1 - w_i^2)}\right] &+ \left(\eta_{sub} - \frac{\gamma - 1}{\gamma}\right)\frac{\ln\left(1 - \dfrac{w^{*4}}{w_i^2}\right)}{\ln(1 - w_i^2)} \end{aligned}\right\} \quad (9\text{-}7)$$

The corresponding values are shown in Fig. B,9e for different values of η_{sub}. Observe that high values of η_{sub} seem difficult to achieve because of the nonuniform conditions after the pseudo-shock, even when this seems to be practically completed before the beginning of the divergent duct. In the figure are shown the experimental values of the efficiency of a diffuser of this type, based on the measurements of [*23*]. They coincide practically with the values calculated for $\eta_{sub} = 0.80$–0.90.

The values of p_f^0/p_i for $\eta_{sub} = 0.80$ are shown as function of M_i in Fig. B,9j.

Continuous supersonic diffuser. The efficiency of the shock-type diffuser decays rapidly for large initial Mach numbers, because of the high losses associated with the shock. Thus a way of improving the efficiency at high Mach numbers is to reduce the strength of the pseudo-shock or to suppress it completely. In principle, in the absence of friction, a shockless diffusion can be obtained simply by inverting the direction of the velocity, without changing its value, in an isentropic nozzle, this being allowed by the reversibility of the processes. The efficiency of the isentropic diffusion would be 1 and the duct should be shaped as a convergent-divergent duct with the ratio of throat area to initial cross-sectional area given as a function of w_i by the same relation (Eq. 6-3) used for nozzles.

The presence of friction and its larger relative importance in diffusion processes modifies this result. The supersonic diffuser cannot be obtained simply by inverting the operation of a nozzle because the gradients of entropy and stagnation pressures do not change sign when the velocity is inverted. For given velocity, stagnation temperature, and flow rate the product p^0A must have a fixed value (Eq. 4-31a), and therefore in the diffuser the area ratio A_t/A_i must be larger than the corresponding isentropic value by the factor $p_i^0/p_t^0 > 1$, while the corresponding value A_t/A_f for the nozzle is smaller than the isentropic value.

The flow in a continuous diffuser with losses can be treated following the same lines of Art. 7. However, as already observed, the treatment of the losses through the skin friction coefficient does not provide meaningful results, and the use of empirical coefficients concerning the flow rate and the velocities is more advisable. One can also introduce the efficiencies and study the flow in the diffuser with the assumption that the efficiency stays constant. If for instance, we assume during the diffusion a constant elemental efficiency (defined through the static values), we get the relation analogous to Eq. 7-16 except for the position of η_{el},

$$\frac{p}{p_i} = \left(\frac{T}{T_i}\right)^{\frac{\gamma\eta_{el}}{\gamma-1}} = \left(\frac{1 - w^2}{1 - w_i^2}\right)^{\frac{\gamma\eta_{el}}{\gamma-1}} \tag{9-8}$$

which corresponds to a polytropic transformation with a polytropic

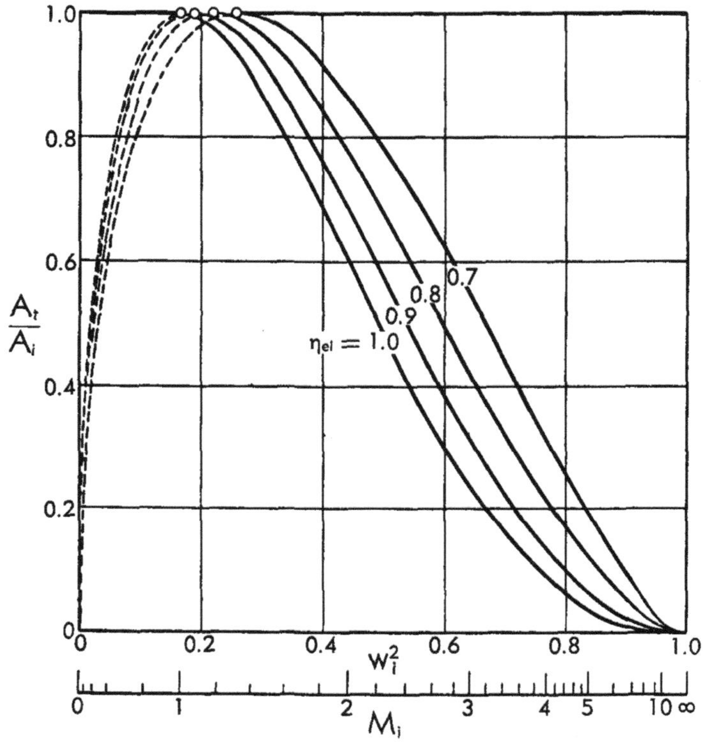

Fig. B,9f. Area ratios of continuous supersonic diffusers in which the elemental efficiency remains constant during the diffusion process. $\gamma = 1.4$.

index n defined by $n/(n - 1) = \gamma\eta_{el}/(\gamma - 1)$, that is by

$$n = \frac{\gamma}{1 - (\gamma - 1)[(1/\eta_{el}) - 1]} \tag{9-9}$$

This relation, similar to Eq. 7-17, shows that for a compression, n is always larger than γ. From Eq. 4-31a one obtains the relation corresponding to Eq. 7-19

$$Aw(1 - w^2)^{\frac{1}{n-1}} = A_i w_i (1 - w_i^2)^{\frac{1}{n-1}} \tag{9-10}$$

from which one finds that the minimum (of the area A_t) is attained for a velocity $w_t > w^*$ given by

$$w_t^2 = \frac{n-1}{n+1}$$

The corresponding value of A_t/A_i can immediately be found and is given for $\gamma = 1.4$ and different values of η_{el} in Fig. B,9f. The dashed part of the curves corresponds to $w_i < w_t$, in which case a simply divergent duct suffices and A_t loses its physical meaning. For large M_i the required throat increases fast with decreasing η_{el}.

The conditions of operation of a continuous supersonic diffuser can be discussed in a similar way to that followed for nozzles. Without entering into details, it suffices to say that if the throat area is smaller than its correct value, the prescribed value of w_i at the entrance of the diffuser cannot be maintained and the conditions in the preceding ducts must undergo a change. We recognize here effects of choking, analogous to

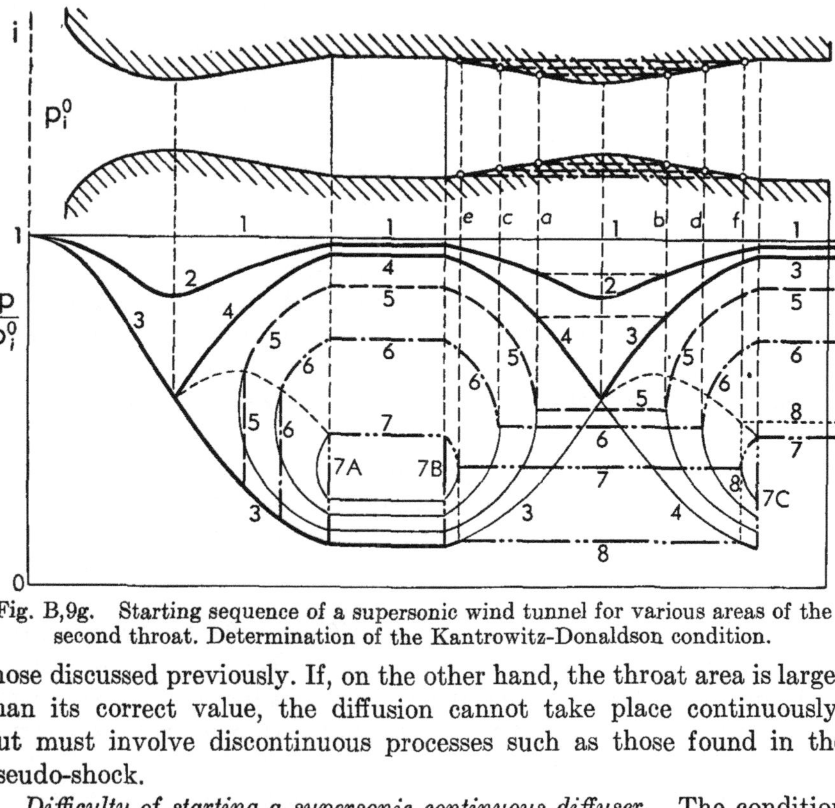

Fig. B,9g. Starting sequence of a supersonic wind tunnel for various areas of the second throat. Determination of the Kantrowitz-Donaldson condition.

those discussed previously. If, on the other hand, the throat area is larger than its correct value, the diffusion cannot take place continuously, but must involve discontinuous processes such as those found in the pseudo-shock.

Difficulty of starting a supersonic continuous diffuser. The condition that the throat area must take a value appreciably larger than its isentropic value is not sufficient. Since the supersonic conditions in any flow system are reached gradually starting from rest, it is important to discuss the conditions in the diffuser during the starting period.

In order to make the discussion more concrete, we take the example of a supersonic wind tunnel in which the supersonic test section is preceded by a Laval nozzle and followed by a continuous diffuser (Fig. B,9g). Assume, for simplicity, that the flow is isentropic everywhere and the areas before and after the diffuser are equal. If the throat areas of both nozzle and diffuser are also equal, a flow configuration represented by curve 3 of the figure can exist in which the flow accelerates steadily in the nozzle up to a final supersonic value, stays equal to this value in the constant area test section, and then decelerates back to subsonic velocities in the diffuser. At both throats the conditions are sonic.

Let us see if this flow configuration, which represents the purpose for which the duct has been so designed, can be obtained starting from a condition where no flow exists, represented in the figure by the horizontal line 1, where $p = p_1^0 = p_b$. Suppose that from this condition of rest, the flow is started by decreasing the back pressure p_b with respect to p_1^0. The flow rate will increase and, after flow configurations such as the one represented by curve 2, reaches its maximum value, with the sonic velocity at the two throats. Between the two throats a subsonic flow is established corresponding to curve 4, and in the remaining regions the subsonic flow corresponding to curve 3.

If we further decrease the back pressure the only change that takes place is in the divergent portion of the diffuser where, after an expansion beyond the sonic velocity, a shock again brings the velocity to subsonic values, as discussed for incorrectly expanded nozzles in Art. 8 and represented by Fig. B,8b. The flow in the rest of the duct, isolated from the variations of p_b by the presence of the sonic velocity at the throat of the diffuser, remains unaltered. We conclude that it is impossible, acting only on the value of the back pressure, to establish the desired supersonic flow in the test section.

This difficulty of starting the supersonic flow at the entrance of a continuous diffuser is obviously also present when the actual flow losses are taken into account, introducing quantitative but not qualitative differences. It is also present for devices other than the supersonic wind tunnel discussed.

Second throat. Kantrowitz-Donaldson diffuser. We have noticed that the impossibility of obtaining supersonic flow between the two throats of the duct considered is connected with the fact that once sonic velocity is established in the second throat, further variations of back pressure are without effect on the flow upstream of this throat.

Obviously the conditions change if one enlarges the second throat, so that the velocity in the whole diffuser is still subsonic when the sonic velocity is attained in the first throat. If, for instance, the portion *ab* of the diffuser, Fig. B,9g, containing the throat, is replaced by a constant area section, as shown in the figure, the corresponding portions of the curves 2 and 4-3 are replaced by the dashed horizontal lines also shown on the figure, the flow being still isentropic. Thus when choking is first produced in the nozzle, the corresponding pressure in the portion *ab* of the diffuser is still above critical and the corresponding velocity subsonic.

Hence a decrease of the back pressure is sensed throughout the subsonic flow region downstream of the first throat and produces, in the nozzle this time, the effects represented by Fig. B,8b. In Fig. B,9g the corresponding conditions are represented first by a shock starting at a certain position of the supersonic part of curve 3 and ending on a curve of constant p^0; then by this curve in the divergent portion of the nozzle, the test section, and the convergent portion of the diffuser, down to section a; by a horizontal segment between section a and b; and finally by the remaining portion of the curve corresponding to the same constant p^0 in the divergent portion of the diffuser.

We have already observed (Fig. B,8b) that the curves p^0 = const have a vertical tangent at the sonic velocity. Thus when, as indicated by curve 5, the slope of the curve p^0 = const becomes vertical at section a, the sonic velocity is attained again in the new throat of the diffuser constituted by the cylindrical portion ab. Once this condition is attained the flow in both the nozzle and the diffuser will stay choked, and further decreases of back pressure will be unable to produce any change between the two throats.

If a larger portion of the diffuser, such as cd, is replaced by a cylindrical section, thus increasing the area of the second throat, the shock moves downstream in the divergent of the nozzle, but again, for a certain p_b, a condition is reached, represented by curve 6, where the flow is also choked in the diffuser. Continuing the increase of the second throat area one finally reaches a value which allows the shock to move down to the entrance of the test section (7A), as represented by curve 7, with sonic velocity in the second throat, constituted now by the cylindrical portion ef. Clearly the shock can now move to any point of the test section without changing the conditions in the rest of the duct. For instance, as shown in the figure, it can sit at the entrance of the diffuser (7B), so that the whole test section is now supersonic, and the primary purpose of reaching the supersonic velocity at the entrance of the diffuser is now attained for the first time.

However, we observe that in these conditions the second throat has no effect on the resulting condition after the diffuser, and that the corresponding convergent-divergent duct does not operate as a diffuser, but rather as a subsonic venturi. Moreover this configuration is unstable, because if the shock moves slightly in the convergent part of the diffuser, this produces a supersonic diffusion, so that the Mach number before the shock and the corresponding dissipation decrease, and the stagnation pressure following the shock increases. Now, whatever the device that regulates the value of the back pressure may be, if the stagnation pressure in front of it increases (the stagnation temperature staying unchanged) it produces an increase of the flow rate through it, and since the flow rate through the nozzle is fixed, there follows an unsteady condition under which gas is evacuated from the intermediate portion of duct. This can only happen at the expense of the gas contained in the subsonic part following the test section, so that the shock must keep moving in the downstream direction, throughout the second throat, until eventually it settles down at the end of the divergent portion of the diffuser (7C) where the final stagnation pressure is reduced again to the undisturbed value. Thus, once the supersonic conditions at the entrance of the diffuser are attained, the shock becomes unstable and must be swallowed through the second throat, after which the configuration is represented by the curve 3, followed by the horizontal line 8, by a portion of curve 4, by the shock (7C), and by the following horizontal line 7. It is immediately checked that the shock (7C) is stable.

Again we observe that in this configuration the presence of the second throat does not bring any improvement in the operation of the diffuser, which results in the same final condition that would be attained in a simple constant section duct. However, if the back pressure is now increased, the shock moves upstream in the divergence of the diffuser. At the same time the shock strength decreases, and the final stagnation pressure increases. The maximum value of the stagnation pressure is obtained when the shock reaches the second throat, as represented by the vertical and horizontal dotted lines 8.

The conclusion of the previous discussion is that we can improve the efficiency of a constant area supersonic diffuser by using a second throat containing a shock, with the condition, however, that the second throat is not smaller than the minimum value allowed by the starting condition. This result was first obtained by Kantrowitz and Donaldson [*32*]. Observe that the diffusion can be extended further by increasing the final area.

When transferring these results to the real case where losses are present even outside of the shock region, the results are quantitatively affected, but present the same qualitative features. The most important difference resides in the replacement of the normal shock with a pseudo-shock. When the length of the pseudo-shock is small, which can happen, following the discussion of Art. 5, at low supersonic velocities and high Reynolds numbers (thin boundary layers), the change may be unessential. It is, however, more important for high Mach numbers and not too high Reynolds numbers, where the length of the pseudo-shock is large. In this case to reach the maximum diffuser efficiency the length of the second throat must be equal to several diameters [*23*], so that most of the pseudo-shock can be completed in the throat. Plate B,9 from [*23*] shows the development of the pseudo-shock in the throat of a diffuser in the presence of thick boundary layers and for $M = 2.55$. Actually here the psuedo-shock begins at the entrance of the convergent portion instead of beginning after it. But this does not produce a substantial difference in the result because most of the pseudo-shock, anyway, is developed in the throat and the part of the dissipation which is increased because it is taking place in the larger sections of the convergent portion is only a small fraction of the total dissipation.

The value of the minimum throat area may also be affected by the substitution of the pseudo-shock for the shock. This is obviously not the case if the pseudo-shock is sufficiently short and the test chamber sufficiently long so that a pseudo-shock beginning at the end of the nozzle may be practically completed before the entrance to the diffuser. However, in the opposite case, which is increasingly likely with increasing Mach number, decreasing Reynolds number, and decreasing length of duct, the pseudo-shock may not be completed before the end of the second throat. In this case the flow through this throat is never subjected to the maximum entropy increase, corresponding to the normal shock, and therefore requires a smaller area for choking. In particular this may be true in hypersonic wind tunnels, which, as was shown experimentally, can be started with a second throat considerably smaller than that determined by Kantrowitz and Donaldson [*34*].

The minimum throat area of the Kantrowitz-Donaldson diffuser, A_{KD}, can be computed according to case 7 of our discussion by assuming that a shock or pseudo-shock occurs in the test section of area A_i entirely before the entrance to the diffuser and from an initial supersonic velocity $w_i \equiv w_1$ to a final subsonic velocity given, following Eq. 5-4, by $w_2 = w^{*2}/w_i$. Since the velocity must be w_2 for the area A_i and w^* for the area A_{KD}, we can apply Eq. 6-2 in the isentropic case and obtain

$$\frac{A_{KD}}{A_i} = \frac{w^*}{w_i}\left(\frac{1 - \frac{w^{*4}}{w_i^2}}{1 - w^{*2}}\right)^{\frac{1}{\gamma - 1}} \tag{9-11}$$

This equation could be corrected for the effects of friction by using, for instance, Eq. 7-19 instead of Eq. 6-2 with n given by Eq. 7-17. However, since the flow from A_i to A_{KD} is expanding, the efficiency can be quite close to 1, and the resulting expression would give values very close to Eq. 9-11.

Also, the efficiency of the Kantrowitz-Donaldson diffuser can be calculated if one assumes constant elemental efficiencies equal to η_{sup} in the supersonic convergence and η_{sub} in the subsonic divergence. The velocity w' at the end of the convergence will be determined in terms of the known area ratio by Eq. 9-10

$$\frac{A_i}{A_{KD}} = \frac{w'}{w_i}\left(\frac{1 - w'^2}{1 - w_i^2}\right)^{\frac{1}{n-1}} \tag{9-12}$$

with n given by Eq. 9-9 in terms of n_{sup}. Thus, because of Eq. 9-11, w' is a function of w_i and η_{sup}. After the pseudo-shock in the throat the velocity w'' is determined by Eq. 5-4. The corresponding shock elemental efficiency is given by Eq. 9-4 as a function of w', and therefore both w'' and η_{sh} are functions only of w_i and η_{sup}. Finally the velocity is reduced from w'' to the final value in the subsonic divergence with efficiency η_{sub}. Using Eq. 9-3 for the subsonic continuous diffusion, and, in terms of static values, for the supersonic continous diffusion and for the shock, one finally has

$$\frac{s_f - s_i}{c_p} = -(1 - \eta_{el}) \ln (1 - w_i^2)$$

$$= (1 - \eta_{sup}) \ln \frac{1 - w'^2}{1 - w_i^2} + (1 - \eta_{sh}) \ln \frac{1 - w''^2}{1 - w'^2}$$

$$- (1 - \eta_{sub}) \ln (1 - w''^2)$$

From this relation one can obtain the elemental efficiency of the diffuser in terms of w_i, η_{sup}, and η_{sub}. In particular if one assumes the same efficiency for the whole continuous part of the diffusion, $\eta_{sup} = \eta_{sub} = \eta_{cont}$, one obtains

$$\frac{\eta_{el} - \eta_{cont}}{\eta_{sh} - \eta_{cont}} = \frac{\ln \dfrac{1 - w'^2}{1 - w''^2}}{\ln (1 - w_i^2)}$$

which can be expanded to give explicitly

$$\eta_{el} = \eta_{cont} - \frac{\dfrac{\gamma - 1}{\gamma} \ln \dfrac{w'^2}{w^{*2}} + \left(\eta_{cont} - \dfrac{\gamma - 1}{\gamma}\right) \left[\ln (1 - w'^2) - \ln \left(1 - \dfrac{w^{*4}}{w'^2}\right)\right]}{\ln (1 - w_i^2)} \tag{9-13}$$

The results obtained from this equation are shown in Fig. B,9h. The curve corresponding to $\eta_{cont} = 1$ covers the whole range $w^* \leqq w_i \leqq 1$, and shows a continuous decrease of η_{el} with increasing w_i.

For other values of η_{cont} a real solution for w' is found only for w_i larger than a certain value w_{i1} shown on the figure. The value of w_{i1} corresponds to the condition where the value of A_{KD} calculated from Eq. 9-11 coincides with the throat area corresponding to a continuous diffusion with the given value of $\eta_{el} = \eta_{cont}$, represented in Fig. B,9f; and, for $w_i < w_{i1}$, A_{KD} becomes smaller than that throat area, and the Kantrowitz-Donaldson condition becomes too restrictive.

Thus for $w_i < w_{i1}$ one has the choice between two possibilities. First, one can use a continuous diffuser, which now satisfies the starting condition. The value of η_{el} for $w_i < w_{i1}$ would then be a constant, equal to η_{cont}. The other possibility is that of establishing a shock at the throat where the velocity w_t is given by Eq. 9-10. The corresponding η_{sh} is then larger than η_{cont}, and so is η_{el}, which is given by the same equation (Eq. 9-13) with w_t substituted for w'. For $w_i < w_t$, as observed previously, the throat disappears, and the rest of the curves down to $w_i = w^*$ correspond to the throatless shock diffuser and are taken from Fig. B,9e. We notice that, within the framework of these calculations, in a definite range of $w_i > w_t$, the throatless shock diffuser is more efficient than one utilizing a throat.

It is interesting to observe that for $w_i \rightarrow 1$ the limit of η_{el} is the same as that for the throatless shock diffuser and is given independently of η_{cont} by $(\gamma - 1)/\gamma$. Thus the improvement due to the Kantrowitz-Donaldson throat vanishes at very high Mach numbers. The reason for

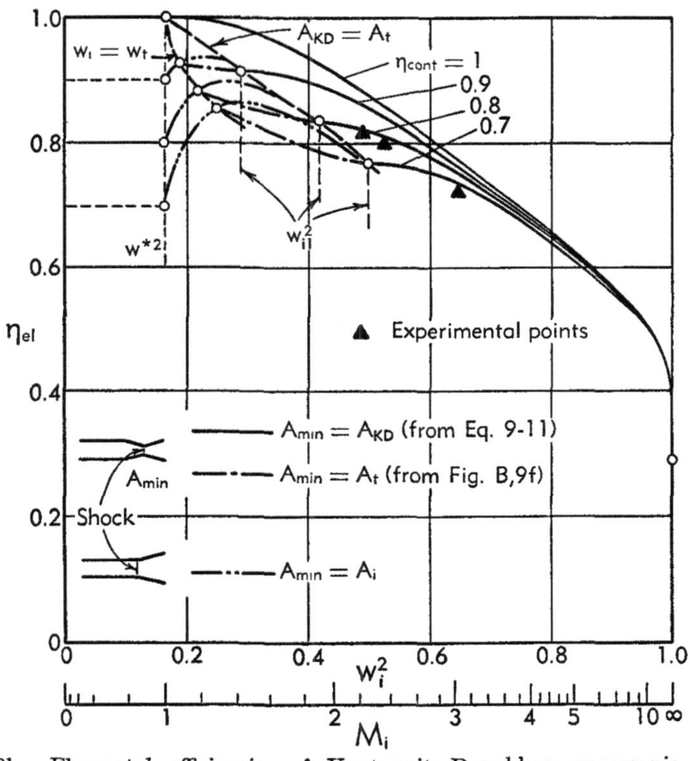

Fig. B,9h. Elemental efficiencies of Kantrowitz-Donaldson supersonic diffusers (full line) and of diffusers apt to replace it below the minimum permitted velocity. $\gamma = 1.4$.

this is that the values of M'/M_i, shown on Fig. B,9i, always stay quite close to unity, and for $w_i \rightarrow 1$ are actually larger than 0.9. Thus the decrease in the shock dissipation due to the Mach number decrease is quite limited, and becomes negligible at very high M_i. On the same figure (Fig. B,9i) the values of the throat areas for the different conditions are also indicated.

On Fig. B,9h are shown some experimental values of the efficiencies for diffusers of the type discussed, taken from [*23*]. The agreement with the curve calculated for $\eta_{cont} = 0.80$ is quite good.

Finally, Fig. B,9j shows, as functions of M_1, the values of the compression ratio p_f^0/p_1 for $\eta_{cont} = 0.80$ in comparison with the values of the same quantity for a throatless shock diffuser with $\eta_{sub} = 0.80$ and for an isentropic diffuser.

Variable geometry diffusers and other types. The results of the previous discussions show that up to Mach numbers of about 2 the throatless and the second throat supersonic diffuser provide good efficiencies, but above Mach number 3 the efficiencies of both types of diffuser drop very fast, and become much worse than the efficiencies that might in principle

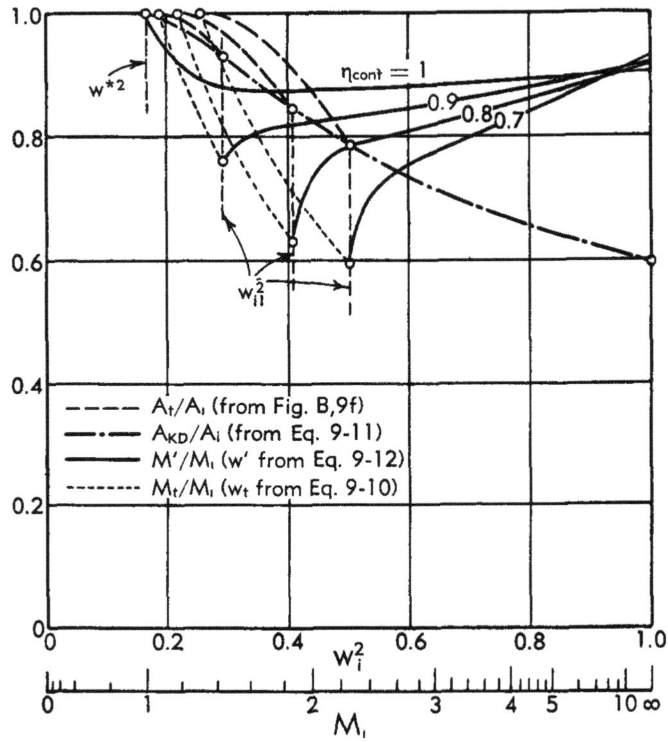

Fig. B,9i. Area ratio and Mach number reduction ratio for the supersonic diffusers of Fig. B,9h. $\gamma = 1.4$.

be obtained from continuous diffusion. Since these limitations are substantially due to the limitation of the minimum throat area deriving from the starting condition, they can be removed if one uses a diffuser of variable geometry, the throat of which may be increased to the necessary value when starting the supersonic flow, and later reduced to a smaller value, such as required for continuous diffusion.

Observe that there are two reasons why it is not actually desirable to reach this condition. First, shocks or pseudo-shocks of small strength being more efficient than any continuous diffusion, the maximum efficiency for the over-all diffusion is obtained before the shock strength is reduced to zero, and therefore for a throat area somewhat larger than the minimum value. This is shown, for instance, by the curves of Fig. B,9h. The optimum value of w' could be calculated without difficulty from Eq. 9-13, and Eq. 9-12 would then give the value of the area ratio, in this case different from Eq. 9-11. Second, a reduction of the throat to the value consistent with continuous diffusion would create an unstable

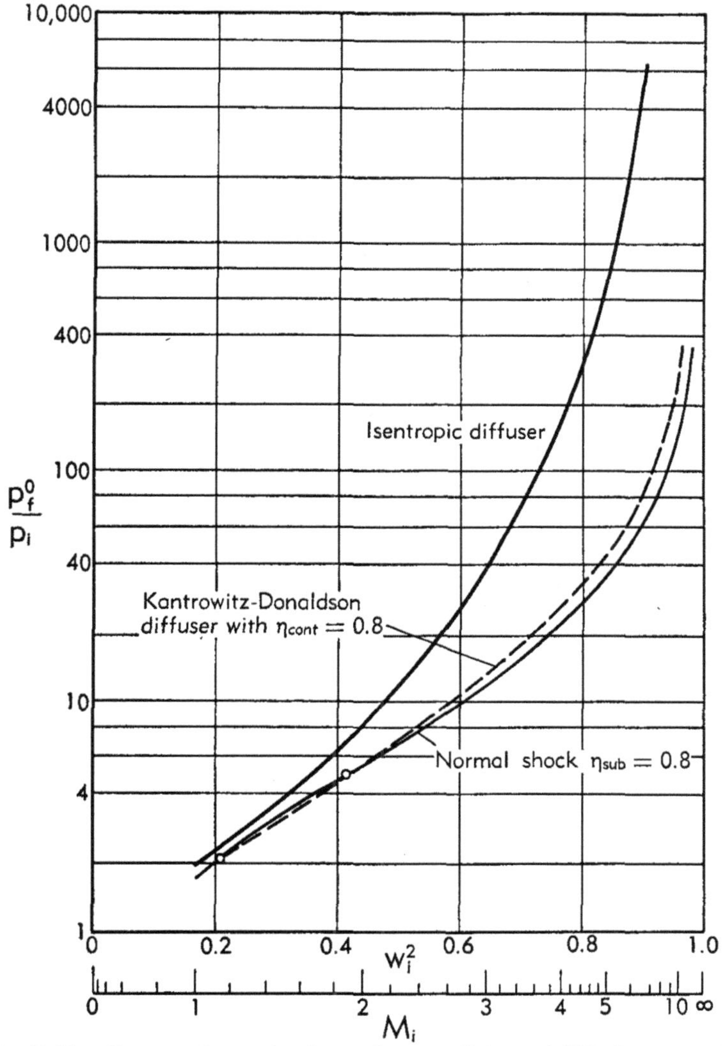

Fig. B,9j. Compression ratios for various conditions of diffusion. $\gamma = 1.4$.

situation because any disturbance producing a shock wave traveling upstream of the throat would increase the entropy in such a way that the throat area would become, temporarily, insufficient for the given flow rate. Hence, fluid should begin to accumulate in front of the throat, downstream of the shock wave which would move upstream, until it would eventually move out of the diffuser, and the velocity would switch back to subsonic in the test section. It is clear that supersonic conditions cannot be restored, except by again enlarging the throat area.

Other procedures, based on nonsteady flow effects, can be used to start a fixed geometry diffuser with a throat smaller than the minimum.

In addition to the one-dimensional diffusers discussed so far, there are several types of diffusers based on oblique shock waves with smaller entropy increases and better efficiencies. These diffusers find a broad field of application in supersonic air intakes, which represent a special use for supersonic diffusers and are discussed in detail in VII,E.

B,10. Long Ducts with Friction. In Art. 7 the effect of friction in nozzles was analyzed, and it was pointed out that for short ducts the one-dimensional treatment of frictional effects does not give satisfactory results because these effects may be overshadowed by those due to the actual development of nonuniformities in the boundary layer. In sufficiently long ducts, however, the initial phase of development of the boundary layers has but little influence on the flow which is essentially determined by what happens after the boundary layers have filled the entire section and the "fully developed pipe flow," mentioned in Art. 5 when dealing with the pseudo-shock, is established.

Incompressible pipe flow. The effect of friction is obtained very simply in the incompressible case. For a constant area duct, in fully developed conditions, the rate of momentum flow remains constant along the duct, and so does the skin friction, which must be balanced by a constant negative pressure gradient. One thus has from Eq. 4-3b, taking into account Eq. 3-5 and 7-20,

$$A(p_1 - p_2) = \frac{4A}{d_h}\tau_w(x_2 - x_1) = 2c_f\frac{A}{d_h}\rho u^2(x_2 - x_1) \tag{10-1}$$

For laminar flow and circular section ($d_h = d$), c_f is given by

$$c_f = \frac{16}{Re}; \qquad Re = \frac{\rho du}{\mu} = \frac{4m}{\pi d\mu} \tag{10-2}$$

where μ is the viscosity coefficient and Re represents the Reynolds number based on the diameter. Laminar flow is present for sufficiently low Reynolds numbers. Practically, below $Re = 1000$, the flow is always laminar.

For turbulent flow with smooth walls and circular section c_f is given by von Kármán's equation

$$\frac{1}{\sqrt{4c_f}} = 2\log\left(Re\sqrt{4c_f}\right) - 0.8 \tag{10-2a}$$

For rough walls c_f is also a function of the relative roughness. See, for instance, [35]. Turbulent flow is established at sufficiently high Reynolds number. In practice, above $Re = 5000$ the flow is always turbulent.

When Re is increased, starting at low values, transition from laminar to turbulent flow must take place. The transition process is very sensitive to the flow conditions, so that the *Re* range in which it is produced varies substantially. In this range the value of c_f cannot be predicted consistently.

Eq. 10-1 is strictly correct only between sections contained in the fully developed region; it is in defect when one or both sections are in the entrance stretch where, as a consequence of the building up of the boundary layers, the rate of momentum flow increases with x, so that the rate of pressure drop is larger than the constant value given by Eq. 10-1 and corresponds to a larger "apparent" skin friction coefficient.

The study of the flow in the entrance stretch cannot be based on purely one-dimensional considerations but requires the detailed study of the boundary layer development [*36,37,38*]. We mention here only a few interesting results for ducts with circular section indicative of the importance of the effects to be expected. If the length of the entrance stretch is called x_{ent} for $x > x_{ent}$ the total pressure drop from the inlet of the duct (assumed to be properly rounded, so that smooth uniform flow can exist in the inlet section) is given by

$$k_p = \frac{p_i - p}{\frac{\rho u^2}{2}} = 4c_f \frac{x}{d} + \Delta k_p = 4c_f \frac{x + \Delta x}{d} \tag{10-3}$$

where k_p represents the pressure drop coefficient, Δk_p its total increase in the entrance stretch, and Δx the equivalent increase of pipe length. Obviously for $x \gg \Delta x$ Eq. 10-3 coincides with Eq. 10-1. For $x < x_{ent}$ Eq. 10-3 can still be used to define Δk_p and Δx which, however, instead of being constant, are functions of x vanishing at $x = 0$ and increasing monotonically with x up to their final constant value.

For laminar flow x_{ent}, Δk_p, and Δx, calculated by Latzko [*37*], are given by

$$\frac{x_{ent}}{d} = 0.029Re = 1.6\frac{\Delta x}{d}; \qquad \Delta k_p = 1.16$$

The agreement with the experimental values is excellent.

For turbulent flow with smooth walls, the results of Pascucci [*38*], based on the logarithmic velocity distribution, are summarized in the following table where c_f is given by Eq. 10-2a:

Table B,10

Re	$4c_f$	$\frac{x_{ent}}{d}$	$\frac{\Delta x}{d}$	Δk_p
2.10^3	0.05	10.40	3.26	0.163
2.10^4	0.0258	14.35	4.03	0.104
2.10^5	0.0155	17.95	5.03	0.078

Observe that x_{ent}/d is almost linear with log Re. Pascucci's results also show that for $x < x_{ent}$, Δx and Δk_p have already practically reached the values given in the table at $x = x_{ent}/2$, so that the length of the entrance stretch for the range of Re considered is between 5 and 9 diameters, and for two sections beyond this length Eq. 10-1 is sufficiently correct.

We must observe, however, that the experimental values are not in complete agreement with Pascucci's result, indicating a larger entrance stretch; they also show a strong sensitivity to the design of the pipe inlet.

When the fluid is compressible, all of the previous results can still be considered as sufficiently accurate if the relative pressure variations are small. For large pressure changes, however, the density cannot be assumed to be constant, and more refined treatments are necessary. Here we shall give only an account of the treatment of the fully developed case.

Fully developed compressible flow. We assume that the flow is adiabatic, so that the stagnation temperature is constant, $T^0 = T^0_i$. Obviously this condition corresponds to well-insulated walls. We shall derive the equations for a general duct of the pressure area power family, for which the momentum equation can be easily expressed in the form, derived from differentiation of Eq. 4-21 with $I_i = 0$,

$$dJ^{(\varepsilon)} = -dX \tag{10-4}$$

The first member is immediately obtained from Eq. 4-34a as

$$dJ^{(\varepsilon)} = m\sqrt{2c_pT^0_i}\left[1 - \varepsilon\frac{\gamma - 1}{2\gamma}\left(1 + \frac{1}{w^2}\right)\right]dw \tag{10-5}$$

The friction force can be expressed from Eq. 3-5 and 7-20 as

$$dX = 4\tau_w\frac{A}{d_h}dx = 2c_fmu\frac{dx}{d_h} = 2c_fm\sqrt{2c_pT^0_i}\,w\frac{dx}{d_h} \tag{10-6}$$

The values of c_f are, for given ε, still functions of Re and of the roughness, as in the incompressible case. In addition they are also affected by the Mach number, an increase of which brings a decrease of c_f which can be substantial at sufficiently high M. This reduction can be predicted theoretically in the laminar case. For the turbulent case, however, although there is strong evidence of this reduction, neither reliable experimental determinations nor completely satisfactory theoretical treatments are available (for cases other than pipe flow, see V,B and F).

Equating Eq. 10-5 to Eq. 10-6, Eq. 10-4 becomes, for $d_h = d$,

$$\left[1 - \varepsilon\frac{\gamma - 1}{2\gamma}\left(1 + \frac{1}{w^2}\right)\right]\frac{dw}{w} = -2c_f\frac{dx}{d} = -\left(\frac{c_f}{c_{f_i}}\frac{d_i}{d}\right)2c_{f_i}\frac{dx}{d_i} \tag{10-7}$$

where c_{f_i} and d_i are constant reference values, for instance the inlet values. The factor in parentheses in the last member of this equation is for smooth walls a function of the Mach number, the Reynolds number, and the diameter. Now observe that for perfect gases μ is a function only of T which, for constant T^0, depends only on w; in addition for a duct of the p-A power family one has from Eq. 4-19

$$d \sim \sqrt{A} \sim p^{\frac{1-\varepsilon}{2\varepsilon}}$$

and, because of Eq. 4-31a, one has

$$pA \sim p^{\frac{1}{\varepsilon}} \sim \frac{1 - w^2}{w}$$

so that d is also only a function of w. Thus, applying Eq. 10-2, the factor in parentheses of the last member of Eq. 10-7 is a function only of w, and the variables can be separated. However, we shall not perform the integration of Eq. 10-7 in its general form. We shall only observe here that the bracketed quantity of the first member vanishes for

$$\frac{1}{w^2} = \frac{2\gamma}{\varepsilon(\gamma - 1)} - 1 = \frac{1}{w_{cr}^2}$$

where w_{cr} is the critical velocity (Eq. 5-20) already introduced before. Thus, since the coefficient of dx in Eq. 10-7 is always positive, we conclude that

$$\frac{dx}{dw} \gtrless 0 \quad \text{for} \quad w \lessgtr w_{cr}$$

Observe that for ducts of the p-A power family the only cause for the changing conditions is the presence of friction, since for $c_f = 0$, Eq. 10-7 gives $w = \text{const}$ (obviously an exception is constituted at supercritical velocities by the pseudo-shock discussed in Art. 5, for which the change, even in the absence of friction, is determined by the internal processes characteristic of pseudo-shocks). Thus we can conclude that the effect of friction in such ducts is to increase the velocity in the subcritical case, and to decrease it in the supercritical case. Moreover the critical velocity represents the maximum value that can be obtained in the subcritical case for a sufficiently long duct with a given ε. Once it is reached we find phenomena analogous to those obtained in Art. 6 when describing the phenomenon of choking in nozzles. Before discussing in detail these effects for the constant section duct ($\varepsilon = 1$), we observe only that for $\varepsilon = 0$ (constant pressure duct) w_{cr} vanishes, and since the flow is always supercritical the velocity gradient is always negative.

Fully developed compressible pipe flow. For constant section ducts, w_{cr} coincides obviously with the reduced sound velocity w^* (Eq. 5-4). Eq. 10-7 in this case becomes somewhat simpler because d is constant and the only variations of c_f are due to the changes of Mach number and Reynolds number, the latter varying only because μ changes, as Eq. 10-2 shows. If, for simplicity, we neglect these effects or choose the proper mean value for c_f given by

$$\bar{c}_f = \frac{1}{x - x_1} \int_{x_1}^{x} c_f dx$$

Eq. 10-7 can be integrated at once, giving

$$-4\bar{c}_f \frac{x - x_1}{d} = \frac{\gamma + 1}{\gamma} \ln \frac{w}{w_1} + \frac{\gamma - 1}{2\gamma} \left(\frac{1}{w^2} - \frac{1}{w_1^2} \right) \qquad (10\text{-}8)$$

It is obvious, from the observations already made on the general form of Eq. 10-7, that the second member of this equation has a minimum at $w = w^*$, and consequently, x takes its maximum value at the sonic velocity. If we call x^* this maximum value we can write the preceding equation in the alternate form, independent of the initial conditions:

$$4\bar{c}_f \frac{x^* - x}{d} = \frac{\gamma + 1}{\gamma} \ln \frac{w}{w^*} + \frac{\gamma - 1}{2\gamma} \left(\frac{1}{w^2} - \frac{1}{w^{*2}} \right) \qquad (10\text{-}8a)$$

For given initial conditions and $\bar{c}_f$, Eq. 10-8 establishes the relation between the velocity and the position along the duct. The pressures can be found immediately from Eq. 4-31a which can be written in the form

$$\sqrt{\frac{\gamma - 1}{2\gamma}} \frac{m \sqrt{\mathcal{R} T_1^0}}{A} = \frac{p_1 w_1}{1 - w_1^2} = \frac{pw}{1 - w^2} = \frac{p^* w^*}{1 - w^{*2}} \qquad (10\text{-}9)$$

where p^* represents the value at $x = x^*$.

Thus the ratio p/p^* is only a function of w and, through Eq. 10-8a, of $4\bar{c}_f(x^* - x)/d$. Instead of relating the pressure to p^* which changes with the flow conditions, we prefer to introduce the stagnation pressure p_1^0 in the initial conditions. The flow rate is then expressed in terms of p_1^0 through Eq. 4-31a as

$$\sqrt{\frac{\gamma - 1}{2\gamma}} \frac{m \sqrt{\mathcal{R} T_1^0}}{A} = p_1^0 w_1 (1 - w_1^2)^{1/(\gamma - 1)} \qquad (10\text{-}10)$$

The flow rate can be related to the maximum possible flow rate that would be obtained in section A after an isentropic expansion to the sonic velocity. In this case one can replace p^* in Eq. 10-9 with

$$p_{is}^* = p_1^0 (1 - w^{*2})^{\gamma/(\gamma - 1)}$$

and obtain

$$\sqrt{\frac{\gamma - 1}{2\gamma}} \frac{m_{is}^* \sqrt{\mathcal{R} T_1^0}}{A} = \frac{p_{is}^* w^*}{1 - w^{*2}} = p_1^0 w^* (1 - w^{*2})^{1/(\gamma - 1)} \qquad (10\text{-}11)$$

This m^*_{is} coincides with the same quantity defined for nozzles (Eq. 6-8). From Eq. 10-9, 10-10, and 10-11 we obtain

$$\frac{m}{m^*_{is}} = \frac{p^*}{p^*_{is}} = \frac{p}{p^0_i} \frac{(1 - w^{*2})^{-1/(\gamma-1)}}{w^*} \frac{w}{1 - w^2} = \frac{w_i(1 - w_i^2)^{1/(\gamma-1)}}{w^*(1 - w^{*2})^{1/(\gamma-1)}} \tag{10-12}$$

Thus the quantity

$$\frac{p}{p^0_i} \frac{m^*_{is}}{m} = (1 - w^{*2})^{\gamma/(\gamma-1)} \frac{p}{p^*} \tag{10-13}$$

simply proportional to p/p^*, is a function only of w or of $4\bar{c}_f(x^* - x)/d$ through Eq. 10-8a. This function is represented, for $\gamma = 1.4$, by the curve of Fig. B,10a. For $w = 0$ both coordinates tend to infinity, while, for

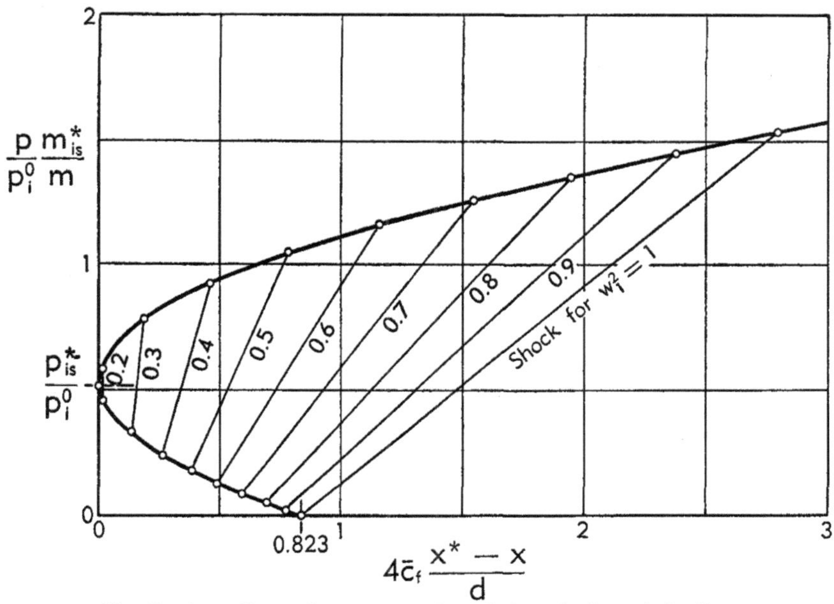

Fig. B,10a. General representation of the solution of the flow equation in pipes with friction. $\gamma = 1.4$.

$w = 1$, the ordinate is zero and the abscissa takes the finite value 0.823.

Pressure distribution in pipes and choking by friction. The curve of Fig. B,10a can be used to determine the values of p/p^0_i at various distances from the initial section for given values of m/m^*_{is}, and therefore of w_i, as Eq. 10-12 shows. The corresponding results are shown in Fig. B,10b, both for subsonic and supersonic values of w_i. Obviously the subsonic case can be obtained by just connecting the pipe through a converging entrance to a tank with pressure p^0_i; in the supersonic case, however, the pipe must be connected to the tank by a Laval nozzle operating in the supercritical range, the throat of which must be such as to produce the desired flow rate m. For each value of m/m^*_{is} there exists a certain maximum length of pipe for which $w = w^*$, $p = p^*$ and beyond which no solution can be found with the given m/m^*_{is}. The corresponding points are joined by the two dashed curves of Fig. B,10b.

In the subsonic case the maximum length increases with decreasing m/m^*_{is} and tends to infinity for $m = 0$. Thus no matter what the length of the pipe is, it is always possible to find a suitable value $m_{max} < m^*_{is}$ of m for which the sonic velocity is established in the final section. To this value of m_{max}/m^*_{is} corresponds, from Eq. 10-12, a well-determined value of $(w_i)_{max}$. For $m < m_{max}$ a flow with $w_i < (w_i)_{max}$ and $w_f < w^*$ is possible.

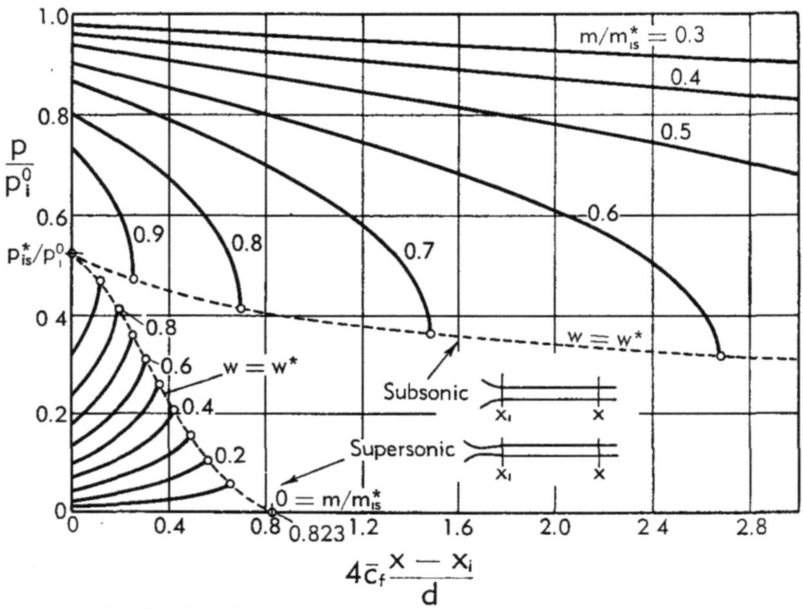

Fig. B,10b. Pressure distributions in pipes with subsonic or supersonic flow at different flow rates. $\gamma = 1.4$.

On the contrary, no solution is allowed for $m > m_{max}$. Therefore a phenomenon of choking analogous to the one discussed for nozzles in Art. 6 is found. The difference between the two cases is that, for nozzles, choking is due to constriction of the area, while here it is due to friction only. We can distinguish the present phenomenon as "choking by friction."

In the supersonic case, the behavior is different. The maximum length increases with increasing w_i (decreasing m/m^*_{is}), and reaches a finite maximum value for $w_i = 1$ ($M_i = \infty$, $m = 0$). This limiting length for a pipe with entirely supersonic flow is given by the relation $4\bar{c}_f l_{lim}/d = 0.823$. Assuming that $\bar{c}_f$ is given by the incompressible von Kármán formula (Eq. 10-2a) and that the Reynolds number is in the range 10^4 to 10^6, then, roughly, $4\bar{c}_f$ stays in the range 0.03–0.01, and therefore l_{lim} in the range $27d$–$82d$. For lower supersonic Mach number the maximum l is always lower. For instance for $M_i = 3$, and the same range of *Re*, the maximum length is in the range 17*d*–52*d*. Observe that these limits would be higher if $\bar{c}_f$ were assumed to decrease with increasing M. We shall see later that, opposite to the subsonic case, solutions with length larger than this maximum are possible for the given initial conditions, but a pseudo-shock is then present in the pipe, which brings the flow from supersonic to subsonic velocities.

Flow rate characteristics of pipes. An interesting cross plot of Fig. B,10b is given in Fig. B,10c where the relation is established between flow rate and pressure ratio p_f/p_i^0 for fixed values of $4\bar{c}_f l/d$, where $l = x - x_i$ is the pipe length and p_f is the pressure in the final section.

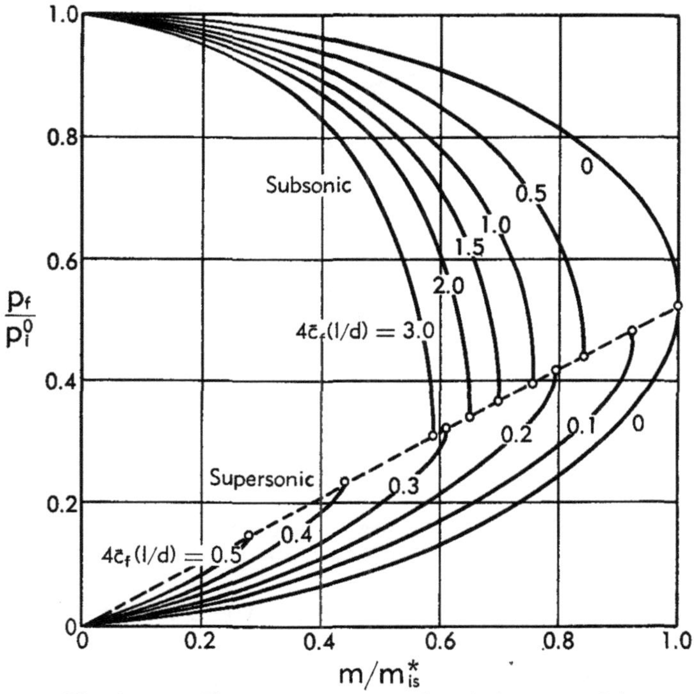

Fig. B,10c. Flow rate characteristics of pipes. $\gamma = 1.4$.

The plane of the figure is divided into a subsonic and a supersonic region by a straight line, the equation of which is obtained by taking $p = p^*$ and $m = m_{max}$ in Eq. 10-13, and is

$$\frac{p^*}{p_i^0} = \frac{m_{max}}{m_{is}^*}\frac{p_{is}^*}{p_i^0}$$

This straight line of course contains the point $p^* = p_{is}^*$, $m_{max} = m_{is}^*$ for $l = 0$, that is for a simple isentropic nozzle. Actually the whole line $l = 0$ coincides with the corresponding curve of Fig. B,6c. We recall that the supersonic portion of this curve does not refer to a nozzle of fixed geometry (as the subsonic portion does), but each point can be obtained with a different throat area.

For $l > 0$ the qualitative behavior of both the subsonic and the supersonic portions of the curves is the same as for $l = 0$. However, the highest value m_{max} of the flow rate, smaller than m^*_{is}, is now obtained at a $p^* < p^*_{is}$. Moreover, the two portions of the curves are not in continuation for the same value of $4\bar{c}_f l/d$.

We see that m_{max}/m^*_{is} is a function of $4\bar{c}_f l/d$ only. This function is shown in Fig. B,10d.

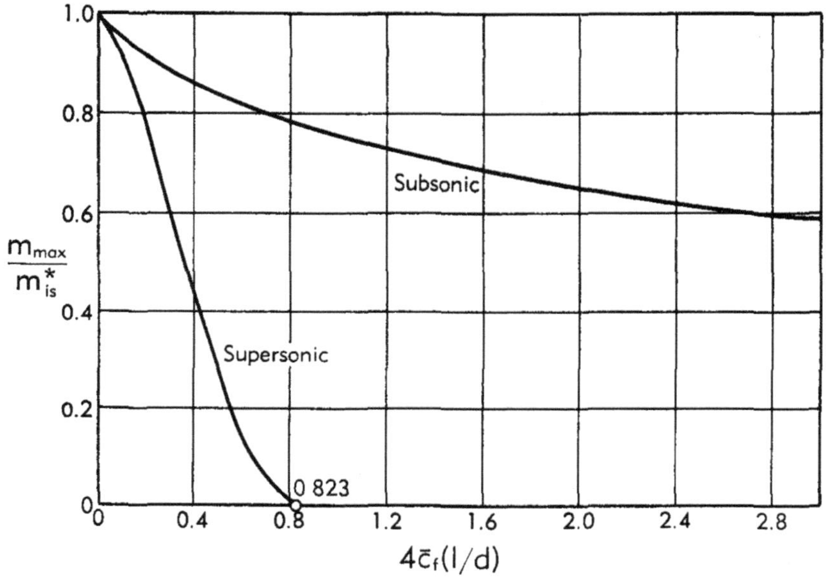

Fig. B,10d. Maximum flow rates in pipes with subsonic or supersonic flow. $\gamma = 1.4$.

Supersonic flow with pseudo-shock. In Art. 5 it was shown that a transition from supersonic to subsonic conditions is always possible in a constant section duct when the proper back pressure is applied. The transition takes place gradually through the complicated process that we have called pseudo-shock. The conditions after the pseudo-shock are very nearly those that would follow a normal shock.

Hence in order to study the effects to be expected when pseudo-shocks are present, we shall replace them by a normal shock, which suddenly brings the velocity from a supersonic value w_1 to a subsonic value $w_2 = w^{*2}/w_1$ (Eq. 5-4). This relation establishes a one-to-one correspondence between the points on the supersonic branch and those on the subsonic branch of Fig. B,10a, which has been indicated on the figure for a few points. Up to the point when the transition takes place the flow proceeds as in the previous case of continuous supersonic flow, with the velocity decreasing and the pressure increasing along the pipe, as indicated by the supersonic curves of Fig. B,10b. After the transition, however, the behavior becomes that of a subsonic flow, with the velocity increasing and the pressure decreasing along the pipe, as for the subsonic curves of Fig. B,10b. The complete behavior is shown in Fig. B,10e, where the values of $(p/p_{\text{i}}^0)(m_{\text{is}}^*/m)$ are plotted vs. $4\bar{c}_f(x - x_{\text{i}})/d$ for a flow with $w_{\text{i}} = 1$. For other values of w_{i} one must start from the proper value of the abscissa. For each w_{i} the curves of p/p_{i}^0 can be obtained from those of Fig. B,10e, because m/m_{is}^* is determined by Eq. 10-12.

It is interesting to observe that for supersonic w_{i} the maximum length of the pipe is limited even in the presence of the shock. The limiting value, corresponding to $w_{\text{i}} = 1$ and a shock situated in the initial section itself,

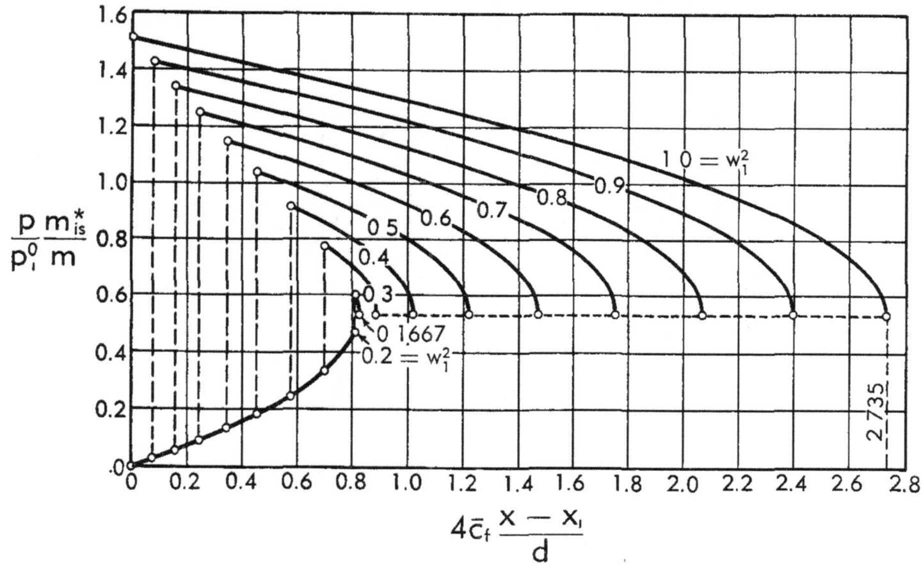

Fig. B,10e. General representation of pressure distribution in pipes in the presence of shocks. $\gamma = 1.4$.

is given by $4\bar{c}_f l_{\text{lim}}/d = 2.735$, that is, for $4\bar{c}_f$ between 0.03 and 0.01, it is in the range $91d$–$273d$. For $M_{\text{i}} = 3$ the limiting length is $40d$–$120d$. These lengths are substantially greater than those calculated previously, corresponding to the same initial velocities in the absence of shocks. While for a length smaller than the lower of the two limiting lengths, and for appropriate values of the back pressure, one has a flow without shocks, shocks must be present if the length is contained between the two values of the limiting length. For lengths larger than the higher value the prescribed initial velocity is impossible and the shock moves into the divergent part of the Laval nozzle preceding the pipe.

Experimental comparisons. The predictions of the theoretical treatment are substantiated by the experimental results of Frössel [*39*]. Fig.

B,10f shows the pressure distributions at subsonic velocities, that is, for a pipe connected to the tank by a convergent duct. The behavior of these curves is very similar to that of the curves of Fig. B,10b. For a quantitative check, however, the abscissa should be altered by a factor $4\bar{c}_f$. This factor will not be the same for all curves because the Reynolds number changes substantially over the full range of tests, though its change along the pipe can be neglected for each test.

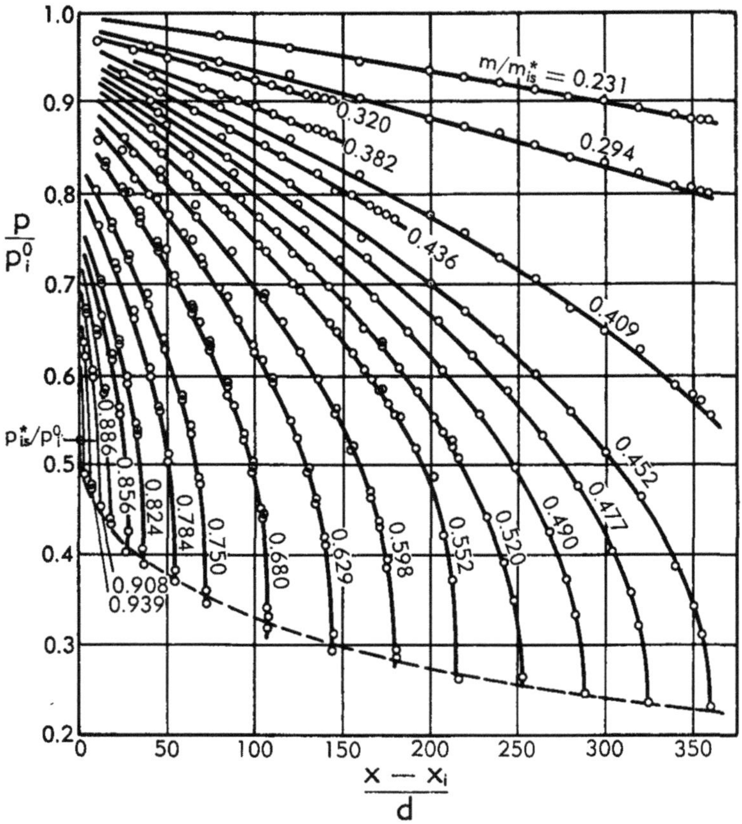

Fig. B,10f. Experimental pressure distributions in pipes with subsonic flow (air).

Fig. B,10g shows the pressure distributions obtained for M_i close to 2 and for $l/d = 26$ by changing the back pressure. Even at the lowest back pressure there is still a shock present, which would indicate that the length of the pipe is above the limiting value without shock. On the other hand, it is certainly smaller than the limiting length with shock. The two theoretical values of the limiting length are given by $4\bar{c}_f l_{\text{lim}}/d$ equal to 0.3 and 0.6, roughly; that is, for $4\bar{c}_f = 0.02$, by $l_{\text{lim}}/d = 15$ and 30. The actual value of 26 lies in fact between these two values. Although these calculations, and others that may be made, prove that a rough agreement can also be expected in the supersonic case, a good quantitative check is not to be expected. Because of the limited length characteristic of the supersonic case, all the measurements are made in the entrance stretch where, especially in the first portion, the theoretical treatment assuming fully developed conditions is certainly in error. Moreover, the departure of the pseudo-shock configuration from that of a simple shock may produce important deviations, particularly when the region behind the pseudo-shock, where the flow presents strong disturbances, is not far from the sonic velocity. Observe that the experimental points in Fig. B,10g are too few to give the actual pressure distribution in the pseudo-

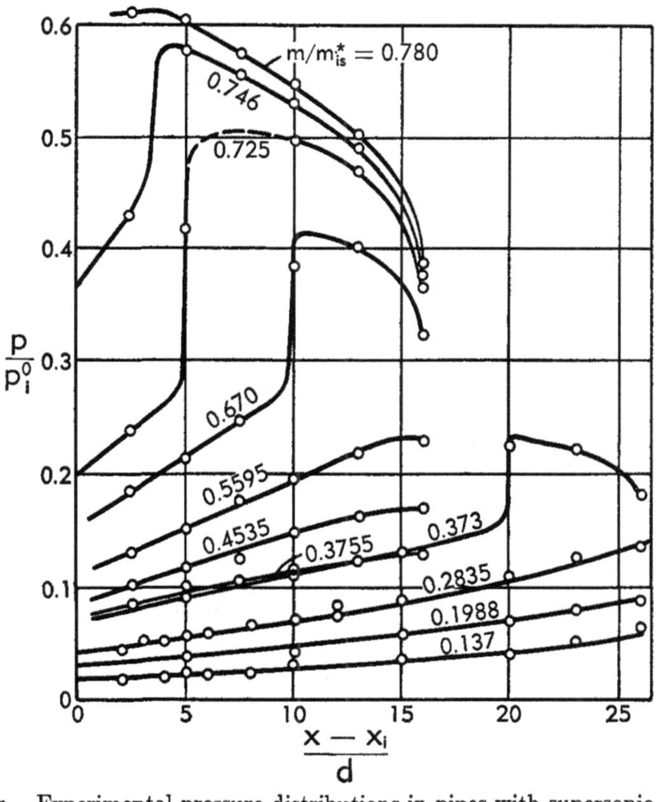

Fig. B,10g. Experimental pressure distributions in pipes with supersonic flow (air).

shock itself, and to give reliability to the actual shape of the curves (reproduced from Frössel) in that region. A similar set of curves, based on the more accurate determinations of [*23*], is given in Fig. B,5d.

Finally, Fig. B,10h shows the experimental relation between p_l/p_i^0 and m/m_{is}^* for constant values of l/d. This figure looks very similar to Fig. B,10c. Again for quantitative checks the parameter of the curves should be multiplied by $4\bar{c}_f$, which, in the whole range of subsonic tests, may change as much as 100 per cent. An observation can be made about the shape of the line dividing the subsonic from the supersonic region, which is somewhat different from the expected straight line, dashed in the figure. Frössel has interpreted this divergence as being due to the nonuniformity of the velocity distribution in fully developed flow. It seems difficult, however, to justify in this way such a noticeable deviation. Moreover, it is possible that this line may be somewhat in error, because it is difficult to determine on the curves of Fig. B,10f the exact value of p/p_i^0 at which m/m_{is}^* is maximum.

Behavior of pipes with varying back pressure. From the results so far described and from the discussion of Art. 8, it is easy to discuss the flow conditions in pipes when the back pressure is changed. The subsonic case, when the pipe is preceded by a purely convergent nozzle, behaves similarly to a convergent nozzle. Decreasing the back pressure p_b in the range $p_{i_1}^0 \geqq p_b \geqq p^*$, the flow rate increases following the lines of Fig. B,10c,

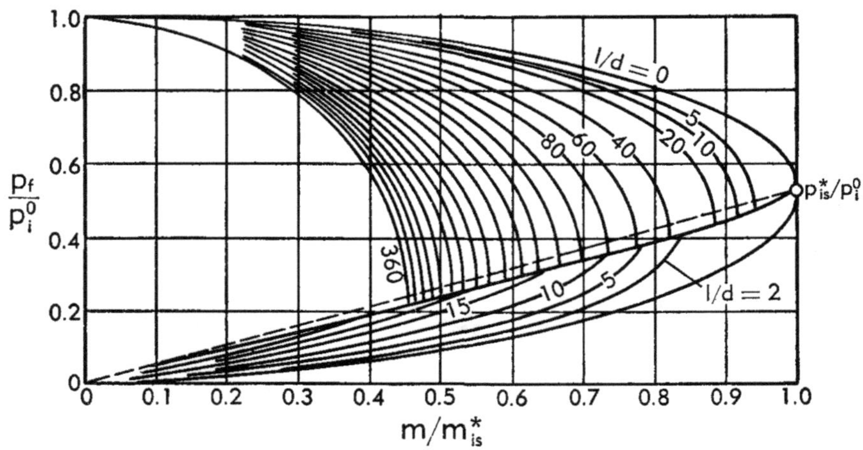

Fig. B,10h. Experimental flow rate characteristics of pipes (air).

where p_f has to be identified with p_b. In the range $p_b < p^*$, however, the flow rate stays constant and equal to m_{max}, the pressure in the final section stays equal to p^*, and the rest of the expansion takes place outside. In other words the flow no longer senses the changes in back pressure (except, perhaps, in a short region close to the exit, where the pressure changes can be propagated upstream in the subsonic part of the boundary layer flow).

When the pipe is preceded by a Laval nozzle, assuming that the length of the pipe is shorter than the limiting length without shocks, for shockless flow one can find on Fig. B,10c the value of the pressure in the final section, in which the velocity is still supersonic. If the back pressure is decreased below this value, the change has no effect on the internal flow which is completed by an external expansion. On the contrary, if the back pressure is increased above this value, a complicated oblique shock pattern will be produced around the exit of the pipe. When the strength of these shocks reaches a certain value, the shock pattern moves definitely inside the pipe until, for sufficiently high back pressure, a complete pseudo-shock followed by a subsonic region will exist inside the pipe. Further increases of back pressure push the pseudo-shock upstream in the pipe and, later, into the divergence of the Laval nozzle. All the changes described so far take place while the flow rate stays constant. However, when the back pressure reaches a sufficiently high value the nozzle unchokes, and from there on the flow rate starts decreasing; in this case the flow stays subsonic throughout the flow process.

If the length of the pipe is included between the limiting length without shocks and that with shocks, a shockless flow is impossible, but there is a flow configuration including a pseudo-shock and a subsonic region terminated by the sonic velocity in the final section. If the back pressure is lower than the final pressure relative to this configuration, it has no effect on the internal flow and is reached through an external expansion. If, on the contrary, it is increased above that value, the pseudo-shock moves upstream and the velocity in the final section becomes subsonic. Again for sufficiently high back pressure, the pseudo-shock moves to the divergence of the nozzle, and for even higher back pressures the nozzle unchokes.

Finally, if the length of the pipe is larger than the limiting length with shocks, a flow with supersonic initial velocity is impossible. In this case, depending on the conditions, the nozzle may be unchoked or choked with a pseudo-shock in the divergent portion. In both eventualities, when the back pressure is decreased below a certain level, the sonic velocity is established in the final section and the internal flow becomes independent of the back pressure itself.

Observe that in all cases in which the nozzle is unchoked the flow does not differ from that obtained in a pipe preceded by a purely convergent nozzle (except for the diffusion losses in the divergent portion).

B,11. Flow with Heat Exchanges or Chemical Reactions. These two types of flow are in reality quite different in principle, the first being essentially nonadiabatic and the second, in general, adiabatic. Also, when chemical reactions are in process, the evolving substance is not in thermochemical equilibrium, and hence, as already observed in Art. 2, certain limitations (concerning the validity of the entropy equations) are present. Despite these differences, the two kinds of processes can be discussed at the same time, as shown by the following considerations. When heat is supplied to, or extracted from, the fluid flowing in a duct one can use the uniformity assumption and suppose that the whole fluid is heated or cooled uniformly on each section (which, as shown in Art. 16, is equivalent to the use of proper mean values). This may be supposed to be true in the case of chemical reactions if the chemical processes are more or less homogeneously distributed in the flow. Both cases have in common the important feature of a variable stagnation temperature, so that, if we make abstraction from the different cause for this variation, and we take it as the fundamental parameter of the transformation, a common treatment is up to a certain extent possible, and forms the object of the following developments.

However, in the case of combustion, the chemical processes may take place in a quasi-homogeneous way only for very high turbulence intensities. When the turbulence is low or absent the homogeneous representation must be replaced by one where both the unburned substances and the products of complete combustion are present at the same time in the same section, the intermediate products being formed only within the flame front, which is a very thin layer separating the two and can, practically, be replaced by a discontinuity (Sec. G). For such cases the consideration of an average temperature on each section leads to appreciable errors and in general the conventional one-dimensional treatment is impossible. However, at the end of this article we shall develop an approximate treatment based on one-dimensional principles of a particularly simple case of combustion through an oblique flame front. Observe also that if one is only interested in the relations between the initial (unburned) and the final (completely burned) conditions; the results are independent of the intermediate stages of combustion, and are the same for a flame-type or a homogeneous-type combustor. Thus, again, the common treatment can be applied.

The variation of stagnation temperature can be an increase for the case of heat supply or exothermic reactions, or a decrease for heat subtraction or endothermic reactions. The first eventuality is the most interesting because it includes the case of combustion. During a combustion process, and in general during a chemical process, the chemical composition of the substance changes, and so do the specific heats (independently of their variation with temperature) and the gas constant. These variations can be taken into account in the definition of the quantity

$$\lambda = \frac{c_{p_\mathrm{f}} T_\mathrm{f}^0}{c_{p_\mathrm{i}} T_\mathrm{i}^0} \tag{11-1}$$

which, according to the discussion of Art. 4, can be introduced to replace the solution of the energy equations (Eq. 4-10 or 4-11).[11] Here c_{p_f} corresponds to the temperature T_f^0 which is intended to be a solution of the energy equation itself. Obviously instead of λ one could use the stagna-

[11] We have indicated here by the subscripts $_\mathrm{i}$ and $_\mathrm{f}$ instead of $_\mathrm{u}$ and $_\mathrm{b}$ the initial and the final conditions of the transformation, so as to include the general case of heat exchange.

tion temperature ratio, or the quantity

$$\Gamma\lambda = \frac{\gamma_i}{\gamma_f}\frac{\gamma_f - 1}{\gamma_i - 1}\frac{c_{p_f}T_f^0}{c_{p_i}T_i^0} = \frac{\mathcal{R}_f T_f^0}{\mathcal{R}_i T_i^0} \tag{11-2}$$

where we have introduced

$$\Gamma = \frac{\gamma_i}{\gamma_f}\frac{\gamma_f - 1}{\gamma_i - 1} \tag{11-3}$$

In the absence of chemical transformations, one can introduce the same quantities, and, obviously, $\Gamma\lambda = T_f^0/T_i^0$. However, for nonpolytropic gases c_{p_f}/c_{p_i} and Γ are different from unity. Only for polytropic gases or for sufficiently small variations of stagnation temperature can one take $\Gamma \cong 1$, and λ coincides with the stagnation temperature ratio.

Observe that in the combustion of ordinary mixtures of hydrocarbons the variation of the gas constant can often be neglected. This is especially true for very diluted mixtures, in which case even the effect of the change of chemical composition on c_p can be disregarded. In this case the process of combustion can be regarded as a simple heat addition process, in which the amount of heat added can be simply computed as indicated in Art. 2 (Eq. 2-20 and 2-21).

In a heat exchange process, the heat is transferred gradually and the conditions at intermediate stages can also be defined through the local stagnation temperature ratio T^0/T_i^0 or through the quantity, similar to λ, $c_pT^0/c_{p_i}T_i^0$. This can also be done for chemical reactions, provided the chemical process takes place homogeneously and the intermediate products are known. For very diluted mixtures we may dispense with the last requirement and consider the substance to be of fixed composition.

In the rest of this article we shall use the simplifying assumption, discussed at the end of Art. 4, that the gases can be approximately taken to be polytropic between each stagnation temperature and the corresponding static temperature. This enables us to make use of the simple expressions derived in Art. 4 for polytropic gases without having to suppose that the specific heats are constant throughout the process.[12]

Frictionless flow in a duct of the p-A power family. From Eq. 4-34a we obtain for a general section the following expression for the generalized impulse function relative to ducts obeying the Eq. 4-19

$$J^{(\epsilon)} = m\sqrt{2c_pT^0}\left(w + \delta\frac{1 - w^2}{2w}\right)$$

$$= m\sqrt{2c_pT^0}\,\frac{\sqrt{\delta(2 - \delta)}}{2}\left(\frac{w}{w_{cr}} + \frac{w_{cr}}{w}\right) \tag{11-4}$$

[12] A treatment without this simplifying assumption is possible. The corresponding equations can be written in differential form, but cannot be integrated, except numerically. See, for instance, the method of Shapiro and Hawthorne (Art. 14).

where for simplicity we have written

$$\delta = \varepsilon \frac{\gamma - 1}{\gamma} \tag{11-5}$$

and where w_{cr} is the critical velocity, defined by Eq. 5-20, i.e.

$$w_{cr}^2 = \frac{\delta}{2 - \delta} \tag{11-6}$$

Both ε and δ can take any positive or negative value; however, w_{cr} is real only in the range $0 \leqq \delta \leqq 2$. For given $J^{(\varepsilon)}$, m, c_pT^0, and δ, one can solve Eq. 11-4 with respect to w. It is immediately checked that for $\delta < 0$ or $\delta > 2$ there is always one and only one positive root w subject only to the limitation of being <1; for $0 \leqq \delta \leqq 2$ and w_{cr} real, w is given by

$$\frac{w}{w_{cr}} = j \pm \sqrt{j^2 - 1} \tag{11-7}$$

with

$$j = \frac{J^{(\varepsilon)}}{m\sqrt{2\delta(2 - \delta)c_pT^0}} = \frac{J^{(\varepsilon)}}{m\sqrt{2\varepsilon(2 - \delta)\mathcal{R}T^0}} \tag{11-8}$$

Obviously there are two real and positive roots for $j > 1$, one supercritical, the other subcritical. For $j = 1$ the two roots coincide together and with w_{cr}, and for $j < 1$ there is no real root.

Neglecting friction the momentum equation is Eq. 4-22, that is

$$J^{(\varepsilon)} = \text{const} = J_1^{(\varepsilon)} = m\sqrt{2c_{p_1}T_1^0}\left(w_1 + \delta_1 \frac{1 - w_1^2}{2w_1}\right) \tag{11-9}$$

where δ_1 is obtained from Eq. 11-5 with $\gamma = \gamma_1$.

Suppose now that the initial conditions and the value of ε are assigned, and that T^0 is made to vary gradually, and with it c_p and γ. Then δ will also change following Eq. 11-5. If δ always stays <0 or >2, for every value of T^0 there will be one, and only one, positive value of w satisfying Eq. 11-9. Suppose, however, that $0 \leqq \delta \leqq 2$ during the whole process, or during only a part of it. Then the quantity j is real and, according to Eq. 11-8 and 11-9 it varies only as a result of the variation of the quantity $\mathcal{R}T^0(2 - \delta)$. If this quantity increases, j decreases and, starting from an initial value, necessarily >1, may become unity if at a certain point the equality

$$\mathcal{R}T^0\left(2 - \varepsilon\frac{\gamma - 1}{\gamma}\right) = \frac{1}{2\varepsilon}\left(\frac{J_1^{(\varepsilon)}}{m}\right)^2 \tag{11-10}$$

is satisfied. A further decrease of j is not compatible with the reality of the solution. This means that if the condition (Eq. 11-10) is attained before the end of the duct, the prescribed initial conditions cannot be maintained and must change so as to make $j \geqq 1$ along the whole duct, and hence the first member of Eq. 11-10 $\leqq$ the second member.

We find here a phenomenon similar to that of choking discussed with reference to nozzles and to pipes with friction. The cause for choking here is the increase of the expression $\mathcal{R}T^0(2 - \delta)$. Since, in general, the variation of T^0 in this expression dominates with respect to the variation of the two other factors, the cause for choking is in general an increase in stagnation temperature. Thus we can call this effect "choking by heat addition," including in this term the case of exothermic reactions. Observe, however, that in principle endothermic reactions may exist in which $\mathcal{R}T^0(2 - \delta)$ increases despite a decrease of T^0, and that choking may be produced by such an endothermic reaction.

The condition for choking is that Eq. 11-10 be satisfied in the final section. Replacing $J_i^{(e)}$ from Eq. 11-9 this condition can be written

$$\lambda\delta_f(2 - \delta_f) = \left[w_i\left(1 - \frac{\delta_i}{2}\right) + \frac{\delta_i}{2w_i}\right]^2 \tag{11-11}$$

This relation can be interpreted in two ways. If w_i is assigned, and δ_i and δ_f are known, Eq. 11-11 gives the maximum value of λ compatible with a real flow. For λ smaller than this value the flow terminates with $w_f \neq w_{cr}$; for λ larger the flow with the given w_i is impossible; finally for λ equal to this maximum value, the flow is choked with $w_f = w_{cr}$. If λ is prescribed and δ_i, δ_f have given values, Eq. 11-11 solved for w_i gives

$$w_i = \frac{\sqrt{\lambda\delta_f(2 - \delta_f)} \pm \sqrt{\lambda\delta_f(2 - \delta_f) - \delta_i(2 - \delta_i)}}{2 - \delta_i} \tag{11-11a}$$

thus establishing two values of w_i, one subcritical and the other supercritical, for which choking is obtained. It is immediately seen that the flow is unchoked for w_i smaller than the first or larger than the second; and that no flow can be realized with values of w_i between the two.

The velocity at each point of the process can be obtained from Eq. 11-7, where, however, one has to choose between two solutions, one subcritical, the other supercritical. It is clear that, by reasons of continuity, the subcritical solution must be chosen everywhere if w_i is subcritical, since a passage to supercritical velocities would only be possible through the critical velocity, where j must be 1. But since $j > 1$ for both $w \lessgtr w_{cr}$, this would indicate that the quantity $\mathcal{R}T^0(2 - \delta)$, after having increased exactly up to the value of Eq. 11-10, should again decrease, a most unlikely coincidence. Thus from the initial subcritical value of w_i, w increases toward w_{cr} or decreases away from it depending on whether $\mathcal{R}T^0(2 - \delta)$ increases or decreases during the process. If w_i is initially supercritical, applying the same line of reasoning it would again seem that w should always remain supercritical, decreasing for increasing $\mathcal{R}T^0(2 - \delta)$ and increasing in the opposite case. However, if we recall the results of Art. 5 about the pseudo-shock, we realize that a transition from super- to subcritical is possible, independent of any change of $\mathcal{R}T^0(2 - \delta)$, this transition being possible, without having to go through the critical velocity, because a "nonuniformizing" process (separation) is present.

Making abstraction from the possible presence of pseudo-shocks, we conclude that in general the effect of an increase of T^0 (heat addition) is an increase of w (and therefore of M) for subcritical flows, and a decrease for supercritical flow. A decrease of T^0 (heat subtraction) has opposite effects.

In reality in the foregoing discussion when we say that w increases or decreases toward w_{cr} we mean that w/w_{cr} increases or decreases toward unity, the two statements being equivalent only for constant w_{cr}, and therefore being nearly equivalent in the actual cases if γ and w_{cr} vary slowly. A more exact determination of the direction in which w changes can be obtained by differentiating Eq. 11-4 with $J^{(\varepsilon)} = \text{const}$, thus obtaining

$$\frac{w^2 - w_{cr}^2}{w^2 + w_{cr}^2}\frac{dw^2}{w^2} = -\frac{d(c_pT^0)}{c_pT^0} - \frac{2w_{cr}^2(1 - w^2)}{w^2 + w_{cr}^2}\frac{d\gamma}{\gamma(\gamma - 1)} \tag{11-12}$$

Constant pressure and constant area ducts. The considerations developed so far may in particular be applied to the two interesting cases of constant pressure and constant area flows. In the constant pressure case ($\varepsilon = \delta = 0$), the solution of Eq. 11-9 is

$$\frac{w}{w_1} = \sqrt{\frac{c_{p_1}T_1^0}{c_pT^0}} \tag{11-13}$$

equivalent to $u = u_1$, i.e. Eq. 4-17. In fact, the velocity cannot change because the total force acting on the fluid vanishes. Thus choking cannot be present, or, more precisely, choking happens for infinite c_pT^0 when the velocity becomes equal to the critical velocity, which is zero in the present case. Of the two solutions (Eq. 11-7), Eq. 11-13 represents the supercritical while the subcritical is always zero.

For the constant area case ($\varepsilon = 1$, $\delta = (\gamma - 1)/\gamma$), w_{cr} is real and coincides with the sonic velocity w^*. Thus there are always two solutions given by

$$\frac{w}{w^*} = \sqrt{\frac{\gamma}{2(\gamma + 1)\mathcal{R}T^0}}\frac{J_1}{m} \pm \sqrt{\frac{\gamma}{2(\gamma + 1)\mathcal{R}T^0}\frac{J_1^2}{m^2} - 1} \tag{11-7a}$$

(one of which is supersonic, the other subsonic), provided the quantity under the second radical is $\geqq 0$ in the whole duct. This condition can be translated into a limitation for T_f^0, or λ, expressed by

$$\lambda \leqq \frac{\gamma_f^2}{\gamma_f^2 - 1}\frac{\gamma_1^2 - 1}{\gamma_1^2}\left(\frac{w_1}{w_1^*} + \frac{w_1^*}{w_1}\right)^2$$

For given λ it corresponds to the condition that w_i must be out of the transonic range determined by the two values

$$w_i = w_i^* \left(\sqrt{\frac{\gamma_f^2 - 1}{\gamma_f^2} \frac{\gamma_i^2}{\gamma_i^2 - 1} \lambda} \pm \sqrt{\frac{\gamma_f^2 - 1}{\gamma_f^2} \frac{\gamma_i^2}{\gamma_i^2 - 1} \lambda - 1} \right)$$

If these conditions are satisfied the flow is possible, and from the previous general discussion we obtain the fact that an increase of T^0 results in an increase or a decrease of Mach number depending on whether the velocity is subsonic or supersonic. In the supersonic case, transition to subsonic conditions through a pseudo-shock can complicate the results. Typical values of the limits for w_i can be obtained from Fig. B,11a, to be discussed later.

All previous equations undergo substantial simplifications for $\gamma =$ const (and therefore $w^* =$ const). If, in addition, no chemical transformation is present, c_p is also constant and one obtains explicit expressions in terms of the heat addition Q, observing that

$$\lambda = 1 + \frac{Q}{mc_pT_i^0}$$

Dependence of choking on duct shape. From the previous discussions it seems clear that the phenomenon of choking by heat addition is strictly connected to the shape of the duct. For instance we have seen that choking can exist only for ducts with $0 < \delta < 2$; and therefore, even staying within the *p-A* power family, ducts can be designed for which a change of T^0, no matter how large, will never produce choking effects. We have seen that, when the conditions of choking are produced, any further increase of T^0 is possible only through a change of initial conditions. However, this is true only if we consider the same type of duct (i.e., in our case, the same value of ε). If, on the contrary, this condition is relieved and the shape of the duct is changed downstream of the point where the critical velocity is attained, further increases of T^0 become possible without modification of the initial conditions. For instance for a constant section duct, once the sonic velocity is attained, further heat addition is possible for fixed initial conditions if one starts enlarging the section.

Let us stress again that the only duct of the *p-A* power family which has a definite geometry is the constant area duct. As observed at the conclusion of Art. 5, for values of $\varepsilon \neq 1$ the resulting shape of the duct depends on the distribution of the processes causing the transformation, that is, in the present case, the rates of heat exchange or of release of chemical energy. Only if these rates are known can one calculate the duct shape corresponding to a prescribed ε.

Static and stagnation pressures. Once w is known, the pressure can be calculated by replacing in Eq. 11-9 the alternate Eq. 4-34 or 4-34a, or else from the continuity equation $m = m_i$ with m given by Eq. 4-31a. Making use of Eq. 4-19 one obtains

$$\left(\frac{p}{p_i}\right)^{\frac{1}{\varepsilon}} = \left(\frac{A}{A_i}\right)^{\frac{1}{1-\varepsilon}} = \frac{pA}{p_iA_i} = \frac{\varepsilon + \gamma_i M_i^2}{\varepsilon + \gamma M^2} = \frac{\gamma_i}{\gamma_i - 1}\,\frac{\gamma - 1}{\gamma}\,\frac{\delta_i + \dfrac{2w_i^2}{1 - w_i^2}}{\delta + \dfrac{2w^2}{1 - w^2}} \tag{11-14}$$

The ratio of the stagnation pressures is then obtained from the equation

$$\left(\frac{p^0}{p_i^0}\right)^{\frac{1}{\varepsilon}} = \left(\frac{p}{p_i^0}\right)^{\frac{1}{\varepsilon}} \frac{(1 - w_i^2)^{1/\delta_i}}{(1 - w^2)^{1/\delta}} \tag{11-15}$$

The consideration of the stagnation pressure is very important for the comparison of the qualities of different ducts when the same irreversible process (such as combustion) takes place, and replaces with advantage the consideration of the entropy increase which is affected strongly by the irreversibility of the process.

If we differentiate Eq. 11-15 for fixed w_i, we obtain, taking Eq. 11-14 into account, the following final expression

$$\frac{\gamma - 1}{\gamma}\frac{dp^0}{p^0} = \frac{w^2 - w_{cr}^2}{w^2 + w_{cr}^2}\frac{dw^2}{1 - w^2} + \left[\frac{2w_{cr}^2 w^2}{w^2 + w_{cr}^2} + (1 - w^2)^{1-\frac{1}{\delta}} \ln(1 - w^2)\right]\frac{d\gamma}{\gamma(\gamma - 1)}$$

from which dw^2 can be eliminated with the help of Eq. 11-12 with the following result

$$\frac{\gamma - 1}{\gamma}\frac{dp^0}{p^0} = -\frac{w^2}{1 - w^2}\frac{d(c_pT^0)}{c_pT^0} + (1 - w^2)^{1-(1/\delta)} \ln(1 - w^2)\frac{d\gamma}{\gamma(\gamma - 1)} \tag{11-16}$$

For $\gamma = \text{const}$ this relation is reduced to

$$\frac{dp^0}{p^0} = -\frac{\gamma}{\gamma - 1}\frac{w^2}{1 - w^2}\frac{d(\mathcal{R}T^0)}{\mathcal{R}T^0} = -\frac{\gamma M^2}{2}\frac{d(\mathcal{R}T^0)}{\mathcal{R}T^0} \tag{11-17}$$

and shows that the stagnation pressure always decreases for increasing $\mathcal{R}T^0$, and increases in the opposite case. However, a γ variation may modify this statement since in general γ changes in the opposite direction of c_pT^0, and the coefficient of $d\gamma$ in Eq. 11-16 is always negative. Observe that the variation of γ is generally of less importance than that of c_pT^0, so that the result of Eq. 11-17 can be applied in practically all cases. We shall see later how friction modifies this result.

As an example of the application of the preceding treatment, Fig. B,11a shows, for a constant section duct ($\varepsilon = 1$) and $\gamma_i = \gamma_f = \gamma = 1.3$, the values of the relative stagnation pressure loss (often called momentum loss) ($|\Delta p^0|/p_i$) $= 1 - (p_f^0/p_i^0)$ divided by $\gamma w_i^2(\lambda - 1)/(\gamma - 1)$ for both subsonic and supersonic w_i and different values of λ (in the supersonic range the value of λ is strongly limited, for larger λ the supersonic flow becoming impossible). In the supersonic range observe the two branches of the λ = const curves. Both branches can have a physical meaning, the supersonic one ($w_f > w^*$) being obtained in the absence of shocks or pseudo-shocks, the other through a pseudo-shock. The two branches terminate together in a cusp on the $w_f = w^*$ line (choked conditions). The cusp itself is of particular interest, corresponding to the "detonation

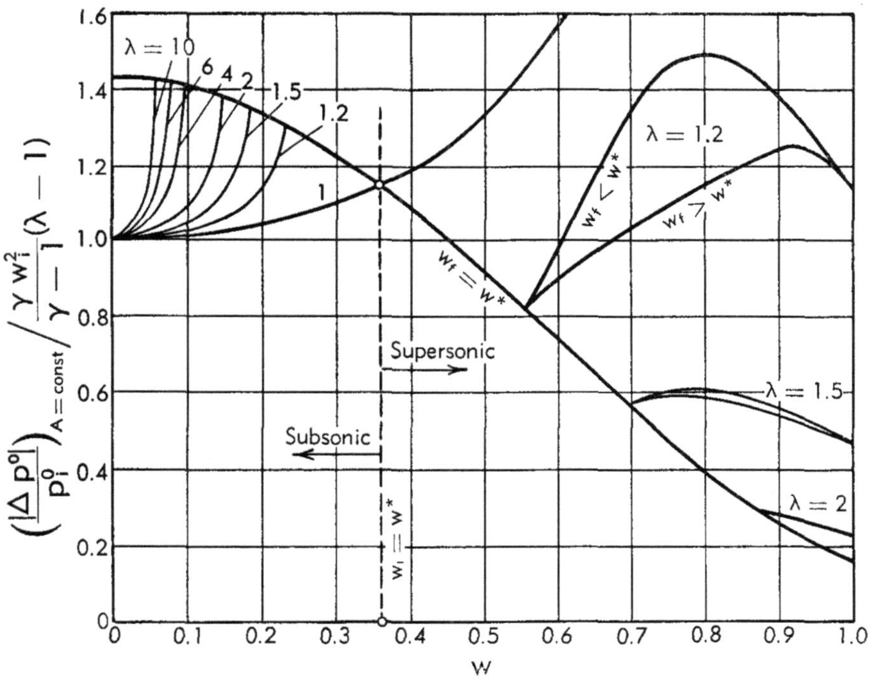

Fig. B,11a. Coefficient of stagnation pressure loss (momentum loss) by heat addition in constant section ducts. $\gamma_i = \gamma_f = \gamma = 1.3$.

wave" and to the Chapman-Jouguet condition (Sec. D and G). In the subsonic range the solution also presents two branches, one subsonic, the other supersonic, and a cusp on the choking line. The supersonic branch, however, has not been indicated in the figure because the transition from subsonic to supersonic is physically impossible.

Comparison of constant pressure and constant area ducts at low velocities. Particularly simple expressions are found when the velocities throughout the process are small enough. It is interesting in this case to compare the results of a process of heat addition or combustion, characterized by a given value of λ, in a constant pressure and a constant area duct. Observe that if $w^2 \ll 1$, Eq. 4-29, expanded in powers of w^2 and stopped at the first power term, can be written approximately as

$$w^2 = \frac{\gamma - 1}{2} M^2$$

In a constant pressure duct, Eq. 11-13, relating final and initial reduced velocities, or the corresponding relation in terms of Mach numbers, can be written in the forms

$$\frac{w_f}{w_i} = \frac{1}{\sqrt{\lambda}}; \quad \frac{\gamma_f M_f^2}{\gamma_i M_i^2} = \frac{1}{\Gamma\lambda} = \frac{\mathcal{R}_i T_i^0}{\mathcal{R}_f T_f^0} \tag{11-18}$$

The static pressure being constant, the relative stagnation pressure drop can be computed, up to terms of the order of w^2 from

$$\left(\frac{|\Delta p^0|}{p_i^0}\right)_{p=\text{const}} = 1 - \frac{p_f^0}{p_i^0} = 1 - \frac{(1 - w_i^2)^{\gamma_i/(\gamma_i - 1)}}{(1 - w_f^2)^{\gamma_f/(\gamma_f - 1)}} \cong \frac{\gamma_i}{\gamma_i - 1} w_i^2 - \frac{\gamma_f}{\gamma_f - 1} w_f^2$$

Thus we obtain

$$\left(\frac{|\Delta p^0|}{p_i^0}\right)_{p=\text{const}} \cong \frac{\gamma_i w_i^2}{\gamma_i - 1}\left(1 - \frac{1}{\Gamma\lambda}\right) \cong \frac{\gamma_f w_f^2}{\gamma_f - 1}(\Gamma\lambda - 1) \tag{11-19}$$

or, in terms of the initial or final Mach number,

$$\left(\frac{|\Delta p^0|}{p_i^0}\right)_{p=\text{const}} \cong \frac{\gamma_i M_i^2}{2}\left(1 - \frac{1}{\Gamma\lambda}\right) \cong \frac{\gamma_f M_f^2}{2}(\Gamma\lambda - 1) \tag{11-19a}$$

The relation between the areas of the final and initial sections of the duct, necessary to give the desired constancy of the pressure, is readily obtained as

$$\frac{A_f}{A_i} \cong \Gamma\lambda$$

For the constant area duct, assuming that $w^2 \ll w_{cr}^2 = w^{*2}$ we can obtain directly from Eq. 11-4 and 11-6

$$\frac{w_f}{w_i} \cong \sqrt{\lambda}\sqrt{\frac{\delta_f(1 - \delta_f)}{\delta_i(1 - \delta_i)}}\frac{w_f^*}{w_i^*} = \frac{\delta_f}{\delta_i}\sqrt{\lambda} = \Gamma\sqrt{\lambda} \tag{11-20}$$

or, in terms of Mach numbers,

$$\frac{\gamma_f M_f^2}{\gamma_i M_i^2} = \frac{1}{\Gamma}\frac{w_f^2}{w_i^2} \cong \Gamma\lambda$$

Comparing with Eq. 11-18 we observe the symmetry of these two relations, and the fact that, for $\Gamma\lambda > 1$, the final Mach number is smaller than the initial in the constant pressure case, and larger in the constant area case. Thus in the former the sufficient condition for the application of the previous approximate expressions is $M_i^2 \ll 1$, but in the latter one has to make certain that $M_f^2 \ll 1$ too. The opposite would be true for heat subtraction ($\Gamma\lambda < 1$).

In the constant area case both the static and the stagnation pressure vary. The corresponding relative pressure drops can be computed from Eq. 11-14 and 11-15 with $\varepsilon = 1$. Up to terms in w^2 we obtain

$$\left(\frac{|\Delta p|}{p_i}\right)_{A=\text{const}} = 1 - \frac{p_f}{p_i} \cong 1 - \frac{1 + \frac{2\gamma_i}{\gamma_i - 1} w_i^2}{1 + \frac{2\gamma_f}{\gamma_f - 1} w_f^2} \cong 2\left(\frac{\gamma_f}{\gamma_f - 1} w_f^2 - \frac{\gamma_i}{\gamma_i - 1} w_i^2\right)$$

$$\left(\frac{|\Delta p^0|}{p_i^0}\right)_{A=\text{const}} = 1 - \frac{p_f}{p_i}\frac{(1 - w_i^2)^{\gamma_i/(\gamma_i - 1)}}{(1 - w_f^2)^{\gamma_f/(\gamma_f - 1)}} \cong \frac{\gamma_f}{\gamma_f - 1} w_f^2 - \frac{\gamma_i}{\gamma_i - 1} w_i^2$$

Thus the static pressure drop is twice as large as the stagnation pressure drop. For the latter we obtain, making use of Eq. 11-20, the following expressions

$$\left(\frac{|\Delta p^0|}{p_i^0}\right)_{A=\text{const}} \cong \frac{\gamma_i w_i^2}{\gamma_i - 1} (\Gamma\lambda - 1) \cong \frac{\gamma_f w_f^2}{\gamma_f - 1}\left(1 - \frac{1}{\Gamma\lambda}\right) \qquad (11\text{-}21)$$

or the corresponding expressions in terms of Mach numbers:

$$\left(\frac{|\Delta p^0|}{p_i^0}\right)_{A=\text{const}} \cong \frac{\gamma_i M_i^2}{2} (\Gamma\lambda - 1) \cong \frac{\gamma_f M_f^2}{2}\left(1 - \frac{1}{\Gamma\lambda}\right) \qquad (11\text{-}21\text{a})$$

In order to establish a comparison between the constant pressure and the constant area case, assuming that in the two cases we have the same process, say of combustion, we note that the values of $\mathcal{R}$, γ, and T^0 in the initial and final sections will be the same in either case. Now from Eq. 4-31a, with $w^2 \ll 1$, we have

$$\frac{m}{A} = \frac{p}{\sqrt{\mathcal{R}T^0}} \sqrt{\frac{2\gamma}{\gamma - 1}}\, w$$

This relation can be applied in both the initial and the final sections. In the constant area case p_f/p_i differs from 1 by an amount of the order of w^2, which can be neglected in the present assumptions, so that p_f and p_i are practically the same. Assuming that their common value is equal to the static pressure of the constant pressure case, we conclude that the values of m/A_i are the same for the two types of duct if w_i (or M_i) is the same, and, similarly, that m/A_f has the same value if w_f (or M_f) is the same.

Therefore, if we compare two ducts with the same initial area per unit flow rate we find from Eq. 11-19 and 11-21

$$\frac{(\Delta p^0)_{p=\text{const}}}{(\Delta p^0)_{A=\text{const}}} = \frac{1}{\Gamma\lambda} \qquad (\text{same } M_i)$$

and the constant pressure duct looks more favorable than the other. But if we compare the same final area per unit flow rate we obtain

$$\frac{(\Delta p^0)_{p=\text{const}}}{(\Delta p^0)_{A=\text{const}}} = \Gamma\lambda \qquad (\text{same } M_f)$$

so that the constant area duct is superior.

Since the final area is the maximum area for the constant pressure duct, the latter result is true if the comparison is made for the same maximum area. Thus in this case, if friction and other losses are neglected, a combustion chamber with a constant area gives the minimum drop of stagnation pressure. This drop, or momentum loss, is given by Eq. 11-21 for small M_f. Of course this relation is not correct for large M_f up to unity (choking). In this case the momentum loss has to be determined more accurately, as discussed previously, and represented in Fig. B,11a. We observe that in the subsonic range and for nonvanishing w_i the values computed from Eq. 11-21 are in defect. However, the difference from the exact values is not exceedingly large even in the limiting case of choking. Thus Eq. 11-21 can be used in the whole subsonic range as a convenient approximate expression for the momentum loss of a constant area combustor, perhaps with a constant numerical factor larger than unity to improve the over-all approximation in the whole range $0 < w_i < w^*$.[13]

Additional considerations on the best shape of combustion duct. The result obtained above about the advantage of the constant area duct can actually be generalized on the following simple grounds. If, in agreement with an observation made at the beginning of this article, we consider the combustion process as one of heat addition, then the minimum entropy increase is obtained when heat is added at every point at the maximum possible static temperature, and, therefore, when the kinetic energy at each point is as low as possible. Obviously this is obtained when the area of each section is equal to the maximum allowed area. Thus for a prescribed maximum area the constant area duct gives the least entropy increase and stagnation pressure drop.

In these comparisons only the drop of stagnation pressure during the combustion process is considered. When other losses are also taken into account, however, the result may be different. Suppose, for instance, that the problem is to find the best shape of duct for the combustion of a mixture supplied at a prescribed (subsonic) velocity and pressure, the maximum cross-sectional area of the duct (larger than the supply area) being also prescribed. There is an infinity of ducts satisfying both conditions. Only to mention a few possibilities, one could first bring the area to the desired maximum value without combustion through a diffuser, and then burn at constant area. Or one could first, through a duct without combus-

[13] The maximum error can be reduced to ± 20 per cent if one multiplies the second and the last member of Eq. 11-21 by a factor 1.2.

tion (which, depending on the circumstances, may be a nozzle or a diffuser) bring the area to such a value that a subsequent constant pressure combustion would terminate with the prescribed maximum value of the area. In both cases the final area equals the prescribed maximum value. These are not, of course, the only possibilities, since ducts can be designed in which the area is first brought, without combustion, to its maximum value in a diffuser, combustion then proceeding in a converging duct.

If we compare these three types of ducts we see that, due to the result established previously, in the first case the stagnation pressure drop during the combustion is a minimum. However, the diffuser preceding the combustion duct corresponds to the maximum possible area increase, and therefore introduces the maximum possible diffusion losses. In the second case, the diffusion losses are certainly decreased or even eliminated; on the other hand the combustion losses are certainly larger. In the third case, the diffusion losses take their maximum value and the combustion losses are larger than for the constant area combustion. Thus we see that the third type of duct is always worse than the first, and must be discarded whenever possible. Between the first two types, however, the choice is undecided, and we see that the best duct should present a certain amount of divergence of the combustion duct which, though introducing an increased combustion loss, will reduce the diffusion loss. The optimum condition, using for instance a combustion duct of the p-A power family, can be easily determined numerically.

In these considerations we have deliberately neglected the friction losses, which may have a certain influence on the results. An even more important effect can derive from the presence of other kinds of losses due to the flame holder (for nearly stoichiometric mixtures) or to the mixing processes (for well-off-stoichiometric mixtures) (see XI, Part 2).

One-dimensional treatment of a particular flame-type combustor. At the beginning of this article it was mentioned that when the combustion takes place through a flame front in such a way that on a given section both unburned and burned gases are present, the treatment given above can only be applied to the final conditions, i.e. when the combustion process is over, and not to the intermediate stages. We want to show now a particular case in which the treatment, though different from the one given previously, can still be based on purely one-dimensional grounds.

For a given combustible mixture with prescribed pressure and temperature, the flame front is characterized by a well-determined value S of the normal velocity of the unburned gases with respect to the flame front, called the "burning velocity" or "flame speed" (Sec. G). When the flow velocity is larger than S, which is the general case in practice because the latter has very low values, it is impossible to maintain a stationary flame, except if this is stabilized by a flame holder downstream of which the flame spreads through the unburned gases as an oblique flame in such a way that the component of the velocity normal to the flame is always equal to S. In the limiting case of a flame holder of a vanishing size the flow process will look, qualitatively, as in the scheme of Fig. B,11b. This flow problem has been investigated in detail in [*40,41,42*] (see also Sec. G) with the common assumptions that the velocity of the gases is always substantially larger than S, the obliquity of the velocities (and that of the flame front) is always quite small, and, as a

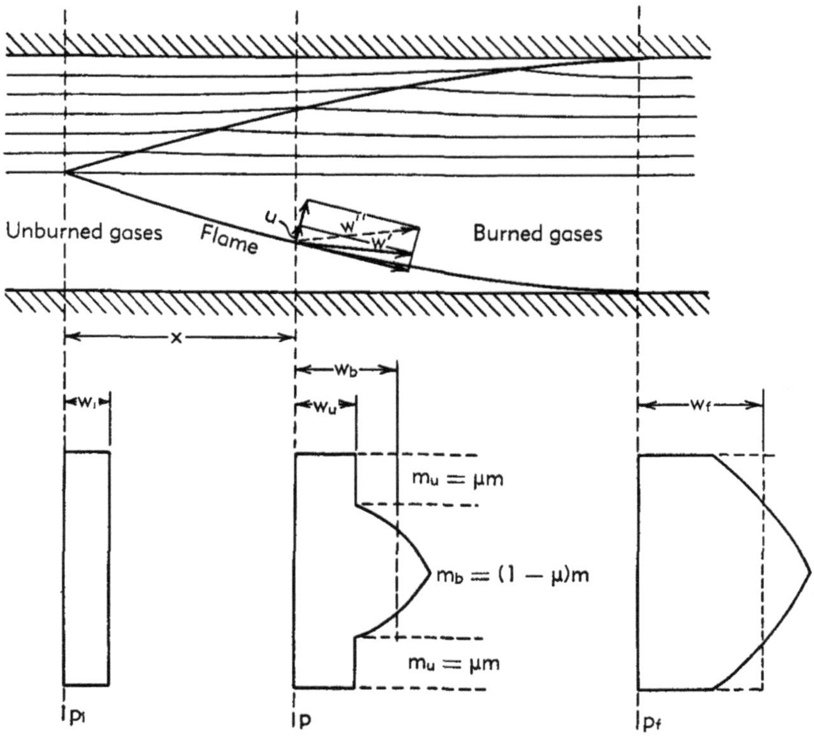

Fig. B,11b. Model of flame-type combustor.

consequence, the static pressure is practically uniform on each section.[14] As a result of these calculations it is found that, while the velocity distribution in the unburned gases is uniform, the velocity distribution in the burned gases has a roughly triangular shape with the vertex on the center line, as shown in the velocity diagrams of Fig. B,11b.

Obviously the derivation of the detailed distributions in this region escapes the purpose of one-dimensional treatments. However, a purely one-dimensional calculation of the pressure distribution along the duct and the shape of the flame front can be obtained if one observes that the assumptions mentioned above are exactly the same as the ordinary one-dimensional assumptions. Therefore, consistent with the usual one-

[14] The flame front discontinuity produces a static pressure drop. However, due to the magnitude of S (Sec. G) this pressure drop can be disregarded.

dimensional simplification, one can replace the triangular velocity distribution in the burned gas region with a uniform mean value of the velocity, as shown by the dashed $w = w_b$ and $w = w_f$ lines of Fig. B,11b. From now on the treatment becomes similar to that of the shockless model of the pseudo-shock (Art. 5).

We assume a duct of constant area (though the treatment can be developed for the general duct of the p-A power family) and consider a section where the total flow rate m consists of the flow rates of unburned gases m_u and burned gases m_b. We define the burned fraction μ, related to the flow rates by

$$\mu = \frac{m_b}{m}, \qquad 1 - \mu = \frac{m_u}{m} \tag{11-22}$$

We choose the initial section at $\mu = 0$ and the final section at $\mu = 1$.

Neglecting friction and heat exchanges, the flow of unburned gases is isoenergetic and isentropic. Thus at any section

$$T_u^0 = \text{const} = T_i^0, \quad \frac{p}{p_i} = \left(\frac{1 - w_u^2}{1 - w_i^2}\right)^{\frac{\gamma_u}{\gamma_u - 1}} \tag{11-23}$$

Consistently with the assumptions of this article, γ_u and c_{p_u} are considered constant because of the constancy of the stagnation temperature. For the same reason, and since the flow is adiabatic, from the energy equation we obtain that the stagnation temperature T_b^0 of the burned gases is constant, and so are γ_b and c_{p_b}. Thus

$$\lambda = \frac{c_{p_b} T_b^0}{c_{p_u} T_u^0} = \frac{c_{p_f} T_f^0}{c_{p_i} T_i^0}$$

is also a constant, the value of which, when known, replaces the energy equation.

The continuity equation, automatically satisfied by Eq. 11-22, is substituted by the equation

$$A = A_u + A_b$$

expressing the constancy of the area. Deriving A_u, A_b, and A from Eq. 4-31a and dividing by $m(\gamma_u - 1)\sqrt{\mathcal{R}_u T_u^0/\gamma_u p}$, this equation becomes

$$(1 - \mu)\frac{1 - w_u^2}{w_u} + \mu\Gamma\sqrt{\lambda}\frac{1 - w_b^2}{w_b} = \frac{p}{p_i}\frac{1 - w_i^2}{w_i} \tag{11-24}$$

where

$$\Gamma = \frac{\gamma_u}{\gamma_b}\frac{\gamma_b - 1}{\gamma_u - 1}$$

coincides with the Γ defined by Eq. 11-3.

The momentum equation,

$$J = J_u + J_b = \text{const} = J_i$$

can also be expressed in terms of the same quantities by replacing for the various J the expressions obtained from Eq. 4-33a, and dividing by $m(\gamma_u - 1)\sqrt{2c_{p_u}T_u^0/\gamma_u}$, with the result

$$(1-\mu)\left(\frac{1}{w_u}+\frac{\gamma_u+1}{\gamma_u-1}w_u\right)+\mu\Gamma\sqrt{\lambda}\left(\frac{1}{w_b}+\frac{\gamma_b+1}{\gamma_b-1}w_b\right) = \frac{1}{w_i}+\frac{\gamma_u+1}{\gamma_u-1}w_i \quad (11\text{-}25)$$

Equating the values of μ obtained from Eq. 11-24 and 11-25 we have

$$\mu = \frac{\dfrac{1-w_u^2}{w_u} - \dfrac{p}{p_i}\dfrac{1-w_i^2}{w_i}}{\dfrac{1-w_u^2}{w_u} - \Gamma\sqrt{\lambda}\dfrac{1-w_b^2}{w_b}} = \frac{(w_u - w_i)\left(\dfrac{\gamma_u+1}{\gamma_u-1} - \dfrac{1}{w_u w_i}\right)}{\dfrac{1}{w_u}+\dfrac{\gamma_u+1}{\gamma_u-1}w_u - \Gamma\sqrt{\lambda}\left(\dfrac{1}{w_b}+\dfrac{\gamma_b+1}{\gamma_b-1}w_b\right)} \quad (11\text{-}25a)$$

Finally we obtain the following equation of the second degree in w_b:

$$\left(1+\frac{\gamma_b+1}{\gamma_b-1}\frac{\gamma_u-1}{\gamma_u+1}F\right)w_b^2 - \frac{1}{\Gamma\sqrt{\lambda}}\left[(1+F)w_u + \left(\frac{\gamma_u-1}{\gamma_u+1}F-1\right)\frac{1}{w_u}\right]w_b + \frac{\gamma_u-1}{\gamma_u+1}F - 1 = 0 \quad (11\text{-}26)$$

with $F(w_i, w_u)$ given by Eq. 5-16 with w_1 and p_1 replaced by w_i and p_i, w' by w_u and w^{*2} by $(\gamma_u - 1)/(\gamma_u + 1)$. Observe that, because of Eq. 11-23, the coefficients of this equation are known functions of w_i and w_u; therefore for prescribed values of γ_u, γ_b, λ, and w_i one can find from Eq. 11-26 w_b as a function of w_u. It can be shown that of the two solutions of Eq. 11-26 one has to choose the smaller, because the other gives a supersonic w_b.

Once $w_b(w_u)$ is known, $\mu(w_u)$ can be derived from Eq. 11-25, and by inverting this relationship one obtains $w_u(\mu)$, $w_b(\mu)$, and $p(\mu)$. Observe that the values of $w_f = w_b(1)$ and $p_f = p(1)$ coincide with the values calculated from Eq. 11-7a and 11-14 for the same $\gamma_i = \gamma_u$, $\gamma_f = \gamma_b$, λ, and w_i. Therefore w_f again becomes sonic for the lower value of w_i (Eq. 11-11a) and choking is produced. It is interesting that it can be shown that, for sonic w_f, $w_u(1)$ is generally supersonic, and that therefore $w_u(1)$ becomes sonic in the unchoked range.

So far only the relations between w_u, w_b, p, and μ have been obtained. The additional information required is the correspondence between μ and x, in order that to any given section one may be able to attribute the proper values of the various quantities. If the burning velocity S is known, and constant, and the duct has a rectangular section of width b and height h $(A = bh)$, the flow rate through an element of flame is

$$dm_b = -dm_u = \rho_u S b dx = m d\mu \tag{11-27}$$

where the axial extent dx of the element can be identified with its length. From this relation one obtains, for constant S,

$$\frac{S}{u_i}\frac{x}{h} = \int_0^\mu \left(\frac{1 - w_i^2}{1 - w_u^2}\right)^{\frac{1}{\gamma_u - 1}} d\mu$$

where the integral can be evaluated once $w_u(\mu)$ is known.

The height of the burned region and therefore the shape of the flame can be evaluated from the additional relation, easily obtained from the previous ones

$$\frac{y}{h} = \frac{A_b}{A} = 1 - \frac{A_u}{A} = 1 - (1 - \mu)\frac{w_i}{w_u}\left(\frac{1 - w_i^2}{1 - w_u^2}\right)^{\frac{1}{\gamma_u - 1}}$$

For a circular duct of diameter d, with a pointlike flame holder, this relation is replaced by

$$\frac{2r}{d} = \sqrt{\frac{A_b}{A}} = \left[1 - (1 - \mu)\frac{w_i}{w_u}\left(\frac{1 - w_i^2}{1 - w_u^2}\right)^{\frac{1}{\gamma_u - 1}}\right]^{\frac{1}{2}} \tag{11-28}$$

where r represents the radius of the burned region at a given section. Eq. 11-27 is replaced by the equation

$$m d\mu = \rho_u S \cdot 2\pi r dx$$

from which one obtains without difficulty

$$\frac{S}{u_i}\frac{x}{d} = \frac{1}{4}\int_0^\mu \sqrt{\frac{A}{A_b}}\left(\frac{1 - w_i^2}{1 - w_u^2}\right)^{\frac{1}{\gamma_u - 1}} d\mu$$

where the integral can be evaluated with the help of Eq. 11-28, when $w_u(\mu)$ is known.

As already observed, these calculations, based on the uniformization of the velocity distribution in the burned gases, are not as accurate as those of Tsien [*41*] and Fabri, Siestrunck, and Fouré [*42*]. However, the resulting pressure distribution and shape of the flame are in excellent coincidence with those calculated by the more accurate theories.

B,12. Flow with Heat Transfer and Friction. In the preceding article frictional effects have been completely disregarded. This is in general well justified for the cases involving combustion, the effects of the latter being of a larger order of magnitude. In the case of heat exchanges through the walls, however, this may be true only when the heat transfer is very intense, that is, when the temperature differences that produce the heat exchanges are large. In the opposite case the effects of friction and heat transfer can be of a comparable order of magnitude and the preceding treatment is not justified. A more refined treatment, based on the close relation existing between the mechanisms responsible for friction and heat exchanges, is developed in this article.

Reynolds' relation. Temperature recovery factor. The Reynolds analogy states that in a purely turbulent process the same mechanism is responsible for the transport of the energy and the momentum. With reference to the wall conditions the energy excess per unit mass of fluid is represented by the total stagnation enthalpy difference between the main flow and the fluid at the wall, $h^0 - h_w$. Similarly the momentum excess is u. The ratio between the heat supplied to the fluid and the friction is equal to the ratio between the energy and the momentum exchanged at the wall, and therefore to the ratio between the energy defect and the momentum excess per unit mass. Thus it can be expressed by

$$\frac{Q}{X} = \frac{h_w - h^0}{u} \tag{12-1}$$

This relation, first derived by Reynolds for low velocity flows, and therefore with the static rather than the stagnation enthalpy, does not apply exactly to actual flows. For instance, in the absence of chemical transformations, Eq. 12-1 predicts that the heat transferred must vanish when the stagnation temperature of the main flow is equal to the wall temperature. In reality, the adiabatic wall temperature, i.e. the temperature T_r on a well-insulated wall is somewhat lower than the main stagnation temperature, and one can define a "recovery factor" in terms of the corresponding enthalpy excesses on the static enthalpy of the main flow, or in terms of the relative temperature excesses. If the specific heats are assumed constant in the range involved, in agreement with the observation at the end of Art. 4, we have for the enthalpy or temperature recovery factor the same value

$$r = \frac{h_r - h}{h^0 - h} = \frac{T_r - T}{T^0 - T} \tag{12-2}$$

For constant pressure flow on a plate and well-developed turbulence, the value of r is around 0.89 for air, while for purely turbulent flows, it should be unity. The reason for this divergence is found in the fact that no actual flow on a wall can be purely turbulent, because the turbulent agitation must vanish at the wall, where the molecular transport of energy and momentum takes over. The relation between these two molecular processes is determined by the magnitude of the Prandtl number $Pr = c_p\mu/k$ (μ = viscosity coefficient, k = conductivity coefficient). Only for $Pr = 1$ would this relation be the same as for turbulent processes and would r be unity. For actual gases, Pr is close but not equal to unity, and $r < 1$ for $Pr < 1$, which is the most general case. The consequent change to Eq. 12-1 would consist in replacing $h_w - h^0$ by $h_w - h_r$. The fact that $Pr \neq 1$, however, also modifies somewhat the numerical relation between the first and the second member of Eq. 12-1. This can be written therefore

$$\frac{Q}{X} = k_q \frac{h_w - h_r}{u} = k_q \frac{c_p(T_w - T_r)}{u} \tag{12-1a}$$

where the factor k_q is only slightly different from unity, its value depending on Pr and other flow conditions (in general $k_q > 1$ because $Pr < 1$). Observe that even after this empirical modification, the Reynolds analogy must be applied with caution, especially when the pressure is not a constant, contrary to what is implicitly assumed in the derivation of Eq. 12-1.

Introducing the friction coefficient,[15] and assuming a circular section with diameter d, one has from Eq. 10-6

$$\frac{dX}{dx} = \frac{2c_f}{d} mu \tag{12-3}$$

and therefore, from Eq. 12-1a one has, for the same element of duct

$$q = \frac{dQ}{dx} = \frac{2c_f k_q}{d} mc_p(T_w - T_r) \tag{12-4}$$

Assuming that the temperature recovery factor is known, the temperature T_r can be expressed explicitly in terms of T^0 and w through Eq. 12-2:

$$T_r = T^0[1 - (1 - r)w^2] \tag{12-5}$$

Flow equations. For constant area ducts, the momentum equation can be written (Eq. 4-5b)

$$dJ = -dX \tag{12-6}$$

Assuming, for simplicity, c_p and $\mathcal{R}$ to be constant, from differentiation of Eq. 4-33a one obtains Eq. 12-6 in the form

$$m\sqrt{2c_pT^0}\left[\left(\frac{\gamma+1}{2\gamma} - \frac{\gamma-1}{2\gamma}\frac{1}{w^2}\right)dw + \left(\frac{\gamma+1}{2\gamma}w + \frac{\gamma-1}{2\gamma}\frac{1}{w}\right)\frac{dT^0}{2T^0}\right] = \frac{2c_f}{d} m\sqrt{2c_pT^0}\, w\,dx \tag{12-7}$$

where dX has been replaced from Eq. 12-3.

The energy equation is given by Eq. 4-4b as

$$dQ = dH^0 = mc_p dT^0 \tag{12-8}$$

[15] Observe that c_f depends not only on Re and M, but also on T_w and T, in a way that is not yet well established. In the following we shall assume that c_f is constant along the pipe.

and becomes, after introduction of Eq. 12-4,

$$\frac{dT^0}{T_w - T_r} = \frac{2c_f k_q}{d}\,dx \tag{12-9}$$

One can eliminate dx from Eq. 12-7 and 12-9, thus obtaining the following equation between w and T^0:

$$\left(\frac{\gamma+1}{2\gamma} - \frac{\gamma-1}{2\gamma}\frac{1}{w^2}\right)\frac{dw}{w} + \left[\left(\frac{\gamma+1}{2\gamma} + \frac{\gamma-1}{2\gamma}\frac{1}{w^2}\right)\frac{1}{2T^0} + \frac{1}{k_q(T_w - T_r)}\right]dT^0 = 0 \tag{12-10}$$

with T_r given by Eq. 12-5.

It is immediately checked that the variables are separable for T_w/T^0 = const. However, this case has no practical interest. More interesting cases, in practice, are the following two.

1. The rate at which heat is supplied to every portion of the pipe is known. This means that $Q(x)$ or $q(x)$ is assigned. Thus $T_w - T_r$ is also known from Eq. 12-4, and $T^0(x)$ can be obtained by integration from Eq. 12-8. Introducing this value of T^0 in Eq. 12-7 we obtain an equation between w and x, the integration of which must be performed numerically.

   Practical examples of this case are found in nuclear reactors, or electrically heated pipes. In the latter case the particularly simple distribution q = const is quite common. Observe that in this case $T_w - T_r$ is constant too, and the solution can be obtained through numerical integration of Eq. 12-10, which contains the variables w and T^0 alone. The connection with x is immediately obtained from Eq. 12-9 as a linear relationship.
2. The outside wall of the pipe exchanges heat with a medium of constant temperature T_e. Assuming that the outside temperature of the pipe is the same as the inside temperature, and assuming a known and constant external coefficient of heat transfer, q must be proportional to $T_e - T_w$, so that, comparing with Eq. 12-4, one has $T_e - T_w = K(T_w - T_r)$, K being a known constant. From this relation one obtains

$$T_w - T_r = \frac{T_e - T_r}{1 + K} \tag{12-11}$$

   Observe that the same relation holds, with a more complicated definition of K, when the temperature drop through the pipe wall is taken into account.

   For very small K, that is, for very large external heat transfer coefficient, T_w practically coincides with T_e and is therefore constant. Such may be the case, for instance, for a pipe immersed in condensing steam or in a molten metal bath. The opposite case of large K, and therefore T_w close to T_r, corresponds to a not-too-well-insulated pipe, or a pipe in a medium with poor conductibility, such as still air.

When Eq. 12-11 is introduced in Eq. 12-10, with T_e = const, this becomes an equation between w and T^0 alone which can be integrated numerically. Once $w(T^0)$ is known, $T_w - T_r$ can also be found from Eq. 12-11 as a function of T^0, and subsequent integration of Eq. 12-9 gives the relation with x.

Thus the exact solution for cases 1 and 2 can only be found through numerical integration. Examples of such integrations were carried out by Shapiro [*15*], in the two particular cases q = const and T_w = const. In the latter case T_r was replaced by T^0. This substitution, exact for $r = 1$, can be accepted as a good approximation when $T_w - T^0 \gg T^0 - T_r$.

Independently of the actual solution of the equations, one can investigate the combined effects of friction and heat transfer on the stagnation pressure [*15*] to see how the presence of friction modifies the result that a stagnation pressure increase can be obtained through a decrease of $\mathcal{R}T^0$ (Eq. 11-17), the point being that friction alone always produces a decrease of stagnation pressure (Eq. 7-22). Following a derivation similar to that of Eq. 7-21, with h^0 variable, one finds for constant c_p and $\mathcal{R}$, (recalling Eq. 4-15a)

$$\frac{\gamma}{\gamma - 1}\frac{dT^0}{T^0} - \frac{dp^0}{p^0} = \frac{ds}{\mathcal{R}} = \frac{dX}{Ap} + \frac{dh^0}{\mathcal{R}T}$$

that is, with $dh^0 = c_p dT^0$ and dX given by Eq. 3-5 and 7-20

$$\begin{aligned}\frac{dp^0}{p^0} &= -\frac{\gamma}{\gamma - 1}dT^0\left(\frac{1}{T} - \frac{1}{T^0}\right) - \frac{2c_f}{d}\gamma M^2 dx\\ &= -\frac{\gamma M^2}{2}\left(\frac{dT^0}{T^0} + \frac{4c_f}{d}dx\right)\end{aligned}$$

We can now use the Reynolds analogy and substitute dT^0 from Eq. 12-9, obtaining

$$\frac{dp^0}{p^0} = -\gamma M^2 \frac{k_q c_f}{d}dx\left[\frac{2}{k_q} + \frac{T_w}{T^0} - 1 + (1 - r)w^2\right]$$

where T_r has been replaced by Eq. 12-5. Since the value of k_q, even when >1, is always considerably smaller than 2, the quantity in brackets is always positive and dp^0 is always negative. Thus the presence of friction prevents the possibility of obtaining an increase in stagnation pressure through heat extraction.

In the rest of this article we shall discuss approximate analytical solutions of the problem, allowed when the wall temperature is in suitable ranges.

Case when friction is preponderant. The coefficient of dT^0 in Eq. 12-10 consists of two terms. It is clearly seen from their derivation that the first of them represents the effect of the heat transfer, and the second the effect of friction. When both terms are of comparable importance, the integration must be carried out by numerical procedures. If, however, one of the two terms is sufficiently larger than the other one, the latter can be neglected and an approximate analytical solution is easily found.

Let us first consider the case when frictional effects are preponderant. The following inequality must be satisfied

$$|T_w - T_r| \ll \frac{2w^2T^0}{k_q\left(\frac{\gamma - 1}{2\gamma} + \frac{\gamma + 1}{2\gamma}\, w^2\right)} \tag{12-12}$$

Obviously this inequality is true for the adiabatic case, $q = T_w - T_r = 0$, $T^0 = \text{const}$, which, in fact, has been treated in Art. 10. It is also true for case 2, when the constant value of T_e is, for instance, equal to the initial stagnation temperature T_i^0 (flow in a pipe from a tank at the ambient temperature), as seen from the following reasoning. The presence of friction makes w change, and with it T_r; however, T_r is initially smaller than T_e, and so it stays during the whole process. Thus $T_r < T_w$, $q > 0$, and T^0 increases continuously above its initial value $T_i^0 = T_e$. Thus we have $T_r < T_w < T_e < T^0$, and we can write

$$T_w - T_r < T^0 - T_r = (1 - r)w^2T^0 \tag{12-13}$$

Comparing the coefficients of w^2T^0 in Eq. 12-12 and 12-13 and remembering that $r \cong 0.89$ for turbulent flows, we see that the latter is always considerably smaller than the first, which, for $\gamma = 1.4$ and $k_q = 1$, takes the values 14 at $w = 0$, 7 at $w = w^*$, and 2 at $w = 1$. Thus, since k_q is always close to unity, the inequality of Eq. 12-12 is always satisfied, particularly well in the subsonic range.

Neglecting in Eq. 12-10 the first term in brackets, and replacing $T_w - T_r$ through Eq. 12-11 and T_r through Eq. 12-5, we easily obtain a first order, linear differential equation for T^0 which can be integrated explicitly, giving $T^0(w)$. Subsequently the integration of Eq. 12-9 provides $x(w)$.

Case when heat transfer is preponderant. This is true when the inequality

$$|T_w - T_r| \gg \frac{2w^2T^0}{k_q\left(\frac{\gamma - 1}{2\gamma} + \frac{\gamma + 1}{2\gamma}\, w^2\right)} \tag{12-14}$$

is satisfied. Replacing from Eq. 12-11 and adding to, or subtracting from, both members the quantity $T^0 - T_r$ since, as already observed, the last member of Eq. 12-13 is much smaller than the right-hand side of Eq. 12-14, this inequality can be converted into the following one

$$\frac{|\Delta T|}{T^0} = \frac{|T_e - T^0|}{T^0} \gg \frac{2w^2}{\frac{\gamma - 1}{2\gamma} + \frac{\gamma + 1}{2\gamma} w^2} \tag{12-14a}$$

where k_q and K have been supposed of order unity. For small velocities this inequality is easily satisfied for not-too-small values of $|\Delta T|$. With increasing w, however, it puts a more severe limitation on $|\Delta T|/T^0$, which, at the sonic velocity, must be much larger than $2\gamma/(\gamma + 1)$, and at the limiting velocity, $w = 1$, much larger than 2.

If the inequality of Eq. 12-14a is satisfied, one can neglect the friction term in Eq. 12-10, which then has the integral $J = \text{const}$, already discussed in this article for frictionless heat addition. However, if the friction term is not completely negligible, i.e. if the inequality of Eq. 12-14a is not satisfied in a decided way, it is possible to take friction into account by the following iterative process. Making use of Eq. 12-3 and 12-9, Eq. 12-6 can be written in integral form as

$$\Delta J = J - J_i = -\int_{x_i}^{x} dX = -\frac{m}{k_q}\int_{T_i^0}^{T^0} w\sqrt{2c_pT^0}\,\frac{dT^0}{T_w - T_r} \tag{12-15}$$

with J and w connected by Eq. 11-7 and 11-8 with $\varepsilon = 1$, that is

$$w(J) = \sqrt{\frac{\gamma - 1}{\gamma + 1}}\,(\sqrt{j} \pm \sqrt{j - 1}); \quad j = \frac{\gamma}{2(\gamma + 1)\mathcal{R}T^0}\frac{J^2}{m^2} \tag{12-16}$$

the upper sign corresponding to supersonic flow, the lower to subsonic flow. Neglecting friction one has in first approximation

$$J_{(1)} = J_i; \qquad w_{(1)} = w[J_{(1)}]$$

and

$$w_{(1)}\sqrt{2c_pT^0} = \sqrt{\frac{2\gamma}{\gamma + 1}\mathcal{R}}\,(\sqrt{a_1} \pm \sqrt{a_1 - T^0}); \quad a_1 = \frac{\gamma}{2(\gamma + 1)\mathcal{R}}\frac{J_i^2}{m^2} \tag{12-17}$$

with constant a_1. A second approximation for J can be calculated using Eq. 12-17 to compute the integral of Eq. 12-15. Explicitly one has the following results.

1. Case $q = \text{const}$. In this case $T_w - T_r$ is also constant and related to q by Eq. 12-4. One obtains

$$\Delta J_{(1)} = -\frac{2c_f m^2 c_p}{qd}\sqrt{\frac{2\gamma}{\gamma + 1}\mathcal{R}}\,\{\sqrt{a_1}\,(T^0 - T_i^0) \mp \tfrac{2}{3}[(a_1 - T^0)^{\frac{3}{2}} - (a_1 - T_i)^{\frac{3}{2}}]\}$$

the upper sign holding for supersonic flow, the lower for subsonic flow.

2. Case T_e = const. Replacing $T_w - T_r$ from Eq. 12-11, and T_r by T^0 (which is allowed because of the inequality of Eq. 12-14, since $T^0 - T_r$ is small compared to $T_e - T_r$), one obtains

$$\Delta J_{(1)} = \frac{(1+K)m}{k_q}\sqrt{\frac{2\gamma}{\gamma+1}}\,\mathcal{R}\left[\sqrt{a_1}\ln\frac{T_w - T^0}{T_w - T_i^0} \mp 2(\sqrt{a - T_i^0} - \sqrt{a - T^0}) \mp f(T^0)\right]$$

where the upper or the lower sign corresponds to supersonic or subsonic flows, and $f(T^0)$ is given by

$$2\sqrt{T_w - a_1}\left(\tan^{-1}\sqrt{\frac{a_1 - T^0}{T_w - a_1}} - \tan^{-1}\sqrt{\frac{a_1 - T_i^0}{T_w - a_1}}\right); \quad \text{for } T_w > a_1$$

$$\sqrt{a_1 - T_w}\left(\ln\frac{\sqrt{a_1 - T_w} + \sqrt{a_1 - T^0}}{\sqrt{a_1 - T_w} + \sqrt{a_1 - T_i^0}} - \ln\frac{\sqrt{a_1 - T_w} - \sqrt{a_1 - T^0}}{\sqrt{a_1 - T_w} - \sqrt{a_1 - T_i^0}}\right); \quad \text{for } T_w < a_1$$

and is zero for $T_w = a_1$. Observe that for small w_i, and therefore large J_i, both eventualities are possible.

Once $\Delta J_{(1)}$ is computed one has $J_{(2)} = J_i + \Delta J_{(1)}$, and replacing it in Eq. 12-16 one can obtain a second approximation $w_2 = w[J_{(2)}]$ for the velocity. In principle $w_{(2)}$ could be used again to calculate a third approximation for J and so on, but the corresponding steps cannot be carried out analytically.

Observe that, as Eq. 12-15 shows, ΔJ is always negative, so that $J_{(2)} < J_i$. Since $w = w^*$ for a fixed value of J^2/T^0 (for which $j = 1$), it results that choking occurs at a lower value of T^0 for the second approximation than for the first (frictionless) approximation. Here, however, it must be noticed that the convergence of the iteration procedure, which is likely to be good for w not too close to w^*, deteriorates rapidly when w^* is approached.

B,13. Effect of Variability of Specific Heats and Gas Imperfections. The developments of the preceding articles have been made with the help of the relations derived in Art. 4 for perfect polytropic gases. At the end of Art. 4 we have indicated that the same relations can be used approximately for perfect nonpolytropic gases when the variations of specific heats in the range between the stagnation and the static temperatures can be disregarded. However, this is not always the case, especially for supersonic flows. Also the gas imperfections may in certain cases have noticeable effects.

In this article we shall leave aside the simple algebraic expressions available for perfect polytropic gases, and consider first the effects of the variability of the specific heats for a perfect gas, and then the effects of the deviations from the perfect gas law for extreme pressure and temperature variations. We shall also briefly discuss the case of two-phase flows, in particular the case of liquid-gas mixtures which may also be considered as a fluid not obeying the perfect gas law. However, let us first review a few general relations and properties, which can be formulated for the most general substance.

Sonic conditions in isoenergetic, isentropic flows. From Eq. 4-2b one has

$$\frac{d\rho}{\rho} + \frac{du}{u} + \frac{dA}{A} = 0 \tag{13-1}$$

On the other hand for an isoenergetic, isentropic frictionless flow one can write the Bernoulli equation (Eq. 4-8) in the following form

$$\frac{d\rho}{\rho} = \frac{1}{a^2}\frac{dp}{\rho} = -M^2\frac{du}{u} \tag{13-2}$$

where use has been made of the definition (Eq. 2-16) of the sound velocity. Combining Eq. 13-1 and 13-2 one obtains

$$(1 - M^2)\frac{dp}{\rho} = u^2\frac{dA}{A}$$

For $dA = 0$ we have either $dp = 0$ or $M = 1$. In other words at a throat we must reach a minimum or maximum of the pressure, except when the sound velocity is established in the throat, in which case the pressure gradient can be different from zero. We see also that

$$\frac{dp}{dA} \gtrless 0 \quad \text{for} \quad M \lessgtr 1$$

that is, a diverging duct acts as a diffuser for subsonic velocities and as an effuser for supersonic velocities; and vice versa for a converging duct.

Thus for physical transformations without state changes of the most general substance, in the absence of heat exchanges or friction, we find the same qualitative results already found for perfect polytropic gases. Obviously the establishment of the sonic conditions in the throat produces the same phenomena of choking discussed in Art. 6.

Adiabatic constant area flow and Fanno line. In this type of flow the changes in the flow characteristics are produced only by the presence of dissipative forces, such as friction. Since

$$h + \frac{u^2}{2} = h^0 = \text{const}, \quad \rho u = \frac{m}{A} = \text{const} \tag{13-3}$$

we see that, for assigned values of the constants, to each value of u correspond well-determined values of h and ρ. Thus if h is given as $h(\rho, T)$, we can determine T and all the other thermodynamic quantities which can also be considered functions of ρ and T.

For instance if the Mollier diagram of the substance (in frozen or equilibrium conditions) is given, for each value of u the intersection between the corresponding ρ = const and h = const lines gives the position of the representative point in the h-s diagram. Varying u one obtains a curve which is called the Fanno line.

Some interesting general properties of the Fanno line can be obtained in the following way. From differentiation of Eq. 13-3 one obtains

$$dh + udu = 0, \quad \frac{d\rho}{\rho} + \frac{du}{u} = 0$$

Eliminating du one has

$$\frac{d\rho}{\rho} = \frac{dh}{u^2} \tag{13-4}$$

On the other hand, considering p a function of ρ and s and using the symbolism of I,A, we have

$$dp = \left(\frac{\partial p}{\partial \rho}\right)_s d\rho + \left(\frac{\partial p}{\partial s}\right)_\rho ds \tag{13-5}$$

Using methods similar to those of I,A,13 one finds

$$\left(\frac{\partial p}{\partial s}\right)_\rho = \rho^2 \left(\frac{\partial T}{\partial \rho}\right)_s \tag{13-6}$$

Also, $(\partial p/\partial \rho)_s$ is by definition the square of the sound velocity a^2. Therefore, making use of Eq. 13-4, Eq. 13-5 becomes

$$dp = \rho \frac{a^2}{u^2} dh + \rho^2 \left(\frac{\partial T}{\partial \rho}\right)_s ds$$

Replacing in

$$Tds = dh - \frac{dp}{\rho}$$

we obtain

$$\left(\frac{dh}{ds}\right)_{\text{Fanno}} = \frac{M^2}{1 - M^2} T \left[1 + \left(\frac{\partial \ln T}{\partial \ln \rho}\right)_s\right] \tag{13-7}$$

For all plausible substances, T varies in the same direction as ρ in an isentropic change. Thus the quantity in brackets is certainly positive (it is γ for a perfect gas). We conclude that $(dh/ds)_{\text{Fanno}}$ is positive for subsonic, negative for supersonic velocities, and infinite at the sonic velocity. The resulting shape of the Fanno line is shown schematically in Fig. B,13a.

Frictionless constant area flow and Rayleigh line. This type is representative of flows in which the conditions change as a result of heat exchanges. Since in reality friction is always present, it must be considered as an ideal limit which may be actually approached when friction is small. Also, the substance considered here should be in frozen or equilibrium conditions. Therefore the application of this line to the discussion of problems where the heat is generated by combustion processes (which are essentially irreversible) is not rigorously correct; it can, however, be useful when the chemical changes are disregarded, as possible especially for diluted mixtures.

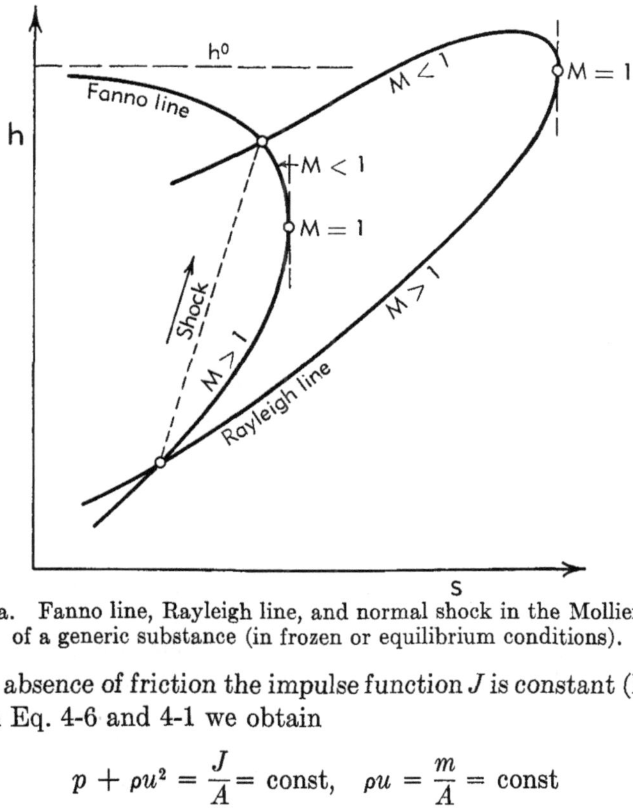

Fig. B,13a. Fanno line, Rayleigh line, and normal shock in the Mollier diagram of a generic substance (in frozen or equilibrium conditions).

In the absence of friction the impulse function J is constant (Eq. 4-18); thus from Eq. 4-6 and 4-1 we obtain

$$p + \rho u^2 = \frac{J}{A} = \text{const}, \quad \rho u = \frac{m}{A} = \text{const} \tag{13-8}$$

For assigned values of the constants and for each value of u, from Eq. 13-8 one can calculate ρ and p, so that the state of the substance (in frozen or equilibrium condition) is determined, and the values of T, h, and s can therefore be obtained. Thus to each value of u corresponds a point of the Mollier diagram. The locus of the points for varying u represents the Rayleigh line.

For the Rayleigh line we can also derive some general properties. From differentiation of the two Eq. 13-8 we have

$$dp + \rho u du = 0, \quad \frac{d\rho}{\rho} + \frac{du}{u} = 0$$

from which du can be eliminated with the result

$$dp = u^2 d\rho \tag{13-9}$$

On the other hand we can rewrite Eq. 13-5 in the form

$$dp = a^2 d\rho + \rho^2 \left(\frac{dT}{d\rho}\right)_s ds \tag{13-10}$$

Elimination of $d\rho$ from Eq. 13-9 and 13-10 produces the relation

$$\frac{M^2 - 1}{M^2} dp = \rho^2 \left(\frac{\partial T}{\partial \rho}\right)_s ds$$

which, introduced in

$$Tds = dh - \frac{dp}{\rho}$$

gives

$$\left(\frac{dh}{ds}\right)_{\text{Rayleigh}} = \frac{T}{M^2 - 1} \left\{\left[1 + \left(\frac{\partial \ln T}{\partial \ln \rho}\right)_s\right] M^2 - 1\right\} \tag{13-11}$$

Since the quantity in brackets is larger than unity, for M^2 sufficiently close to unity $(dh/ds)_{\text{Rayleigh}}$ has the same sign as $M^2 - 1$ and is, therefore, positive or negative for $M \gtrless 1$, and infinite for $M = 1$. Moreover, it vanishes for a value of $M < 1$, below which it is again positive. The resulting shape of the Rayleigh line is thus the one shown in Fig. B,13a. The Rayleigh line is intersected by the Fanno line at two points, where the Fanno line is always steeper than the Rayleigh line, as clearly shown by the comparison of Eq. 13-11 and 13-7.

Normal shock. Since the conservation equations for a normal shock are (Eq. 13-3 and 13-8), the final point must lie on both the Fanno and Rayleigh lines passing through the initial point and is therefore represented by their other intersection. This does not mean that the representative point of the continuous transition from the initial to the final condition follows either one of the two lines, since within the shock both dissipative forces and internal heat exchanges are present.

Examples of the utilization of the Fanno and Rayleigh lines to determine the results of high Mach number shocks in air are given in Fig. B,13b using the Mollier diagram of Fig. B,2t. Three initial pressures have been considered ($p_1 = 10^{-1}$, 10^{-3}, and 10^{-5} atm), and one initial temperature ($T_1 = 216.5°K$). For the lowest Mach number, $M_1 = 10$, both the Rayleigh and the Fanno lines have been drawn completely between the two intersections corresponding to the initial and the final points. At higher Mach numbers only the parts of the two lines close to the second intersection have been indicated. Observe that the Rayleigh line, if drawn completely, would extend beyond the limits of the Mollier diagram in its higher part. It is interesting to observe that the final points are well beyond the limits where the dissociations can be neglected. Actually for the highest M_1 and the lowest p_1 we are not far from the conditions where the content of N_2 and O_2 in the air becomes negligibly small. For even higher M_1 and not too high p_1 simpler and sufficiently accurate calculations could be based on the assumption that only atomic nitrogen and oxygen are present after the shock.

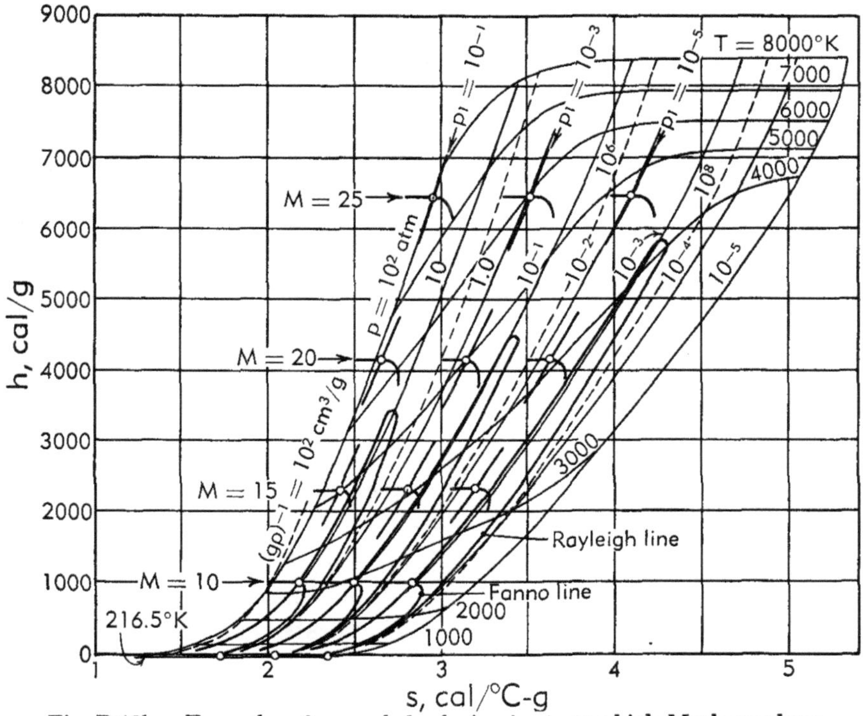

Fig. B,13b. Examples of normal shocks in air at very high Mach numbers.

We observe that these calculations based on the achievement of thermochemical equilibrium in the final gases are subject to caution when the processes are not of a one-dimensional nature and the flow following the shock is not indefinitely uniform. The reason for this is that the dissociations of the molecular species are only reached after a sufficient number of collisions, with the consequence that the shock thickness necessary to achieve the assumed thermochemical equilibrium may be considerably larger than in the absence of dissociations, and may become comparable with the radius of curvature of the shock or other characteristic dimensions of the problem. Calculations on "relaxation" effects of this kind can only be performed by taking into account the chemical kinetics of the processes.

The main results of Fig. B,13b are reproduced in Table B,13a, as obtained from large scale plots. For comparison we give the values of the temperature, pressure, and density ratios calculated for a perfect, polytropic reference gas with $\gamma = 1.4$. They are indicated with a subscript $_0$ on the left of the final quantity. It is obvious that for such high Mach numbers the polytropic approximation becomes very poor. This is due essentially to the high temperature level reached, which enhances the chemical dissociations. The approximation would be better for lower T_1 or higher p_1 (as happens in wind tunnels); in which case for $M_1 = 10$ it may be possible to neglect the dissociations completely, and only take into account the variability of the specific heats, as will be done in the following article.

Table B,13a

M_1	p_1,atm	T_2,°K	T_2/T_1	$_0T_2/T_1$	p_2/p_1	$_0p_2/p_1$	ρ_2/ρ_1	$_0\rho_2/\rho_1$
	10^{-1}	3335	15.40		121.9		7.78	
10	10^{-3}	2925	13.51	20.4	125.3	116.5	8.79	5.71
	10^{-5}	2530	11.69		130.0		10.26	
	10^{-1}	4980	23.00		281.8		10.45	
15	10^{-3}	4095	18.91	44.7	287.1	262.3	12.16	5.87
	10^{-5}	3450	15.93		291.7		14.26	
	10^{-1}	6380	29.47		537.1		13 03	
20	10^{-3}	5015	23.16	78.8	539.5	466.5	15.52	5.93
	10^{-5}	4055	18.73		547.0		18.75	
	10^{-1}	7780	35.93		831.8		13.68	
25	10^{-3}	5905	27.27	122.6	839.5	729.	16.90	5.95
	10^{-5}	4730	21.85		847.2		20.51	

Efficiencies. In handling the problem of flow in real ducts it is useful to make use of the concept of efficiency, following the same lines we did for perfect polytropic gases. We shall discuss here the definitions of both the adiabatic and the elemental efficiency for the most general gas, and then specify the definition for the important case of perfect nonpolytropic gas. For η_{ad}, the definitions are still Eq. 7-8 or the first Eq. 7-9 for nozzles, and Eq. 9-1 for diffusers. Therefore we can write

$$\eta_{ad} = \frac{u_f^2}{2[h_i^0 - (h_f)_{is}]} = \frac{h_i^0 - h_f}{h_i^0 - (h_f)_{is}} = \frac{h(s_i, p_i^0) - h(s_f, p_f)}{h(s_i, p_i^0) - h(s_i, p_f)} \tag{13-12}$$

in the case of nozzles and

$$\eta_{ad} = \frac{2[(h_f^0)_{is} - h_i]}{u_i^2} = \frac{(h_f^0)_{is} - h_i}{h_f^0 - h_i} = \frac{h(s_i, p_f^0) - h(s_i, p_i)}{h(s_f, p_f^0) - h(s_i, p_i)} \tag{13-13}$$

for diffusers.

The graphical interpretation of Eq. 13-12 is obviously (Fig. B,13c):

$$\eta_{ad} = \frac{A^0B_1}{A^0C} \tag{13-12a}$$

with the line AB representing the actual transformation during the expansion process. If the initial conditions and the final pressure are assigned one can determine the points A^0 and C on the diagram, and, for known η_{ad}, obtain the point B_1 from Eq. 13-12a, after which B is found at the intersection with the line of p_f and the horizontal line through B_1.

Similarly the graphical interpretation of Eq. 13-13 is given on Fig. B,13d by

$$\eta_{ad} = \frac{AC}{AB_1}$$

and the determination of B is immediate when η_{ad} and u_f are known.

For the elemental efficiency let us assume momentarily, to simplify the matter, that the initial kinetic energy in an expansion process can be disregarded, so that the points A and A^0 coincide (Fig. B,13c). For an intermediate point P of the transformation, and for an elementary expansion from p to $p - dp$, let us define the elemental efficiency η_{el} by the equation corresponding to Eq. 13-12

$$\eta_{el} = \frac{dh}{dh_{is}} \tag{13-14}$$

Now from Eq. 2-12 and from the definition of dh_{is} we have

$$dh_{is} - \frac{dp}{\rho} = dh_{is} - dh + Tds = (1 - \eta_{el})dh_{is} + Tds = 0 \tag{13-15}$$

so that we obtain

$$ds = -(1 - \eta_{el})\frac{dp}{\rho T}$$

Thus, assuming that η_{el} keeps a constant value during the whole transformation we find

$$1 - \eta_{el} = \frac{s_f - s_i}{\displaystyle\int_{p_f}^{p_i} \frac{dp}{\rho T}} \tag{13-16}$$

Similarly for a process of compression we define

$$\eta_{el} = \frac{dh_{is}}{dh} \tag{13-17}$$

and obtain

$$\frac{1}{\eta_{el}} - 1 = \frac{s_f - s_i}{\displaystyle\int_{p_i}^{p_f} \frac{dp}{\rho T}} \tag{13-18}$$

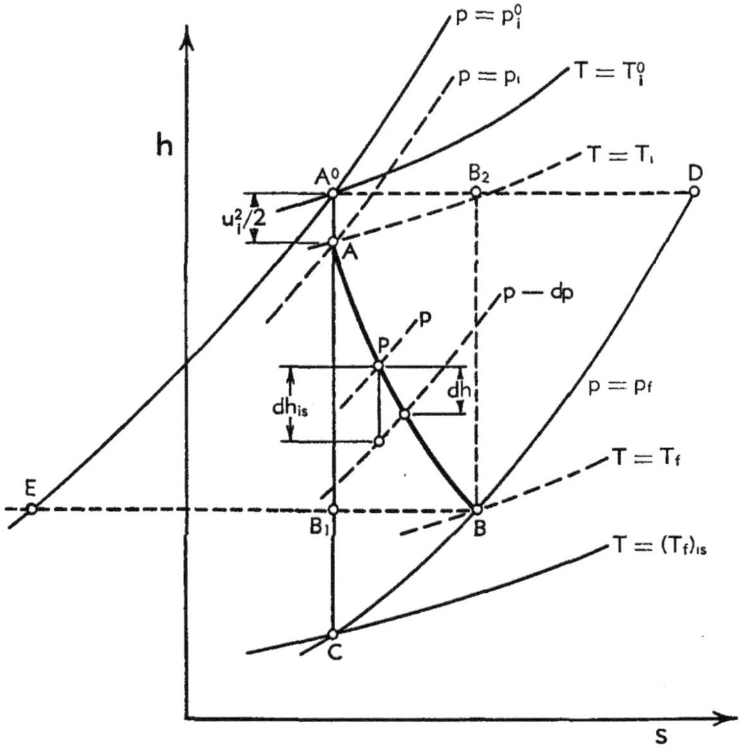

Fig. B,13c. Representation of the expansion of a generic substance in a nozzle, illustrating the graphical definition of the efficiencies.

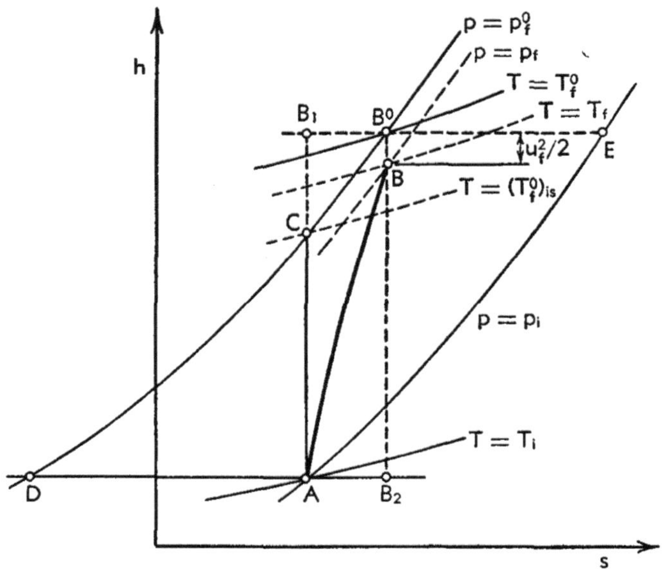

Fig. B,13d. Representation of the compression of a generic substance in a diffuser, illustrating the graphical definition of the efficiencies.

Obviously if η_{el} is not a constant during the transformation, one can still make use of Eq. 13-16 and 13-18 to define an average value of it, which we shall call the elemental efficiency of a finite transformation.

The importance of this definition of the efficiency is immediately realized if one applies it to a series of N stages of, say, an expansion process, all characterized by the same value of η_{el}. We can then write for each stage

$$(1 - \eta_{el}) \int_{(p_f)_r}^{(p_i)_r} \frac{dp}{\rho T} = (s_f)_r - (s_i)_r \qquad (r = 1, 2, \ldots, N)$$

Adding these N equations together, and observing that $(p_i)_r = (p_f)_{r-1}$ and $(s_i)_r = (s_f)_{r-1}$, we obtain the relation

$$(1 - \eta_{el}) \int_{(p_f)_N}^{(p_i)_1} \frac{dp}{\rho T} = (s_f)_N - (s_i)_1$$

which shows that the elemental efficiency of the N stages considered as a single unit is identical with that of the individual stages and is therefore fit to represent the quality of the transformation independently of its extent. The result is obviously also true for compression processes, but it is not true, as easily recognized, for η_{ad}. In fact it can be shown that a combination of stages having all the same η_{ad} has an η_{ad} larger than that of the individual stages for an expansion process, and smaller for a compression process. The definitions of the elemental efficiency (Eq. 13-16 or 13-18) have been expressly derived for the case when the initial kinetic energy of the expansion process or the final kinetic energy of the compression process is negligibly small. If this is not the case we can obviously modify the definitions by replacing p_i by p_i^0 in Eq. 13-16 and p_f by p_f^0 in Eq. 13-18. For the sake of symmetry, one can also replace s_i by $s_i^0 = s_i$ in Eq. 13-16 and s_f by $s_f^0 = s_f$ in Eq. 13-18.[16]

A disadvantage of these definitions of the elemental efficiency is that, for a general gas, they do not depend exclusively on the initial and final conditions as, on the contrary, do Eq. 13-12 and 13-13. This is an inconvenience because in most instances the actual path of the transformation is not known and one wants to calculate the final conditions from the assigned efficiency and initial conditions or, conversely, determine the efficiency from the experimentally determined conditions before and after the transformation.

[16] These modifications apply to devices, such as nozzles and diffusers, directed to the production or utilization of kinetic energy. For devices such as turbines and compressors where, instead of kinetic energy, the object is mechanical work, the stagnation conditions must replace the static conditions at both the initial and the final point. This is also the usual practice for the definitions of η_{ad}.

This inconvenience, however, disappears for a gas following the perfect gas law, no matter how the specific heats vary. In this case Eq. 13-16, corrected for the initial kinetic energy, becomes

$$1 - \eta_{el} = \frac{s_f - s_i^0}{\mathcal{R} \ln \dfrac{p_i^0}{p_f}} \tag{13-16a}$$

or, splitting the entropy into its temperature- and pressure-dependent terms following Eq. 2-13,

$$\eta_{el} = \frac{\sigma_i^0 - \sigma_f}{\mathcal{R} \ln \dfrac{p_i^0}{p_f}} = \frac{\sigma_i^0 - \sigma_f}{\sigma_i^0 - (\sigma_f)_{is}} \tag{13-16b}$$

where σ is a function of the temperature alone given by Eq. 2-12 and, from Eq. 13-15, one has replaced

$$\mathcal{R} \ln \frac{p_i^0}{p_f} = \int_{(T_f)_{is}}^{T_i^0} \frac{dh_{is}}{T} = \int_{(T_f)_{is}}^{T_i^0} d\sigma_{is} = \sigma_i^0 - (\sigma_f)_{is} \tag{13-19}$$

It is interesting to observe the analogy between the definition (Eq. 13-16b) of η_{el} and the definition (Eq. 13-12) of η_{ad}, the only difference being the substitution of $\sigma(T)$ to $h(T)$.

Similarly for a compression process one obtains from Eq. 13-18, taking into account the final kinetic energy,

$$\frac{1}{\eta_{el}} - 1 = \frac{s_f^0 - s_i}{\mathcal{R} \ln \dfrac{p_f^0}{p_i}}; \quad \eta_{el} = \frac{\mathcal{R} \ln \dfrac{p_f^0}{p_i}}{\sigma_f^0 - \sigma_i} = \frac{(\sigma_f^0)_{is} - \sigma_i}{\sigma_f^0 - \sigma_i} \tag{13-18a}$$

The analogy with the definition (Eq. 13-13) of η_{ad} is obvious. Clearly, for a polytropic gas, Eq. 13-16b and 13-18a coincide with the definitions (Eq. 7-10 and 9-2) used previously. Eq. 13-16b has a simple graphical interpretation which is immediately obtained by rewriting it in the equivalent forms

$$\eta_{el} = \frac{s(h_i^0, p_i^0) - s(h_f, p_i^0)}{s(h_f, p_f) - s(h_f, p_i^0)} = \frac{s(h_i^0, p_f) - s(h_f, p_f)}{s(h_i^0, p_f) - s(h_i^0, p_i^0)} \tag{13-20}$$

Thus in Fig. B,13c we have for expansion processes

$$\eta_{el} = \frac{EB_1}{EB} = \frac{B_2D}{A^0D} \tag{13-20a}$$

the two ratios being identical because, for a perfect gas, $EB = A^0D$.

Similarly for a compression process, we obtain from Eq. 13-18a

$$\eta_{el} = \frac{s(h_f^0, p_i) - s(h_f^0, p_f^0)}{s(h_f^0, p_i) - s(h_i, p_i)} = \frac{s(h_i, p_i) - s(h_i, p_f^0)}{s(h_f^0, p_f^0) - s(h_i, p_f^0)} \qquad (13\text{-}21)$$

These relations are interpreted in Fig. B,13d as

$$\eta_{el} = \frac{B^0E}{B_1E} = \frac{DA}{DB_2} \qquad (13\text{-}21a)$$

For a perfect gas the expressions of Eq. 13-20 are both identical with Eq. 13-16, and Eq. 13-21 with Eq. 13-18. For a real gas, however, each one of these expressions corresponds to a different definition of η_{el}, because, as easily checked for the first Eq. 13-20, this is obtained from Eq. 13-16 by performing the integration at the denominator along the path EB instead of the actual path of the transformation; only for perfect gases (or, more generally, if ρT is a function of p alone) do the two paths give the same result. Similarly, the second definition (Eq. 13-20) corresponds to the path of integration A^0D, and therefore gives different results from the first and from the correct definition of Eq. 13-16. The same is true for compression processes. Nevertheless it is suggested that the use of either Eq. 13-20 or 13-21, and the corresponding graphical constructions (Eq. 13-20a and 13-21a), may provide a better indication of the "quality" of the transformation than that indicated by η_{ad}.

Flow of perfect nonpolytropic gases. If the gas obeys the perfect gas equation, transformations with known adiabatic or elemental efficiency can be easily calculated with only the help of the functions $h(T)$ and $\sigma(T)$ in graphical or tabular form. Suppose, for instance, that one has to compute the final results of the adiabatic expansion in a nozzle from the initial conditions p_i^0, T_i^0 to the final pressure p_f. Since h_i^0 and σ_i^0 corresponding to T_i^0 are known, one can, if η_{ad} is assigned, calculate $(\sigma_f)_{is}$ from Eq. 13-19, determine the corresponding values of $(T_f)_{is}$ and $(h_f)_{is}$, and derive from Eq. 13-12 the value of h_f. Then T_f can be determined, ρ_f can be derived from the equation of state, and u_f from the energy equation (Eq. 4-9). Similarly, if η_{el} is given, one can, from Eq. 13-16b, calculate σ_f and the corresponding values of T_f and h_f from which ρ_f and u_f can be obtained.

The complementary problem, in which the value of u_f to be obtained is assigned and p_f has to be determined, can be treated in a similar way. If η_{ad} is given, h_f can be computed from the energy equation, the corresponding T_f determined, the value of $(h_f)_{is}$ obtained from Eq. 13-12, and the pressure p_f calculated using in Eq. 13-19 the corresponding value of $(\sigma_f)_{is}$. If η_{el} is known, p_f can be obtained introducing in Eq. 13-16b the value of σ_f corresponding to h_f. The preceding methods can be applied, the only change being the elimination of the subscript $_f$, to the calculation of the flow conditions at sections other than the final section, provided the distribution of η_{ad} or η_{el} along the duct is supposed to be known.

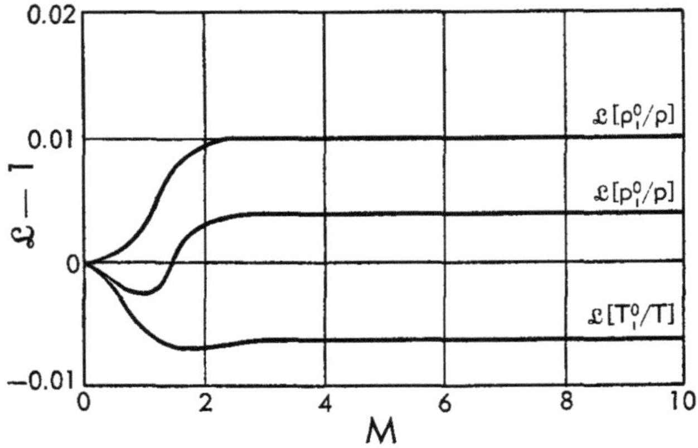

Fig. B,13e. Correction factors for isentropic expansions of air with respect to the polytropic case with $\gamma = 1.4$. $T_i^0 = 1000°R$.

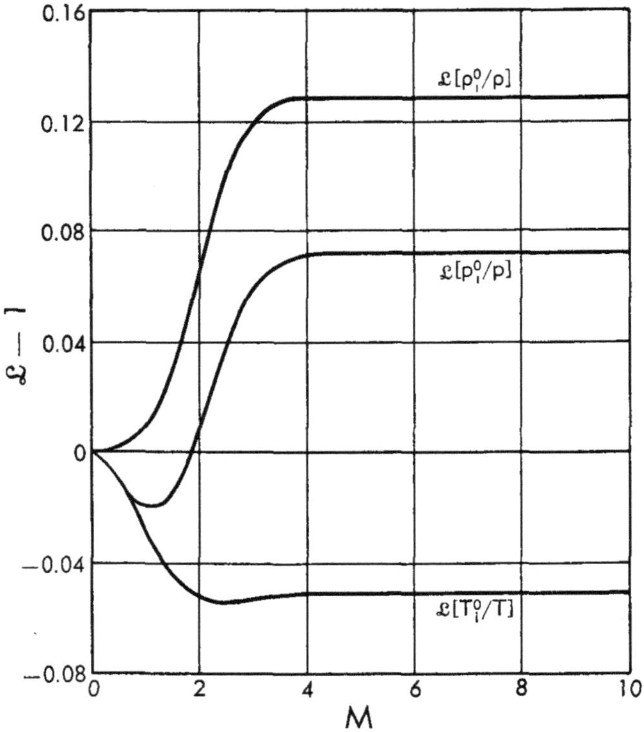

Fig. B,13f. Correction factors for isentropic expansions of air with respect to the polytropic case with $\gamma = 1.4$. $T_i^0 = 2000°R$.

Calculations of this type have been made by several authors. Here we shall only report in Fig. B,13e and B,13f the results of Donaldson [4] for the isentropic expansions of air ($\eta_{ad} = \eta_{el} = 1$) from two different initial temperatures T_i^0 (the pressure level is unessential since the gas is considered perfect). The results are expressed for better accuracy in the form of correction factors to be applied to the relations giving the various quantities for a polytropic gas with $\gamma = 1.4$, and are plotted in function of $M = M_i$ obtained from u_i dividing by the sound velocity relative to $T = T_i$. For instance one can write[17]

$$\frac{T_i^0}{T} = \mathcal{L}\left[\frac{T_i^0}{T}\right](1 + 0.2M^2); \quad \frac{p_i^0}{p} = \mathcal{L}\left[\frac{p_i^0}{p}\right](1 + 0.2M^2)^{3.5};$$

$$\frac{\rho_i^0}{\rho} = \mathcal{L}\left[\frac{\rho_i^0}{\rho}\right](1 + 0.2M^2)^{2.5}$$

It is interesting to observe that all $\mathcal{L}$ curves level off beyond a certain value (depending on T_i^0) of M. The reason for this behavior is found in the fact that below ambient temperature the air is actually polytropic with $\gamma = 1.4$. The correction factors vary only when T is in the range where c_p is variable, and stay constant below this range. The higher the initial temperature, the wider becomes the range of variability and the larger become the departures of the correction factors from unity. This is also clear from the figures, which show that the polytropic approximation can be used with confidence at all M for $T_i^0 \leqq 1000°\text{R}$, the errors being smaller than 1 per cent. However, the errors involved may become larger than 10 per cent for $T_i^0 \geqq 2000°\text{R}$ (except for small M).

The flow in diffusers can be handled in a very similar manner through the use of the η_{ad} or η_{el} defined for the case of compression. Here we shall mention only the results of the calculations for isentropic flows, which have an important application to the determination of the stagnation conditions at high Mach number $M = M_i$. The results of Fig. B,13g are reproduced from [*43*,*44*], both pertaining approximately to the same initial temperature $T = T_i$ (500°R for [*43*]; 520°R for [*44*]), and are given, again, in the form of correction factors to be applied to the various quantities calculated from the polytropic relations.

The errors are here considerably larger than in the expansion case, because the temperature range is displaced to higher values where the approximation $\gamma = 1.4$ becomes worse. Observe that the temperature attained after the compression process in the range of M envisaged here, though high, is still below the level where the dissociations of oxygen and nitrogen become important, and the results can be used with confidence.

Another interesting example of a compression process is the compression through a normal shock wave. This case can be treated starting from the same equations (Eq. 5-1) used in the case of a polytropic gas. They can be rewritten here in the form

$$\rho_2 u_2 = \rho_1 u_1, \quad p_2 + \rho_2 u_2^2 = p_1 + \rho_1 u_1^2, \quad h_2 + \frac{u_2^2}{2} = h_1 + \frac{u_1^2}{2} = h^0 \qquad (13\text{-}22)$$

[17] These results are presented in a slightly different form from the reference, where, for instance, $1 - \mathcal{L}[T_i^0/T]$ is considered as a fractional error of the constant c_p expression of T/T_i^0 with respect to the exact value.

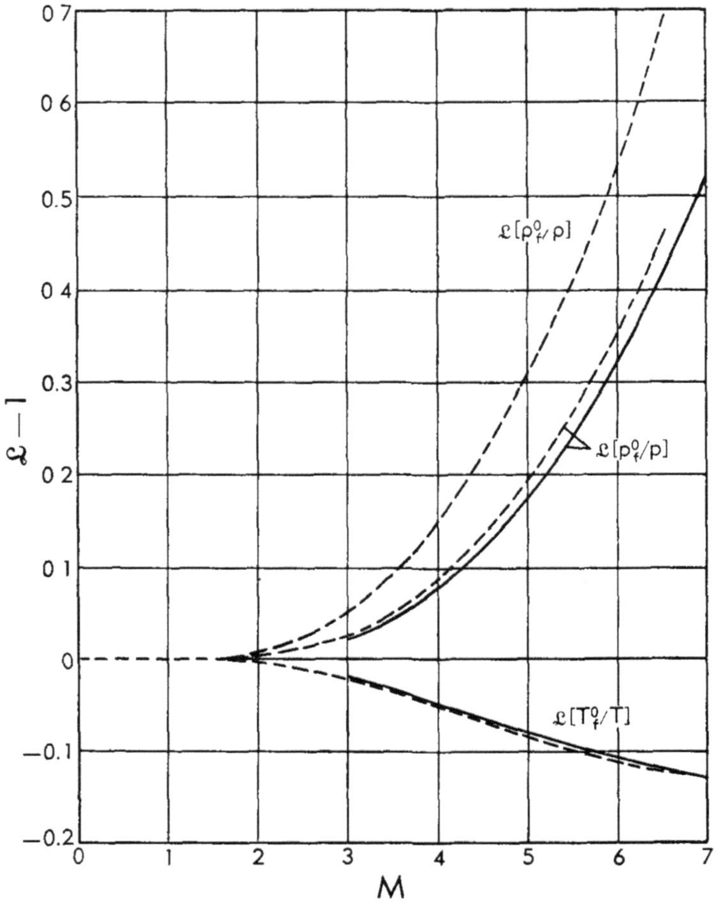

Fig. B,13g. Correction factors for isentropic compressions of air with respect to the polytropic case with $\gamma = 1.4$. ——— from [43] with $T_i = 500°R$; -------- from [44] with $T_i = 520°R$.

Replacing p_1 and p_2 (from the equation of state) in the second equation and making use of the first equation, we obtain

$$u_2 + \frac{\mathfrak{R}T_2}{u_2} = u_1 + \frac{\mathfrak{R}T_1}{u_1} \tag{13-23}$$

Substituting here the value of u_2 obtained from the third equation (Eq. 13-22), we obtain

$$\sqrt{2(h^0 - h_2)} + \frac{\mathfrak{R}T_2}{\sqrt{2(h^0 - h_2)}} = u_1 + \frac{\mathfrak{R}T_1}{u_1}$$

which for given values of u_1 and T_1, and therefore of h^0, represents an equation for T_2, the solution of which can be obtained using the graphs or the tables of $h(T)$. Once T_2 has been determined, u_2 can be obtained

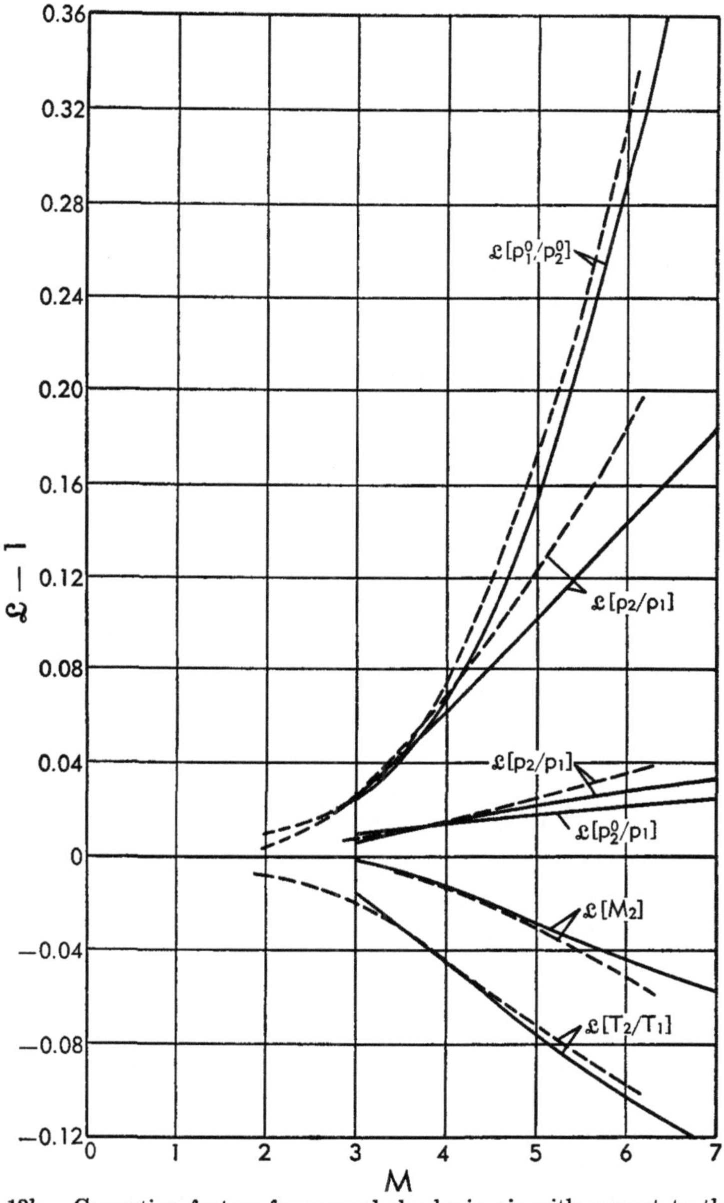

Fig. B,13h. Correction factors for normal shocks in air with respect to the polytropic case with $\gamma = 1.4$. ——— from [*43*] with $T_1 = 500°\text{R}$; -------- from [*44*] with $T_1 = 520°\text{R}$.

from Eq. 13-23, and ρ_2, p_2 from Eq. 13-22 and the assigned values of ρ_1, p_1. The results of the calculations of [*43,44*] are shown in Fig. B,13h in the form of correction factors to be applied to the values of M_2, p_2/p_1, and ρ_2/ρ_1 given by Eq. 5-4a and 5-5a and to the values of the other quantities that can be obtained from these, with $\gamma = 1.4$. Again we see that very important corrections are required at the higher M, except for the values of p_2/p_1 and p_2^0/p_1, which depart from those based on the polytropic approximation only by a few per cent at the most. These results apply to an initial temperature T_1 of 500°R or 520°R. The corrections would be smaller for lower T_1 and larger for higher T_1.

Approximate relations for expansion processes. The approximate thermodynamic relations developed in Art. 2 can be used for the calculation of flows in expansion. The resulting accuracy may be sufficient for many engineering purposes when the flowing gas is air or the products of combustion of hydrocarbon fuels with air, and when T_i^0 is in the proper ranges discussed in Art. 2.

Two kinds of problems can be treated, each of them for assigned values of p_i^0, T_i^0 and of η_{ad} or η_{el}.

Problem 1. Calculate the values of p_f necessary to obtain a given velocity u_f, and the corresponding value of T_f. From these, ρ_f can also be obtained. Using Eq. 13-12 and 13-16b and the approximate relations of Eq. 2-33 and 2-37 one obtains the following explicit expressions for p_f and T_f:

$$\left.\begin{aligned}\frac{\gamma_i^0-1}{\gamma_i^0}\ln\frac{p_f}{p_i^0} &= \frac{(\sigma_f)_{is}-\sigma_i^0}{c_{p_i}^0} = \mathfrak{D}\left(\frac{u_f^2}{2\eta_{ad}c_p^0 T_i^0}\right)\ln\left(1-\frac{u_f^2}{2\eta_{ad}c_p^0 T_i^0}\right)\\ &= \frac{\sigma_f-\sigma_i^0}{\eta_{el}c_{p_i}^0} = \frac{1}{\eta_{el}}\mathfrak{D}\left(\frac{u_f^2}{2c_{p_i}^0 T_i^0}\right)\ln\left(1-\frac{u_f^2}{2c_{p_i}^0 T_i^0}\right)\end{aligned}\right\} \tag{13-24}$$

$$\frac{T_f}{T_i^0} = \mathfrak{B}\left(\frac{u_f^2}{2c_{p_i}^0 T_i^0}\right)\left(1-\frac{u_f^2}{2c_{p_i}^0 T_i^0}\right) \tag{13-25}$$

where $c_{p_i}^0$ and γ_i^0 are the values corresponding to T_i^0, and $\mathfrak{B}$ and $\mathfrak{D}$ can be given the values represented by Eq. 2-34 and 2-38, that is, for instance, if the expansion range is not very large,

$$\mathfrak{B}(x) = 1 - 0.565x^{3.2}, \quad \mathfrak{D}(x) = 1 + 0.29x^4$$

An alternate expression of T_f/T_i^0 can be found from Eq. 2-31 using the value of $\sigma_f - \sigma_i^0$ given by Eq. 13-24. Neither expression contains the values of the efficiencies, both being fundamentally obtained from the energy equation where they do not appear.

Problem 2. Determine the values of u_f and T_f obtained in an expansion to p_f. Again using Eq. 13-12 and 13-16b and the approximate relations of Eq. 2-31, 2-33, and 2-35 we find

$$\left.\begin{aligned}\frac{u_f^2}{2c_{p_i}^0 T_i^0} = \frac{h_i^0-h_f}{c_{p_i}^0 T_i^0} &= \mathfrak{C}\left(\eta_{el}\frac{\gamma_i^0-1}{\gamma_i^0}\ln\frac{p_i^0}{p_f}\right)\left[1-\left(\frac{p_f}{p_i^0}\right)^{\eta_{el}\frac{\gamma_i^0-1}{\gamma_i^0}}\right]\\ = \eta_{ad}\frac{h_i^0-(h_f)_{is}}{c_{p_i}^0 T_i^0} &= \eta_{ad}\mathfrak{C}\left(\frac{\gamma_i^0-1}{\gamma_i^0}\ln\frac{p_i^0}{p_f}\right)\left[1-\left(\frac{p_f}{p_i^0}\right)^{\frac{\gamma_i^0-1}{\gamma_i^0}}\right]\end{aligned}\right\} \tag{13-26}$$

$$\frac{T_f'}{T_i^0} = \mathfrak{A}\left(\eta_{el} \ln \frac{p_i^0}{p_f}\right)\left(\frac{p_f}{p_i^0}\right)^{\eta_{el}\frac{\gamma_i^0 - 1}{\gamma_i^0}} = \mathfrak{B}\left(\frac{h_i^0 - h_f}{c_{p_i}^0 T_i^0}\right)\left(1 - \frac{h_i^0 - h_f}{c_{p_i}^0 T_i^0}\right) \tag{13-27}$$

with $h_i^0 - h_f$ in the last expression given by the above Eq. 13-26 in terms of η_{ad} and p_f/p_i^0, and $\mathfrak{A}$, $\mathfrak{C}$ given by Eq. 2-32 and 2-34,

$$\mathfrak{A}(x) = 1 + 0.0041x^{2.1}; \quad \mathfrak{C} = 1 - 0.022x^{1.8}$$

As already observed in Art. 2 the values of $\mathfrak{C}$ and $\mathfrak{D}$, for a given transformation, are substantially closer to unity than those of $\mathfrak{A}$ and $\mathfrak{B}$. Thus for not-too-large expansion ratios one can be justified in taking $\mathfrak{C} = \mathfrak{D} = 1$ in Eq. 13-24 and 13-26 (which are reduced in this case to the expressions obtained for polytropic gases with the initial values of c_p and γ). On the other hand, the use of the polytropic expressions for the calculation of T_f is subject to considerably larger errors.

Using the above expressions one can calculate the flow properties at each point of a nozzle, assuming that either η_{ad} or η_{el} is constant for the whole expansion. The continuity equation can be written in the form

$$\frac{m}{A} = \rho u = \frac{p_i^0}{\sqrt{\mathfrak{R}T_i^0}}\sqrt{\frac{2\gamma_i^0}{\gamma_i^0 - 1}}\sqrt{\frac{u^2}{2c_{p_i}^0 T_i^0}}\frac{p}{p_i^0}\frac{T_i^0}{T} \tag{13-28}$$

where $u^2/2c_{p_i}^0 T_i^0$, p/p_i^0, and T/T_i^0 can be connected together by the above relations, written without the subscript $_f$. Thus, for given m, Eq. 13-28 represents an equation between A and one of these three quantities, the solution of which provides the values of the flow parameters at each section.

Flow of real gases. Equations. For large pressure or temperature variations, such as those which are likely to be encountered when the flow involves very high Mach numbers, the gas imperfections may produce appreciable deviations from the perfect gas law. Their effects, combined with those of the specific heat variations, have been studied in a number of papers. Tsien [*45*] using an approximate method, and Donaldson [*4*] rigorously, have investigated the isentropic one-dimensional flow of a gas following the van der Waals equation of state. Their methods consist of deriving the expressions for the enthalpy, entropy, and sound velocity in terms of pressure and temperature, and applying laborious numerical procedures. Here we shall rather describe the method of Crown [*2*] based on the use of the correction factors defined by Eq. 2-41 and 2-42 and given for the air by Fig. B,2q, B,2r, B,2s, B,2t, B,2u, and B,2v. Making use of these correction factors we can write the following equations. First we can write Eq. 2-8 in the form

$$\frac{h}{h^0} = \frac{\mathfrak{F}}{\mathfrak{F}^0}\frac{T}{T^0} = 1 - \frac{u^2}{2h^0} = 1 - w^2 \tag{13-29}$$

where the reduced velocity $w = u/u_{max}$ is based on the u_{max} given by Eq. 4-24 rather than Eq. 4-24a (observe that for real gases the constancy of h^0 does not involve that of T^0); and $\mathfrak{F} = \mathfrak{F}(p, T)$, $\mathfrak{F}^0 = \mathfrak{F}(p^0, T^0)$ can be read on Fig. B,2r.

Similarly, one finds

$$M^2 = \frac{u^2}{a^2} = \frac{2(h^0 - h)}{{}_0\gamma \mathfrak{R} T \mathcal{G}} = \frac{2}{{}_0\gamma - 1} \frac{\mathfrak{F}}{\mathcal{G}} \frac{h^0 - h}{h} = \frac{2}{{}_0\gamma - 1} \frac{\mathfrak{F}}{\mathcal{G}} \frac{w^2}{1 - w^2} \tag{13-30}$$

or

$$w^2 = \frac{M^2}{M^2 + \dfrac{2}{{}_0\gamma - 1} \dfrac{\mathfrak{F}}{\mathcal{G}}} \tag{13-30a}$$

with $\mathcal{G} = \mathcal{G}(p, T)$ given by Fig. B,2s and ${}_0\gamma = 1.4$ for air.

Proceeding in a similar way we obtain the equations corresponding to Eq. 4-30a, 4-31a, 4-32a, and 4-33a in the form

$$u = \sqrt{2h^0}\, w \tag{13-31}$$

$$m = \frac{1}{\mathscr{g}} \frac{2\,{}_0\gamma}{{}_0\gamma - 1} \frac{pA}{\sqrt{2h^0}} \frac{w}{1 - w^2} \tag{13-32}$$

$$I = m\sqrt{2h^0}\, w = \frac{1}{\mathscr{g}} \frac{2\,{}_0\gamma}{{}_0\gamma - 1} pA \frac{w^2}{1 - w^2} \tag{13-33}$$

$$J = m\sqrt{2h^0}\left(w + \mathscr{g}\, \frac{{}_0\gamma - 1}{2\,{}_0\gamma} \frac{1 - w^2}{w}\right)$$

$$= pA\left(1 + \frac{1}{\mathscr{g}} \frac{2\,{}_0\gamma}{{}_0\gamma - 1} \frac{w^2}{1 - w^2}\right) \tag{13-34}$$

with $\mathscr{g}$ given by Fig. B,2t in function of p and T. Similarly one could obtain the expression of $J^{(\epsilon)}$ for ducts of the p-A power family. For the entropy one obtains from Eq. 2-40 and 2-42:

$$s = s_r' + \mathfrak{R} \ln\left(\frac{\mathcal{J}}{p} T^{\frac{{}_0\gamma}{{}_0\gamma - 1}}\right) \tag{13-35}$$

with $\mathcal{J}$ given by Fig. B,2u. Here the additive constant is related to the one defined previously by $s_r' = s_r - {}_0c_p \ln T_r$. Thus for an isentropic transformation we have

$$p \sim \mathcal{J} T^{\frac{{}_0\gamma}{{}_0\gamma - 1}} \tag{13-35a}$$

Flow of real gases in nozzles or diffusers. With the help of the above expressions one can treat the same problems analyzed in the preceding articles for a perfect polytropic gas. However, due to the presence of the correction factors, the solution has to be found by iteration following the procedure outlined by Crown for isentropic flow. As an example we shall consider here the flow in a nozzle with known adiabatic efficiency, given initial conditions p_i^0, T_i^0, and assigned final pressure p_f.

Combining Eq. 2-41 and 13-35 we have

$$h = {}_0c_p\mathfrak{F}\left(\frac{p}{\mathfrak{G}}\right)^{\frac{{}_0\gamma-1}{{}_0\gamma}} e^{\frac{s-s_r'}{{}_0c_p}}$$

Thus, using Eq. 13-29 and the definition (Eq. 13-12) of η_{ad}, we obtain successively

$$\left.\begin{aligned} w_f^2 &= \frac{u_f^2}{2h_i^0} = 1 - \frac{\mathfrak{F}_f}{\mathfrak{F}_i^0}\frac{T_f}{T_i^0} \\ &= \eta_{ad}\left[1 - \frac{h(s_i, p_f)}{h(s_i, p_i^0)}\right] = \eta_{ad}\left[1 - \frac{\mathfrak{F}_f}{\mathfrak{F}_i^0}\left(\frac{\mathfrak{G}_i^0}{\mathfrak{G}_f}\frac{p_f}{p_i^0}\right)^{\frac{{}_0\gamma-1}{{}_0\gamma}}\right] \end{aligned}\right\} \qquad (13\text{-}36)$$

From these relations w_f, u_f, and T_f can be obtained by the following procedure. Determine $\mathfrak{F}_i^0 = \mathfrak{F}(p_i^0, T_i^0)$ from Fig. B,2r and similarly for $\mathfrak{G}_i^0$. Compute a first approximation of T_f by assuming $\mathfrak{F}_f = \mathfrak{F}_i^0$ and $\mathfrak{G}_f = \mathfrak{G}_i^0$. Corresponding to this T_f and the assigned p_f calculate a better value for $\mathfrak{F}_f$ and $\mathfrak{G}_f$, and a second approximation for T_f. Iterate as necessary (generally the second approximation is sufficiently accurate). Finally the correct values of w_f and u_f can be computed, h_i^0 being given by ${}_0c_pT_i^0\mathfrak{F}_i^0$. Then, the area A_f/m per unit flow rate can be obtained from Eq. 13-32, and the thrust, if required, from Eq. 8-3 and 13-34.

If the values of η_{ad} are known at all points of the nozzle (for instance, η_{ad} = const), the relations obtained from Eq. 13-36 by removing the subscript f can be used to determine the quantities at each section. For instance one can write

$$\frac{T}{T_i^0} = \frac{\mathfrak{F}_i^0}{\mathfrak{F}}(1 - w^2), \quad \frac{p}{p_i^0} = \frac{\mathfrak{G}}{\mathfrak{G}_i^0}\left[\frac{\mathfrak{F}_i^0}{\mathfrak{F}}\left(1 - \frac{w^2}{\eta_{ad}}\right)\right]^{\frac{{}_0\gamma}{{}_0\gamma-1}} \qquad (13\text{-}37)$$

and, replacing in Eq. 13-32, obtain the relation

$$\frac{{}_0\gamma - 1}{2{}_0\gamma}\,\mathfrak{G}_i^0\,\frac{\sqrt{2h_i^0}}{p_i^0}\,\frac{m}{A} = \frac{\mathfrak{G}_i^0}{\mathfrak{G}}\,\frac{\mathfrak{G}}{\mathfrak{G}_i^0}\,\frac{w}{1 - w^2}\left[\frac{\mathfrak{F}_i^0}{\mathfrak{F}}\left(1 - \frac{w^2}{\eta_{ad}}\right)\right]^{\frac{{}_0\gamma}{{}_0\gamma-1}} \qquad (13\text{-}38)$$

which together with Eq. 13-37 allows the determination of A/m for any given value of w, following an iteration procedure similar to that described above. The minimum value (if any) of A, resulting from these computations, represents the throat area.

These relations become simpler for isentropic flow ($\eta_{ad} = 1$), in which case we know from the beginning of this article that the throat corresponds to $M = 1$, and therefore (Eq. 13-30a and 13-37) to the following values of w, p, and T:

$$w^{*2} = \frac{1}{1 + \dfrac{2}{_0\gamma - 1}\dfrac{\mathfrak{F}^*}{\mathcal{G}^*}}, \quad \frac{T^*}{T_i^0} = \frac{\mathfrak{F}_i^0}{\mathfrak{F}^* + \dfrac{_0\gamma - 1}{2}\mathcal{G}^*},$$

$$\frac{p^*}{p_i^0} = \frac{\mathcal{J}^*}{\mathcal{J}_i^0}\left(\frac{\mathfrak{F}_i^0}{\mathfrak{F}^* + \dfrac{_0\gamma - 1}{2}\mathcal{G}^*}\right)^{\frac{_0\gamma}{_0\gamma - 1}}$$

Once these values are determined by iteration, from Eq. 13-38 we obtain for the isentropic nozzle, after easy manipulations,

$$\left(\frac{A^*}{A}\right)_{is} = \frac{\mathcal{K}}{\mathcal{K}^*}\frac{w}{w^*}\left(\frac{\mathfrak{F}^*}{\mathfrak{F}}\frac{1 - w^2}{1 - w^{*2}}\right)^{\frac{1}{_0\gamma - 1}} \tag{13-39}$$

where $\mathcal{K}$, defined by Eq. 2-42, can be found from Fig. B,2v. This and the previous relations can be expressed in terms of M, substituting w^2 from Eq. 13-30a.

For a diffuser, analogous relations can be found using the definition (Eq. 13-13) of η_{ad}. For assigned p_i, T_i, and u_i, the final stagnation temperature and pressure are found to be

$$\frac{T_f^0}{T_i} = \frac{\mathfrak{F}_i}{\mathfrak{F}_{f_i}^0}\left(1 + \frac{u_i^2}{2h_i}\right), \quad \frac{p_f^0}{p_i} = \frac{\mathcal{J}_{fi}^0}{\mathcal{J}_i}\left[\frac{\mathfrak{F}_i}{\mathfrak{F}_f^0}\left(1 + \eta_{ad}\frac{u_i^2}{2h_i}\right)\right]^{\frac{_0\gamma}{_0\gamma - 1}}$$

If preferred, one can easily develop an iterative procedure for the calculation of nozzle and diffuser flows in terms of the elemental efficiency, using either one of the two definitions given by Eq. 13-20 for nozzles and Eq. 13-21 for diffusers.

Shocks in real gases. The conservation equations can be written

$$m_2 = m_1, \quad J_2 = J_1, \quad h_2^0 = h_1^0 = h_1 + \frac{u_1^2}{2} \tag{13-40}$$

For assigned initial p_1, T_1, and u_1, one can calculate h_1, h_1^0, ρ_1, w_1, and M_1 from the values of the correction factors and the above relations. Similarly, if M_1, instead of u_1, is assigned one can determine the remaining quantities.

From the second equation (Eq. 13-40), taking into account the first and the third, and using Eq. 13-33, we obtain the following two equations:

$$w_2\left(\frac{2_0\gamma}{_0\gamma - 1} - \mathcal{J}_2\right) + \frac{\mathcal{J}_2}{w_2} = w_1\left(\frac{2_0\gamma}{_0\gamma - 1} - \mathcal{J}_1\right) + \frac{\mathcal{J}_1}{w_1} \tag{13-41}$$

$$\frac{p_2}{p_1} = \frac{1 + \dfrac{1}{\mathcal{J}_1}\dfrac{2_0\gamma}{_0\gamma - 1}\dfrac{w_1^2}{1 - w_1^2}}{1 + \dfrac{1}{\mathcal{J}_2}\dfrac{2_0\gamma}{_0\gamma - 1}\dfrac{w_2^2}{1 - w_2^2}} \tag{13-42}$$

which, in addition to the relation

$$\frac{T_2}{T_1} = \frac{\mathfrak{F}_1}{\mathfrak{F}_2}\frac{1 - w_2^2}{1 - w_1^2} \tag{13-43}$$

obtained from Eq. 13-29 are sufficient for the evaluation of the conditions after the shock with the following iterative procedure. Assume in first approximation $\mathcal{G}_2 = \mathcal{G}_1 = 1$ and $\mathfrak{F}_2 = \mathfrak{F}_1$. Then the equations coincide with those of Art. 5, the significant solution of which is given by Eq. 5-2 and 5-5. Obtain from the figures the values of $\mathcal{G}_2$ and $\mathfrak{F}_2$ corresponding to these p_2 and T_2, and those of $\mathcal{G}_1$ and $\mathfrak{F}_1$ corresponding to the known initial conditions, and derive a second approximation for w_2 by solving Eq. 13-41. Obviously the smaller of the two solutions has to be chosen, the other one corresponding again to the trivial solution $w_2 = w_1$, $p_2 = p_1$, and $T_2 = T_1$. Now a second approximation for p_2 and T_2 can be obtained from Eq. 13-42 and 13-43 and, if necessary, the whole procedure can be repeated to convergence. From the first of Eq. 13-40 one obtains $\rho_2/\rho_1 = w_1/w_2$. On the other hand, one has from Eq. 2-41

$$\frac{\rho_2}{\rho_1} = \frac{\mathcal{E}_2}{\mathcal{E}_1}\frac{p_2}{p_1}\frac{T_1}{T_2}$$

One of these two equations is not independent of the others. It can be used either to check the calculations or to replace one of the equations (Eq. 13-42 and 13-43).

Finally, the stagnation values after the shock are obtained from the equations, immediately obtained from $h_2^0 = h_1^0$ and $s_2^0 = s_2$:

$$\frac{T_2^0}{T_1^0} = \frac{\mathfrak{F}_1^0}{\mathfrak{F}_2^0}, \quad \frac{p_2^0}{p_2} = \frac{\mathcal{G}_2^0}{\mathcal{G}_2}\left(\frac{T_2^0}{T_2}\right)^{\frac{{}_0\gamma}{{}_0\gamma - 1}}, \quad \frac{\rho_2^0}{\rho_2} = \frac{\mathcal{K}_2^0}{\mathcal{K}_2}\left(\frac{T_2^0}{T_2}\right)^{\frac{1}{{}_0\gamma - 1}}$$

Here, too, one calculates a first approximation for T_2^0 and p_2^0 with $\mathfrak{F}_2^0 = \mathfrak{F}_1^0$ and $\mathcal{G}_2^0 = \mathcal{G}_2$; with this value a better estimate of $\mathfrak{F}_2^0$ and $\mathcal{G}_2^0$ and a second approximation of T_2^0 and p_2^0 is obtained. Then the procedure is repeated as necessary.

Numerical examples of flow of real gases. The following results of calculations based on the procedures described above are taken from [*2*]. We shall indicate with a subscript zero placed on the left of a symbol the result of calculations made for the perfect polytropic reference gas.

1. Isentropic expansion. Expand from $p_i = 50$ atm, $T_i = 300°$K to $T_f = 50°$K. We obtain ${}_0p_f = 0.0945$ atm. Corresponding to T_f and ${}_0p_f$ we find that $\mathcal{G}_i = 0.8925$ and $\mathcal{G}_f = 0.9650$.[18] Thus we obtain from Eq. 13-35a] $p_f = 0.1022$ atm. A second iteration is not necessary. The fractional error $({}_0p_f - p_f)/p_f$ which results from neglecting the gas imperfections and the specific heat variability is here -7.5 per cent, and is to be attributed entirely to deviations from the perfect gas law.

[18] These values are obtained from large scale plots. However, those obtained from Fig. B,2u are sufficiently accurate for practical purposes.

2. Isentropic compression. Compress from $p_i = 0.1$ atm, $T_i = 300°$K to $T_f = 1800°$K. We have $\mathcal{J}_i = 1.001$ and, corresponding to T_f and ${}_0p_f = 52.9$ atm, $\mathcal{J}_f = 1.77$. Thus $p_f = 93.4$ atm with a fractional error of -43.3 per cent. Despite this large error it is not necessary to iterate, since in this range of temperature $\mathcal{J}_f$ is not substantially affected by the pressure. In other words the deviations from the perfect gas law are unimportant in this compression process and only the variability of the specific heats is the cause for the large correction.

For the same reason, if we assign the final pressure, as in the following example of isentropic compression from $p_i = 1$ atm, $T_i = 1000°$K, to p_f

Table B,13b

M_f	p_i^0,atm	T_i^0,°K	$\frac{{}_0p_f - p_f}{p_f}$, per cent	$\frac{{}_0T_f - T_f}{T_f}$, per cent	$\frac{{}_0\left(\frac{A_f}{A^*}\right) - \frac{A_f}{A^*}}{\frac{A_f}{A^*}}$, per cent
5	50	300	−3.9	2.2	0.8
5	50	500	−2.1	−0.02	3.5
5	50	700	−0.8	−0.18	0.5
5	10	400	−0.5	0.03	0.6
5	50	400	−2.0	0.6	2.6
5	100	400	−4.7	1.1	4.6
10	50	700	−1.0	−2.0	1.0

$= 10$ atm, several iterations are necessary, because the computed ${}_0T_f = 1931°$K is appreciably different from the final correct value $T_f = 1750°$K, and $\mathcal{J}_f$ changes at each iteration. The fractional error $({}_0T_f - T_f)/T_f$ is here 10.2 per cent, practically due entirely to the specific heat variability.

3. Isentropic flow in a nozzle. A nozzle is required to produce an assigned Mach number M_f in the exit section for given initial conditions p_i^0 and T_i^0. Neglecting the presence of the correction factors one calculates ${}_0w_f$ from Eq. 13-30a, and ${}_0p_f$ and ${}_0T_f$ from Eq. 13-37 (with $\eta_{ad} = 1$). With these values one can calculate a value of $\mathfrak{F}_f$, $\mathcal{G}_f$, and $\mathcal{J}_f$, and a second approximation for w_f, p_f, and T_f, which is generally found to be sufficiently accurate. From Eq. 13-39 one can calculate the values of ${}_0(A_f/A^*)$ disregarding the correction factors, or those of the correct A_f/A^* using the value of $\mathcal{K}_f$. The values of the starred quantities are obtained as described above. Table B,13b summarizes the errors due to neglecting the gas imperfection and the variability of the specific heats for a few representative cases.

Inspection of this table shows that despite the wide range of variation of pressure and temperature in the nozzle the errors never exceed a few per cent, a result that is substantially due to the fact that at the lowest temperatures reached, at which the deviation from the perfect gas law could be high, high deviations are actually prevented by the corresponding low level of pressure.

4. Shock waves. For assigned p_1 and T_1, and a given M_1 one can compare the approximate results obtained for the perfect polytropic reference gas from the equations of Art. 5, with the accurate values calculated as discussed above in this article. For $M_1 = 7$ and $T_1 = 275°K$ the following approximate results are obtained:

$$ {}_0T_2 = 2879°K, \quad \frac{{}_0p_2}{p_1} = 57, \quad \frac{{}_0\rho_2}{\rho_1} = 5.444, \quad {}_0T_2^0 = 2970°K $$

The fractional errors of these results with respect to the correct ones are given in Table B,13c for three values of p_1.

Table B,13c

p_1,atm	$\frac{{}_0T_2 - T_2}{T_2}$, per cent	$\frac{{}_0p_2 - p_2}{p_2}$, per cent	$\frac{{}_0\rho_2 - \rho_2}{\rho_2}$, per cent	$\frac{{}_0T_2^0 - T_2^0}{T_2^0}$, per cent
0	13.6	−3.1	14.7	14.7
1	13.8	−3.0	13.8	14.7
5	14.9	−2.7	10.4	16.3

It is seen that considerable errors are present, due essentially to the variability of the specific heats as a result of the high temperatures involved. It is interesting to notice, however, that the error on the pressure ratio is relatively small, in agreement with the results of Fig. B,13h. Calculations of this type can be useful when the initial conditions are assigned as happens in free flight and pressurized air ranges. For the case of wind tunnels it is also useful to examine a combined process consisting of expansion in a nozzle to a given Mach number, followed by a shock wave. If p_i^0 and M_t for the expansion are assigned, and T_t is also assigned (in order to avoid too low values at which air condensation would result) the values of p_t and T_i^0 can be obtained and compared with their approximate values for the perfect polytropic reference gas. For the shock, one has to take $p_1 = p_t$, $T_1 = T_t$, and $M_1 = M_t$, and proceed as indicated above. The final results in terms of fractional errors are given for $p_i^0 =$ 100 atm, $T_t = 50°K$, and two values of M_t in Table B,13d. Observe that some of the errors are of opposite signs at the two values of M_t considered. This is due to the fact that at $M_t = 5$ the correct initial temperature is $T_i^0 = 309°K$, so that the errors are entirely due to the gas imperfections, while for $M_t = 10$ we have $T_i^0 = 1002°K$, and the specific heat variability has an important effect in a direction opposite to that of the gas imperfections. It is also interesting to observe that at the higher Mach numbers the correction of the indication of a Pitot tube placed after the nozzle (identical with p_2^0) is negligible.

Table B,13d

M_f	$\frac{{}_0T_i^0 - T_i^0}{T_i^0}$, per cent	$\frac{{}_0p_f - p_f}{p_f}$, per cent	$\frac{{}_0p_2 - p_2}{p_2}$, per cent	$\frac{{}_0T_2 - T_2}{T_2}$, per cent	$\frac{{}_0p_2^0 - p_2^0}{p_2^0}$, per cent	$\frac{{}_0T_2^0 - T_2^0}{T_2^0}$, per cent
5	−3.0	−3.0	−2.8	2.8	−2.5	3.1
10	4.8	2.0	1.1	3.8	0.2	4.2

Two-phase flow. There are essentially two types of two-phase flows, that in which the two phases flow distinctly, separated by a simple shape of surface, and that in which one of the two phases is intimately dispersed into the other. In general a pure thermodynamic treatment of these flows, as if the flowing substance were a single substance with well-defined thermodynamic properties, is impossible. The two phases have different temperatures and velocities and must be considered separately in the evaluation of the conservation equations. An example of this kind of treatment for a well-dispersed mixture of air and droplets is given in Art. 14, where it is shown that the problem can be solved when the laws governing the rates of momentum, heat, and matter exchange between the two phases are known. The treatment of the case when the phases are not mixed can be made following almost exactly the same lines, again provided that the laws governing the exchanges between phases are known. This problem is of particular interest to chemical engineers, and for more detail we refer the reader to some of the specialized publications on the subject [*46,47*].

However, as already observed in Art. 2, a pure thermodynamic treatment of the two-phase flow is possible for mixtures in which the rate of exchanges is either so low that it can be neglected or so high that the two phases can be considered at each instant in a state of mechanical, thermal, or physical equilibrium. Examples of this kind are considered in II,B,5.

Here we shall analyze only the flow of the mixture of a liquid with small gas bubbles for which the thermodynamic relations have been derived in Art. 2. These relations (Eq. 2-46 to 2-50) show a particular type of deviation from the perfect gas law, due to the presence of the covolume of the incompressible liquid. The specific heats can be considered constant and it is possible to define a constant adiabatic index Γ, defined by Eq. 2-52, which is always smaller than that of the gaseous phase alone and for practical values of the mixture ratio μ is generally very close to unity [*14*].

The flow of this type of mixture can be treated following the same lines used for the cases treated previously. One can define a stagnation enthalpy and proper values of the stagnation temperature and pressure, using Eq. 2-8 and the expression (Eq. 2-48) for the enthalpy. We obtain

$$h^0 = c_p T^0 + \frac{p^0}{(1+\mu)\rho_{\text{liq}}} = c_p T + \frac{p}{(1+\mu)\rho_{\text{liq}}} + \frac{u^2}{2} \qquad (13\text{-}44)$$

Obviously this relation defines h^0 but not T^0 and p^0, for which we must also use the condition that, by definition, the entropy must be the same in static and stagnation conditions, i.e. from Eq. 2-51

$$s = \mathcal{R} \ln \frac{(T^0)^{\Gamma/(\Gamma-1)}}{p^0} + \text{const} = \mathcal{R} \ln \frac{T^{\Gamma/(\Gamma-1)}}{p} + \text{const}$$

or

$$\frac{T}{T^0} = \left(\frac{p}{p^0}\right)^{(\Gamma-1)/\Gamma} = y^{(\Gamma-1)/\Gamma} \qquad (13\text{-}45)$$

where, for simplicity, we have indicated by y the static-to-stagnation pressure ratio.

It follows from Eq. 13-44 that for adiabatic flows the fact that the process stays isoenergetic ($h^0 = \text{const}$) does not imply the constancy of T^0. However, if the flow is also isentropic, then both T^0 and p^0 must remain constant during the whole process.

It may be useful [*14*] to introduce another quantity, which represents the ratio of the total gas bubble volume to the volume of the liquid and which is given by

$$\delta = \frac{\mu\rho_{\text{liq}}}{\rho_g} = \frac{\mu\rho_{\text{liq}}\mathcal{R}_g T}{p} = \frac{(1+\mu)\rho_{\text{liq}}\mathcal{R}T}{p}$$

This "volume ratio" is obviously variable. In terms of the value δ^0 which it takes under stagnation conditions, and of the stagnation enthalpy, we find the following expressions for T^0 and p^0:

$$c_p T^0 = \frac{h^0}{1+z}, \quad p^0 = \frac{z(1+\mu)\rho_{\text{liq}}h^0}{1+z} \qquad (13\text{-}46)$$

with

$$z = \frac{\Gamma - 1}{\Gamma\delta^0} \qquad (13\text{-}47)$$

From Eq. 13-44, 13-45, and 13-46 we obtain the equation

$$u^2 = 2h^0 \frac{1 - y^{(\Gamma-1)/\Gamma} + z(1-y)}{1+z} \qquad (13\text{-}48)$$

from which the relation between the reduced velocity $w = u/\sqrt{2h^0}$ and the pressure ratio y is immediately obtained for any given Γ (fixed for a given mixture) and stagnation volume ratio δ^0. For a pure gas we have $\mu = \infty$, $\delta^0 = \infty$, $\Gamma = \gamma$, and Eq. 13-45 and 13-48 are reduced to the corresponding Eq. 4-28. In the present case, however, instead of using these relations to express all the relevant quantities in terms of T^0 and w, as in Art. 4, it is more convenient to express them in terms of h^0, y, and z (or δ^0). One obtains without difficulty

$$m = \rho u A = (1 + \mu)\rho_{\text{liq}} A \sqrt{2h^0} \frac{[(1 - y)^{(\Gamma-1)/\Gamma} + z(1 - y)]^{\frac{1}{2}}}{(1 + \delta^0 y^{-1/\Gamma})(1 + z)^{\frac{1}{2}}} \quad (13\text{-}49)$$

$$I = (1 + \mu)\rho_{\text{liq}} A \cdot 2h^0 \frac{1 - y^{(\Gamma-1)/\Gamma} + z(1 - y)}{(1 + \delta^0 y^{-1/\Gamma})(1 + z)}$$

$$J = \frac{(1 + \mu)\rho_{\text{liq}} A h^0}{1 + z}\left[zy + 2\frac{1 - y^{(\Gamma-1)/\Gamma} + z(1 - y)}{1 + \delta^0 y^{-1/\Gamma}}\right] \quad (13\text{-}50)$$

Similarly one can obtain the expression of $J^{(s)}$ for ducts of the p-A power family.

The equations derived above can be used for the solution of a large number of flow problems. For instance, for the expansion from assigned initial values p_i^0, T_i^0 to a final pressure p_f in a nozzle with assigned adiabatic efficiency, we obtain the value of u_f and T_f from Eq. 13-12 by introducing the values obtained from Eq. 2-48 for h_i^0 and h_f and the value

$$(h_f)_{\text{is}} = c_p T_i^0 \left(\frac{p_f}{p_i^0}\right)^{(\Gamma-1)/\Gamma} - \frac{p_f}{(1 + \mu)\rho_{\text{liq}}}$$

Then the value of ρ_f can be obtained from the equation of state (Eq. 2-46) and the area A_f from $m/\rho_f u_f$.

Analogous calculations can be performed for diffusers. Moreover, by assuming a known distribution of η_{ad} along the duct one can find the pressure distributions in nozzles or diffusers of assigned geometry. The relations for isentropic flows ($\eta_{\text{ad}} = 1$) are substantially simpler because both p^0 and T^0 are constant along the transformation. This case is developed in [*14*] which contains numerical tables and graphs and a few experimental comparisons.

Phenomena similar to those discussed in Art. 6 for perfect polytropic gases are obtained in the present case. Choking occurs when the velocity in the throat reaches the sonic value given by Eq. 2-53. For choked flow, supersonic velocities are obtained in the divergent portion of the nozzle. Approximate relations can be derived for Γ close to 1 and for not-too-large values of p_i^0/p_f. For additional details the reader is referred to [*14*].

When the flow is supersonic, shocks may be produced. The relations between the conditions (p_1, T_1, and u_1) before the shock and those (p_2, T_2, and u_2) after the shock are found from the conservation equations (Eq. 13-40). After introduction of Eq. 13-50, the equation $J_2 = J_1$ provides an equation of the second degree for the determination of z_2 (or δ^0 from Eq. 13-47) in terms of y_2 and J_1. Replacing the proper root of this equation in the equation obtained from $m_2 = m_1$ and Eq. 13-49, one obtains a complicated equation for y_2. After numerical determination of the root of this equation, one can calculate z_2 and determine the values of T_2^0, p_2^0, and u_2 from Eq. 13-46 and 13-48. The values of T_2 and p_2 follow from Eq. 13-45.

B,14. More General Types of Flow. It is clear from the results of the preceding articles that the only types of flow for which the exact solutions can be expressed in closed form are those of polytropic gases in which there is only one independent cause for the transformation, such as area variation, friction, and heat addition. We include in these types of flows those in ducts of the p-A power family, for which the area variation is not independent, but is determined by the independent cause of the transformation (friction, heat transfer, etc.). When the specific heats are variable or the flow includes the combined effects of two independent causes (area variation and friction, friction and heat transfer), only approximate or numerical treatments are generally available.

This is of course true for more general types of flow containing several independent causes, such as those corresponding to the general formulation of Art. 3. The solution for these flows can be achieved only through numerical calculations, starting from the general equations of conservation, written in differential form, that is, from Eq. 4-2a, 4-3a, and 4-4a which we shall rewrite here

$$dm = dm_i \tag{14-1}$$

$$dI = -Adp - dX_w - dX_b - dI_i \tag{14-2}$$

$$dH^0 = dQ_w + dQ_b + dW_b + dH_i^0 \tag{14-3}$$

In the energy equation (Eq. 14-3) we have, for more generality, kept the term dW_b representing the work, or mechanical energy in any form, supplied from the outside to the flowing fluid. If the injected substance consists of a discrete number N of streams, each with a uniform axial injection velocity and stagnation enthalpy (indicating with this term the sum of the enthalpy and kinetic energy even for nongaseous substances), we have

$$dm_i = \sum_1^N (dm_i)_j, \quad dI_i = \sum_1^N (u_i dm_i)_j, \quad dH_i^0 = \sum_1^N (h_i^0 dm_i)_j \tag{14-4}$$

where the subscript $_j$ refers to any of the N streams. Observe that, in accordance with Art. 2, if changes of state or chemical composition are present, the enthalpy for each substance, and for the main stream, must include the latent heat and the heat of formation. Obviously the use of Eq. 14-1, 14-2, and 14-3 is based on the assumption that the changes of all quantities are continuous. In the case of discontinuous changes, the equations connecting the quantities on the two sides of the discontinuity are the original conservation equations.

Method of Shapiro and Hawthorne. A considerable clarification of the numerical calculation and of the qualitative interpretation is made possible if one manipulates the conservation equations and the other relevant equations, following a method devised by Shapiro and Hawthorne [*15,48*]. The method consists in expressing the differential of each relevant dependent quantity (such as velocity, pressure, density, etc.) through a linear combination of the independent elementary causes of the transformation (such as friction, heat addition, area change, etc.), the coefficients of these linear combinations, called the "influence coefficients," being expressed as functions of a single variable (the Mach number).[19] We observe here that the first introduction in the published literature of the fertile idea of influence coefficients, depending only on M, can be credited to Bailey [*49*], who used it for the treatment of simpler flow problems. Although the method of Shapiro and Hawthorne can be easily studied in the mentioned references, because of its importance we shall give a brief account of it here.

The type of flow considered is quite general since it includes the effects of area change, wall friction, drag of immersed bodies, chemical reactions, mixing and change of phase of injected substances, and changes in molecular weight (caused by chemical reactions or by mixing). The effect of mass forces is not considered, but it might easily be included. The single constituents of the flowing gas are assumed to obey the perfect gas law. Thus one can write for the flowing gas the equation of state

$$p = \rho \mathcal{R} T \tag{14-5}$$

where $\mathcal{R}$, however, is not constant if the molecular weight $\mathfrak{M}$ varies, being related to it and to the universal gas constant R by the equation

$$\mathcal{R} = \frac{R}{\mathfrak{M}} \tag{14-6}$$

Of course, the variability of $\mathcal{R}$ does not alter the expression of the sound velocity, which is still given by Eq. 2-17:

$$a^2 = \gamma \mathcal{R} T \tag{14-7}$$

From logarithmic differentiation of Eq. 14-5 and 14-6 one obtains

$$\frac{dp}{p} = \frac{d\rho}{\rho} + \frac{dT}{T} - \frac{d\mathfrak{M}}{\mathfrak{M}} \tag{14-8}$$

which can therefore represent, in differential form, the equation of state.

[19] We shall follow here the same procedure although it would be easy to express the influence coefficients in terms of w, which is the variable that we have more often used. The transformation can be made quite easily, recalling the relation (Eq. 4-29) between M and w.

The continuity equation, in differential form, is obtained from the logarithmic differentiation of $m = \rho u A$ and substitution from Eq. 14-1:

$$\frac{d\rho}{\rho} + \frac{du}{u} + \frac{dA}{A} = \frac{dm_i}{m} = \sum_1^N \frac{(dm_i)_j}{m} \tag{14-9}$$

The momentum equation (Eq. 14-2) can be elaborated in the following way. We have first

$$dI = d(mu) = mdu + udm_i$$

Also, we can express dX_w through Eq. 3-5 and 7-20 as

$$dX_w = \gamma p A M^2 \cdot \frac{2c_f}{d_h} dx$$

Replacing in Eq. 14-2, dividing by pA, and observing that with the help of Eq. 14-5 and 14-7 one gets

$$m = \rho u A = \frac{\gamma p A M^2}{u} \tag{14-10}$$

the momentum equation is expressed in the form

$$\frac{dp}{p} + \gamma M^2 \frac{du}{u} = -\gamma M^2 \left\{ 2c_f \frac{dx}{d} + \frac{dX_b}{\gamma A p M^2} + \sum_1^N \left[\left(1 - \frac{u_i}{u}\right) \frac{dm_i}{m} \right]_j \right\} \tag{14-11}$$

where use has been made of Eq. 14-4. Similarly, for the energy equation, one has

$$dH^0 = d(mh^0) = mdh^0 + h^0 dm_i = m(dh + udu) + h^0 dm_i \tag{14-12}$$

where dh can be obtained from differentiation of Eq. 2-9 in the form

$$dh = dh_r + d\left(\int_{T_r}^T c_p dT\right) = dh_r + c_p dT + \int_{T_r}^T dc_p dt \tag{14-13}$$

Here h_r and c_p change as a result of changes in gas composition due to both chemical changes and physical mixing with the gaseous or gasified injected substances. Hence we can write

$$dh_r = (dh_r)_{ch} + (dh_r)_{ph}; \quad dc_p = (dc_p)_{ch} + (dc_p)_{ph} \tag{14-14}$$

Consequently we have in Eq. 14-13 two terms which can be related to the thermal effect dq_{th} of the elementary chemical reaction at temperature T by the relation, similar to Eq. 2-19,

$$(dh_r)_{ch} + \int_{T_r}^T (dc_p)_{ch} dT = -dq_{th} \tag{14-15}$$

and two other terms which can be evaluated, observing that in a mixing process both h_r and c_p follow the additive rule. One can thus write

$$(m + dm_i)[h_r + (dh_r)_{ph}] = mh_r + \sum_1^N (h_{ri}dm_i)_j$$

where $(h_{ri})_j$ is the enthalpy of the jth injected substance, gaseous or gasified, at the reference temperature. A similar relation can be written for the specific heats. Hence we obtain

$$(dh_r)_{ph} = \sum_1^N [(h_{ri})_j - h_r] \frac{(dm_i)_j}{m}, \quad (dc_p)_{ph} = \sum_1^N [c_{p_i})_j - c_p] \frac{(dm_i)_j}{m} \qquad (14\text{-}16)$$

From Eq. 14-14, 14-15, and 14-16 we obtain

$$dh_r + \int_{T_r}^T dc_p dT = -dq_{th} + \sum_1^N [(h_{iT})_j - h] \frac{(dm_i)_j}{m} \qquad (14\text{-}17)$$

where

$$(h_{iT})_j = (h_{ri})_j + \int_{T_r}^T (c_{p_i})_j dT$$

represents the enthalpy of the jth injected substance, in gaseous form, at the temperature T of the main stream.

Inserting Eq. 14-17 in Eq. 14-13, this in Eq. 14-12 and the latter in Eq. 14-3 divided by c_pT, and making use of Eq. 14-4, we finally obtain the energy equation in the form

$$\frac{dT}{T} + (\gamma - 1)M^2 \frac{du}{u} = \frac{dQ_w + dQ_b + dW_b + d\mathcal{H}}{mc_pT} \qquad (14\text{-}18)$$

where for the sake of brevity we have written

$$d\mathcal{H} = mdq_{th} + \sum_1^N (h_i^0 - h_{iT}^0)_j (dm_i)_j \qquad (14\text{-}19)$$

with the quantity

$$(h_{iT}^0)_j = (h_{iT})_j + \frac{u^2}{2}$$

representing the stagnation enthalpy of the jth injected substance in gaseous form at the temperature T and velocity u of the main stream.

Observe that the conservation equations (Eq. 14-9, 14-11, and 14-19) are in a somewhat more general form than that given to them by Shapiro and Hawthorne, who considered only two injected streams, one of a gaseous substance and the other of a liquid which vaporizes as soon as it is injected. For the injected liquid, then, the value of $(h_i^0)_j$ is computed for the substance in liquid form, and that of $(h^0_{iT})_j$ in vapor form, the difference between the two corresponding values of h_r being the latent heat of vaporization at the reference temperature.

If to the conservation equations and the state equation (Eq. 14-8), we add the following equation obtained from logarithmic differentiation of Eq. 14-7 and 14-6

$$\frac{da}{a} = \frac{1}{2}\left(\frac{d\gamma}{\gamma} + \frac{dT}{T} - \frac{d\mathfrak{M}}{\mathfrak{M}}\right) \tag{14-7a}$$

and the one derived from the definition of M:

$$\frac{dM^2}{M^2} = 2\left(\frac{du}{u} - \frac{da}{a}\right) \tag{14-6a}$$

we have at our disposal six linear homogeneous equations relating the logarithmic differentials of the dependent quantities M^2, u, a, T, ρ, and p to the elementary processes producing the transformation, as represented by the second members of Eq. 14-9, 14-11, and 14-19 and by the logarithmic differentials of A, γ, and $\mathfrak{M}$. Observe that the coefficients of the equations are all constants or simple functions of M^2.

These six equations can be solved with respect to the dependent quantities, so that finally the logarithmic differential of each of these is expressed as a linear combination of the already mentioned expressions representing the independent elementary processes. The coefficients of the linear combinations, that is the influence coefficients, functions of M^2 alone, are shown in the first six rows of Table B,14a. The last two rows give the values of the influence coefficients for two additional important dependent quantities, J and s. For the impulse function we obtain from logarithmic differentiation of Eq. 4-33

$$\frac{dJ}{J} = \frac{dp}{p} + \frac{dA}{A} + \frac{\gamma M^2}{1 + \gamma M^2}\left(\frac{d\gamma}{\gamma} + \frac{dM^2}{M^2}\right)$$

For the second, Shapiro and Hawthorne have made use of the equation

$$\frac{ds}{c_p} = \frac{dT}{T} - \frac{\gamma - 1}{\gamma}\frac{dp}{p} \tag{14-20}$$

which, however, does not represent the total entropy change appearing in the expression (Eq. 4-7a) of the second law, or even the entropy change of the total amount of fluid flowing, expressed by

$$m d\bar{s} = d(ms) - \sum_1^N (s_i dm_i)_j = m ds + \sum_1^N [s - (s_i)_j](dm_i)_j \tag{14-21}$$

The expression (Eq. 14-20) does not even represent the value to replace for the specific entropy variation ds of the main stream alone in

Table B,14a

	$\frac{dA}{A}$	$\frac{dQ + dW_b + d\mathcal{H}}{mc_pT}$	$4c_f \frac{dx}{d_h} + \frac{2dX_b}{\gamma ApM^2} + \sum^N \left[\left(1 - \frac{u_i}{u}\right)\frac{dm_i}{m}\right]_i$	$\frac{dm_i}{m}$	$\frac{d\mathfrak{M}}{\mathfrak{M}}$	$\frac{d\gamma}{\gamma}$
$\frac{dM^2}{M^2}$	$-\frac{2\left(1 + \frac{\gamma - 1}{2}M^2\right)}{1 - M^2}$	$\frac{1 + \gamma M^2}{1 - M^2}$	$\frac{\gamma M^2\left(1 + \frac{\gamma - 1}{2}M^2\right)}{1 - M^2}$	$\frac{2(1 + \gamma M^2)\left(1 + \frac{\gamma - 1}{2}M^2\right)}{1 - M^2}$	$-\frac{1 + \gamma M^2}{1 - M^2}$	-1
$\frac{du}{u}$	$-\frac{1}{1 - M^2}$	$\frac{1}{1 - M^2}$	$\frac{\gamma M^2}{2(1 - M^2)}$	$\frac{1 + \gamma M^2}{1 - M^2}$	$-\frac{1}{1 - M^2}$	0
$\frac{da}{a}$	$\frac{\frac{\gamma - 1}{2}M^2}{1 - M^2}$	$\frac{1 - \gamma M^2}{2(1 - M^2)}$	$-\frac{\gamma(\gamma - 1)M^2}{4(1 - M^2)}$	$-\frac{\frac{\gamma - 1}{2}M^2(1 + \gamma M^2)}{1 - M^2}$	$\frac{\gamma M^2 - 1}{2(1 - M^2)}$	$\frac{1}{2}$
$\frac{dT}{T}$	$\frac{(\gamma - 1)M^2}{1 - M^2}$	$\frac{1 - \gamma M^2}{1 - M^2}$	$-\frac{\gamma(\gamma - 1)M^2}{2(1 - M^2)}$	$-\frac{(\gamma - 1)M^2(1 + \gamma M^2)}{1 - M^2}$	$\frac{(\gamma - 1)M^2}{1 - M^2}$	0
$\frac{d\rho}{\rho}$	$\frac{M^2}{1 - M^2}$	$-\frac{1}{1 - M^2}$	$-\frac{\gamma M^2}{2(1 - M^2)}$	$-\frac{(\gamma + 1)M^2}{1 - M^2}$	$\frac{1}{1 - M^2}$	0
$\frac{dp}{p}$	$\frac{\gamma M^2}{1 - M^2}$	$-\frac{\gamma M^2}{1 - M^2}$	$-\frac{\gamma M^2[1 + (\gamma - 1)M^2]}{2(1 - M^2)}$	$-\frac{2\gamma M^2\left(1 + \frac{\gamma - 1}{2}M^2\right)}{1 - M^2}$	$\frac{\gamma M^2}{1 - M^2}$	0
$\frac{dJ}{J}$	$\frac{1}{1 + \gamma M^2}$	0	$-\frac{\gamma M^2}{2(1 + \gamma M^2)}$	0	0	0
$\frac{ds}{c_p}$	0	1	$\frac{(\gamma - 1)M^2}{2}$	$(\gamma - 1)M^2$ [?]	0	0

Eq. 14-21, except in the special case when the chemical constitutions of all injected substances are identical and the same as those of the main stream, and when no chemical reaction is present. In the general case a much more complicated expression for ds should replace Eq. 14-20, involving the variation of the reference entropy s_r (Eq. 2-11) due to the chemical reactions or to the addition of the injected substances, the variations of γ and $\mathfrak{M}$, and finally the entropy increase following the homogenization of the gas mixture.

Nevertheless the inclusion of the last row of the table is interesting because it allows the immediate derivation of some of the equations obtained in the preceding articles when calculating the entropy change due to friction and heat exchanges. However, we have indicated with a question mark the influence coefficient of dm_i/m, which must be applied with caution because Eq. 14-20 provides the correct expression to introduce in Eq. 14-21 only for no changes in chemical composition and when $dm_i = 0$ or when the injected substances are identical with the main stream.

Flows with constant molecular weight and specific heat. This case may also include the case where the variations of both quantities, and consequently of γ, are small enough so that the gas can be regarded as approximately of fixed composition and polytropic. Such may be the case for the combustion of diluted mixtures in accordance with the observations made at the beginning of Art. 11. Obviously the assumption of constant $\mathfrak{M}$ and c_p implies that the injected substances, if present, have the same values of $\mathfrak{M}$ and c_p, a fact that with rare exceptions may be interpreted as chemical identity of the injected substances with the main stream.

Use can be made, in this case, of the relations (Eq. 4-28) to introduce the stagnation temperature and pressure. By logarithmic differentiation of these relations one obtains

$$\frac{dT^0}{T^0} = \frac{dT}{T} + \frac{\frac{\gamma - 1}{2} M^2}{1 + \frac{\gamma - 1}{2} M^2} \frac{dM^2}{M^2}$$

$$\frac{dp^0}{p^0} = \frac{dp}{p} + \frac{\frac{\gamma}{2} M^2}{1 + \frac{\gamma - 1}{2} M^2} \frac{dM^2}{M^2}$$

Also, Eq. 14-18 can be rewritten as

$$\frac{dT^0}{T} = \left(1 + \frac{\gamma - 1}{2} M^2\right) \frac{dT^0}{T^0} = \frac{dQ_w + dQ_b + dW_b + d\mathcal{H}}{mc_pT}$$

and shows that the logarithmic differential of T^0 can be used instead of the second member of Eq. 14-18 to characterize the corresponding group of elementary causes of the transformation.

Again one can obtain the influence coefficients for the dependent relevant quantities, among which one can include p^0, in terms of the independent elementary causes. The results are given in Table B,14b. The last row must still be considered as concerning the variation of specific entropy of the main stream alone. However, if chemical reactions are present, this represents, instead of the correct entropy variation, the entropy variation of a fictitious flow where the effect of the chemical reactions is replaced by the mere heat addition or substraction corresponding to the thermal effect. Numerical tables for the influence coefficient are given in [*50*] for different values of γ.

Applications of the method. Many of the results of the previous articles can be obtained very easily from the systematic application of Table B,14a and B,14b, and for their derivation the reader is referred to the cited references. Here we give only a brief account of what can be considered one of the most refined and complicated applications of one-dimensional methods.

We have seen in Art. 11 that, if friction is neglected, the stagnation pressure of a flow can be increased by a process of heat subtraction. This opens the attractive possibility of improving the performance of gas turbines by lowering the back pressure below the atmospheric level and reconducting the exhaust pressure to atmospheric through subtraction of heat. In practice, however, this subtraction cannot be performed through ordinary heat exchanges, because in this case the heat exchange is bound to the wall friction by the Reynolds analogy (Eq. 12-1a) and, as shown in Art. 12, the effect of friction always prevails, with a consequent decrease of stagnation pressure.

Shapiro and coworkers [*51*] looked into the alternate possibility of absorbing heat through the evaporation of a liquid. For this purpose water, readily and cheaply available in many types of gas turbine plants, is injected into the hot stream of exhaust gases in a state of fine atomization, so as to increase the contact surface and to produce fast acceleration of the droplets and high rates of exchange.

It is clear that, as in the case of heat exchangers, the transfer of heat cannot be obtained without a certain parasitic frictional effect, represented by the drag of the droplets. We can in this case, as in the case of turbulent flow discussed in Art. 12, write the ratio between the heat exchanged and the drag in the form of the Reynolds analogy (Eq. 12-1a), the coefficient k_q being adjusted so as to make it true. However, the velocity of the main stream u, in the denominator, must be replaced by the relative velocity $|u - u_L|$ between the droplets and the main stream. It will be seen later that k_q always stays around unity, despite the fact that the assumptions on which the Reynolds analogy is based are not verified for the flow around droplets. However, the fact that $|u - u_L|$ can be considerably smaller than u produces a displacement in favor of the heat

Table B,14b

	$\frac{dA}{A}$	$\frac{dT^\circ}{T^\circ}$	$4c_f\frac{dx}{d_h}+\frac{2dX_b}{\gamma ApM^2}+\sum_1^N\left(\left[1-\frac{u_i}{u}\right)\frac{dm_i}{m}\right]_i$	$\frac{dm_i}{m}$
$\frac{dM^2}{M^2}$	$-\frac{2\left(1+\frac{\gamma-1}{2}M^2\right)}{1-M^2}$	$\frac{(1+\gamma M^2)\left(1+\frac{\gamma-1}{2}M^2\right)}{1-M^2}$	$\frac{\gamma M^2\left(1+\frac{\gamma-1}{2}M^2\right)}{1-M^2}$	$\frac{2(1+\gamma M^2)\left(1+\frac{\gamma-1}{2}M^2\right)}{1-M^2}$
$\frac{du}{u}$	$-\frac{1}{1-M^2}$	$\frac{1+\frac{\gamma-1}{2}M^2}{1-M^2}$	$\frac{\gamma M^2}{2(1-M^2)}$	$\frac{1+\gamma M^2}{1-M^2}$
$\frac{da}{a}$	$\frac{\frac{\gamma-1}{2}M^2}{1-M^2}$	$\frac{\frac{1-\gamma M^2}{2}\left(1+\frac{\gamma-1}{2}M^2\right)}{1-M^2}$	$-\frac{\gamma(\gamma-1)M^2}{4(1-M^2)}$	$-\frac{\frac{\gamma-1}{2}M^2(1+\gamma M^2)}{1-M^2}$
$\frac{dT}{T}$	$\frac{(\gamma-1)M^2}{1-M^2}$	$\frac{(1-\gamma M^2)\left(1+\frac{\gamma-1}{2}M^2\right)}{1-M^2}$	$-\frac{\gamma(\gamma-1)M^2}{2(1-M^2)}$	$-\frac{(\gamma-1)M^2(1+\gamma M^2)}{1-M^2}$
$\frac{d\rho}{\rho}$	$\frac{M^2}{1-M^2}$	$-\frac{1+\frac{\gamma-1}{2}M^2}{1-M^2}$	$-\frac{\gamma M^2}{2(1-M^2)}$	$-\frac{(\gamma+1)M^2}{1-M^2}$
$\frac{dp}{p}$	$\frac{\gamma M^2}{1-M^2}$	$-\frac{\gamma M^2\left(1+\frac{\gamma-1}{2}M^2\right)}{1-M^2}$	$-\frac{\gamma M^2[1+(\gamma-1)M^2]}{2(1-M^2)}$	$-\frac{2\gamma M^2\left(1+\frac{\gamma-1}{2}M^2\right)}{1-M^2}$
$\frac{dp^0}{p^0}$	0	$-\frac{\gamma M^2}{2}$	$-\frac{\gamma M^2}{2}$	$-\gamma M^2$
$\frac{dJ}{J}$	$\frac{1}{1+\gamma M^2}$	0	$-\frac{\gamma M^2}{2(1+\gamma M^2)}$	0
$\frac{ds}{cp}$	0	$1+\frac{\gamma-1}{2}M^2$	$\frac{(\gamma-1)M^2}{2}$	$(\gamma-1)M^2$

exchanges, which may upset the unfavorable balance of Art. 12 and result in an increase of stagnation pressure. It is clear that any reduction of $|u - u_L|$ tends to make this possibility more likely, a fact that already shows the advantage of reducing the mean diameter of the droplets and avoiding large droplets even if their number is relatively small.

The problem of the two-phase flow of gases and evaporating droplets can be studied by one-dimensional methods if one assumes that not only the characteristics of the flowing gas, but also those of the droplet stream are uniformly distributed throughout each section. The problem is further simplified by the assumption that only droplets of one size are present at the beginning of the process, and therefore during the whole process.

One can consider the gaseous and the droplet stream as two separate flowing systems, and apply to each of them the equations of conservation. The two systems are connected through the mass, momentum, and heat exchanges between droplets and gases. In order to solve the flow problem, one has to formulate the rates of these exchanges in terms of the characteristics of the two systems.

The equations of conservation can be obtained directly for this particular problem, following the procedure of the reference. Here, without going through the detailed derivation, we shall briefly indicate how the conservation equations can be obtained from the general equations of the Shapiro-Hawthorne method.

Let us indicate with $m + m_L$ the total flow rate of the two systems, m_L being that of the droplet stream. Obviously the total flow rate stays constant, so that

$$m + m_L = \text{const} = m_0 + m_{L0} \tag{14-22}$$

where the subscript $_0$ indicates the conditions in the initial section. Thus at each point the flow in the gaseous system is constituted by a flow rate m_0 of air and $m_{L0} - m_L$ of vapor. Its molecular weight $\mathfrak{M}$ is therefore given by

$$\frac{m}{\mathfrak{M}} = \frac{m_0}{\mathfrak{M}_A} + \frac{m_{L0} - m_L}{\mathfrak{M}_V}$$

where the subscripts A and V indicate air and vapor and $\mathfrak{M}$ is a function only of m_L. The gas constant is then obtained from Eq. 14-6. Obviously such thermodynamic functions as c_p and h, obeying the additive rule, can be obtained from the relation

$$mh = m_0 h_A + (m_{L0} - m_L) h_{VT}$$

or the analogous one for c_p. Here the subscript T indicates that the relative value of h_V must be taken at the temperature of the gaseous stream. Once $\mathcal{R}$ and c_p are known, one can obtain the value of γ.

From Eq. 14-22 one finds that the elementary mass dm_i, injected per unit time into the gaseous system and extracted from the liquid system, which appears in the continuity equation (Eq. 14-9), is

$$dm_i = dm = -dm_L \tag{14-23}$$

This injected mass carries with it the velocity u_L of the droplets (which therefore can replace u_i in the equations) and the enthalpy content of the vapor at the temperature T_L of the droplets, which we shall indicate simply by h_V. Thus we can write, in place of Eq. 14-4,

$$dI_i = -u_L dm_L, \quad dH_i^0 = -\left(h_V + \frac{u_L^2}{2}\right) dm_L$$

The elementary drag dX_b of the droplets is the force linking the two systems. It can be considered as acting on the liquid system in the positive x direction. Applying to this system the law of conservation of momentum (law 2 of Art. 3) we can write $d(m_L u_L) = dX_b + u_L dm_L$, the last term representing the (negative) momentum injected in the system within the control surface. In the previous equation the force due to the pressure gradient acting on the very small cross-sectional area of the droplets has been disregarded. One can therefore replace dX_b by the equivalent expression $dX_b = m_L du_L$, so that, taking into account Eq. 14-10, the corresponding term in the momentum equation (Eq. 14-11) can be replaced by

$$\frac{dX_b}{\gamma A p M^2} = \frac{m_L}{m} \frac{du_L}{u} \tag{14-24}$$

The heat $-dQ_b$ supplied by the gaseous to the liquid system is partly used to increase the enthalpy of the droplets, partly to evaporate the amount $-dm_L$. Thus we can write

$$-dQ_b = m_L dh_L - (h_V - h_L) dm_L \tag{14-25}$$

where h_L is the enthalpy of the liquid corresponding (like h_V for the vapor) to the temperature T_L. Obviously $h_V - h_L$ represents the latent heat of vaporization at temperature T_L.

The term $-dW_b$ stands for the energy extracted in mechanical form from the gaseous system. Now, the work performed by the drag of the droplets is not entirely extracted from the system, since part of it is dissipated within the gaseous system itself. The only mechanical energy extracted is the increase of kinetic energy of the liquid system. Therefore we can write

$$-dW_b = m_L u_L du_L \tag{14-26}$$

Finally, since $dq_{th} = 0$, the last term (Eq. 14-19) appearing in the energy equation (Eq. 14-18) is given by

$$d\mathcal{H} = -\left(h_V - h_{VT} + \frac{u_L^2}{2} - \frac{u^2}{2}\right) dm_L \tag{14-27}$$

We must still write the relations governing the mass, momentum, and heat exchanges between the two systems. These relations can easily be written for the single droplets, observing that the diameter δ of these is related to the mass m_L by the simple relation

$$\left(\frac{\delta}{\delta_0}\right)^3 = \frac{m_L}{m_{L0}} \tag{14-28}$$

δ_0 representing the (uniform) diameter of the droplets in the initial section. Observing also that for the droplet system $dx = u_L dt$, one can write the evaporation rate for a droplet as [52]

$$-\frac{d}{dt}\left(\frac{\pi\delta^3}{6}\rho_L\right) = -u_L\frac{d}{dx}\left(\frac{\pi\delta^3}{6}\rho_L\right) = \alpha_D\pi\delta^2(\rho_V - \rho_{VT}) \tag{14-29}$$

where ρ_V represents the density of the saturated vapor at temperature T_L, ρ_{VT} the vapor density at temperature T and at its partial pressure in the main stream, and the coefficient α_D is related to the diffusivity D of the vapor in the air, to the viscosity μ, and to the other quantities by the relation

$$\alpha_D = [2 + 0.6(Sc)^{\frac{1}{3}}(Re)^{\frac{1}{2}}]\frac{D}{\delta}, \quad Sc = \frac{\mu}{\rho D}, \quad Re = \frac{\rho\delta|u - u_L|}{\mu}$$

The drag of a droplet, balancing its inertia force, is given by

$$\rho_L\frac{\pi\delta^3}{6}\frac{du_L}{dt} = \rho_L\frac{\pi\delta^3}{6}u_L\frac{du_L}{dx} = C_D\frac{\pi\delta^2}{4}\frac{\rho}{2}(u - u_L)|u - u_L| \tag{14-30}$$

where the drag coefficient C_D can be approximated by

$$C_D = \frac{24}{Re} \quad \text{for} \quad 0 < Re \leqq 1$$

$$C_D = \frac{24}{(Re)^{\frac{2}{3}}} \quad \text{for} \quad 1 \leqq Re \leqq 390$$

$$C_D = 0.45 \quad \text{for} \quad 390 \leqq Re$$

Finally, the heat supplied to the droplet, related to the rate of increase of the droplet enthalpy and to the evaporation rate by the relation corresponding to Eq. 14-25, is given by

$$\alpha_T\pi\delta^2(T - T_L) = \frac{\pi\delta^3}{6}\rho_L u_L\frac{dh_L}{dx} - (h_V - h_L)u_L\frac{d}{dx}\left(\frac{\pi\delta^3}{6}\rho_L\right) \tag{14-31}$$

with the heat transfer coefficient α_L expressed in terms of the conductivity k, the viscosity, and the other quantities by the relation [52]

$$\alpha_T = [2 + 0.6(Pr)^{\frac{1}{3}}(Re)^{\frac{1}{2}}]\frac{k}{\delta}, \quad Pr = \frac{c_p\mu}{k}$$

Summarizing the results, we still have at our disposal the same six equations of the generalized Shapiro-Hawthorne method, containing now, in addition to the same six dependent variables, m_L, δ, T_L, and u_L because of the introduction of Eq. 14-24, 14-25, 14-26, and 14-27. In addition we

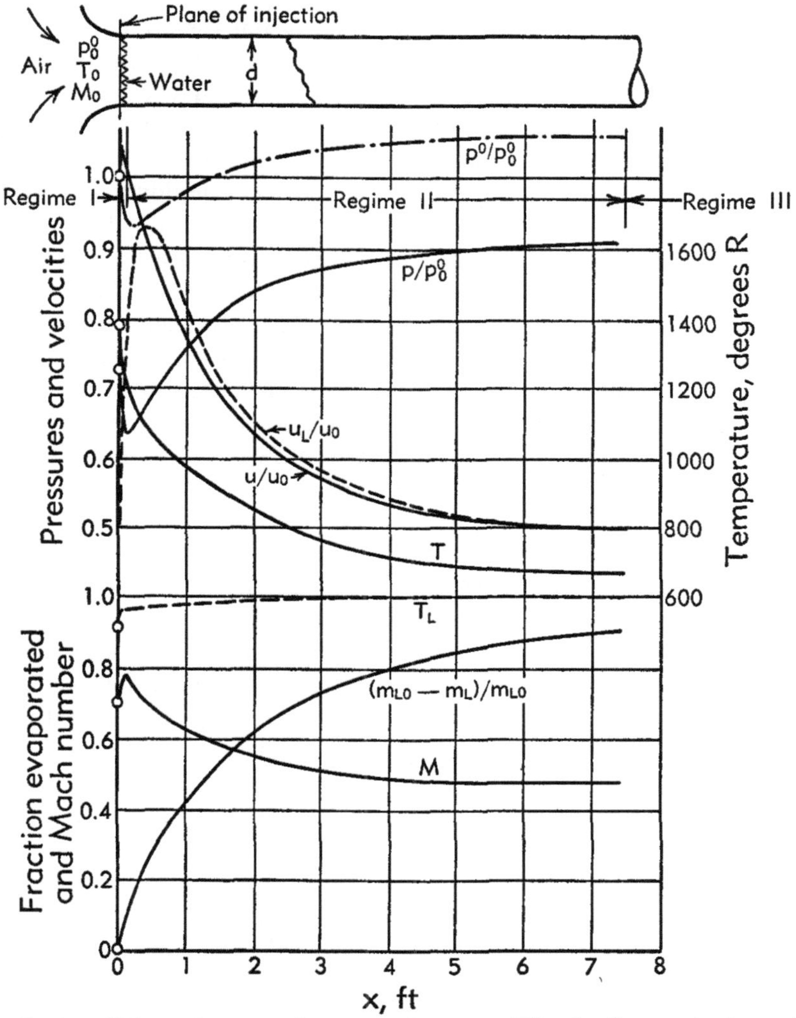

Fig. B,14a. Effect of evaporation on various quantities in the constant area flow of air with water droplets. $M_0 = 0.7$; $T_0^0 = 1500°R$; $p_0^0 = 16.7\,lb/in.^2$; $c_f/d = 0.004\,ft^{-1}$; $m_{L0}/m_0 = 0.2$; $T_{L0} = 530°R$; $u_{L0} = 0$; $\delta_0 = 13.8\mu$. The flow stays entirely subsonic.

have the four extra Eq. 14-28, 14-29, 14-30, and 14-31. Thus we have a complete, though complicated, differential system with x as independent variable, which can be numerically integrated for given initial conditions and given duct geometry as represented by $A(x)$.

In [*52*] a certain number of examples are given. Here we shall report only a few characteristic results for the case of a constant area duct.

Fig. B,14a shows the variations of the relevant quantities for a particular set of values and subsonic velocity throughout the process. One can distinguish three regimes. In regime I, the relative velocity between gas and droplets is large (initially $u_{L0} = 0$). Thus the effect of the droplet drag is predominant, and, despite the fact that the evaporation rate is maximum here, the stagnation pressure falls. In the meantime the droplets are being heated up and very rapidly attain a temperature close to the "wet bulb temperature," where the heat supplied practically balances the evaporation heat, and after which T_L stays nearly constant. In regime II the relative velocity, and therefore the drag, has dropped to a sufficiently low value and the effect of droplet evaporation becomes predominant, so that p^0 increases and becomes appreciably larger than its initial value. Observe that since u decreases during this regime, the droplets after a period of acceleration reach $u_L = u$, after which they stay faster than the gas stream, and the corresponding drag, becoming negative, helps in producing a p^0 increase. This effect, however, is quite small since u_L stays very close to u. In regime III the gas temperature, which has been steadily decreasing, is sufficiently close to T_L, and the droplet diameter is sufficiently small, so as to reduce the evaporation rate to a very low value. Then the effect of wall friction becomes predominant and p^0, after reaching a maximum p^0_{max}, starts decreasing again.

By varying the initial conditions it is possible to show the influence of different parameters on the results. One of the most important parameters, in agreement with our previous statement, seems to be the initial diameter of the droplet; by varying this diameter the length of duct required to evaporate a certain fraction of the liquid varies approximately as the square of the diameter. For too large δ_0 the evaporation rate may become so small that regime II disappears completely, while even a small decrease of δ_0 produces a substantial improvement of p^0_{max}. The diameter of the duct directly affects the relative importance of the wall friction, and therefore the transition from regime II to regime III and the value of p^0_{max}. Below a certain duct size, regime II disappears. However, increasing the duct diameter has only an asymptotic effect, and above a certain size no further appreciable gain can be realized. The effect of the initial water-to-air ratio beyond a certain limit is not very critical. It becomes critical, however, below this limit, a result consistent with the obvious fact that for zero water content no favorable effect can be present.

The initial Mach number has a very important influence on the development of the process, as shown in Fig. B,14b, where the variations of M and p^0 are given as functions of the fraction evaporated for a set of fixed initial conditions, except for varying M_0 (δ_0 varies too, since it is related to M_0 during the atomization process). The most important effect of the increase of M_0 is to accentuate a phenomenon taking place during regime I. Due to the effect of droplet drag, the Mach number increases during most of regime I, and starts decreasing because of the effects of evaporation only after having reached a maximum. This maximum becomes sharper and higher with increasing M_0, until for a well-determined critical value of M_0 it reaches unity. Correspondingly the same phenomena observed in choking due to constriction (Art. 6) can be observed. Beyond the point where $M = 1$ one can again have a subsonic flow, or a supersonic flow followed by a transition to subsonic through a particular type of

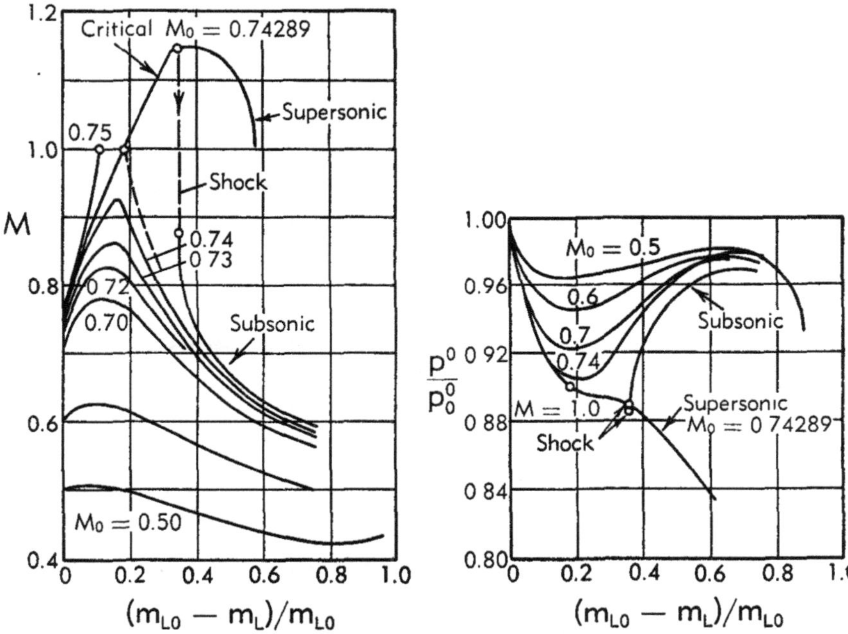

Fig. B,14b. Effect of initial Mach number on the constant area flow of air with water droplets ($T_0^0 = 1500°R$; $p_0^0 = 14.7$ lb/in.^{2}; $c_f/d = 0.0226$ ft^{-1}; $m_{L0}/m_0 = 0.2$; $T_{L0} = 530°R$; $u_{L0} = 0$; $\delta_0 = 19\mu$ at $M_0 = 0.5$; $\delta_0 = 14.9\mu$ at $M_0 = 0.65$; $\delta_0 = 13.1\mu$ at $M_0 = 0.743$). Corresponding to the critical value of M_0 choking is produced, the velocity may become supersonic and shock phenomena may appear.

discontinuity similar to a shock, the position of which is determined by the back pressure. Only one such discontinuity is shown in Fig. B,14b, the values of the various quantities after the discontinuity being related to those before by the conservation equations, as derived in [*52*]. Obviously no flow is physically possible for M_0 larger than the critical value. In Fig. B,14c the variations of the static pressure as a function of x are given for the same initial conditions, for three values of M_0 and for various back pressures. They are compared with experimental values obtained in conditions as close as possible to those assumed in the calculations, showing an agreement which must be considered quite good for such a complicated process. Another result, obtained experimentally, but likely to be obtainable also from calculations, is that, so far as p^0_{max} is concerned, the effect of increasing T_0^0 and p_0^0 is beneficial.

The results given here are for A = const. Considerable improvements of the maximum increase of p^0 can result from the relief of this condition, as shown in the reference for a particular example. Values up to 20 per cent above the entrance stagnation pressure, including the losses in the final exhaust diffuser, seem to be possible.

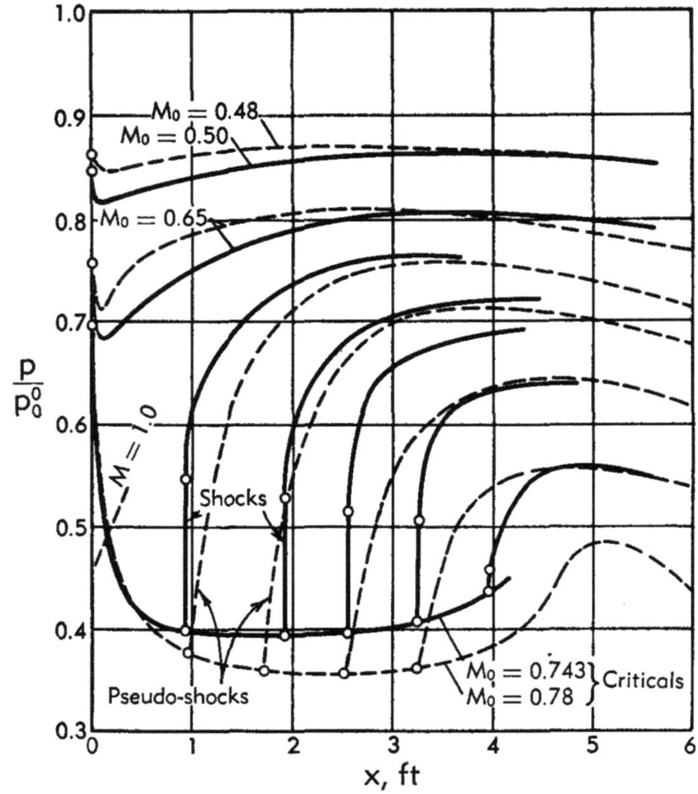

Fig. B,14c. Pressure distributions along the duct. Solid line: theoretical (T_0^0 = 1500°R; p_0^0 = 14.7 lb/in.^{2}; c_f/d = 0.0226 ft^{-1}; m_{L0}/m_0 = 0.2; T_{L0} = 530°R; u_{L0} = 0; δ_0 = 19μ at M_0 = 0.5; δ_0 = 14.9μ at M_0 = 0.65; δ_0 = 13.1μ at M_0 = 0.743). Dashed line: experimental.

Application of the second law of thermodynamics. When material is being extracted from the main stream, one must be careful in the selection of the values of certain quantities if absurd results in contradiction with the second law of thermodynamics are to be avoided.

This can be shown easily by writing down explicitly Eq. 4-7a for the particular case where chemical reactions are absent and the injected substance is gaseous and identical with the main flow (which is in particular true for the case of mass extraction). The value of ds to introduce in Eq. 4-7a is now correctly given by Eq. 14-20. Using the influence coefficients of Table B,14a, and observing that in the absence of immersed bodies, $dQ_b = dW_b = dX_b = 0$, ds is given by

$$\frac{ds}{c_p} = \frac{dQ_w}{mc_pT} + (\gamma - 1)M^2\frac{dX_w}{mu} + \left[\frac{h_i^0 - h^0}{c_pT} + \left(1 - \frac{u_i}{u}\right)(\gamma - 1)M^2\right]\frac{dm_i}{m}$$

Here, because of Eq. 12-3, $dX_w/mu = 2c_f dx/d$.

Therefore we obtain Eq. 4-7a in the form

$$\frac{dS}{mc_p} = \left(\frac{1}{T} - \frac{1}{T_w}\right)\frac{dQ_w}{mc_p} + (\gamma - 1)M^2\frac{dX_w}{mu} + \left[\frac{1}{c_p}\left(s - s_i + \frac{h_i^0 - h^0}{T}\right)\right.$$
$$\left. + \left(1 - \frac{u_i}{u}\right)(\gamma - 1)M^2\right]\frac{dm_i}{m} \geqq 0$$

We can now express dQ_w in terms of dX_w using the Reynolds analogy, (Eq. 12-1a). Introducing the enthalpy recovery factor, defined by the first Eq. 12-2, we obtain

$$\frac{dS}{mc_p} = \left\{\frac{k_q}{c_p}(h_w - h)\left(\frac{1}{T} - \frac{1}{T_w}\right) + \left[1 - \frac{k_q r}{2}\left(1 - \frac{T}{T_w}\right)\right](\gamma - 1)M^2\right\}\frac{dX_w}{mu}$$
$$+ \left[\frac{1}{c_p}\left(s - s_i - \frac{h - h_i}{T}\right) + \frac{1}{2}\left(1 - \frac{u_i}{u}\right)^2(\gamma - 1)M^2\right]\frac{dm_i}{m} \quad (14\text{-}32)$$

Now observe that the coefficient of dX_w is certainly positive because so are its first term, both for $T > T_w$ or $T < T_w$ (it is zero for $T = T_w$) and its second term, if the condition $k_q r < 2$ is satisfied (as it is in general). Similarly, the coefficient of dm_i is certainly positive because

$$s - s_i - \frac{h - h_i}{T} = \int_{T_i}^{T} c_p\left(\frac{1}{T'} - \frac{1}{T}\right)dT'$$

is larger than zero for $T_i \gtrless T$ and zero for $T = T_i$. Thus if $dX_w \geqq 0$ and $dm_i \geqq 0$ (injection), the second law is certainly satisfied.

For $dm_i < 0$ (extraction) the question is more delicate. It is true that if one assumes that the material is extracted at the conditions of the main stream ($T_i = T$, $u_i = u$) the coefficient of dm_i vanishes and the second law is satisfied for positive or zero friction. However, this assumption is certainly not a realistic one, because we know that the temperature of the material to be extracted must reach the wall temperature and its velocity must vanish when it reaches the wall. With $T_i = T_w$, $u_i = 0$ the coefficient of dm_i is always positive and the corresponding contribution to dS is negative. It is therefore clear that a certain relation must be satisfied between dm_i and dX_w if the results are not to contradict the second law. For instance, the neglect of friction ($dX_w = 0$) would certainly lead to absurd results. A simdle, sufficient form of the condition to be satisfied between dX_w and dm_i is obtained if one rewrites Eq. 14-32, with $T_i = T_w$ and $u_i = 0$, in the form

$$\frac{dS}{mc_p} = \left\{\frac{k_q}{c_p}(h_w - h)\left(\frac{1}{T} - \frac{1}{T_w}\right) + \left[1 - \frac{k_q r}{2}\left(1 - \frac{T}{T_w}\right)\right](\gamma - 1)M^2\right\}$$
$$\left(\frac{dX_w}{mu} + \frac{dm_i}{m}\right) + \left\{\frac{1}{c_p}\left[s - s_w - \frac{h - h_w}{T} - k_q(h_w - h)\left(\frac{1}{T} - \frac{1}{T_w}\right)\right]\right.$$
$$\left. - \frac{1}{2}\left[1 - k_q r\left(1 - \frac{T}{T_w}\right)\right](\gamma - 1)M^2\right\}\frac{dm_i}{m}$$

Now the coefficient of dm_i is certainly negative if $1 \leqq k_q \leqq 1/r$ because

$$s - s_w - \frac{h - h_w}{T} - (h_w - h)\left(\frac{1}{T} - \frac{1}{T_w}\right) = \int_{T_w}^{T} c_p\left(\frac{1}{T'} - \frac{1}{T_w}\right)dT' \leqq 0$$

and it is likely to be negative in all practical cases even if the above condition for k_q is not strictly verified, $k_q r$ being, however, as is generally the case, close to 1.

Hence we see that the condition $dX_w \geqq 0$, which is sufficient to insure the fulfillment of the second law for $dm_i \geqq 0$, can be replaced by the sufficient condition

$$dX_w + u\,dm_i \geqq 0$$

when $dm_i < 0$. The interpretation of this condition is obvious. In order to assure the fulfillment of the second law in the case of mass extraction one must include in the equations a wall friction at least equal to the loss of momentum of the material extracted when it reaches the wall. Clearly a friction coefficient constantly equal to that which characterizes the flow without injection cannot satisfy this condition, and the friction coefficient must increase with increasing extraction rate. This is actually what happens in reality, accompanied by a simultaneous decrease of the boundary layer thickness. Similar effects, in the opposite direction, are produced when mass is injected, in which case the skin friction decreases without, however, being bound by particular limitations in relation to the second law.

B,15. Flows with Piecewise Uniformity. In Art. 5 and 11 we have given the one-dimensional treatment of two particular types of flow (shockless model of pseudo-shock and flame-type combustor), characterized by the fact that the velocity and temperature distributions, initially uniform on the whole section, become nonuniform under the effect of a nonuniformizing process (separation, in the first case; combustion, in the second). The treatment is made possible by the approximate assumption that the flow can be divided into two regions with uniform velocity and temperature distributions, so that on each section the distributions are "piecewise uniform." In the cases considered the pressure distribution still satisfies the condition of uniformity on each section.

Following substantially the same lines, it is possible to develop a one-dimensional treatment of other types of flow characterized by the piecewise uniformity of the various distributions including, possibly, the pressure distribution. A quite general class of flows, treated in this article, is that in which the nonuniformity is deliberately introduced in the initial section of a duct, for some particular purpose. Important examples of devices of this kind are the injectors or ejectors widely used in engineering, the twin-flow (by-pass) turbojet with mixing before the expansion in the nozzle, and the ducted rocket. In these devices a higher velocity flow (primary, inducing) and a lower velocity flow (secondary, induced) are brought through separate inlet ductings to the initial section of a mixing duct in which the flows are free to mix. The mixing proceeds in principle until the flow reaches a state of practical uniformity at the end of the mixing duct. The flow is then brought to the additional outlet ducting (such as diffuser or nozzle) required to attain the particular purpose of the device.

All these devices may differ in the nature and the disposition of the inlet and outlet ductings, but have in common the presence and the purpose of the mixing duct. There are, however, other possible distinctions that can be made concerning the processes taking place in the latter. In the simplest case the primary and the secondary flows convey the same (or practically the same) substance at identical stagnation temperatures (for instance, ejector to increase the vacuum at the downstream end of a blowdown supersonic wind tunnel, using the same source of compressed air). In other cases the two substances may exhibit more or less different physicochemical properties, and be at more or less different stagnation temperatures (ejectors for evacuation of chemicals; twin-flow turbojet). Finally, in addition to the above differences, the difference of chemical nature may result in a chemical reaction (for instance, combustion) when the two flows are mixed together (ducted rocket).

Other diversities concerning the mechanical conditions in which the two flows are brought together are the following. If the velocities of the two flows in the initial section of the mixing chamber are subsonic (which happens for unchoked flows, but also when one or both flows are discharged into the mixing duct through Laval nozzles operating in choked conditions with very strong overexpansion (see Art. 8), the static pressures of the two flows must be the same, so that the pressure, uniform in the initial section, may remain uniform throughout the mixing duct. This pressure uniformity may of course also be attained in the case when one or both flows are discharged at supersonic velocities, the corresponding Laval nozzles being designed so as to produce correct expansion; but this is not true in general, and the Laval nozzles are likely to be incorrectly expanded except for one particular operating condition, with the result that the static pressures of the two flows are different and the initial pressure distribution in the mixing duct is piecewise uniform. In these cases, as already indicated in Art. 8, complicated recurrent patterns of oblique shock waves and expansion waves must appear in the mixing region, where the pressure distribution, certainly nonuniform, may not be correctly represented even by a piecewise uniform distribution.

The geometry of the system may have considerable influence on the actual development of the processes in the mixing duct. This is obviously true for the shape of the mixing duct itself or for the angle between the directions of the inlet ductings when they merge together (this angle should be as close to zero as possible). A more subtle effect results from the relative extent of the interface between the two flows. For instance, if the two flows issue from concentric circular nozzles, the resulting interface is smaller than that obtained discharging one of the two flows through several circular nozzles more or less uniformly distributed in the single circular nozzle carrying the other flow. The rate at which mixing takes place increases with the extent of interface, and the minimum length of mixing duct required for practically uniform mixing decreases. Also, the pressure nonuniformities deriving from incorrect expansion of the nozzles are likely to decay faster.

That the mixing rate may be a primary factor in the operation of a mixing duct is proved by the fact that in certain instances the minimum length of duct required for practically complete mixing is well above the maximum length compatible with other requirements of the device under consideration. In such cases the mixing duct cannot perform its function properly unless the necessary increase of mixing rate is achieved, for instance by acting on the interface.

The mixing rate, and therefore the minimum length required for the mixing duct, also plays a role through the frictional effects which obviously decrease with decreasing length. Other geometric conditions can also affect the frictional effects, which are minimized if the primary flow with its higher velocity is imbedded in the secondary flow and does not come in contact with the walls of the mixing duct.

Finally observe that an interesting particular type of mixing duct is obtained when the velocity of the secondary flow is reduced to zero. In this case the corresponding discharge section can be replaced by a solid wall and the mixing duct takes the aspect of an ordinary duct with a sudden enlargement of the section.

One-dimensional treatment of mixing ducts. Let us first consider the case when the pressure is uniformly distributed in the initial section, and therefore also at any other section of the mixing duct. This case presents some similarity with the pseudo-shock model discussed in Art. 5 and can be treated along similar lines. If, for simplicity, we neglect the wall friction, instead of the isentropic core and the dissipative region of the pseudo-shock, there are in the present case two isentropic regions with practically uniform velocity and temperature distributions corresponding to the fractions of the two original flows (assumed to satisfy, separately, the condition of practical uniformity in the initial section) which have not yet been reached by the spreading of the third region, in which mixing and dissipation take place. Obviously for a given section the velocity, temperature, and other properties in the dissipative region vary continuously between those of the two isentropic regions and are by no means uniformly distributed. However, here again we can make the same approximation used in Art. 5 by replacing the actual dissipative region with a fictitious one where the distributions are uniform, with proper mean values of velocity, temperature, and other properties (Art. 16). With this approximation, assuming the knowledge of the spreading rates of the dissipative region in the two isentropic regions, and with proper assumptions about the chemical reactions if present (for instance, that the reactions take place instantly at mixing), the one-dimensional treatment of the mixing process becomes possible. It is clear that at a certain distance from the initial section the isentropic regions will disappear, which happens in general at different distances for the two regions. After both regions have disappeared, only the dissipative region is left and, as a result of the approximation of uniformity in this region, the mixing process is over. In practice this is not exactly true, and it will take some additional length of duct before the process of uniformization in the dissipative region is practically through.

Here we shall not carry out in detail this one-dimensional treatment of the mixing process, not only because it represents only an extension of that of Art. 5, but also because it does not apply to the other case in which pressure nonuniformities are also present. It is actually doubtful that a satisfactory one-dimensional model can be developed for this case. We shall instead concentrate our attention only on the determination of the final conditions attained when the mixing and the chemical processes are practically over and the uniformity conditions are satisfied. This separate determination is not possible for the most general geometry of mixing duct, because the final conditions are not independent of the intermediate stages; but it is possible for ducts of the p-A power family (including the constant area and the constant pressure case) for which this independence is realized. Moreover this determination can also cover the case when pressure nonuniformities are present and the area is constant. Since for this particular duct shape the resultant of the pressures on the wall vanishes, the fact that the mean wall pressure on each section (defined by Eq. 3-8) does not coincide with the mean pressure on the section (defined by Eq. 3-7) does not introduce any difficulty. However, if one attempts to extend to the case of nonuniform pressure distributions the concept of the p-A power family, some difficulties result from the difference of these two mean pressures. The difficulties can be overcome in principle if one concentrates only on the initial and final conditions and on the relation

$$\frac{\ln A_f - \ln A_i}{\ln (\bar{p}_f A_f) - \ln (\bar{p}_i A_i)} = \frac{\int_{A_i}^{A_f} \bar{p}_w dA}{\bar{p}_f A_f - \bar{p}_i A_i} = 1 - \varepsilon \tag{15-1}$$

into which Eq. 4-19 and 4-19a should be generalized if the momentum equation has to be reduced to the form of Eq. 4-21 with $I_i = 0$, as shown later. If, by proper shaping of the intermediate geometry of the duct between two fixed sections, the two members of this equation can be actually made equal, their common value represents the proper value of $1 - \varepsilon$ to introduce in the equations that follow. Observe that $\varepsilon = 1$ still represents the constant area case, but $\varepsilon = 0$ does not correspond now to the condition of constant wall pressure, because, as immediately seen, the mean value of $\bar{p}_w$ must now be equal to $\bar{p}_f = \bar{p}_i$ while, due to the pressure nonuniformity in the initial section, the initial wall pressure is different from $\bar{p}_i$.

Another observation is fundamental when pressure nonuniformities are present. The result of the determination of the final conditions makes sense only if in the final section not only the velocity, temperature, and composition are sufficiently uniform, but also the pressure nonuniformities have decayed to a sufficiently low level. In the opposite case, which may be true for supersonic final conditions, the pressure itself, as well as the velocities and the temperatures, must go on oscillating in nonuniform patterns, and the result of the determination only represents an average value of these patterns along the duct. These oscillating patterns cannot be present when the final velocity is subsonic.

Determination of the final conditions in mixing ducts. Assume, for more generality, that $N \geqq 2$ streams are brought together in the initial section with individual flow rates m_j ($j = 1, 2, \ldots, N$). Then, since no further mass injection is effected between the initial and final sections, the conservation of mass is expressed by

$$m_i = \sum_1^N m_j = m_f = m \tag{15-2}$$

m being the constant flow rate common to all sections.

For each stream we assume uniform velocity u_{ij}, temperature T_{ij}, pressure p_{ij}, and composition; and we call A_{ij} the corresponding portion of the total cross-sectional area A_i, and h_{ij} the corresponding stagnation enthalpy. We have the following relations

$$\begin{aligned} A_i &= \sum_1^N A_{ij}, & \bar{p}_i A_i &= \sum_1^N p_{ij} A_{ij} \\ I_i &= \sum_1^N I_{ij} = \sum_1^N m_j u_{ij}, & H_i^0 &= \sum_1^N H_{ij}^0 = \sum_1^N m_{ij} h_{ij}^0 \end{aligned} \tag{15-3}$$

the second defining, according to Eq. 3-7, the mean value of the pressure, and the third and fourth the total flux of momentum and stagnation enthalpy, all referred to the initial section.

For each stream one can define, in accordance with Eq. 4-20, a generalized impulse function $J_{ij}^{(\varepsilon)} = I_{ij} + \varepsilon p_{ij} A_{ij}$ with a given value of ε; one then has

$$J_i^{(\varepsilon)} = \sum_1^N J_{ij}^{(\varepsilon)} = I_i + \varepsilon \bar{p}_i A_i \tag{15-4}$$

We assume now that in the final section the mixing is practically complete, so that the flow can be treated as uniform. The conservation of momentum is then expressed by Eq. 3-6 with $I_i = 0$,[20] the pressure p_1 replaced by $\bar{p}_i$; and p replaced by $\bar{p}_w$ (Eq. 3-8) in the integrand, with the following result

$$I_f - I_i = \bar{p}_i A_i - p_f A_f - \int_{A_i}^{A_f} \bar{p}_w dA - X \tag{15-5}$$

where $X = X_w + X_b + X_m$; in general $X_m = X_b = 0$ and $X = X_w$.

If the value of ε has been chosen so as to satisfy Eq. 15-1, Eq. 15-5 can be rewritten in the compact form

$$J_f^{(\varepsilon)} - J_i^{(\varepsilon)} = -X \tag{15-6}$$

that is, in the same form as Eq. 4-21, despite the pressure nonuniformities. Finally the conservation of energy, in the absence of mass forces and work exchanges, is obtained from Eq. 3-13 and 15-3 in the form

$$mh_f^0 - \sum_1^N m_{ij} h_{ij}^0 = Q \tag{15-7}$$

where Q represents the total heat received from external sources. If chemical reactions take place between the various streams the enthalpies should be referred to a reference chemical system (such as the elements) so as to include the heat of reaction, in the way discussed in Art. 2.

If the conditions in the initial section are known and X, Q, and ε are prescribed, from Eq. 15-2 and 15-7 one can obtain h_f^0 and from this derive the value of T_f^0 and c_{p_f}, γ_f corresponding to the final gases and the

[20] In Eq. 3-6, the subscript $_i$ has a meaning different from that used in this article.

relative chemical composition. The values of w_f, p_f, and A_f are then obtained from the value of $J_f^{(e)}$ determined by Eq. 15-6, because from Eq. 4-34a we have

$$J_f^{(e)} = m\sqrt{2c_{p_f}T_f^0}\left(w_f + \delta_f\frac{1 - w_f^2}{2w_f}\right) = p_f A_f\left(\varepsilon + \frac{2\gamma_f}{\gamma_f - 1}\frac{w_f^2}{1 - w_f^2}\right) \quad (15\text{-}8)$$

where, as in Art. 11, we have used the abbreviation of Eq. 11-5, that is

$$\delta_f = \varepsilon\frac{\gamma_f - 1}{\gamma_f}$$

Solving the first of these equations with respect to w_f we have

$$w_f = \frac{J_f^{(e)}}{m\sqrt{2c_{p_f}T_f^0}\,(2 - \delta_f)} \pm \sqrt{\left[\frac{J_f^{(e)}}{m\sqrt{2c_pT_f^0}\,(2 - \delta_f)}\right]^2 - \frac{\delta_f}{2 - \delta_f}} \quad (15\text{-}9)$$

This expression for w_f shows that for $\delta_f < 0$ ($\varepsilon < 0$) or $\delta_f > 2$ ($\varepsilon > 2\gamma_f/(\gamma_f - 1)$) only the upper sign gives positive values of w_f and the other solution has no physical meaning. Thus in this range of δ_f there is always one and only one positive solution w_f which is, however, limited by the condition $w_f \leqq 1$. This condition can be translated into $J_f^{(e)} \leqq m\sqrt{2c_pT_f^0}$ for $\delta_f < 0$ or $J_f^{(e)} \geqq m\sqrt{2c_pT_f^0}$ for $\delta_f > 2$, and in turn one obtains the corresponding limitations for $J_i^{(e)}$ from Eq. 15-6. We shall discuss these limitations further in Art. 17.

The situation is different when δ_f is in the range

$$0 \leqq \delta_f \leqq 2 \quad (15\text{-}10)$$

which is seen by writing the solution (Eq. 15-9) in the form, similar to Eq. 11-7,

$$w_f = (w_f)_{cr}(j_f \pm \sqrt{j_f^2 - 1}) \quad (15\text{-}11)$$

with

$$(w_f^2)_{cr} = \frac{\delta_f}{2 - \delta_f}, \quad j_f = \frac{J_f^{(e)}}{m\sqrt{2\varepsilon(2 - \delta_f)\mathcal{R}_f T_f^0}} \quad (15\text{-}12)$$

In the range of Eq. 15-10 both $(w_f)_{cr}$ and j_f are real and positive, and hence there are two positive solutions w_f for $j_f > 1$, one supercritical ($>(w_f)_{cr}$), the other subcritical ($<(w_f)_{cr}$). For $j_f = 1$ the two solutions coincide with $(w_f)_{cr}$. Finally for $j_f < 1$ there is no real solution.

These results are similar to those obtained in Art. 11 for the case of heat addition and can be discussed in a similar way. If the initial conditions and the prescribed value of ε are such that with the value of $J_f^{(e)}$ obtained from Eq. 15-6 the resulting value of j_f is <1, the flow is impossible. Thus we find again a phenomenon of choking when the initial conditions or the heat addition are gradually altered in such a way that the value $j_f = 1$ is attained, starting from values larger than one. The final velocity at choking equals the critical value defined by Eq. 15-12. Observe, however, that the cause for choking can in this case be unrelated with heat addition, since both heat addition and chemical reactions can be absent and choking may be produced only by the mixing of the streams if the initial value of $J_i^{(\varepsilon)}$ is below a well-determined value. For instance, if one mixes a supersonic and a subsonic air stream in a constant section duct ($\varepsilon = 1$), and if the velocity of the subsonic stream is gradually increased, choking is produced when a certain subsonic value is achieved. This kind of choking could be distinguished from the others discussed in preceding articles as "choking by mixing." Obviously, when both mixing and heat addition or chemical reactions are present, the cause for choking resides in the combination of the various effects.

An interesting difference between the case of Art. 11 and the present case concerns the problem of choosing between the two solutions given by Eq. 15-11 when δ_f is in the range of Eq. 15-10 and the flow is unchoked. In Art. 11 the choice was made on the grounds of the continuity of the solutions between the initial and the final sections. In the present case, however, we are unable to determine on purely one-dimensional grounds the intermediate solutions and to apply the principle of continuity. The only thing we can say at this point is that, of the two values of w_f, the subcritical solution must always be possible because it can always be obtained from the supercritical through a pseudo-shock; on the contrary the physical possibility of the supercritical solution is subject to limitations. An attempt to express these limitations for the case when the initial pressure distribution is uniform is discussed in Art. 17.

Once the value of w_f is determined, application of the second Eq. 15-8 and of the first Eq. 15-1 gives

$$\left(\frac{p_f}{p_i}\right)^{\frac{1}{\varepsilon}} = \left(\frac{A_f}{A_i}\right)^{\frac{1}{1-\varepsilon}} = \frac{p_f A_f}{\bar{p}_i A_i} = \frac{J_f^{(\varepsilon)}}{\bar{p}_i A_i \left(\varepsilon + \dfrac{2\gamma_f}{\gamma_f - 1}\dfrac{w_f^2}{1 - w_f^2}\right)} \qquad (15\text{-}13)$$

from which p_f and A_f can be obtained. The value of p_f^0 also follows.

Pure mixing of two streams. One of the most common, and one of the simplest, applications of a mixing duct is that in which only two nonreactive streams are brought together in the initial section, and adiabatically mixed in the duct. Ejector pumps represent the most usual devices using this application: the primary gas may be generally different from the secondary, it is supplied at known stagnation pressure $p^{0\prime}$ and temperature $T^{0\prime}$, and its velocity w_i' at the initial section of the mixing duct is, in general, supersonic, and supplied by a Laval nozzle. The secondary stagnation pressure $p^{0\prime\prime}$ and temperature $T^{0\prime\prime}$ may also be supposed to be assigned, although they can vary in certain ranges during the operation.

The secondary velocity w_i'' at the initial section is generally subsonic and is produced by expansion in a simply convergent nozzle. In ejectors, the mixing duct is followed by a subsonic diffuser bringing the pressure of the mixed flow up to the back pressure p_b. The shape of the mixing duct may have considerable importance for the results.

The problem of calculating the performance and geometry of an ejector pump for a given design condition has been treated by several authors for the case of constant pressure mixing [*53,54,55,56*] and of constant area mixing [*54,55,56,57*]. Substantially, the problem can be stated so as to include both cases as follows: determine the flow rate m' of the primary gas of known $p^{0'} > p_b$ and $T^{0'}$, necessary to bring the secondary gas of known $p^{0''} < p_b$ and $T^{0''}$ up to the pressure p_b with a flow rate m''. The pressure $p_i < p^{0''}$ at the entrance of the mixing duct is chosen arbitrarily, and the mixing duct can be assumed to be of the p-A power family with an arbitrary ε. Pressure nonuniformities need not be taken into consideration if the supersonic primary nozzle is correctly expanded for the pressure p_i.

The problem can be treated through the general equations already developed. Neglecting friction and considering for simplicity the case when the two gases are identical with a gas constant $\mathcal{R}$ and an adiabatic index γ uniform throughout the mixing process, δ will also be a constant and the conservation equations can be written in the form

$$m' + m'' = m \tag{15-14}$$

$$m' \sqrt{T^{0'}} \left[w_i' \left(1 - \frac{\delta}{2}\right) + \frac{\delta}{2} \frac{1}{w_i'} \right] + m'' \sqrt{T^{0''}} \left[w_i'' \left(1 - \frac{\delta}{2}\right) + \frac{\delta}{2} \frac{1}{w_i''} \right]$$
$$= m \sqrt{T_f^0} \left[w_f \left(1 - \frac{\delta}{2}\right) + \frac{\delta}{2} \frac{1}{w_f} \right] \tag{15-15}$$

$$m'T^{0'} + m''T^{0''} = mT_f^0 \tag{15-16}$$

Here, assuming approximately isentropic expansion for both streams down to the pressure p_i, the two velocities at the entrance of the mixing duct are determined by

$$\frac{p_i}{p^{0'}} = (1 - w_i'^2)^{\gamma/(\gamma-1)}, \quad \frac{p_i}{p^{0''}} = (1 - w_i''^2)^{\gamma/(\gamma-1)} \tag{15-17}$$

The pressure p_f at the end of the mixing duct can be determined from Eq. 15-13 or from the equivalent relation obtained by repeated application of Eq. 4-31a,

$$\left(\frac{p_i}{p_f}\right)^{\frac{1}{\varepsilon}} = \left(\frac{A_i}{A_f}\right)^{\frac{1}{1-\varepsilon}} = \frac{p_i A_i}{p_f A_f} = \frac{p_i A_i' + p_i A_i''}{p_f A_f}$$
$$= \frac{m' \sqrt{T^{0'}} \dfrac{1 - w_i'^2}{w_i'} + m'' \sqrt{T^{0''}} \dfrac{1 - w_i''^2}{w_i''}}{m \sqrt{T_f^0} \dfrac{1 - w_f^2}{w_f}} \tag{15-18}$$

Finally assuming a complete diffusion process to the pressure p_b with elemental efficiency η we have

$$\frac{p_f}{p_b} = (1 - w_f^2)^{\eta\gamma/(\gamma-1)} \tag{15-19}$$

For given values of $p^{0\prime}$, $p^{0\prime\prime}$, p_i, $T^{0\prime}$, and $T^{0\prime\prime}$, w_i' and w_i'' are determined by Eq. 15-17, and T_f^0 can be expressed in terms of the ratio $m''/m' = \mu$ through Eq. 15-16 and 15-14. Thus Eq. 15-15 represents an equation between the two unknowns μ and w_f. Another equation between the same variables is obtained by eliminating p_f from Eq. 15-18 and 15-19. Both equations are of second degree in μ, so that μ can be explicitly eliminated, thus producing a single complicated equation for the determination of

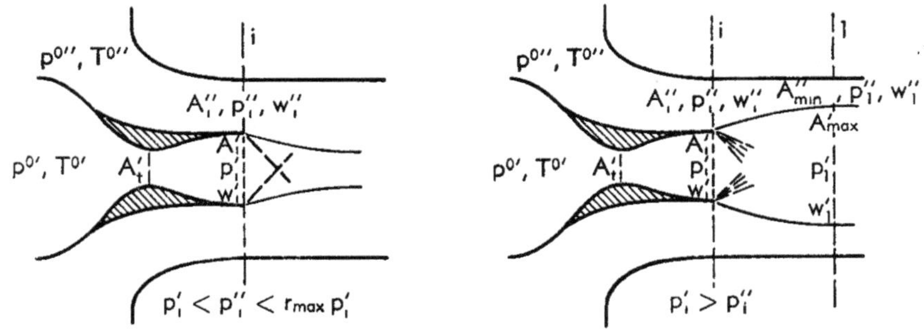

Fig. B,15a. Two conditions of choking of the secondary flow: Left, choking at the nozzle exit; right, choking in the free jet region. The primary nozzle is choked.

w_f. Once this equation is solved the values of the other unknowns μ, T_f^0, p_f, and A_f/A_i are immediately calculated. The primary and the secondary area, A_i' and A_i'', and hence A_i and A_f, can be calculated from Eq. 4-31a.

Observe that the solution of the system becomes easier if, following the usual practice, rather than prescribing p_b, one assigns μ. Then the value of w_f is found from the second degree equation (Eq. 15-15) and p_f and p_b are obtained from Eq. 15-18 and 15-19. Again, as observed before, there will be a supersonic value and a subsonic value of w_f, the latter being always physically possible, the former not always.

Thus we have the solution for the given values of p_i and ε. The problem of the optimization of the performance through proper choice of p_i has been investigated in [*56*] for the constant pressure and the constant area case ($\varepsilon = 0$ and $\varepsilon = 1$), with the result that the former gives better results than the latter. This does not mean, however, that the constant pressure mixing duct represents the best possibility; actually the contrary can be proved by optimizing with respect to ε, as will be seen in Art. 17.

An alternate problem to that briefly summarized above concerns the off-design operation of an ejector of given geometry. The problem is discussed in [*57*] for the constant area mixing duct. Actually, this represents the only well-defined duct shape of the whole *p-A* power family, and the only one for which keeping ε constant produces the same duct shape independently of the operating conditions. Here we shall briefly discuss some interesting peculiarities of the results.

We assume again that the two gases are identical. The areas A'_i and A''_i are now assigned, and so are the constant area $A = A'_i + A''_i$ of the mixing duct and the throat area A'_t of the primary nozzle (Fig. B,15a). For given $p^{0'}$, if p_b is sufficiently low and $p^{0''}$ is in the appropriate range, both the primary and the secondary flows are choked. The values of p'_i, w'_i, and m' then have fixed values, satisfying (if one assumes approximately isentropic expansion) the relations

$$m' = (m'_{max})_{is} = \frac{p^{0'}A'_t}{\sqrt{\mathcal{R}T^{0'}}}\sqrt{\frac{2\gamma}{\gamma-1}}\,w^*(1 - w^{*2})^{1/(\gamma-1)}$$

$$= \frac{p^{0'}A'_i}{\sqrt{\mathcal{R}T^{0'}}}\sqrt{\frac{2\gamma}{\gamma-1}}\,w'_i(1 - w'^2_i)^{1/(\gamma-1)}$$

$$= \frac{p'_iA'_i}{\sqrt{\mathcal{R}T^{0'}}}\sqrt{\frac{2\gamma}{\gamma-1}}\,\frac{w'_i}{1 - w'^2_i} \quad (15\text{-}20)$$

the last two relations being true for $p''_i < r_{max}p'_i$, since in this case no separation occurs in the primary nozzle (see Art. 8). For higher values of p_b and $p^{0''}$ the opposite condition may be realized, and the flow in the primary may still be choked and supersonic but separated before the nozzle exit, or choked and subsonic but unseparated and containing a pseudo-shock, or finally unchoked and subsonic without pseudo-shock. The corresponding cases could be discussed but do not present much practical interest.

The choked secondary flow may still present two different conditions, depending on whether $p''_i \geqq p'_i$ or $p''_i < p'_i$. In the first eventuality (Fig. B,15a, left) the primary jet undergoes a contraction within a short distance from the nozzle, and choking in the secondary flow takes place at the exit of the secondary nozzle where one has to take $w''_i = w^*$. Therefore,

$$m'' = \frac{p^{0''}A''_i}{\sqrt{\mathcal{R}T^{0''}}}\sqrt{\frac{2\gamma}{\gamma-1}}\,w^*(1 - w^{*2})^{1/(\gamma-1)}, \quad \mu = \frac{m''}{m'} = \frac{A''_i}{A'_t}\frac{p^{0''}}{p^{0'}}\sqrt{\frac{T^{0'}}{T^{0''}}} \quad (15\text{-}21)$$

and hence for given $T^{0'}$ and $T^{0''}$ the flow rate ratio μ is proportional to $p^{0''}$, as shown by the corresponding straight line through the origin in Fig. B,15b.

Consider now the other eventuality, $p''_i < p'_i$, which is realized when $p^{0''}$ is sufficiently decreased (Fig. B,15a, right). In this case the primary jet must undergo an increase in area through a complicated pattern of expansion waves of the kind discussed in Art. 8 for underexpanded nozzles. If we neglect, for a short distance from the nozzle exit, the mixing of the two streams, we can consider a well-defined, cross-sectional area of the primary jet. This area will reach a maximum A'_{max} at a certain distance downstream of the nozzle exit, say at station 1. Correspondingly the secondary flow will occupy the minimum area $A''_{min} = A - A'_{max}$ at station 1. It is clear that in these conditions choking of the secondary flow occurs at station 1, and the corresponding flow rate ratio is now

$$\mu = \frac{m''}{m'} = \frac{A''_{min}}{A'_t}\frac{p^{0''}}{p^{0'}}\sqrt{\frac{T^{0'}}{T^{0''}}} \tag{15-22}$$

smaller than that expressed by Eq. 15-21, since $A''_{min} < A''_i$. It is also clear that for decreasing $p^{0''}$ (and p''_i), A'_{max} will go on increasing, and that for a certain value of $p^{0''} = p^{0''}_{min}$, it may become equal to A. For this value of $p^{0''}$, m'' and μ must vanish.

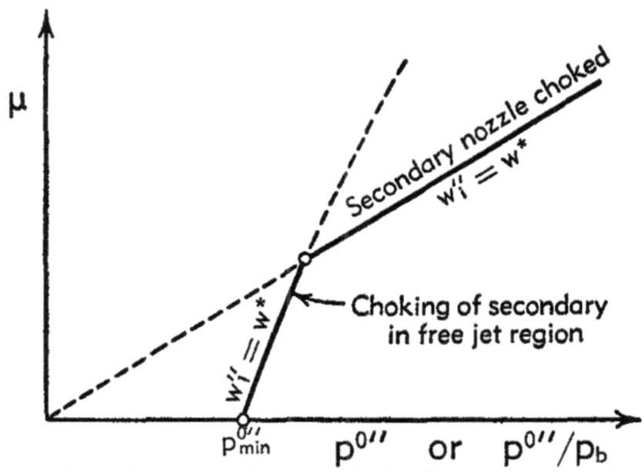

Fig. B,15b. Secondary flow rate characteristics for assigned $p^{0'}$ and very low p_b.

The calculation of A'_{max} should in principle be made using the method of characteristics which would also give details about the nonuniform velocity and pressure distributions in the primary jet. A rough calculation, following [*57*], can be based on the fact, already mentioned several times and analyzed in detail in Art. 16, that for not too large degrees of nonuniformity, one can replace the actual flow by an equivalent uniform flow. Obviously the value of the mean pressure at a given station of the jet is now different from the actual pressure at the interface of the primary and secondary flow, and therefore from the mean pressure of the latter at the same station.

Assuming approximately isentropic expansions in both flows, and $w''_1 = w^*$ at station 1, one has

$$\frac{A'_{max}}{A'_i} = \frac{w'_i}{w'_1}\left(\frac{1 - w'^2_i}{1 - w'^2_1}\right)^{\frac{1}{\gamma-1}}, \quad \frac{A''_{max}}{A''_i} = \frac{w''_i}{w^*}\left(\frac{1 - w''^2_i}{1 - w^{*2}}\right)^{\frac{1}{\gamma-1}} \tag{15-23}$$

While w''_i goes from w^* to zero, A''_{min} goes from A''_i to zero, A'_{max} goes from

A'_i to A, and w'_1 goes from w'_i to a maximum value $(w'_1)_{max} > w'_i$. Thus the relations (Eq. 15-23) establish a correspondence between w'_1 and w''_i. From the momentum equation, which can now be written, neglecting friction, as

$$J'_i + J''_i = J'_1 + J''_1$$

or using Eq. 4-33a, as

$$m' \sqrt{T^{0'}} \left(w'_i + \frac{w^{*2}}{w'_i} \right) + m'' \sqrt{T^{0''}} \left(w''_i + \frac{w^{*2}}{w''_i} \right) = m' \sqrt{T^{0'}} \left(w'_1 + \frac{w^{*2}}{w'_1} \right) + 2m'' \sqrt{T^{0''}}\, w^*$$

one obtains, taking into account Eq. 15-22 and 15-23:

$$\mu \sqrt{\frac{T^{0''}}{T^{0'}}} = \frac{w''_i (w'_1 - w'_i)(w'_i w'_1 - w^{*2})}{w'_i w'_1 (w^* - w''_i)^2} = \frac{A''_{min} p^{0''}}{A'_t p^{0'}} = \frac{w''_i}{w^*} \left(\frac{1 - w''^2_i}{1 - w^{*2}} \right)^{\frac{1}{\gamma - 1}} \frac{A''_i p^{0''}}{A'_t p^{0'}} \qquad (15\text{-}24)$$

Thus, for every value of w''_i in the range 0–w^*, Eq. 15-23 gives $w'_1(w''_i)$ and Eq. 15-24 gives $\mu(w''_i)$ and $p^{0''}(w''_i)$. The curve corresponding to these results will appear schematically as shown in Fig. B,15b. This curve intersects the previous straight line at the point where $p''_i = p'_i$. Obviously only those parts of each line offering the smaller value of μ for a given $p^{0''}$ have a physical meaning.

The value of $p^{0''}_{min}$ corresponding to $\mu = 0$ is found from Eq. 15-24 by letting $w''_i = 0$ and $w'_1 = (w'_1)_{max}$:

$$\frac{p^{0''}_{min}}{p^{0'}} = \frac{A'_t}{A''_i} \frac{(1 - w^{*2})^{1/(\gamma-1)} [w'_i (w'_1)_{max} - w^{*2}]}{w^* w'_i} \qquad (15\text{-}25)$$

and is a function only of A'_i/A'_t and A/A'_t. We shall see later that the actual values of $p^{0''}_{min}$ lie well below the value calculated from Eq. 15-25.

So far the value of p_b has not appeared in our discussion, the only assumption being that it is sufficiently low. This is reasonable, since there is no doubt that if p_b is very low so that no counter pressure is exerted on the mixing duct, the velocity resulting from mixing the supersonic primary jet with the sonic (or supersonic) secondary flow will be supersonic. Supposing that the final mixed velocity is sufficiently uniform (that is the mixing duct is sufficiently long) and assuming, for simplicity, that no divergence is added to the duct, the conditions at the exit of the mixing duct are similar to those at the exit of a supersonic nozzle. Hence they are independent of p_b so far as $p_b \leq r_{max} p_f$, the final pressure p_f of the mixed flow being, in the absence of friction, given by Eq. 15-18 with $\varepsilon = 1$ (corrections for friction can be carried out easily). Clearly in this case p_f must be calculated using for w_f the supersonic root of Eq. 15-15.

In the opposite case, $p_b > r_{max}p_f$, separation is produced before the final section of the duct. Suppose, for the sake of discussion, that the mixing duct is long enough to produce practically complete mixing, and to contain, in addition, a complete pseudo-shock; then the pressure can be increased until the flow again fills the final section with w_f given by the subsonic root of Eq. 15-15 and p_f by the corresponding value obtained from Eq. 15-18. Increasing p_b in this range has no effect on the flow pattern preceding the pseudo-shock.

Further increases of p_b will force the pseudo-shock to move in the mixing region, thus producing complicated flow patterns, until finally for a sufficiently high value of p_b the effects of the back pressure will reach the initial section and the secondary flow will unchoke. From this point on, the flow condition at the initial section will be determined by the value of p_b and can be calculated by using the conservation equations (Eq. 15-14, 15-15, and 15-16), which, however, can be more conveniently written so as to have A_i' and A_i'' appearing explicitly in the equations, rather than m' and m''. Taking $\delta = (\gamma - 1)/\gamma$ and making use of Eq. 4-31a and 15-19 (to cover the general case where a subsonic diffuser is present at the end of the mixing duct; the case when this is absent being immediately obtained by putting $\eta = 0$, $p_f = p_b$), the conservation equations can be written in the form

$$\frac{A_i'p^{0'}}{Ap_b} w_i'(1 - w_i'^2)^{1/(\gamma-1)} + \frac{A_i''p^{0''}}{Ap_b}\sqrt{\frac{T^{0''}}{T^{0'}}}\, w_i''(1 - w_i''^2)^{1/(\gamma-1)} = \sqrt{\frac{T_f^0}{T^{0'}}}\, \frac{w_f}{(1 - w_f^2)^{1-[\gamma\eta/(\gamma-1)]}} \tag{15-26}$$

$$\frac{A_i'p^{0'}}{Ap_b}\left(1 + \frac{w_i'^2}{w^{*2}}\right)(1 - w_i'^2)^{1/(\gamma-1)} + \frac{A_i''p^{0''}}{Ap_b}\left(1 + \frac{w_i''^2}{w^{*2}}\right)(1 - w_i''^2)^{1/(\gamma-1)} = \frac{1 + \dfrac{w_f^2}{w^{*2}}}{(1 - w_{f}^2)^{1-[\gamma\eta/(\gamma-1)]}} \tag{15-27}$$

$$\frac{A_i'p^{0'}}{Ap_b} w_i'(1 - w_i'^2)^{1/(\gamma-1)} + \frac{A_i''p^{0''}}{Ap_b}\sqrt{\frac{T^{0'}}{T^{0''}}}\, w_i''(1 - w_i''^2)^{1/(\gamma-1)} = \sqrt{\frac{T^{0'}}{T_f^0}}\, \frac{w_f}{(1 - w_f^2)^{1-[\gamma\eta/(\gamma-1)]}} \tag{15-28}$$

The solution of this system can be found for given values of $p^{0'}/p_b$, $T^{0''}/T^{0'}$, and w_i' by assigning the value of $w_i'' < w^*$ and eliminating $T_f^0/T^{0'}$ and $p^{0''}/p_b$, thus obtaining an equation for w_f. Choosing the subsonic solution for w_f, the resulting values of $p^{0''}/p_b$, and, if required, those of $T_f^0/T^{0'}$ and p_f/p_b can be found. Finally, the value of μ is obtained as the ratio of the second to the first term of Eq. 15-26. If the curve connecting μ to $p^{0''}/p_b$ is calculated and drawn on the figures, two cases are possible as shown in Fig. B,15c and B,15d, depending on the position of the intersections of the computed curve with the previously calculated lines. In both cases the minimum secondary pressure $p^{0''}_{\min}$ is reached at $\mu = 0$.

It is interesting to notice that while the value of $p^{0''}_{\min}$ calculated from

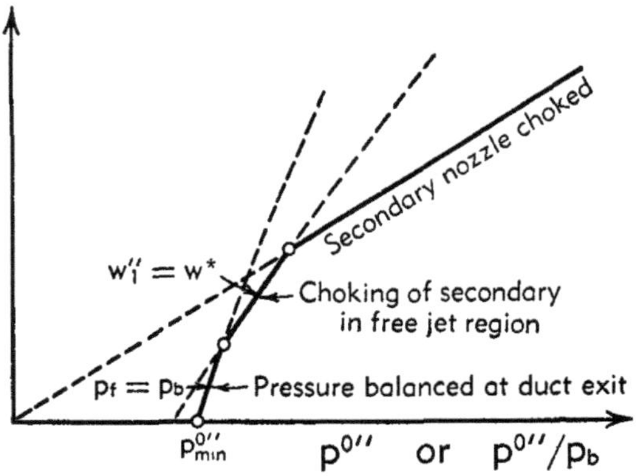

Fig. B,15c. Secondary flow rate characteristics for assigned $p^{0'}$ in the intermediate range of p_b.

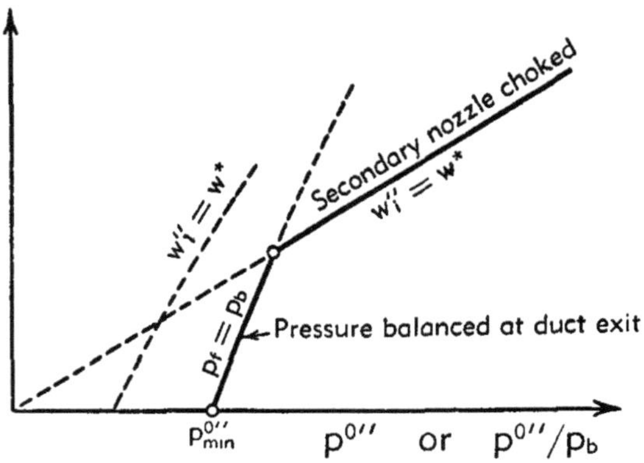

Fig. B,15d. Secondary flow rate characteristics for assigned $p^{0'}$ and sufficiently high p_b.

Eq. 15-25 is proportional to $p^{0'}$, the latter value of $p^{0''}_{\min}$, on the contrary, varies in the opposite direction as $p^{0'}$, as is seen later (Eq. 15-32). Thus the situation shown by Fig. B,15c is possible only in a limited range of $p^{0'}/p_b$, above which only the conditions of Fig. B,15b are produced and below which only those of Fig. B,15d are present. If the value of $p^{0''}_{\min}/p_b$ is plotted for given geometrical conditions as a function of $p^{0'}/p_b$, the resulting behavior is shown in Fig. B,15e. We shall see later that the behavior indicated by these figures is only qualitatively correct, since, as already mentioned, Eq. 15-25 gives values larger than those observed. However, if one excludes the corresponding regions of the curves, the agreement with the experiments is quite good (see [*57*]).

We finally observe that the above discussion has been based on the assumption of a mixing duct sufficiently long for complete mixing to occur, followed by a pseudo-shock. If the mixing duct is shorter, the limiting value of p_b above which the flow unchokes will be lower and will decrease steadily with the duct length. Thus the performance of the ejector will be worsened for p_b above this limiting value. The effect of shortening the mixing duct may be felt not only as a result of the incompleteness of mixing, but also as a result of the recurrent wave patterns

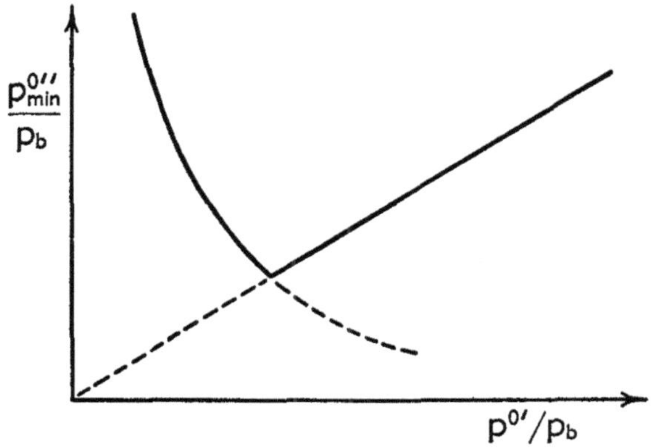

Fig. B,15e. Relation between secondary tank pressure at zero flow rate and primary pressure ratio.

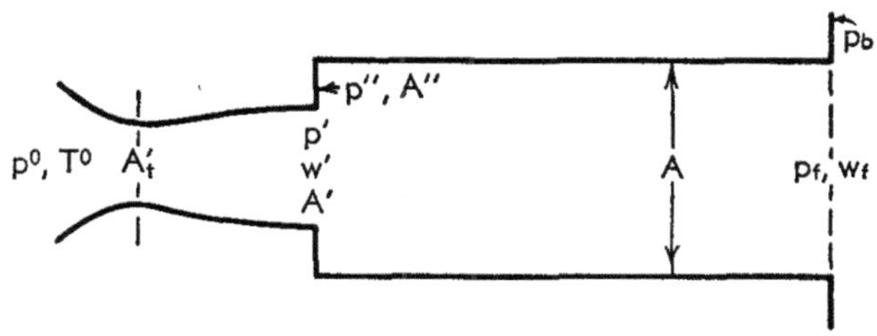

Fig. B,15f. Duct with sudden enlargement followed by a constant section duct.

present in the region of supersonic velocities when the primary nozzle is incorrectly expanded. The effect of these wave patterns on the performance varies cyclically with decreasing length of the duct.

Sudden enlargement of the section. It has already been noted that if the secondary flow rate vanishes the problem of the mixing duct is reduced to that of the effect of a section discontinuity. Such a problem was treated first by Nusselt [*58*], who also checked the results with tests, and more recently by Fabri and Siestrunck [*59*], who carried out an exhaustive theoretical and experimental survey. An interesting practical application of this problem is found in wind tunnels with a free jet test section. This particular problem was treated by Hermann [*60*].

The problem to be analyzed is that shown in Fig. B,15f. The corresponding equations can be obtained from those of the mixing duct by taking $m'' = 0$, $m' = m$, $p^{0\prime} = p^0$, $p'_i = p'$, $w'_i = w'$, $A'_i = A'$, $A''_i = A''$, and $p^{0\prime\prime} = p''_i = p''$. Moreover, if no additional divergent duct is present, η can be taken as zero and $p_f = p_b$, except for supersonic final velocity (see below). For more completeness we take into account the friction in the constant section duct of length l. The energy equation (Eq. 15-16) immediately gives $T^0_f = T^0$. The other two conservation equations are obtained from Eq. 15-26 and 15-27 (modified for the friction term) in the form:

$$\frac{A'p^0}{Ap_f} w'(1 - w'^2)^{1/(\gamma-1)} = \frac{w_f}{1 - w_f^2} \tag{15-29}$$

$$\frac{A'p^0}{Ap_f}\left(1 + \frac{w'^2}{w^{*2}}\right)(1 - w'^2)^{1/(\gamma-1)} + \frac{A''p''}{Ap_f} = \frac{1 + \dfrac{w_f^2}{w^{*2}}}{1 - w_f^2} + \frac{X_w}{Ap_f} \tag{15-30}$$

The friction force X_w can be calculated approximately from Eq. 3-5 and 7-20 by assuming constant values of the velocity and pressure, coinciding with those in the final section. The result is

$$\frac{X_w}{Ap_f} = \frac{4c_fl}{d_h}\frac{\gamma}{\gamma - 1}\frac{w_f^2}{1 - w_f^2} \tag{15-31}$$

The solution can be found explicitly by solving Eq. 15-29 for w_f and substituting in Eq. 15-30 to obtain p''/p_f.

Using the abbreviations

$$Y = \frac{Ap_f}{A'p'} = \frac{Ap_f}{A'p^0(1 - w'^2)^{\frac{\gamma}{\gamma-1}}}, \quad Z = \frac{1 + w'^2}{1 - w'^2}, \quad f = \frac{2c_fl}{d_h}$$

we obtain

$$\frac{A''p''}{Ap_f} = \frac{1}{(\gamma + 1)Y}\left[\gamma(1 + f)\sqrt{Y^2 + Z^2 - 1} - (1 + \gamma f)Y - \gamma Z + 1\right] \tag{15-32}$$

The behavior of this solution is easily checked for $f = 0$. In this case the second member vanishes (and $p'' = 0$) for the two values of Y

$$Y_1 = 1, \quad Y_2 = \frac{2}{\gamma^2 - 1}(\gamma Z - 1) - 1$$

If $w' \geqq w^*$, one has $Z \geqq \gamma$, and Y_2 is $\geqq 1$. Only for $Y > Y_2$ or $Y < Y_1$ is p'' positive and the solution physically possible. For $w' = w^*$, Y_1 and Y_2 coincide. Actually, as is shown later, only the solutions with $Y > Y_2$

can exist in practice, although not in the whole range. Fig. B,15g shows the behavior of the solution for $f = 0$ and different values of M'. For each value of the area enlargement ratio A/A' the corresponding curve of p''/p_f vs. p'/p_f or p^0/p_f can be drawn. This is done in Fig. B,15h, taken from [*59*], where the values calculated from Eq. 15-32 for $w' = 0.770$ ($M' = 1.886$) and $A/A' = 2.82$ with $4l/d_h = 48$ and $c_f = 0.0053$ are shown by curve (a).

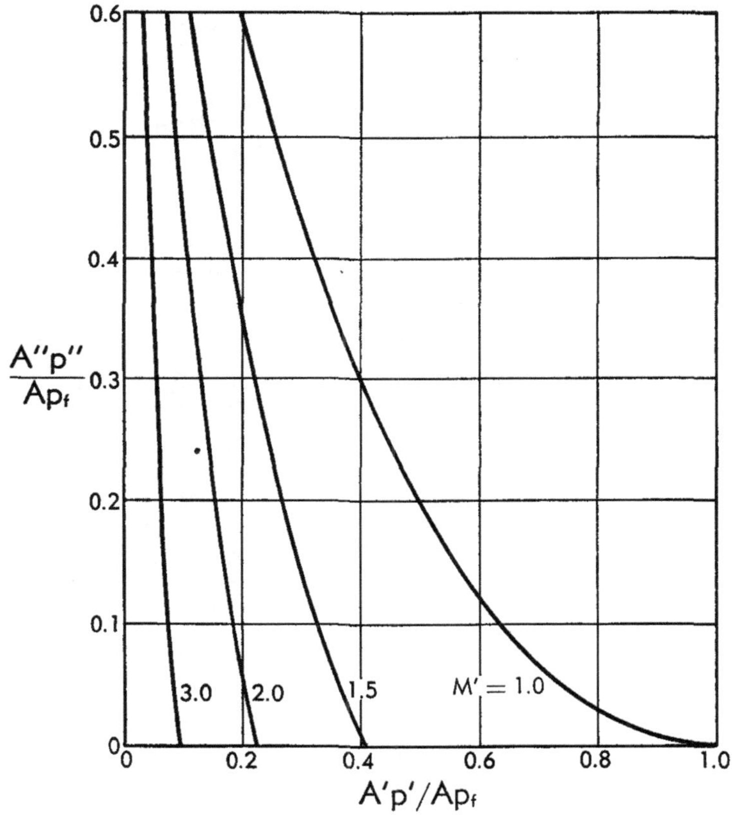

Fig. B,15g. Representation of Eq. 15-32 for $f = 0$, $\gamma = 1.4$.

This curve is fit to represent the real results only in a certain range. If p'' is $> r_{max}p'$ the separation of the supersonic jet upstream of the nozzle exit will result in larger p' and smaller A'. Assuming, in agreement with the discussion of Art. 8 for overexpanded nozzles, that the pressure in the nozzle following the separation section stays constant and equal to p'', A'' increases as much as A' decreases, and calculations can be performed with the same equation (Eq. 15-32). This was done in [*59*] with the constant value 2.5 for r_{max}, and the results are shown by curve (b). Experimental points in this range show that, for a nozzle with a conical divergence, curve (b) fits the actual values very well. However, for a nozzle with parallel exit velocity the jet has a tendency to separate only for larger values of p''/p' ($\cong 3.8$). When p^0/p_f approaches unity, other regimes, including pseudo-shocks and eventually unchoking of the nozzle, take place for which analytical estimates have not been given, although this would not involve fundamental difficulties.

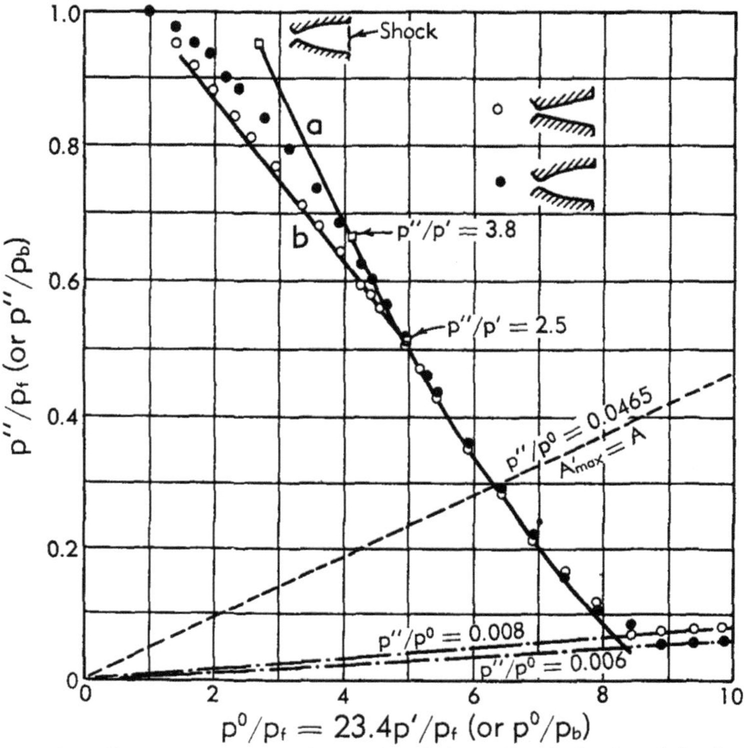

Fig. B,15h. Comparison of experimental and theoretical behavior of the duct of Fig. B,15f. ($M' = 1.886$; $A/A' = 2.82$; $\gamma = 1.4$; $4l/d_h = 48$; $c_f = 0.0053$).

Another limit for the validity of Eq. 15-32 is found in the range $p'' < p'$. Clearly if p'' is sufficiently low the maximum section of the free jet downstream of the nozzle exit may become equal to A. Within the assumptions mentioned earlier, the corresponding value of p''/p^0 is given by Eq. 15-25 and is equal to 0.0465. When this situation is reached, further decreases of p''/p_f should become impossible and the value of p''/p' should stay constant. This condition is represented by the dashed straight line through the origin of Fig. B,15h. Actually, however, this limitation appears to be too restrictive, and further decreases of p''/p_f with increasing p'/p_f can take place, closely following Eq. 15-32. Only at a much lower value of p''/p_f do deviations from Eq. 15-32 become apparent and, after a short transition, the experimental points fall on a straight line through the origin. The slope of this straight line (and therefore the value of p''_{min}/p^0) is considerably lower than predicted by Eq. 15-25, and seems to be entirely determined for a given nozzle by the enlargement ratio.

Observe that the condition $p_f = p_b$ is not maintained along the whole extension of the straight line, because, once supersonic flow is established on the entire final section, the value of p_f remains constant regardless of further decreases of p_b. On the portion of the straight line on which p_f is still equal to p_b, the solution can deviate from Eq. 15-32 because another of the basic assumptions on which this equation was obtained is not satisfied, namely that of uniformity. In fact in the corresponding region a pseudo-shock will start emerging gradually from the final section of the duct and will allow a gradual transition from the subsonic w_f corresponding to Eq. 15-32 to a uniform supersonic w_f.

The physical illustration of these effects is provided by the series of schlieren pictures of Plate B,15, also from [*59*]. In the first picture the nozzle is too much underexpanded and separation occurs within the nozzle itself. In the second no separation takes place, though the nozzle is still underexpanded. The third picture shows the case when the nozzle is only moderately underexpanded. In all three cases nonsymmetric behavior of the jet is observed. The fourth picture shows a case when the jet almost fills the nozzle with its maximum section. The dead water region around the jet is not yet isolated from the effects of the back pressure. In the fifth and sixth picture, on the contrary, the dead water region is isolated from the back pressure, and the jet pattern in the region close to the nozzle becomes independent of the downstream conditions. In the fifth picture a pseudo-shock clearly starts at a distance of 2–2.5 duct widths from the enlargement, and in the sixth the flow is supersonic everywhere. As already observed p_f becomes independent of p_b after this condition is established.

It is clear, by observing the last two pictures, that the jet is issued from the nozzle as if it should attain a maximum section substantially larger than the duct area, and that this expansion is stopped by two strong oblique shock waves when the jet reaches the walls. The compression due to these shock waves allows the pressure in the dead water region to go well below that predicted by Eq. 15-25. This phenomenon can be related [*61,62*] to the base pressure phenomenon on blunt-based bodies, and its treatment is very closely connected with the conditions of reattachment of the jet to produce a boundary layer, which escapes the realm of one-dimensional theory (see VI,B). Observe that, as the pictures show clearly, the shocks do not reach the wall because of the presence of subsonic flows in the boundary layer and in the zone where the jet mixes with the gases recirculating in the dead water region. However, these subsonic regions are thin, and if their longitudinal extent is sufficient, they cannot prevent the isolation of the dead water region from the downstream effects.

As a last example of application of the theory developed above, we give in Fig. B,15i, in a different representation, the curves calculated for the case of a purely convergent nozzle followed by a sudden enlargement, for different values of the reciprocal of the enlargement ratio A'/A and for $f = 0$.

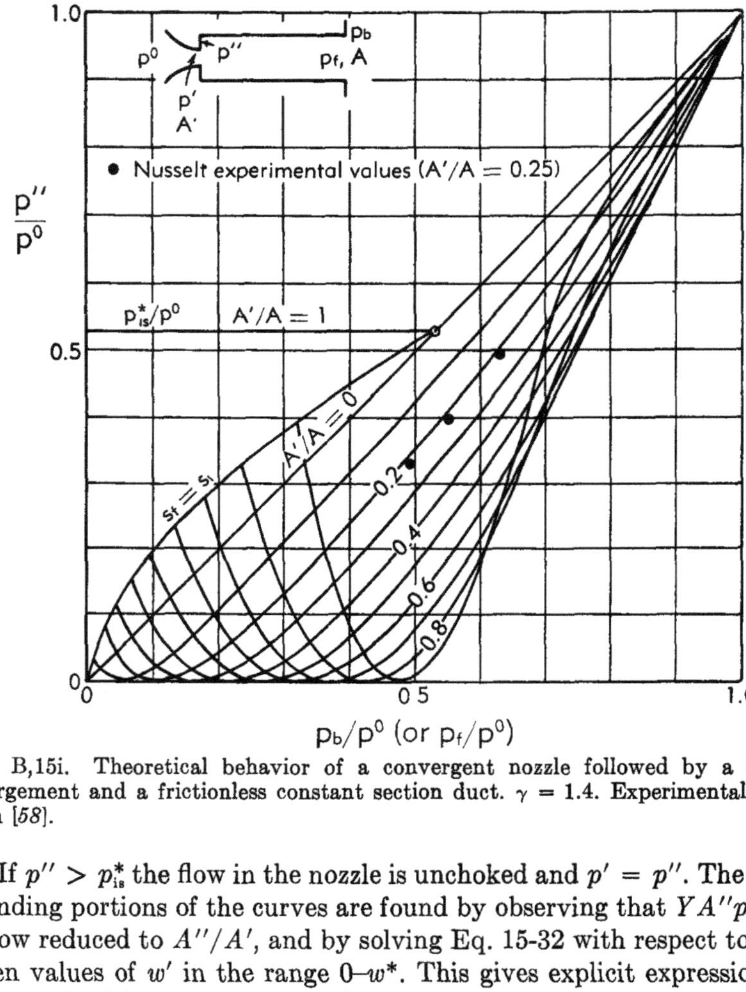

Fig. B,15i. Theoretical behavior of a convergent nozzle followed by a sudden enlargement and a frictionless constant section duct. $\gamma = 1.4$. Experimental values from [*58*].

If $p'' > p^*_{is}$ the flow in the nozzle is unchoked and $p' = p''$. The corresponding portions of the curves are found by observing that $YA''p''/Ap_f$ is now reduced to A''/A', and by solving Eq. 15-32 with respect to Y for given values of w' in the range 0–w^*. This gives explicit expressions for p_f/p^0 and p''/p^0 as functions of the geometry, f (if $\neq 0$) and w'. For $p'' \leqq p^*_{is}$ the flow is choked and $w' = w^*$, $p' = p^*_{si}$. For this condition, as observed previously, the values Y_1 and Y_2 for which Eq. 15-32 gives $p'' = 0$ both coincide with unity. Thus the total pressure force $A'p^*_{is}$, acting at station i, is equal to the pressure force Ap_f on the final section, and the final velocity must be equal to the initial velocity $w_f = w^*$. To the right of the point where $p'' = 0$, w_f is subsonic, and to the left, supersonic. The supersonic portions of the curves have been extended, although deprived of physical significance, up to the line where $s_f = s_i$; beyond that line $s_f - s_i$ would be negative in violation of the second law of thermodynamics. It is interesting to notice that the conditions on the line $s_f = s_i$ correspond to those assumed in the calculation of Eq. 15-25. Thus if a horizontal line is drawn through the intersection with this line of the curve corresponding to a given A'/A, only that part of the subsonic branch of this curve which lies above the horizontal line should have a physical meaning, and the rest of the curve should be replaced by the horizontal line.

We know already that this condition is too restrictive and that the position of the horizontal line must be shifted to substantially lower levels in order to agree with the experiments. On the figure a few experimental values from [*58*] show good agreement with the calculated curves.

B,16. Validity of the Uniformity Assumption. The assumption of uniform, or piecewise uniform, velocity, temperature, and composition on each section allows the explicit formulations of Art. 4 and the treatments of the following articles. However, as already noted, this assumption presents some contradiction with the actual physical situation. It is therefore necessary to discuss its meaning and to check its limitations.

Suppose that u, T, $\mathcal{R}$, and c_p have arbitrary distributions. There are several ways in which one can define the mean values of the corresponding quantities, and express m, I, H, E_k, and H^0 in terms of these mean values. However, the effect of the nonuniformity cannot be completely eliminated, and in general appears in the form of "nonuniformity coefficients." These coefficients are unity for uniform distributions, and therefore their deviation from unity is a measure of the importance of the effects of nonuniformity.

Substantially, the results of these calculations do not depend on the particular way in which the mean quantities are defined. A certain amount of arbitrariness, contained in the definitions that follow, has therefore no effect on the conclusions. First let us consider the flux of stagnation enthalpy (Eq. 3-11). For every element $d\mu$ of the flow, h^0 is different because T^0 is different and because $h(T)$ is different. We can imagine adiabatically mixing all the elements together in two steps. In the first we uniformize the stagnation temperature to a mean value $\overline{T}^0$ without making the composition uniform, so that each element now has the stagnation enthalpy $h(\overline{T}^0)$. In the second we also make the composition uniform, while the temperature stays unchanged, and obtain a uniform value $\bar{h}(\overline{T}^0) = \bar{h}^0$. The flux of stagnation enthalpy must not change during the whole process. Therefore we can write

$$H^0 = \int_m h(T^0)d\mu = \int_m h(\overline{T}^0)d\mu = \int_m \bar{h}(\overline{T}^0)d\mu = m\bar{h}^0 \qquad (16\text{-}1)$$

Now we assume that for each element of gas the specific heat can be taken constant between T^0 and $\overline{T}^0$. As a consequence of Eq. 16-1 we obtain

$$\int_m [h(T^0) - h(\overline{T}^0)]d\mu \cong \int_m c_p(T^0 - \overline{T}^0)d\mu = 0$$

This equation can be written in the form

$$\hat{H}^0 = \int_m c_pT^0 d\mu = \overline{T}^0 \int_m c_p d\mu = m\bar{c}_p\overline{T}^0 \tag{16-2}$$

where we have indicated with $\hat{H}^0$ the flux of c_pT^0, and we have introduced the average specific heat

$$\bar{c}_p = \frac{1}{m}\int_m c_p d\mu \tag{16-3}$$

Similarly the flux of c_pT can be designated by $\hat{H}$. A mean static temperature $\overline{T}$ can be defined through $\hat{H}$ and is given by

$$\hat{H} = \int_m c_pT d\mu = \overline{T}\int_m c_p d\mu = m\bar{c}_p\overline{T} \tag{16-4}$$

If we assume, in agreement with the final observations of Art. 4, that for each element c_p is constant in the range T^0–T we can write Eq. 3-11 as

$$E_k = H^0 - H \cong \int_m c_p(T^0 - T)d\mu = \hat{H}^0 - \hat{H} = m\bar{c}_p(\overline{T}^0 - \overline{T})$$

where E_k is the flux of kinetic energy.

Defining a mean square velocity $\bar{u}$ through the relation

$$E_k = \frac{1}{2}\int_m u^2 d\mu = m\frac{\bar{u}^2}{2} \tag{16-5}$$

we obtain in analogy to Eq. 4-27 the following relation between mean quantities

$$\bar{c}_p\overline{T} + \frac{\bar{u}^2}{2} = \bar{c}_p\overline{T}^0 \tag{16-6}$$

An average gas constant and an average γ can be defined by the relations:

$$\overline{\mathcal{R}} = \frac{1}{m}\int_m \mathcal{R}d\mu, \quad \frac{\bar{\gamma}}{\bar{\gamma} - 1} = \frac{\bar{c}_p}{\overline{\mathcal{R}}} \tag{16-7}$$

A different definition of mean velocity and temperature is obtained in the following way. Define a mean velocity as

$$\tilde{u} = \frac{I}{m} \tag{16-8}$$

One can then define a mean density, related to a mean temperature by the equation of state, as

$$\tilde{\rho} = \frac{p}{\overline{\mathcal{R}}\tilde{T}} = \frac{m}{A\tilde{u}}$$

so that

$$\overline{\mathcal{R}}\tilde{T} = \frac{pAI}{m^2} \tag{16-9}$$

With these definitions of $\tilde{u}$ and $\tilde{T}$ the expressions of m and I in terms of the mean quantities keep the same form as for the uniform case, (Eq. 4-1) in the same way that, with the definitions of $\bar{u}$ and $\overline{T}$, Eq. 16-6 keeps the same form as Eq. 4-27.

Coefficients of nonuniformity. For uniform flow $\bar{u} = \tilde{u}$ and $\overline{T} = \tilde{T}$ coincide. However, for nonuniform flows in general $\bar{u} \neq \tilde{u}$ and $\overline{T} \neq \tilde{T}$. We introduce two nonuniformity coefficients

$$\alpha = \frac{\overline{T}}{\tilde{T}}, \quad \beta = \left(\frac{\bar{u}}{\tilde{u}}\right)^2 \tag{16-10}$$

or, explicitly,

$$\alpha = \frac{\bar{\gamma} - 1}{\bar{\gamma}}\frac{m\hat{H}}{pAI} = \frac{m\int_m \mathcal{R}d\mu \int_m c_pTd\mu}{\int_m c_pd\mu \int_m ud\mu \int_m \frac{\mathcal{R}T}{u}d\mu} \tag{16-11}$$

$$\beta = \frac{2mE_k}{I^2} = \frac{m\int_m u^2d\mu}{\left(\int_m ud\mu\right)^2} \tag{16-12}$$

In Eq. 16-11 we have used the relation, immediately obtained from Eq. 3-2,

$$pA = \int_m \frac{\mathcal{R}T}{u}d\mu \tag{16-13}$$

Based on the above quantities we can define a mean Mach number and a mean reduced velocity as

$$\tilde{M} = \frac{\tilde{u}}{\sqrt{\bar{\gamma}\overline{\mathcal{R}}\tilde{T}}}, \quad \tilde{w} = \frac{\tilde{u}}{\sqrt{2\bar{c}_p\overline{T}^0}} \tag{16-14}$$

Then one has

$$\bar{\gamma}\tilde{M}^2 = \frac{I}{pA} = \frac{1}{A}\int_A \gamma M^2 d\eta \tag{16-15}$$

Here use has been made of Eq. 3-4, and M indicates the local Mach number. For uniform γ, the $\tilde{M}$ defined by Eq. 16-15 coincides with the mean square value, the average being taken with respect to the area.

With the use of Eq. 16-10 and 16-13, we obtain from Eq. 16-6

$$\frac{T^0}{\tilde{T}} = \alpha + \beta\frac{\tilde{\gamma}-1}{2}\tilde{M}^2 = \frac{\alpha}{1-\beta\tilde{w}^2} \tag{16-16}$$

Thus the relation between $\tilde{M}$ and $\tilde{w}$ is

$$\tilde{M}^2 = \frac{2}{\tilde{\gamma}-1}\frac{\alpha\tilde{w}^2}{1-\beta\tilde{w}^2}, \quad \tilde{w}^2 = \frac{\dfrac{\tilde{\gamma}-1}{2}\tilde{M}^2}{\alpha+\beta\dfrac{\tilde{\gamma}-1}{2}\tilde{M}^2} \tag{16-17}$$

Eq. 16-13, 16-16, and 16-17 are generalizations of Eq. 4-23, 4-25, 4-28, and 4-29 from which they differ only by the presence of the nonuniformity coefficients. Similarly one can derive the relations equivalent to Eq. 4-30, 4-31, 4-32, 4-33, 4-34, 4-30a, 4-31a, 4-32a, 4-33a, and 4-34a in the form

$$\tilde{u} = \sqrt{\mathcal{R}\tilde{T}^0}\left[\sqrt{\tilde{\gamma}}\,\tilde{M}\sqrt{\frac{\tilde{T}}{\tilde{T}^0}}\right] = \sqrt{\mathcal{R}\tilde{T}^0}\left[\sqrt{\frac{2\tilde{\gamma}}{\tilde{\gamma}-1}}\,\tilde{w}\right] \tag{16-18}$$

$$m = \frac{pA}{\sqrt{\mathcal{R}\tilde{T}^0}}\left[\sqrt{\tilde{\gamma}}\,\tilde{M}\sqrt{\frac{\tilde{T}^0}{\tilde{T}}}\right] = \frac{pA}{\sqrt{\mathcal{R}\tilde{T}^0}}\left[\sqrt{\frac{2\tilde{\gamma}}{\tilde{\gamma}-1}}\frac{\alpha\tilde{w}}{1-\beta\tilde{w}^2}\right] \tag{16-19}$$

$$\left.\begin{aligned} I &= m\sqrt{\mathcal{R}\tilde{T}^0}\left[\sqrt{\tilde{\gamma}}\,\tilde{M}\sqrt{\frac{\tilde{T}}{\tilde{T}^0}}\right] = pA\tilde{\gamma}\tilde{M}^2 \\ &= m\sqrt{\mathcal{R}\tilde{T}^0}\left[\sqrt{\frac{2\tilde{\gamma}}{\tilde{\gamma}-1}}\,\tilde{w}\right] = pA\left[\frac{2\tilde{\gamma}}{\tilde{\gamma}-1}\frac{\alpha\tilde{w}^2}{1-\beta\tilde{w}^2}\right] \end{aligned}\right\} \tag{16-20}$$

$$\left.\begin{aligned} J &= m\sqrt{\mathcal{R}\tilde{T}^0}\left[\frac{1+\tilde{\gamma}\tilde{M}^2}{\sqrt{\tilde{\gamma}}\,\tilde{M}}\sqrt{\frac{\tilde{T}}{\tilde{T}^0}}\right] = pA[1+\tilde{\gamma}\tilde{M}^2] \\ &= m\sqrt{\mathcal{R}\tilde{T}^0}\left[\sqrt{\frac{2\tilde{\gamma}}{\tilde{\gamma}-1}}\left(\tilde{w}+\frac{\tilde{\gamma}-1}{2\tilde{\gamma}}\frac{1-\beta\tilde{w}^2}{\alpha\tilde{w}}\right)\right] \\ &= pA\left[1+\frac{2\tilde{\gamma}}{\tilde{\gamma}-1}\frac{\alpha\tilde{w}^2}{1-\beta\tilde{w}^2}\right] \end{aligned}\right\} \tag{16-21}$$

$$\left.\begin{aligned} J^{(\varepsilon)} &= m\sqrt{\mathcal{R}\tilde{T}^0}\left[\frac{\varepsilon+\tilde{\gamma}\tilde{M}^2}{\sqrt{\tilde{\gamma}}\,\tilde{M}}\sqrt{\frac{\tilde{T}}{\tilde{T}^0}}\right] = pA[\varepsilon+\tilde{\gamma}\tilde{M}^2] \\ &= m\sqrt{\mathcal{R}\tilde{T}^0}\left[\sqrt{\frac{2\tilde{\gamma}}{\tilde{\gamma}-1}}\left(\tilde{w}+\varepsilon\frac{\tilde{\gamma}-1}{2\tilde{\gamma}}\frac{1-\beta\tilde{w}^2}{\alpha\tilde{w}}\right)\right] \\ &= pA\left[\varepsilon+\frac{2\tilde{\gamma}}{\tilde{\gamma}-1}\frac{\alpha\tilde{w}^2}{1-\beta\tilde{w}^2}\right] \end{aligned}\right\} \tag{16-22}$$

where the effect of the nonuniformities is all concentrated in the coefficients α and β. Here $T^0/\tilde{T}$ must be considered to be expressed by Eq. 16-16. It is clear that the less α and β deviate from unity, the smaller will be the error resulting from taking $\alpha = \beta = 1$ as if the flow were actually uniform.

Effect of pressure nonuniformities. It is interesting to observe that even if we abandon the condition of pressure uniformity, the effects of the nonuniformity of the velocity, temperature, composition, and pressure can still be concentrated in the same two nonuniformity coefficients defined by Eq. 16-11 and 16-12, provided that in all previous equations p is replaced by the mean pressure

$$\bar{p} = \frac{1}{A}\int_A p d\eta = \frac{1}{A}\int_m \frac{\mathcal{R}T}{u}\, d\mu \tag{16-23}$$

previously introduced (Eq. 3-7). For instance, $\tilde{T}$ must now be defined by the relations

$$\tilde{\rho} = \frac{\bar{p}}{\bar{\mathcal{R}}\tilde{T}} = \frac{m}{A\tilde{u}} \tag{16-24}$$

and $\tilde{M}$ by

$$\tilde{\gamma}\tilde{M}^2 = \frac{I}{\bar{p}A} = \frac{\int_A \gamma p M^2 d\eta}{\int_A p d\eta} \tag{16-25}$$

The expressions for m, I, and J are unchanged except for the presence of $\bar{p}$ instead of p. Thus the existence of pressure nonuniformities does not alter the form of the equations, provided one uses the proper definitions for the mean quantities.

A third nonuniformity coefficient. For flows in which entropy considerations are important, it may be useful to introduce a third coefficient related to the entropy flux. The most interesting case is that of a nonuniform, isoenergetic, and isentropic flow, to which, for simplicity, we shall restrain our derivations. For such flows the values of p^0 and T^0 are constant everywhere, and their distributions on each section are uniform. Hence, if the velocity is not uniformly distributed, both p and T are nonuniform on each section with mean values $\bar{p}$ and $\overline{T}$. Obviously p^0 and T^0 can here be interpreted as the pressure and temperature in a tank in which the unmixed gas elements are originally contained.

Let us now consider the uniform flow having at each section the pressure $\bar{p}$, the temperature $\overline{T}$, the gas constant $\overline{\mathcal{R}}$, and the specific heat $\bar{c}_p$. Clearly in the tank $\bar{p} = p^0$ and $\overline{T} = T^0$. Thus the variation of entropy flux from the tank to an arbitrary station is

$$\overline{S} - S^0 = m\left(\int_{T^0}^{\overline{T}} \bar{c}_p \frac{dT}{T} - \overline{\mathcal{R}} \ln \frac{\bar{p}}{p^0}\right) = m\left(\bar{c}_p \ln \frac{\overline{T}}{T^0} - \mathcal{R} \ln \frac{\bar{p}}{p^0}\right) \tag{16-26}$$

if, consistent with the assumption already used, $\bar{c}_p$ can be approximately considered constant in the range T^0–$\overline{T}$. In general $\overline{S} - S^0$ is not zero, and therefore the mean uniform flow is not exactly isentropic. Eq. 16-26 can be put in the alternate form

$$\frac{\bar{p}}{p^0}\left(\frac{T^0}{\overline{T}}\right)^{\frac{\bar{\gamma}}{\bar{\gamma}-1}} = e^{\frac{S^0-\overline{S}}{\overline{\mathfrak{R}}}} = \Xi \tag{16-27}$$

where Ξ represents a third nonuniformity coefficient. Clearly the less Ξ deviates from unity, the smaller is the error resulting from assuming isentropic expansion for the mean flow.

If, instead of $\overline{T}$ one uses the other definition, $\tilde{T}$, of the mean temperature, the corresponding relation, similar to Eq. 16-27, will be expressed in terms of the two nonuniformity coefficients Ξ and α.

Values of the nonuniformity coefficients. Pipe flow. The deviations of α, β, and Ξ from unity represent a measure of the error resulting from the uniformity assumption. For certain intrinsically nonuniform flows (such as those with piecewise uniformity or those with large separation effects) the divergence from unity may be substantial. For more regular flows, however, the deviation can be quite small, as we are going to show for three types of flow, and the assumption of uniform flow can produce a very good approximation.

We first consider the effects of the boundary layers on the values of the nonuniformity coefficients in a pipe. For this evaluation we choose the worst case, that of fully developed pipe flow, for which the boundary layer fills the entire section. In these conditions a well-determined velocity distribution is developed and remains approximately stationary along the pipe. In view of the complications of the exact analysis of the compressible viscous laminar flow and of the lack of a sure basis of analysis for the turbulent flow, and in view of the approximate character of the results we are looking for, we shall perform the calculations on the basis of some plausible approximate relations which result in simple expressions for the nonuniformity coefficients.

We assume a round pipe with inner radius R. At a certain radius $r < R$ the quantity

$$\frac{\eta}{A} = 1 - \frac{r^2}{R^2} \tag{16-28}$$

expresses the surface of the annulus between r and R as a fraction of the cross-sectional area. For low Mach number and isothermal flow, the density is practically constant and the velocity distribution can be represented with good approximation by a power law

$$\frac{u}{u_a} = \left(\frac{\eta}{A}\right)^{\frac{1}{n}} \tag{16-29}$$

where the subscript $_a$ represents the value on the axis and n can be taken as 1 for laminar flows and 7 for ordinary turbulent flows (larger for larger Reynolds numbers). Observe that near the wall, for small $R - r$, Eq. 16-29 gives a velocity proportional to the $1/n$ power of the distance from the wall. The mass flow rate through the annulus of area η is given by

$$\mu = \rho \int_r^R u d\eta = \frac{n}{n+1} \rho u_a A \left(\frac{\eta}{A}\right)^{\frac{n+1}{n}} = m \left(\frac{\eta}{A}\right)^{\frac{n+1}{n}} \tag{16-30}$$

Therefore the relation

$$\frac{u}{u_a} = \left(\frac{\mu}{m}\right)^{\frac{1}{n+1}} \tag{16-31}$$

is equivalent to Eq. 16-29.

If ρ is not constant (high Mach numbers or nonisothermal, low Mach number flows), we define a new variable

$$y = \frac{\int_0^\eta \rho d\eta}{\int_0^A \rho d\eta}$$

and we make the assumption that the velocity distribution can be expressed by

$$\frac{u}{u_a} = y^{\frac{1}{n}} \tag{16-32}$$

which coincides with Eq. 16-29 for constant ρ. The variable y can be regarded as a kind of Howarth variable (IV,B) generalized for axially symmetric flows. Certainly the assumption of Eq. 16-32 does not correspond exactly to the actual distribution. However, there is sufficient evidence in boundary layer theory that it may represent a good approximation to the actual distribution for both laminar and turbulent flows.

With the assumption of Eq. 16-32 one finds, instead of Eq. 16-30,

$$\mu = \int_r^R \rho u d\eta = \frac{n}{n+1} u_a \int_0^A \rho d\eta \cdot y^{\frac{n+1}{n}} = m y^{\frac{n+1}{n}}$$

As a result Eq. 16-31 holds even in the case of variable density. Once the velocity distribution is known, the temperature distribution can be obtained from the energy integral ([*63*]; also IV,B,4)

$$\frac{h_0 - h_w}{h_a^0 - h_w} \cong \frac{T^0 - T_w}{T_a^0 - T_w} = \frac{u}{u_a} \tag{16-33}$$

where the subscripts $_a$ and $_w$ indicate the axis and wall conditions. For simplicity c_p has been assumed to be uniform throughout the section and, in agreement with previous assumptions, constant between the static and the stagnation temperatures. Similarly, $\mathcal{R}$ and γ are assumed uniform in the following developments. The energy integral, Eq. 16-33, gives a good approximation of the actual distribution for laminar flows, and an even better approximation for turbulent flows in the general case when heat is being exchanged through the wall. The case of the insulated wall is included as a particular case for $T^0_a = T_w$ and Eq. 16-33 is reduced to $T^0 = \text{const}$. For $T^0_a \gtrless T_w$ heat is being transferred from or to the fluid.

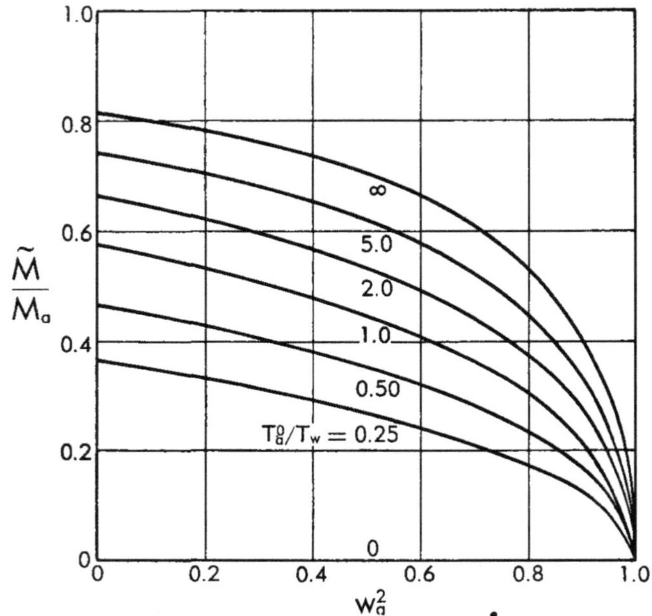

Fig. B,16a. Mean-to-axial Mach number ratio in laminar pipe flow. $n = 1$.

From Eq. 16-33 and 16-31 one obtains the following relation between the static temperature and μ

$$\frac{T}{T_w} = 1 + \left(\frac{T^0_a}{T_w} - 1\right)\left(\frac{\mu}{m}\right)^{\frac{1}{n+1}} - \frac{u_a^2}{2c_pT_w}\left(\frac{\mu}{m}\right)^{\frac{2}{n+1}} \tag{16-34}$$

We can now proceed to the calculation of some of the mean quantities and of α and β. Introducing Eq. 16-31 we obtain from Eq. 16-8 and 16-12

$$\frac{\tilde{u}}{u_a} = \frac{n+1}{n+2}, \quad \beta = 1 + \frac{1}{(n+1)(n+3)} \tag{16-35}$$

With the additional help of Eq. 16-34 we can evaluate pA from Eq. 16-13 and $\tilde{M}^2$ from Eq. 16-15. The reduced velocity and the Mach number on the axis of the pipe are given by

$$w_a^2 = \frac{T_w}{T^0_a}\frac{u_a^2}{2c_pT_w}, \quad M_a^2 = \frac{2}{\gamma - 1}\frac{w_a^2}{1 - w_a^2} \tag{16-36}$$

It is interesting to evaluate the quantity

$$\frac{M_a^2}{\tilde{M}^2} = 1 + \frac{1}{n+1} \frac{1 + \frac{n+2}{n} \frac{T_w}{T_a^0}}{1 - w_a^2} \tag{16-37}$$

For α we derive from Eq. 16-11 the following relation

$$\frac{1}{\alpha} = 1 + \frac{1}{n} \frac{\frac{T_w}{T_a^0} + \frac{n(n+1)}{(n+2)(n+3)} w_a^2}{\frac{T_w}{T_a^0} + (n+1)\left(1 - \frac{n+2}{n+3} w_a^2\right)} \tag{16-38}$$

The corresponding values of $\tilde{M}/M_a$ and α are shown in Fig. B,16a and B,16b for $n = 1$ and in Fig. B,16c and B,16d for $n = 7$ in the whole range of variation of w_a and T_w/T_a^0.

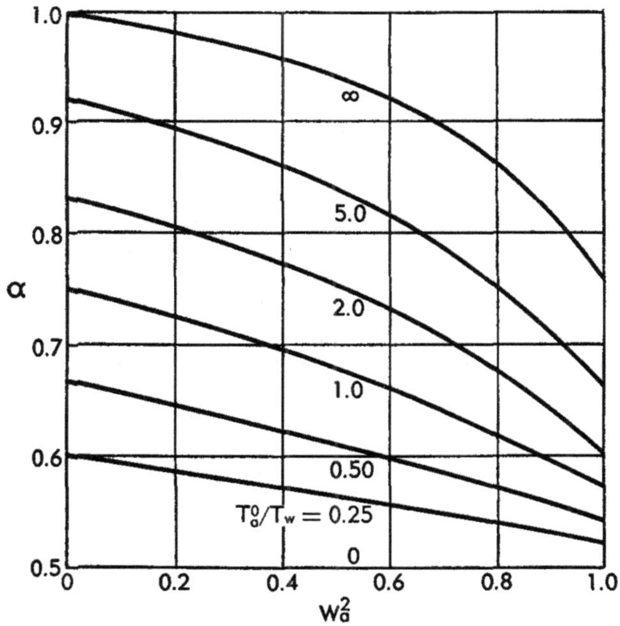

Fig. B,16b. Nonuniformity coefficient α for laminar pipe flow. $n = 1$.

For the case of insulated walls ($T_a^0 = T_w$) and subsonic velocities the values are always quite close to those corresponding to $w_a = 0$, that is

$$\alpha = 1 - \frac{1}{(n+1)^2}, \quad \frac{\tilde{M}}{M_a} = \sqrt{\frac{n}{n+2}}$$

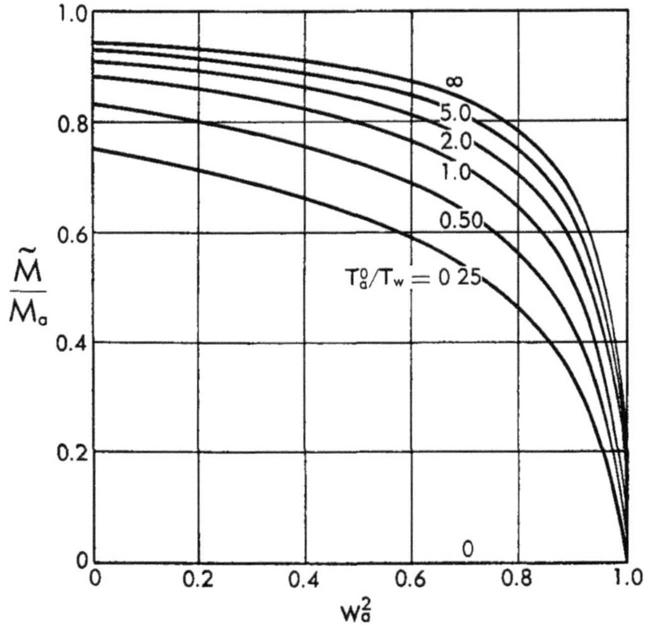

Fig. B,16c. Mean-to-axial Mach number ratio in turbulent pipe flow. $n = 7$

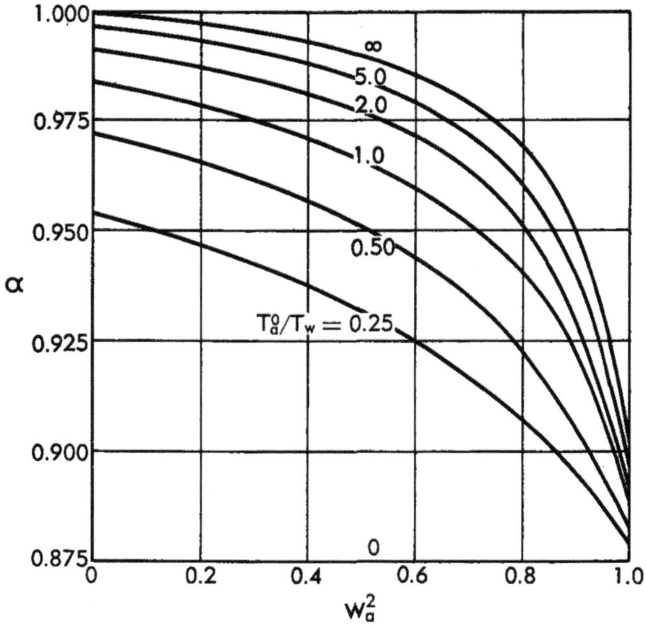

Fig. B,16d. Nonuniformity coefficient α for turbulent pipe flow. $n = 7$.

The following table gives for $n = 1$ and $n = 7$ the values obtained from these two expressions and from Eq. 16-35 for β, which is independent of the flow conditions.

Table B,16a

n	α	β	$\tilde{M}/M_a$
1	0.75	1.125	0.578
7	0.9843	1.0125	0.883

From this table and the figures, and from consideration of the nonuniformity coefficients, we see that for $n = 1$ (laminar flow) large deviations from the uniformity are still present, but the deviations for $n = 7$ (turbulent flow) and low Mach number are of the order of 1 per cent (except for low values of T_a^0/T_w). In all cases the values of $\tilde{M}/M_a$ deviate substantially more from unity than those of α and β. For $n = 7$ and low Mach number the values of the maximum Mach number exceed by more than 10 per cent the mean value. This fact may have effects in the transonic range.

Vortex flow in a two-dimensional annular duct. As a second example consider the isentropic plane flow between two concentric cylinders of radius R and $\nu R (\nu < 1)$. Due to the centrifugal forces, the pressure, and therefore the velocity and the temperature, must vary with the distance r from the center of curvature. The particular type of flow for which

$$wr = \text{const} = w_{\max}\nu R \tag{16-39}$$

($w_{\max}$ representing the velocity on the inner cylindrical surface) is of special significance and is called "vortex flow." The evaluation of the values of the nonuniformity coefficients for vortex flow can give a rough idea of the effects of the curvature for more general types of ducts and of the possibility of approximating the flow in a curved channel by a uniform flow.

Assuming constant c_p, $\mathcal{R}$, T^0, p^0, and ρ^0, the temperature and density for each w are given by Eq. 6-4a and 6-4c. The flow rate $d\mu = \rho u d\eta$ is obtained by taking, according to Eq. 16-39, 4-25, and 6-4c,

$$\frac{d\eta}{A} = \frac{1}{1-\nu}\frac{dr}{R} = \frac{\nu}{1-\nu} w_{\max} d\left(\frac{1}{w}\right),$$
$$u = w\sqrt{2c_pT^0}, \quad \rho = \rho^0(1 - w^2)^{1/(\gamma-1)}$$

where A once again represents the cross-sectional area of the duct.

Introducing these expressions in Eq. 16-11 and 16-12, and taking into account Eq. 6-4a, $T = T^0(1 - w^2)$, one obtains

$$\alpha = \frac{\int_{\nu w_{\max}}^{w_{\max}} (1 - w^2)^{1/(\gamma-1)} \frac{dw}{w} \int_{\nu w_{\max}}^{w_{\max}} (1 - w^2)^{\gamma/(\gamma-1)} \frac{dw}{w}}{\int_{\nu w_{\max}}^{w_{\max}} (1 - w^2)^{1/(\gamma-1)} dw \int_{\nu w_{\max}}^{w_{\max}} (1 - w^2)^{\gamma/(\gamma-1)} \frac{dw}{w^2}}$$

$$\beta = \frac{\int_{\nu w_{\max}}^{w_{\max}} (1 - w^2)^{1/(\gamma-1)} \frac{dw}{w} \int_{\nu w_{\max}}^{w_{\max}} (1 - w^2)^{1/(\gamma-1)} w dw}{\left[\int_{\nu w_{\max}}^{w_{\max}} (1 - w^2)^{1/(\gamma-1)} dw\right]^2}$$

Similarly from the definition (Eq. 16-4) of $\overline{T}$ we get

$$\frac{\overline{T}}{T^0} = \frac{\int_{\nu w_{\max}}^{w_{\max}} (1 - w^2)^{\gamma/(\gamma-1)} \frac{dw}{w}}{\int_{\nu w_{\max}}^{w_{\max}} (1 - w^2)^{1/(\gamma-1)} \frac{dw}{w}}$$

and from that of $\bar{p}$ (Eq. 3-7) and Eq. 6-4b, we obtain

$$\frac{\bar{p}}{p^0} = \frac{\nu w_{\max}}{1 - \nu} \int_{\nu w_{\max}}^{w_{\max}} (1 - w^2)^{\gamma/(\gamma-1)} \frac{dw}{w^2} \tag{16-40}$$

The value of Ξ can thus be calculated from Eq. 16-27. Fig. B,16e and B,16f show for several values of the Mach number $M_{\max}$ corresponding to $w_{\max}$ and for ν between 0.4 and 1 the values of the three nonuniformity coefficients. It clearly appears that if the value of ν is sufficiently close to 1 (that is, if the radius of curvature of the duct is sufficiently large compared with its width) the effects of nonuniformities can be approximately neglected for all $M_{\max}$ as far as α and β are concerned, and may be neglected for not too large $M_{\max}$ when considering Ξ. Observe that in the same range of ν the pressure nonuniformities are comparatively large, as shown by the curves of Fig. B,16g obtained from the relation

$$\frac{p_{\max} - p_{\min}}{p^0} = (1 - w_{\max}^2)^{\gamma/(\gamma-1)} - (1 - \nu^2 w_{\max}^2)^{\gamma/(\gamma-1)}$$

divided by Eq. 16-40.

Choked flow in an axially symmetric throat. As a last example let us consider the isentropic flow of a polytropic gas in the throat of an axially symmetric nozzle. Due to the curvature of the streamlines and the resulting centrifugal forces, the pressure must increase from the walls to the axis. The velocity must have the opposite behavior, as schematically shown in Fig. B,16h, so that at the throat it will be supersonic near the walls and subsonic on the axis.

The exact calculation of this flow pattern presents analytical diffi-

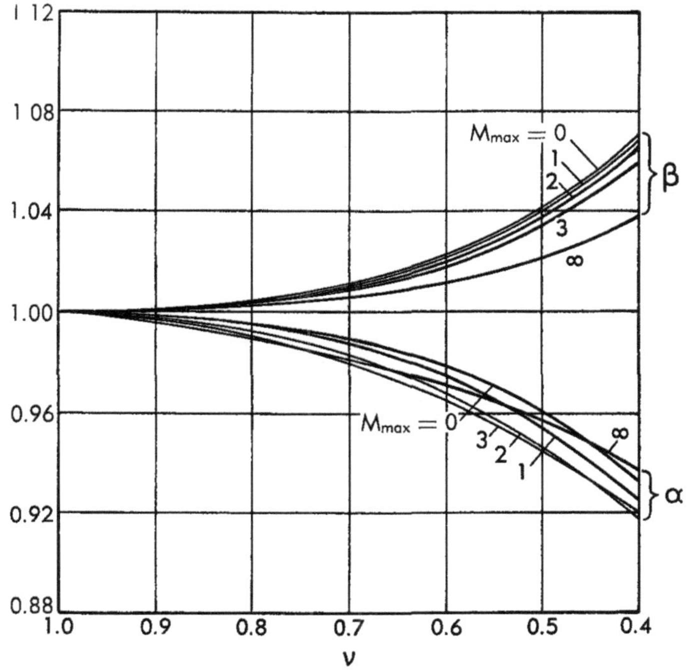

Fig. B,16e. Nonuniformity coefficients α and β for vortex flow. $\gamma = 1.4$.

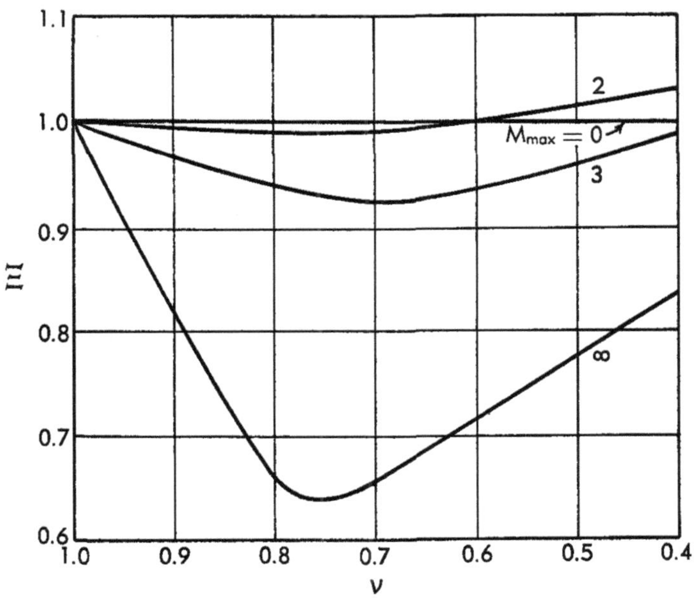

Fig. B,16f. Nonuniformity coefficient Ξ for vortex flow. $\gamma = 1.4$.

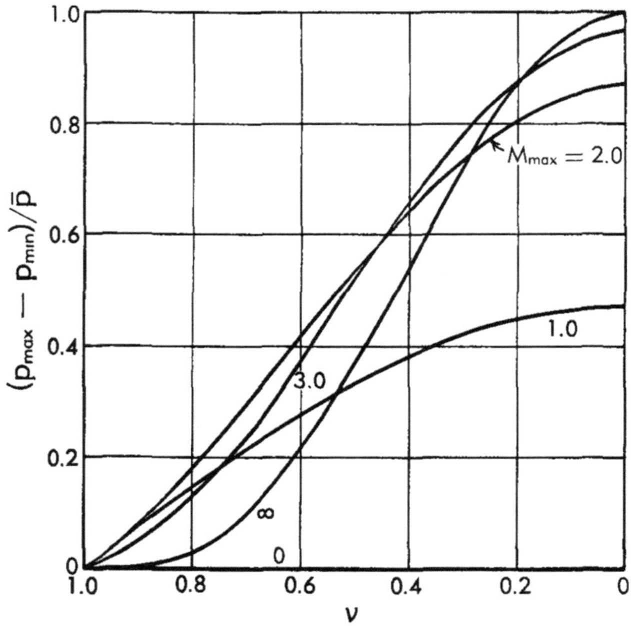

Fig. B,16g. Pressure nonuniformities in vortex flow. $\gamma = 1.4$.

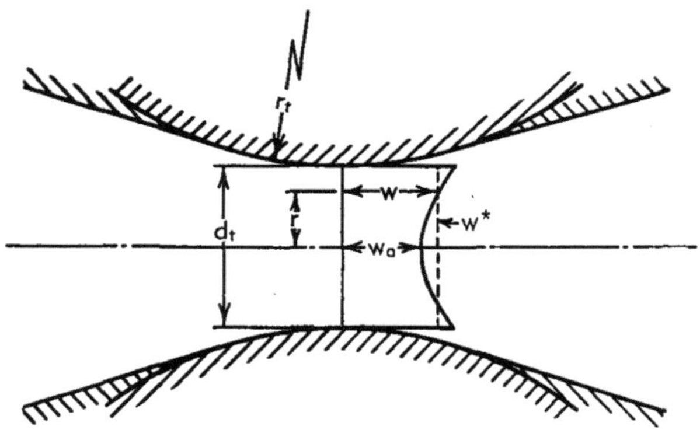

Fig. B,16h. Velocity distribution at the throat of a supercritical axially symmetric nozzle.

culties. Here, in view of the nature of the desired calculations, we shall use the approximate results of [64].[21] The velocity distribution can be expanded in a series of powers of μ/m, where μ represents the flow rate between the axis and the radius r in the form

$$w = w_a\left(1 + g\frac{\mu}{m} + h\left(\frac{\mu}{m}\right)^2 + \cdots\right)$$

Here from [64] we have

$$g = \frac{d_t}{4r_t}$$

[21] Identical results can be obtained from the treatments of [65,66] and others, based on the series expansions originally suggested by Meyer and Taylor-Hooker.

Moreover h can be expected to be of order g^2. Its actual value could be evaluated, but it does not appear in the results of these calculations.

The value of w_a can also be obtained from [64], and is given by

$$w_a = w^*\left(1 - \frac{g}{2} + k\right) \tag{16-41}$$

where k is of the order g^2 and does not need to be calculated for the purpose of this discussion. With $T/T^0 = 1 - w^2$, we find from Eq. 16-11 and 16-12

$$\alpha = \frac{\int_0^1 (1 - w^2)d\left(\frac{\mu}{m}\right)}{\int_0^1 wd\left(\frac{\mu}{m}\right)\int_0^1 \frac{1 - w^2}{w} d\left(\frac{\mu}{m}\right)}, \quad \beta = \frac{\int_0^1 w^2 d\left(\frac{\mu}{m}\right)}{\left[\int_0^1 wd\left(\frac{\mu}{m}\right)\right]^2} \tag{16-42}$$

Moreover, we have

$$\frac{\overline{T}}{T^0} = 1 - \int_0^1 w^2 d\left(\frac{\mu}{m}\right) \tag{16-43}$$

The mean pressure can be calculated from

$$\begin{aligned}\frac{\bar{p}}{p^0} &= \frac{1}{p^0 A}\int_A p d\eta = \frac{1}{p^0 A}\int_m \frac{p}{\rho u} d\mu \\ &= \frac{m\mathcal{R}T^0}{p^0 A\sqrt{2c_p T^0}}\int_0^1 \frac{1 - w^2}{w} d\left(\frac{\mu}{m}\right) \\ &= w^*(1 - w^{*2})^{1/(\gamma-1)}\int_0^1 \frac{1 - w^2}{w} d\left(\frac{\mu}{m}\right)\end{aligned} \tag{16-44}$$

since, within the accuracy of the present approximation we can still write (this can be checked from the results of [64]):

$$d\mu = \rho u d\eta, \quad m = \rho_{is}^* A a^* = \rho^0 A\sqrt{2c_p T^0}\, w^*(1 - w^{*2})^{1/(\gamma-1)},$$

The above expressions can be reduced to the evaluation of three integrals, which are calculated below up to terms of the order g^2:

$$\int_0^1 \frac{1}{w} d\left(\frac{\mu}{m}\right) = \frac{1}{w_a}\left(1 - \frac{g}{2} + \frac{g^2 - h}{3}\right)$$

$$\bar{w} = \int_0^1 wd\left(\frac{\mu}{m}\right) = w_a\left(1 + \frac{g}{2} + \frac{h}{3}\right)$$

$$\bar{w}^2 = \int_0^1 w^2 d\left(\frac{\mu}{m}\right) = w_a^2\left(1 + \frac{g}{2} + \frac{g^2 + 2h}{3}\right)$$

Introducing these expressions and that of w_a (Eq. 16-41) into Eq. 16-42, 16-43, and 16-44 we find, up to terms of order g^2,

$$\alpha = 1 - \frac{\gamma}{12} g^2 = 1 - \frac{\gamma}{192}\left(\frac{d_t}{r_t}\right)^2$$

$$\beta = 1 + \frac{1}{12} g^2 = 1 + \frac{1}{192}\left(\frac{d_t}{r_t}\right)^2$$

$$\frac{\bar{p}}{p^0} = (1 - w^{*2})^{\gamma/(\gamma-1)}\left[1 + \frac{7\gamma+1}{24} g^2 - \gamma\left(k + \frac{h}{3}\right)\right]$$

$$\frac{\bar{T}}{T^0} = (1 - w^{*2})\left[1 + \frac{5}{24}(\gamma-1)g^2 - (\gamma-1)\left(k + \frac{h}{3}\right)\right]$$

Finally we obtain from Eq. 16-27

$$\Xi = 1 + \frac{2\gamma+1}{24} g^2 = 1 + \frac{2\gamma+1}{384}\left(\frac{d_t}{r_t}\right)^2$$

It is interesting to notice that the h and k contribution to the nonuniformity coefficients is of an order higher than g^2. Therefore the result is that, while the velocity (and the pressure) nonuniformities are of order g, the deviations of the nonuniformity coefficients from unity are of order g^2. In the following table are shown, for $\gamma = 1.4$, the values of α, β, Ξ, and those of $(w_{max} - w_{min})/\bar{w}$ and $(p_{max} - p_{min})/\bar{p}$ (the last two calculated only up to terms of order g) for three values of d_t/r_t. We see that for values

Table B,16b

$\frac{d_t}{r_t}$	α	β	Ξ	$\frac{w_{max} - w_{min}}{\bar{w}}$	$\frac{p_{max} - p_{min}}{\bar{p}}$
0.5	0.998	1.001	1.002	0.124	0.174
1.0	0.993	1.005	1.010	0.250	0.350
2.0	0.972	1 020	1.040	0.500	0.700

of $d_t \leq r_t$ the deviations of α, β, and Ξ from unity are within 1 per cent, and the one-dimensional approximation can be applied with confidence even though the variation of velocity across the channel amounts to as much as one fourth of the mean velocity, and that of the pressure to 35 per cent of the mean value.

Meaning of the uniformity assumption. When the nonuniformity coefficients are sufficiently close to unity, they can be taken approximately equal to 1, with the result that all the conservation equations for the nonuniform flow come to coincide with those of a uniform flow possessing the mean properties. This is the actual meaning of the uniformity assumption. The errors deriving from this approximation are of the same order as the deviations from unity of the nonuniformity coefficients.

Observe that, when this approximation is used, the way the mean quantities are defined becomes immaterial, since within the same order of error all the definitions lead to the same value, as is clear from Eq. 16-10.

B,17. Some General Properties of Nonuniform Flows. In Art. 16 the nonuniformity coefficients were introduced with the purpose of showing that for many typical flows they can be approximately identified with unity, thus justifying the uniformity assumption. However, when the deviations of the nonuniformity coefficients from unity are of the order of 1, the uniformity assumption has to be rejected. Such is the case, for instance, for flows with piecewise uniformity, in which the nonuniformity may be deliberately introduced for some useful purpose. Clearly the same is true for more general types of flow with strong nonuniformities. The purpose of this article is to present some interesting results that can be derived for flows of this type without any restrictive assumption on the deviation from uniformity.

Two inequalities. We shall discuss first a few general properties of the coefficients α and β. For this purpose we recall the Schwartz inequality

$$\left(\int_{x_1}^{x_2} f(x)g(x)dx\right)^2 \leqq \int_{x_1}^{x_2} [f(x)]^2dx \int_{x_1}^{x_2} [g(x)]^2dx \tag{17-1}$$

which holds for arbitrary functions $f(x)$ and $g(x)$. The equal sign only applies when $f/g = \text{const}$, as can be immediately checked. For a derivation of this inequality, see [*67*, p. 49].

In the more restrictive, but still quite broad, case in which df/dx and dg/dx are either of the same sign or of opposite signs in the whole interval x_1, x_2 in which the functions are defined, we can write the following additional inequality

$$\int_{x_1}^{x_2} dx \int_{x_1}^{x_2} f(x)g(x)dx \gtrless \int_{x_1}^{x_2} f(x)dx \int_{x_1}^{x_2} g(x)dx \tag{17-2}$$

where the upper or lower sign has to be taken depending on whether

$$\frac{df/dx}{dg/dx} \gtrless 0 \tag{17-3}$$

that is whether f and g vary in the same direction or in opposite directions. In a somewhat unconventional way, we shall say here, to indicate these two cases, that f and g are covariant in the first case and contravariant in the second.[22] When the left-hand side of Eq. 17-3 is either zero or infinity, that is for constant f or g, the inequality of Eq. 17-2 is transformed into an equality.

The proof of the inequality of Eq. 17-2 can be given as follows. In performing the integrations appearing in Eq. 17-2, the order of the elements

[22] This definition can cover the case when f and g are discontinuous, and hence is more general than Eq. 17-3.

dx can be arbitrarily changed. Thus without any restriction, one can assume that $f(x)$ is a monotonically increasing function of x in the interval x_1, x_2. Then $g(x)$ will also be a monotonically increasing function of x if f and g are covariant, and a monotonically decreasing function of x if they are contravariant. Let us define the mean value of f in the interval x_1, x_2

$$f_m = \frac{\int_{x_1}^{x_2} f dx}{\int_{x_1}^{x_2} dx} \tag{17-4}$$

and the corresponding value of x_m such that[23]

$$f(x_m) = f_m$$

Then if f and g are covariant we have[24]

$$f(x) - f_m \gtrless 0; \quad g(x) - g(x_m) \gtrless 0 \quad \text{for} \quad x \gtrless x_m$$

and if they are contravariant

$$f(x) - f_m \gtrless 0; \quad g(x) - g(x_m) \lessgtr 0 \quad \text{for} \quad x \gtrless x_m$$

Hence we can write

$$\int_{x_1}^{x_2} [f(x) - f_m][g(x) - g(x_m)]dx \gtrless 0$$

depending on whether f and g are covariant or contravariant. Adding to this expression the quantity

$$g(x_m) \int_{x_1}^{x_2} [f(x) - f_m]dx = 0$$

which vanishes identically because of Eq. 17-4, we obtain

$$\int_{x_1}^{x_2} [f(x) - f_m]g(x)dx \gtrless 0$$

or

$$\int_{x_1}^{x_2} f(x)g(x)dx \gtrless f_m \int_{x_1}^{x_2} g(x)dx$$

depending on whether f and g are covariant or contravariant.

Introducing for f_m the expression of Eq. 17-4, we find the inequality of Eq. 17-2.

General properties of α and β. From the definition of β (Eq. 16-12) and the Schwartz inequality (Eq. 17-1) we see immediately, taking $dx = dm$, $f = u$, and $g = 1$, that

$$\beta \geqq 1 \tag{17-5}$$

[23] If $f(x)$ presents discontinuities and f_m falls in the range of f corresponding to one such discontinuity, x_m has to be taken equal to the value of x where that discontinuity takes place.

[24] If g is discontinuous at x_m, $g(x_m)$ can be assigned any value in the range of g corresponding to the discontinuity.

the sign of equality being correct only for uniform u, no matter how nonuniform the distributions of the other quantities are. This result corresponds to the well-known property that the mean square value of a quantity is larger than the square of its mean value, except, of course, when the quantity is constant.

The properties of α are easily determined when γ is uniform on each section. In this case c_p and $\mathcal{R}$ are proportional to each other and Eq. 16-11 is reduced to

$$\alpha = \frac{\int_m d\mu \int_m \mathcal{R}T d\mu}{\int_m u d\mu \int_m \frac{\mathcal{R}T}{u} d\mu} \tag{17-6}$$

Thus taking $dx = d\mu$, $f = u$, and $g = \mathcal{R}T/u$ we find from the inequality of Eq. 17-2 that $\alpha \gtrless 1$, depending on whether u and $1/g = u/\mathcal{R}T$ are contravariant or covariant, and $\alpha = 1$ if u is either uniform or proportional to $\mathcal{R}T$.

Thus for uniform velocity and γ distributions both α and β are unity, regardless of the distribution of the other quantities.

Observe that the fact the temperature has nonvanishing values everywhere, while u, and therefore $u/\mathcal{R}T$ vanishes at the wall, suggests that for certain types of flow the latter quantity is likely to be covariant with u. This can be easily checked to be true for flows where the distribution of stagnation temperature is uniform or obeys the energy integral (Eq. 16-29). For flows of this type α is certainly <1, as happens, for instance, for the examples of Art. 16. However, other types of flow, such as those present in mixing ducts, may have values of α larger than unity, although for practical devices the opposite is more likely. These statements can be easily checked by considering the simple case of two polytropic gases with identical c_p and γ, brought together at the same static pressure in a mixing duct with different stagnation temperatures $T^{0\prime}$ and $T^{0\prime\prime} < T^{0\prime}$, and different Mach numbers M' and M'', or velocities u' and u''.[25] The following relations are immediately found:

$$\frac{u'}{\sqrt{2c_pT^{0\prime\prime}}} = \sqrt{\frac{T^{0\prime}}{T^{0\prime\prime}}\left(1 - \frac{T'}{T^{0\prime}}\right)}, \quad \frac{u''}{\sqrt{2c_pT^{0\prime\prime}}} = \sqrt{1 - \frac{T''}{T^{0\prime\prime}}}$$

$$\sqrt{\frac{T^{0\prime\prime}}{2c_p}}\frac{u'}{T'} = \frac{T^{0\prime}}{T'}\sqrt{\frac{T^{0\prime\prime}}{T^{0\prime}}\left(1 - \frac{T'}{T^{0\prime}}\right)}, \quad \sqrt{\frac{T^{0\prime\prime}}{2c_p}}\frac{u''}{T''} = \frac{T^{0\prime\prime}}{T''}\sqrt{1 - \frac{T''}{T^{0\prime\prime}}}$$

with $T^{0\prime}/T'$ and $T^{0\prime\prime}/T''$ related to M' and M'' by Eq. 4-28. These rela-

[25] Here the slight nonuniformities due to the presence of boundary layers are disregarded, in view of the strong nonuniformity deliberately introduced in the flow. This is necessary in order to be able to apply the inequality of Eq. 17-2.

tions are represented in Fig. B,17a for different values of $T^{0\prime}/T^{0\prime\prime}$ and of T^0/T (equal to $T^{0\prime}/T^\prime$ or $T^{0\prime\prime}/T^{\prime\prime}$). The curve corresponding to $T^{0\prime}/T^{0\prime\prime} = 1$ represents not only the relation for the first stream when the two stagnation temperatures are identical, but also the relation for the second

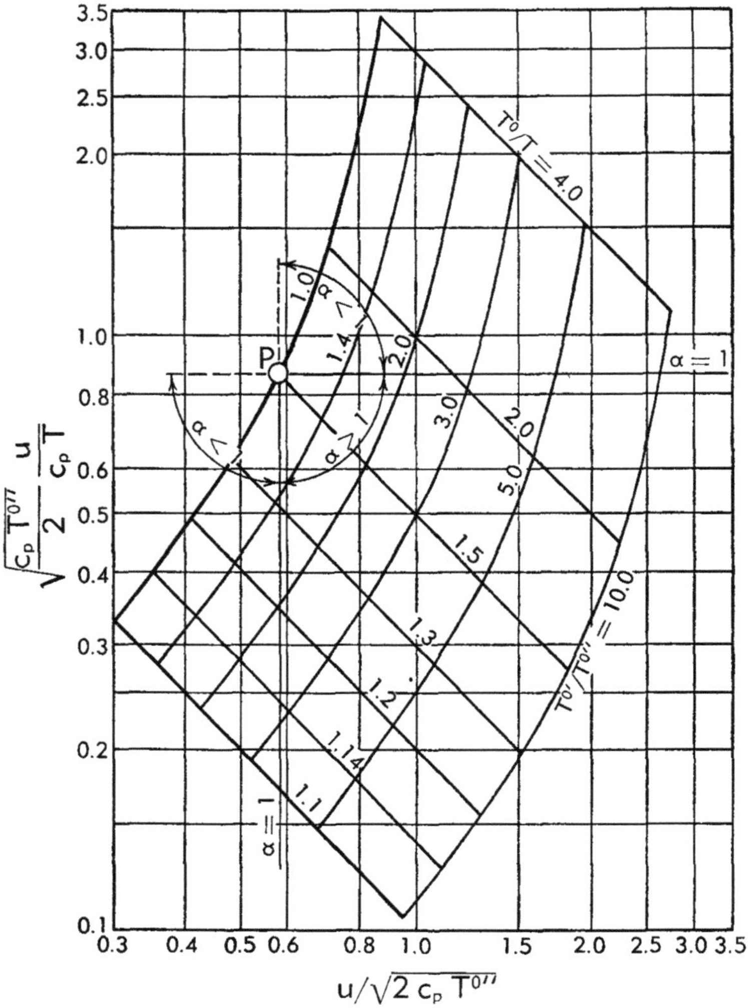

Fig. B,17a. Diagram for the determination of the qualitative behavior of the nonuniformity coefficient α in ducts where two streams (with equal c_p and γ) are mixed.

stream in all cases, parametered by the values of $T^{0\prime\prime}/T^{\prime\prime}$. If the latter quantity, and therefore $M^{\prime\prime}$, has an assigned value (for instance $T^{0\prime\prime}/T^{\prime\prime} = 2$), the point representative of the second stream has a fixed position (for instance the point P of the figure). The point representing the first stream can be immediately found if $T^{0\prime}/T^{0\prime\prime}$ and $T^{0\prime}/T^\prime$ are given. It is clear that depending on which quadrant around the point P the point representing the first stream lies, u and u/T will be covariant or contravariant and, hence, $\alpha \gtrless 1$ as shown in the figure. It is apparent that if the imbalance between the two Mach numbers (i.e. between the values of $T^{0\prime}/T'$ and $T^{0\prime\prime}/T''$) is large, large values of $T^{0\prime}/T^{0\prime\prime}$ are required to make $\alpha > 1$; but this condition can easily be achieved if M' and M'' are close to each other, and is always achieved when they are equal. In most practical devices the two Mach numbers are substantially different from each other. For these devices, therefore, α is likely to be <1, and only for exceptionally high differences in stagnation temperature can the opposite happen. Thus the most common case in practice is $\alpha < 1$. However, examples of practical applications where M' and M'' are quite close to each other can be found (for instance a twin-flow (by-pass) turbojet with mixing before the expansion in the nozzle); and in this case the most likely condition is $\alpha > 1$. When this happens α may be smaller or larger than β. The corresponding condition is found from the relation (still relative to the case of uniform γ):

$$\frac{\alpha}{\beta} = \frac{\int_m u d\mu \int_m \mathcal{R} T d\mu}{\int_m u^2 d\mu \int_m \frac{\mathcal{R}T}{u} d\mu} = \frac{\int_m u d\mu \int_m a^2 d\mu}{\int_m u^2 d\mu \int_m \frac{a^2}{u} d\mu}$$

immediately obtained from Eq. 16-12 and 17-6. Applying again the inequality of Eq. 17-2 with $dx = ud\mu$, $f = u$, and $g = a^2/u^2 = 1/M^2$, we see that $\alpha \gtrless \beta$ depending on whether u and M are contravariant or covariant. If u is uniform $\alpha = \beta = 1$; but we also have $\alpha = \beta$ for uniform M. For practical flows, including the case of mixing ducts, u and M are likely to vary in the same directions, faster portions of the flow also having the higher Mach number. Hence the most common condition is $\alpha < \beta$.

A quantity whose behavior is important for the considerations of this article is (again for uniform γ)

$$\Delta' = \frac{m \int_m c_p T^0 d\mu}{\int_m u d\mu \int_m \frac{c_p T^0}{u} d\mu} \tag{17-7}$$

Δ' can also be considered as a nonuniformity coefficient, analogous to Eq. 17-6, with the stagnation temperature in place of the static temperature. Applying the inequality of Eq. 17-2 with $dx = d\mu$, $f = u$, and $g = c_pT^0/u$, we obtain $\Delta' \gtrless 1$ depending on whether u and u/c_pT^0 are contravariant or covariant. Also, we have $\Delta' = 1$ both for uniform u or for u proportional to c_pT^0.

Again, for well-behaved flows, such as the case when T^0 is uniform or obeys the energy integral (Eq. 16-29), u and u/c_pT^0 vary in the same direction and hence $\Delta' < 1$. However, the opposite may be true for the case of mixing ducts, as can be shown from the consideration of the simple

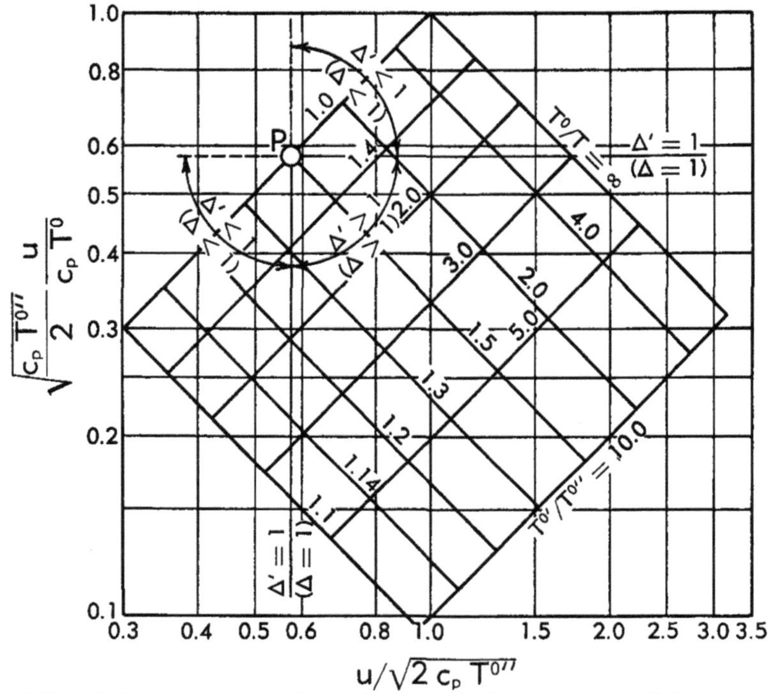

Fig. B,17b. Diagram for the determination of the qualitative behavior of the nonuniformity coefficients Δ' and Δ in ducts where two streams (with equal c_p and γ) are mixed.

case already discussed for α. We can write immediately

$$\sqrt{\frac{T_0''}{2c_p}}\frac{u'}{T^{0'}} = \sqrt{\frac{T^{0''}}{T^{0'}}\left(1 - \frac{T'}{T^{0'}}\right)}, \quad \sqrt{\frac{T^{0''}}{2c_p}}\frac{u''}{T^{0''}} = \sqrt{1 - \frac{T''}{T^{0''}}}$$

and draw Fig. B,17b, similar to Fig. B,17a. The discussion is here identical to that developed for α. Again depending on which quadrant around the point P (representing the second stream) the point that represents the first stream is located, we have $\Delta' \gtrless 1$, as indicated in the figure. Qualitatively, one can make the same considerations about this figure and the conditions under which Δ' may be >1, as in the case of α. Quantitatively, however, it is easily checked that it is more likely for Δ' to be >1 than for $\alpha > 1$. This is in agreement with the following relations (Eq. 17-11 and 17-11a).

The quantity Δ' can be expressed through α and β in the following way. We have from Eq. 16-2, 16-6, and 16-10

$$\int_m c_p T^0 d\mu = \hat{H}^0 = m\left(\alpha\tilde{c}_p\tilde{T} + \frac{\beta}{2}\tilde{u}^2\right)$$

Moreover, for uniform γ, making use of Eq. 16-23 we find

$$\int_m \frac{c_p T^0}{u}\, d\mu = \int \frac{c_p T}{u}\, d\mu + \tfrac{1}{2}\int u\, d\mu$$

$$= \frac{\gamma}{\gamma - 1}\bar{p}A + \tfrac{1}{2}I$$

Thus, recalling the definition (Eq. 16-14) of $\tilde{M}$ and Eq. 16-24 and 16-25 we obtain from Eq. 17-7

$$\Delta' = \frac{\alpha + \beta\dfrac{\gamma - 1}{2}\tilde{M}^2}{1 + \dfrac{\gamma - 1}{2}\tilde{M}^2}$$

or, in terms of $\tilde{w}$, making use of Eq. 16-16 and 16-17

$$\Delta' = \frac{\dfrac{\alpha}{1 - \beta\tilde{w}^2}}{1 + \dfrac{\alpha\tilde{w}^2}{1 - \beta\tilde{w}^2}} = 1 + \frac{\Delta - 1}{1 + \dfrac{\gamma - 1}{2}\tilde{M}^2} \tag{17-8}$$

where we have used the abbreviation

$$\Delta = \frac{\alpha(1 - \tilde{w}^2)}{1 - \beta\tilde{w}^2} \tag{17-9}$$

The quantity Δ may also be considered as a nonuniformity coefficient, and in fact it is unity for $\alpha = \beta = 1$. Since the denominator of Eq. 17-8 is always positive, the conditions discussed above for $\Delta' \gtrless 1$ can be translated into the other condition that

$$\Delta \gtrless 1 \tag{17-10}$$

depending on whether u and u/c_pT^0 are contravariant or covariant. Again, as for α and Δ', Δ is <1 for well-behaved flows, and is likely to be such also for mixing ducts when the Mach number imbalance is large and the stagnation temperature imbalance not too large, as happens in the most common practical applications. However, in some practical applications the opposite may be true and Δ may be larger than unity.

Other useful properties of these nonuniformity coefficients are the following. It is easily checked that, by virtue of the inequality of Eq. 17-5, one always has

$$\Delta' \gtrless \alpha \quad \text{for} \quad \beta \gtrless \alpha$$

and that

$$\Delta' > 1 \ (\Delta > 1) \quad \text{for} \quad \alpha \geqq 1 \tag{17-11}$$

Obviously, if $\alpha < 1$, since $\beta \geqq 1$, we always have $\Delta' > \alpha$. In the latter case, however, the sign of $\Delta' - 1$, and hence of $\Delta - 1$, is not determined and one can have

$$\Delta' \gtrless 1 \ (\Delta \gtrless 1) \quad \text{for} \quad \alpha < 1 \tag{17-11a}$$

All previous derivations, except the one concerning β, were obtained for γ uniformly distributed on each section. No simple exact conclusions can be reached when γ itself is nonuniform. However if its nonuniformity is contained within narrow limits, so that the effect of the other nonuniformities is still predominant, the above relations can be approximately applied even for nonuniform γ, which will actually be done throughout the following discussion. For instance, if u and u/c_pT^0 are strongly nonuniform the condition represented by Eq. 17-10 is likely to be true, independent of the nonuniformity of γ. Only in marginal cases, when u is nearly uniform or nearly proportional to c_pT^0, and therefore Δ is very close to one, the nonuniformity of γ may be predominant in determining the sign of $\Delta - 1$.

Flow equations for ducts of the p-A power family. We shall now derive the relations between the mean quantities at two different stations (for instance, the initial and the final sections) of a duct of the p-A power family for the general nonuniform flow. We shall assume that injection of matter and mass forces are absent.

Let us first consider the case when the pressure is uniform on each section, and related to the area by the power law (Eq. 4-19) characteristic of the p-A power family, i.e. by the relation

$$pA^{\frac{\varepsilon}{1-\varepsilon}} = \text{const} \tag{17-12}$$

A quantity $J^{(\varepsilon)}$ can be defined, similar to the generalized thrust function introduced in Art. 4, given, for the generic section, by

$$J^{(\varepsilon)} = I + \varepsilon pA \tag{17-13}$$

in terms of which the conservation of momentum can be expressed by the same Eq. 4-21 derived for uniform flows (with $I_i = 0$)

$$J^{(\varepsilon)} = J_{\mathrm{i}}^{(\varepsilon)} - X \tag{17-14}$$

where $J_{\mathrm{i}}^{(\varepsilon)}$ and $J_{\mathrm{f}}^{(\varepsilon)}$ represent the values corresponding to the initial and final sections, and X the total friction and body force acting between the two sections.

The conservation of mass is simply expressed by

$$m = \text{const}$$

and the conservation of energy by

$$H_f^0 = H_i^0 + Q \tag{17-15}$$

Following the same procedure already used several times, we can suppose that Eq. 17-15 has been solved and that the value of the ratio

$$\frac{\hat{H}_f^0}{\hat{H}_i^0} = \frac{\bar{c}_{p_f}\bar{T}_f^0}{\bar{c}_{p_i}\bar{T}_i^0} = \lambda \tag{17-16}$$

has been calculated.

Similarly, for known X and $J_i^{(\varepsilon)}$, the ratio

$$\frac{J_f^{(\varepsilon)}}{J_i^{(\varepsilon)}} = 1 - \frac{X}{J_i^{(\varepsilon)}} = \nu \tag{17-17}$$

is known. Now eliminate the variable $\bar{T}^0$ from Eq. 16-2 and 16-22, forming the expression

$$\Phi^{(\varepsilon)}(\alpha, \beta, \bar{\gamma}, \tilde{M} \text{ or } \varpi) = \frac{J^{(\varepsilon)}}{\sqrt{2m\hat{H}^0}}$$

$$= \frac{1}{\bar{\gamma}}\sqrt{\frac{\bar{\gamma}-1}{2}}\,\frac{\varepsilon + \bar{\gamma}\tilde{M}^2}{\tilde{M}\sqrt{\alpha + \beta\dfrac{\bar{\gamma}-1}{2}\tilde{M}^2}} = \varpi + \varepsilon\frac{\bar{\gamma}-1}{2\bar{\gamma}}\frac{1-\beta\varpi^2}{\alpha\varpi} \tag{17-18}$$

In terms of $\Phi^{(\varepsilon)}$, Eq. 17-16 and 17-17 can be combined in the form

$$\frac{\Phi_f^{(\varepsilon)}}{\Phi_i^{(\varepsilon)}} = \frac{\Phi^{(\varepsilon)}(\alpha_f, \beta_f, \bar{\gamma}_f, \tilde{M}_f \text{ or } \varpi_f)}{\Phi^{(\varepsilon)}(\alpha_i, \beta_i, \bar{\gamma}_i, \tilde{M}_i \text{ or } \varpi_i)} = \frac{\nu}{\sqrt{\lambda}} \tag{17-19}$$

From this equation, for given α_i, β_i, $\bar{\gamma}_i$, and $\tilde{M}_i$ or ϖ_i, ν, λ, and $\bar{\gamma}_f$, and if α_f and β_f are known, one can calculate $\tilde{M}_f$ or ϖ_f. The value of p_f and the corresponding A_f can then be obtained from the relations

$$\left.\begin{aligned} \left(\frac{p_f}{p_i}\right)^{\frac{1}{\varepsilon}} = \left(\frac{A_f}{A_i}\right)^{\frac{1}{1-\varepsilon}} = \frac{p_fA_f}{p_iA_i} = \nu\frac{\varepsilon + \bar{\gamma}_i\tilde{M}_i^2}{\varepsilon + \bar{\gamma}_f\tilde{M}_f^2} \\ = \nu\frac{\varepsilon + \dfrac{2\bar{\gamma}_i}{\bar{\gamma}_i - 1}\dfrac{\alpha_i\varpi_i^2}{1-\beta_i\varpi_i^2}}{\varepsilon + \dfrac{2\bar{\gamma}_f}{\bar{\gamma}_f - 1}\dfrac{\alpha_f\varpi_f^2}{1-\beta_f\varpi_f^2}} = \sqrt{\lambda}\frac{\dfrac{\bar{\gamma}_i}{\bar{\gamma}_i - 1}\dfrac{\alpha_i\varpi_i}{1-\beta_i\varpi_i^2}}{\dfrac{\bar{\gamma}_f}{\bar{\gamma}_f - 1}\dfrac{\alpha_f\varpi_f}{1-\beta_f\varpi_f^2}} \end{aligned}\right\} \tag{17-20}$$

which are immediately derived from Eq. 17-12, 16-22, and 17-17, or from Eq. 16-19 and 17-16.

Clearly, similar equations could be written for two generic stations, different from the initial and final, provided that the quantities λ and ν are modified accordingly. For instance one could obtain the value of ϖ at any intermediate station, when the initial conditions are known and the values of $\bar{\gamma}$, λ, ν, α, and β are assigned the proper values, by using the same equation (Eq. 17-17) with the subscript $_f$ removed. Similarly the corresponding p and A would be obtained from Eq. 17-18 written without the subscript $_f$.

The preceding relations have been obtained under the assumption of uniform pressure distribution on each section. However, this restrictive assumption can be waived, in a way very similar to that followed in Art. 15, if one concentrates only on the relations between the initial and final stations of a duct which is properly shaped so that the equality between the first two members of Eq. 15-1 is satisfied; then ε takes the value given by this equation. If this condition is satisfied, all of the previous equations remain correct, provided that $p_\imath$ and p_f are replaced everywhere by $\bar{p}_\imath$ and $\bar{p}_f$. However, the application to intermediate stations is, in this case, incorrect.

Properties of the function $\Phi^{(\varepsilon)}(\alpha, \beta, \bar{\gamma}, \varpi)$. If α, β, and $\bar{\gamma}$ are known and ε is assigned, $\Phi^{(\varepsilon)}$ can be considered to be a function only of ϖ. In reality the separation of the effects of α and β from those of ϖ is artificial, since the values of these three quantities are strictly interconnected. However, from a purely mathematical standpoint, the separation is allowed and leads to the derivation of interesting properties.

The range in which ϖ can vary is actually restricted by the condition that $\tilde{M}^2$ is a definitely positive quantity (as shown by Eq. 16-15 and 16-24), and therefore from Eq. 16-16 we have the limitation

$$\varpi \leqq \frac{1}{\sqrt{\beta}} < 1 \tag{17-21}$$

This is not, however, the only limitation on ϖ. Eq. 17-18, defining $\Phi^{(\varepsilon)}$ as a function of ϖ, can be written in the form:

$$\Phi^{(\varepsilon)}(\alpha, \beta, \bar{\gamma}, \varpi) = \varpi + \frac{\delta}{2\Delta}\,\frac{1 - \varpi^2}{\varpi}$$

where we have used the abbreviation, similar to Eq. 11-5,

$$\delta = \varepsilon\,\frac{\bar{\gamma} - 1}{\bar{\gamma}} \tag{17-22}$$

and Δ is the quantity defined by Eq. 17-9. On the other hand

$$\Phi^{(\varepsilon)}(1, 1, \bar{\gamma}, \varpi) = \varpi + \frac{\delta}{2}\,\frac{1 - \varpi^2}{\varpi} \tag{17-23}$$

Therefore we have the limitation

$$\left.\begin{array}{l}\Phi^{(\varepsilon)}(\alpha, \beta, \bar{\gamma}, \bar{w}) \lesseqgtr \Phi^{(\varepsilon)}(1, 1, \bar{\gamma}, \bar{w}) \\ \text{for } \Delta \gtrless 1 \text{ if } \varepsilon > 0, \text{ or } \Delta \lessgtr 1 \text{ if } \varepsilon < 0\end{array}\right\} \tag{17-24}$$

On the other hand, from Eq. 17-9 one has

$$\Delta \gtrless 1 \quad \text{for} \quad (\beta - \alpha)\bar{w}^2 \gtrless 1 - \alpha \tag{17-25}$$

Now in view of the conditions stated by Eq. 17-11 and 17-11a we can consider three cases:

Case I. $\alpha < 1 < \beta$; $\Delta < 1$. The inequality of Eq. 17-25 requires that

$$\bar{w}^2 < \frac{1 - \alpha}{\beta - \alpha} < \frac{1}{\beta} \tag{17-26}$$

Case II. $\alpha < 1 < \beta$; $\Delta > 1$. The range of $\bar{w}^2$ is

$$\frac{1 - \alpha}{\beta - \alpha} < \bar{w}^2 < \frac{1}{\beta} \tag{17-26a}$$

Case III. $\alpha > 1$; $\Delta > 1$. The whole range of Eq. 17-21 is possible.

The significance of the above restrictions is that, in view of the actual meaning of the α and β coefficients, values of $\bar{w}$ out of the ranges of Eq. 17-26 and 17-26a are not physically compatible with the assigned values of α and β. Thus, although $\Phi^{(\varepsilon)}$ is mathematically defined for all $\bar{w}$ values, it has no physical meaning out of the above ranges.

The behavior of $\Phi^{(\varepsilon)}$ as a function of $\bar{w}$ depends strongly on the value of ε (or δ). At $\bar{w} = 0$, $\Phi^{(\varepsilon)}$ is $\pm\infty$ for $\delta \gtrless 0$ and is 0 for $\delta = 0$. At $\bar{w} = 1/\sqrt{\beta}$, $\Phi^{(\varepsilon)}$ is independent of δ and equal to $1/\sqrt{\beta}$. $\Phi^{(\varepsilon)}$ has a minimum when the mean reduced velocity takes its critical value, defined by

$$w_{cr}^2 = \frac{1}{\dfrac{2\alpha}{\delta} - \beta} \tag{17-27}$$

The critical velocity exists and falls in the range of Eq. 17-21, only if

$$0 \leqq \delta \leqq \frac{\alpha}{\beta} \tag{17-28}$$

Out of this range of δ, $\Phi^{(\varepsilon)}$ goes monotonically from $\pm\infty$ at $\bar{w} = 0$ (depending on whether $\delta \gtrless 0$) to $1/\sqrt{\beta}$ at $\bar{w} = 1/\sqrt{\beta}$. Thus for $\delta < 0$, $\Phi^{(\varepsilon)}$ goes from negative to positive values when $\bar{w}$ increases; it is zero for $\bar{w} = w_0$ with

$$w_0^2 = \frac{1}{\beta - \dfrac{2\alpha}{\delta}} > 0 \tag{17-29}$$

The value of $\Phi_{min}^{(\varepsilon)}$, when it does exist, is

$$\Phi_{min}^{(\varepsilon)} = \sqrt{\frac{\delta}{\alpha}\left(2 - \frac{\delta\beta}{\alpha}\right)} \tag{17-30}$$

We notice that the expression (Eq. 17-27) for w_{cr} coincides for $\alpha = \beta = 1$ with the expressions (Eq. 5-20 or 11-6) already used in Art. 5, 9, and 11 for uniform flows, and hence represents a generalization of this quantity to the case of nonuniform flows. For $\alpha = \beta = \varepsilon = 1$, w_{cr} coincides with w^*; and it is always zero for $\varepsilon = 0$ when the $\Phi^{(\varepsilon)}$ curve degenerates into the

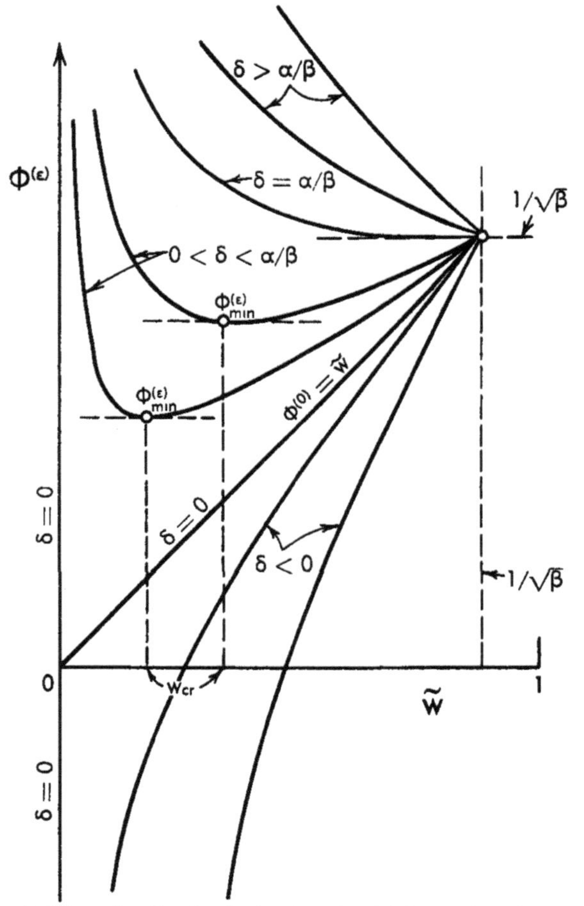

Fig. B,17c. Qualitative behavior of $\Phi(\varepsilon)$ for assigned α, β, $\bar{\gamma}$ and for different values of $\delta = \varepsilon(\bar{\gamma} - 1)/\bar{\gamma}$.

two straight lines $\tilde{w} = 0$ and $\Phi^{(0)} = \tilde{w}$. We observe also that, for $\varepsilon = 1$, the inequality of Eq. 17-28 is satisfied even for considerable departures of α and β from unity. Thus for a constant area duct the critical velocity in general exists in the range of Eq. 17-21. Whether or not it falls in the allowed ranges of Eq. 17-26 or 17-26a when $\alpha < 1$ and $\Delta \lesseqgtr 1$ may be easily verified. The results of the above discussion are schematically shown in Fig. B,17c.

From now on, in agreement with our definitions in Art. 5, we shall call subcritical the range $\tilde{w} < w_{cr}$ and supercritical the range $\tilde{w} > w_{cr}$. These denominations can be extended to cover the case when w_{cr} is out of the physical range, or imaginary, by calling subcritical all cases for which $\partial\Phi^{(\epsilon)}/\partial\tilde{w}$ is negative (as always happens for $\delta > \alpha/\beta$) and calling supercritical all cases for which it is positive (such as all cases with negative δ).

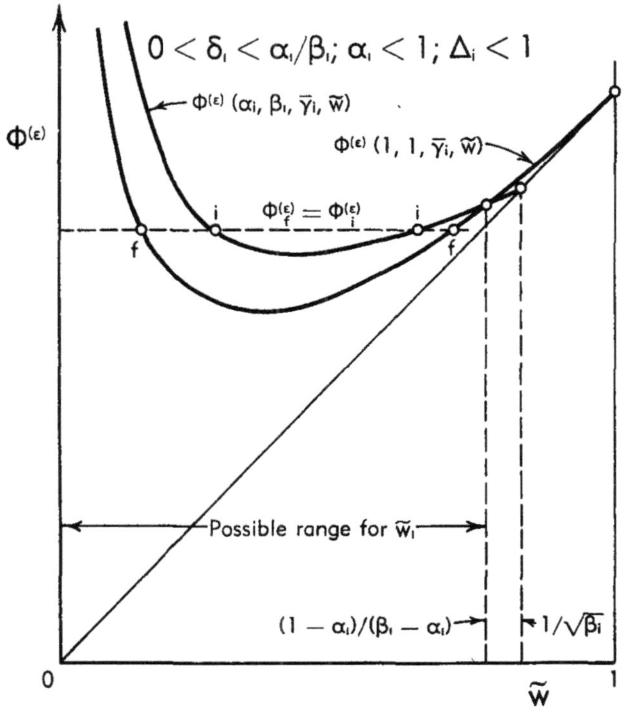

Fig. B,17d. Pure mixing. Qualitative representation of the solutions of the flow equation in case I.

Discussion of the solutions. Flows with uniformization. The problem of determining the mean final conditions of the flow in a duct of the *p*-*A* power family when the initial conditions are assigned and the values of λ, ν, $\bar{\gamma}_f$, α_f, and β_f are known can be solved through the use of Eq. 17-19 (which provides $\tilde{w}_f$) and Eq. 17-20 (which gives p_f). As already noticed, the same equations with p_i and p_f replaced by $\bar{p}_i$ and $\bar{p}_f$ can be used if pressure nonuniformities are present.

From the properties of the α, β, and Δ coefficients and the behavior of the $\Phi^{(\epsilon)}$ function it is possible to discuss certain general properties of the solutions. Let us discuss first the case of flows in which, due to mixing and other uniformizing processes, a final quasi-uniform state is reached starting from an initially nonuniform state. The term "quasi" has been used here to indicate the fact, already observed several times, that although the flow in a duct can never be rigorously uniform, practical uniformity can be assumed when the nonuniformity coefficients have reached values sufficiently close to unity, which happens for many of the examples considered in Art. 16. Thus we shall assume that, starting with values of α_i and β_i substantially different from unity, the flow reaches a final condition where the departure of both α and β from unity can practically be disregarded, and one can take α_f and $\beta_f \cong 1$. Despite this fact, we shall

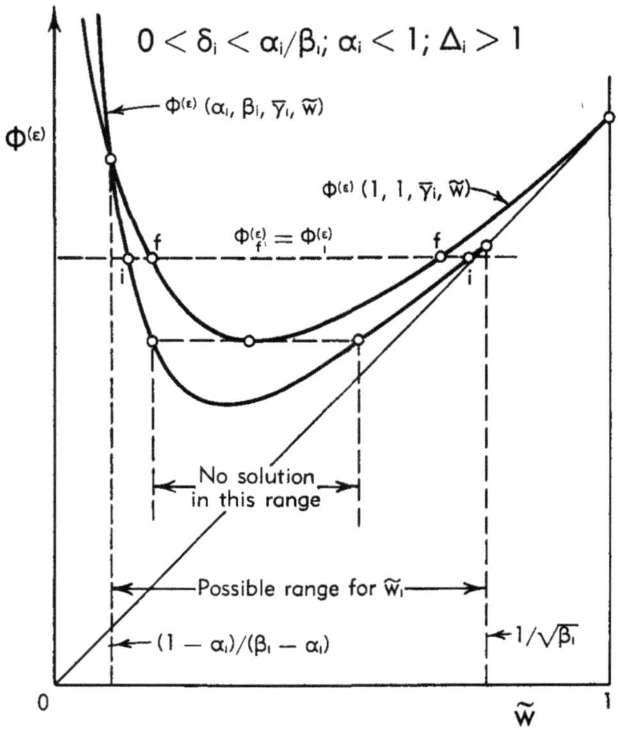

Fig. B,17e. Pure mixing. Qualitative representation of the solutions of the flow equations in case II.

go on using mean values for the various quantities in the final section, because there can still be quite an amount of nonuniformity present, not only in the velocities but also in the other quantities. We shall start considering a positive value of ε such that, for the given values of α_i, β_i, and $\bar{\gamma}_i$, δ_i falls in the range of Eq. 17-28. In this case due to the fact that α/β is likely (in the great majority of practical cases) to be <1, δ_f also falls certainly in the range of Eq. 17-28 if $\bar{\gamma}_f \leqq \bar{\gamma}_i$ (a condition which is also likely to cover the great majority of practical cases, with or without combustion).

Depending on whether $\alpha_i \lesseqgtr 1$ and $\Delta_i \lesseqgtr 1$ we can again distinguish the three cases discussed previously (assuming that the case $\alpha_i > 1$ and $\Delta_i < 1$, forbidden for uniform γ, can also be practically excluded for nonuniform γ). Accordingly the curves of $\Phi^{(\epsilon)}(\alpha_i, \beta_i, \bar{\gamma}_i, \tilde{w})$ and $\Phi^{(\epsilon)}(1, 1, \bar{\gamma}_i, \tilde{w})$ appear as shown in Fig. B,17d (case I), B,17e (case II), and B,17f (case III). On the figures we have also indicated, according to Eq. 17-26, 17-26a and 17-21, the ranges of $\tilde{w}_i$ that are physically allowed.

The simplest problem to be solved is that of a flow with pure mixing, that is, without heat addition or chemical changes. In this case we have

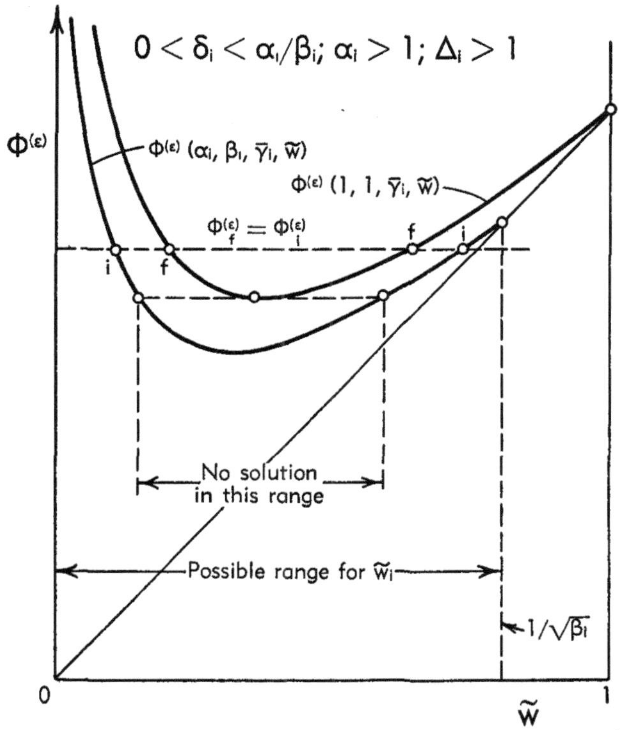

Fig. B,17f. Pure mixing. Qualitative representation of the solutions of the flow equations in case III.

$\lambda = 1$ and $\bar{\gamma}_f = \bar{\gamma}_i = \bar{\gamma}$. Therefore $\delta_f = \delta_i = \delta$. If, in addition, we disregard friction, we also have $\nu = 1$ and hence Eq. 17-19 and 17-20 can be written in the form

$$\Phi_f^{(\epsilon)} = \Phi_i^{(\epsilon)}, \quad \left(\frac{\bar{p}_f}{\bar{p}_i}\right)^{\frac{1}{\epsilon}} = \left(\frac{A_f}{A_i}\right)^{\frac{1}{1-\epsilon}} = \frac{\bar{p}_f A_f}{\bar{p}_i A_i} = \frac{\alpha_i \tilde{w}_i}{1 - \beta_i \tilde{w}_i^2} \frac{1 - \tilde{w}_f^2}{\tilde{w}_f} \qquad (17\text{-}31)$$

The first Eq. 17-31 is solved graphically for $\tilde{w}_f$ on the figures, which shows that for subcritical $\tilde{w}_i$ and $\tilde{w}_f$ one has $\tilde{w}_f \lessgtr \tilde{w}_i$ for $\Delta_i \lessgtr 1$, and for supercritical $\tilde{w}_i$ and $\tilde{w}_f$ one has $\tilde{w}_f \gtrless \tilde{w}_i$ for $\Delta_i \lessgtr 1$. From the first Eq. 17-31, recalling the expression (Eq. 17-18) of $\Phi^{(\epsilon)}$ we see that, since ϵ is positive, the quantity

$$\frac{\bar{\gamma}}{\bar{\gamma} - 1} \frac{\alpha \tilde{w}}{1 - \beta \tilde{w}^2} \qquad (17\text{-}32)$$

must vary in the same direction as $\bar{w}$ itself. Thus from the second Eq. 17-31 we obtain that, depending on whether $\Delta_i \lesseqgtr 1$, we have $\bar{p}_f/\bar{p}_i \gtreqless 1$ for subcritical velocities and $\bar{p}_f/\bar{p}_i \lesseqgtr 1$ for supercritical velocities. Hence recalling the inequality of Eq. 17-10, a process of pure mixing and uniformization in a subcritical flow is accompanied by an increase or decrease of mean pressure, depending on whether u and u/c_pT^0 are co-variant or contravariant; and the opposite for supercritical flows.

From the second relation of Eq. 17-31 we obtain also that the cross-sectional area of the duct varies in the same direction as the mean pressure for $\varepsilon < 1$ and in the opposite direction for $\varepsilon > 1$.

The above results have been obtained with the assumption that a solution for $\bar{w}_f$ exists. Now it is clear that for $\Delta_i < 1$, due to the relative positions of the two curves of Fig. B,17d, two solutions for $\bar{w}_f$, one subcritical and the other supercritical, exist regardless of the initial conditions. The same is not true, however, for $\Delta_i > 1$ because, as the two corresponding figures show clearly, for a well-defined range around the value of w_{cr} relative to the initial conditions, no real solution can be obtained. Hence there exists an upper limit for $\bar{w}_i$ in the subcritical range and a lower limit for $\bar{w}_i$ in the supercritical range, for which the two solutions for $\bar{w}_f$ come in coincidence and coincide with the corresponding critical velocity. This effect is nothing but a case of choking, due in this case to mixing only and can exist only for $\Delta_i > 1$, that is, recalling Eq. 17-10, if u and u/c_pT^0 are contravariant.

Let us now remove the restriction that δ falls in the range of Eq. 17-28 (δ may also be out of the range 0 to 1 or, in marginal cases, it may still fall in that range). It can be easily checked that if $\delta > \alpha_i/\beta_i$ the same conclusions derived above for subcritical flows still hold. In the other alternative, $\delta < 0$, we have, in view of the inequality of Eq. 17-24, $\bar{w}_f \gtreqless \bar{w}_i$ depending on whether $\Delta_i \gtreqless 1$. Since ε is negative, from the first Eq. 17-31 one obtains that the quantity of Eq. 17-32 must now vary in the opposite direction from $\bar{w}$; so that, from Eq. 17-31, $\bar{p}_f/\bar{p}_i \gtreqless 1$ and $A_f/A_i \lesseqgtr 1$ for $\Delta_i \gtreqless 1$. When δ is out of the range of Eq. 17-28 and the range 0 to 1, the critical velocities are either imaginary or out of the physical range, and choking cannot exist. It can exist, however, in marginal cases, if $\alpha_i/\beta_i < \delta < 1$, provided that the value of $\Phi^{(e)}_{\min}$ for $\alpha_f = \beta_f = 1$ falls above the minimum value of $\Phi^{(e)}(\alpha_i, \beta_i, \bar{\gamma}, \bar{w})$, i.e. $1/\sqrt{\beta_i}$. It is immediately found that the corresponding condition is

$$1 - \sqrt{1 - \frac{1}{\beta_i}} < \delta < 1$$

For $\delta > 1$ and $\Delta_i > 1$ it is easily checked that there also exists an upper limit to $\bar{w}_i$ corresponding to $\Phi^{(e)}(\alpha_i, \beta_i, \bar{\gamma}, \bar{w}_i) = 1$.

Hence very neat general results are found for the case of pure mixing without friction and heat addition or chemical changes. When one of these factors is taken into account the results cannot always be formulated so clearly. For instance the case when only friction is present can be discussed by taking $\lambda = 1$, $\bar{\gamma}_f = \bar{\gamma}_i$, but $\nu < 1$. The only change to Eq. 17-31 is that now $\Phi_f^{(\varepsilon)} = \nu\Phi_1^{(\varepsilon)} < \Phi_i^{(\varepsilon)}$. Fig. B,17g shows the graphical solution of this equation for $0 < \delta < \alpha_i/\beta_i$ and $\Delta_i > 1$ (the cases II ($\alpha_i < 1$) and III ($\alpha_i > 1$) are here represented together, the only difference residing in the existence or absence of a lower limitation to the values

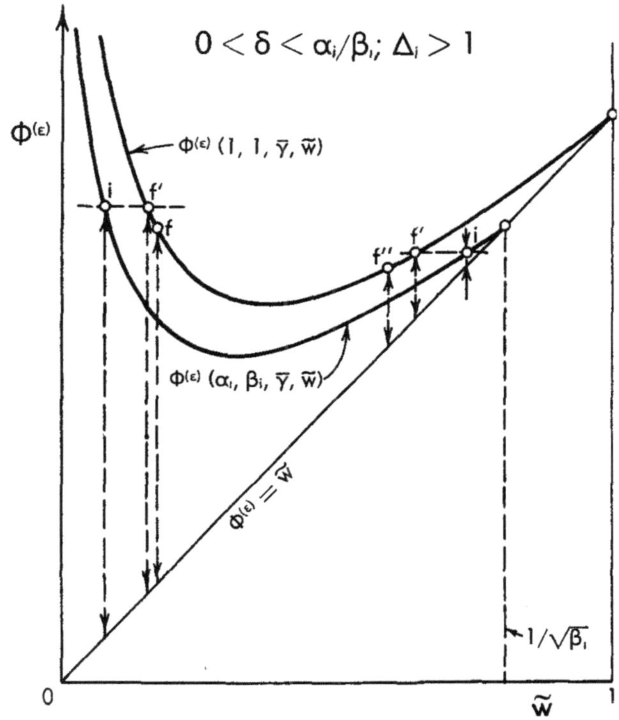

Fig. B,17g. Mixing and friction. Qualitative representation of the solutions of the flow equations in cases II and III.

of $\tilde{w}$). If the flow is subcritical the point f, representing the final condition, lies on the curve $\Phi^{(\varepsilon)}(1, 1, \bar{\gamma}, \tilde{w})$ at a lower level than the point f′ corresponding to $\nu = 1$. It is clear that the vertical distance of f from the line $\Phi^{(\varepsilon)} = \tilde{w}$, which, in view of Eq. 17-18 represents the reciprocal of Eq. 17-32 times a positive constant, is smaller than the vertical distance of f′ from the same line, which in turn is smaller than that of the point i. The opposite is true if the flow is supercritical. Thus the final value of Eq. 17-32 is larger than its initial value for subcritical flow and smaller for supercritical, so that, due to the second Eq. 17-31, $\bar{p}_f/\bar{p}_i$ is $\lesseqgtr 1$ (and, as immediately seen from the figure, $\tilde{w}_f \gtreqless \tilde{w}_i$) depending on whether the flow is sub- or supercritical. For $\delta > \alpha_i/\beta_i$ one finds that the same results apply as for supercritical flow. It can be immediately verified that no clearcut answer can be obtained for $\Delta_i < 1$.

For $\varepsilon < 0$, using the same line of reasoning on Fig. B,17h, which represents the two cases $\Delta_i < 1$ and $\Delta_i > 1$, we find that one has certainly $\tilde{w}_f < \tilde{w}_i$ for $\Delta_i < 1$ and $\tilde{w}_i > w_0$ (w_0 being defined by Eq. 17-29), and $\tilde{w}_f > \tilde{w}_i$ for $\Delta_i > 1$ and $\tilde{w}_i < w_0$. However, in these two cases no definite answer can be given about the direction in which the mean pressure varies. On the other hand, one has $\bar{p}_f/\bar{p}_i < 1$ for $\Delta_i < 1$ and $\tilde{w}_i < w_0$, and $\bar{p}_f/\bar{p}_i > 1$ for $\Delta_i > 1$ and $\tilde{w}_i > w_0$, but no definite statement can be made in these two cases about $\tilde{w}$.

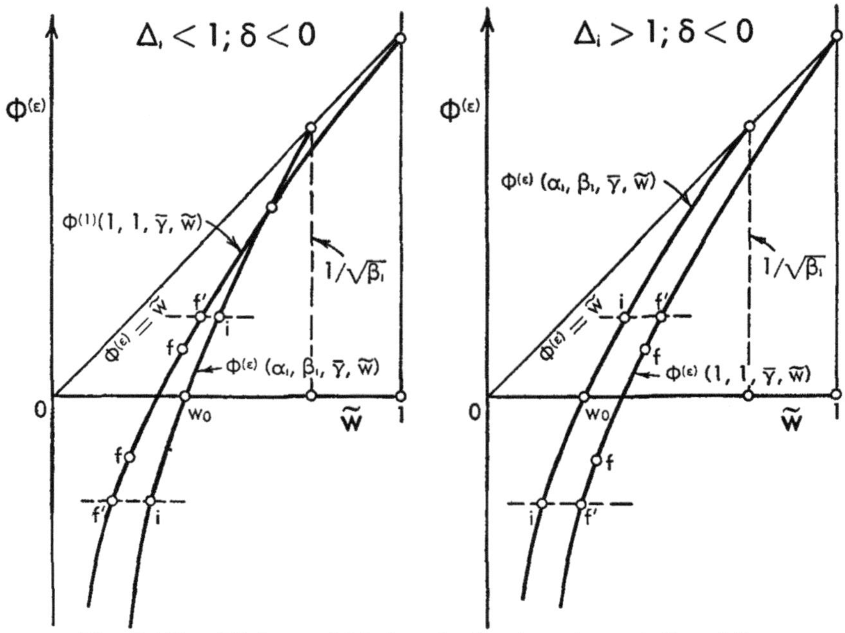

Fig. B,17h. Mixing and friction. Qualitative representation of the solution of the flow equations for $\varepsilon < 0$.

Choking can occur under the combined effect of mixing and friction if δ is in the range of Eq. 17-28. It is more likely to occur for $\Delta_i > 1$, in which case the effect of friction and mixing are concurrent, but can also occur in the case $\Delta_i < 1$, if the effect of friction prevails over that of mixing. Choking may also occur, more exceptionally, when δ is out of the range expressed by Eq. 17-28.

Another simple case to consider is that of simultaneous mixing and heat transfer. If friction is neglected, one can take $\nu = 1$ and $\lambda \neq 1$. If the range of temperatures during the whole process is moderate or if the gases are polytropic, one can still take $\bar{\gamma}_f = \bar{\gamma}_i = \bar{\gamma}$ and therefore $\delta_f = \delta_i = \delta$. Then Eq. 17-19 and 17-20 give the following relations

$$\Phi_f^{(\varepsilon)} = \frac{\Phi_i^{(\varepsilon)}}{\sqrt{\lambda}}, \quad \left(\frac{\bar{p}_f}{\bar{p}_i}\right)^{\frac{1}{\varepsilon}} = \left(\frac{A_f}{A_i}\right)^{\frac{1}{1-\varepsilon}} = \frac{\bar{p}_f A_f}{\bar{p}_i A_i} = \frac{\dfrac{\delta}{2} + \dfrac{\alpha_i \tilde{w}_i^2}{1 - \beta_i \tilde{w}_i^2}}{\dfrac{\delta}{2} + \dfrac{\tilde{w}_f^2}{1 - \tilde{w}_f^2}} \qquad (17\text{-}33)$$

For $\lambda > 1$ (heat addition) the solution of the first of Eq. 17-33 is similar to that discussed for $\lambda = 1$, $\nu < 1$, and is shown schematically on Fig. B,17i for δ in the range of Eq. 17-28 and $\Delta_i > 1$. It is clear that $\tilde{w}_f \gtrless \tilde{w}_i$ depending on whether the flow is subcritical or supercritical. It is also clear from the figure that the slope $\Phi^{(\epsilon)}/\tilde{w}$ of the lines connecting the representative points to the origin decreases going from i to f′ and to f if

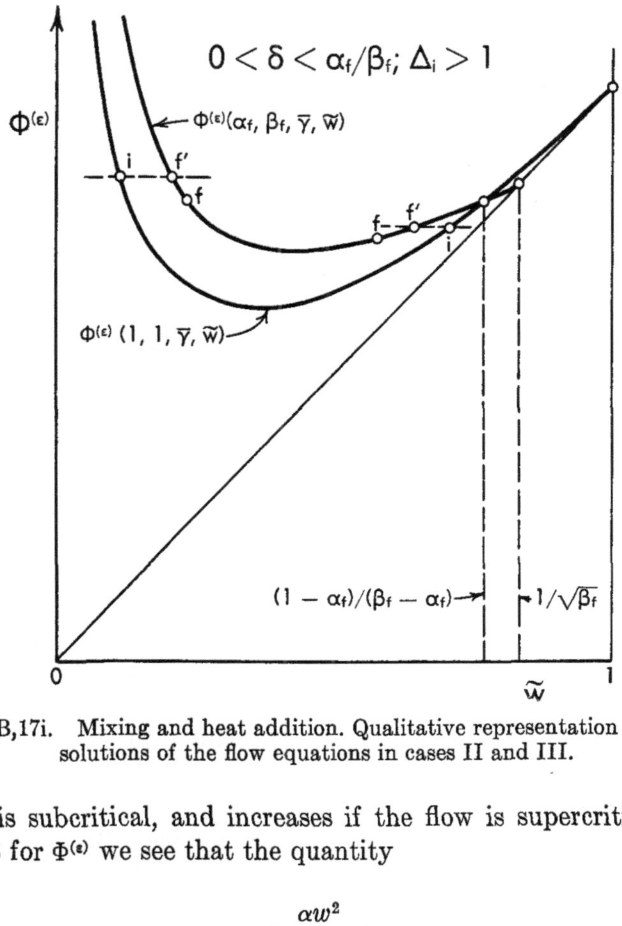

Fig. B,17i. Mixing and heat addition. Qualitative representation of the solutions of the flow equations in cases II and III.

the flow is subcritical, and increases if the flow is supercritical. From Eq. 17-18 for $\Phi^{(\epsilon)}$ we see that the quantity

$$\frac{\alpha w^2}{1 - \beta w^2}$$

must vary in the direction opposite to the slope, and from Eq. 17-33 that the quantity $\bar{p}_f A_f/\bar{p}_i A_i$ must vary in the same direction as the slope itself. Thus we have $\bar{p}_f/\bar{p}_i \lesseqgtr 1$ depending on whether the flow is sub- or supercritical.

For $\delta > \alpha_i/\beta_i$ and $\Delta_i > 1$, it can easily be checked that we may apply the same results as for subcritical flows. For $\delta < 0$ and $\Delta_i > 1$, making use of the construction of Fig. B,17h and considering again the variation of the slopes, we find that $\bar{p}_f/\bar{p}_i < 1$ for all values of $\tilde{w}_i$.

No definite answer as to the direction in which the mean pressure varies can be given if $\Delta_i < 1$, because the effects of mixing and heat addition are opposite to each other.

Proceeding along similar lines for the case $\lambda < 1$ (heat extraction) and $\Delta_i < 1$ we find that for $\varepsilon > 0$, $\bar{p}_f/\bar{p}_i$ is $\gtrless 1$ depending on whether the flow is sub- or supercritical, and it is always >1 for $\varepsilon < 0$. No definite conclusion can be reached for $\lambda < 1$ and $\Delta_i > 1$.

Obviously the area always varies in the same direction as the mean pressure for $0 < \varepsilon < 1$ and in the opposite direction for $\varepsilon < 0$ or $\varepsilon > 1$.

Choking under the combined action of mixing and heat addition is definitely possible if δ is in the range of Eq. 17-28, and also, in marginal cases, out of this range. It is more likely to happen when $\Delta_i > 1$ and $\lambda > 1$, but can also be present for $\Delta_i < 1$ and $\lambda < 1$.

Finally we discuss briefly the case when chemical changes are present. The effects are here complicated by the variation of $\bar{\gamma}$ between the initial and the final conditions. However, in the most interesting practical case, that of combustion, the value of λ can be substantially larger than unity, and consequently the effect of λ is likely to dominate over the effects of changing $\bar{\gamma}$, mixing, and friction. Thus the qualitative behavior of the solutions is likely to be the same as that discussed in Art. 11. Here we summarize briefly those results and a few others that can immediately be obtained from the graphical procedure outlined above.

For δ_i in the range of Eq. 17-28 we have $\bar{w}_f \gtrless \bar{w}_i$ depending on whether the flow is sub- or supercritical, but $\bar{p}_f A_f/\bar{p}_i A_i$ and therefore $\bar{p}_f/\bar{p}_i$ are always <1. For $\delta_i > \alpha_i/\beta_i$ we find the same result as for subcritical flows. For $\delta_i < 0$ we have $\bar{w}_f \gtrless \bar{w}_i$ depending on whether $\bar{w}_i \lesseqgtr w_0$. However, we always obtain $\bar{p}_f A_f/\bar{p}_i A_i > 1$ and $\bar{p}_f/\bar{p}_i < 1$. Thus the effect of combustion is always in the direction of decreasing the mean pressure. For the reasons mentioned above, when choking is produced under the combined action of combustion, mixing, and friction, the main cause for choking is the combustion process.

In the above discussion, we have considered only flows that are entirely subcritical or supercritical. Obviously one must also take into account the possibility that a flow, initially subcritical, results in a supercritical final condition, and vice versa, in which case the previous results have to be modified. Later we shall discuss a tentative criterion concerning this possibility.

Flows with nonuniformization. Boundary layer formation. We have so far assumed that the only appreciable nonuniformities are those deliberately introduced, and that when the mixing process is complete both α_f and β_f can be identified with unity. This is of course only an approximation and, due to the presence of boundary layers, α_f and β_f are always somewhat different from unity. If, therefore, a flow starts without preexisting boundary layers and the boundary layers are gradually built up, as happens in the entrance stretch of a pipe, the corresponding values of α and β, which are unity at the initial section, will gradually deviate from unity. The maximum deviation is reached when the boundary layers fill the whole section; the relative values of α and β have been calculated under certain simplifying assumptions in Art. 16, where it has been shown that the deviations from unity of both α and β may be quite small for regular turbulent flows, so that when other effects predominate as in the preceding developments of this article, they can be replaced by unity.

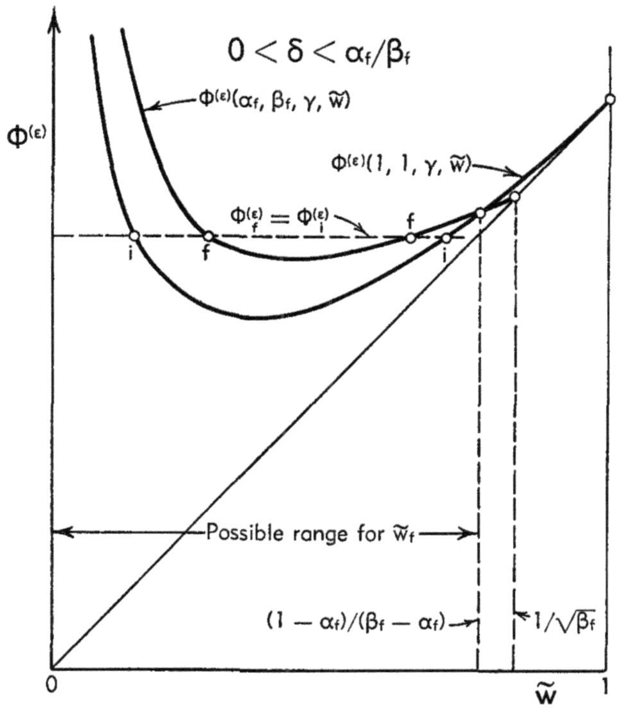

Fig. B,17j. Nonuniformization due to boundary layer formation. Qualitative representation of the solutions of the flow equations.

However, when more delicate effects are investigated, such as the effect of friction in short ducts (or in the entrance stretch of a pipe), the variation of α and β has to be taken into account, and is actually responsible for the inadequacy, already noticed in Art. 7, of a treatment of frictional effects in short ducts based on the uniformity assumption.

The nonuniformizing effects due to the boundary layer formation can be treated with the same procedure used for the case of uniformization with invariable $\bar{\gamma}$ and δ, the only difference being that now one has to take $\alpha_i = \beta_i = 1$, and α_f and β_f different from unity. In addition, since u and u/c_pT^0 are certainly covariant in the final section, (because T^0 is either nearly uniform or close to satisfying the energy integral (Eq. 16-29)), we also have $\Delta_f < 1$, so that the curve $\Phi^{(\epsilon)}(\alpha_f, \beta_f, \bar{\gamma}, \tilde{w})$ is all above (in the allowed range of $\bar{w}$) the curve $\Phi^{(\varepsilon)}(1, 1, \bar{\gamma}, \bar{w})$ as shown for $0 < \delta < \alpha_f/\beta_f$ in Fig. B,17j. In these conditions it is immediately seen from the figure that $\bar{w}_f \gtrless w_i$, depending on whether the flow is sub- or supercritical, and, following the same arguments as in the previous discussion, it is found that, correspondingly, $\bar{p}_f/p_i \lessgtr 1$. It is also clear that in both kinds of flow the effect of nonuniformization is concurrent with that of friction. It is easily checked that this is also true for the case $\delta > \alpha_f/\beta_f$ which behaves as a subcritical flow. The case $\delta < 0$ must again be considered separately. It is found that the nonuniformization and the friction act in the same direction on the velocity if $w_i < w_0$, and on the pressure if $w_i > w_0$.

Quantitatively one can obtain simple expressions for α and β close to unity by expanding $\Phi^{(\varepsilon)}$ in a Taylor series around $\alpha = 1$, $\beta = 1$, and $\bar{w} = w_i$

$$\Phi_f^{(\varepsilon)} - \Phi_i^{(\varepsilon)} = \left(\frac{\partial\Phi^{(\varepsilon)}}{\partial w}\right)_i (\bar{w}_f - w_i) + \left(\frac{\partial\Phi^{(\varepsilon)}}{\partial \alpha}\right)_i (\alpha_f - 1) + \left(\frac{\partial\Phi^{(\varepsilon)}}{\partial \beta}\right)_i (\beta_f - 1) + \cdots$$

and neglecting the higher order terms. The partial derivatives of $\Phi^{(\varepsilon)}$ must be calculated by first differentiating Eq. 17-18 and then taking $\alpha = \beta = 1$, $\bar{w} = w_i$. Equating this expression to the following one obtained from Eq. 17-19,

$$\Phi_f^{(\varepsilon)} - \Phi_i^{(\varepsilon)} = -(1 - \nu)\Phi_i^{(\varepsilon)} = -(1 - \nu)\frac{\delta}{2}\left(\frac{w_i}{w_{cr}^2} + \frac{1}{w_i}\right)$$

where w_{cr} is given by Eq. 17-27 with $\alpha = \beta = 1$, and solving for $\bar{w}_f - w_i$, we obtain

$$\left(\frac{1}{w_{cr}^2} - \frac{1}{w_i^2}\right)(\bar{w}_f - w_i) = \frac{1 - w_i^2}{w_i}(\alpha_f - 1) + w_i(\beta_f - 1) - \left(\frac{w_i}{w_{cr}^2} + \frac{1}{w_i}\right)(1 - \nu) \quad (17\text{-}34)$$

Similarly from Eq. 17-31 one obtains the Taylor expansion

$$\frac{p_f A_f}{p_i A_i} = 1 + \frac{1 + w_i^2}{w_i^2}(\bar{w}_f - w_i) - \frac{1 - w_i^2}{w_i}(\alpha_f - 1) - w_i(\beta_f - 1) + \cdots$$

Neglecting again the higher order terms and using Eq. 17-34, the values of $p_f A_f/p_i A_i$, p_f/p_i, and A_f/A_i can be found in terms of $\alpha_f - 1$, $\beta_f - 1$, and $1 - \nu$.

The above equations can be applied to the entrance stretch of a pipe by using, for instance, for α_f and β_f, the values from Eq. 16-38 and 16-35 and calculating ν in terms of the friction coefficient. It can be checked that the contributions of the terms in $\alpha_f - 1$ and $\beta_f - 1$ are of the same order as those of the term in $1 - \nu$. As already observed before, this is the reason for the inadequacy of the conventional one-dimensional treatment in studying the effects of friction in short ducts.

Pseudo-shock and exclusion rule. In Art. 5 we have shown that transition from a uniform supercritical flow to a uniform subcritical flow is not only allowed by the second law of thermodynamics, but is also

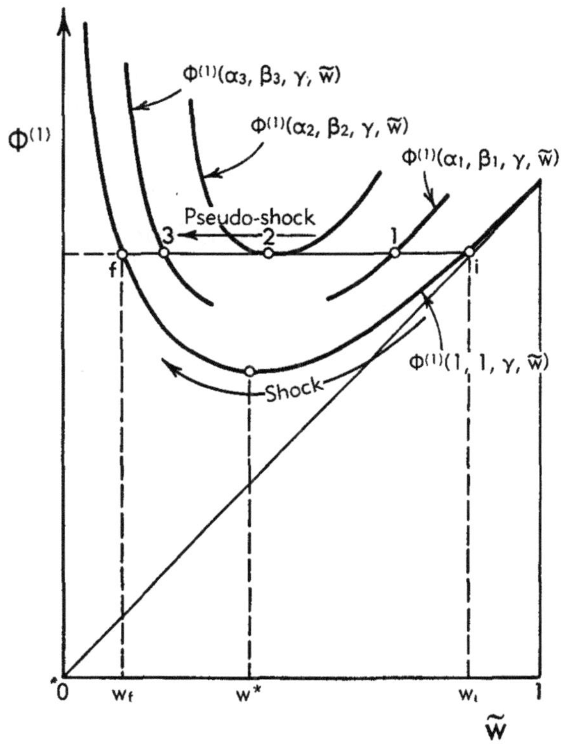

Fig. B,17k. Qualitative representation of the intermediate stages of a shock and a pseudo-shock in a constant section duct.

physically possible, and we have also discussed two possible mechanisms by which the transition is produced. However, the opposite transition, from uniform subcritical to uniform supercritical, must be excluded by virtue of the second law. Before we discuss the extension of this exclusion rule to nonuniform flows, it is interesting to illustrate on the $\Phi^{(e)}$, $\tilde{w}$ plane the difference between the two mechanisms which may produce the transition.

The shock process always takes place at constant area, and therefore we can take $\varepsilon = 1$. On Fig. B,17k we have schematically shown the curve $\Phi^{(1)}(1, 1, \gamma, \tilde{w})$ and the initial and final points of a shock which, following Eq. 17-19, must lie on the same horizontal line because $\lambda = \nu = 1$ (absence of over-all heat exchanges and friction). Within the structure of the shock, however, internal heat exchanges and viscous stresses are present and are responsible for the transition from the initial uniform w_i to the final uniform w_f. Hence the local intermediate values of λ and ν are different from unity. On the other hand, since the flow remains uniform at all intermediate stations, all points representing the intermediate conditions must stay on the line $\Phi^{(1)}(1, 1, \gamma, \mathfrak{w})$. Therefore the conditions within the shock structure are represented by the points of this line between the initial and the final point, the variation of $\Phi^{(1)}$ being possible because at the intermediate points λ and ν are different from 1.

For a pseudo-shock the situation is different. If, in analyzing its structure, we disregard the microstructure of the oblique shocks that are present and regard these as discontinuities, and if, in addition, we disregard the friction on the walls, the values of λ and ν are now unity at all intermediate stations. Therefore, by virtue of Eq. 17-19, the value of $\Phi^{(1)}$ must be constant and the point representing the transition must move on the same horizontal line containing the initial and the final point. This transition is made possible by the presence of a strong nonuniformizing process such as separation. As a result of this process the values of α and β will deviate from their initial value which is unity, in such a way that the curve of $\Phi^{(1)}(\alpha, \beta, \gamma, \mathfrak{w})$ will be above that of $\Phi^{(1)}(1, 1, \gamma, \mathfrak{w})$ (because T^0 remains uniform and therefore $\Delta < 1$), as shown for three consecutive points on the figure. Observe that for point 1 the value of $\mathfrak{w}$ is still supercritical, while it is subcritical for point 3. The transition from super- to subcritical takes place in a continuous fashion, going through the critical velocity, as shown by point 2. Obviously the value of $\Phi^{(1)}_{\min}$ must first increase, reach a maximum equal to $\Phi^{(1)}_i$, and then decrease again. It is also obvious that α and β must vary at the beginning in a certain direction (α decreasing, because T^0 is uniform, and β increasing) and at the end in the opposite direction in order to go back to unity.

It can be shown that, due to the presence of the shock pattern characteristic of the pseudo-shock, the variation of α and β can invert its direction several times. However, on the average, α must first decrease and then increase, and the opposite for β. It can also be checked that for the shockless model discussed in Art. 5 the variations of α and β are monotonic both when increasing and when decreasing, with a single intermediate inversion.

The same conclusions reached for $\varepsilon = 1$ can be applied to pseudo-shocks in the general duct of the p-A power family, the transition curve being represented by a line $\Phi^{(\varepsilon)} = \text{const}$, and being possible because of suitable variations of α and β.

For nonuniform processes involving mixing, friction, and heat addition or combustion, it is clear that if we suppose, as we did before, that the process takes place completely in the supercritical or in the subcritical field, a pseudo-shock can always be produced in the final quasi-uniform flow if it is supercritical, thus ending with a subcritical condition; but the opposite, with transition from subcritical to supercritical, is impossible, thus resulting in the same exclusion rule as for uniform flows.

However, this separation of the pseudo-shock from the rest of the process does not necessarily represent the actual situation, where the phenomena that characterize the pseudo-shock may take place simultaneously with the other processes. Obviously the application of the second law fails to give a conclusive criterion for a complicated process of this kind because the presence of mixing, friction, and combustion must be accompanied by a finite entropy increase.

A supporting argument for the correctness of the exclusion rule in the case when pressure nonuniformities are absent can be found under the assumption that a simple uniformization process must correspond to a monotonic variation of α and β from their initial values to unity. It is immediately found from Eq. 17-30 that $\Phi^{(e)}_{\min}$ increases or decreases along the duct depending on whether

$$-d\beta - 2\left(\frac{1}{\delta} - \frac{\beta}{\alpha}\right)d\alpha = -d\beta\left[1 + 2\left(\frac{1}{\delta} - \frac{\beta}{\alpha}\right)\frac{d\alpha}{d\beta}\right] \gtrless 0 \quad (17\text{-}35)$$

Since $\beta_i > 1$, if β varies toward the final value, which is unity, monotonically, $d\beta$ is always negative. The same is true for $d\alpha$ for $\alpha_i > 1$; in which case $\Phi^{(e)}_{\min}$, whenever it exists in the range of Eq. 17-21, is certainly, by virtue of Eq. 17-28, a monotonically increasing quantity. For $\alpha_i < 1$, $d\alpha/d\beta$ is negative, and in principle $\Phi^{(e)}_{\min}$ may not only decrease but can also invert the direction in which it varies. However, this seems unlikely. For instance, if $\beta - 1$ and $1 - \alpha$ decrease following the same decay factor, it is immediately seen that $d\alpha/d\beta$ stays constant during the whole process. Hence, if the relative variations of the factor $(1/\delta) - (\beta/\alpha)$ in Eq. 17-35 are small, which is certainly true if δ is sufficiently distant from its limit α/β, the quantity of Eq. 17-35 remains of the same sign and $\Phi^{(e)}_{\min}$ varies monotonically during the whole process.

If we accept as a probable but unproved fact that $\Phi^{(e)}_{\min}$ varies monotonically when α and β are monotonic, and that the last statement is correct when the pressure distributions are uniform and only processes of a uniformizing character are present, it is immediately clear from the consideration of the process in the $\Phi^{(e)}$, ϖ plane that, in the absence of friction and heat exchanges or combustion, a transition from subcritical to supercritical, or vice versa, is impossible, since for this transition $\Phi^{(e)}_{\min}$ should first increase, reach a maximum, and then decrease again. The only possibility for such a transition is offered by a nonmonotonic variation of α and β due to some kind of nonuniformizing process, such as the process of separation which is characteristic of the pseudo-shock.

Observe that the previous argument fails if pressure nonuniformities are present, because these are accompanied by the characteristic recurrent shock patterns which produce a nonmonotonic variation of α and β.

In conclusion we can say that when the pressure distributions are uniform, a certain number of arguments can be advanced to support the following tentative exclusion rule:

1. If the flow is initially subcritical, only the final subcritical solution is permitted, and the supercritical must be excluded.
2. If the flow is initially supercritical, both the supercritical and the subcritical final conditions are allowed, the latter being made possible by the presence of nonuniformizing processes of the same kind as those present in a pseudo-shock.

When pressure nonuniformities are present this exclusion rule seems more difficult to justify, not only in view of the nonmonotonic behavior of α and β, but also because the attainment of a final uniform condition may require ducts of an excessive length.

Finally, we observe that in the following discussion of the optimization of the duct shape we shall find another argument in favor of the exclusion rule.

Optimization of the duct shape. Transformation of the equations. We shall now investigate the problem of determining the best duct of the p-A power family for a flow with mixing and combustion when the initial nonuniformities are assigned. This is not the most general problem of optimization that should also allow the nonuniformities to vary, which can be clearly seen by considering, for instance, the case of an ejector, where the initial stagnation pressure of the primary and secondary flows are assigned and the conditions at the entrance of the mixing duct are desired for which a given mass flow ratio results in the maximum back pressure, or a given back pressure requires the minimum mass flow ratio. These optimum conditions may be determined by allowing the static pressure at the entrance of the mixing duct to vary and by looking for its optimum value. This variation results in different relative variations of velocity and temperature of the two streams, and therefore also in changes of the initial nonuniformity coefficients. Calculations of this kind can be performed numerically, after the proper assumptions have been made about the characteristics of the mixing duct.

In the rest of this article, however, we only consider the first optimization problem mentioned above, which can be investigated analytically and in a very general way. We shall therefore assume that the nonuniformity coefficients, α_i and β_i, at the entrance of the mixing duct are assigned, and so are the mean velocity $\bar{w}_i$, the mean pressure $\bar{p}_i$, and the mean adiabatic index $\bar{\gamma}_i$. Moreover, when combustion processes are present and are allowed to go to completion, we assume a given value of λ and $\bar{\gamma}_f$, since, within the approximation of our treatment, these quantities can be considered to be independent of the shape of the mixing duct. Finally, we shall neglect friction.

We define the quantity, similar to Eq. 11-3,

$$\Gamma = \frac{\bar{\gamma}_i}{\bar{\gamma}_f}\frac{\bar{\gamma}_f - 1}{\bar{\gamma}_i - 1}$$

and, for simplicity, we write

$$\delta = \delta_f = \frac{\bar{\gamma}_f - 1}{\bar{\gamma}_f}\,\varepsilon \tag{17-36}$$

so that

$$\delta_i = \frac{\delta}{\Gamma} \tag{17-37}$$

For the sake of brevity we introduce the following two quantities which can be considered to have assigned values:

$$\varphi = \frac{\Gamma\alpha_i\varpi_i^2}{1 - \beta_i\varpi_i^2}, \quad \psi = \frac{\Gamma\lambda\alpha_i}{1 - \beta_i\varpi_i^2} = \frac{\lambda\varphi}{\varpi_i^2} \tag{17-38}$$

and the ratio of the final to initial mean (dimensional) velocity (still unknown):

$$\xi = \frac{u_f}{\tilde{u}_i} = \sqrt{\lambda}\,\frac{\varpi_f}{\varpi_i} \tag{17-39}$$

In terms of these quantities we have

$$\varpi_i^2 = \lambda\frac{\varphi}{\psi}, \quad \varpi_f^2 = \xi^2\frac{\varphi}{\psi} \tag{17-40}$$

With the use of the above relations we obtain from Eq. 17-18 the following expressions:

$$\Phi_i^{(s)} = \sqrt{\frac{\lambda\varphi}{\psi}}\left(1 + \frac{\delta}{2\varphi}\right), \quad \Phi_f^{(s)} = \xi\sqrt{\frac{\varphi}{\psi}}\left(1 - \frac{\delta}{2}\right) + \frac{\delta}{2\xi}\sqrt{\frac{\psi}{\varphi}}$$

where the mixing process has been supposed to be practically completed at the final section, so that $\alpha_f = \beta_f = 1$. Introducing these expressions in Eq. 17-19 we obtain the following equation for the determination of ξ

$$\left(1 - \frac{\delta}{2}\right)\xi^2 - \left(1 + \frac{\delta}{2\varphi}\right)\xi + \frac{\delta\psi}{2\varphi} = 0 \tag{17-41}$$

This equation has, in agreement with the previous discussions of this article, either two roots or none. However, the roots have a physical meaning only if $0 \leqq \varpi_f \leqq 1$, that is, from Eq. 17-40, if ξ is in the range

$$0 \leqq \xi \leqq \xi_{max} = \sqrt{\frac{\psi}{\varphi}} \tag{17-42}$$

Since φ and ψ have fixed values, the smaller root corresponds to the smaller, i.e. the subcritical, ϖ_f and the larger root to the supercritical.

When the two roots coincide, $\bar{w}_f$ takes the critical value $(w_{cr})_f = \delta/(2 - \delta)$.

In order to be able to apply the exclusion rule in the selection of the proper root, we must check if $\bar{w}_i$ is sub- or supercritical. From Eq. 17-27, 17-37, and 17-38 we have

$$\frac{1}{(w_{cr})_i^2} = 2\frac{\Gamma}{\delta}\alpha_i - \beta_i, \quad \frac{1}{\bar{w}_i^2} = \frac{\Gamma}{\varphi}\alpha_i + \beta_i$$

Hence $\bar{w}_i$ is $\lesseqgtr (w_{cr})_i$ depending on whether

$$\delta \gtrless \delta_{cr} = \frac{1}{\dfrac{1}{2\varphi} + \dfrac{\beta_i}{\Gamma\alpha_i}} \tag{17-43}$$

Thus, if δ is smaller than the value δ_{cr} defined by this relation, both solutions are possible; however, if δ is larger than δ_{cr}, only the smaller root of Eq. 17-41 is allowed.

The values of $\bar{p}_f$ and A_f can be found from Eq. 17-20, which can be written by virtue of Eq. 17-38 and 17-40 and the fact that $\nu = 1$, in the form

$$\left(\frac{\bar{p}_f}{\bar{p}_i}\right)^{\frac{1}{\epsilon}} = \left(\frac{A_f}{A_i}\right)^{\frac{1}{1-\epsilon}} = \frac{\delta + 2\varphi}{\delta + \dfrac{2\varphi\xi^2}{\psi - \varphi\xi^2}} \tag{17-44}$$

The mean final stagnation pressure can also be calculated as if the flow were uniform, from the relation

$$\frac{\bar{p}_f^0}{\bar{p}_f} = (1 - \bar{w}_f^2)^{-\frac{\bar{\gamma}_f}{\bar{\gamma}_f - 1}} = \left(1 - \frac{\varphi\xi^2}{\psi}\right)^{-\frac{\bar{\gamma}_f}{\bar{\gamma}_f - 1}} \tag{17-45}$$

We now want to determine the value to be selected for δ in order to obtain the maximum possible $\bar{p}_f^0$ for given φ, ψ, $\bar{w}_i$, $\bar{p}_i$, λ, and Γ. Rather than follow the direct procedure of solving Eq. 17-41 for ξ, and replacing this value in Eq. 17-44 and 17-45, the discussion is much simpler if we solve Eq. 17-41 with respect to δ, thus avoiding radicals, and introduce this expression of $\delta(\xi)$ into Eq. 17-44 and 17-45. In this way we obtain

$$\delta(\xi) = \frac{2\varphi\xi(1 - \xi)}{\psi - \xi - \varphi\xi^2} \tag{17-46}$$

Eq. 17-44 can then be written in the form

$$\left(\frac{\bar{p}_f}{\bar{p}_i}\right)^{\frac{\bar{\gamma}_f - 1}{\bar{\gamma}_f}} = \left(\frac{\psi - \varphi\xi^2}{\xi}\right)^{\delta(\xi)} \tag{17-47}$$

and Eq. 17-45 can be combined with Eq. 17-47 to give

$$P(\xi) = \left(\frac{\bar{p}_f^0}{\bar{p}_i}\right)^{\frac{\bar{\gamma}_f - 1}{\bar{\gamma}_f}} = \frac{\psi}{\psi - \varphi\xi^2}\left(\frac{\psi - \varphi\xi^2}{\xi}\right)^{\delta(\xi)} \tag{17-48}$$

For given φ and ψ both δ and P can be considered functions of ξ only. Clearly $\bar{p}_f^0$ is maximum when P is maximum, so that our optimization problem is reduced to that of finding from Eq. 17-48 the value of ξ which makes P maximum, and of determining from Eq. 17-46 the corresponding value of δ.

Behavior of $\delta(\xi)$ *and* $P(\xi)$. Before proceeding with the process of optimization we shall discuss the properties of the functions defined by Eq. 17-46 and 17-48. Obviously $\delta(\xi)$ vanishes at $\xi = 0$ and $\xi = 1$, and is $\pm\infty$ for $\psi - \xi - \varphi\xi^2 = 0$, the only positive root of this equation being

$$\xi_1 = \frac{\sqrt{1 + 4\varphi\psi} - 1}{2\varphi} = \frac{\sqrt{(1 + 2\varphi)^2 + 4\varphi[\psi - (1 + \varphi)]} - 1}{2\varphi} \tag{17-49}$$

Clearly we have $\xi_1 \lesseqgtr 1$ depending on whether $\psi \lesseqgtr 1 + \varphi$.

At the upper limit of the range of Eq. 17-42, $\xi = \xi_{max}$, we have

$$\delta(\xi_{max}) = 2\varphi\left(\sqrt{\frac{\psi}{\varphi}} - 1\right) \tag{17-50}$$

Now from the definitions of Eq. 17-38, and since $\lambda \geqq 1$ and $\beta_i \geqq 1$, we find the series of inequalities:

$$\frac{\psi}{\varphi} = \lambda\left(\beta_i + \frac{\Gamma\alpha_i}{\varphi}\right) \geqq 1 + \frac{\Gamma\alpha_i}{\varphi\beta_i}$$

$$> \left(1 + \frac{\Gamma\alpha_i}{\varphi\beta_i}\right)\left[1 - \frac{1}{\left(1 + \frac{2\varphi\beta_i}{\Gamma\alpha_i}\right)^2}\right] = 4\left(\frac{1 + \frac{\varphi\beta_i}{\Gamma\alpha_i}}{1 + \frac{2\varphi\beta_i}{\Gamma\alpha_i}}\right)^2$$

Hence, not only is $\psi/\varphi > 1$ and therefore ξ_{max} is >1 and $\delta(\xi_{max})$ is positive, but also, comparing with Eq. 17-43, we have

$$\delta(\xi_{max}) > 2\varphi\left(2\frac{1 + \frac{\varphi\beta_i}{\Gamma\alpha_i}}{1 + \frac{2\varphi\beta_i}{\Gamma\alpha_i}} - 1\right) = \delta_{cr} \tag{17-51}$$

Differentiating Eq. 17-46 we find

$$\frac{d\delta}{d\xi} = 2\varphi\frac{(1 + \varphi)\xi^2 - 2\psi\xi + \xi}{(\psi - \xi - \varphi\xi^2)^2} \tag{17-52}$$

Thus $d\delta/d\xi$ can be zero only when the numerator vanishes, that is for

$$\xi_2 = \frac{\psi}{1 + \varphi}\left(1 - \sqrt{1 - \frac{1 + \varphi}{\psi}}\right), \quad \xi_3 = \frac{\psi}{1 + \varphi}\left(1 + \sqrt{1 - \frac{1 + \varphi}{\psi}}\right) \tag{17-53}$$

When ξ_2 and ξ_3 are real, that is for $\psi > 1 + \varphi$, we have $\xi_2 < 1$ and $\xi_3 > 1$. In this case $d\delta/d\xi$ is negative in the range $\xi_2 < \xi < \xi_3$ and positive out of that range, so that δ presents a relative maximum at ξ_2 and a relative minimum at ξ_3, given by

$$\frac{1}{\delta(\xi_2)} = 1 + \frac{1}{2\varphi} + \frac{1+\varphi}{\varphi}(\xi_3 - 1); \quad \frac{1}{\delta(\xi_3)} = 1 + \frac{1}{2\varphi} - \frac{1+\varphi}{\varphi}(1 - \xi_2) \tag{17-54}$$

Thus

$$\delta(\xi_2) < \frac{1}{1 + \dfrac{1}{2\varphi}} < \delta(\xi_3)$$

Observe that if $\bar{\gamma}_f \leqq \bar{\gamma}_i$ (which is generally true for mixing and combustion processes) and $\alpha_i < \beta_i$ (which, as shown previously, is true if M and u are covariant, i.e. for the most likely condition) we have from Eq. 17-43

$$\delta_{cr} < \frac{1}{1 + \dfrac{1}{2\varphi}} < \delta(\xi_3) \tag{17-55}$$

This inequality is therefore very likely to be satisfied in all practical cases; however, the opposite might exceptionally be true.

The smaller root ξ_2 always falls in the range of Eq. 17-42. However, we have $\xi_3 \lesseqgtr \xi_{max}$ depending on whether

$$\psi \lesseqgtr \varphi\left(1 + \frac{1}{2\varphi}\right)^2 = 1 + \varphi + \frac{1}{4\varphi} \tag{17-56}$$

Correspondingly, as immediately derived from Eq. 17-50,

$$\delta(\xi_{max}) \lesseqgtr 1 \tag{17-57}$$

It is easily checked from Eq. 17-40 that

$$\varpi_f(\xi_2) = \frac{\delta(\xi_2)}{2 - \delta(\xi_2)}, \quad \varpi_f(\xi_3) = \frac{\delta(\xi_3)}{2 - \delta(\xi_3)}$$

and, therefore, at both ξ_2 and ξ_3 the final conditions are critical. For $\psi < 1 + \varphi$, ξ_2 and ξ_3 have no real existence and $d\delta/d\xi$ is always positive, so that δ is a monotonically increasing function of ξ.

Let us now investigate the behavior of $P(\xi)$. Near $\xi = 0$ we have from Eq. 17-46 $\delta(\xi) \cong 2\varphi\xi/\psi$. Thus from Eq. 17-48 we find

$$P(0) = \lim_{\xi \to 0}\left(\frac{\psi}{\xi}\right)^{\frac{2\varphi\xi}{\psi}} = 1$$

At $\xi = \xi_1$ ($\delta = \pm\infty$) we have

$$P(\xi_1) = \lim_{\xi \to 1} \frac{\psi}{\xi_1}[\zeta(\xi_1)]^{\frac{2\varphi(1-\xi_1)}{\zeta(\xi_1)-1}}$$

where for brevity we have written

$$\zeta(\xi) = \frac{\psi - \varphi\xi^2}{\xi} \tag{17-58}$$

Evaluating the limit we see that $P(\xi_1)$ is finite and given by

$$P(\xi_1) = \frac{\psi}{\xi_1} e^{2\varphi(1-\xi_1)}$$

Thus the fact that δ is infinite does not correspond to a physical singularity. For infinite δ, the p-A power law defining the duct is reduced to $\bar{p}_i A_i = \bar{p}_f A_f$.

At $\xi = \xi_{max}$ we have

$$P(\xi_{max}) = \lim_{\xi \to \xi_{max}} \frac{\psi}{\xi_{max}^{\delta(\xi_{max})}} [\varphi(\xi_{max}^2 - \xi^2)]^{\delta(\xi_{max})-1}$$

We see that $P(\xi_{max})$ is 0 or ∞ depending on whether $\delta(\xi_{max}) \gtrless 1$, that is whether $\psi \gtrless 1 + \varphi + (1/4\varphi)$, by virtue of the inequalities of Eq. 17-56 and 17-57.

Finally, differentiating Eq. 17-48 we obtain

$$\frac{1}{P}\frac{dP}{d\xi} = \frac{\dfrac{1}{\zeta(\xi)} - 1 + \ln \zeta(\xi)}{[1 - \zeta(\xi)]^2} \frac{2\varphi}{\xi^2} [(1 + \varphi)\xi^2 - 2\psi\xi + \psi] \tag{17-59}$$

where $\zeta(\xi)$ is defined by Eq. 17-58; and since the first factor of this expression is always positive, we see by comparison with Eq. 17-52 that $dP/d\xi$ always has the same sign as $d\delta/d\xi$. It is also checked immediately that $dP/d\xi$ is infinite at $\xi = 0$.

Thus P has a relative maximum, like δ, at $\xi = \xi_2$, and a relative minimum at $\xi = \xi_3$. These extrema exist only for $\psi < 1 + \varphi$. If, on the contrary, $\psi < 1 + \varphi$, P is a monotonically increasing function of ξ.

Optimization of the duct shape. Discussion. As a result of the properties of $\delta(\xi)$ and $P(\xi)$ we can distinguish four cases:

Case I: $\psi < 1 + \varphi$. This case is possible in the absence of combustion, when the quantity Δ' defined by Eq. 17-7 is <1. In fact from Eq. 17-8 and 17-38 we obtain $\psi = \lambda\Delta'(\Gamma + \varphi)$, a relation that in the absence of combustion or heat exchanges ($\lambda = \Gamma = 1$) is reduced to $\psi = \Delta'(1 + \varphi)$, thus proving the above statement. The shape of the $\delta(\xi)$ and $P(\xi)$ curves, with $\xi_1 < 1$, is shown schematically by Fig. B,17l. We see that $P(\xi)$ increases without limit when $\xi \to \xi_{max}$. However, a limitation is obtained from the application of the exclusion rule. The line $\delta = \delta_{cr}$ is indicated in the figure; this line necessarily cuts both branches of the $\delta(\xi)$ curve because of Eq. 17-55. The branch to the left represents subcritical final flow conditions, that to the right, supercritical final conditions. On the other hand, by virtue of the inequality of Eq. 17-43 the points below the line $\delta = \delta_{cr}$ correspond to supercritical initial conditions, while those above that line correspond to subcritical initial conditions. The exclusion rule, therefore, excludes the points of the supercritical branch that are situated above the line $\delta = \delta_{cr}$. This portion of the curve $\delta(\xi)$ and the corresponding portion of the curve $P(\xi)$ are shown dashed on the figure. Thus the maximum value of $P(\xi)$ is reached at $\xi = \xi_{opt}$ and corresponds to the value of δ which renders the flow, in the initial section, critical.

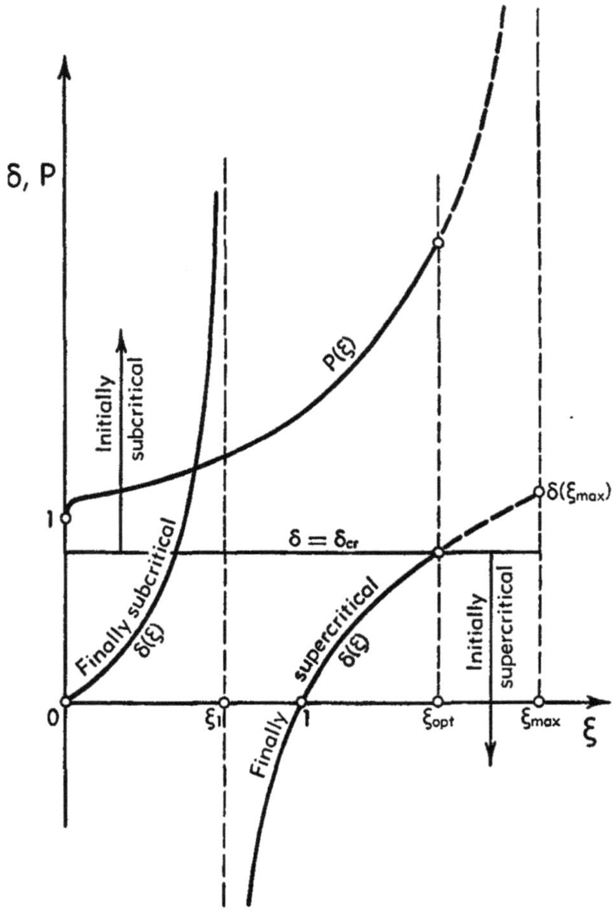

Fig. B,171. Optimization of the duct shape. Case I: $\psi < 1 + \varphi$.

Observe that, as one can see immediately from Eq. 17-47, $\bar{p}_f \gtrless \bar{p}_i$ for $\xi \lessgtr 1$. Therefore, the optimum conditions are obtained with a decrease of the mean pressure. Observe finally that all values of δ are allowed.

Case II. $\psi = 1 + \varphi$. This case can be obtained when combustion is absent and $\Delta' = 1$, in particular when the flow is uniform, so that α_i and β_i are also unity. The curve of $\delta(\xi)$ degenerates into the curve

$$\delta = \frac{2\varphi\xi}{1 + \varphi(1 + \xi)} \tag{17-60}$$

and the vertical line $\xi = \xi_1 \equiv 1$ (δ arbitrary). The two lines intersect each other at $\delta(\xi_1) = 2\varphi/(1 + 2\varphi)$. Depending on the position of δ_{cr} with respect to $\delta(\xi_1)$ we can distinguish the three subcases shown in Fig. B,17m, B,17n, and B,17o. The subcase of Fig. B,17n corresponds to $\alpha_i = \beta_i$ ($\Gamma = 1$), and, in particular, to the case of uniform flow. Applying the exclusion rule we obtain that if the initial flow is subcritical, only the solution $\xi = 1$ is possible, that is $u_2 = u_1$, and, as immediately checked

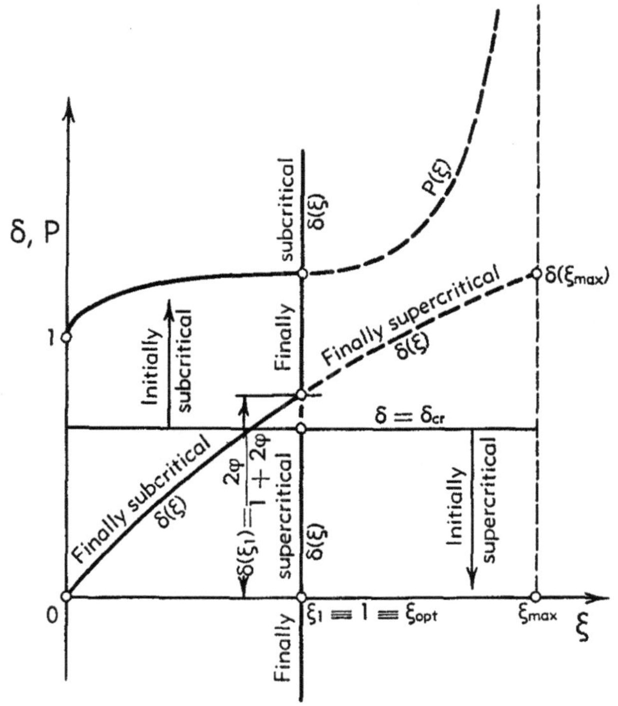

Fig. B,17m. Optimization of the duct shape. Case II: $\psi = 1 + \varphi$. First subcase: $\delta_{cr} < \delta(\xi_1)$.

from Eq. 17-47, $p_2 = p_1$. This is therefore the trivial solution implying no change in the flow conditions. If the flow is initially supercritical, in addition to the trivial supercritical solution, we also have a subcritical solution which is reached through a pseudo-shock. These are the same results presented in Art. 5 for uniform flows. As we see, they apply equally well to the case when $\alpha_i = \beta_i$, that is, when the local Mach number is proportional to the local velocity.

The most common subcase, where M and u are covariant, gives the inequality of Eq. 17-55 and is represented in Fig. B,17m. Here the exclusion rule also indicates the impossibility of the segment of the line $\xi = \xi_1$ between δ_{cr} and $\delta(\xi_1)$. As Eq. 17-47 shows, the isobaric solution is now possible only for $\delta > \delta(\xi_1)$ or $\delta < \delta_{cr}$, but between the two $\bar{p}_f$ must be different from $\bar{p}_i$.

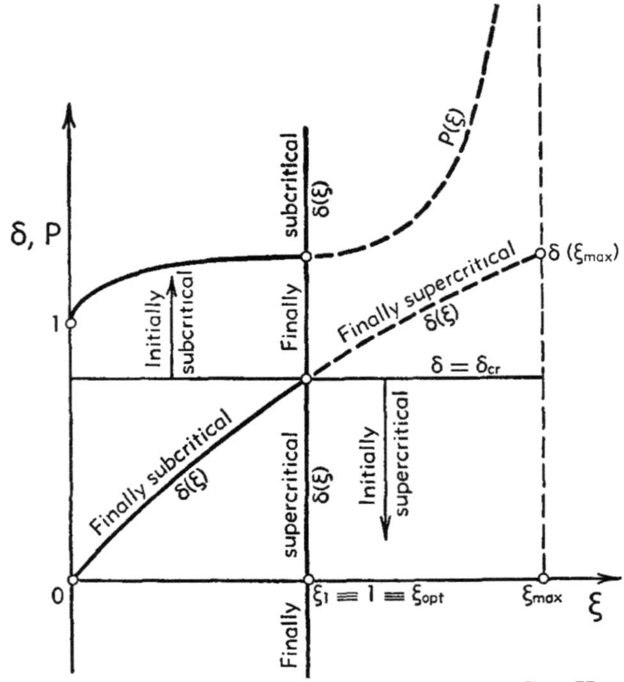

Fig. B,17n. Optimization of the duct shape. Case II: $\psi = 1 + \varphi$. Second subcase: $\delta_{cr} = \delta(\xi_1)$.

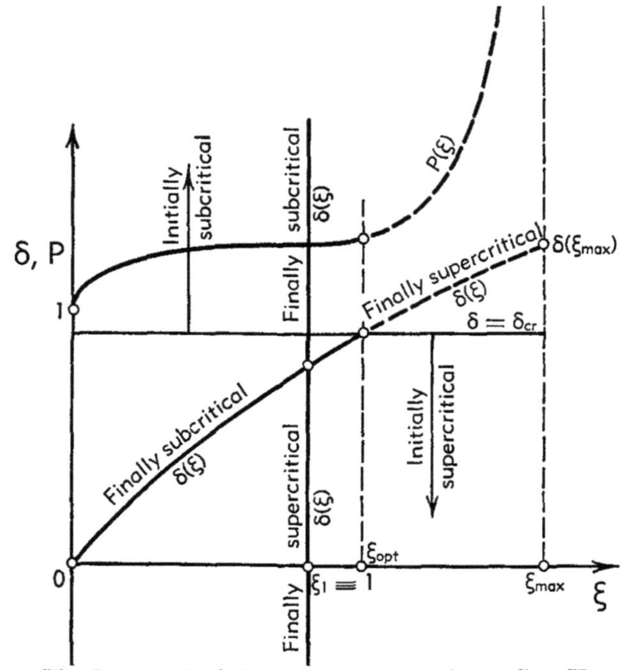

Fig. B,17o. Optimization of the duct shape. Case II: $\psi = 1 + \varphi$. Third subcase: $\delta_{cr} > \delta(\xi_1)$.

The subcase of Fig. B,17o, though possible, is most unlikely, because it requires that M and u be contravariant. The supercritical segment of the curve of Eq. 17-60 between $\delta(\xi_1)$ and δ_{cr} is now allowed. We see clearly that the optimum value of P, equal to $1 + \varphi$, is provided in the first two cases by the isobaric solution $\xi = 1$. In the last case, however, the maximum P, larger than $1 + \varphi$, is attained for $\xi_{opt} > \xi_1$, at the value of δ that makes the initial flow critical.

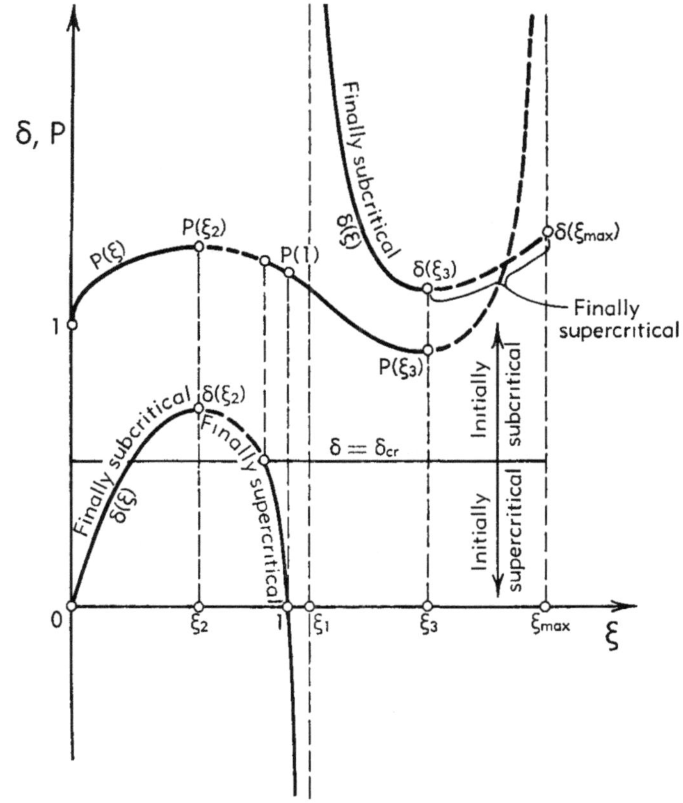

Fig. B,17p. Optimization of the duct shape. Case III: $1 + \varphi < \psi < 1 + \varphi + (\frac{1}{4}\varphi)$. First subcase: $\delta_{cr} < \delta(\xi_2)$.

As the curves show, and as can be checked from Eq. 17-59, the curve $P(\xi)$ now has, in case II, an inflection point with horizontal tangent at $\xi = 1$.

Case III. $1 + \varphi < \psi < 1 + \varphi + (1/4\varphi)$. This case is possible in the absence of combustion for $\Delta' > 1$, or, in the presence of combustion processes, regardless of the value of Δ'. The value of ξ_1 (Eq. 17-49) is now >1, and the values of ξ_2 and ξ_3 (Eq. 17-53) are real and <1. The resulting shape of the $\delta(\xi)$ curve is indicated schematically in Fig. B,17p, B,17q, and B,17r. The curve $\delta(\xi)$ is now split into a lower branch with a relative maximum $\delta(\xi_2)$ and an upper branch with a relative minimum

$\delta(\xi_3) > \delta(\xi_2)$. Thus not all values of δ are allowed, and choking is present for all values of δ between $\delta(\xi_2)$ and $\delta(\xi_3)$.

We can distinguish three subcases. In Fig. B,17p δ_{cr} is $<\delta(\xi_2)$ and a segment of both branches is forbidden by the exclusion rule. In Fig. B,17q δ_{cr} is between $\delta(\xi_1)$ and $\delta(\xi_2)$, and only the segment of the upper branch beyond the minimum is forbidden. Finally, in Fig. B,17r, δ_{cr} is

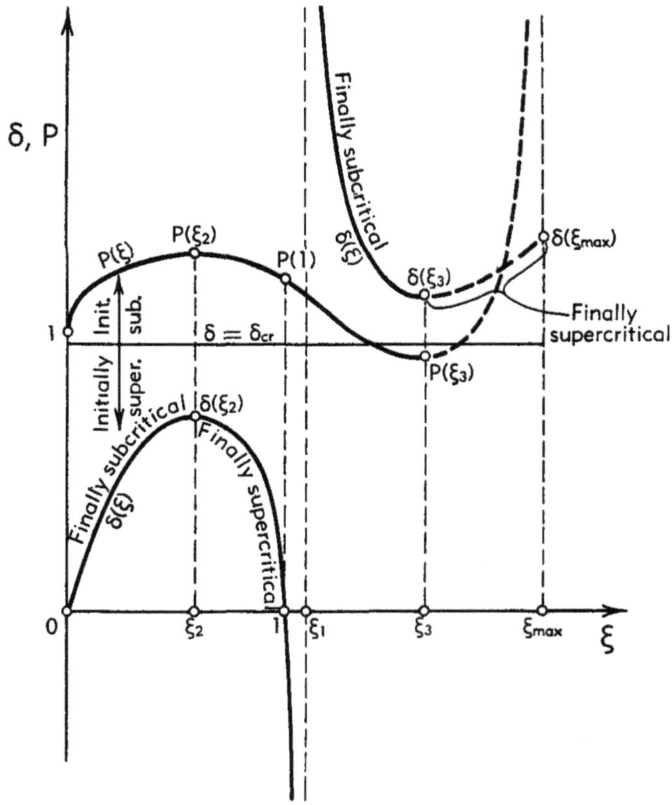

Fig. B,17q. Optimization of the duct shape. Case III: $1 + \varphi < \psi < 1 + \varphi + (1/4\varphi)$. Second subcase: $\delta(\xi_2) < \delta_{cr} < \delta(\xi_3)$.

$> \delta(\xi_3)$. This is a very unlikely condition, as observed when discussing the inequality represented by Eq. 17-55. In this case, only a smaller segment, $\xi_4 < \xi < \xi_{max}$, of the upper branch is forbidden. Considering the curves of $P(\xi)$, it is clear that for $\delta < \delta(\xi_3)$, P is maximum at ξ_2, for $\delta = \delta(\xi_2)$, that is, as previously observed, when δ is such as to make the final flow critical. The value of $P(\xi_2)$ can be easily calculated. When $\delta_{cr} > \delta(\xi_3)$, $P(\xi_2)$ is still likely to be the maximum value of P that can be obtained except in the improbable case when $P(\xi_4) > P(\xi_2)$. Considering that this is a very unlikely case, we obtain the simple result that in case III the final flow must be critical in order to reach the highest stagnation pressure. The value $P(\xi_2)$, as shown by the figures, is always higher than the value $P(1)$ corresponding to the isobaric case. From Eq. 17-47 it is actually shown easily that at ξ_2 we always have $\bar{p}_f > \bar{p}_i$, so that, contrary to case I, we need a certain amount of compression for best results. Observe that the value of δ corresponding to a constant section duct, $\delta = (\bar{\gamma}_f - 1)/\bar{\gamma}_f$, may be above or below $\delta(\xi_2)$, and only exceptionally coincides with it. Thus the constant section duct has no special properties with respect to the optimization of P.

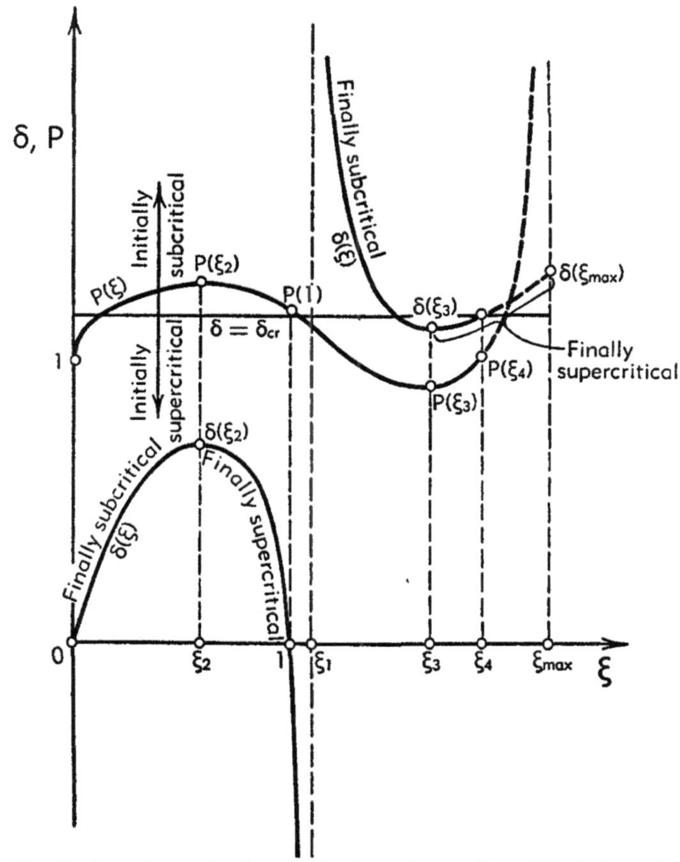

Fig. B,17r. Optimization of the duct shape. Case III: $1 + \varphi < \psi < 1 + \varphi + (1/4\varphi)$. Third subcase: $\delta(\xi_3) < \delta_{cr}$.

Case IV. $\psi > 1 + \varphi + (1/4\varphi)$. In this case, which may be realized when combustion is present, we have $\xi_3 > \xi_{max}$, and $\delta(\xi_{max}) > 1$ represents the lowest possible value on the upper branch of the $\delta(\xi)$ curve. Fig. B,17s and B,17t illustrate the two subcases $\delta_{cr} < \delta(\xi_2)$ and $\delta(\xi_2) < \delta_{cr}$. In the first subcase, a segment of the lower branch is still forbidden; in the latter subcase, all the points of the $\delta(\xi)$ curves for $0 \leqq \xi \leqq \xi_{max}$ are allowed. The value of $P(\xi_{max})$ is now zero. The absolute maximum of the $P(\xi)$ curve is again obtained for $\xi = \xi_2$, i.e. for critical final conditions.

We observe that case IV is the only one in which values of ξ up to ξ_{max} are possible, and it is the only case in which $P(\xi_{max}) = 0$. In the other three cases we had $P(\xi_{max}) = \infty$, that is, an infinite increase in stagnation pressure, and therefore an infinite entropy decrease, which is certainly contrary to the second law of thermodynamics. The fact that the attainment of ξ_{max} is forbidden by the exclusion rule in all the cases in which

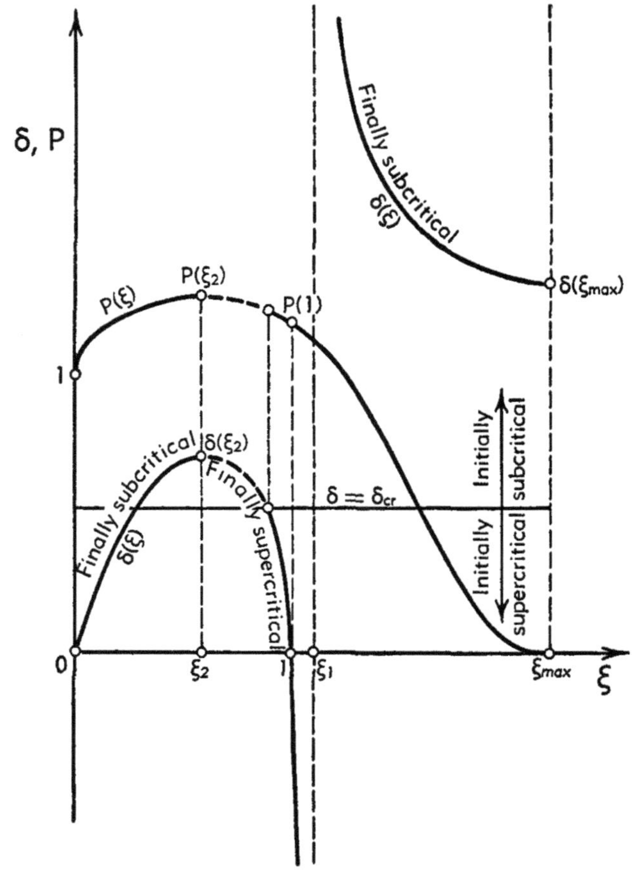

Fig. B,17s. Optimization of the duct shape. Case IV: $1 + \varphi + < \psi$.
First subcase: $\delta_{cr} < \delta(\xi_2)$.

$P(\xi_{max}) = \infty$, represents a link between the exclusion rule and the second law of thermodynamics and an argument in favor of the validity of the exclusion rule.

Final considerations on the duct optimization. The attainment for given initial conditions of the maximum stagnation pressure at the end of the mixing duct does not necessarily represent for all practical devices the correct criterion of optimization. In ejectors, for instance, the optimization should rather be operated on the stagnation pressure at the exit of the diffuser that follows the mixing duct. As a result of the introduction of a diffuser efficiency, which must be a function of the Mach number, this optimization can only be carried out numerically. Due to the fact that the diffuser efficiency decreases with the Mach number, the optimum value of ξ moves toward lower values. For instance, in case I one would find that, instead of obtaining best results with an expansion in

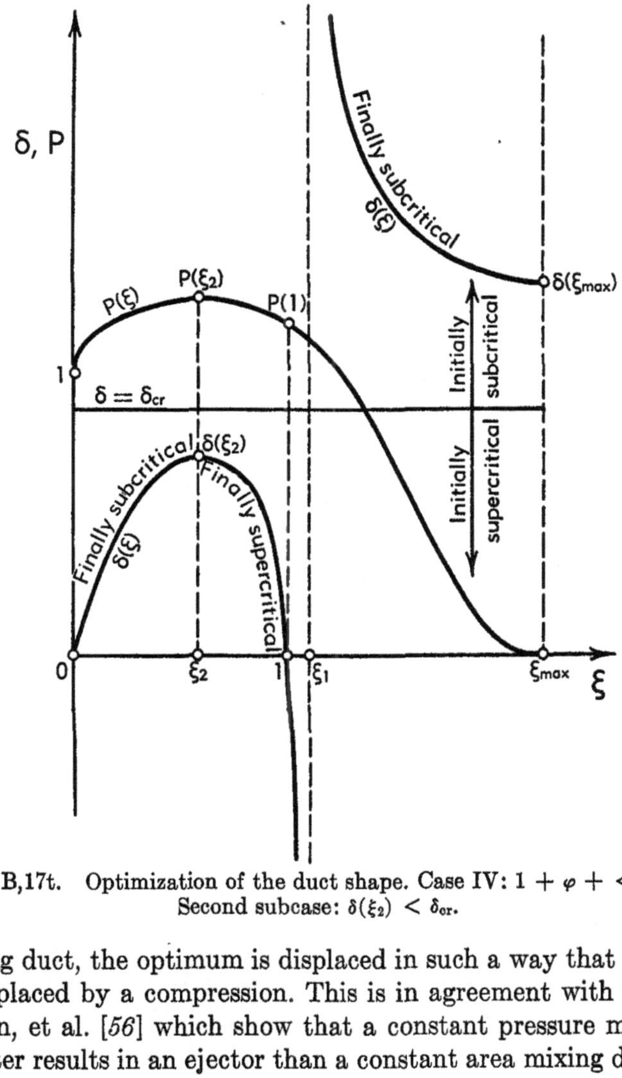

Fig. B,17t. Optimization of the duct shape. Case IV: $1 + \varphi + < \psi$. Second subcase: $\delta(\xi_2) < \delta_{cr}$.

the mixing duct, the optimum is displaced in such a way that the expansion is replaced by a compression. This is in agreement with the results of Keenan, et al. [*56*] which show that a constant pressure mixing duct gives better results in an ejector than a constant area mixing duct, where in the present case the uniformization process is accompanied by a pressure decrease.

In the preceding discussion no restraint has been imposed on the geometry of the duct, which has been left free to take whatever shape is required in order to maximize the stagnation pressure, with the only restriction that of satisfying the proper relation (Eq. 15-1). If, however, conditions of a geometric nature are introduced, such as determining the best duct for given maximum cross-sectional area, the results can change substantially. This problem would be analogous to that discussed in Art. 11, when looking for the most favorable shape of combustion chamber, and actually the case of uniform flow with combustion is included as a particular case in the more general flow problem considered in this article.

B,18. Cited References.

1. Dodge, B. F. *Chemical Engineering Thermodynamics.* McGraw-Hill, 1944.
2. Crown, J. C. *Nav. Ord. Lab. Rept. 2148,* White Oak, Maryland, 1951.
3. Keenan, J. H., and Kaye, J. *Gas Tables.* Wiley, 1945.
4. Donaldson, C. duP. *NACA Research Mem. L8J14,* 1948.
5. Schmidt, E. *Jahrbuch deut. Luftfahrtforschung,* 314–319 (1938).
6. Noeggerath, W. *Jet Propulsion 25,* 454–462 (1955).
7. Crocco, L. *Monografie sci. aeronaut. 6,* 1947.
8. Hershey, R. L., Eberhardt, J. E., and Hottel, H. C. *S.A.E. Quart. Trans. 39,* 409 (1936).
9. Hottel, H. C., and Williams, G. C. *NACA Tech. Note 1026,* 1946.
10. Crocco, L. *Supplementi Tecnici, Monografie sci. aeronaut. 2,* 1947.
11. Hottel, H. C., Williams, G. C., and Satterfield, C. N. *Thermodynamic Charts for Combustion Processes.* Wiley, 1949.
12. Krieger, F. J., and White, W. B. *Rand Corp., Project Rand Rept. R-149,* Santa Monica, Calif., 1949.
13. Heinrich, G. *Z. angew. Math. u. Mech. 22,* 117–118 (1942).
14. Tangren, R. F., Dodge, C. H., and Seifert, H. S. *J. Appl. Phys. 20,* 637–649 (1949).
15. Shapiro, A. H. *The Dynamics and Thermodynamics of Compressible Fluid Flow.* Ronald Press, 1953.
16. Foa, J. V. Mach number functions for ideal diatomic gases. *Cornell Aeronaut. Lab.,* 1949.
17. *A.R.C. Compressible Flow Tables Panel, A Selection of Tables for Use in Calculations of Compressible Airflow.* Clarendon Press, Oxford, 1952.
18. Weise, A. *Tech. Berichte 10,* 59–61 (1943).
19. Moeller, J. H., Vas, I. E., and Bogdonoff, S. M. *Princeton Univ. Aeronaut. Eng. Dept. Rept. 298,* 1956.
20. Bogdonoff, S. M. *Heat Transfer and Fluid Mech. Inst. Conference,* Palo Alto, Calif., June 1955.
21. Crocco, L., and Probstein, R. F. *Princeton Univ. Aeronaut. Eng. Dept. Rept. 254,* 1954.
22. Lustwerk, F. *Mass. Inst. Technol. Gas Turbine Lab. Meteor Rept. 61,* 1950.
23. Neumann, E. P., and Lustwerk, F. *J. Appl. Mech. 71,* 195–202 (1949).
24. Prandtl, L., and Proell, A. *Z. Ver. deut. Ing.,* 348 (1904).
25. Stodola, A. *Steam and Gas Turbines.* Peter Smith, New York, 1945.
26. Kestin, J., and Zaremba, S. K. *Aircraft Eng. 25,* 172–175 (1953).
27. Fabri, J. *Office natl. Etudes et Recherches aéronaut. 17,* 1950.
28. Malina, F. J. *J. Franklin Inst. 230,* 433 (1940).
29. Crocco, L. *Proc. of Conference on High Speed Aeronautics, Polytech. Inst. of Brooklyn,* 1955.
30. Prandtl, L. *Reale Accad. d'Italia, Proceeding of the Volta Congress on High Speed Aerodynamics 449,* Rome, 1935.
31. Summerfield, M., Foster, C. R., and Swan, W. C. *Jet Propulsion 24,* 319–321 (1954).
32. Kantrowitz, A., and Donaldson, C. duP. *NACA Advance Restricted Rept. L5D20,* 1945.
33. Kuchemann, D., and Weber, J. *Aerodynamics of Propulsion.* McGraw-Hill, 1953.

34. Wegener, P. P., and Lobb, R. K. *J. Aeronaut. Sci. 20,* 105–110 (1953).
35. Moody, L. F. *Trans. Am. Soc. Mech. Engrs. 66,* 671 (1948).
36. Schiller, L. *Z. angew. Math. u. Mech. 2,* 96 (1922).
37. Latzko, H. *Z. angew. Math. u. Mech. 1,* 268 (1921).
38. Pascucci, L. *Atti Guidonia 87–88,* 1943.
39. Frössel, W. *Ver. deut. Ing. Forschungsheft A7,* 75 (1936).
40. Scurlock, A. C. *Mass. Inst. Technol. Fuel Research Lab. Meteor. Rept. 19,* 1948.
41. Tsien, H. S. *J. Appl. Mech. 18,* 188–194 (1951).
42. Fabri, J., Siestrunck, R., and Fouré, C. *Fourth Symposium on Combustion,* pp. 443–450. Williams & Wilkins, 1953.
43. Durham, F. P. *J. Appl. Mech. 19,* 57–62 (1952).
44. Schetzer, J. D. *Univ. Mich. Dept. of Eng. Research External Mem. 6,* 1947.
45. Tsien, H. S. *J. Math. and Phys. 25,* 301–324 (1947).
46. Lockhardt, R. W., and Martinelli, R. C. *Chem. Eng. Progr. 45,* 39–45 (1949).
47. Bergelin, O. P. *Chem. Eng. News 56,* 104–106 (1949).
48. Shapiro, A. H., and Hawthorne, W. R. *J. Appl. Mech. 14,* A317–A336 (1947).
49. Bailey, N. P. *J. Aeronaut. Sci. 11,* 227–238 (1944).
50. Edelman, G. M., and Shapiro, A. H. *J. Appl. Mech. 14,* A344–A351 (1947); *15,* A169–A175 (1948).
51. Shapiro, A. H., Wadleigh, K. R., Gavril, B. D., and Fowle, A. A. *Trans. Am. Soc. Mech. Engrs. 78,* 617–653 (1956).
52. Ranz, W. E., and Marshall, W. R. *Chem. Eng. Progr. 48,* 141–146, 173–180 (1952).
53. Crocco, L. *Aerotecnica 15,* 237–275, 735–778 (1935).
54. Flugel, G. *Ver. deut. Ing. Berlin Forschungsheft 395,* 1939.
55. Roy, M. *Publs. sci. et tech. Ministère air, France, 203,* 1947.
56. Keenan, J. H., Neumann, E. P., and Lustwerk, F. *J. Appl. Mech. 17,* 299–309 (1950).
57. Fabri, J., LeGrives, E., and Siestrunck, R. *Jahrbuch 1953 der W.G.L.,* pp. 106–110. Braunschweig, 1954.
58. Nusselt, W. *Ver. deut. Ing. Berlin Forschungsheft A11,* 250–255 (1940).
59. Fabri, J. and Siestrunck, R. *Rev. Générale des Sciences Appliquées 2,* 229–237 (1955).
60. Hermann, R. *J. Aeronaut. Sci. 19,* 375–384 (1952).
61. Wick, R. S. *J. Aeronaut. Sci. 20,* 675–682 (1953).
62. Korst, H. M., and Wick, R. S. *J. Aeronaut. Sci. 21,* 568–569 (1954); *22,* 135–137 (1955).
63. Crocco, L. *Aerotecnica 12,* 181–197 (1932).
64. *New York Univ. Appl. Math. Group Rept. 43,* 1944.
65. Sauer, R. *Deut. Luftfahrtforschung, Forschungsbericht 1992.* See also *NACA Tech. Mem. 1147,* 1947.
66. Oswatitsch, K., and Rothstein, W. *Jahrbuch Luftfahrtforschung* I, 91–102 (1942). See also *NACA Tech. Mem. 1215,* 1949.
67. Courant, R., and Hilbert, D. *Methods of Mathematical Physics,* Vol. I. Interscience, 1953.

SECTION C

ONE-DIMENSIONAL TREATMENT OF NONSTEADY GAS DYNAMICS

A. KANTROWITZ

C,1. Introduction. This section treats those gas flow problems in which the velocity and the thermodynamic state of the gas depend to a sufficient degree of approximation on only one space coordinate, x, and the time. One-dimensional treatment gives useful results in those channel flow problems for which the nonuniformities (in the y and z directions) initially present, or introduced by divergence or curvature of the channel walls, by friction or by heat transfer, can be neglected or handled by corrections. Explosion (and its reverse, implosion) processes in which we have approximately spherical or cylindrical symmetry can also be treated one-dimensionally.

In this section we shall assume the perfect gas equation of state. The most important problems which are excluded in this way are explosive phenomena in liquids and solids. In the development of the fundamental equations (Eq. 2-8 and 2-9), it is also assumed that the number of gas molecules and the heat capacity are constant. This assumption is dropped where necessary for the treatment of special problems.

Historically, the development of this subject followed the early development of the theory of sound. Poisson [*1*] and Earnshaw [*2*] first extended the theory of sound to the case of finite amplitude disturbances propagating in one direction. Riemann [*3*] presented the theory in a form suitable for calculating the propagation of plane waves of finite amplitude proceeding in both directions. He observed a steepening of compression waves which eventually leads to the formation of discontinuities in perfect fluids. The conservation-law treatment of these discontinuities and their relation to shock waves produced by explosions by Rankine [*4*] and Hugoniot [*5*] completed the essential principles of the subject.

During World War II a considerable impetus was given to this subject from two sources. First, there was considerable interest in explosions and their resultant pressure waves. Wartime research in this direction by the group working with Courant and Friedrichs has been summarized in their book [*6*]. This work contains a considerable elaboration of Riemann's treatment and its application to various examples. This book also contains a considerable discussion of these problems from a mathematical point of view, and those interested in this side of the subject will find there much fuller discussions of the mathematical theory of characteristics than is presented herein. (See also VI,B.)

A second source of interest in waves of finite amplitude was the wartime German development of pulse-jet engines (Schmidt tubes) which were developed to power the V-1 missile. For this purpose the method of characteristics was developed to permit numerical integration as had previously been done by Prandtl and Busemann [*7*, pp. 499–509] for steady two-dimensional supersonic flow.

In this numerical integration procedure (given in Art. 9) it is possible to take account of many effects of practical importance which cannot be handled by analytical procedures. In common with other numerical procedures, however, results are only numerical answers for special cases rather than the highly surveyable results which sometimes can be obtained from problems treated analytically.

The general plan of this section is to develop the general equations in Art. 2, to drastically simplify these so as to exhibit their wave propagation properties in Art. 3, and to exhibit the other properties of one-dimensional flows by including gradually more and more physical effects.

A good bibliography of this field has been prepared by the United Aircraft Corporation and is to be found in [*8*].

C,2. Fundamental Equations. Our basic assumption is that the velocity u and the thermodynamic state of the gas depend only on a single space coordinate x and the time t. This means that surfaces normal to the direction of x (normal surfaces) will be surfaces of constant velocity and state. This assumption also implies that any velocity changes produced by pressure forces will be parallel to the x direction. We shall therefore make the additional postulate that the flow velocity is everywhere parallel to x.[1]

The equation of continuity (conservation of mass) can now be written. Considering a volume element bounded by streamlines and normal surfaces, the statement that the difference between mass inflow and outflow rates equals the rate of accumulation of mass in the element can be written

$$A\frac{\partial\rho}{\partial t} + \frac{\partial}{\partial x}(\rho u A) = 0 \tag{2-1}$$

where ρ and u are the gas density and velocity and A is the area of the

[1] By this postulate we exclude a group of one-dimensional flows in which there is a velocity normal to x, which is a function of x and t only. For example, we have excluded the case of a cylindrical wave propagating outward from the center of a vortex.

normal surface of the element. Note that we are considering the streamlines fixed in time so that A is a function of x only.[2]

The equation of motion can also be written simply if we consider that only pressure forces act on the fluid (i.e. if we neglect viscous and body forces). Following a particle of fluid, its rate of change of velocity with time is

$$\frac{Du}{Dt} = \frac{\partial u}{\partial t} + \frac{\partial u}{\partial x}\frac{Dx}{Dt} \tag{2-2}$$

(D denotes derivatives taken following a particle and Dx/Dt is just the particle velocity.) The rate of change of the momentum of an element of volume dV is $\rho(Du/Dt)dV$. The net pressure force on the particle is $-(\partial p/\partial x)dV$. Equating the force to the rate of change of momentum we obtain

$$\frac{\partial u}{\partial t} + u\frac{\partial u}{\partial x} + \frac{1}{\rho}\frac{\partial p}{\partial x} = 0 \tag{2-3}$$

Note that viscous forces, which have been neglected, will be present in general (see Sec. D and H). However, these forces are very small except when the velocity gradient changes appreciably in a molecular mean free path. Such rapid changes occur, for example, inside shock waves and in rarefied gas flow problems. Shock waves will be treated separately in this analysis and we shall in other cases restrict ourselves to those cases where the mean free path is very small compared to the other dimensions in the problem.

The entropy change per unit mass in a thermodynamically reversible elementary process is given by

$$ds = \frac{dq}{T} = \frac{de + pdv}{T} \tag{2-4}$$

where dq is the heat added, de is the change in internal energy, dv is the change in specific volume, and T is the absolute temperature. For a perfect gas this can be written

$$\begin{aligned} ds &= c_v d \ln T + \mathcal{R} d \ln v = 2c_v d \ln a - \mathcal{R} d \ln \rho \\ &= 2c_p d \ln a - \mathcal{R} d \ln p = c_v d \ln p - c_p d \ln \rho \end{aligned} \tag{2-5}$$

where $a = \sqrt{(c_p/c_v)\mathcal{R}T} = \sqrt{\gamma \mathcal{R} T} = \sqrt{\gamma p/\rho}$ (a will later be identified with the local speed of sound) and c_v, c_p, and $\mathcal{R}$ are the heat capacities and gas constant per unit mass. Since entropy is a function of the thermodynamic state alone, Eq. 2-5 (or an integrated form of it) will apply even though the entropy change was produced irreversibly. It is necessary

[2] Problems such as the propagation of pressure waves in fluids bounded by elastic tubes are thereby excluded. The propagation of waves in elastic tubes can be treated by numerical integration along characteristics (see Guderley [9]).

only that the initial and final states be at equilibrium for the entropy change to have meaning. (Otherwise the irreversible process cannot be replaced by a reversible one to calculate the entropy change.)

It will be convenient to rewrite the equation of continuity (Eq. 2-1) and the equation of motion (Eq. 2-3) in terms of a and s as thermodynamic state variables in place of p and ρ. Using the relations (Eq. 2-5), Eq. 2-1 becomes

$$n \frac{\partial \ln a}{\partial t} + un \frac{\partial \ln a}{\partial x} + u \frac{\partial u}{\partial x} + \frac{d \ln A}{dx} - \frac{1}{\mathcal{R}}\left(\frac{\partial s}{\partial t} + u \frac{\partial s}{\partial x}\right) = 0 \quad (2\text{-}6)$$

where $n = 2c_v/\mathcal{R}$, the classical number of degrees of freedom of the gas molecules. Eq. 2-3 becomes, using the perfect gas law $p/\rho = \mathcal{R}T = a^2/\gamma$,

$$\frac{\partial u}{\partial t} + u \frac{\partial u}{\partial x} + a^2 n \frac{\partial \ln a}{\partial x} - \frac{a^2}{\gamma \mathcal{R}} \frac{\partial s}{\partial x} = 0 \quad (2\text{-}7)$$

We assume now that the c_v and c_p are constant, i.e. independent of temperature.[3] These equations can then be put into a more useful form by making a transformation introduced by Riemann [*3*]. Multiplying Eq. 2-6 by a and adding it to Eq. 2-7, we obtain

$$\frac{\partial}{\partial t}(na + u) + (u + a)\frac{\partial}{\partial x}(na + u) + ua \frac{d \ln A}{dx} - \frac{a^2}{\gamma \mathcal{R}} \frac{\partial s}{\partial x} - \frac{a}{\mathcal{R}}\left(\frac{\partial s}{\partial t} + u \frac{\partial s}{\partial x}\right) = 0 \quad (2\text{-}8)$$

Similarly, subtracting from Eq. 2-7 yields

$$\frac{\partial}{\partial t}(na - u) + (u - a)\frac{\partial}{\partial x}(na - u) + ua \frac{d \ln A}{dx} + \frac{a^2}{\gamma \mathcal{R}} \frac{\partial s}{\partial x} - \frac{a}{\mathcal{R}}\left(\frac{\partial s}{\partial t} + u \frac{\partial s}{\partial x}\right) = 0 \quad (2\text{-}9)$$

These equations are sufficiently general to cover most of the problems treated in this section. However, it will be easier to exhibit their properties by first simplifying them greatly (to the case of plane, small amplitude, isentropic disturbances) to exhibit their essential wave propagation properties (Art. 3). Following this we shall reinstate the nonlinear terms which become important for large amplitude disturbances and exhibit the characteristic nonlinear wave motion (Art. 4). The shock waves (Art. 5) which form from the large amplitude disturbances will be treated in Art. 6 and 7, first by approximate analytic methods valid for weak and very strong (Art. 8) shocks, and later by numerical methods (Art. 9) for shocks of moderate strength.

[3] For the case of a gas with variable heat capacity, see Art. 9.

C,3. Acoustic Waves.

Plane acoustic waves. The most striking properties of Eq. 2-8 and 2-9 are that they represent the propagation of waves.[4] These can best be exhibited by considering the case of plane waves of very small amplitude in an isentropic gas. For this case A and s are not functions of x or t, which fact removes the last three terms from Eq. 2-8 and 2-9. The restriction to small amplitude means that the velocity disturbances everywhere in the fluid are considered small compared to $a_0 = \sqrt{\gamma \mathcal{R} T_0}$.

It can be seen that variations in na are of the same order of magnitude as variations in u, since variations of these variables play symmetrical roles in Eq. 2-8 and 2-9 (for s = const). We denote the temperature of the undisturbed gas by T_0 and use a coordinate system moving with the velocity of the undisturbed gas. Then neglecting u and variations of a compared to $a_0 \equiv \sqrt{\gamma \mathcal{R} T_0}$, Eq. 2-8 and 2-9 become[5]

$$\frac{\partial P}{\partial t} + a_0 \frac{\partial P}{\partial x} = 0 \tag{3-1}$$

and

$$\frac{\partial Q}{\partial t} - a_0 \frac{\partial Q}{\partial x} = 0 \tag{3-2}$$

Here we have used the symbols $P = na + u$ and $Q = na - u$.

The general solutions of Eq. 3-1 and 3-2 can be found by noticing that for a point moving with the velocity $+a_0$ Eq. 3-1 yields the result that P = const. Thus

$$P = P(x - a_0 t) \tag{3-4}$$

Similarly, for a point moving with the velocity $-a_0$, Eq. 3-2 yields Q = const. Thus

$$Q = Q(x + a_0 t) \tag{3-5}$$

This behavior is illustrated in Fig. C,3. Here we consider a small region of initially disturbed gas (at $t = 0$), surrounded by undisturbed medium.

[4] This article is not intended as a treatment of acoustics but merely as an illustration of wave propagation. For treatments of acoustics, consult Rayleigh [*10*] or Morse [*11*].

[5] These equations can be reduced to the usual form of the wave equation by differentiating each of them with respect to x (and multiplying by a) and with respect to t. If the resultant equations are added and subtracted, so as to eliminate the mixed partial derivatives, we obtain

$$\begin{aligned} \frac{\partial^2 P}{\partial t^2} - a_0^2 \frac{\partial^2 P}{\partial x^2} &= 0 \\ \frac{\partial^2 Q}{\partial t^2} - a_0^2 \frac{\partial^2 Q}{\partial x^2} &= 0 \end{aligned} \tag{3-3}$$

which are the wave equations. Similar equations can also be found for u, a, or ρ.

Note that we need a specification of the velocity and one thermodynamic variable. The latter is readily transformed into a variation of a using the relations in Eq. 2-5, since s = const. Now from the definitions of P and Q these quantities can be found at $t = 0$ in the disturbed region (outside the disturbed region $P = Q = na_0$) for the given initial conditions. Then P and Q are known everywhere for all later times from our solutions (Eq. 3-4 and 3-5). From these u and a can be found at all x and t since

$$u = \frac{P - Q}{2} \quad \text{and} \quad a = \frac{P + Q}{2n}$$

General disturbances of u and a result in disturbed values of both P and Q. However, if $u = n(a - a_0)$ then Q is undisturbed and, similarly,

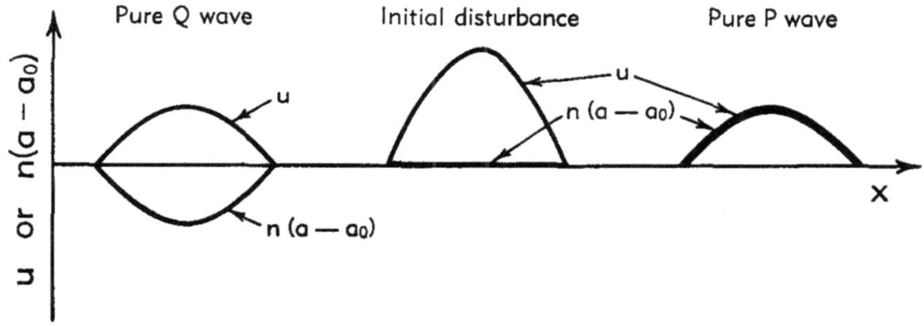

Fig. C,3. Illustrating the propagation of sound from a finite initially disturbed region (center). The initial disturbance is chosen in this case to be a simple velocity u disturbance with no temperature, a, disturbance. The conditions after a time t are shown at the sides. At this time there is no disturbance in the center.

if $-u = n(a - a_0)$ then P is undisturbed. In these cases we say that we have *simple waves* or pure P and Q waves respectively. Generalizing from Fig. C,3, it is seen that a finite arbitrarily disturbed region eventually splits into a pure P pulse[6] moving with the velocity of sound a_0 in the $+x$ direction and a pure Q pulse moving in the $-x$ direction.

In simple waves the pressure changes are uniquely related to the velocity since $u = \pm n(a - a_0) \cong \pm(a_0/\gamma p_0)(p - p_0) = \pm(p - p_0)/\rho_0 a_0$ (where $\gamma = c_p/c_v$) from the thermodynamic relations (Eq. 2-5). The + signs refer to pure P waves and the − signs to pure Q waves. Note that even in the solution of simple wave problems it is necessary to use both Eq. 3-4 and 3-5. However, in such cases, either Eq. 3-5 becomes $Q = Q_0$ (pure P wave) or Eq. 3-4 becomes $P = P_0$ (pure Q wave).

The utility of the quantities P and Q (which are sometimes called *Riemann invariants*) rests in the fact that they represent the disturbances moving in the positive and negative x directions even in an arbitrary nonsimple wave case. Thus, quite generally, P and Q are propagated un-

[6] The word *pulse* is used to denote a propagating disturbance which is confined to a finite region of space with neighboring regions undisturbed.

changed but this is not true for u or a except for the case of a simple wave. Extensive use will be made of these quantities throughout this section.

The power transmitted by a sound wave can be readily calculated. The work done per unit time by a particle on a neighboring particle is the product of the force and the velocity. Then the power transmitted in the $+x$ direction is pu per unit area. We divide this power into two parts $(p - p_0)u$ and p_0u, where p_0 is undisturbed pressure. The term p_0u represents the net flow work transmitted. This quantity will vanish for processes in which there is no net displacement of the fluid particles (e.g. cyclic phenomena). The term $(p - p_0)u$ is called the wave energy transmission. For a small amplitude isentropic wave,

$$\frac{(p - p_0)u}{p_0} \cong (n + 2)\,\frac{(a - a_0)u}{a_0}$$

and

$$(p - p_0)u = \frac{\gamma}{4}\frac{p_0}{a_0}[(P - P_0)^2 - (Q - Q_0)^2]$$

It is thus clear that the net wave energy transmission can be resolved into parts

$$\frac{\gamma}{4}\frac{p_0}{a_0}(P - P_0)^2 \quad \text{and} \quad \frac{\gamma}{4}\frac{p_0}{a_0}(Q - Q_0)^2$$

moving in the x and $-x$ directions, even in a region of generalized disturbance.

Spherical sound waves. The acoustic approximation for a spherical wave is easily worked out by the use of the potential. Starting with the equation for the potential (derived in Sec. A, Eq. 9-8)

$$\nabla^2\phi = \frac{1}{a^2}\frac{\partial^2\phi}{\partial t^2} + \frac{2}{a^2}u_i\frac{\partial^2\phi}{\partial t\partial x_i} + \frac{1}{a^2}u_iu_j\frac{\partial^2\phi}{\partial x_j\partial x_i} \qquad \text{(A,9-8)}$$

where ϕ is the velocity potential. Neglecting $u_i^2 \ll a^2$, this becomes the potential equation for sound waves

$$\nabla^2\phi = \frac{1}{a^2}\frac{\partial^2\phi}{\partial t^2}$$

For the case of spherical symmetry,

$$\nabla^2\phi = \frac{\partial^2\phi}{\partial r^2} + \frac{2}{r}\frac{\partial\phi}{\partial r} = \frac{1}{r}\frac{\partial^2(r\phi)}{\partial r^2}$$

where r is the radius. The general solution of this equation can be written as

$$r\phi = f(at - r) + F(at + r)$$

and hence we obtain for outgoing waves

$$u = -\frac{f(at - r)}{r^2} - \frac{f'(at - r)}{r}$$

The pressure is found from the potential, using

$$\frac{\partial \phi}{\partial t} + \frac{u^2}{2} + \frac{\gamma}{\gamma - 1}\frac{p}{\rho} = \frac{\gamma}{\gamma - 1}\frac{p_0}{\rho_0} \qquad \text{(A,9-3)}$$

If we neglect $u^2/2$ compared to the other terms, we obtain

$$\frac{\partial \phi}{\partial t} = \frac{af'(at - r)}{r} = \frac{\gamma}{\gamma - 1}\left(\frac{p_0}{\rho_0} - \frac{p}{\rho}\right)$$

Neglecting the term r^2 in the denominator of the above equation for u, it will be noted that at large distances from a spherically symmetrical source, a simple progressive wave is obtained in which the pressures and velocities produced in the wave decay with $1/r$. It is clear that this approximation can only be used for very weak explosive waves. A better approximation has been made by Whitham [*12*] who expands the velocity and the pressure along each characteristic in a series of terms with ascending negative powers of r. In this way he obtains a solution which is valid for explosion waves in which the entropy increases in the shocks can be neglected. Since this treatment is analogous to that given in Art. 6, it is not covered in detail here.

Amplitude of sound. In concluding our consideration of small amplitude disturbances it may be well to give some numerical data on sound amplitudes. The smallest periodic pressure amplitude detectable by the ear is of the order of 10^{-3} dyne/cm^2 (at about 2000 cycles). On the other hand, the threshold of feeling corresponds to a pressure amplitude roughly 10^6 times greater. Thus these limits of what is ordinarily called sound correspond to pressure fluctuations of 10^{-9} to 10^{-3} atmospheres, which justifies the assumption of small amplitudes in the theory of sound.

C,4. Plane Isentropic Waves of Large Amplitude. We now consider the case where the velocity is not negligible compared to the velocity of sound. At the same time, variations of the thermodynamic parameters (p, T, a, s) in general become significant. We continue for the present the restriction to plane isentropic disturbances. Eq. 2-8 and 2-9 become

$$\frac{\partial P}{\partial t} + (u + a)\frac{\partial P}{\partial x} = 0 \qquad \text{(4-1)}$$

and

$$\frac{\partial Q}{\partial t} + (u - a)\frac{\partial Q}{\partial x} = 0 \qquad \text{(4-2)}$$

These equations differ from the acoustic case, Eq. 3-1 and 3-2, in that we have not neglected u and variations of a compared to the sound velocity. Eq. 4-1 and 4-2 are nonlinear in that the second term in each is second degree in the dependent variables. The solutions of these nonlinear equations also correspond to the propagation of waves, but these differ from the acoustic case in that (even for simple waves) a continuous change in shape occurs.

Since P and Q are functions of x and t only, we can write

$$dP = \frac{\partial P}{\partial x}\, dx + \frac{\partial P}{\partial t}\, dt \tag{4-3}$$

and

$$dQ = \frac{\partial Q}{\partial x}\, dx + \frac{\partial Q}{\partial t}\, dt \tag{4-4}$$

If now we consider a point which moves with the velocity $u + a$ so that the ratio of dx to dt in Eq. 4-3 is $u + a$, we find by Eq. 4-1 that the increment in P vanishes. Similarly, moving with the velocity $u - a$, Q is invariant.

These results can be understood physically by considering an advancing P wave. Consider a main P wave to which we have added small marker "pips," i.e. small discontinuities which can be used to follow the propagation of the wave. Since these marker pips are small disturbances, they propagate with the local velocity of sound (which is appreciably different from the undisturbed velocity of sound) relative to the local fluid. They find themselves immersed in a fluid which is moving with a non-negligible velocity u, i.e. the velocity due to the main P wave in the neighborhood of the pip. The final propagation velocity is the sum $u + a$. Note that we have assumed that the marker pips are pure P disturbances so that they move with the main wave. The result of Eq. 4-1 and 4-2 can therefore be stated: When we move with the local propagation velocity the values of the Riemann invariants are unchanged.

The symbol $\delta^+/\delta t$ is used to represent variations with time when considering a point moving with the velocity $u + a$ in applying Eq. 4-1 (or more generally, Eq. 2-8). The symbol $\delta^-/\delta t$ is used to represent variations following a point moving with velocity $u - a$ in applying Eq. 4-2 (or Eq. 2-9). Then Eq. 4-1 and 4-2 can be written

$$\frac{\delta^+ P}{\partial t} = 0 \tag{4-5}$$

and

$$\frac{\delta^- Q}{\partial t} = 0 \tag{4-6}$$

The treatment of wave propagation problems is most conveniently represented on an x, t diagram in which x is plotted as abscissa and t as ordinate, as in Fig. C,4a. In the x, t diagram, curves with the slope $dt/dx = 1/(u + a)$ connect points where $P =$ const. These lines are called P waves. Similarly curves with slope $1/(u - a)$ connect points of constant Q and are called Q waves.

Notice that

$$u + a = \frac{P}{2}\left(1 + \frac{1}{n}\right) - \frac{Q}{2}\left(1 - \frac{1}{n}\right) \tag{4-7}$$

$$u - a = \frac{P}{2}\left(1 - \frac{1}{n}\right) - \frac{Q}{2}\left(1 + \frac{1}{n}\right) \tag{4-8}$$

With the aid of Fig. C,4a we apply the relations (Eq. 4-5, 4-6, and 4-7) to solve the problem of a simple advancing P pulse. Starting with an

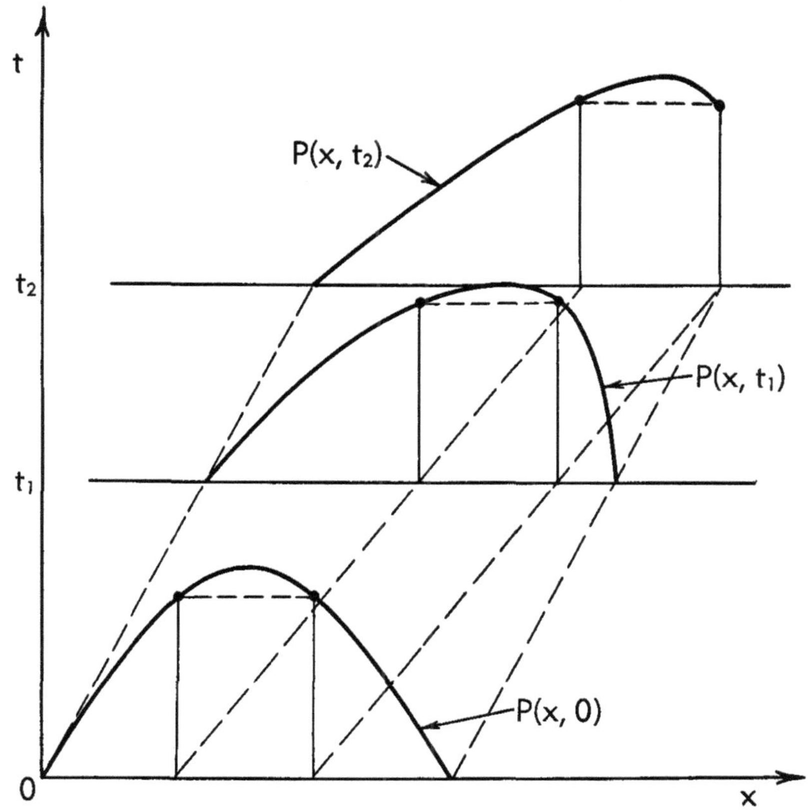

Fig. C,4a. Illustrating the propagation of a pure isentropic P pulse of finite amplitude. The P wave whose shape is determined by Eq. 4-5 and 4-7 is shown at times $t = 0$, $t = t_1$, and $t = t_2$. The distance traveled by each marker pip (indicated by the dashed lines) is $(u + a)t$ where u and a are the local values at each pip. Note that the initial and final markers travel at the undisturbed sound speed a_0, and thus remain a constant distance apart.

initial given distribution of P at time $t = 0$ and the fact that $Q = Q_0$ everywhere, we can trace the history of the pulse. The problem can be solved by merely drawing the paths traveled by portions of the P wave. These slopes are determined from Eq. 4-7 with the given initial data. Thus P and Q $(=Q_0)$ are determined everywhere from Eq. 4-5 and 4-6 and the problem is solved. P is also plotted against x at the later times t_1 and t_2 in Fig. C,4a.

It is seen from this plot that the given P pulse has suffered a considerable change in shape. Note that one portion of the P pulse has become very much steeper while the other has appreciably flattened. If we consider conditions at a constant x station, then the portion of the P pulse in which P was rising with the time has become steeper, while the portion in which P was falling has flattened. A rising P means an increasing u and an increasing a (since Q = const) and pressure. Thus the portion which steepens is a compression and the portion which flattens is an expansion. Since Q and P interchange places if we interchange the $+x$ and $-x$ directions, it is also true for pure Q waves that compressions become continually steeper and expansions become continually flatter. This change in shape is the most characteristic result of the nonlinear nature of our differential equations (Eq. 4-1 and 4-2).

If the propagation of the P pulse is considered at later times than t_1 shown in Fig. C,4a, the P waves in the compression region will eventually meet. This means that at the point where two waves carrying different values of P meet, a discontinuity in P will be found. If the diagram were extended further, P would become multiple-valued so that a simple extension of this diagram beyond the point where a discontinuity first appears would lead to physically meaningless results.

We must note, however, that when the gradient of P first becomes very steep (just before a discontinuity appears), our neglect of viscosity and heat conduction must be modified. A steep gradient in P means that the velocity and the temperature are changing rapidly so that viscous forces appreciably modify the equation of motion (Eq. 2-3). In addition, heat conduction and the viscous dissipation of mechanical energy cause appreciable entropy changes which must be taken into account. These regions of rapidly changing velocity and pressure are known as shock waves. In Sec. D, shock waves are treated in detail and it is shown that the regions where the viscosity and heat conduction are important are only a few mean free paths in thickness. Thus, for many cases, the thickness of the shocks is small compared to other dimensions in the problem. In this section (see Art. 5), we shall take advantage of their small thickness to treat the shocks separately (directly from the conservation laws) and will continue to neglect viscosity and heat conduction outside of shock waves.

The change in shape of a pure P wave can be considered quantitatively as follows. We differentiate Eq. 4-1 with respect to x and obtain

$$\frac{\partial \epsilon}{\partial t} + (u + a)\frac{\partial \epsilon}{\partial x} = -\epsilon\left(\frac{\partial u}{\partial x} + \frac{\partial a}{\partial x}\right) = -\frac{\epsilon^2}{2}\left(1 + \frac{1}{n}\right) \tag{4-9}$$

where $\epsilon = \partial P/\partial x$ and we have used the fact that $\partial Q/\partial x = 0$. We fix our attention on a point which moves with the propagation velocity $u + a$ and consider the change in ϵ with time for this point. If we denote this change by $\delta^+\epsilon/\delta t$, Eq. 4-9 can then be written

$$\frac{\delta^+\epsilon}{\delta t} = -\frac{\epsilon^2}{2}\left(1 + \frac{1}{n}\right) \tag{4-10}$$

which can be immediately integrated to give

$$\frac{\epsilon}{\bar{\epsilon}} = \frac{1}{\left(1 + \frac{1}{n}\right)\frac{t\bar{\epsilon}}{2} + 1} \tag{4-11}$$

where $\bar{\epsilon}$ is the value of ϵ at $t = 0$. Now if $\bar{\epsilon}$ is positive, ϵ decreases toward 0 as the wave propagates. On the other hand, if $\bar{\epsilon}$ is negative, $|\epsilon|$ increases indefinitely, becoming infinite at $t = -2/[1 + (1/n)]\bar{\epsilon}$. This is the time required for a shock to form when we are given the initial gradient of P.

It is possible to prove a useful theorem concerning the propagation of a simple P pulse. In such a pulse, points with a given value of P have the same propagation velocity. Thus markers placed at points with equal values of P remain a constant distance apart as the pulse propagates. From this fact it can be seen that quantities of the form

$$\int f(P)dx$$

do not change as the pulse propagates. This can be shown formally as follows:

Let $f(P)$ be an arbitrary well-behaved function of P. Multiply Eq. 4-1 by $df(P)/dP$, and integrate with respect to x to obtain

$$\frac{\partial}{\partial t}\int f(P)dx + \int \left[\frac{P}{2}\left(1 + \frac{1}{n}\right) - \frac{Q}{2}\left(1 - \frac{1}{n}\right)\right] df(P) = 0 \tag{4-12}$$

Here we have introduced the value of the propagation velocity from Eq. 4-7. The limits of integration are to be taken over the whole of a finite pulse so that P vanishes at both limits. Since $Q = \text{const}$, the second integral then vanishes and we obtain

$$\frac{\partial}{\partial t}\int_{\text{pulse}} f(P)dx = 0 \tag{4-13}$$

As obvious special cases of this theorem, we might note that the mass, energy, and momentum carried with a simple pulse are preserved as the pulse propagates, since the quantities can clearly be expressed as functions of P (Q and s are constant).

The special case where $f(P) = P - P_0$ is of interest in connection with our consideration of the propagation of pulses containing weak shocks and in connection with the study of disturbances superposed on a steady channel flow. We shall call $\int(P - P_0)dx$ the pulse area. The quantity pulse area is used because it is simpler in this formulation than other similar quantities such as the mass, energy, or the momentum carried by the pulse. The flow of pulse area past a point can readily be found by first transforming Eq. 4-1 (with the aid of Eq. 4-7) into

$$\frac{\partial}{\partial t}(P - P_0) + \frac{\partial}{\partial x}\left\{(P - P_0)\left[\frac{P - P_0}{4}\left(1 + \frac{1}{n}\right) + \frac{P_0}{n}\right]\right\} = 0 \quad (4\text{-}14)$$

where $P_0 = Q_0 = na_0$. Integrating Eq. 4-14 with respect to x we obtain

$$\frac{\partial}{\partial t}\int_A^B Pdx + \left\{(P - P_0)\left[\frac{P - P_0}{4}\left(1 + \frac{1}{n}\right) + \frac{P_0}{n}\right]\right\}_a^b = 0 \quad (4\text{-}15)$$

If we take the point A in the undisturbed region, then we see from Eq. 4-13 and 4-15 that the expression

$$\left\{(P - P_0)\left[\frac{P - P_0}{4}\left(1 + \frac{1}{n}\right) + \frac{P_0}{n}\right]\right\} \quad (4\text{-}16)$$

represents the rate at which the pulse area flows past a point B, since Eq. 4-16 is the rate of accumulation of pulse area between the points A and B.

To conclude this article we offer some examples of simple waves which occur in practice. Consider (see Fig. C,4b) that we have an infinite straight tube containing gas at rest and that the negative end is bounded by a piston. If the piston is accelerated gradually to a velocity u, then a simple P wave will propagate in the plus x direction. The wave is readily calculated from the fact that $Q = Q_0$ everywhere. We know the piston velocity as a function of time, and since $Q_0 = na_0 = na - u$ we can find P at the surface of the piston as a function of t. Remembering that $\delta^+ P/\delta t = 0$, P is determined everywhere. Note that eventually a shock is formed as the wave propagates. The shock produces variations in Q (Q waves). However, until the Q waves produced by the shock reach the piston, we can readily determine the pressure on the piston as a function of its velocity. From $Q = Q_0$ we get $\Delta a = (1/n)u$. Since the flow is isentropic, we have

$$\frac{p}{p_0} = \left(\frac{a}{a_0}\right)^{n+2}$$

and finally:

$$\frac{p}{p_0} = \left(1 + \frac{1}{n}\frac{u}{a_0}\right)^{n+2} \quad (4\text{-}17)$$

In Fig. C,4b is plotted the pressure ratio obtained in this way versus the ratio of the piston velocity to the undisturbed velocity of sound. If we consider a case in which we have a gas flowing steadily through a long tube at a velocity u, an interesting comparison can be made between the pressure rise obtained when the flow is stopped by a simple wave, such as would be created by gradually shutting off the tube with a valve, and the pressure rise obtained in an impact tube inserted in the flow. In the case

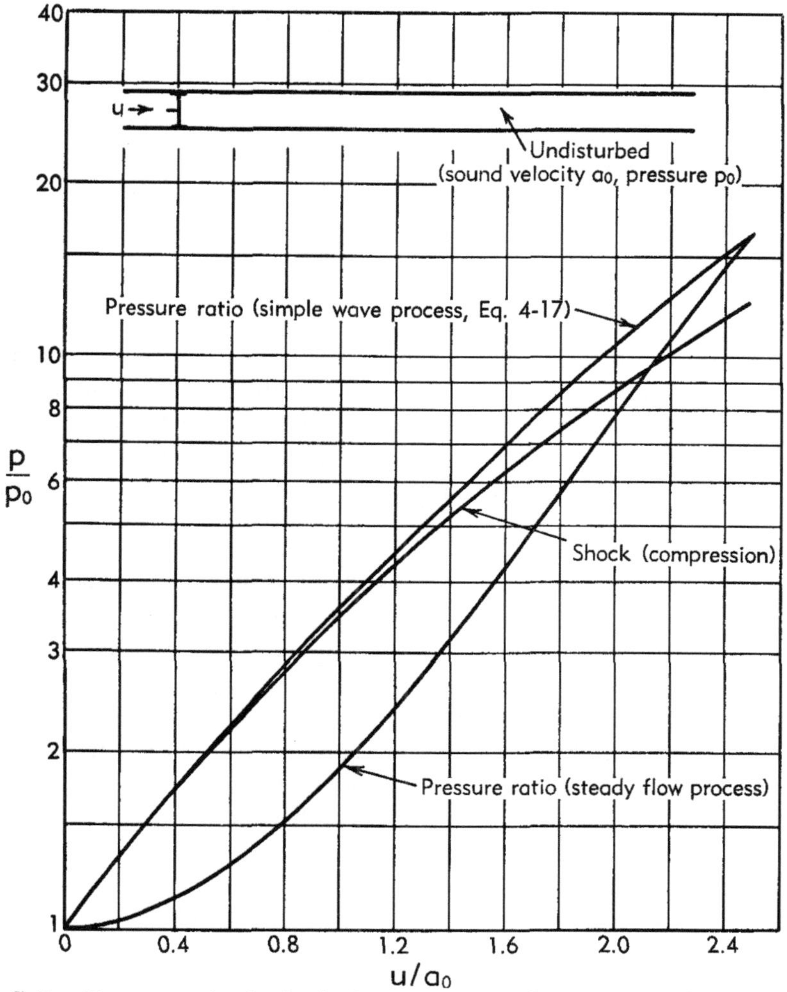

Fig. C,4b. Pressure ratio obtained when a gas is accelerated to a velocity u by a compressive simple wave. This is compared with the ratio of stagnation to static pressures for a steady flow process in which the transition is accomplished using both P and Q waves. For comparison the pressure ratio produced by a sudden acceleration to a velocity u, which produces a shock, is also plotted. $\gamma = 1.40$.

of the impact tube, we have assumed that the flow in the immediate vicinity of the tube is steady and have employed the standard steady flow relationship between the total head and the static pressure. Notice that for small velocities the pressure rise can be obtained from the acoustic approximation $p - p_0 = \rho u a_0$ and is therefore proportional to the first power of the velocity. On the other hand, the pressure rise in the impact tube is $\frac{1}{2}\rho u^2$ for low velocities and is proportional to the second power of the velocity. Thus for small velocities the pressure obtained by shutting off the tube suddenly is much larger than that obtained by stagnating the gas in a steady flow.

The relation between velocity and pressure changes for a steady flow is

$$dp = -\rho u du \tag{4-18}$$

as can be seen from Eq. 2-3 by neglecting the variation of u with time. On the other hand, in the unsteady flow, considering now that a positive u is altered by a simple Q wave so that $P = P_0$, the variation of pressure with velocity is given by

$$dp = (n + 2)p \frac{da}{a} = -\frac{\gamma p}{a} du = -\rho a du \tag{4-19}$$

Comparison of Eq. 4-18 and 4-19 shows that when the velocity (relative to the coordinate system used in the steady flow) is subsonic then the pressure change per unit velocity change is larger for the simple wave. On the other hand, for supersonic steady flows, the pressure change per unit velocity change is larger than for simple waves. This behavior can be seen in Fig. C,4b.

It is possible that the relatively high pressures which can be obtained by stopping a subsonic gas flow could be useful in practical aeronautics. In cases where it is desired to obtain a pressure higher than stagnation pressure, this could be done by (1) allowing air to flow through a tube at essentially the flight velocity, (2) interrupting the air flow periodically with a valve at the downstream end, and (3) coordinating with this valve a mechanism to draw off the high pressure air during the time that the wave is propagating in the upstream region of the tube. When the stopping wave reaches the upstream end, it is reflected as a rarefaction which reduces the pressure. Before this rarefaction reaches the downstream end of the tube, the tube should be shut off from the high pressure air. Then the downstream valve should be opened, which will allow the flow in the tube to return to its initial velocity, whereupon the process can be repeated.

C,5. Shock Waves. Approximations for Weak Shocks. We have seen in Art. 4 that shock waves develop rapidly in the propagation of waves of finite amplitude. In this article we set forth the Rankine-Hugoniot theory of shock waves and derive forms of the shock wave equations suitable for our later applications. We shall depend upon the assumption that the shock is very thin compared to the other dimensions in the problem as mentioned earlier. In this case we may choose planes on both sides of the shock far enough from the shock so that heat conduction and viscosity can be neglected at these planes, which are at the same time close enough together so that any accumulation of mass, energy, or momentum between them can be neglected. Designating planes so chosen by the subscripts 1 and 2, referring to fluid before and after passage through the shock, we may write the equation of conservation of mass as

$$\rho_1(u_1 - U) = \rho_2(u_2 - U) \tag{5-1}$$

where U is the shock velocity. The equation of the conservation of energy crossing the planes 1 and 2 may be written as

$$na_1^2 + (u_1 - U)^2 = na_2^2 + (u_2 - U)^2 \tag{5-2}$$

Finally, the equation of the conservation of momentum is

$$p_1 + \rho_1(u_1 - U)^2 = p_2 + \rho_2(u_2 - U)^2 \tag{5-3}$$

These equations together with the perfect gas equation $\gamma p/\rho = a^2$ are sufficient to determine the conditions behind the shock (plane 2) in terms of the conditions ahead of the shock and the shock velocity.

More generally, there are seven quantities p_1, ρ_1, u_1, p_2, ρ_2, u_2, and U connected by three relations. Thus given any four quantities or independent relations among these we may determine the remaining three. These equations can be combined to give

$$(u_1 - U)(u_2 - U) = \frac{(u_1 - U)^2 \frac{1}{n} + a_1^2}{1 + \frac{1}{n}} = a^{*2} \tag{5-4}$$

where a^* is known as the critical speed of sound (shock coordinates). Eq. 5-4 is known as Prandtl's relation.

Solving Eq. 5-1, 5-2, and 5-3 simultaneously, we can also obtain (see Courant and Friedrichs [*6*, Sec. 67]):

$$\left(1 + \frac{1}{n}\right)\frac{u_2 - u_1}{a_1} = \frac{U - u_1}{a_1} - \frac{a_1}{U - u_1} \tag{5-5}$$

Eq. 5-5 can be written

$$\frac{U - u_1}{a_1} = \left(1 + \frac{1}{n}\right)\frac{u_2 - u_1}{2a_1} \pm \sqrt{1 + \left(1 + \frac{1}{n}\right)^2 \frac{(u_2 - u_1)^2}{4a_1^2}} \tag{5-6}$$

Note that the $\pm$ signs correspond in the limit where $u_2 = u_1$ to $U = u_1 \pm a_1$, i.e. to the propagation velocities of P and Q waves. We therefore refer to the shocks obtained by using the upper or lower signs in Eq. 5-6 as P shocks and Q shocks respectively.

In Art. 6 we shall require an approximation for the velocity of weak shocks. Expanding Eq. 5-6 with the use of the binomial theorem we obtain for P shocks

$$U = u_1 + a_1 + \left(1 + \frac{1}{n}\right)\left(\frac{u_2 - u_1}{2}\right) + \left(1 + \frac{1}{n}\right)^2 \frac{(u_2 - u_1)^2}{8a_1} + \cdots \quad (5\text{-}7)$$

which is valid to the second order in the shock strength $(u_2 - u_1)$.

From Eq. 2-5 we may readily obtain the entropy increase across a shock wave as

$$s_2 - s_1 = c_v \ln \left(\frac{p_2}{p_1}\right)\left(\frac{\rho_1}{\rho_2}\right)^\gamma \quad (5\text{-}8)$$

Following Liepmann and Puckett [*13*, p. 41] it can readily be shown that if the entropy increase is expanded in a series with ascending powers of the shock strength, the first nonvanishing term is the third power term. The pressure ratio across the shock can be written as

$$\frac{p_2}{p_1} = \frac{2\gamma}{\gamma + 1} M_1^2 - \frac{\gamma - 1}{\gamma + 1} \quad (5\text{-}9)$$

where

$$M_1^2 = \left(\frac{U - u_1}{a_1}\right)^2 = 1 + m$$

Note that from Eq. 5-7,

$$m = \left(1 + \frac{1}{n}\right)(u_2 - u_1)$$

to first order accuracy, so that m may be used as a measure of shock strength also. Then the entropy increase can be written

$$\frac{s_2 - s_1}{c_v} = \ln \left(1 + \frac{2\gamma}{\gamma + 1} m\right)\left(1 + \frac{\gamma - 1}{\gamma + 1} m\right)^\gamma (1 + m)^{-\gamma} \quad (5\text{-}10)$$

Expanding Eq. 5-10 in a power series in m we obtain

$$\frac{s_2 - s_1}{c_v} = \frac{2\gamma(\gamma - 1)}{(\gamma + 1)^2} \frac{m^3}{3} + \cdots \quad (5\text{-}11)$$

Writing Eq. 5-2 in terms of the Riemann invariants before and after the shock wave, we obtain

$$n\left(\frac{P_1 + Q_1}{2n}\right)^2 + \left(\frac{P_1 - Q_1}{2} - U\right)^2 = n\left(\frac{P_2 + Q_2}{2n}\right)^2 + \left(\frac{P_2 - Q_2}{2} - U\right)^2 \quad (5\text{-}12)$$

This can be written as

$$(\Delta P)^2 - \frac{2n-2}{n+1}\Delta P\Delta Q + \Delta P\left(2P_1 - \frac{2n-2}{n+1}Q_1 - \frac{4}{1+\frac{1}{n}}U\right)$$

$$+ \Delta Q\left(2Q_1 - \frac{2n-2}{n+1}P_1 + \frac{4}{1+\frac{1}{n}}U\right) + (\Delta Q)^2 = 0 \quad (5\text{-}13)$$

where $\Delta P = P_2 - P_1$ and $\Delta Q = Q_2 - Q_1$. This is a quadratic equation in ΔP and may be put in the form

$$(\Delta P)^2 + b\Delta P + c = 0 \quad (5\text{-}14)$$

where

$$b = 2P_1 - \frac{2n-2}{n+1}Q_1 - \frac{4}{1+\frac{1}{n}}U - \frac{2n-2}{n+1}\Delta Q$$

$$= \frac{4a_1}{1+\frac{1}{n}}(M_1+1) - \frac{2n-2}{n+1}\Delta Q$$

and

$$c = \Delta Q\left[2Q_1 - \frac{2n-2}{n+1}P_1 + \frac{4U}{1+\frac{1}{n}} + \Delta Q\right]$$

$$= \Delta Q\left[\frac{4a_1}{1+\frac{1}{n}}(1-M_1) + \Delta Q\right]$$

With the aid of Eq. 5-14 it is possible to show that, for weak Q shocks, ΔP is third order in ΔQ, and vice versa. The solution to Eq. 5-14 can be written

$$\frac{2\Delta P}{b} = -1 \pm \sqrt{1 - \frac{4c}{b^2}} \quad (5\text{-}15)$$

Thus, using the $-$ sign, ΔP is of larger order than ΔQ. On the other hand, using the $+$ sign, ΔQ is of larger order than ΔP. These cases therefore represent P shocks and Q shocks respectively. Taking the $+$ sign and expanding the radical by the use of the binomial theorem we obtain

$$\frac{2\Delta P}{b} = -\frac{2c}{b^2} - \frac{2c^2}{b^4} + \cdots \quad (5\text{-}16)$$

From its definition, b is a zero order quantity. To obtain the order of magnitude of c we expand Eq. 5-2, and for weak Q shocks we obtain to first order (at least)

$$n\Delta a + \Delta u = 0 \tag{5-17}$$

With the aid of Eq. 5-17 and 5-7 it is seen that c is third order in ΔQ. Therefore ΔP is third order in ΔQ for Q shocks, and vice versa. This fact,

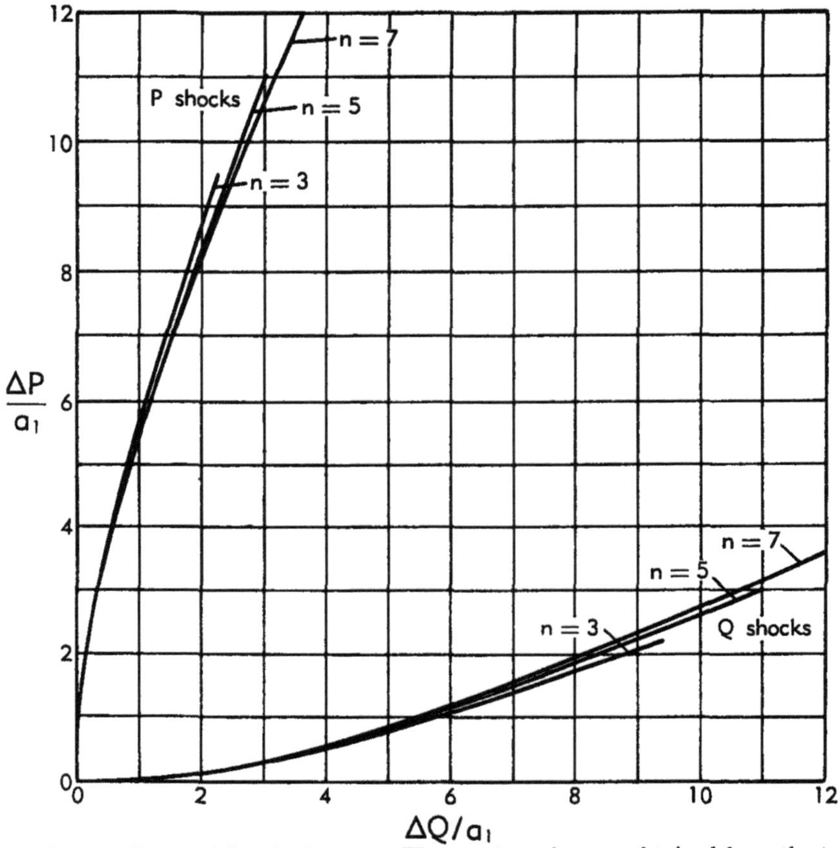

Fig. C,5a. ΔP vs. ΔQ for shock waves. The two branches are obtained from the two solutions of Eq. 5-14. The solutions go over in the limit of vanishing shock strength to P and Q waves; hence for finite strength they are called P and Q shocks. This graph is intended for use in connection with numerical integration along characteristics (see Art. 9).

together with the fact that the entropy increase is third order in the shock strength, makes it possible to treat simple waves analytically even in the presence of small shocks (see Art. 6).

With the aid of the fact that ΔQ is third order in ΔP (so that to third order $u_2 - u_1 = n(a_2 - a_1)$) Eq. 5-7 can be written in the alternative forms, for P shocks,

$$\begin{aligned} U &= \tfrac{1}{2}(u_1 + a_1 + u_2 + a_2) + \frac{1}{8a_1}(u_2 + a_2 - u_1 - a_1)^2 + \cdots \\ &= u_1 + a_1 + \frac{1 + (1/n)}{4}\Delta P + \frac{[1 + (1/n)]^2}{32a_1}(\Delta P)^2 + \cdots \end{aligned} \tag{5-18}$$

with a similar form for Q shocks. Notice that to first order the shock velocity is an average of the propagation velocities ahead and behind.

In Art. 9 we shall discuss a numerical method for integrating the equations of motion in the presence of shock waves of moderate strength,

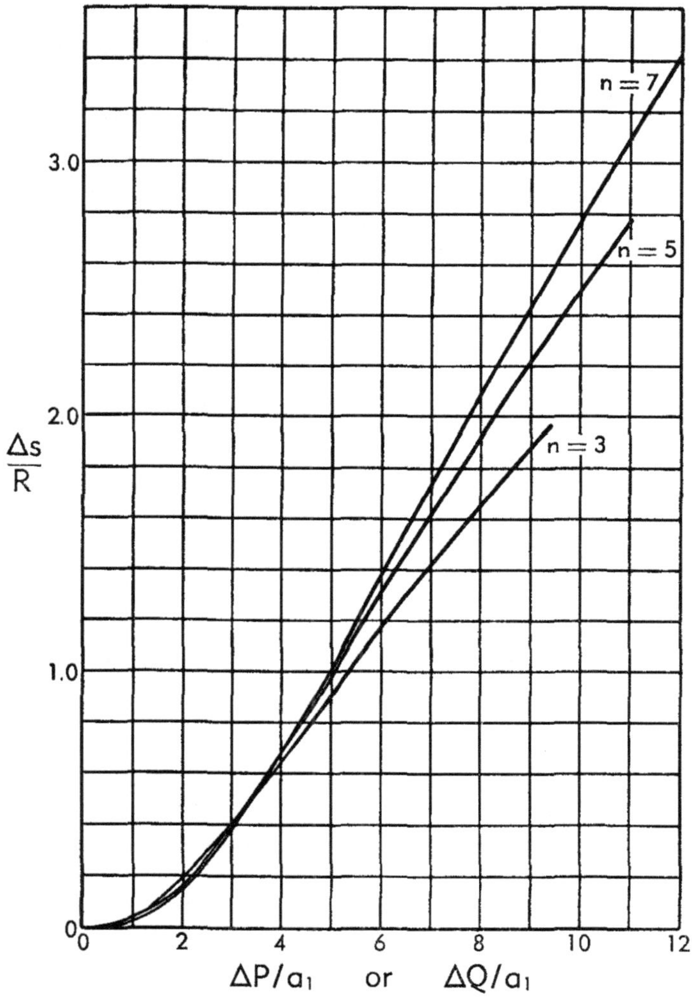

Fig. C,5b. The entropy increase as a function of shock strength from Eq. 5-8. (For use with Art. 9.)

etc. For this purpose it is convenient to have graphical representations of Eq. 5-6, 5-8, and 5-14.

If we use a coordinate system fixed with respect to the shock wave, we notice that shock waves form a single parameter (e.g. M_1) family. This simplicity is not lost if we use, in our graphical representation, quantities which are invariant to changes in the coordinate system velocity. Examples of such quantities are ΔP, ΔQ, $(P - U)$, $(Q + U)$, Mach number relative to the shock (i.e. M_1), velocity of sound, and other thermodynamic quantities. For use in connection with Art. 9, plots of the nondimensional invariant quantities $\Delta P/a_1$ versus $\Delta Q/a_1$ from Eq. 5-14 are given in Fig. C,5a. Similarly, Fig. C,5b is a plot of $\Delta S/R$ versus

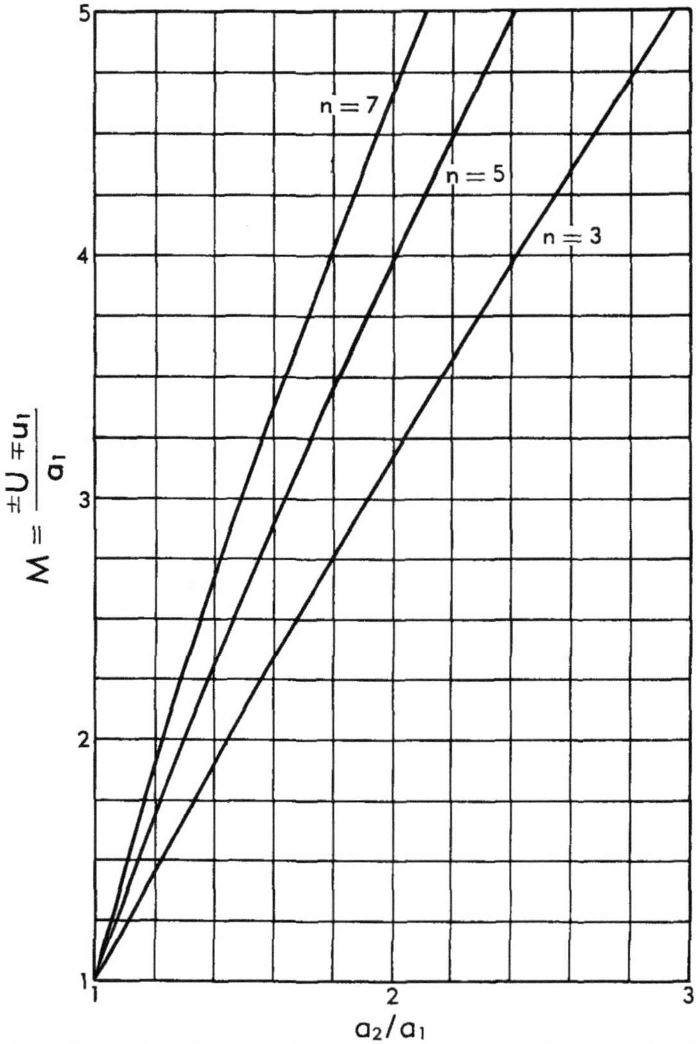

Fig. C,5c. The shock velocity or Mach number vs. a_2/a_1. (For use with Art. 9.)

$\Delta P/a_1$ or $\Delta Q/a_1$ (for P and Q shocks respectively) made from Eq. 5-8. Similarly Fig. C,5c is a plot of $(\pm U \mp u_1)/a_1$, i.e. M_1 versus a_2/a_1.

C,6. Propagation of Simple Waves Containing Weak Shocks.

Growth and decay of weak shocks. In Art. 4 (see Eq. 4-11) it was shown that the compression phases of simple waves steepen to form shocks in finite time. In this article an expression will first be derived for the rate of growth or decay of the shocks formed from the simple wave. The growth behavior becomes especially simple for the case of weak shocks.

In general the presence of a shock formed from a P pulse, for example, will initiate disturbances in Q and entropy. Thus, an initially simple disturbance will become an interacting wave after the formation of a shock. However, it was shown in Art. 5 that the Q and entropy disturbances emitted by a weak shock were third order in the shock strength. For this reason there is a considerable range of shock strengths for which quite accurate results can be obtained without departure from the simple wave theory.

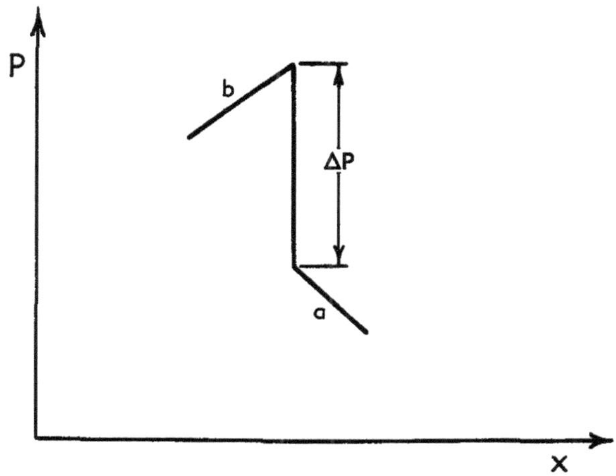

Fig. C,6a. Illustrating the change in strength of a weak shock, forming part of a pure P wave. In time the point b overtakes the shock and simultaneously the shock overtakes the point a. The shock strength thus changes to $P_b - P_a$ (in a constant area channel).

Consider that at time t we have a distribution of P in the neighborhood of a P shock such as is given in Fig. C,6a. It can be shown from Prandtl's relation (Eq. 5-4) that for a P shock

$$u_1 + a_1 < U < u_2 + a_2 \tag{6-1}$$

Therefore the shock velocity lies between the propagation velocity ahead and behind the shock. In fact from Eq. 5-18 the velocity of weak shocks is halfway between the propagation velocities on the two sides of the shock. Referring now to Fig. C,6a it is seen that a P wave originating at point b overtakes the shock with a relative velocity $u_2 + a_2 - U$. At the time when a P wave from the point b has overtaken a shock, the value of P_2 will have changed to the value carried by a P wave through b. In the absence of changes in entropy or area (i.e. $d(\ln A)/dx = 0$ in Eq. 2-8) this is simply the value of P at b. Similarly, the shock overtakes a P wave starting at point a; therefore the value of P_1 changes to that carried by a P wave through a. The changes in P_1 and P_2 in a time dt (in the absence of entropy or area changes) are

$$dP_2 = -\left(\frac{\partial P}{\partial x}\right)_2 (u_2 + a_2 - U)dt \tag{6-2}$$

$$dP_1 = \left(\frac{\partial P}{\partial x}\right)_1 (U - u_1 - a_1)dt \tag{6-3}$$

and the rate of change of shock strength with time is

$$\frac{d\Delta P}{dt} = \frac{dP_2 - dP_1}{dt} \tag{6-4}$$

Using Eq. 5-18 for the propagation velocity of weak shocks, Eq. 6-4 may be written to second order in ΔP:

$$\frac{d\Delta P}{dt} = -\left[\left(\frac{\partial P}{\partial x}\right)_2 + \left(\frac{\partial P}{\partial x}\right)_1\right]\frac{[1 + (1/n)]}{4}\Delta P + \left[\left(\frac{\partial P}{\partial x}\right)_2 - \left(\frac{\partial P}{\partial x}\right)_1\right]\frac{[1 + (1/n)]^2(\Delta P)^2}{32a_1} \tag{6-5}$$

Consider now the later history of the pulse shown in Fig. C,4a. This pulse was drawn so that the shock first forms at the leading edge (head) of the pulse so that $(\partial P/\partial x)_1 = 0$. The P waves from the remainder of the pulse then gradually overtake the shock, at first increasing and, later, decreasing its strength. Note that finally the pulse consists of the shock followed by the P waves which originated very close to the trailing edge of the pulse.

The pulse area. Our understanding of the propagation of simple pulses containing weak shocks can be improved by study of the pulse area $\int(P - P_0)dx$, which was introduced previously. In Art. 4 it was shown that quantities of this type are exactly preserved for simple isentropic pulses. We will now show that the rate of change of pulse area caused by the presence of weak shocks in the pulse is third order in the shock strength.

The pulse area consumed by the advancing shock in region 1 is made up of that overtaken by the advancing shock $U(P - P_0)$ minus the amount of pulse area which flows through a point just ahead of the shock (Eq. 4-16), which is

$$U(P_1 - P_0) - (P_1 - P_0)\left[\frac{P_1 - P_0}{4}\left(1 + \frac{1}{n}\right) + \frac{P_0}{n}\right] \tag{6-6}$$

Similarly, the pulse produced at the rear of the shock (region 2) is equal to $U(P - P_0)$ minus the pulse which flows into the shock as found from

Eq. 4-16. The net production of pulse area is

$$U\Delta P - (P_2 - P_0)\left[\frac{P_2 - P_0}{4}\left(1 + \frac{1}{n}\right) + \frac{P_0}{n}\right] + (P_1 - P_0)\left[\frac{P_1 - P_0}{4}\left(1 + \frac{1}{n}\right) + \frac{P_0}{n}\right] \quad (6\text{-}7)$$

Now inserting the expression for U (Eq. 5-18) in Eq. 6-7, we find that the net rate of creation of pulse area by the shock is

$$\frac{[1 + (1/n)]^2}{32a_1} (\Delta P)^3 + \cdots \quad (6\text{-}8)$$

It is thus seen that the rate of change of pulse area is third order in the shock strength.

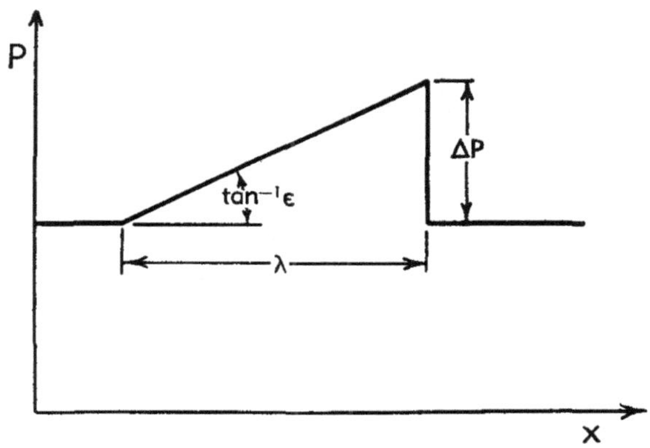

Fig. C,6b. The final triangular shape approached by a decaying pulse whose initial form is given by Fig. C,4a. As the pulse propagates, λ increases and ΔP decreases in proportion to $\sqrt{t}$. Thus the pulse area $\frac{1}{2}\Delta P\lambda$ remains constant.

The progress of a simple pulse can now be easily sketched with the use of Eq. 4-11 and the rule of conservation of pulse area. Returning now to the later history of the pulse shown in Fig. C,4a, all of the P waves except those from the trailing edge of the pulse will have caught up to the shock formed near the leading edge. Thus the pulse will in its later stages have a triangular shape as shown in Fig. C,6b, where the slope $\partial P/\partial x = \epsilon$ of the entire expansion region is to be determined from Eq. 4-11, using a value of $\bar{\epsilon}$ appropriate to the trailing edge of the pulses. The shock position and magnitude can then be found from the pulse area rule. The pulse area for a triangular pulse,

$$\frac{(\Delta P)^2}{2\epsilon} = \frac{(\Delta P)^2}{2}\left[\left(1 + \frac{1}{n}\right)\frac{t}{2} + \frac{1}{\bar{\epsilon}}\right] \quad (6\text{-}9)$$

is equal to the original pulse area plus the integral of the rate of creation (Eq. 6-8) over the period following the formation of the shock. Neglecting this latter quantity, we find that the shock strength is after $t \gg 2/[1 + (1/n)]\bar{\epsilon}$,

$$(\Delta P)^2 t = \frac{4}{1 + (1/n)} \int_{\text{pulse}} (P - P_0) dx \tag{6-10}$$

where the integral is taken over the whole of the initial pulse. The extension of the disturbed regions, λ (Fig. C,6b) is, to the first order,

$$\lambda = \frac{\Delta P}{\epsilon} = \left[\left(1 + \frac{1}{n}\right)\frac{t}{2} + \frac{1}{\bar{\epsilon}}\right]\left[\frac{4}{[1 + (1/n)]t} \int_{\text{pulse}} (P - P_0) dx\right]^{\frac{1}{2}} \tag{6-11}$$

Thus after a long time $\Delta P \sim t^{-\frac{1}{2}}$ and $\lambda \sim t^{\frac{1}{2}}$.

We can examine the accuracy of this procedure by calculating the time τ_1 required (in a triangular pulse) for the contribution (Eq. 6-8) of the shock to the pulse area to become comparable with the pulse area. This time can then be compared with the time τ_2 required for the shock to decay to half amplitude. The rate (Eq. 6-8) creates a pulse area $\frac{1}{2}\Delta P\lambda$ in a time of the order of magnitude

$$\tau_1 \sim \frac{32\lambda a_1}{(\Delta P)^2[1 + (1/n)]^2}$$

On the other hand from Eq. 6-5,

$$\tau_2 \sim \frac{4\lambda}{\Delta P} \frac{1}{1 + (1/n)}$$

Hence for

$$\frac{\Delta P}{a_1} \ll \frac{8}{1 + \frac{1}{n}}$$

$\tau_1 \gg \tau_2$ and the pulse area rule gives an accurate history of the decay of pulses.

N waves. We shall examine briefly the later history of pulses of two other shapes. First consider the case of an expansion between two compressions such as is shown in Fig. C,6c (N wave). Such a pulse is produced by a finite obstacle passing through the air or by moving a piston into a tube, decelerating it, and finally retracting it and again bringing it to rest. After some time all of the expansion waves run into one or the other of the shocks formed except those formed very near to the point where the piston was at rest. This type of pulse can readily be divided into two triangular pulses, a positive one such as that previously treated, and a negative one which can be treated by almost the same method.

These are divided by a point of no disturbance (to third order) across which there is no propagation of pulse area. The history of such pulses is similar to the history of the triangular pulses considered earlier.

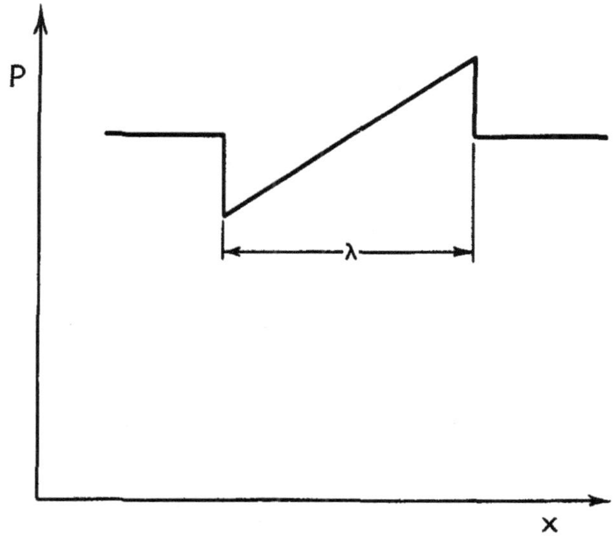

Fig. C,6c. N wave. Such a wave will develop if a piston is accelerated into a tube, then decelerated, retracted, and finally brought to rest at its initial position. The shock amplitudes decay and λ grows as $\sqrt{t}$, as in Fig. C,6b.

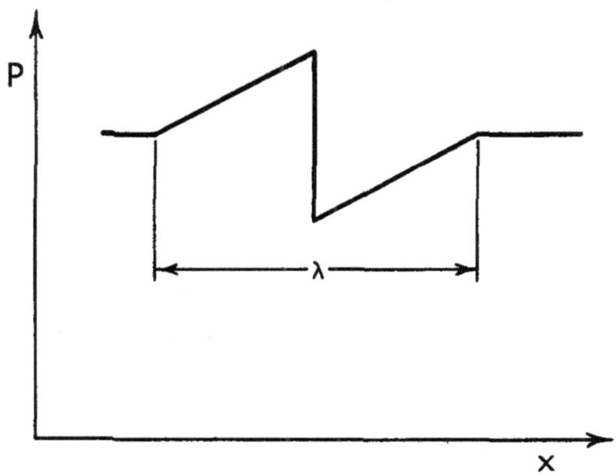

Fig. C,6d. Another type of N wave. A wave of this type can be produced by initially retracting a piston, bringing it to rest, and advancing it to its initial position. The shock amplitude here decays as t, and λ remains constant.

A somewhat different type (Fig. C,6d) can be produced by initially retracting a piston, bringing it to rest, and advancing it to its original position. In this way it is possible to produce a pulse in which a compression lies between two expansions. After some time such a pulse assumes another N shape (Fig. C,6d). Note, however, that in this case the length of the disturbed region remains constant since the leading and trailing waves propagate with the undisturbed speed of sound. The shock strength can then readily be found by computing $\partial P/\partial x$ from Eq. 4-11. Considering the case where there is no net pulse area we see that the shock strength finally decays inversely with the first power of the time, rather than with the square root of the time as in the other cases.

Other treatments of pulses containing weak shocks. It was first pointed out by Chandrasekhar [14] that the progress of certain (P) pulses containing weak shocks can be treated analytically (by a method differing from the above) if we neglect the third order disturbances in entropy and Q and use the approximation (Eq. 5-18) to second order terms for the shock velocity. The integration was carried out somewhat more generally by Friedrichs [15] and his treatment is followed here. We assume that a shock heads a P pulse so that $P_1 = P_0$. At $t = 0$ let P be a given function $P(x)$ of x. Write

$$\xi = x - (u + a)t \tag{6-12}$$

Now $P = P(\xi)$ for all t as long as we consider only points behind the shock. Note that

$$u + a = \left(1 + \frac{1}{n}\right)\left(\frac{P - P_0}{2}\right) + a_0$$

is also a function of ξ only. Also note that

$$\frac{d\Delta P}{dt} = \frac{dP_2}{dt} = \left(\frac{dP}{d\xi}\right)_2 \frac{d\xi_2}{dt} \tag{6-13}$$

where ξ_2 is the value of ξ at the rear of the shock. Finally from Eq. 6-12 (or from Eq. 4-11)

$$\left(\frac{\partial P}{\partial x}\right)_2 = \frac{\left(\dfrac{dP}{d\xi}\right)_2}{1 + \dfrac{[1 + (1/n)]}{2}\left(\dfrac{dP}{d\xi}\right)_2 t} \tag{6-14}$$

Noting that here $a_1 = a_0$, we can write Eq. 6-5 as

$$\left[\left(1 + \frac{1}{n}\right)\frac{\Delta P}{4} - \left(1 + \frac{1}{n}\right)^2 \frac{(\Delta P)^2}{32a_0}\right]\frac{dt}{d\xi} + \left(1 + \frac{1}{n}\right)\frac{1}{2}\left(\frac{dP}{d\xi}\right)_2 t = -1 \tag{6-15}$$

Eq. 6-15 can be integrated to give

$$a_0 t = 8\left[\frac{4 - \left(1 + \dfrac{1}{n}\right)\dfrac{\Delta P}{2a_0}}{\left(1 + \dfrac{1}{n}\right)\dfrac{\Delta P}{2a_0}}\right]^2 \int_{\xi_2}^{\xi_0} \frac{\dfrac{1}{2a_0}\,(P(\mu) - P_0)\left(1 + \dfrac{1}{n}\right)}{\left[4 - \dfrac{1}{2a_0}\left(1 + \dfrac{1}{n}\right)(P(\mu) - P_0)\right]^3}\,d\mu \tag{6-16}$$

where ξ_0 is the value of ξ_2 at $t = 0$.

From Eq. 6-16 the time required for the shock to reach a strength ΔP can be found (obtaining ξ_2 from $\Delta P = P(\xi_2) - P_0$), accurate to the second order in ΔP for a pulse for which $P(\xi)$ is known.

Eq. 6-16 can be reduced to a more surveyable form, accurate to the first order, by neglecting $\Delta P/a_1$ and $(P(\mu) - P_0)/a_1$ compared to $4/[1 + (1/n)]$. We obtain

$$t(\Delta P)^2 = \frac{4}{[1 + (1/n)]} \int_{\xi_2}^{\xi_0} (P(\mu) - P_0)d\mu \tag{6-17}$$

After the pulse is propagated for a long time ξ_2 closely corresponds to the value of ξ for the trailing edge of the pulse. Thus the integral in Eq. 6-17 becomes constant and simply represents the initial pulse area. This then is in complete agreement with Eq. 6-10.

Lighthill [*16*] has made a very careful study of the accuracy of the Friedrichs theory. He shows that, although only third order terms have been neglected, these can over a long period of time ($\sim\tau_2$) integrate to produce a second order effect. This has already been illustrated from a different point of view by considering the expression (Eq. 6-8) for the rate of creation of pulse area by the shock. Thus the error in these analyses is eventually second order.

Lighthill has considered the third order effects, i.e. the entropy increase and the Q disturbance produced at the P shock. He shows that the entropy terms can be more conveniently treated by the use of a function of the pressure in place of na in the characteristic quantities P and Q. He also showed that the Q disturbances can be accounted for as the reflection of the original pulse from the shock wave.

C,7. Formation and Stability of Normal Shock Waves in Steady Channel Flows.

Introduction. There is experimental evidence that steady channel flows involving shock-free deceleration through the speed of sound are unstable. This article presents an analysis of nonviscous unsteady channel flow which was made by Kantrowitz [*17*] to gain some insight into this apparent instability, to provide some information on the factors determining the minimum shock intensity for a stable flow, and to study in general the formation and motion of shocks in channel flows.

Experimental investigation of transonic channel flows has established a striking difference between the processes of acceleration and deceleration through the speed of sound. In a de Laval nozzle, a gas is accelerated to the velocity of sound in the converging (i.e. converging downstream) part of the nozzle and is accelerated further to supersonic velocities in the diverging part. Presuming that the nozzle is shaped to avoid the condensation of compression waves, the acceleration through the speed of sound can be made very smoothly (without shock waves).

As far as is known, the reverse flow, smooth deceleration through the speed of sound in a channel flow, has never been observed. The experimental situation can be summed up briefly as follows: In starting the supersonic flow, a normal shock always forms ahead of the converging part of the channel (supersonic diffuser). By suitably changing the velocities and pressures at the ends of the diffuser, the normal shock can be made to jump to the diverging part. With further changes in velocities, pressures, and/or geometry, the normal shock can be pushed up the diverging channel and its intensity reduced somewhat. However, before it can be made to disappear, it suddenly (to visual observations) jumps to a position ahead of the converging part of the diffuser and the starting process must begin again. The experiments strongly indicate, therefore, that smooth deceleration through the speed of sound in a channel flow is unstable.

The analysis of this apparent instability presented here neglects boundary layer effects entirely. It is well known, however, that these effects are important in steady channel flows involving shock waves. Therefore the results obtained can at best be only qualitatively similar to experiment. Perhaps it will be possible later to include boundary layer effects where they are important following a procedure frequently used in gas dynamics.

The general plan of this treatment can best be explained by considering the central problem, the stability of smooth deceleration through the speed of sound, from a physical point of view. In such a flow, disturbances starting in the rear of the channel will be trapped in the sonic region. The presence of this disturbance trap constitutes a distinct difference between this flow and smooth flows known to be stable.

Two types of processes which govern the behavior of pulses in the sonic region are considered separately in this article. First, there are isentropic processes similar to the propagation of waves of finite amplitude considered in Art. 4. Shocks form in the channel flow as before. If these shock waves are assumed to be of very small intensity, it is shown from first order theory that a pulse approaches a stationary state. Thus the amplitude, shape, and position of a pulse become stationary in the channel. The motion of the shock waves in these "trapped" pulses is considered more accurately later, and it is found that the shock moves, causing the pulse to be consumed or to grow depending on its sign. In cases where the intensity of the shocks involved is not too large, the consumption process is much slower than the trapping process. In these cases, it is possible to obtain a fairly complete history of the behavior of a pulse without considering the interaction of these two processes.

A similar analysis is made for the case where a shock wave forms part of the initial channel flow. In this case, disturbances from the rear serve to move the shock to a new position, and if *changes* in shock strength are neglected, it is found that the pulses in this case are also trapped as shock displacements. Here again, the slower effects on this shock displacement can now be studied, including the changes in shock strength. In this way, the stability of shock waves in channel flows is studied.

Finite plane disturbances superposed on a channel flow. We shall assume the existence of an equilibrium flow in the channel with quantities denoted by subscript $_0$. Primed quantities denote departures from this equilibrium. Thus

$$\begin{aligned} u &= u_0 + u' \qquad & a &= a_0 + a' \\ P &= P_0 + P' \qquad & Q &= Q_0 + Q' \end{aligned} \tag{7-1}$$

Note that for steady channel flows

$$\frac{d \ln A}{dx} = \frac{1}{u_0} (M_0^2 - 1) \frac{du_0}{dx} \tag{7-2}$$

where $M_0 \equiv u_0/a_0$. Also

$$n \frac{da_0}{dx} = -M_0 \frac{du_0}{dx} \tag{7-3}$$

Introducing Eq. 7-1 and 7-2, Eq. 2-8 and 2-9 can be written (for the isentropic case),

$$\begin{aligned} \frac{\delta^+ P'}{\delta t} &\equiv \frac{\partial P'}{\partial t} + (u + a) \frac{\partial P'}{\partial x} \\ &= (M_0 - 1) \left[\frac{P'}{2}\left(1 + \frac{1}{n}\right) - \frac{Q'}{2}\left(1 - \frac{1}{n}\right)\right] \frac{du_0}{dx} \\ &\quad - (M_0^2 - 1) \left[\frac{P'^2 - Q'^2}{4nu_0} + \frac{P' - Q'}{2M_0} + \frac{P' + Q'}{2n}\right] \frac{du_0}{dx} \end{aligned} \tag{7-4}$$

and

$$\begin{aligned} \frac{\delta^- Q'}{\delta t} &\equiv \frac{\partial Q'}{\partial t} + (u - a) \frac{\partial Q'}{\partial x} \\ &= (M_0 + 1) \left[\frac{P'}{2}\left(1 - \frac{1}{n}\right) - \frac{Q'}{2}\left(1 + \frac{1}{n}\right)\right] \frac{du_0}{dx} \\ &\quad - (M_0^2 - 1) \left[\frac{P'^2 - Q'^2}{4nu_0} + \frac{P' - Q'}{2M_0} + \frac{P' + Q'}{2n}\right] \frac{du_0}{dx} \end{aligned} \tag{7-5}$$

If we now consider transonic flow in the $+x$ direction, $M_0 \cong 1$, so that we may regard $M_0 - 1$ as a small quantity of first order. We may then obtain a first order description of wave processes in the transonic region by neglecting the right-hand side of Eq. 7-4 and the last term on the right-hand side of Eq. 7-5 and replacing $M_0 + 1$ by 2, which yields

$$\frac{\delta^+ P'}{\delta t} \cong 0 \tag{7-6}$$

$$\frac{\delta^- Q'}{\delta t} \cong 2 \left[\frac{P'}{2}\left(1 - \frac{1}{n}\right) - \frac{Q'}{2}\left(1 + \frac{1}{n}\right)\right] \frac{du_0}{dx} \tag{7-7}$$

These equations are now sufficiently simple so that they may be treated by the pulse area methods used in Art. 6. It is evident from these equations that the behavior of Q waves (upstream-moving waves) is different in first order from the behavior of P waves, which are substantially unaffected by the transonic steady flows. It should also be remarked that P waves spend little time in the transonic region because of their large propagation velocity, while Q waves propagate only very slowly against the stream and eventually become trapped. We shall therefore concentrate our attention on upstream-moving waves and assume that no downstream P disturbances are present.

Conservation of pulse area. The propagation of an upstream-moving pulse can be studied with the aid of Eq. 7-7. In Art. 6 it was shown that the quantity $\int Q'dx$ (pulse area) is constant in the isentropic propagation of plane pulses in a homogeneous gas. By the use of the first order equation (Eq. 7-7) it can easily be shown that the same rule is approximately true for propagation in a flow field. Eq. 7-7 can also be written (cf. Eq. 4-14)

$$\frac{\partial Q'}{\partial t} + \frac{\partial}{\partial x}\left\{Q'\left[u_0 - a_0 - \frac{Q'}{4}\left(1 + \frac{1}{n}\right)\right]\right\} = 0 \tag{7-8}$$

as can readily be verified by differentiation.

Integrating this equation with respect to x over the range x_a to x_b,

$$\frac{\partial}{\partial t}\int_a^b Q'dx = -\left\{Q'\left[u_0 - a_0 - \frac{Q'}{4}\left(1 + \frac{1}{n}\right)\right]\right\}_a^b$$

If the integration is taken over the whole of a finite pulse, the right-hand term vanishes and the pulse area remains constant to the order of accuracy of Eq. 7-7. It can also be seen from the above equation that, at an arbitrary point, Q pulse area flows through at the rate

$$Q'\left[u_0 - a_0 - \frac{Q'}{4}\left(1 + \frac{1}{n}\right)\right]$$

It was shown in Art. 6 that the rule of conservation of pulse area is valid to third order quantities in the shock strength. This is, of course, also true for the shocks which develop out of the propagation of a pulse in regions where the steady flow is initially shock free. Since the existence of a continuous velocity gradient cannot affect relations between quantities measured immediately ahead of and behind the shock, the proof of Eq. 6-8 can be carried over unchanged for this case.

Another similar case which is of interest in connection with later considerations occurs when a small amplitude, upstream-moving pulse interacts with a shock which is part of the initial equilibrium flow. This case is illustrated in Fig. C,7a. Here it is necessary to assume that the shock intensity is low (equilibrium shock Mach number close to 1) to get a simple result. It is also necessary to neglect the variations in local equilibrium velocities, etc., with position compared to the disturbance velocities. These changes, which are due to the changing channel area, are responsible for the stability of position of the shock wave. Thus, to put it differently, the stability of the shock wave is neglected compared to the

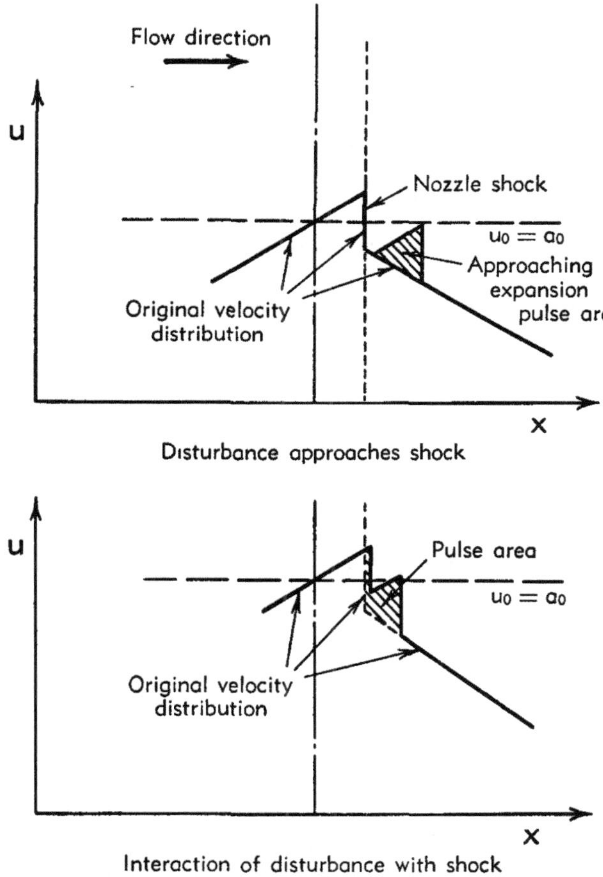

Fig. C,7a. The interaction of an expansion pulse with a nozzle shock. It is shown in the text that the pulse area which disappears on the downstream side of the shock is equal to the pulse area which appears on the upstream side of the shock. Note that the pulse areas plotted here are $(-\frac{1}{2}\int Q'dx)$ since here $u' = -Q'/2$. From [*17*].

disturbance. The circumstances under which this assumption is permissible will become clearer later. Neglecting the stability of the shock, the velocity with which it moves under the influence of a small upstream-moving disturbance can be found from Eq. 5-18 (properly modified for Q shocks) to first order quantities to be

$$U = -\frac{1}{4}\left(1 + \frac{1}{n}\right)(Q_1' + Q_2') \tag{7-9}$$

where we have used the fact that the shock is at rest in the steady flow.

Now, in addition to the contributions to the pulse area which were present in the previous problem, pulse area is contributed by the fact that the shock is merely out of position. Thus the equation which must now be satisfied to demonstrate conservation of pulse area is

$$UQ_1' - Q_1'\left[u_{10} - a_{10} - \frac{Q_1'}{4}\left(1 + \frac{1}{n}\right)\right] - 2U(u_{10} - u_{20})$$
$$= UQ_2' - Q_2'\left[u_{20} - a_{20} - \frac{Q_2'}{4}\left(1 + \frac{1}{n}\right)\right] \quad (7\text{-}10)$$

where the subscripts $_{10}$ and $_{20}$ denote equilibrium conditions before and after the shock. Noting that to first order accuracy from Eq. 5-2 and 5-4,

$$\frac{u_{20} - a_{20}}{u_{10} - u_{20}} = -\frac{1}{2}\left(1 + \frac{1}{n}\right) \quad \text{and} \quad \frac{u_{10} - a_{10}}{u_{10} - u_{20}} = +\frac{1}{2}\left(1 + \frac{1}{n}\right)$$

for Mach numbers close to 1 and, introducing the above shock velocity from Eq. 7-9, it is found that Eq. 7-10 is identically satisfied. Thus it seems that the rule of conservation of pulse area is a useful approximation with considerable generality.

The pulse shape. The shape of a pulse (that is, Q' as a function of x) advancing upstream in a velocity field, in the absence of downstream-moving disturbances, can be studied in a flow field by considering its x derivative $\partial Q'/\partial x \equiv \epsilon'$. The variation of this quantity with time can be found by differentiating Eq. 7-7 with respect to x, obtaining[7]

$$\frac{\partial \epsilon'}{\partial t} + \left[u_0 - a_0 - \frac{Q'}{2}\left(1 + \frac{1}{n}\right)\right]\frac{\partial \epsilon'}{\partial x}$$
$$+ \frac{\epsilon'}{2}\left(1 + \frac{1}{n}\right)\left(4\frac{du_0}{dx} - \epsilon'\right) + \left(1 + \frac{1}{n}\right)Q'\frac{d^2u_0}{dx^2} = 0 \quad (7\text{-}11)$$

For the case where the equilibrium velocity gradient is constant, that is $d^2u_0/dx^2 = 0$, the behavior of the pulse shape can be studied simply from this differential equation. Consider the point under observation as moving with the wave, that is, as moving with the local propagation velocity. Let $\delta^-\epsilon'/\delta t$ be the rate of change of ϵ' when the point under observation moves with the propagation velocity; then Eq. 7-11 becomes

$$\frac{\delta^-\epsilon'}{\delta t} = -\frac{\epsilon'}{2}\left(1 + \frac{1}{n}\right)\left(4\frac{du_0}{dx} - \epsilon'\right) \quad (7\text{-}12)$$

The variables in Eq. 7-12 can be separated. Introducing the dimensionless parameters

$$y \equiv -\frac{\epsilon'}{4(du_0/dx)} \quad \text{and} \quad t^* \equiv \frac{du_0}{dx}t$$

[7] Note that if we differentiate Eq. 7-5 and then neglect $M_0^2 - 1$ compared to 1 we get another term which, however, can be neglected for $Q'/a_0 \ll 1$.

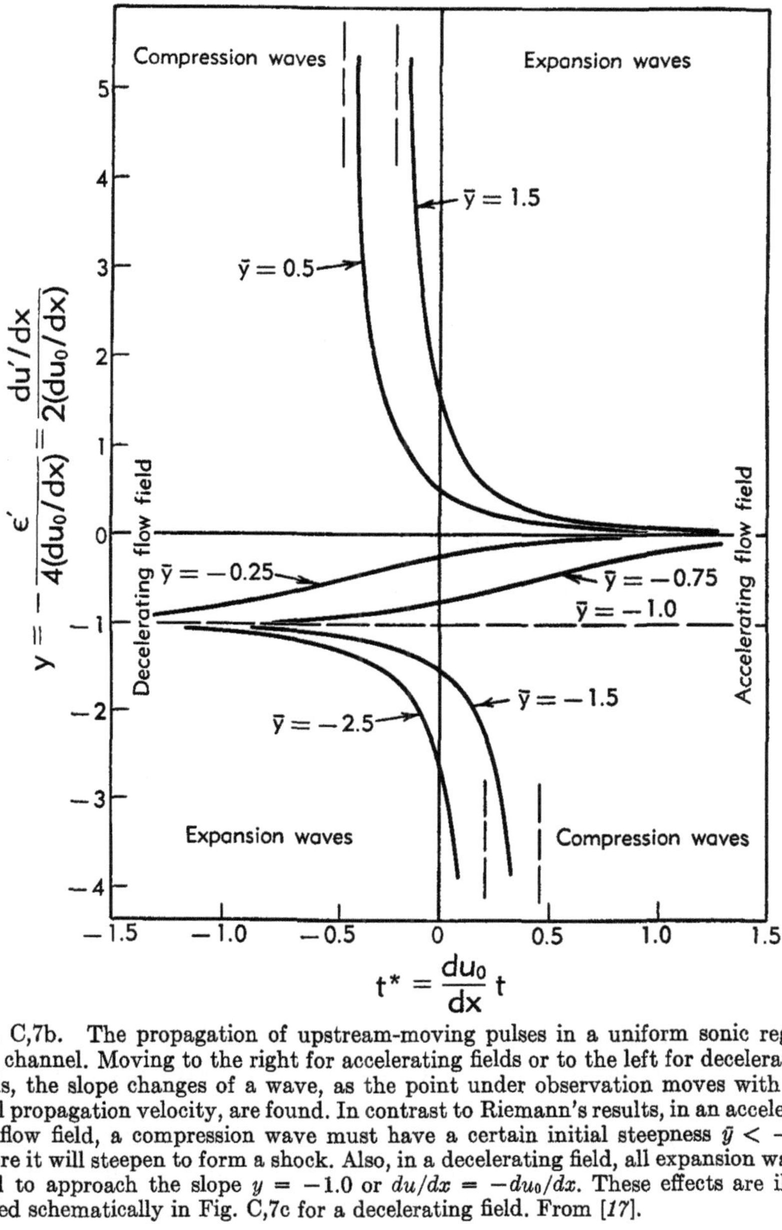

Fig. C,7b. The propagation of upstream-moving pulses in a uniform sonic region of a channel. Moving to the right for accelerating fields or to the left for decelerating fields, the slope changes of a wave, as the point under observation moves with the local propagation velocity, are found. In contrast to Riemann's results, in an accelerating flow field, a compression wave must have a certain initial steepness $\bar{y} < -1.0$ before it will steepen to form a shock. Also, in a decelerating field, all expansion waves tend to approach the slope $y = -1.0$ or $du/dx = -du_0/dx$. These effects are illustrated schematically in Fig. C,7c for a decelerating field. From [*17*].

this equation can be integrated to give

$$y = -\frac{1}{1 - [1 + (1/\bar{y})]e^{2[1+(1/n)]t^*}} \tag{7-13}$$

where $\bar{y}$ is the dimensionless slope of the wave at the time $t = 0$. Positive values of t^* in Eq. 7-13 represent accelerating equilibrium velocity fields, i.e. $du_0/dx > 0$, and negative values of t^* represent decelerating fields.

Also, in decelerating fields, positive values of y represent compressions and negative values represent expansions, the opposite being true for accelerating fields. Eq. 7-13 is plotted in Fig. C,7b for various values of $\bar{y}$.

It can be seen from this figure that the results are considerably different from those obtained in Art. 4 for a fluid with zero velocity gradient in the undisturbed state. Thus, in an accelerating field, compression waves must have a certain initial steepness, $\bar{y} < -1$, or they will flatten out in time. On the other hand, in a decelerating field, expansion waves instead

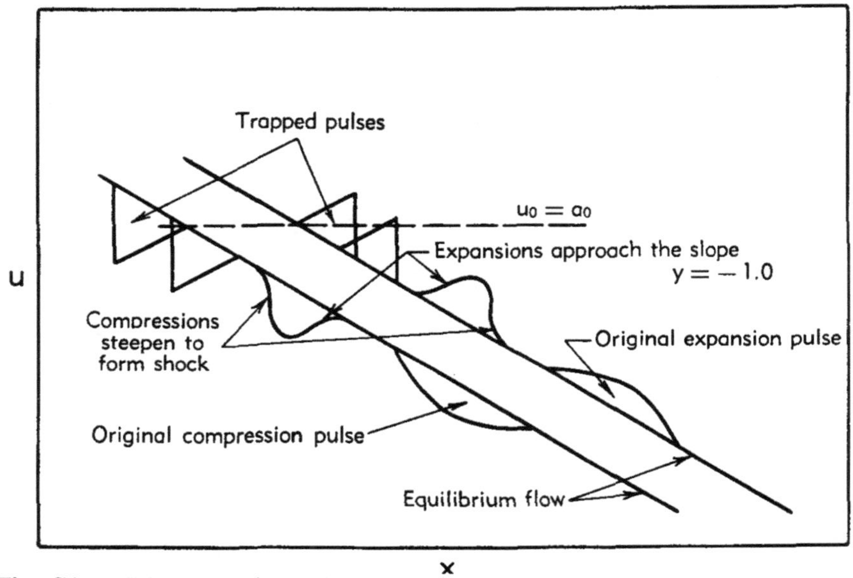

Fig. C,7c. Schematic illustration of four successive positions in the progress of expansion and compression pulses in a uniformly decelerating field. The pulse area is conserved and pulse shape approaches an isosceles triangle. When the leading portion of the pulse approaches the local velocity of sound, the pulse approaches a "trapped" state. From [*17*].

of flattening out (as occurs in Riemann's case) approach a slope given by $y = -1$. It should be noted that $y = -1$ corresponds to a total velocity gradient $du/dx = -du_0/dx$.

Trapped pulses. Using the approximate relations which have been developed in the preceding sections, it can be shown that a pulse moving upstream in a uniformly decelerating flow approaches a stationary state. Thus the pulse area stays constant, the expansion phases approach the slope given by $y = -1$, and the compression phases become shocks. As these developments take place, the pulse advances to a position where the propagation velocities of the shocks approach zero. When this has happened, the pulse is said to be "trapped." The trapping of compression and expansion pulses in a decelerating transonic channel flow is shown schematically in Fig. C,7c.

The asymptotic result for expansion phases of trapped pulses is compared in Fig. C,7d with the exact accelerating equilibrium channel flow. It can be seen from this figure that if all the velocities involved are close to the local speed of sound, the expansion phases are very close to the

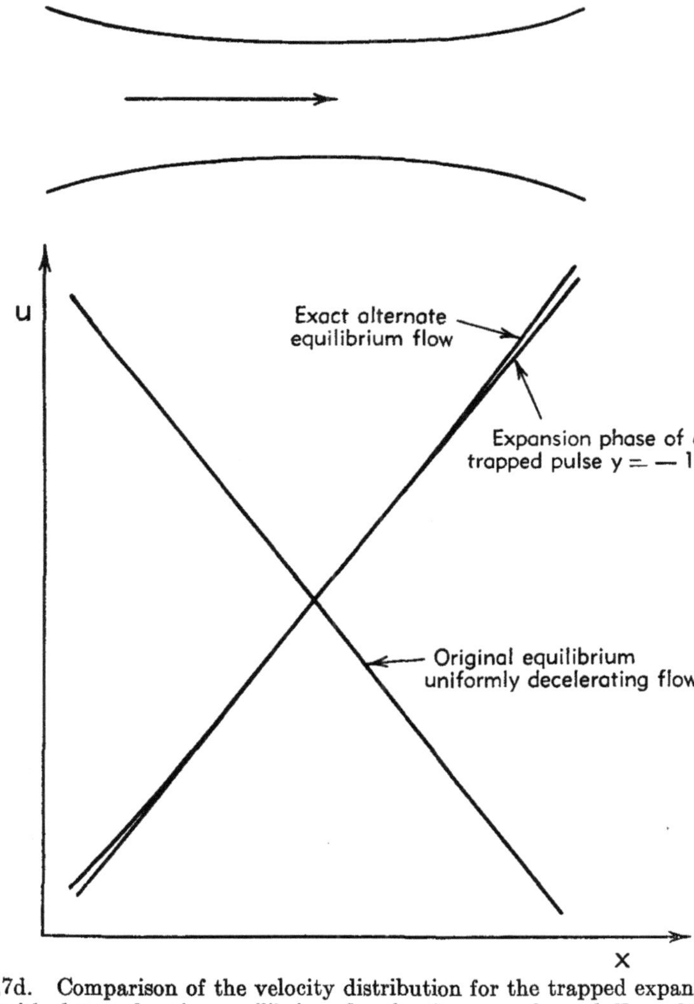

Fig. C,7d. Comparison of the velocity distribution for the trapped expansion wave with the accelerating equilibrium flow for the same channel. From [*17*].

accelerating steady flow. It seems reasonable to infer from Fig. C,7d that the differences between the two curves are due to the approximations introduced in deriving Eq. 7-7 from the exact equations (Eq. 7-4 and 7-5), by neglecting reflections and $M_0^2 - 1$ terms. To put it a little differently, it seems reasonable to infer that an exact calculation from Eq. 7-4 and 7-5 would yield the exact accelerating steady flow. This inference will be adopted later and the exact accelerating flow will be used in place of the $y = -1$ result in calculations of the shock velocities.

It will be valuable to calculate the order of magnitude of the times in which the first order trapping processes take place. These times will later be compared to times in which the slower (higher order) processes take place. For example, the leading edge of an expansion wave propagates upstream with the velocity $u_0 - a_0$. If x is the distance of this leading edge from the sonic point of a channel, then in a uniformly decelerating field (for M close to 1)

$$u_0 - a_0 = \left(1 + \frac{1}{n}\right) x \frac{du_0}{dx}$$

Thus the equation for the approach of the leading edge to the sonic point is

$$\frac{dx}{dt} = \left(1 + \frac{1}{n}\right) x \frac{du_0}{dx}$$

The relaxation time for the approach is therefore

$$\frac{1}{\left(1 + \frac{1}{n}\right)\left(-\frac{du_0}{dx}\right)}$$

A relaxation time for the slope of an expansion wave to approach its equilibrium slope can similarly be found from Eq. 7-12. If the slope is close to the equilibrium value $\epsilon' = 4(du_0/dx)$, then Eq. 7-12 can be written

$$\frac{\delta^- \left(\epsilon' - 4\frac{du_0}{dx}\right)}{\delta t} = 2\left(1 + \frac{1}{n}\right)\frac{du_0}{dx}\left(\epsilon' - 4\frac{du_0}{dx}\right)$$

Hence the relaxation time for this process is

$$\frac{1}{2\left(1 + \frac{1}{n}\right)\left(-\frac{du_0}{dx}\right)}$$

or half as large as that for the approach of the leading edge to the sonic point. It is clear from Fig. C,7b that the formation of shocks from compression waves is a faster process and a shock is formed in a finite time.

Application to the formation of a normal shock in a de Laval nozzle. The theory just developed can be applied to trace the formation of shock waves in a de Laval nozzle. Consider the converging-diverging nozzle shown schematically in Fig. C,7e. Suppose the back pressure to be adjusted to such a value that the local speed of sound is reached but not exceeded at the throat. The form of velocity distribution that would be obtained is also shown schematically in Fig. C,7e. The small curved region near the local speed of sound in the velocity curve which would be obtained in a practical case has been neglected to simplify the argument.

It is known from many experimental results that if the back pressure is lowered further, a normal shock will form in the nozzle. The formation

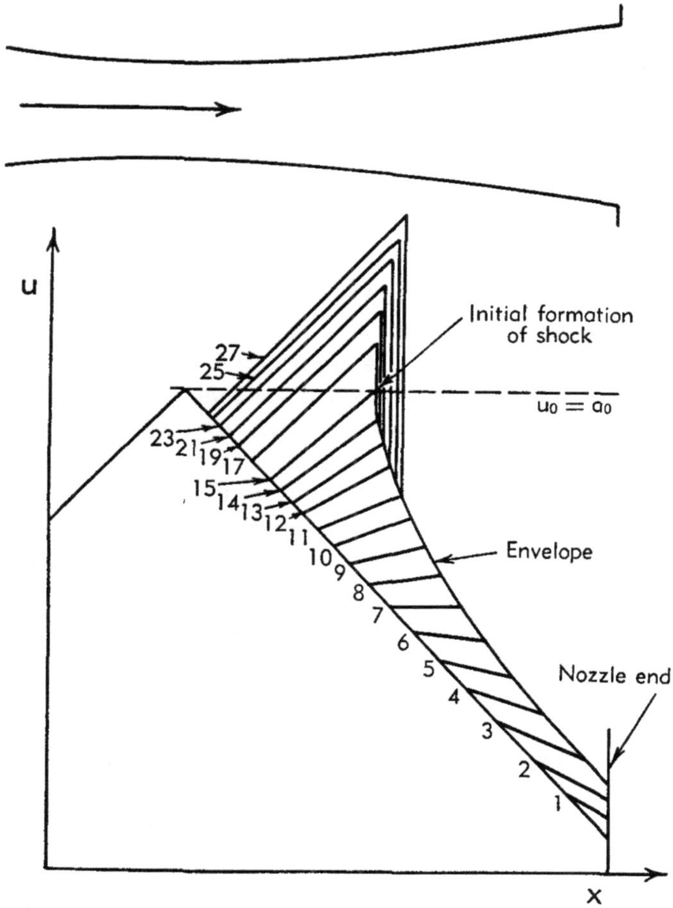

Fig. C,7e. The development of a weak normal shock in a de Laval nozzle as the back pressure is lowered at the nozzle end. The envelope curve is calculated in Appendix C of [*17*]. The successive expansion wave positions are calculated by finding the propagation velocities of the end points. The time intervals are equal up to the formation of shock, and double thereafter. From [*17*].

of this shock can be readily described by use of the foregoing theory if the velocity at the end of the nozzle is sufficiently close to the speed of sound. It will be presumed that the back pressure will be lowered continuously for a time and then held constant. (It will also be presumed that the back pressure is not lowered enough to produce supersonic flow at the end of the nozzle.) While the back pressure is being lowered, the velocity at the end of the nozzle increases. The space derivative of the velocity disturbance created can be found from Eq. 7-7 if downstream-moving disturbances are neglected. If this equation is solved for

$$\frac{\partial u'}{\partial x} = -\frac{1}{2}\frac{\partial Q'}{\partial x}$$

then

$$\frac{\partial u'}{\partial x} = -\frac{\left(1+\frac{1}{n}\right)u'\frac{du_0}{dx}}{u_0 - a_0 + \left(1+\frac{1}{n}\right)u'} - \frac{\frac{\partial u'}{\partial t}}{u_0 - a_0 + \left(1+\frac{1}{n}\right)u'} \tag{7-14}$$

It can be seen from Eq. 7-14 that some time before the expansion stops, $\partial u'/\partial x$ becomes negative. This occurs when

$$\left(1+\frac{1}{n}\right)u'\frac{du_0}{dx} + \frac{\partial u'}{\partial t} < 0$$

since the denominators in Eq. 7-14 are negative for subsonic velocities.

A negative $\partial u'/\partial x$ means that a compression wave travels upstream and steepens, as previously shown, to form a compression shock. The compression shock and the wedge-shaped area ahead of it grow by the addition of expansion pulse area coming from the rear of the nozzle. The shock grows until its intensity and its dissipation are large enough so that the pressure at the rear of the nozzle is again at equilibrium.

The calculations on which Fig. C,7e is based are given in [*17*]. The leading phase which originated while the nozzle back pressure was falling can be calculated from Eq. 7-12 and the known propagation velocity of its leading edge. The steady phase which originates after the nozzle back pressure has again become steady can be found by integrating Eq. 7-7, making use of the fact that $\partial u'/\partial t = 0$ for this phase. It is found that, for the case considered, an infinite decelerating velocity gradient occurs at the point where the velocity first reaches the local speed of sound. The shock that follows is calculated step by step using Eq. 5-7. The calculations used in drawing Fig. C,7e are thus only a first approximation to the shock motion. A more accurate shock velocity calculation would show the shock to settle at an equilibrium position.

Considering the unsteady aspect of a channel flow problem, it is thus easy to see how shocks arise in the transonic flow. It is quite possible that an analysis of this kind could be extended to two or three dimensions and that, if this extension were made, much light could be shed on the formation of normal shocks in these more complicated problems.

The motion and stability of shock waves in channel flow. The trapping of pulses and the formation of shock waves in decelerating channel flows has been considered to a first approximation. In the case of the central problem, the stability of smooth transonic deceleration, these calculations indicate that a trapped pulse converts a part of the channel flow to the accelerating steady flow and that the entire pulse becomes stationary in the sonic region of the channel. The occurrence of these trapped pulses in transonic decelerating channel flow means that to the order of accuracy of these calculations, this flow has a kind of neutral stability to pulses coming from the rear of the channel. Therefore it will be necessary to make more accurate shock velocity calculations before it can be decided whether this flow is stable or unstable.

In [*17*] more accurate calculations of the shock velocity are given for special cases. It is presumed that conditions at the ends of the channel are steady; that is, the shock velocity is calculated for the case where the channel flow is out of equilibrium but where no disturbances are originating at the end of the channel. Thus disturbances are treated by the first order methods already given; these more cumbersome calculations will be used to answer stability questions which cannot be answered by the first order methods.

Consider that a shock exists in a transonic channel flow and that its motion, if any, is so slow that only small shock displacements can occur during the times required for the first order trapping processes to take place. Thus it is assumed that the first order trapping processes have time to reach completion. Also, the conditions immediately upstream from the shock are assumed to be the original equilibrium supersonic flow or the alternate acceleration flow assumed to result from a trapped pulse.

To find the shock velocity, a single relation between the parameters immediately downstream must be known. This relation could of course be found in principle by finding the waves propagating between the shock and the downstream channel end. In view of the difficulty of this procedure, a somewhat similar simple problem is solved instead. Several types of boundary conditions are assumed to occur immediately downstream from the shock, and the shock velocity can then be readily calculated. The boundary conditions considered are: (1) that the velocity behind the shock is constant and equal to its equilibrium value, (2) that the pressure is similarly constant, and (3) that downstream-moving waves are not reflected, i.e. that the shock is followed by an infinite straight tube.

For low Mach numbers it is found that the shock velocity changes only slightly when these various boundary conditions are adopted. This insensitivity to the particular condition downstream from the shock wave makes it seem likely that the results found here will have general significance in view of the fact that these boundary conditions are widely different.

The stability of smooth deceleration through the speed of sound. These shock velocity results can now be applied to discuss the later stages of development of the trapped pulses which were found earlier.

For the case where the equilibrium flow is smooth deceleration through the speed of sound, the shock velocity can be easily calculated. It will be seen that the shock velocity depends only on local conditions at the shock position. It is found that the shock velocity is always negative; that is, the shock always moves upstream. Thus in the case of the trapped expansion pulse (Fig. C,7c) the shock moves upstream and eventually consumes the pulse. Therefore, smooth deceleration through the speed of sound is stable to expansion pulses coming from the rear. On the other hand, in the case of the trapped compression pulse, the shock wave again moves upstream, and in this case the pulse grows continuously. Therefore smooth deceleration through the speed of sound is unstable to compression pulses coming from the rear.

Stability of shock position in channel flows. It appears then that in stable channel flows, deceleration through the speed of sound is accomplished in a shock wave. Consider now a converging-diverging channel: there are, in general, two possible shock positions which will yield equilibrium flows, i.e. the shock can be either in the converging or diverging part of the channel. Experimentally, it is found that the shock is stable in the diverging part of the channel and unstable in the converging part of the channel. A theory of the shock motion similar to the above has been used in [*17*] to study this question theoretically and agrees with the experimental findings. Quantitative relaxation times for the return to equilibrium for the diverging channel are found.

Application to supersonic diffusers. It is interesting to consider the maximum pulse which a supersonic diffuser with a shock in a diverging channel can absorb and still return to its initial configuration (that is, its initial shock position). Consider first the effect of expansion pulses which move the shock downstream. As long as the shock is not moved beyond the end of the diverging part of the channel, it would be expected that the stability would return it to its original position.

The effect of compression pulses strong enough to push the shock through the diffuser throat is more interesting. After the pulse has completely interacted with the shock wave (that is, when the trailing edge of the pulse has just reached the shock), the shock will be displaced by an amount which can be readily estimated from the pulse area conservation rule. The first order results indicate that the flow behind the shock assumes the alternate (subsonic) steady flow possible for this channel. If the pulse is sufficiently strong, the displaced position of the shock will be in the converging part of the channel where the shock position is unstable. There is, of course, an unstable equilibrium position for the shock in the converging part of the channel, and the displaced shock will move away from this unstable equilibrium position. If now the displaced shock position is downstream from the unstable equilibrium position, it will move farther downstream and eventually return to the diverging part of the channel where it will again assume its stable equilibrium position. On the other hand, if the displaced position is upstream from the unstable equilibrium shock position, it will continue to move upstream and eventually convert the supersonic flow in the converging part of the channel to a subsonic flow. Thus the supersonic flow with a shock in the diverging part of the channel is stable to compression pulses which are

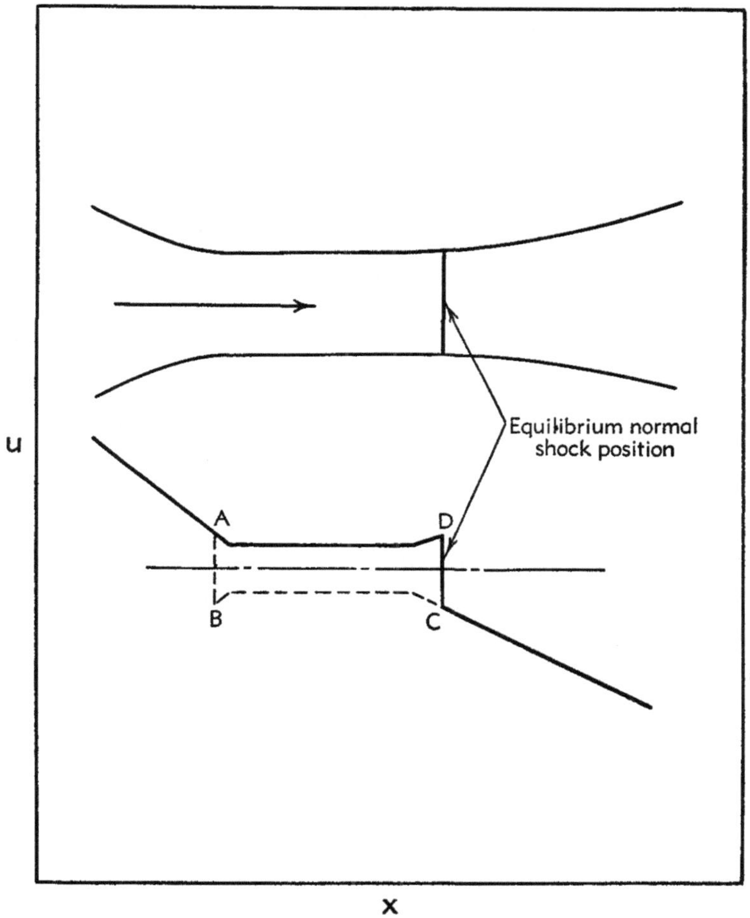

Fig. C,7f. Long throat supersonic diffuser. The limiting compression pulse that could be absorbed by this diffuser would have the pulse area, *ABCD*. Note that the pulse areas plotted here are $(-\frac{1}{2}\int Q'dx)$ since here $u' = -Q'/2$. From [*17*].

not sufficiently strong to displace the shock beyond the unstable equilibrium position in the converging part of the channel.

In the practical design of supersonic diffusers, it is desirable to insure that disturbances of a given magnitude (that is, a given pulse area) do not force the shock beyond this limiting position. Of course, it is also desirable that the equilibrium shock intensity be kept as low as possible. It appears, therefore, that diffusers with a long throat region, which produce a velocity distribution such as that shown in Fig. C,7f, should be considered. In this case, the compression pulse area that can be absorbed by the diffuser, i.e. the area $ABCD$, is increased without increasing the shock intensity. However, it may be noticed that the skin friction would also be increased in this way. Further investigation is necessary to determine the optimum throat length and shape.

C,8. Very Intense Explosions and Implosions. (See also Art. 12.)

Strong wave similarity. For very strong spherical (and cylindrical) shock waves the flow problem can under certain conditions be reduced essentially from two independent variables t and r to a problem involving only one dependent variable which is a combination of r and t. This reduction can be made when (1) the shock is sufficiently strong so that the density ratio across it and the Mach number of the flow behind it (in shock coordinates) become constant and (2) when the boundary and initial conditions can be expressed in terms of the combination variable. Under these conditions the flow problem can be reduced to the solution of an ordinary nonlinear differential equation. Numerical solutions of such ordinary differential equations for various assumed conditions have been given by Taylor [*18*], Guderley [*19*], Brinkley and Kirkwood [*20*], and others.

Explosions. The treatment given by Taylor for a sudden release of energy concentrated at a single point is followed here. In distinction to an ordinary explosion it is assumed that no gas release occurs. This treatment was made in conjunction with the atomic bomb development and his comparison with the first atomic explosion will be discussed.

Denote the undisturbed pressure and density by p_0 and ρ_0, the radius of the shock wave by R, and the outward radial velocity by u. Taylor assumes from energy argument that the following similarity conditions prevail:

$$\frac{p}{p_0} = y = R^{-3}f_1(\eta) \tag{8-1}$$

$$\frac{\rho}{\rho_0} = \psi(\eta) \tag{8-2}$$

$$u = R^{-\frac{3}{2}}\omega_1(\eta) \tag{8-3}$$

where $\eta = r/R$ and $f_1(\eta)$, $\psi(\eta)$, $\omega_1(\eta)$ are to be determined. Using the equations of motion, continuity, and the equation of state for a perfect gas with constant γ, Taylor obtains

$$\frac{\partial f}{\partial \eta}\left[(\eta - \omega)^2 - \frac{f}{\psi}\right] = f\left[-3\eta + \omega\left(3 + \frac{\gamma}{2}\right) - 2\gamma\frac{\omega^2}{\eta}\right] \tag{8-4}$$

where f and ω are nondimensional forms of f_1 and ω_1. The boundary conditions at the shock wave are consistent with the similarity assumptions (Eq. 8-1, 8-2, and 8-3) only when the shock wave is very strong.

For the case of very strong shocks Taylor obtains a numerical solution of Eq. 8-4 in which he finds, for the inner half of the radius, that the pressure is constant and the velocity is proportional to the radius. For the outer half the pressure rises steeply to a peak behind the shock wave and the velocity rises somewhat more rapidly than linearly with the radius. He also derives approximate formulas to describe his numerical results and concludes that the pressure immediately behind the shock wave is given by

$$P_{max} = 0.155R^{-3}E \tag{8-5}$$

where E is the energy released by the explosion.

Taylor compares the results of this theory with several experimental chemical explosions. The range of shock wave radii in which comparison can be made is severely limited, because the shock radius must be large enough so that the mass of material in which the energy is originally concentrated is small compared with the mass of air involved in the disturbance. On the other hand, the shock wave must be considered strong enough so that the density behind it is essentially constant to validate the similarity conditions. Therefore, comparisons can be made only over a restricted radius range (less than a factor of 2). The agreement is at least as good as could be expected.

Taylor also compares this theory with the (New Mexico) atomic explosion of 1945. Fig. C,8a, taken from his paper, shows that for shock radii between 20 and 185 meters, the radius varies with $t^{\frac{2}{5}}$. This result is in agreement with the similarity condition (Eq. 8-3), since the radial velocity at the shock wave is directly proportional to the shock velocity and the shock radius is simply an integral of the shock velocity over time. Taylor points out, however, that the agreement is surprising in view of the assumption of constant heat capacity which is certainly not valid at the high temperatures which would occur. He points out that important radiation effects may contribute to the apparent constancy of γ. The measured shock velocities can be used together with Eq. 8-5 to calculate the energy released in the explosion. In this way, assuming $\gamma = 1.4$, he finds that the TNT equivalent of the energy of the New Mexico explosion, or more strictly that energy which was not radiated from the "ball of fire," was 16,800 tons. Using the calculated value of the heat capacity for a radius of 100 meters, which gives $\gamma = 1.29$, he finds the TNT equivalent to be 23,700 tons.

Brinkley and Kirkwood have given a method of treatment of the propagation of plane, cylindrical, and spherical shock waves in which it is possible to use the exact equations of motions, the exact Rankine-Hugoniot relations, and the exact fluid properties in a numerical integration procedure. It is assumed, however, that the pressure peak following the shock wave has a certain simple shape which reduces to the appropriate approximate shape for weak shock waves and is exponential for strong shock waves. This amounts to the imposition of similarity constraints on the explosions.

Implosions. Guderley has treated the cases of cylindrical and spherical shock waves converging toward the center. He starts from the

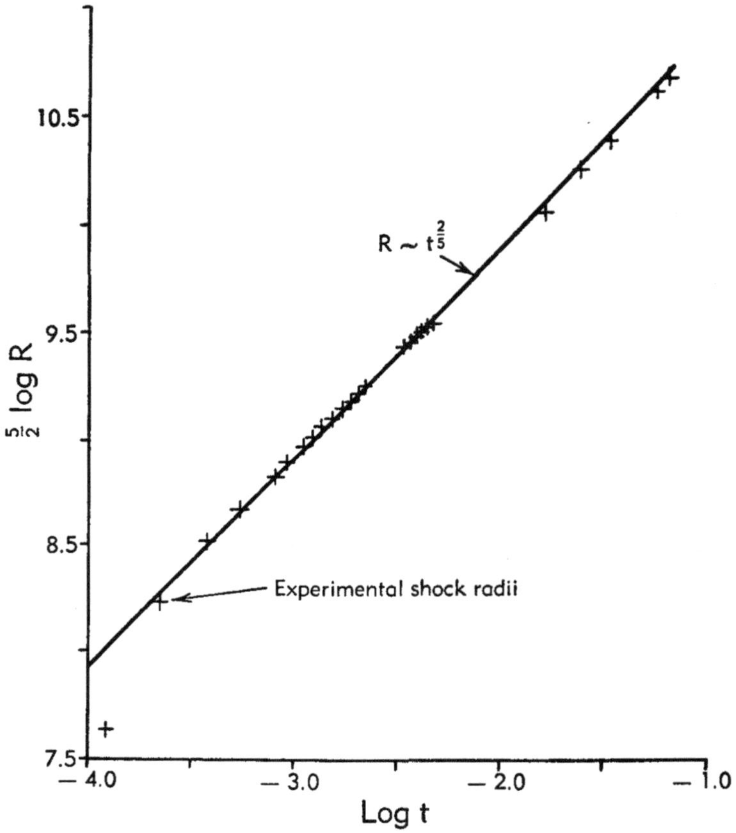

Fig. C,8a. Comparison of Taylor's similarity assumption (Eq. 8-3) with the New Mexico atomic explosion of 1945. From [*18*].

assumption that the radius of a shock as a function of time can be represented by an equation of the form

$$R = \sum a_m(-t)^m \tag{8-6}$$

where the time scale is chosen so that $t = 0$ for the shock at the center. If now we are sufficiently close to the point $t = 0$, then the term with the smallest value of m, i.e. m', will dominate Eq. 8-6 which then can be represented by

$$R = a_{m'}(-t)^{m'} \tag{8-7}$$

If now the strong shock equations are introduced, he then finds it possible to again reduce the system to an ordinary differential equation. The differential equation found has solutions for various values of m'. In general these solutions correspond to various disturbances coming from the

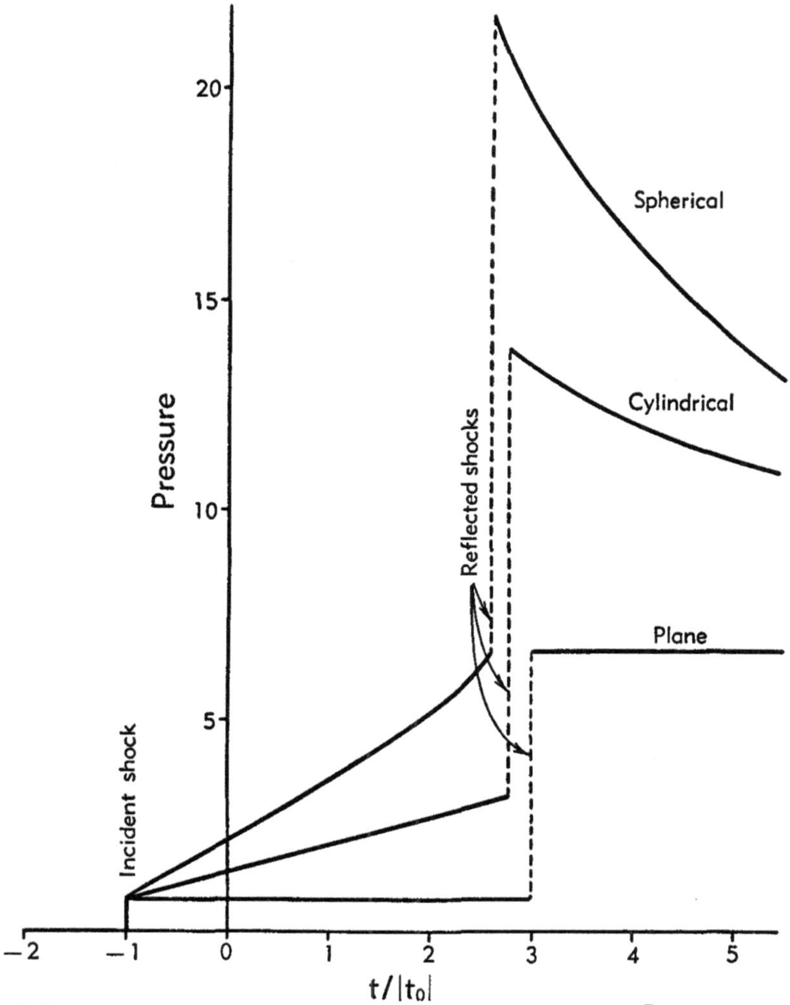

Fig. C,8b. Pressure at a fixed station as a function of time for a strong implosion. Pressure is measured in units of the pressure behind the incident shock and time in units of the interval required for the shock to propagate from the fixed station to the center. The shock arrives at the center at $t = 0$. Taken from Guderley [*19*].

region external to the shock wave, and the disturbances necessary to preserve a given value of m' become infinite as the shock approaches the center. However, for $m' = 0.834$ for cylindrical waves, and $m' = 0.717$ for spherical waves, infinite external disturbances are not required. Guderley therefore concludes that these are the proper values of m' to associate with these waves. From these results he concludes that the pressure immediately behind the shock is proportional to $(1/R)^{0.396}$ for cylindrical waves, and proportional to $(1/R)^{0.792}$ for spherical waves. The numerical values were obtained for $\gamma = 1.4$.

Using the method of characteristics, Guderley calculates the shock which is reflected from the center of convergence. In Fig. C,8b, taken from Guderley's paper, the pressure at a fixed radius is plotted as a function of time for plane, cylindrical, and spherical shock waves.

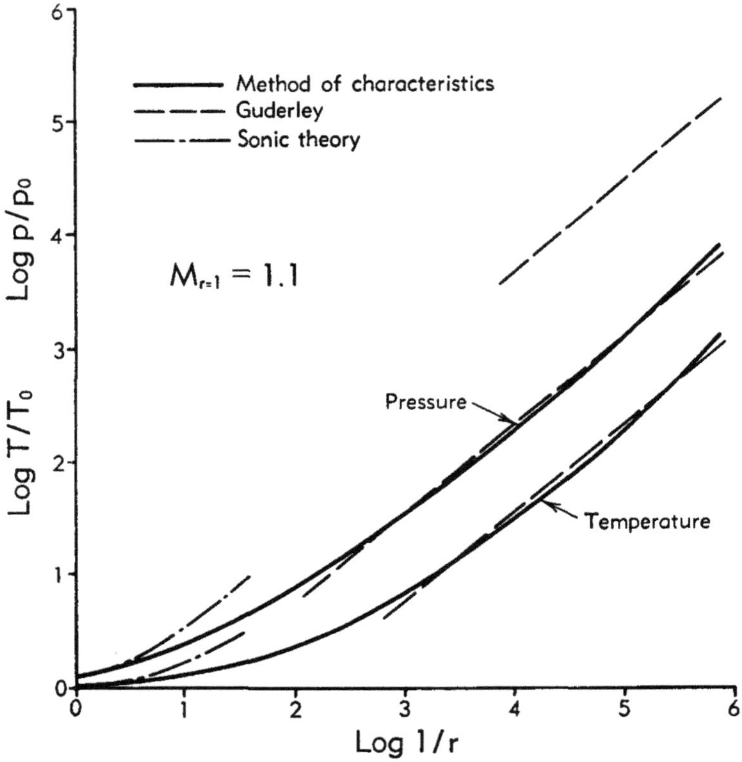

Fig. C,8c. Theory of converging spherical shocks. Comparison between sonic theory, Guderley's solution, and the method of characteristics for spherical converging shock waves ($\gamma = 1.40$). p/p_0 and T/T_0 are the ratios of pressure and temperature immediately behind the shock to the initial values. The short upper curve corresponds to the pressure after reflection from the center. From [21].

The whole history of an imploding spherical wave can be traced by means of the method of characteristics. This calculation was carried out by Perry and Kantrowitz and the results are reproduced in Fig. C,8c in which the pressure and temperature immediately behind the shock are plotted as functions of the radius. It is seen that Guderley's results are confirmed for pressure ratios greater than 100.

Perry and Kantrowitz [21] investigated the stability of these converging shock waves by considering the smoothing-out of curvature concentrations. A considerable amount of study has been given to corners (which are concentrations of a finite amount of curvature in an infinitesimal region). Such corners are produced when plane shocks intersect a wedge (see Fig. C,8d). In the resultant Mach reflection, part of the original shock curvature θ is smoothed out over the Mach shock and a part ϕ remains at the triple point. Thus, in the Mach reflection, a fraction $1 - (\phi/\theta)$ (the attenuation) of the original curvature concentration has been smoothed out.

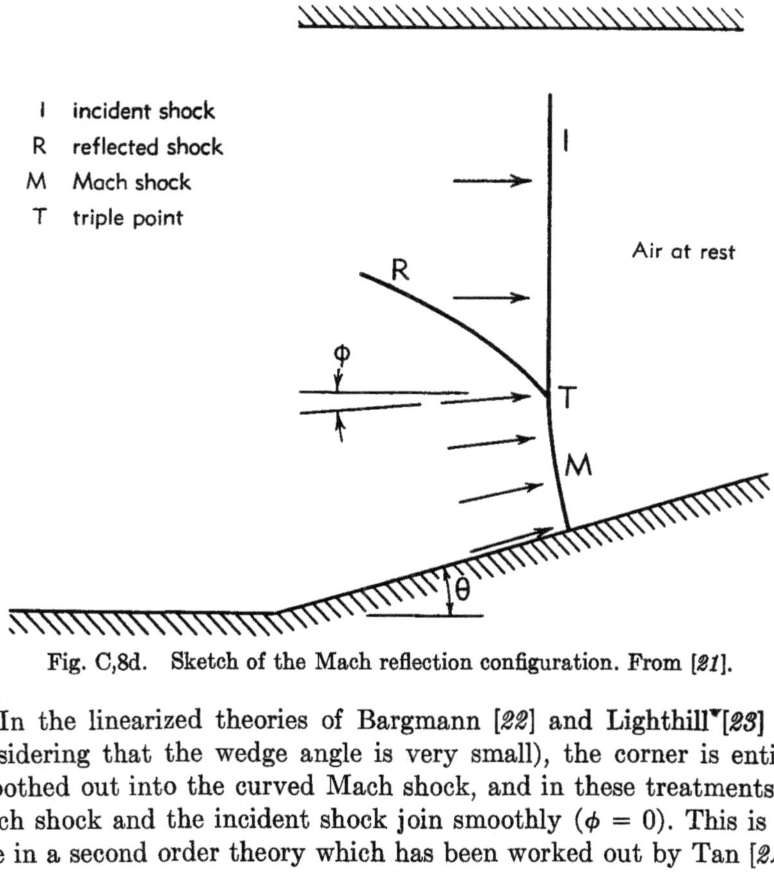

Fig. C,8d. Sketch of the Mach reflection configuration. From [*21*].

In the linearized theories of Bargmann [*22*] and Lighthill [*23*] (i.e. considering that the wedge angle is very small), the corner is entirely smoothed out into the curved Mach shock, and in these treatments the Mach shock and the incident shock join smoothly ($\phi = 0$). This is also true in a second order theory which has been worked out by Tan [*24*].

Information on the attenuation of corners must therefore be obtained from experiment. Analyzing the experiments of L. G. Smith [*25,26*], it is found that for small wedge angles and for small shock intensities, the attenuation is very large. However, it is found that for wedge angles approaching those at which regular reflection occurs, very little attenuation is obtained. Also, for strong shocks the region of high attenuation becomes restricted to small corners, i.e. small departures from a straight shock. This analysis thus indicates that shock waves of weak and moderate strength should have an appreciable tendency to form shapes of constant curvature. This tendency is very well illustrated in shock tube experiments where it is found that shocks propagating in a straight tube become very flat, even though they are produced with an irregular shape as a result of irregular diaphragm bursts.

Using the apparatus shown in Fig. C,8e it was found possible to produce cylindrical implosions. Schlieren photographs of incoming and outgoing shocks are shown in Plate C,8. These photographs were made starting with quite weak plane shocks. When the pressure ratios across the diaphragm are adjusted to produce stronger plane shocks (i.e. about $M = 2$), the resultant cylindrical shocks are not nearly so regular in agreement with the analysis based on Mach reflection experiments.

It would be expected that near the center of convergence of cylindrical shock waves, very high temperatures and pressures would be produced. Thus it would be expected that if good convergence were obtained,

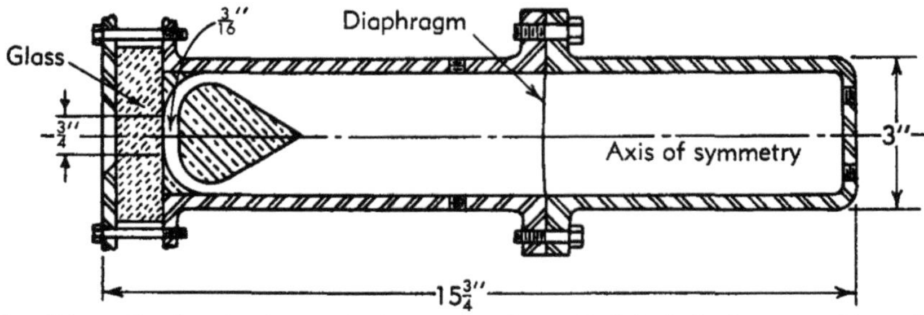

Fig. C,8e. Shock tube for producing converging cylindrical shock waves. From [*21*].

the center of convergence would become luminous. This luminosity was observed and photographs of it are reproduced in [*21*].

C,9. General Problems. Numerical Integration Along Characteristics.

Introduction. In the preceding articles we have discussed those problems which can be treated either analytically or can be reduced to the numerical integration of an ordinary differential equation. General problems involving disturbances in both P and Q cannot be handled by these methods.[8] In these cases we must resort to numerical integration of the equations of motion. When high speed computing machines are

[8] For the case of plane, isentropic, interacting waves a method of solution was originally given by Riemann and is discussed by Courant and Friedrichs [*6*, Sec.82]. Riemann transforms the differential equation by taking P and Q as the independent variables, and x and t as the dependent variables. In this way the equations become linear and the solutions can be given in terms of hypergeometric functions. According to Courant and Friedrichs these solutions have been found simpler than numerical integration only for the case of interacting, centered, rarefaction waves, i.e. for the case where we have a tube bounded by two pistons each of which suddenly starts moving away from the gas with a constant velocity. In most real problems, however, shock waves and resultant entropy variations appear, making this type of analytical procedure impossible, and it is to this larger class of problems that the methods of this article are directed.

available these can be used to integrate the equations of motion directly. However, numerical integration along characteristics, e.g. using Eq. 2-8 and 2-9, is far more convenient where hand computing techniques are to be employed. (Integration along characteristics, employing high speed machines, involves a complex programming associated with any shock waves that may arise during the calculation.) Using numerical integration along characteristics, it is possible to include various physical properties of the medium which have been thus far excluded. For example, we shall

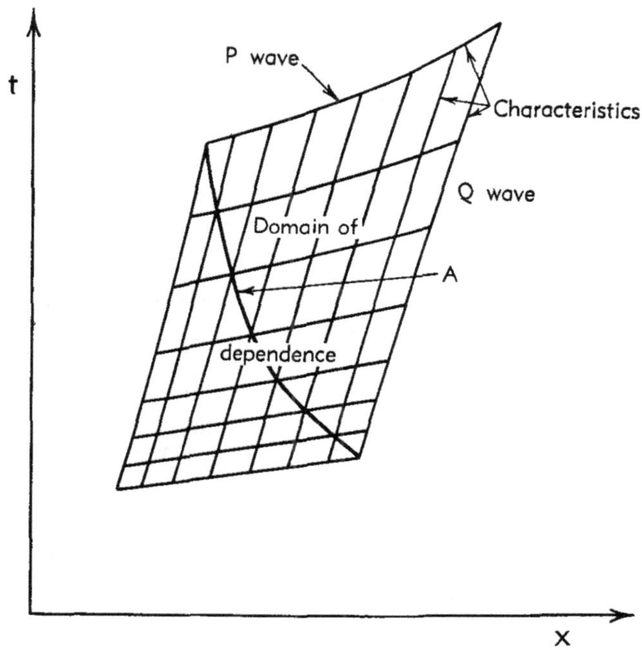

Fig. C,9a. Illustrating the domain of dependence for an isentropic wave propagation problem. Complete data (P and Q) given along A determine P and Q and thus u and a everywhere within the domain of dependence.

give equations for the cases where the heat capacity of the gas is a function of temperature and where the various parts of the heat capacity may be out of equilibrium.

Characteristics. The essential simplification which occurs when we employ these equations is best illustrated by considering the case of isentropic plane waves. In this case, Eq. 2-8 and 2-9 reduce to Eq. 4-5 and 4-6, i.e. to the statements that P and Q are unchanged along P and Q waves respectively. If now we are given values of P and Q along some curve A in the x, t plane (see Fig. C,9a), then the values of P and Q can be found everywhere in a roughly diamond-shaped region formed by the extreme waves passing through this curve. The region so determined has been called, e.g. by Courant and Friedrichs, the domain of dependence. Clearly, if the curve A is the path of a P or Q wave the domain of dependence vanishes. This property is usually used as the mathematical definition of a characteristic. Presuming now that the curve A is not a P or Q wave, the problem of solving the flow in the domain of dependence reduces to the problem of drawing the P and Q waves, i.e. of drawing curves, which have the local slope $1/(u \pm a)$. All points on such curves have the same value of P or Q which is known to be equal to the values at the intersection of these waves with the curve A.

The procedure for determining the characteristics is to start from the curve A, drawing short straight lines with a slope $1/(u \pm a)$. At the intersections of waves of different families, P and Q—and hence everything—are known. From such points new short extensions of the characteristics can be drawn and new intersections, at which everything is known, determined. In this way the complete network of characteristics can be constructed step by step.

The accuracy of this procedure depends only on the magnitude of the steps used. Inaccuracies are introduced due to the fact that the characteristics have been replaced by a series of short straight lines having slopes appropriate to the point from which these lines are drawn. Thus the errors in the location of the intersection points will be second order in the length of the linear steps. These errors can be reduced to third order by the following procedure. Instead of drawing lines with the slope appropriate to the point from which they come, the value of $u \pm a$ is computed from an average of the values of P and Q at the initial and final points. Note that the values of P and Q at the intersection point are known before the intersection point has been located. This averaging procedure generally results in more rapid calculation to a given order of accuracy, since the labor per step is increased by something less than a factor of 2 and the magnitude of the steps used may in many cases be increased by something like a factor of two. Since the number of line segments to be drawn to cover the domain of dependence is inversely proportional to the square of their length, the labor of the computation is reduced by a factor of the order of two in this way.

A convenient aid in drawing lines with the slopes $1/(u \pm a)$ or $1/u$ (the latter will be used in connection with computations involving entropy variations) was devised by Schmidt and is reported by Guderley [9]. In Fig. C,9b values of $P/2$ and $Q/2$, $(P/2)[1 + (1/n)]$ and $(Q/2)[1 - (1/n)]$, $(P/2)[1 - (1/n)]$ and $(Q/2)[1 + (1/n)]$ are plotted horizontally on lines 1 unit apart. If now lines are drawn connecting the values of P and Q at a point in each of these diagrams, then the slope of the particle paths $1/u$, and the P and Q waves, $1/(u \pm a)$, are equal to the slopes of the connecting lines on these diagrams. We assume, of course, that the vertical axis in these diagrams is made parallel to the t axis on the x, t diagram.

Variable channel area and entropy. We next consider the case of the propagation of waves in a variable area channel. We also consider that the entropy of the gas varies smoothly in space and time. The cases of shock waves and finite discontinuities in entropy or in the other properties of a gas are treated separately below.

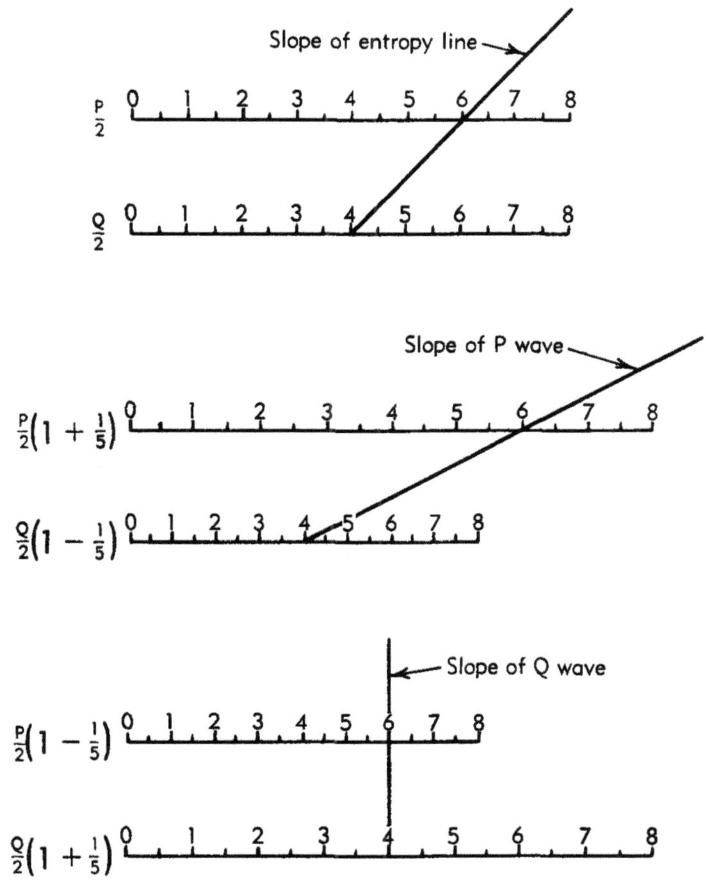

Fig. C,9b. Schmidt diagram [9] for finding the slopes of entropy lines, P and Q waves in drawing characteristic on an x, t diagram where a_0t is the ordinate net. The figure is drawn for $n = 5$ and the examples are drawn for $P = 6a_0$ and $Q = 4a_0$.

Eq. 2-8 and 2-9 can be written

$$\frac{\delta^+ P}{\delta t} = \frac{a^2}{\gamma R}\frac{\partial s}{\partial x} + \frac{a}{R}\left(\frac{\partial s}{\partial t} + u\frac{\partial s}{\partial x}\right) - ua\frac{d \ln A}{dx} \tag{9-1}$$

$$\frac{\delta^- Q}{\delta t} = -\frac{a^2}{\gamma R}\frac{\partial s}{\partial x} + \frac{a}{R}\left(\frac{\partial s}{\partial t} + u\frac{\partial s}{\partial x}\right) - ua\frac{d \ln A}{dx} \tag{9-2}$$

Since

$$\frac{\delta^+}{\delta t} = \frac{\partial}{\partial t} + (u + a)\frac{\partial}{\partial x}$$

and writing

$$\frac{D}{Dt} = \frac{\partial}{\partial t} + u\frac{\partial}{\partial x}$$

we can write

$$\frac{\partial s}{\partial x} = \frac{1}{a}\frac{\delta^+ s}{\delta t} - \frac{1}{a}\frac{Ds}{Dt}$$

and

$$\frac{d\ln A}{dx} = \frac{1}{u+a}\frac{\delta^+ \ln A}{\delta t}$$

Now Eq. 9-1 can be written

$$\delta^+ P = -\frac{ua}{u+a}\delta^+ \ln A + \frac{a}{\gamma R}\delta^+ s + \frac{a}{c_p}\left(\frac{Ds}{Dt}\right)\delta t \tag{9-3}$$

and similarly,

$$\delta^- Q = -\frac{ua}{u-a}\delta^- \ln A + \frac{a}{\gamma R}\delta^- s + \frac{a}{c_p}\left(\frac{Ds}{Dt}\right)\delta t \tag{9-4}$$

Consider for the moment that the entropy is constant when we move with a particle so that $Ds/Dt = 0$. We proceed now to construct a network of characteristics similar to that shown in Fig. C,9a, the P and Q waves being drawn as before. It is now necessary to draw the particle paths, i.e. the curves[9] with the slope $1/u$. Since $Ds/Dt = 0$ these are lines of constant entropy and will sometimes be referred to as entropy lines. Now to determine the value of the entropy at an intersection it is in general necessary to interpolate between entropy lines. When the location of the intersection point is determined approximately, we can determine the variations of the entropy and the channel area between the initial points and the intersection. With the use of Eq. 9-3 and 9-4 the values of P and Q can then be found.

It is necessary to use average values of u and a in applying Eq. 9-3 and 9-4. We see that this process must essentially be iterative, since the intersection point cannot be determined accurately until fairly accurate values of P and Q at this point are known. On the other hand, until the intersection point is known, we do not know the values of P and Q in distinction to the previous (isentropic, constant area) problem. In performing the calculation a preliminary estimate of P and Q at the intersection must first be made, the characteristics (including the particle paths) drawn according to these estimates, and the indicated values of P and Q found. The iteration must be repeated until the calculated values

[9] Note that these curves are also characteristics in that if we are given information only on one particle path, the entropy elsewhere is unknown and the domain of dependence vanishes.

of P and Q agree with the preliminary estimates to the required accuracy. This iterative process may become very slowly convergent or actually divergent if the steps are made too large. This is especially true when u is close to $\pm a$ or when there is a rapid variation of channel area. After some experience has been gained in making close preliminary estimates larger steps can be used.

When Ds/Dt is not zero, that is, in cases where heat is being added to the gas or where irreversible processes are taking place in the body of the gas, the last term in Eq. 9-3 and 9-4 must also be included. The method of handling this is not essentially different from the methods treating

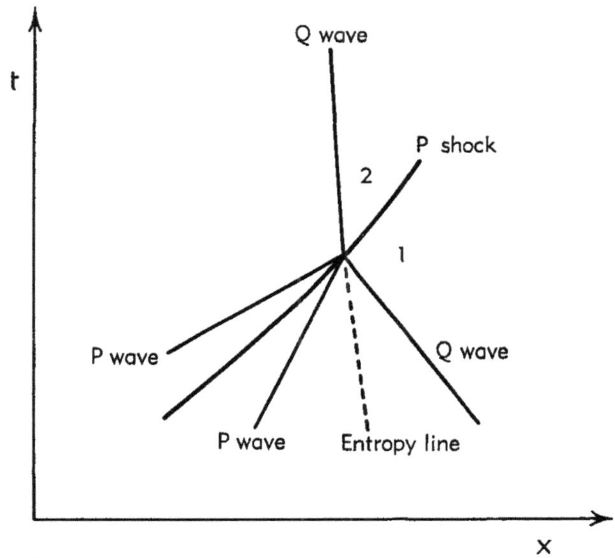

Fig. C,9c. Illustrating the calculation of shock problems by numerical methods.

the other quantities in these equations. The increments in P and Q caused by the Ds/Dt term are for a finite increment δt equal to

$$\int \frac{a}{c_p}\frac{Ds}{Dt}\,dt = \frac{1}{c_p}\overline{\left(a\,\frac{Ds}{Dt}\right)}\,\delta t$$

where the bar indicates a time average at points along the characteristic. The evaluation of this term is somewhat more tedious than the previous terms in that the average value of a rate of variation must be found, whereas the previous calculation required only averages of simple functions of u and a. In other respects the calculation proceeds as previously.

Shock waves. Many unsteady flow problems involve shock waves, which are most conveniently treated by the use of graphical representations of results obtained from the Rankine-Hugoniot relations presented for $n = 3$, 5, or 7 in Fig. C,5a, C,5b, and C,5c. Considering that we are dealing with a P shock (see Fig. C,9c), we in general know values of P_1,

Q_1, and s_1. Thus we have a complete statement of conditions ahead of the shock. Also, since the shock is overtaken by waves from behind, the value of P_2 would be known if the entropy behind the shock were known. Presuming for the moment that a preliminary estimate of the value of P_2 is made we can find Q_2 from Fig. C,5a, and the entropy behind the shock from Fig. C,5b. The accuracy of the preliminary estimate can then be checked by a computation of the value of P_2 from Eq. 9-3. It may be necessary to perform further iterations to obtain the shock strength with the required accuracy. However, unless the shocks are very strong, the variation in entropy produced by variations in shock strength results in such small changes in P that the process converges very rapidly. Note

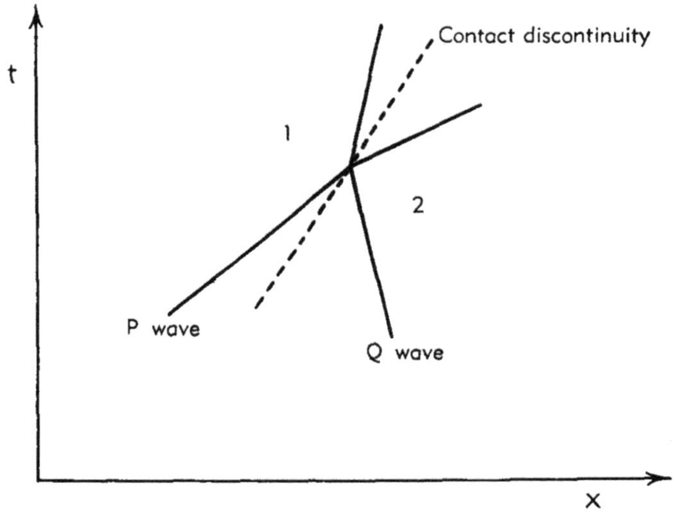

Fig. C,9d. Illustrating the technique of crossing a contact discontinuity.

that, in general, interpolations are necessary at every point in order to find values of Q_1, P_1, s_1, and P_2 applicable to a single point on the shock wave. The calculations would become very tedious if many shock waves were present. However, in many problems there are only a small number of shock waves in which the total labor involved is not a large part of the total labor involved in constructing the two-dimensional network. Note that for weak P shocks the change in Q and entropy is negligible and these may be represented in a characteristic network simply as a line with slope halfway between that of the adjacent P waves.

After the shock has been determined at one point a line can be drawn in the local direction of the shock motion with the use of Fig. C,5c, and the process can be repeated to find the shock strength at the next point.

Contact discontinuities. In many problems contact discontinuities occur, i.e. discontinuities between gases of greatly different properties. For example, two dissimilar gases may exist side by side as in a shock tube. Again, a portion of a gas may have been greatly heated so that its entropy is considerably different from the neighboring gas. In these cases (see Fig. C,9d) we know the values of P and Q arriving at any point of the discontinuity since, of course, the discontinuity moves with the local particle velocity. Also, assume that the entropy is known on both sides of the contact discontinuity. The two sides of the contact discontinuity are labeled 1 and 2 where $x_2 > x_1$. We know then P_1, Q_2, s_1, and s_2, and we need to determine P_2 and Q_1.

For this purpose we have first the fact that the particle velocity is the same on both sides of the discontinuity, which yields

$$2u_1 = P_1 - Q_1 = 2u_2 = P_2 - Q_2 \quad \text{or} \quad \Delta P = \Delta Q \tag{9-5}$$

Secondly, the pressure must be the same on both sides of the discontinuity. From Eq. 2-5 the pressure may be written as

$$p_1 = \left(\frac{a_1}{a_{A_1}}\right)^{\frac{2c_{p_1}}{R_1}} e^{-\frac{(s_1 - s_{A_1})}{R_1}} = p_2 = \left(\frac{a_2}{a_{A_2}}\right)^{\frac{2c_{p_2}}{R_2}} e^{-\frac{(s_2 - s_{A_2})}{R_2}} \tag{9-6}$$

where the subscript $_A$ refers to some standard state. We may choose $s_{A_1} = s_{A_2} = 0$ without loss of generality. Using Eq. 9-5,

$$a_1 = \frac{P_1 + Q_1}{2n_1} = \frac{P_1 + Q_2 - \Delta P}{2n_1}$$

and

$$a_2 = \frac{P_2 + Q_2}{2n_2} = \frac{P_1 + Q_2 + \Delta P}{2n_2}$$

Now Eq. 9-6 may be written

$$\left(\frac{P_1 + Q_2 - \Delta P}{2n_1 a_{A_1}}\right)^{\frac{2c_{p_1}}{R_1}} e^{-\frac{s_1}{R_1}} = \left(\frac{P_1 + Q_2 + \Delta P}{2n_2 a_{A_2}}\right)^{\frac{2c_{p_2}}{R_2}} e^{-\frac{s_2}{R_2}} \tag{9-7}$$

from which ΔP (and using Eq. 9-5, ΔQ) may be determined from the known quantities.

If we now have merely an entropy discontinuity between gases of the same chemical nature, $a_{A_1} = a_{A_2}$, $c_{p_1} = c_{p_2}$, $R_1 = R_2$, and we can solve Eq. 9-7 for ΔP, explicitly obtaining after some simple algebra:

$$\Delta P = (P_1 + Q_2) \tanh \frac{\Delta s}{4c_p} = \Delta Q \tag{9-8}$$

Collision of shocks with shocks or contact discontinuities. The collision of two shock waves (Fig. C,9e) can be treated with the aid of the Rankine-Hugoniot equation. This, however, leads to the necessity of solving equations of fourth degree. It is much more convenient to solve this problem by a trial and error procedure. For example, we estimate the strength of the shock $4 \rightarrow 3$. This gives us the pressure and velocity in region 3, which in turn are equal to the pressure and velocity in region 2. Since everything is assumed known in region 1, we may easily determine the strength of the shock $1 \rightarrow 2$ so as to make the velocities equal in 2 and 3 (for example, from Fig. C,5a). We then may check the pressure in region 2 against that in region 3. If these pressures are not equal, our original

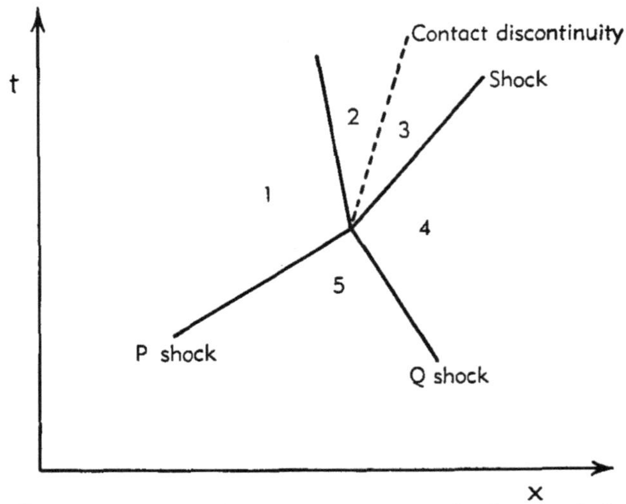

Fig. C,9e. Illustrating the calculation of the collision of a *P* and a *Q* shock. The case of the collision of two *P* or two *Q* shocks is handled in an analogous manner.

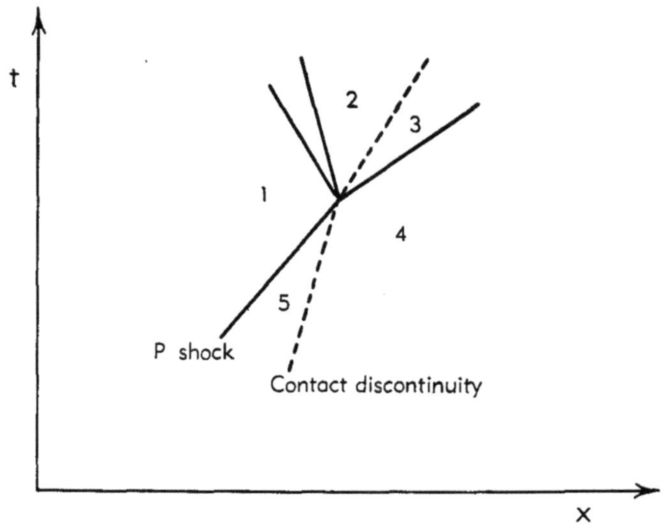

Fig. C,9f. Illustrating the calculation of the collision of a shock and a contact discontinuity.

estimate of the strength of the shock $4 \rightarrow 3$ must be revised, and the procedure repeated until satisfactory accuracy has been obtained.

The recommended procedure when a shock collides with a contact discontinuity (Fig. C,9f) is of the same sort. In this case we again assume a value for the strength of the shock $4 \rightarrow 3$, calculate the velocity in regions 3 and 2, find the wave $1 \rightarrow 2$ which yields the proper velocity at 2, and compare the pressures obtained in regions 2 and 3. Note that the wave reflected from the collision (the wave $1 \rightarrow 2$) may be a rarefaction or a shock. The necessity for treating these collision problems by a trial and error procedure or by the use of the fourth degree equation is not particularly onerous, since there are usually very few such collisions in a flow problem.

Boundary conditions. When an end of a tube is approached, an additional relation between P and Q is required and is obtained from the boundary condition. Three examples of these boundary conditions illustrate this point.

1. Consider the case of a tube with a closed end where $u = 0$. In this case we know that at the end $P = Q$.
2. Consider a tube discharging air directly into a chamber at a known pressure p. When a disturbance reaches the end of such a tube, spherical waves are radiated. These spherical outgoing waves result in incoming waves of opposite kind, for example, compression waves are reflected as expansion waves. However, if the typical lengths in the disturbance (i.e. the distances in which considerable changes in one of the Riemann invariants take place) are large compared to the tube diameter, then the discharge from the end of the tube will be much like a steady flow discharge, and the spherical waves will produce a small effect. Some indication of the errors made when we assume that the discharge is like a steady flow discharge, i.e. that the pressure in the end of the tube is the same as the chamber pressure, can be found from the acoustical theory of organ pipes. According to Rayleigh [*10*, Vol. II, Chap. 16], the pitch of organ pipes can be calculated correctly if one uses a length greater than the geometrical length by a value somewhat less than the radius. The length correction to be added is of the order of one half the tube radius and depends somewhat on the detailed geometry just outside the tube (flanges, etc.).

   Taking into account then that we may be making an error equivalent to that which we would make if we used a tube length in error by about half a radius, we may set the pressure at the end of the tube equal to the chamber pressure. In this case we know that $a = e^{s/2c_p}p^{R/2c_p}$, where s and a are the entropy and the sonic speed of the air being discharged. Thus

$$\frac{P+Q}{2n} = e^{\frac{s}{2c_p}}p^{\frac{R}{2c_p}} \tag{9-9}$$

   Under certain conditions the flow out of a tube may become sonic. Under these conditions no boundary condition need be used since no waves are able to propagate into the tube. The conditions when the flow out of a tube becomes sonic are easily determined by the use of the boundary condition (Eq. 9-9).

3. Consider a faired nozzle discharging air into a tube from a chamber where the stagnation pressure and temperature are known. If the nozzle is very short compared to the typical length of the disturbances, we may again in this case assume that the nozzle flow is steady, and that the pressure and velocity at the tube end are related to the pressure

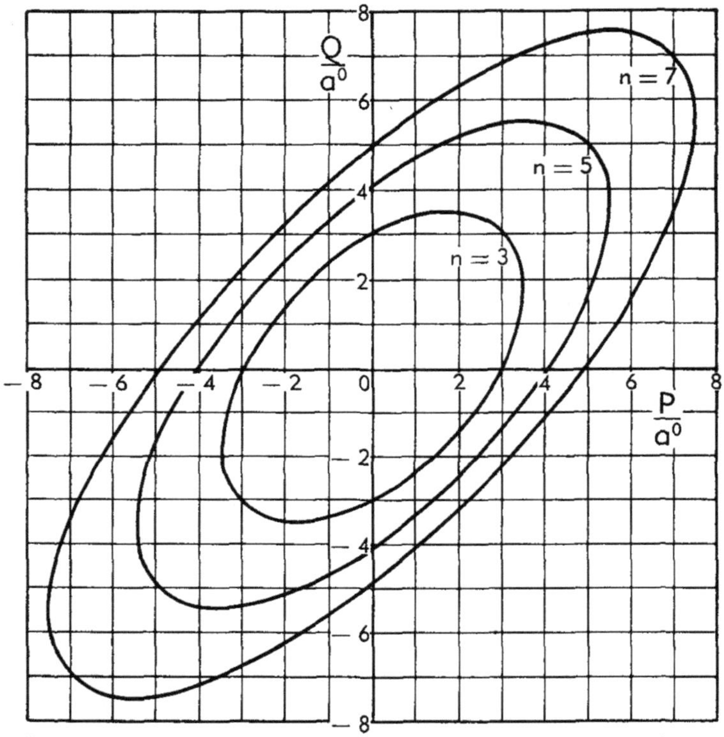

Fig. C,9g. The relation between P and Q when gas with a stagnation speed of sound a^0 is discharged into a long tube. (From Eq. 9-10.)

and velocity in the chamber by steady flow relations. Using the steady flow relation

$$u^2 + na^2 = n(a^0)^2$$

or

$$\left(\frac{1}{n} + 1\right) P^2 + 2\left(\frac{1}{n} - 1\right) PQ + \left(\frac{1}{n} + 1\right) Q^2 = 4n(a^0)^2 \qquad (9\text{-}10)$$

where a^0 is the speed of sound in the chamber. Eq. 9-10 is presented graphically in Fig. C,9g.

Variable and lagging heat capacity. In the previous theory of this section we have considered only the case of perfect gases with heat capacity independent of the temperature. In polyatomic gases there is in general a heat capacity due to molecular vibration, which varies with the temperature. Also, in hydrogen and in gases such as methane consisting of one heavy atom and hydrogen atoms, the heat capacity due to molecular rotation varies at temperatures below room temperature.

In these cases where the heat capacity depends on temperature, it has also been found [*27*] that the vibrational (or rotational, in the case of hydrogen) heat capacity requires an appreciable time to adjust to changes in state. This relaxation time is especially large in the case of combustion gases which contain large amounts of nitrogen.

Hence there is little point in including heat capacities variable with temperature unless we simultaneously include the lag effects, since in most cases the part of the heat capacity which varies with temperature also lags behind the rapid temperature changes which occur in wave processes. This can be illustrated from sound propagation in these gases. At very low frequencies the speed of sound can be calculated correctly for polyatomic gases by using the total heat capacity to compute γ. On the other hand at very large frequencies (for gases other than hydrogen, etc.), the velocity of sound can be calculated correctly if we use the heat capacity due to translation and rotation only. In intermediate frequency ranges, the velocity of sound is intermediate but there is a very large absorption of sound. It is clearly necessary to include these heat capacity lag effects in studying the propagation of finite amplitude disturbances in gases with vibrational heat capacity, except when the relaxation time is very short compared to the typical times in the problem. If the relaxation time is very long compared to all the typical times in the problem, it would be possible to neglect the vibrational energy entirely, proceeding as before.

We shall accordingly develop a procedure for dealing with variable and lagging heat capacity simultaneously. Divide the heat capacity of the gas into two parts: first a part c_v', which is independent of the temperature and which reaches equilibrium in a negligible time. This includes the translational heat capacity $\frac{3}{2}R$ and, for gases other than hydrogen, etc., the rotational heat capacity which is R for linear molecules (diatomic molecules CO_2, etc.) and $\frac{3}{2}R$ for nonlinear molecules. The second category c_v'' includes those parts of the heat capacity which vary with the temperature, and we assume that for this portion of the heat capacity a finite adjustment time is required. We restrict ourselves to cases where there is only one lagging heat capacity (with a single relaxation time).

We can adopt Eq. 9-3 and 9-4 to this case if we make the following modifications. We take the entropy as used in Eq. 2-5 to be the entropy s' associated with the heat capacity c_v'. Thus when there is a change in temperature, heat flowing from c_v'' to c_v' causes increases in the entropy s', even though there are no irreversible or external heat addition processes taking place. If now we write

$$a' = \sqrt{\frac{c_v' + R}{c_v'} \frac{p}{\rho}}, \qquad n' = \frac{2c_v'}{R}$$

$$P' = n'a' + u, \qquad Q' = n'a' - u$$

we can reproduce the derivations of Eq. 9-3 and 9-4 and obtain

$$\delta^+ P' = -\frac{ua'}{u + a'} \delta^+ \ln A + \frac{a'}{\gamma' R} \delta^+ s' + \frac{a'}{c_p'} \left(\frac{Ds'}{Dt}\right) \delta t \tag{9-11}$$

and

$$\delta^- Q' = -\frac{ua'}{u - a'} \delta^- \ln A + \frac{a'}{\gamma' R} \delta^- s' + \frac{a'}{c_p'} \left(\frac{Ds'}{Dt}\right) \delta t \tag{9-12}$$

Assume that the flow of heat dq from c_v'' to c_v' is proportional to the difference between a temperature T' associated with c_v' and a temperature T'' associated with c_v''. This involves the following assumptions: (1) that it is proper to assign temperatures to the various parts of the heat capacity, thus implying equilibria associated with each of the two parts of the heat capacity; (2) the assumption that the heat flow is linearly proportional to the temperature difference is clearly only a first approximation valid for small temperature differences. This assumption for the rate of heat flow has been used in all previous treatments of heat capacity lag in acoustics and gas dynamics. We have then

$$dq = k(T'' - T')dt \tag{9-13}$$

The meaning of the constant can be seen by integrating Eq. 9-13 under the condition where the total enthalpy of the gas is constant. Constant enthalpy means

$$c_p'T' + \int_0^{T''} c_v'' dT'' = \text{const} \tag{9-14}$$

Since we have already restricted this analysis to cases where T' is not greatly different from T'' by the use of Eq. 9-13, it entails no further loss of generality to use an appropriate average value of c_v'' in Eq. 9-14 and this equation becomes

$$c_p'T' + c_v''T'' = c_pT^* \tag{9-15}$$

where $c_p = c_p' + c_v''$ and T^* is the temperature reached when the gas attains equilibrium. Eq. 9-13 can now be written for the constant enthalpy case as

$$\frac{dT'}{T' - T^*} = -k\left(\frac{1}{c_p'} + \frac{1}{c_v''}\right) dt = -\frac{dt}{\tau} \tag{9-16}$$

where $\tau = c_p'c_v''/kc_p$ is called the relaxation time of the gas. Integrating Eq. 9-16 gives

$$T' - T^* = (\text{const})\, e^{-\frac{t}{\tau}}$$

The values of relaxation times to be used for actual gases can be found in [*28*] and in I,H.

Numerical calculations from Eq. 9-11 and 9-12 can be most easily carried out for the case where there is no heat addition from external sources and where there are no irreversible processes taking place locally in the gas other than the flow of heat from c_v'' to c_v'. In this case $dq = T'ds'$. We can find T'' from $dq = -c_v''dT''$. The entropy variations of the gas particles can then be found using Eq. 9-13, noting that $T' = a'^2/\gamma'R$. Now using Eq. 9-11 and 9-12, we can construct the characteristic network as before.

When heat is added to the gas we must distinguish the cases where it is added to c_v' from those cases where it is added to c_v'' (e.g. by radiation or by chemical processes). In many cases, e.g. combustion, the information on which to base such a division of the heat input is not now available (see Vol. II).

C,10. Application of Pressure Waves in Heat Engines.

Introduction. In this article a brief survey is made of the application of one-dimensional wave propagation theory to processes occurring in heat engines. Perhaps the most important application of this kind is to the pulse jet ("buzz-bomb" engine). This engine is discussed in detail in XII,F. An important modification of this engine, which omits the upstream check valve, has recently been developed and is also described in the section referred to above. In the analysis of the performance of these engines, the techniques given in Art. 9 are generally applied. An essential difficulty appears, however, in that it is frequently very difficult to specify the heat release rate produced by the combustion. Until more is known about turbulent combustion, it is necessary to continue to design these engines by semiempirical procedures.

In the operation of "steady flow" ramjets, a phenomenon has been discovered in which the diffuser shock periodically moves upstream of the diffuser inlet. As a result, the diffuser swallows a smaller air flow which in turn affects the combustion processes. The diffuser shock is then again swallowed and the process repeats periodically. An unsteady flow analysis of this phenomenon appears in XII,E,3.

The chugging of rockets is another wave propagation phenomenon (see XII,G and H).

Internal combustion engines. In internal combustion engines, pressure waves of considerable amplitude are created in the inlet and exhaust piping by the rapid flow of gases into and out of the cylinder. Thus, for example, if the intake valve opens quickly at a time such that the cylinder pressure is lower than the pressure in the intake manifold, an expansion wave propagates along the intake manifold. A considerable effort has been made to employ these unsteady flow phenomena in the intake and exhaust manifolds to improve the scavenging and supercharging of internal combustion engines [*29,30*].

An example of this type of effect is the so-called "Kadenacy effect." In utilizing this effect the exhaust port opens rapidly into a long exhaust pipe. During the early high pressure portion of the discharge, a compression wave moves down the pipe. This wave is then reflected from the end of the exhaust pipe and the type of reflection is governed by the exhaust pipe termination. It has been found that when a long pipe is terminated with a diverging diffuser, a considerable reflected expansion wave can be obtained in practice. A momentary but considerable pressure reduction at the exhaust port can be achieved in this way, and Kadenacy has employed this pressure reduction to achieve improved scavenging especially in two cycle engines without the use of a blower. Jenny [*31*] has made a considerable study of the pressure at the exhaust valve which is obtained when we have various terminations, pipe lengths, etc. He has compared his experimental results with calculations made from the method of characteristics and finds good agreement, provided that effects of wall friction are included.

Wave engines. A series of engines appeared, starting about the year 1900, in which one gas (A) was used directly to compress another gas (B). In these machines a compressed gas (A) is introduced at one end of a tube which contains the gas (B). After being compressed, gas (B) is withdrawn from the opposite end of the tube. Machines of this type can be made to operate with valves at the two ends of the tube or continuously by mounting a bank of such tubes on a rotor. In the latter case the passage of the tube ends past nozzles, gas pickups, etc. provides valving action.

Machines of this type have been designed to serve several purposes. First, as gas generators, they have been used to provide a supply of high pressure gas. In this application, use is made of the fact that a given mass of high temperature gas acting as gas A can displace a larger mass of the gas at low temperature acting as gas B. In this way a net output of compressed gas can be achieved and the machine becomes a gas generator. It has been proposed to use this generated high pressure gas for various purposes. For example, the Brown Boveri Company [*32,33*], in their machine called the "Comprex," proposed to use a machine of this type to drive an auxiliary turbine. In their designs this machine was used in conjunction with a conventional gas turbine with a compressor producing a pressure ratio of four. The "Comprex" operated with a pressure ratio of 2.5, thus generating gas at a final pressure of 10 atmospheres. They reported considerable improvement in the efficiency and power output of the gas turbine upon the addition of the "Comprex."

The author and his associates, employing the method of characteristics, have designed machines of this type for considerably higher pressure ratios [*34*]. Thus, for example, gas generators giving ratios in excess of 3 have been constructed.

In these machines, intended for large pressure ratios, important wave propagation phenomena appear. Thus the sudden introduction of a high pressure gas A into a tube filled with low pressure gas produces a strong shock wave. In various other porting phenomena, strong expansion and shock waves are produced. The early inventors of machines of this type apparently had an inadequate understanding of some of these phenomena. For example, in the Brown Boveri design as described in [*32,33*], all the wave phenomena are described as sound waves which lead to important inaccuracies in timing. More recently, the method of characteristics has become available in the engineering literature and it is to be hoped that application of this knowledge of the propagation of waves of finite amplitude will lead to improved designs of these machines.

C,11. Cited References and Bibliography.

Cited References

1. Poisson, S. D. *J. école polytech. 14*, 319–392 (1808).
2. Earnshaw, S. *Trans. Roy. Soc. London 150*, 133–148 (1860).
3. Riemann, B. *Göttingen Abhandl. 8*, 1860.
4. Rankine, W. J. M. *Trans. Roy. Soc. London A160*, 277–288 (1870).
5. Hugoniot, H. *J. école polytech. 58*, 1–125 (1889).
6. Courant, R., and Friedrichs, K. O. *Supersonic Flow and Shock Waves.* Interscience, 1948.
7. Prandtl, L., and Busemann, A. *Näherungsverfahren zur Zeichnerischen Ermittlung von Ebenen Strömungen mit Uberschallgeschwindigkeit.* Stodola Festschrift, Zurich, 1929.
8. Hall, N. A. An introduction to the analysis of one-dimensional non-steady flow. *United Aircraft Corp. Meteor Rept. UAC-44*, Jan. 1950.
9. Guderley, G. Nonstationary gas flow in thin pipes of variable cross section. *NACA Tech. Mem. 1196*, 1948. Transl. from *Z. W. B. Forschungsbericht 1744*, 1942.
10. Rayleigh, Lord (J. W. Strutt). *The Theory of Sound*, 2nd ed. Reprinted by Dover, 1945.
11. Morse, P. M. *Vibration and Sound*, 2nd ed. McGraw-Hill, 1948.
12. Whitham, G. B. *Proc. Roy. Soc. London A203*, 571–581 (1950).
13. Liepmann, H. W., and Puckett, A. E. *Aerodynamics of a Compressible Fluid.* Wiley, 1947.
14. Chandrasekhar, S. On the decay of plane shock waves. *Ballistic Research Lab. Rept. 423*, Nov. 1943.
15. Friedrichs, K. O. *Commun. on Pure and Appl. Math. 1*, 211–245 (1948).
16. Lighthill, M. J. *Phil. Mag.* 7, 1101–1128 (1950).
17. Kantrowitz, A. The formation and stability of normal shock waves in channel flow. *NACA Tech. Note 1225*, 1947.
18. Taylor, G. I. *Proc. Roy. Soc. London A201*, 159–186 (1950).
19. Guderley, G. *Luftfahrtforschung 19*, 302–312 (1942).
20. Brinkley, S. R., and Kirkwood, J. G. *Phys. Rev. 71*, 606–611 (1947).
21. Perry, R. W., and Kantrowitz, A. *J. Appl. Phys. 22*, 878–886 (1951).
22. Bargmann, V. On nearly glancing reflection of shocks. *NDRC Appl. Math. Panel Rept. 108. 2R*, 1945.
23. Lighthill, M. J. *Proc. Roy. Soc. London A198*, 454–470 (1949).
24. Tan, H. S. *J. Aeronaut. Sci. 18*, 768–769 (1951).
25. Smith, L. G. Photographic investigation of the reflection of plane shocks in air. *NDRC A350, Office Sci. Research and Develop. Rept. 6271*, Nov. 1945.
26. Harrison, F. B., and Bleakney, W. Remeasurement of reflection angles in regular and Mach reflection of shock waves. *Private communication*, Mar. 1947.
27. Richards, W. T. *Revs. Mod. Phys. 11*, 36–64 (1939).

28. Kantrowitz, A. *J. Chem. Phys. 14*, 150–164 (1946).
29. Buchi, A. *Motortech. Z. 1*, 198–199 (1939).
30. Carter, H. D. *Proc. Inst. Mech. Engrs. 154*, 386–396, 1946.
31. Jenny, E. *Brown Boveri Mitt. 37*, 1950.
32. Seipple, C. *U.S. Patent 2,399,394*, Apr. 1946.
33. Meyer, A. *Electrical World 127*, 38–40 (1947).
34. Kantrowitz, A. Heat engines based on wave processes. Preprint of paper delivered at *ASME Annual Meeting in New York City*, Nov. 29, 1948.

Bibliography

Acoustical oscillations in burners. *Battelle Memorial Inst. Progress and Summary Rept.*, Columbus, Ohio, 1947.

Aschenbrenner, J. *Forsch. Gebiete Ingenieurw. 8*, 118–130 (1937). Transl. by *Cornell Univ. Aeronaut. Lab.*, 1946.

Bolle, and Seitz. *Einfuehrung in die Innere Ballistik*, pp. 114–135. Vieweg, Braunschweig, 1941.

Busemann, A. Probleme der Detonation. *Schriften Deut. Akad. Luftfahrtforschung 1023/41G*, 59–89 (1941).

Damköhler, G., and Schmidt, A. Contributions of fluid dynamics to the evaluation of flame experiments in tubes. *Z. Elektrochem. 47*, 547–567 (1941).

de Haller, P. The application of a graphic method to some dynamic problems in gases. *Sulzer Tech. Rev. Switz. 1*, 6–24 (1945). Also *Ministry of Supply RTP Transl. 2555*, 1945.

de Haller, P. Une solution graphique du problème de Langrange en balistique intérieure. *Bull. tech. Suisse Romande 3*, Jan. 1948.

Diedrich, G. The aero-resonator power plant of the V-1 flying bomb. Transl. by *Princeton Univ. Project Squid Tech. Mem. Pr-4*, June 1948.

Elias, P. Flame and particle motions in small pulse jet engine. *New York Univ. Project Squid Tech. Rept. 16*, July 1948.

Hall, N. A. An introduction to the analysis of one-dimensional non-steady flow. *United Aircraft Corp. Meteor Rept. UAC-44*, Jan. 1950.

Hett, J. H. Report on conference on the "Measurement of Temperatures of Pulsating Burning Gases." *New York Univ. Project Squid Tech. Mem. NYU-2*, Apr. 1948.

Huber, P. W., Fitton, C. E., and Delpino, F. Experimental investigation of moving pressure disturbances and shock waves and correlation with one-dimensional unsteady-flow theory. *NACA Tech. Note 1903*, 1949.

Kahane, A., and Lees, L. Unsteady one-dimensional flows with heat addition on entropy gradients. *Princeton Univ. Project Squid Tech. Mem. Pr-2*, Nov. 1947.

Kamrass, M. Two-dimensional supersonic wind tunnel investigation of flow in a duct with fluctuating exit pressure. *Cornell Univ. Aeronaut. Lab. Project Squid Tech. Mem. CAL-7*, June 1947.

Kobes. *Z. Osterr. Ing. u. Architekt-Ver. 35*, 1910.

Kuo, Y. H. *Quart. Appl. Math. 4*, 349–360 (1947).

Lewy, H. Motion of gas in half-open pipes. *New York Univ. Appl. Math. Group Rept. 73*, Oct. 1944.

Logan, J. G., Jr. On the possibility of representing one-dimensional gas motion by means of an electrical analogy. *Cornell Univ. Aeronaut. Lab. Project Squid Tech. Mem. CAL-2*, June 1947.

Logan, J. G., Jr. Suggested forms of air duct motors utilizing intermittent combustion. Part I: Intermittent jet engines. Part II: Intermittent combustion turbines. *Cornell Univ. Aeronaut. Lab. Project Squid Tech. Mem. CAL-15*, Feb. 1948. Part III: Modification of combustion cycle to obtain wave reinforcements. *CAL-16*, Mar. 1948.

MacDonald, J. K. L. A gas dynamical formulation for waves and combustion in pulse-jets. *New York Univ. Appl. Math. Group Rept. 151*, June 1946.

McFadden, J. A. *Phys. Rev. 84*, 611 (1951).

Pfriem, H. *Akust. Z. 6*, 222–241 (1941).

Pfriem, H. *Forsch. Gebiete Ingenieurw. 12*, 51–64, 244–256 (1941).

Pfriem, H. *Forsch. Gebiete Ingenieurw. 13*, 112–124 (1942).

Pfriem, H. Transl. from *Akust. Z. 7,* 56–65 (1942). Abstract in *Roy. Aeronaut. Soc. J. 1,* 306–307, 1944.

Pischinger, A. Oscillations in gas columns, with special reference to introduction and scavenging processes in 2-stroke engines. *Brit. Aeronaut. Research Council RTP. Transl. 2494.* Transl. from *Forsch. Gebiete Ingenieurw. 6,* 273–280 (1935).

Rayleigh, Lord (J. W. Strutt). *Scientific Papers 5,* pp. 573–610. Cambridge Univ. Press, 1912.

Roberts, R. C. The method of characteristics in compressible flow, Part II. *Air Materiel Command Tech. Rept. F-TR-1173D-ND,* Dec. 1947.

Rothstein, W. Unsteady compressible flow. Transl. from *AVA Monographs by Brit. RTP 993,* Apr. 1948.

Rozzell, G. M. A summary of pulse jet theory. *Curtiss-Wright Corp. Airplane Division, Columbus, Ohio, Rept. E-47-2,* Mar. 1947.

Rudinger, G., and Logan, J. G., Jr. Study of gas oscillations in half-open pipes of various shapes. *Cornell Univ. Aeronaut. Lab. Project Squid. Part I. Tech. Rept. 3,* June 1947. Also *Squid Tech. Mem. CAL-4,* June 1947.

Rudinger, G., Logan, J. G., Jr., and Dashifsky, W. Investigation of acoustic jets, Part I. *Cornell Univ. Aeronaut. Lab. Project Squid Tech. Mem. CAL-13,* Feb. 1948.

Rudinger, G., Logan, J. G., Jr., and Finamore, O. B. Investigation of acoustic jets, Part II. *Cornell Univ. Aeronaut. Lab. Project Squid Tech. Mem. CAL-18,* Apr. 1948.

Rudinger, G., and Rinaldi, L. D. The construction of wave diagrams for the study of one-dimensional non-steady gas flow. *Cornell Univ. Aeronaut. Lab. Project Squid Tech. Mem. CAL-14,* Mar. 1948.

Sauer, R. On the theory of the jet tube. *Goett. AVA, Zent. Wissensch. Berichtsw. über Luftfahrtforsch. Forschungsbericht 1864-1-2,* Oct. 1943. (On microfilm in German.)

Sauer, R. The method of characteristics for the one-dimensional unsteady flow of a gas. Transl. from *Ing.-Arch. 13,* 78–89 (1942); also *Brown Univ. Transl. A9-T-2,* 1948, and *Brit. Aeronaut. Research Council RTP Transl. 2021.*

Sauer, R. Theory of the nonstationary gas flow. I: Laminar steady gas waves. *Deut. luft. FB 1675/1, Army Air Force Transl. F-TS-763-RE,* July 1946. II: Plane gas waves with compression shocks. *Deut. luft. FB 1675/2, Army Air Force Transl. F-TS-758-RE,* Aug. 1946. III: Laminar flow in tubes of variable cross-section. *Deut. luft. FB 1675/3, Army Air Force Transl. F-TS-770-RE,* Feb. 1947. IV: Plane gas flow with friction heat transfer and temperature differences. *Deut. luft. FB 1675/4, Army Air Force AAF Transl. F-TS-949-RE,* Mar. 1947.

Sauer, R. Zur Theorie des nichtstationären ebenen Verdichtungsstosse. *Ing.-Arch. 14,* 1943.

Sauer, R. Zur Theorie des nichtstationären Verdichtungsstosse. *Ing.-Arch. 13,* 1942.

Schardin, H. *Physik. Z. Jahrb. 33,* 60–64 (1932).

Schmidt, P. The Schmidt tube. *Zent. Wissensch. Berichtsw. über Luftfahrtforsch. THM/Schmidt 44,* June 1944. Transl. by Univ. Mich. for Bur. of Aeronautics, Dept. of the Navy.

Schultz-Grunow, F. Flow phenomena in an impulsive duct. Transl. by *Brit. Ministry of Supply, Volkenrode RTP 371,* Feb. 1947.

Schultz-Grunow, F. *Forsch. Gebiete Ingenieurw. 11,* 170–187, 1942.

Schultz-Grunow, F. Gas-dynamic investigations of the pulse-jet tube, Parts I and II. *NACA Tech. Mem. 1131,* 1947. Transl. from *FB 2015-1,-2 Tech. Hochschule,* Aachen.

Schultz-Grunow, F. *Ing.-Arch. 14,* 21–29 (1943). Transl. by *Brit. Ministry of Supply RTP 2071.*

Schultz-Grunow, F. Three contributions on non-stationary one-dimensional gas flow. *Brit. Ministry of Supply, Volkenrode RTP 223,* Mar. 1947.

Torda, P. Approximate theory of compressible air inflow through reed valves for pulse jet engines. *Polytech. Inst. Brooklyn Project Squid Tech. Rept. 12,* Sept. 1948.

Torda, P., Villalba, I. P., and Brick, J. H. Compressible flow through reed valves for pulse jet engines. Part I: Hinged reed valves. *Polytech. Inst. Brooklyn Project Squid Tech. Rept. 9,* June 1948. Part II: Clamped reed valves. *10,* July 1948.

Whitham, G. B. *Proc. Roy. Soc. London A203,* 572 (1950).

SECTION D

THE BASIC THEORY OF GASDYNAMIC DISCONTINUITIES

WALLACE D. HAYES

It is the aim of this section to present a theoretical introduction to the subject of gasdynamic discontinuities, from as general a point of view as is practicable. A gasdynamic discontinuity is a surface in a fluid field across which various properties of the fluid appear from a macroscopic point of view to change discontinuously and across which there is some flow of the fluid. A gasdynamic discontinuity is thus distinguished from a contact discontinuity across which there is no flow of fluid and across which the pressure is continuous. Hydrodynamic discontinuities occur in many fields of fluid mechanics; in the field of gas dynamics or high speed aerodynamics they appear in essentially all flows of practical importance. Their existence was first recognized in the middle of the nineteenth century, though it was not until the work of Hugoniot (1889) [*1*, p. 80] that the presently accepted formulation of the discontinuity relations was established. For a discussion of the early history of the study of discontinuities in hydrodynamics the reader is referred to the book of Courant and Friedrichs [*2*].

The most important gasdynamic discontinuity is the shock wave, in which a gas or other material undergoes a sudden increase in pressure, density, temperature, and entropy. In an ordinary shock wave the gas behind the shock obeys the same equation of state as does the gas in front of the shock. Many other types of gasdynamic discontinuity do not have this property; perhaps the most important type is the combustion wave, in which a chemical reaction takes place in the discontinuity proper so that the materials on the two sides of the discontinuity obey quite different equations of state. Another important type which may appear in aerodynamic or wind tunnel problems is the condensation shock, in which the gas in front of the discontinuity contains a vapor in a supersaturated state which partially condenses in the discontinuity proper.

Important in very strong discontinuities or shocks are phenomena which heretofore have been of interest particularly to physicists. Among these effects may be cited ionization, dissociation, and relaxation. With large amounts of radiation the energy effects of the radiation may strongly affect the discontinuity, and with sufficiently high velocities relativistic effects might have to be taken into account.

It is true, of course, that these gasdynamic discontinuities are not discontinuities in the strict sense; a shock, combustion wave, or condensation shock has a finite thickness across which the physical properties change continuously. If this thickness is small compared with some appropriate macroscopic dimension of the flow field, such as the radius of curvature of a curved shock, the physical relationships may be obtained by an analysis which treats the discontinuity as strict. The assumption that the discontinuity thickness is small compared with a macroscopic dimension is a fundamental one for this section.

The term "structure" as applied to a gasdynamic discontinuity refers to the values of the physical properties of the fluid within the small but finite thickness of the discontinuity. If thermodynamic equilibrium in a substance is disturbed, a characteristic time must elapse before equilibrium can be approximately reestablished; this time times the velocity of the fluid defines a characteristic distance which is of the order of a molecular mean free path or greater. If the physical and chemical changes occurring in the discontinuity are sufficiently slow, so that the thickness of the discontinuity is large compared with this characteristic distance, the concept of thermodynamic quasi-equilibrium may be considered to apply. In this case the Navier-Stokes equations are applicable, as are the classical chemical kinetic laws for slow reactions. If the discontinuity is thin, with the physical and chemical changes occurring rapidly, the essential absence of thermodynamic equilibrium must be taken into account. This may involve abandoning the continuum concept and taking the point of view of kinetic theory. Certain molecular processes such as diffusion and nucleation must be taken into account in investigating the structure of discontinuities involving, for example, chemical reactions or condensation. (See Sec. F, G, and H.)

It is clear that the thorough study of gasdynamic discontinuities and their structures combines in an essential way the fields of hydrodynamics, physics, and chemistry, and that there is no lack of problems which deserve attention. At the present time there is a strong need for additional concepts which properly describe phenomena taking place in discontinuities and which will permit some simplification of the complicated laws governing the attendant phenomena, so that more suitable theoretical approaches to the problems may be made.

D,1. Basic Relations in a Normal Discontinuity. The basic laws governing normal discontinuities in hydrodynamics are now given and discussed. The investigation is made on the basis of a steady state process; the flow is considered to be one-dimensional and the discontinuity surface is assumed to be perpendicular to the direction of the flow. Although most discontinuities as they occur in nature do not satisfy these restrictions, they do not introduce any essential lack of generality. The motion of the discontinuity may be eliminated by having the observer move so as to always lie on the discontinuity. Unsteadiness and nonuniformity of the flow field may be eliminated by having the observer take a sufficiently microscopic point of view, with small space and time scales. Obliquity of the discontinuity may be eliminated by having the observer move along the discontinuity so that the tangential velocity components are zero.

In the establishment of the basic laws certain assumptions are necessary, on which the validity of the results depends. The principal basic assumption which must be made is that the fluid on each side of the discontinuity obeys a known equation of state[1] and has unique definable values of the velocity and of the various intensive and specific extensive thermodynamic variables. This assumption may pose difficulties in certain cases, such as in a mixture of solid and gaseous combustion products, or a case where there is a marked deviation from thermal equilibrium. Energy exchange through radiation is neglected, and the laws of Newtonian or nonrelativistic mechanics are assumed.

The basic laws of discontinuities are derived by use of the basic conservation principles of mechanics. Since these laws, or very similar ones, have already been derived in previous sections, no detailed derivation is given here. The standard notation is used, with p the pressure, ρ the density, u the velocity, h the specific enthalpy, and e the specific internal energy. For convenience the additional symbol v is used for the inverse of the density, i.e. for the specific volume.

Referring to Fig. D,1a, the principle of the conservation of mass gives the basic law

$$m = \rho_1 u_1 = \rho_2 u_2 = \frac{u_1}{v_1} = \frac{u_2}{v_2} \tag{1-1}$$

The quantity m thus defined is termed the mass flow. The principle of the conservation of momentum gives the law

$$p_0 = p_1 + \rho_1 u_1^2 = p_2 + \rho_2 u_2^2 \tag{1-2}$$

Finally, the principle of the conservation of energy gives the law

$$h_0 = h_1 + \tfrac{1}{2}u_1^2 = h_2 + \tfrac{1}{2}u_2^2 \tag{1-3}$$

These three are termed the conservation laws.

[1] The term "equation of state" here and subsequently in this section is used in a sense encompassing all the usual thermodynamic variables, and not in the engineering sense specifying only the relationship connecting pressure, volume, and temperature. The specification of enthalpy as a function of entropy and pressure, for example, may be considered the equation of state in the sense used here, because such a relation gives complete thermodynamic information. Also, the term "an equation of state" is occasionally used to refer to any equation connecting state variables.

A convenient result, valid for all types of gasdynamic discontinuity, may be obtained immediately from the first two conservation laws (Eq. 1-1 and 1-2). It is

$$p_2 - p_1 = m(u_1 - u_2) = m^2(v_1 - v_2) \tag{1-4}$$

and shows that a decrease of the flow velocity across the discontinuity is associated with an increase in the pressure and an increase in the density. Conversely, an increase in the flow velocity is associated with a decrease in the pressure and a decrease in the density. This permits the immediate classification of all discontinuities into those of compression (deceleration) types and those of expansion (acceleration) types. If the three quantities equated in Eq. 1-4 are zero, there is no discontinuity. If they are small the

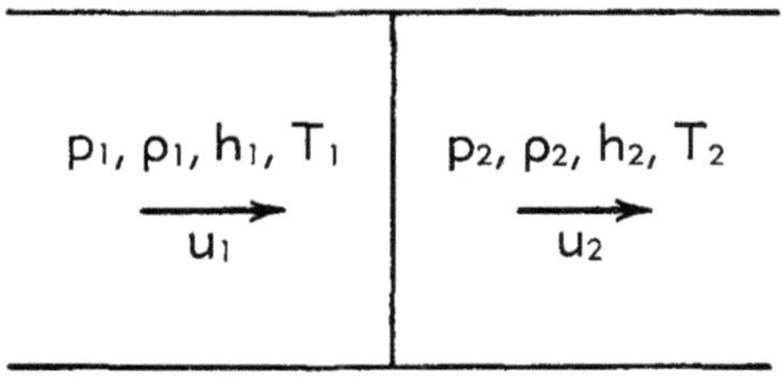

Fig. D,1a. Normal discontinuity.

discontinuity is termed weak. Alternate forms obtainable from Eq. 1-4 are

$$\frac{p_2 - p_1}{\rho_2 - \rho_1} = u_1 u_2 \tag{1-5}$$

$$(u_1 - u_2)^2 = (p_2 - p_1)(v_1 - v_2) \tag{1-6}$$

The energy function in Eq. 1-3 is the specific enthalpy, related to the specific internal energy by the relation

$$h = e + pv \tag{1-7}$$

An alternate expression for the energy equation may be obtained as

$$h_0 = e_1 - \tfrac{1}{2}u_1^2 + \frac{p_0 u_1}{m} = e_2 - \tfrac{1}{2}u_2^2 + \frac{p_0 u_2}{m} \tag{1-8}$$

From the conservation laws may be derived the so-called Hugoniot relation, which may be expressed in two alternate forms:

$$h_2 - h_1 = \tfrac{1}{2}(p_2 - p_1)(v_1 + v_2) \tag{1-9}$$

$$e_2 - e_1 = \tfrac{1}{2}(p_2 + p_1)(v_1 - v_2) \tag{1-10}$$

This relation is of great importance in considerations of detonations and deflagrations and has the particular property that the flow velocities do not appear, so that the equation is a purely thermodynamic one. This relation leads to the Hugoniot diagram, in which p_2 is plotted vs. v_2 for a given choice of p_1 and v_1. If the fluid on both sides of the discontinuity obeys the same equation of state, the point p_1, v_1 lies on the curve. The cases in which a reaction or change of state occurs, such that the point p_1, v_1 does not lie on the curve, may conveniently be divided into those for which the point lies below the curve, termed exothermic, and those for which the point lies above the curve, termed endothermic.[2] Only the exothermic case is generally encountered; this is the case for which the Hugoniot diagram is illustrated in Fig. D,1b. Other diagrams are useful in the study of discontinuity phenomena, though they do not have the property of the Hugoniot diagram of involving only thermodynamic variables.

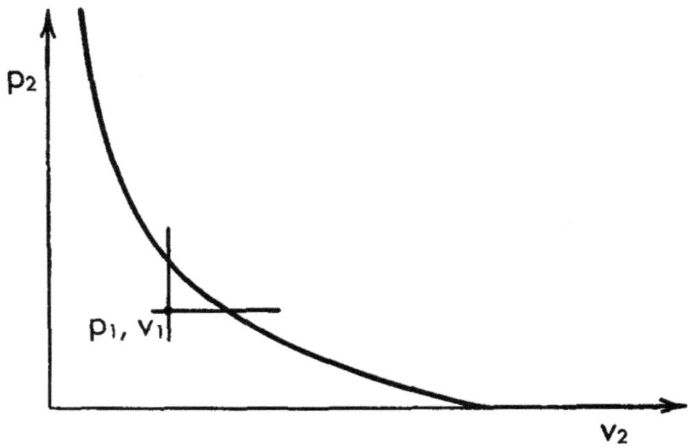

Fig. D,1b. Hugoniot diagram.

One of the most important questions that arises in the study of gasdynamic discontinuities is the question of existence. Some conditions for the existence of such a discontinuity may be given:

1. The conservation laws given above must be satisfied. This condition is a necessary one but is far from being sufficient. Many hypothetical discontinuities which satisfy the discontinuity laws do not exist.
2. The specific entropy of the material must increase. This condition is imposed to satisfy the second law of thermodynamics and is a necessary one. Together with condition 1 it is still not sufficient to ensure the existence of the discontinuity, but may be useful in eliminating

[2] This use of the words exothermic and endothermic is unconventional, and perhaps some such terms as expansive and contractive might be more appropriate. However, this usage would lead to such confusing entities as a compression expansive discontinuity, and the author prefers the terms exothermic and endothermic despite the lack of a clear-cut definitional connection with the release or absorption of heat. In general a reaction that gives off heat would give a discontinuity that is exothermic in the sense used here, and one that absorbs heat would give a discontinuity that is endothermic; hence the terminology does have some connection with conventional usage.

certain nonexistent discontinuities which would be permitted by condition 1.

3. The discontinuity must correspond in its structure to a physically realizable process. Here the term "physically" includes considerations of the chemistry and chemical kinetics of any reactions taking place. This condition is necessary and sufficient for the existence of the discontinuity in the small provided the discontinuity is internally stable. Since the appropriate physical and thermodynamic laws must be satisfied within the structure of the discontinuity, conditions 1 and 2 are automatically satisfied. The precise physical process must in most cases be approximated by a somewhat idealized process for purposes of theoretical treatment. For example, the Navier-Stokes equations are not strictly valid within a shock of finite strength and if they are used in an analysis the resultant picture of the physical process is an approximation. The pertinent point here is that the validity of an existence proof based on the demonstration of a physical process depends upon the accuracy with which the assumed process approximates the actual one.
4. The discontinuity must be internally stable. This means that if an equilibrium solution undergoes a disturbance allowed by the local hydrodynamic conditions the solution must return to the equilibrium one. The theory for such internal or local stability has not as yet been greatly developed. An attempt is given in Art. 4 below to outline such a theory and to present a few results.
5. The discontinuity must be stable in the large. A demonstration of the satisfaction of conditions 3 and 4 given above can at most prove the possible existence of a discontinuity in the small. A condition for the existence of a discontinuity in any particular case is that it must be stable with respect to possible changes in the configuration and hydrodynamic solution of the entire flow. The question of this type of stability lies outside the scope of this section.

Gasdynamic discontinuities must be classified according to whether or not the equations of state governing the material change as the material crosses the discontinuity. If the equations of state are the same for the material on both sides of the discontinuity it is termed a shock. A special case sometimes encountered is that in which the equation of state for a material has different forms in different thermodynamic regions because of different phases, so that the equations of state for the same material on the two sides of a discontinuity might be of different form. An example would be one in which a partially condensed gas changed through the discontinuity to a state with no condensed phase; such a discontinuity might be called an evaporation shock and would occur, for example, if a sufficiently strong shock were passed through a mist. However, in most cases in which the form of the equations of state changes, the material on the front side of the discontinuity is in a metastable state. Examples of such discontinuities are condensation shocks in which the vapor before the shock is supersaturated, and detonations or deflagrations, in which a definite chemical change takes place. In all practical cases known to the author the change is exothermic.

The next two articles depend to a large extent on detailed considerations of the Hugoniot diagram. To serve as a basis for these two articles, some general results are presented here. In the Hugoniot diagram the original state of the fluid (subscript $_1$) is assumed known and fixed. The subscript $_2$ is now dropped from the unknown quantities behind the discontinuity for the purposes of the next two articles and the remainder of this one. Except briefly at the end of this article, phase changes are not considered in this section, so that the thermodynamic variables are generally considered to have continuous first derivatives.

As a preliminary, some basic inequalities from thermodynamics are recalled [*3*, pp. 79–81]. Two of these, based on the thermal stability of a system, may be expressed

$$c_p > c_v > 0 \tag{1-11}$$

The other, based on the hydrostatic stability of a system, may be expressed

$$\left(\frac{\partial p}{\partial v}\right)_s = \frac{c_p}{c_v}\left(\frac{\partial p}{\partial v}\right)_T < 0 \tag{1-12}$$

A fundamental physical condition is that the mass flow m is real, or from Eq. 1-4 that $p - p_1$ and $v_1 - v$ must have the same sign. This limits the parts of the Hugoniot curve which are of physical interest to those lying in the upper left quadrant from the point p_1, v_1, corresponding to compression solutions, and in the lower right quadrant, corresponding to expansion solutions. Any portion of the curve in either of the other two quadrants would be of interest only for considerations of mathematical continuity.

Of particular interest are the points on the Hugoniot curve for which various quantities are stationary. One type of stationary point is of especial significance, that for which the specific entropy s and the mass flow m are stationary. The relation

$$dh_0 = \tfrac{1}{2}v_1 dp_0 = v_1^2 m dm \tag{1-13}$$

shows that the quantities h_0 and p_0 have extrema at the same points m does. From the Hugoniot relation Eq. 1-9 or 1-10, together with Eq. 1-4, may be obtained the relations

$$Tds = \tfrac{1}{2}(p - p_1)dv + \tfrac{1}{2}(v_1 - v)dp \tag{1-14}$$

$$Tds = (v_1 - v)^2 m dm = (p - p_1)(v_1 - v)\frac{dm}{m} \tag{1-15}$$

Thus it is seen that the quantities s, m, p_0, h_0, and u_1 have extrema of the same kind (maxima or minima) simultaneously, and that this must occur on a point for which

$$-\frac{dp}{dv} = -\left(\frac{\partial p}{\partial v}\right)_s = \rho^2 a^2 = \frac{p - p_1}{v_1 - v} = m^2 = \rho^2 u^2 \qquad (1\text{-}16)$$

or for which the velocity behind the discontinuity is sonic, with

$$u^2 = a^2 \qquad (1\text{-}17)$$

Such a point is termed a Chapman-Jouguet point, after the important discoveries of Chapman [*4*] and of Jouguet [*5*]. The designation Ch-J will be used henceforth for these points. If the point p_1, v_1 lies on the Hugoniot curve, it is a Ch-J point for which only s need be stationary.

In considering the occurrence of extrema of other quantities, recourse is made to the differentiated forms of Eq. 1-1, 1-4, 1-6, 1-9, and 1-10, together with the law relating internal energy and enthalpy with the entropy

$$de = Tds - pdv \qquad (1\text{-}18)$$

$$dh = Tds + vdp \qquad (1\text{-}19)$$

A couple of the less obvious relations may be expressed:

$$\tfrac{1}{2}d(u_1 - u)^2 = de + p_1 dv = -dh + v_1 dp \qquad (1\text{-}20)$$

$$(v_1 - v)^2 m du = (2v_1 - v)Tds - (v_1 - v)^2 dp \qquad (1\text{-}21)$$

$$(v_1 - v)^2 m du = (2v_1 - v)dh - v_1^2 dp \qquad (1\text{-}22)$$

In the upper left quadrant of the Hugoniot diagram corresponding to compression solutions the extrema must appear in the following cyclic order or its reverse:

p	minimum
h	minimum
Ch-J	minimum
u	minimum
v	minimum
e	maximum
$u_1 - u$	maximum
p	maximum

The designations of maximum or minimum are reversed if the cycle is repeated or if the order becomes reversed. Thus, for example, the extremum which would precede "p minimum" in the list above would be "$u_1 - u$ minimum." Also, if the order of the list were to become reversed after "v minimum" but before "e maximum," the next extremum would be "v maximum," followed by "u maximum." Two of these extrema can only coincide at a singular or nonanalytic point of the curve. The extrema on the compression branch are illustrated on Fig. D,1c, on an exaggerated version of the Hugoniot curve.

To illustrate the method by which this sequence is derived the three quantities p, h, and s are considered. The Hugoniot curve is considered to be continuous and the investigator passes from a minimum of p along the curve to a minimum of s (or Ch-J). At the minimum of p, dp is zero and ds is negative (because the minimum of s is being approached); from Eq. 1-19 dh must be negative there. At the minimum of s, ds is zero and dp is positive; there dh must be positive. The conclusion is drawn that between the minimums of p and of s, the enthalpy must have a minimum. This argument, repeated for various choices of the three variables involved, justifies the construction of the entire sequence.

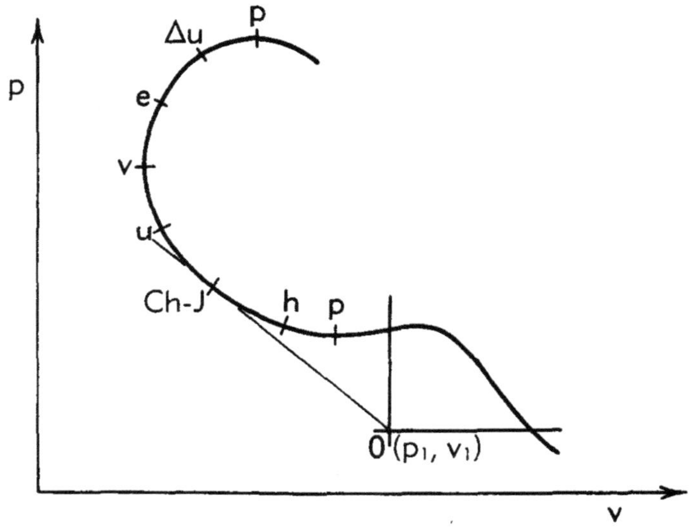

Fig. D,1c. Extrema on the compression branch of the Hugoniot curve.

For the lower right quadrant corresponding to expansion solutions the extrema must appear according to the following corresponding sequence:

v	minimum
e	maximum
Ch-J	maximum
u	maximum (if $v < 2v_1$)
p	minimum
u	maximum (if $v > 2v_1$)
h	minimum
$u - u_1$	maximum
v	maximum

The Ch-J points corresponding to extrema of the slope of the line from the point p_1, v_1 are lateral extrema (of u_1) on a velocity plot, of u vs. u_1.

This relation between the two plots has a reciprocal property, as extrema of the slope of the line through the origin on the velocity plot, u/u_1, are also lateral extrema (of v) on the Hugoniot curve. The point p_1, v_1 on the Hugoniot diagram plays an important part in interpreting the properties of the Hugoniot curve, while the point 0, 0 does not. The point p_1, v_1 is termed the "origin" of the Hugoniot diagram, for ease in reference. The reader may note that if the Hugoniot diagram is considered as a plot of $p - p_1$ vs. $v - v_1$ this point is the origin in the conventional sense.

Some subsidiary formulas involving derivatives of the thermodynamic quantities are of interest, and are expressed in the form

$$d\eta = \frac{dv}{1 - \frac{1}{2}(v_1 - v)\left(\frac{\partial p}{\partial e}\right)_v} \tag{1-23a}$$

$$= \frac{Tds}{\frac{1}{2}(p - p_1) + \frac{1}{2}(v_1 - v)\left(\frac{\partial p}{\partial v}\right)_s} \tag{1-23b}$$

$$= \frac{-de}{\frac{1}{2}(p + p_1) - \frac{1}{2}(v_1 - v)\left(\frac{\partial p}{\partial v}\right)_e} \tag{1-23c}$$

$$= \left(\frac{\partial v}{\partial p}\right)_s \frac{dp}{1 - \frac{1}{2}(p - p_1)\left(\frac{\partial v}{\partial h}\right)_p} \tag{1-23d}$$

$$= \left(\frac{\partial v}{\partial p}\right)_s \frac{dh}{\frac{1}{2}(v_1 + v) + \frac{1}{2}(p - p_1)\left(\frac{\partial v}{\partial p}\right)_h} \tag{1-23e}$$

$$= \left(\frac{\partial v}{\partial p}\right)_s \frac{\frac{1}{2}d(u_1 - u)^2}{\frac{1}{2}(v_1 - v)\left[1 - (p - p_1)\left(\frac{\partial v}{\partial h}\right)_p\right] - \frac{1}{2}(p - p_1)\left(\frac{\partial v}{\partial p}\right)_s} \tag{1-23f}$$

The derivation of these equations is not presented here, except to indicate the method used. Equations derivable from the Hugoniot relation of the type of Eq. 1-14 involve three first differentials. But between three first differentials there is a thermodynamic, or state, relation, such as

$$dp = \left(\frac{\partial p}{\partial s}\right)_v ds + \left(\frac{\partial p}{\partial v}\right)_s dv \tag{1-24}$$

Elimination of one of the variables gives an equation connecting two first differentials; eliminating dp from Eq. 1-14 and 1-24 gives the relation connecting Eq. 1-23a and 1-23b. Suitable repetition of the method yields the complete set of Eq. 1-23. The differential appearing in various forms equated in Eq. 1-23 may not vanish as long as there is a change in any one of the variables; this may be considered the differential of a variable η defined along the Hugoniot curve, termed the "Hugoniot curve variable," which has no stationary points. If one of the denominators in Eq. 1-23 is zero at a point on the Hugoniot curve, the corresponding variable is stationary there, and conversely.

From Eq. 1-14, 1-23a, and 1-23b the following expression may be obtained

$$m^2 - \rho^2 a^2 = \left[1 - \tfrac{1}{2}(v_1 - v)\left(\frac{\partial p}{\partial e}\right)_v\right]\left(m^2 + \frac{dp}{dv}\right) \tag{1-25}$$

which may be used to relate the geometry of the Hugoniot curve to the question of whether the velocity behind the discontinuity is subsonic or supersonic.

Finally, a last property of the Ch-J points is given, with regard to the second derivative of the Hugoniot curve there:

$$\left[1 - \tfrac{1}{2}(v_1 - v)\left(\frac{\partial p}{\partial e}\right)_v\right]\frac{d^2p}{dv^2} = \left(\frac{\partial^2 p}{\partial v^2}\right)_s = 2\rho^3 a^2 \Gamma \tag{1-26}$$

where

$$\Gamma = \frac{1}{a}\left(\frac{\partial \rho a}{\partial \rho}\right)_s \tag{1-27}$$

This dimensionless quantity Γ is equal to $\frac{1}{2}(\gamma + 1)$ for a perfect gas with constant specific heats and it is the appropriate replacement for $\frac{1}{2}(\gamma + 1)$ in the theory of transonic similitude in an imperfect gas. Eq. 1-26 may be derived by differentiating Eq. 1-25 with respect to v and applying the extremal properties of the Ch-J point and those given by Eq. 1-16.

In addition to the fundamental thermodynamic inequalities of Eq. 1-11 and 1-12 other inequalities are considered as conditions which must be satisfied in order that the Hugoniot curve satisfy certain conditions of proper behavior. These are conditions which cannot be shown to be satisfied by any general thermodynamic argument; they are, on the other hand, conditions which are satisfied by most materials but which may not all be satisfied by certain materials. A symbol α for the coefficient of thermal expansion is introduced

$$\alpha = \frac{1}{v}\left(\frac{\partial v}{\partial T}\right)_p \tag{1-28}$$

and the following classical thermodynamic relations are presented

$$\frac{pv}{c_p}\alpha = p\left(\frac{\partial v}{\partial h}\right)_p = \frac{p}{T}\left(\frac{\partial T}{\partial p}\right)_s = 1 + \frac{v^2}{a^2}\left(\frac{\partial p}{\partial v}\right)_e \tag{1-29a}$$

$$\frac{a^2}{c_p}\alpha = v\left(\frac{\partial p}{\partial e}\right)_v = -\frac{v}{T}\left(\frac{\partial T}{\partial v}\right)_s = -1 - \frac{a^2}{v^2}\left(\frac{\partial v}{\partial p}\right)_h \tag{1-29b}$$

The conditions which are of interest are:

Condition I $$\left(\frac{\partial^2 p}{\partial v^2}\right)_s = 2\rho^3 a^2 \Gamma > 0 \tag{1-30}$$

Condition II-weak $$\frac{pv}{c_p}\alpha = \frac{p}{T}\left(\frac{\partial T}{\partial p}\right)_s > -\frac{2pv}{a^2} \tag{1-31}$$

Condition II-strong $$\frac{pv}{c_p}\alpha = \frac{p}{T}\left(\frac{\partial T}{\partial p}\right)_s > -\frac{pv}{a^2} \tag{1-32}$$

Condition III-weak $$\frac{pv}{c_p}\alpha = \frac{p}{T}\left(\frac{\partial T}{\partial p}\right)_s < 2 \tag{1-33}$$

Condition III-strong $$\frac{pv}{c_p}\alpha = \frac{p}{T}\left(\frac{\partial T}{\partial p}\right)_s < 1 \tag{1-34}$$

It may be noted that condition II-strong implies condition II-weak, and condition III-strong implies condition III-weak.

With phase changes the conditions of Eq. 1-31, 1-32, 1-33, and 1-34 are unchanged, except that they may be simplified by replacing $(\partial T/\partial p)_s$ by dT/dp for a two-phase mixture. For a simple two-phase mixture, condition I is also unchanged, but the quantity Γ or $(\partial^2 p/\partial v^2)_s$ is undefined at the point where a substance changes from one-phase to two-phase or vice versa, i.e. under saturation conditions. At such a point the quantity

$$-\left(\frac{\partial p}{\partial v}\right)_s = \rho^2\left(\frac{\partial p}{\partial \rho}\right)_s = \rho^2 a^2 \tag{1-35}$$

is discontinuous, and condition I may be replaced by the condition that ρa increase with an isentropic increase in the density. Note that the quantity a^2 as used here always refers to a thermodynamic derivative which is necessarily identified with the square of the propagation velocity of sound only for sound waves of sufficiently low frequency.

Bethe [*6*] has carried out an investigation of shock waves in an arbitrary fluid following an approach which has certain features in common with the approach of the first two articles of this section, and specifying certain conditions which are the same as some of those above. Condition I above is Bethe's condition I; condition II-weak is Bethe's condition II; and condition III-strong is Bethe's condition III. He investigates in detail the possibility of violations of his three conditions by specific substances; such considerations are not included in the present contribution. It may be said that common materials satisfy all of Bethe's conditions and also condition II-strong. With respect to condition I Bethe notes that at the (thermodynamic state) boundary between a one-phase region and a two-phase region the quantity ρa is always greater for the single phase than for the mixture of two phases. Thus condition I at a phase boundary may be expressed: isentropic expansion must lead to a phase transition from a one-phase state to a two-phase state. Bethe notes that this is always true if one of the phases is a vapor but that it may be untrue for certain liquid-solid and solid-solid transitions. Of course, the basic theory must be modified in general if any of the material involved is a solid under stress.

These various conditions are applied in the next two articles. In general, it may be seen from Eq. 1-23 that in compression discontinuities condition III-strong prevents an extremum of e and condition III-weak prevents an extremum of p; analogously, in expansion discontinuities condition II-strong prevents an extremum of h and condition II-weak prevents an extremum of v.

D,2. The Normal Shock Wave. A shock wave is a gasdynamic discontinuity in which no chemical reaction is involved, and for which the material on both sides may be considered to obey the same equation of state (allowing certain discontinuous behavior if a phase change should be involved). In the analysis of this article no phase changes are considered, although in general the conclusions reached would be valid with such changes present. It is found that conditions I and III-strong are sufficient to guarantee the good behavior of compression shocks, by ensuring their existence and uniqueness under appropriate boundary conditions, and by ensuring that the velocity of the fluid is supersonic with respect to the shock in front of it and is subsonic behind it. The point of view which is taken is that the state of the material in front of the shock is known, and that behind the shock is unknown.

Weak shocks are considered first. Since the Hugoniot curve passes through the "origin" (p_1, v_1) and is presumed to have continuous slope, it satisfies the approximate equation

$$\frac{dp}{dv} \cong \frac{p - p_1}{v - v_1} \tag{2-1}$$

to first order accuracy. From Eq. 1-14 and 1-26 it may be seen that the origin is not only a stationary point for s, or a Ch-J point, but also a point for which the Hugoniot curve osculates the isentrope. Condition I requires that the curvature of the Hugoniot curve there be positive. This indicates that the entropy change across a weak shock can be no larger than of the third order in the pressure or volume change.

In order to obtain a quantitative expression for the entropy change across a weak shock, Eq. 1-14 is reexpressed and differentiated twice with respect to v

$$T\frac{ds}{dv} = \tfrac{1}{2}(v_1 - v)\frac{dp}{dv} + \tfrac{1}{2}(p - p_1) \tag{2-2a}$$

$$\frac{dT}{dv}\frac{ds}{dv} + T\frac{d^2s}{dv^2} = \tfrac{1}{2}(v_1 - v)\frac{d^2p}{dv^2} \tag{2-2b}$$

$$\frac{d^2T}{dv^2}\frac{ds}{dv} + 2\frac{dT}{dv}\frac{d^2s}{dv^2} + T\frac{d^3s}{dv^3} = \tfrac{1}{2}(v_1 - v)\frac{d^3p}{dv^3} - \frac{1}{2}\frac{d^2p}{dv^2} \tag{2-2c}$$

Evaluation of the derivatives of s at the origin gives

$$T_1\left(\frac{ds}{dv}\right)_1 = 0 \tag{2-3a}$$

$$T_1\left(\frac{d^2s}{dv^2}\right)_1 = 0 \tag{2-3b}$$

$$T_1\left(\frac{d^3s}{dv^3}\right)_1 = -\frac{1}{2}\left(\frac{d^2p}{dv^2}\right)_1 = -\frac{1}{2}\left(\frac{\partial^2p}{\partial v^2}\right)_s = -\rho^3a^2\Gamma \tag{2-3c}$$

where the thermodynamic quantities are evaluated at the origin and Γ is defined in Eq. 1-27. Using the symbol Δ to indicate change across the shock, expanding Δs in a Taylor series gives

$$T_1\Delta s = -\frac{1}{12}\left(\frac{\partial^2p}{\partial v^2}\right)_s \Delta v^3 = -\frac{\Gamma}{6a}\Delta u^3 \tag{2-4}$$

with an error of fourth order.

Another method involves the use of a variant of Taylor's expansion of a function of two variables, which has been made symmetric with respect to the two argument points. In terms of the function $e(v, s)$ this expansion is expressed

$$\begin{aligned}\Delta e(v, s) = &\left(\frac{\partial e}{\partial v}\right)_{\text{av}}\Delta v + \left(\frac{\partial e}{\partial s}\right)_{\text{av}}\Delta s\\ &- \frac{1}{12}\left(\frac{\partial^3e}{\partial v^3}\right)_{\text{av}}\Delta v^3 - \frac{1}{4}\left(\frac{\partial^3e}{\partial v^2\partial s}\right)_{\text{av}}\Delta v^2\Delta s\\ &- \frac{1}{4}\left(\frac{\partial^3e}{\partial v\partial s^2}\right)_{\text{av}}\Delta v\Delta s^2 - \frac{1}{12}\left(\frac{\partial^3e}{\partial s^3}\right)_{\text{av}}\Delta s^3\\ &+ 0(\Delta v^5, \Delta s^5)\end{aligned} \tag{2-5}$$

where the subscript av denotes the arithmetic mean over the interval. Anticipating the result that Δs is of the order of Δv^3 this may be rewritten

$$\Delta e = -p_{\text{av}}\Delta v + T_{\text{av}}\Delta s + \frac{1}{12}\left(\frac{\partial^2p}{\partial v^2}\right)_{\text{av}}\Delta v^3 + 0(\Delta v^5) \tag{2-6}$$

The Hugoniot relation (Eq. 1-10) may be rewritten

$$\Delta e = -p_{\text{av}}\Delta v \tag{2-7}$$

whence Eq. 2-6 becomes

$$T_{\text{av}}\Delta s = -\frac{1}{12}\left(\frac{\partial^2p}{\partial v^2}\right)_{\text{av}}\Delta v^3 + 0(\Delta v^5) \tag{2-8}$$

This derivation is essentially the same as that of Bethe [*6*]. Noting that to the same approximation $u_{\text{av}} = a_{\text{av}}$, Eq. 2-8 may be reexpressed

$$T_{\text{av}}\Delta s = \frac{\Gamma_{\text{av}}}{6a_{\text{av}}}(u_1 - u)^3 + 0[(u_1 - u)^5] \tag{2-9}$$

in agreement with Eq. 2-4. Since condition I is that $\Gamma > 0$ and the second law of thermodynamics requires that $\Delta s > 0$, this condition requires that a weak shock must be compressive. The behavior of s at the origin is of the third order, hence s has neither a maximum or minimum and the origin is what might be termed a double Ch-J point.

For the Hugoniot curve as a whole, the importance of the behavior near the origin is that the position on the series of extrema discussed in Art. 1 is established. The origin is itself a Ch-J point, and condition I requires that s increase with decreasing v, for compression solutions, and that s decrease with increasing v, for expansion solutions. Any Ch-J solution must occur with dp/dv negative (see Eq. 1-16), extrema of p for compression solutions may be excluded with condition III-weak, and extrema of v for expansion solutions may be excluded with condition II-weak. Under these conditions, if there were one or more Ch-J points besides the double one at the origin, the one which lies nearest the origin (p_1, v_1) on either side must have negative curvature on the basis of geometrical considerations. Eq. 1-26 shows that this would violate condition I; thus no maximum or minimum for s may exist on the Hugoniot curve. This gives the conclusion: *Conditions I, II-weak, and III-weak are sufficient to establish the monotonicity of the entropy for shock waves over the entire Hugoniot curve.* If only the compression branch is being considered, conditions I and III-weak are sufficient, and all compression shocks correspond to an increase in entropy. With conditions I and II-weak, all expansion shocks correspond to a decrease in entropy and are precluded by the second law of thermodynamics. It should be noted that these conditions are far from necessary. Thus, in general, expansion shocks are forbidden and compression shocks permitted; in order for expansion shocks to exist or for compression shocks to be forbidden by the second law of thermodynamics, it is necessary that condition I be violated.

Attention is henceforth confined to the compression branch of the Hugoniot curve for shock waves and is now directed to the question of uniqueness. The boundary conditions which determine a shock wave are generally applied to the material behind the shock in one of a variety of ways. Two types of boundary conditions may be considered as representative extremes, that in which p is given, and that in which $u_1 - u$ is given. Condition III-weak assures that p varies monotonically on the compression branch of the Hugoniot curve and that, if p is given, the solution is unique. Condition III-weak is not generally sufficient for the monotonicity of $u_1 - u$, but condition III-strong is sufficient and assures a unique solution if $u_1 - u$ is given; condition III-strong also precludes an extremum of e. Thus *condition III-strong, together with condition I, serves to guarantee uniqueness of compression shock waves.*

For very strong shocks, the pressure and internal energy of the fluid before the shock may be neglected, and the Hugoniot relation may be expressed in the approximate form

$$e = \tfrac{1}{2}p(v_1 - v) \tag{2-10}$$

To illustrate the strong shock relations the van der Waals model is considered, in the approximate form valid at very high pressures

$$p(v - b) = \mathcal{R}T \tag{2-11}$$

It may be shown that for any material for which p/T is a function of

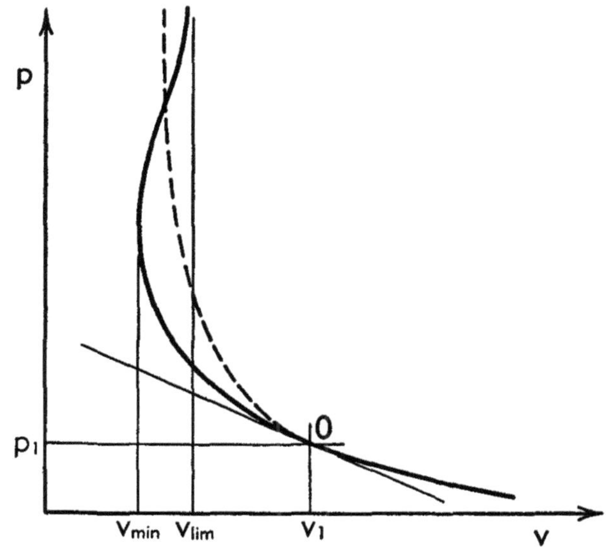

Fig. D,2. Strong shock asymptotic behavior of the Hugoniot curve.

v alone, the internal energy e is a function of T alone. In the case considered e may be expressed, for constant γ

$$e = \int c_v dT = \frac{\mathcal{R}T}{\gamma - 1} = \frac{1}{\gamma - 1}p(v - b) \tag{2-12}$$

Eq. 2-10 and 2-12 may be combined to yield the asymptotic relation

$$v_{\text{lim}} = \frac{\gamma - 1}{\gamma + 1}v_1 + \frac{2}{\gamma + 1}b \tag{2-13a}$$

or, for example,

$$v_{\text{lim}} = \tfrac{1}{4}v_1 + \tfrac{3}{4}b \tag{2-13b}$$

if $\gamma = \frac{5}{3}$. In most gases v has at least one minimum less than v_{lim} and greater than zero. A typical Hugoniot curve for a gas is illustrated in Fig. D,2. The dashed line represents where the Hugoniot curve would lie if there were no dissociation or ionization taking place at higher temperatures. For a further discussion of the shape of the Hugoniot curve at high temperature and pressures, the reader is referred to Bethe [*6*].

The asymptotic behavior of the Hugoniot curve shows that the quantities $u_1 - u$ and p are unbounded. Existence of a solution is thus ensured for the two types of boundary conditions for which uniqueness has been considered.

The velocity behind the shock wave may be compared with the speed of sound there through a consideration of Eq. 1-25. Up to the point for which v has its first extremum (a minimum), $m^2 - \rho^2 a^2$ and $m^2 + dp/dv$ must have the same sign. The geometrical properties dependent upon condition I which have already been established ensure that the sign of $m^2 - \rho^2 a^2$ is negative, or that *the flow behind a shock wave is subsonic.* Since $1 - \frac{1}{2}(v_1 - v)(\partial p/\partial e)_v$ changes sign with dv on passing an extremum of v this conclusion remains valid on the remainder of the compression Hugoniot curve, subject only to geometric limitations ensured by the sufficient condition III-weak.

The geometric properties of the Hugoniot curve imposed by condition I also ensure that the entire curve lies above the tangent to the curve at the origin, which has slope $-\rho_1^2 a_1^2$. Thus m^2 must be always greater than $\rho_1^2 a_1^2$ on the compression branch of the curve, and *the flow in front of the shock wave is supersonic.*

Although phase changes are strictly excluded from the considerations given above, the conclusions remain valid with appropriate translation of the thermodynamic conditions. It is of interest to quote Bethe's conclusions [*6*, Sec. 14] regarding the situation in which condition I is violated at a phase boundary: If such material is compressed so that no single shock solution is permitted, the solution appears to involve two shock waves, traveling at different speeds. The first and faster shock takes the material in front of the shock (single phase) to the phase boundary; the material between the shocks is still single phase but on the phase boundary; the material behind the second and slower shock is two phase. If such material is expanded under an analogous condition, with the material in front of the wave in the two-phase region but near the phase boundary, the result must be a wave of finite amplitude, an expansion shock.

The results of this article are not in conflict with those of other investigators when account is taken of their assumptions or method of approach. Courant and Friedrichs [*2*, Sec. 64, 65] assume in their investigation of uniqueness that dp/dv is negative everywhere on the Hugoniot curve, thus allowing for no minimum of v. In addition to the explicit assumption that the coefficient of thermal expansion is positive, the assumption of no minimum of v is also made by them in discussing other features of shock waves in a general fluid. Kline and Shapiro [*7*] attach significant importance to the sign of the coefficient of thermal expansion; their conclusions are controlled by the fact that they investigate the shock wave on the basis of a Rayleigh line process, for which the sign of the coefficient of thermal expansion strongly affects the solution. The assumptions of Courant and Friedrichs and of Kline and Shapiro are very stringent, and are violated by known materials. The author believes that the sufficient conditions stated by him are the least stringent ones possible which depend only upon the thermodynamic state behind the shock.

Finally, a word about the so-far discarded portion of the Hugoniot diagram corresponding to the impossible expansion shocks. This portion of the curve has a useful physical interpretation in terms of the problem inverse to that heretofore considered: Given the state of the material *behind* a shock, what are the possible states of the material in front of the shock? In an inverse shock wave Hugoniot diagram, for which p_1 and v_1 are the variables and p and v are held fixed, it is the lower right portion of the curve which has physical meaning and the upper left portion which must be discarded, provided, of course, that condition I is satisfied.

D,3. Exothermic Discontinuities. The Hugoniot diagram is now used to investigate the nature of discontinuities in which there is a change in the equation of state, with primary considerations given to those discontinuities which are embraced by the term "exothermic" in the sense defined below. The compression part of the Hugoniot curve lying in the upper left quadrant from the "origin" (p_1, v_1) is termed the detonation branch for exothermic discontinuities; the expansion part lying in the lower right quadrant is termed the deflagration branch. Each of these two branches is divided by Ch-J points into two parts, termed "weak" and "strong." The main purpose of this article is to express the conditions under which this classification is valid, and to describe the characteristics of each branch with regard to whether the velocities involved are subsonic or supersonic and with regard to the question of determinacy. For an alternate and thorough treatment the reader is referred to Courant and Friedrichs [*2*, Sec. 84–92].

Attention must now be directed again to the geometry of the Hugoniot curve. Except for the shock wave case treated in the previous article the case for which the Hugoniot curve passes through the origin (p_1, v_1) is exceptional and is not considered here. The state of a material, under the assumption of the existence of an equation of state, is presumed to be determined uniquely by p and h or by v and e. Consideration of the Hugoniot relations (Eq. 1-9 and 1-10) shows that the Hugoniot curve must have a single point on the axis $p = p_1$ and a single point on the axis $v = v_1$. Also, it may be shown that if the point on the axis $p = p_1$ lies to the right of the origin, the point on the axis $v = v_1$ must lie above it, and inversely. These conclusions are obtainable from the monotonicity of the Hugoniot-curve variable η on the basis of the continuity of the Hugoniot curve. Eq. 1-23a and 1-23d show that on the axis $p = p_1$ the quantity η must decrease with increasing p, and on the axis $v = v_1$ the quantity η must increase with increasing v; the Hugoniot curve may not cross either axis more than once and must pass the origin either to the right and above or below and to the left.

A Hugoniot curve which passes above and to the right of the origin (p_1, v_1) will be termed "exothermic." The term "endothermic" is defined as the opposite of "exothermic." Although such a definition is not in accord with the usual concept of "exothermic" as involving release of heat, heat-producing reactions are generally of the type here defined as exothermic, and the terminology may be considered reasonably apt. Thus, exothermic as here used means that the pressure increases in a constant volume, constant internal energy reaction (with no heat transfer if quasi-reversible); alternatively, it means that the volume increases in a constant pressure, constant enthalpy reaction.

Such a definition only applies to the unreacted material at the origin. In order to generalize the concept somewhat, it is assumed that there exists a single parameter ϵ which characterizes states of the material intermediate between those for the unreacted material and those for the fully reacted material. The parameter ϵ is considered as a new additional state variable which increases as the reaction proceeds. It is assumed to have the value 0 in the unreacted material, and another constant value (conveniently 1) in the fully reacted material. The basic formulas of thermodynamics are assumed to apply for the hypothetical intermediate material for which ϵ is constant.

In terms of this parameter ϵ the condition for a reaction to be exothermic may be stated

$$\left(\frac{\partial p}{\partial \epsilon}\right)_{v,s} > 0 \tag{3-1a}$$

$$\left(\frac{\partial v}{\partial \epsilon}\right)_{p,s} > 0 \tag{3-1b}$$

Because the quantity $(\partial p/\partial v)_{\epsilon,s}$ must be negative (see Eq. 1-12) these two statements are equivalent. For an endothermic reaction the inequalities are reversed. Using the same approach by which Eq. 1-23 were obtained but with consideration of the additional parameter ϵ, the equation

$$\left[1 - \tfrac{1}{2}(v_1 - v)\left(\frac{\partial p}{\partial e}\right)_{\epsilon,v}\right]\frac{dp}{d\epsilon} - \left(\frac{\partial p}{\partial v}\right)_{\epsilon,s}\left[1 - \tfrac{1}{2}(p - p_1)\left(\frac{\partial v}{\partial h}\right)_{\epsilon,p}\right]\frac{dv}{d\epsilon} = \left(\frac{\partial p}{\partial \epsilon}\right)_{v,s} \tag{3-2}$$

may be obtained. With an exothermic reaction, from the inequality Eq. 3-1a, the left-hand side of Eq. 3-2 is positive.

If the material at and near the origin satisfies the condition of Eq. 3-1, the Hugoniot curve for the fully reacted material must lie above and to the right of the origin, and the two statements of the concept are consistent. If the material satisfies Eq. 3-1 everywhere, Eq. 3-2 shows that the Hugoniot curve for the fully reacted material must always lie on the same side of the Hugoniot curve for the unreacted material (the shock wave Hugoniot curve, which passes through the origin and which satisfies Eq. 1-23a and 1-23d). In this case the discontinuity is termed "always exothermic." The term "always endothermic" may be defined analogously.

With the condition "always exothermic" and with condition I in the material in front of the discontinuity, the Hugoniot curve (for the reacted material) lies definitely above the tangent to the shock curve at the origin. As is concluded for the shock wave in the previous article, *the velocity in front of the discontinuity is supersonic for the compression or detonation branch of the Hugoniot curve.* Similarly, it may be shown that *the velocity in front is subsonic for the expansion or deflagration branch.* The two conditions expressed are only sufficient ones for the validity of these conclusions.

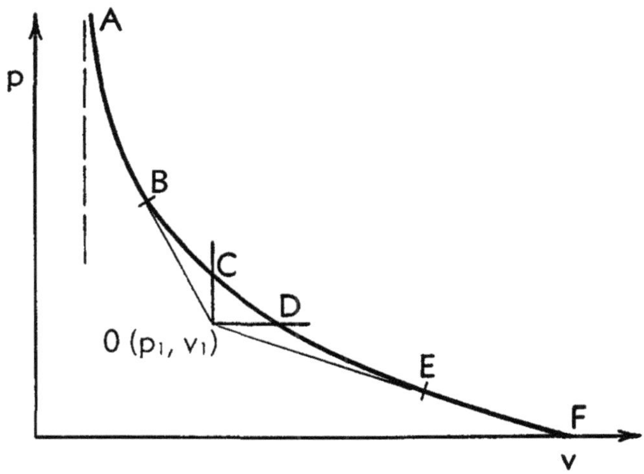

Fig. D,3a. Exothermic Hugoniot diagram.

The Hugoniot curve for an exothermic discontinuity is represented in Fig. D,3a, in which certain specific points have been lettered. The point 0 is the origin (p_1, v_1) corresponding to the state of the material in front of the discontinuity. On the curve, the point for which $p = \infty$ is denoted A, that for which $v = v_1$ is denoted C, that for which $p = p_1$ is denoted D, and that for which $p = 0$ is denoted F. A ray from 0 to the curve is tangent to it at one point B lying between A and C, and at one point E between D and F. These two points B and E are Ch-J points.

The physically interesting sections of the curve are thus divided into four distinct parts, into the arc A-B corresponding to "strong detonation" solutions, the arc BC corresponding to "weak detonation" solutions, the arc DE corresponding to "weak deflagration" solutions, and the arc EF

corresponding to "strong deflagration" solutions. The Hugoniot curve usually cuts the axis $p = 0$ at a finite angle, and it is not necessary that the point E and the strong deflagration branch EF exist.

Consideration of Eq. 1-23 shows that point C must lie between $u_1 - u$, or p if condition III-weak is satisfied, and Ch-J on the compression sequence of extrema developed in Art. 1. The Ch-J point nearest point C must be one for which the curve is convex and for which condition I is satisfied. As for the shock wave case treated in Art. 2, conditions III-weak and I preclude the possibility of another compression Ch-J point, and the Ch-J point found is unique and is the point B of Fig. D,3a. Analogously, point D must lie between $u_1 - u$, or v if condition II-weak is satisfied, and Ch-J on the expansion sequence of extrema. If conditions II-weak and I hold, there may be a single Ch-J point, the point E of Fig. D,3a.

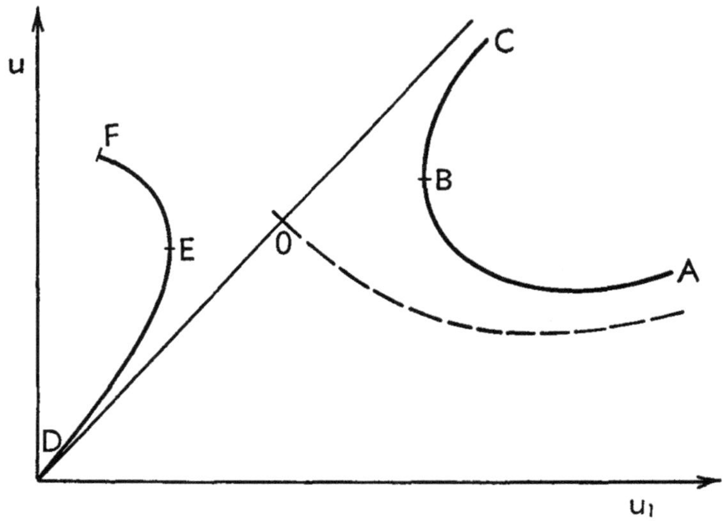

Fig. D,3b. Velocity plot.

As in Art. 2, the sign of $m^2 - \rho^2 a^2$ may be investigated through Eq. 1-24. The resulting conclusions are that *the velocity u behind the discontinuity is subsonic for a strong detonation or a weak deflagration* (arc AB or DE) *and is supersonic for a weak detonation or strong deflagration* (arc BC or EF). At the two Ch-J points (B and E) the velocity behind the discontinuity is sonic, in agreement with Eq. 1-17.

There is a physical significance to certain of the characteristic properties of the Ch-J points. If the flow behind a discontinuity is sonic it can support a sound (infinitesimal) wave which lies a small but finite distance behind the discontinuity. This sound wave may be conceptually included with the discontinuity as a perturbation of the downstream conditions. Since it does not affect the discontinuity ahead of it the quantity u_1 is unchanged. Since the sound wave is itself isentropic the quantity s is unchanged. Thus, from physical considerations, a finite discontinuity with respect to which the downstream flow is sonic must be at a stationary point for u_1 and for s. Sometimes useful conceptually is the related consideration of a discontinuity as two distinct discontinuities each with the same mass flow m, moving at a fixed small but finite distance apart. For example, a strong detonation may be considered conceptually as a shock wave followed by a weak deflagration, or as a weak detonation followed by a shock wave.

It is instructive to consider also a plot of u vs. u_1 for an exothermic discontinuity. Such a plot is illustrated in Fig. D,3b, in which the points A to F of Fig. D,3a are indicated. The dashed line indicates the solution curve for an ordinary shock wave in the original material.

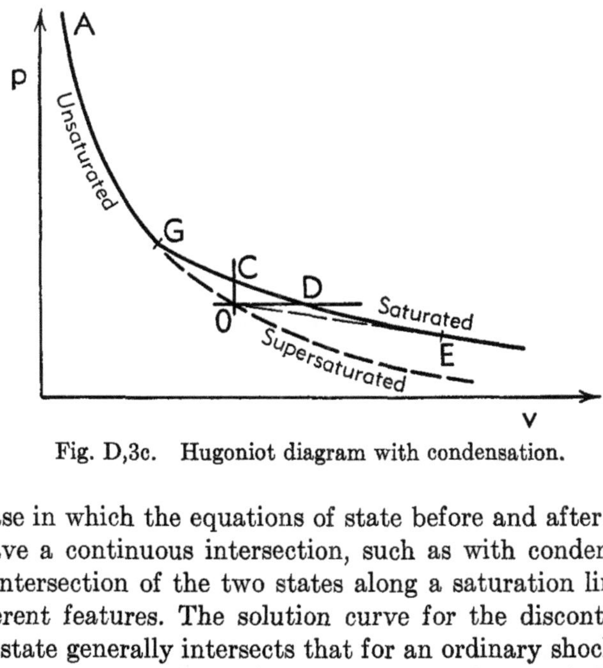

Fig. D,3c. Hugoniot diagram with condensation.

The case in which the equations of state before and after the discontinuity have a continuous intersection, such as with condensation in a gas with intersection of the two states along a saturation line, presents some different features. The solution curve for the discontinuity with change of state generally intersects that for an ordinary shock wave; the solution curve may thus have discontinuities in slope and, for example, the Ch-J detonation point may not exist. Fig. D,3c presents a possible configuration of the Hugoniot diagram for a case with condensation. In this diagram the point 0 lies on the supersaturated curve and G is the point corresponding to pure saturated vapor. The arc AG corresponds to an ordinary compression shock, the arc GC to a detonation-type (here depicted as all weak) condensation shock, and the arc DE to a weak deflagration-type condensation discontinuity. The corresponding velocity plot is depicted in Fig. D,3d. The dashed line again indicates the solution curve for an ordinary shock wave in the original (supersaturated) vapor.

In such cases in which the equations of state have a continuous intersection, it is possible to use the second law of thermodynamics to exclude certain solutions. In the case of simple condensation of a vapor investigated by Hayes [*8*], part of the strong deflagration branch corresponds to a decrease in the entropy and is therefore forbidden. A discontinuity in which condensation takes place but the vapor in front is not supersaturated is also forbidden by the second law of thermodynamics.

For the question of determinacy it is necessary to consider the number of essential unknowns in determining which of the solutions possible would occur in a particular case, and the number of specified quantities available to determine the unknowns. The formulation of determinacy

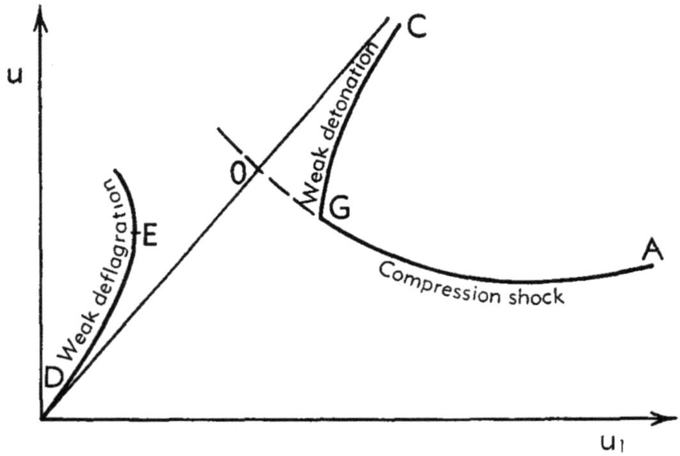

Fig. D,3d. Velocity plot with condensation.

conditions depends quite a bit on the point of view taken. The point of view which perhaps shows the differences between the various discontinuity types most clearly is that of an experimenter who initiates the disturbance at one end of a tube of unreacted gas, and this is the point of view taken here. The state of the gas far ahead of the discontinuity is given. If the flow in front of the discontinuity is subsonic (deflagration case) the disturbance caused by the discontinuity may cause a pressure wave to be sent out into the flow ahead and thus alter the state immediately ahead. The magnitude of this change of state is one unknown in the deflagration case. The other unknown, with the state immediately ahead of the discontinuity known, is the choice of the possible solutions offered by the Hugoniot diagram. If the flow behind the discontinuity is subsonic (strong detonation or weak deflagration) then downstream boundary conditions affect the solution and an appropriate statement of these conditions gives one specified quantity to determine the unknowns. Thermodynamic and chemical considerations of the actual reaction process may be expected to furnish an additional quantity to determine the unknowns. In order to clarify the unknowns and the quantities needed to be specified for determinacy, Table D,3 is given. In a strong detonation the hydrodynamic boundary conditions downstream determine the solution completely, which then must be independent of the reaction process. In a weak detonation the one unknown must be determined by the reaction process. In a weak deflagration both the reaction process involved and the downstream boundary conditions determine the solution.

In the case of a strong deflagration there are two unknowns and downstream boundary conditions do not affect the solution. This means that considerations of the reaction process must be used to determine both unknowns. In Art. 6 it is shown that it is very difficult to find a physically realizable process for a strong deflagration and that, effectively, reaction considerations can determine only one unknown.

In practical cases involving combustion the velocity of a true combustion wave (deflagration) is largely controlled by diffusion processes and is generally slow, slow enough so that conditions approximate those at

Table D,3

	Strong detonation	Weak detonation	Weak deflagration	Strong deflagration
Unknowns				
Strength of pressure wave in front	0	0	1	1
Choice of solution on Hugoniot curve	1	1	1	1
Quantities specified				
Downstream boundary conditions	1	0	1	0
Others needed	0	1	1	2

point D in Fig. D,3a and D,3b. Solutions near the Ch-J deflagration point and in the weak detonation regime require a high velocity of propagation for the chemical reaction itself and are not observed in practice. Detonations observed in practice are of the Ch-J type because of the boundary conditions usually established on the burned side of the discontinuity, but it is recognized that with suitable boundary conditions, viz. with a piston following the front with a high enough velocity, a strong detonation would be found. With the assumption that the chemical reaction itself can only support a very low velocity of propagation, a Ch-J detonation or strong detonation can only occur if the normal shock traveling at the same speed in the unburned medium is strong enough to raise the temperature to the ignition point. This corresponds to the interpretation of Courant and Friedrichs [*2*] of a usual detonation as a deflagration initiated by shock.

If the ignition point of a combustible mixture were very sharp, with a negligible rate of reaction at lower temperatures and an appreciable one at higher temperatures, it would be possible to obtain a weak detonation by preheating the mixture to a temperature sufficiently close to but below the ignition point. The fact that no weak detonation has been found in practice might conceivably be explainable by the fact that no experimenter has used this approach so far. It is probable, however, that the fact that the ignition point is not sharp may effectively prevent the production of a weak detonation-type combustion wave.

In the case of condensation waves the situation is much different. In practice condensation waves are not controlled by diffusion processes but by flow processes which decrease the temperature below the saturation point to an extent that nucleation and condensation take place spontaneously, as in the expansion section of a supersonic or hypersonic wind tunnel or in the accelerated flow over an airplane wing under humid atmospheric conditions. In condensation phenomena weak detonations may occur.

Approximate equations for solutions in the vicinity of the special points D, C, A, and F of Fig. D,3a are now given. The point D corresponds to a constant pressure process and neighboring solutions are slow deflagrations. The appropriate approximate equations are

$$\left.\begin{aligned} p &= p_1 \\ h &= h_1 \end{aligned}\right\} \tag{3-3}$$

which determine the quantity v, and the velocity u may be obtained from Eq. 1-1 if u_1 is known. The mass flow is of the first order of magnitude in the velocities and (see Eq. 1-4) the actual pressure difference is of the second. This limiting case is the one considered in conventional laminar flame studies; a critical discussion of the assumptions underlying this case may be found in [*9*, p. 684].

The point C corresponds to a constant density process and neighboring solutions are very fast weak detonations. The appropriate approximate equations are

$$\left.\begin{aligned} v &= v_1 \\ e &= e_1 \end{aligned}\right\} \tag{3-4}$$

whence the pressure p behind the discontinuity may be determined. From Eq. 1-4 it may be seen that the inverse mass flow is of the first order of magnitude in the velocity difference and that the actual specific volume difference is of the second. This limiting case is not approached in practice.

The point A corresponds to a strong detonation process at unbounded high velocities. The appropriate equations are obtained by neglecting p_1 and h_1 in the conservation equations and are

$$\left.\begin{aligned} p &= \rho_1^2 u_1^2 (v_1 - v) \\ h &= \tfrac{1}{2}\rho_1^2 u_1^2 (v_1^2 - v^2) \end{aligned}\right\} \tag{3-5}$$

The quantity u_1 may be eliminated, whereby the equation

$$h = \tfrac{1}{2} p (v_1 + v) \tag{3-6}$$

is obtained, equivalent to Eq. 2-10. There is no essential difference between this case and that of a very strong shock, and the remarks in Art. 2 and the analysis for a van der Waals gas apply equally here. This limiting case is important in the theory of hypersonic flow.

For the point F the quantity p is zero and $h = h(0, v)$, and an equation similar to Eq. 3-6 holds with the subscripts changed. This permits the expression of the limiting value of v

$$v = 2\frac{h_1 - h(0, v)}{p_1} - v_1 \tag{3-7}$$

which is implicit if $h(0, v)$ is not constant. Although the point F in a

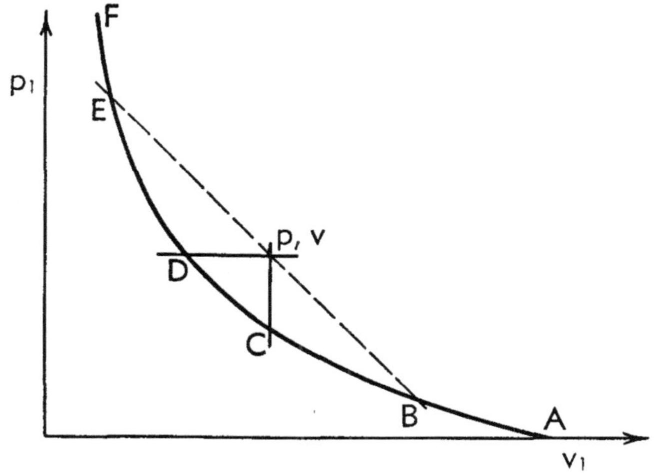

Fig. D,3e. Inverse exothermic (or endothermic) Hugoniot diagram.

regular Hugoniot diagram has no physical significance, this case is of interest for the inverse Hugoniot diagram discussed below.

An occasionally helpful plot is the inverse Hugoniot diagram, showing the possible states of the material in front of the discontinuity consistent with a given state of the material behind. Such a plot is depicted in Fig. D,3e for a point p, v with respect to which the reaction is exothermic, and with points lettered to correspond with the lettering on Fig. D,3a. The dotted line has slope $-\rho^2a^2$ and intersects the curve in the two Ch-J points B and E. With condition I satisfied for the material in front of the discontinuity, there is no ray from the origin tangent to the curve, no inverse Ch-J point. This diagram may also be used with Eq. 1-25 to show that the velocity in front of a detonation is supersonic and that in front of a deflagration is subsonic.

A word may be said here about the endothermic case, although no cases of gasdynamic discontinuities of this type are known to the writer. The Hugoniot curve passes below the origin, and Fig. D,3e, with p, v and

p_1, v_1 interchanged, may be used to consider the endothermic case. The ray of slope $-\rho_1^2 a_1^2$ intersects the curve in two inverse Ch-J points B and E; these points do not have the Ch-J properties of being stationary points for s, or for u_1 or u. The branch AB is essentially the same as the strong deflagration branch of an exothermic discontinuity and the branch EF the same as the strong detonation branch, but here the similarity ceases. The branch BC has the velocity properties of a weak detonation but is, of course, of the expansion type; it might be termed the "endothermic fast deflagration" branch. Similarly, the branch DE has the velocity properties of a weak deflagration but is of the compression type; it might be termed the "endothermic slow detonation" branch. With regard to the question of determinacy, on both branches the solution would require one quantity to be specified from the reaction process in order to determine the solution.

D,4. Internal Stability Considerations. The concept of internal stability here discussed is perhaps an unconventional one in that it involves no partial differential equations or roots of determinantal equations. The principal question asked is this: If a discontinuity of the type considered in this section is established, is it possible for it to jump to another solution permitted by the basic hydrodynamic laws? The question is obviously related to the questions of determinacy discussed in Art. 3. Thus a strong detonation is determined only by hydrodynamic boundary conditions; once these are established in a steady state solution no change in solution is permitted by the basic hydrodynamic laws and it will be found that the concept of internal stability does not apply. The principal question may only be asked about those discontinuities which are partly determined by other than hydrodynamic laws. The only case for which the concept may have practical significance appears to be that of the weak deflagration, for which solutions near the Ch-J point should be internally unstable.

Internal stability as here discussed refers to simple one-dimensional stability in which the disturbances considered leave the discontinuity normal and involve simple plane waves emitted into the material in front of or behind the discontinuity. Presented in this article is a brief treatment of the internal stability of combustion waves, or of exothermic discontinuities with reaction rate increasing with temperature. One of the main results obtained is that a discontinuity of the strong deflagration type, if such can exist, must be internally unstable. This result was found by Landau [*10*, p. 83] in 1944 but appears to have been overlooked by many investigators.

The assumption is now made that the velocity of the front with respect to the material in front of it depends on the temperature in front of it, the velocity increasing with increasing temperature because of an increased reaction rate. In the case of a detonation the temperature in front of the discontinuity cannot be affected by an internal disturbance and only the temperature behind may be considered. The dependence of the wave velocity on the temperature behind the wave may be expected to be extremely weak but of the same kind.

Two types of plots are found to be useful in these considerations. One of these is the familiar (Sec. C) track of the wave on a plot of distance versus time. The other is a plot of pressure versus the velocity change, p vs. $u_1 - u$. With given initial conditions this figure has the appearance of Fig. D,4a. In this figure the various points are labeled according to the letters assigned in Fig. D,3a. The slope of either branch of the curve at the corresponding Ch-J point is $1/\rho a$.

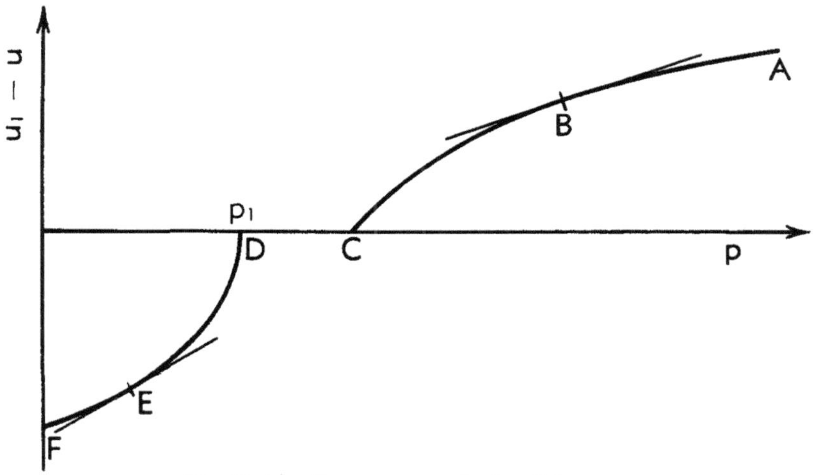

Fig. D,4a. Pressure vs. velocity difference.

The case of the strong detonation is considered first. If an infinitesimal change is to take place in the conditions behind the front, it must follow the equation

$$du = \frac{dp}{\rho a} \tag{4-1}$$

because the adjustment in the velocity and pressure must be made across a sound wave traveling downstream from the front. This relation corresponds to a line of negative slope in the AB arc of the curve of Fig. D,4a. Since the flow after the change must also correspond to a point on the curve and the line indicated by the relation Eq. 4-1 does not intersect the curve, no such adjustment is possible. This means that the conclusion of Art. 3, that a strong detonation is fully determined by the downstream boundary conditions, also holds in the small. A strong detonation is always stable internally since no perturbation is permitted by the hydrodynamic conditions.

The case of a weak detonation is depicted in a distance-time plot in Fig. D,4b, with the wave going from right to left in order to maintain the orientation of Fig. D,1a. The corresponding pressure-velocity change plot is given in Fig. D,4c. The particle path separating regions 4 and 4′ of Fig. D,4b is a contact discontinuity or slipstream, across which pressure and velocity are continuous. An increase in the velocity u_1 may be shown with the aid of the Hugoniot diagram to correspond to a decrease

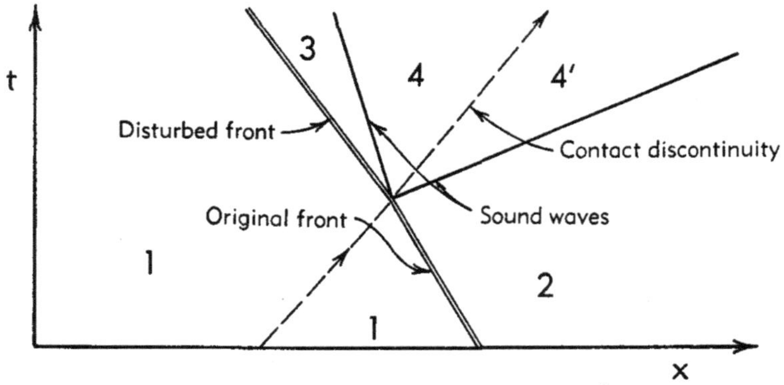

Fig. D,4b. Disturbance of a weak detonation.

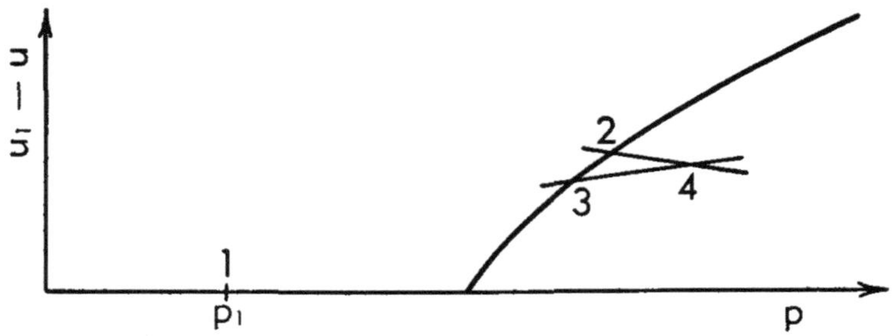

Fig. D,4c. Weak detonation.

in internal energy, which means a decrease in temperature in perfect gases and in most other cases. With the general assumptions of this article a weak detonation is certainly internally stable if the material behind the front behaves as a perfect gas, since with a temperature decrease an increase in u_1 cannot be supported. In view of the expected very weak dependence of u_1 on T in general, essentially any weak detonations which otherwise can exist may be expected to be stable.

In the case of deflagrations the analysis is more complicated, because conditions in front of the discontinuity may change. This change in front of the discontinuity is an isentropic one, for which

$$dp_1 = a_1^2 d\rho_1 \tag{4-2}$$

The Hugoniot curve is shifted, to the left with an increase of pressure and density, or conversely. Differentiation of the Hugoniot relation (Eq. 1-9) yields the result

$$dp_1 - u_1^2 d\rho_1 + dp - u^2 d\rho + \frac{1}{\beta}(dp - a^2 d\rho) = 0 \qquad (4\text{-}3)$$

where

$$\beta = \tfrac{1}{2}(v - v_1)\left(\frac{\partial p}{\partial e}\right)_\rho \qquad (4\text{-}4)$$

The connection with Eq. 1-25 is evident here. Differentiation of the

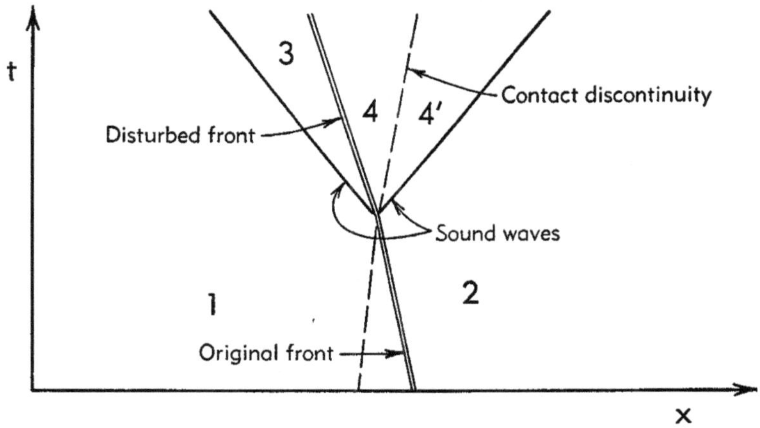

Fig. D,4d. Disturbance of a weak deflagration.

expressions for m, Eq. 1-1 and 1-4, gives

$$\begin{aligned} 2(v - v_1)m\,dm &= 2u_1(u - u_1)d\rho_1 + 2\rho_1(u - u_1)du_1 \\ &= (dp_1 - u_1^2 d\rho_1) - (dp - u^2 d\rho) \end{aligned}$$

which may be reexpressed as

$$2\rho_1(u - u_1)du_1 = (a_1^2 + u_1^2 - 2u_1 u)d\rho_1 - (dp - u^2 d\rho) \qquad (4\text{-}5)$$

Finally, the pressure equation (Eq. 1-2) gives the relation

$$2m\,d(u - u_1) = (dp_1 + u_1^2 d\rho_1) - (dp + u^2 d\rho) \qquad (4\text{-}6)$$

These equations alone are insufficient to determine the perturbation solution.

In the case of a weak deflagration there is a simple wave in front of the discontinuity and one behind it, as depicted in Fig. D,4d. The pressure-velocity difference plot corresponding to Fig. D,4d is shown in Fig. D,4e. In order to make the changes from points 1 to 3 and 2 to 4 follow the slopes $1/\rho_1 a_1$ and $-1/\rho a$ appropriate to the corresponding simple waves, the change from point 1 to point 3 is indicated with a suitable shift of the p axis.

Denoting the change in the absolute velocity of the front by δu, the differentials of the relative velocities u_1 and u may be expressed

$$du_1 = \delta u - \frac{dp_1}{\rho_1 a_1}$$

$$du = \delta u + \frac{dp}{\rho a}$$

giving the relation

$$d(u - u_1) = \frac{dp_1}{\rho_1 a_1} + \frac{dp}{\rho a} \tag{4-7}$$

Combination of this equation with Eq. 4-6 gives

$$(a_1 - u_1)^2 d\rho_1 = \left(1 + 2\frac{u}{a}\right) dp + u^2 d\rho \tag{4-8}$$

This equation provides the additional information needed to solve Eq.

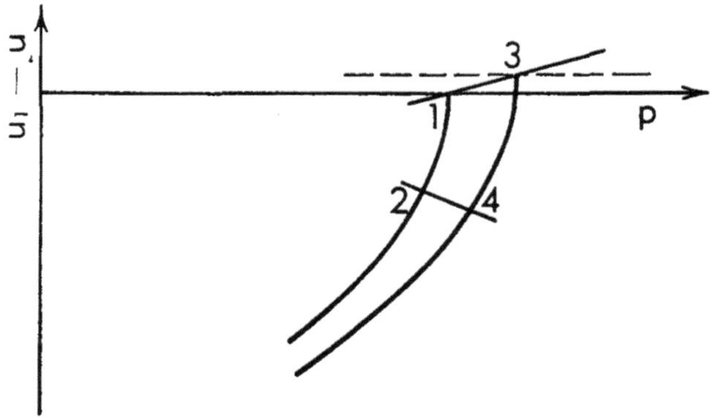

Fig. D,4e. Weak deflagration.

4-2 to 4-5. The result for $du_1/d\rho_1$ is

$$\rho_1(u - u_1)\left(a + u + 2\frac{u^2}{a}\beta\right)\frac{du_1}{d\rho_1} = (a_1 u + u_1 a)(a_1 - u)$$

$$+ u(u - u_1)(a_1 - u_1) + 2\frac{u^2}{a}(a_1^2 - u_1 u)\beta \tag{4-9}$$

With β positive and with $a_1 > u$ then the quantity $du_1/d\rho_1$ is certainly positive, and it may be generally expected to be positive over the entire range of weak deflagrations to the Ch-J point.

To interpret this result, the assumption made earlier about the increase of u_1 with increasing temperature (itself increasing with p_1 and ρ_1) is recalled. If, with a given increase in ρ_1, the velocity increase given by Eq. 4-9 is greater than that which the increased temperature can support, the deflagration is internally stable. If the velocity increase given by Eq. 4-9 is less than that which the increased temperature can support, the deflagration is internally unstable. For slow combustion waves the propagation velocity is governed primarily by heat conduction and diffusion, rather than by how near the temperature is to the ignition temperature, and such waves may be expected to be internally stable. For rapid combustion waves, at or near the Ch-J point, the opposite is plausible and such waves may be expected to be internally unstable. Because of the here-undiscussed important question of stability in the large this does not indicate that slow normal combustions should exist, but only that very rapid ones should not.

The case for which the propagation velocity increases with decreased temperature is quite different. Condensation shocks, the propagation of which depends critically on the supersaturation, may be expected to fall into this category. Here, as long as $du_1/d\rho_1$ is positive, there is always internal stability in a discontinuity of weak deflagration type.

With a strong deflagration the flow behind the front is supersonic and adjustment there to a perturbation may be made through two wave systems. This leaves an unresolved indeterminacy in the perturbation solution. This means that it is always possible to find a perturbation solution with a suitably small change in the perturbation velocity such that the thermodynamic change in front of the wave will be able to support the velocity change. The conclusion may be drawn that all strong deflagrations are internally unstable, based on the assumption that the solution still exists after the perturbation.

It should be noted that the point of view of this section may be used for the question of determinacy. In this respect the results are completely consistent with those of the preceding section, indicating that the number of thermodynamic or chemical quantities necessary to determine the solution is 0 for a strong detonation, 1 for a weak detonation or a weak deflagration, and 2 for a strong deflagration. The instability conclusion for strong deflagrations may be considered to be related to the assumption that the propagation velocity depends on the state in front of the wave and the implicit assumption that only this one quantity is provided by thermodynamic and chemical considerations.

The approach of this section may be readily extended to take into consideration finite perturbations. In the case of the strong deflagration, a one-parameter family of such finite perturbations can change the discontinuity to a weak or Ch-J deflagration, or to a weak or Ch-J detonation. A weak deflagration may also be changed to a discontinuity of the detonation type. Accordingly the question of stability against finite perturbations should also be raised.

The question of hydrodynamic stability in the large is not treated here, except for a brief mention of a result of Landau [*10*]. (See also G, Chap. 2.) He investigated the problem of the stability of two-dimensional disturbances of a normal discontinuity under certain simplifying assumptions, primarily that the fluids on both sides of the front may be considered incompressible. The result of this investigation is that a discontinuity through which the density decreases, i.e. a deflagration, is unstable for two-dimensional disturbances of lateral wavelength large compared with the discontinuity thickness. Since this result applies to slow combustions, which it was indicated should be internally stable, the net conclusion is that normal deflagrations are never stable except for very particular cases. Such cases might be a combustion in a tube whose diameter is of the same order as the thickness of the reaction zone, or an oblique combustion wave at small distances from the edge which serves as a flame holder. In practice this conclusion is borne out, and normal slow combustions are not observed. What are observed are flame fronts with a cellular structure, fronts which have a thick turbulent mixing-reaction zone, and fronts which exhibit marked nonstationary behavior. These phenomena are discussed elsewhere in this series, and a comprehensive description of them may be found, for example, in the book of Lewis and von Elbe [*11*]. Landau considers the two-dimensional instability found by him as leading to turbulence. He concludes that the mixing-reaction zone thickness should be of the same order of magnitude as the tube diameter in which the combustion takes place and that the propagation velocity of the front should be of the order of magnitude of this distance divided by the characteristic time of the reaction, within the limits of the incompressibility assumption.

D,5. Navier-Stokes Shock Structure. The actual physical process which occurs within a shock wave or other gasdynamic discontinuity involves the transfer of both energy and momentum. In considering these transfer phenomena for the purpose of obtaining a theory for the structure of a shock, two distinct points of view may be taken. The most fundamental point of view is that the transfer is accomplished through atomic and molecular collisions, and leads to the problem of solving the Boltzmann equation (I,B and D). The other point of view is that the material may be considered to be a continuum with stresses and heat transfer dependent upon the spatial derivatives of the velocity and temperature, and requires the determination of the laws governing this dependence.

With moderately strong shocks the effective thickness appears to be of the order of magnitude of a few molecular mean free paths. Thus it would appear at first glance that the first point of view, that of the kinetic theory, was necessarily the correct one and that the continuum point of view was necessarily in error. This might be true if complete and precise solutions to the Boltzmann equation were available and if no experimentally based extension of the realm of validity of the continuum theory beyond that provable by the kinetic theory were to be considered acceptable. In fact, present studies of the structure of moderately strong shocks using kinetic theory have all used assumptions of varying arbitrariness as to the nature of the solutions to the Boltzmann equation yielding approximations to the shock wave structure. And these studies are mostly limited to the case of a monatomic gas with particular postulated laws of interaction. On the other hand, the continuum theory has shown evidence of an empirical type indicating a range of validity appreciably wider than that granted to it by most kinetic theorists.

Most of the kinetic theory approaches and a sufficiently general continuum theory approach (e.g. Truesdell [*12*]) arrive at a set of tensorial relations connecting the stress tensor, energy (or heat) flow vector, velocity gradient tensor, temperature gradient vector, and the total time and the distance derivatives of these quantities. The Navier-Stokes relations (here taken to include the Fourier heat conduction relation) appear naturally as the lowest dissipative approximation with either approach. The essential difference between a "classical" kinetic theory approach of this type and a continuum theory lies in the determination of the coefficients involved: with a kinetic theory they are to be determined by calculation from a knowledge of the molecular collisions involved, while with a continuum theory they are to be determined by experimental means as are other material properties. Either way of determining these coefficients (besides the shear viscosity and the heat conduction coefficient) is beset with important difficulties. In addition, either of these approaches involves the assignment of a temperature under nonequilibrium conditions for which no temperature may be defined strictly thermodynamically; this difficulty may be obviated by the concept of quasi-equilibrium only if the Navier-Stokes relations are involved.

The classical kinetic theory approaches stem primarily from the approximation schemes associated with the names of Hilbert, Chapman, Enskog, and Burnett; the classical exposition of the subject is to be found in [*13*]. Analyses of a shock wave by this approach have been made by Wang-Chang [*14*], Zoller [*15*], Herpin [*16*], Broer [*17*], and Travers [*18*]. Grad [*19*] has made a study of shock structure based on a different approximation scheme. These analyses result in what appear to be reasonable pictures of shock wave structure for moderately weak shocks in a perfect monatomic gas obeying a Maxwellian or elastic-sphere force law.

Mott-Smith [*20*] has used a completely different kind of hypothesis, assuming that the velocity distribution function is made up within the shock of a linear combination of that before, and that after, the shock. The state of the gas changes through an assumed law of interaction between the two Maxwellian distributions. Mott-Smith's method is open to a number of criticisms and his results are admittedly in error for weak shocks, but his has been the only published attempt on the problem of the structure of a strong shock. The strong shock problem is one for which neither continuum methods nor the classical kinetic approximation theory can give even an approximate solution. The approach of Mott-Smith may well point the way to a satisfactory attack on this problem.

The writer is among those of the opinion that for weak and moderately weak shocks a continuum approach is clearly preferable to a kinetic theory approach. To summarize the reasons for this stand, it may be noted that (cf. Truesdell [*21*, Sec. 5A]):

1. The validity of the Boltzmann equation for nonequilibrium states, though generally accepted, remains somewhat in doubt. This point has been discussed by Grad [*22*], who points to a "molecular chaos" assumption as the critical one underlying the Boltzmann equation, and also by Jeffreys [*23*].
2. The validity of the approximation schemes employed has never been established in any case even in an asymptotic sense. On the contrary, Truesdell [*24*], in studying his exact solution for shear flow with Maxwellian molecules and his mathematical model for studying the classical approximation scheme by analogy has shown that in certain cases the Navier-Stokes solution may be a correct asymptotic solution but that further approximations are worse ones. The disagreement among investigators using various versions of the classical approximation scheme may well be a symptom of this basic difficulty.
3. The kinetic theory as presently developed is limited in application to gases which are monatomic and which obey a specified force law. Thus, a polyatomic perfect gas, or even a monatomic gas with a general dependence of viscosity on temperature, cannot be treated. These limitations do not appear in continuum theory.
4. In the only case for which a comparison of experiment with theory is available (Sherman [*25*]) agreement was obtained with the appropriate continuum theory and not with kinetic theory.

The classical Navier-Stokes treatment of the shock wave is due to Becker [*26*], who obtained the first solutions for nonweak shocks involving both viscosity and heat conduction. He made the correct but meaningless note that his solution with constant coefficient of viscosity gave shock thicknesses smaller than a mean free path in the gas ahead of the shock. Of course, if a correct continuum theory gave large changes in physical quantities in a distance small compared with the local mean free path the theory could be dismissed as inapplicable. The point of importance here is that the temperature variation of the viscosity coefficient and the temperature variation of the mean free path at constant volume are closely connected, and the proper comparison to be made is between the shock thickness according to a continuum theory, using the correct temperature variation of viscosity coefficient and an average value of the correct mean free path measured within the shock. Although Becker's remark may not be taken literally, it is on precisely this point that the Navier-Stokes solutions are most open to criticism for moderately strong shocks, from the continuum point of view as well as from the kinetic theory point of view. Thus for moderately strong shocks a correction is indicated, involving an altered relation between stress and velocity derivatives. It has been claimed by various authors (e.g. Travers [*18*]) that the correction term proportional to the square of the velocity gradient is the most important term to include, while terms in high derivatives of the velocity and temperature are less important. Gilbarg and Paolucci [*27*] have given a proof of the existence of the mathematical solution with a general nonlinear relation between stress and velocity gradient. However, Broer [*17*] and others have pointed out that in an expansion development in a parameter proportional to shock strength, the linear second derivative terms are of larger order of magnitude than the quadratic first derivative terms.

Most classical presentations of Navier-Stokes shock structure have assumed the Stokes relation between shear and second coefficients of viscosity, and have thus assumed zero bulk viscosity. The Stokes relation has not even been shown to be correct for all monatomic gases and is completely false for most other gases and liquids (see Truesdell [*28*]). Moreover, for the purpose of the theory of shock structure it is unnecessary and unhelpful. The additional viscous stress in a one-dimensional flow is $\mu''(\partial u/\partial x)$, where μ'' is a longitudinal coefficient of viscosity. This coefficient is related to the usual shear coefficient of viscosity μ by the relation

$$\mu'' = \tfrac{4}{3}\mu + \mu' = 2\mu + \lambda \tag{5-1}$$

where μ' is the dilational or bulk coefficient of viscosity which must be equal to or greater than zero, and λ is the classical second coefficient of viscosity. Analyses of shock structure in which the Stokes relation has been assumed may be corrected by simply replacing $\frac{4}{3}\mu$ by μ''. The pressure is defined as the thermodynamic pressure in terms of which the equation of state is expressed.

The subscript 2 is now restored to quantities behind the discontinuity in this and succeeding articles, and unsubscripted quantities refer to conditions within the structure of the discontinuity. For the case of a shock, the steady state one-dimensional equations of motion with the Navier-Stokes viscous stress and the linear heat conduction relations may be expressed, in correspondence with Eq. 1-1, 1-2, and 1-3,

$$\rho u = m \tag{5-2}$$

$$p + \rho u^2 - \mu'' \frac{du}{dx} = p_0 \tag{5-3}$$

$$h + \frac{u^2}{2} - \frac{\mu''}{\rho}\frac{du}{dx} - \frac{k}{\rho u}\frac{dT}{dx} = h_0 \tag{5-4}$$

where k is the coefficient of heat conduction. The quantities m, p_0, and h_0 are assumed fixed and known. If the quantities p, h, e, μ'', and k are presumed known as functions of T and ρ, then T and u (equal to m/ρ) may be used as the independent variables. In this sense the equations may be rewritten

$$\frac{\mu'' u}{m}\frac{du}{dx} = \frac{p}{\rho} + u^2 - \frac{p_0 u}{m} \tag{5-5}$$

$$\frac{k}{m}\frac{dT}{dx} = e - \frac{u^2}{2} + \frac{p_0 u}{m} - h_0 \tag{5-6}$$

The procedure of solving this pair of differential equations is to first eliminate the variable x by dividing one equation by the other. For the purposes of solving the resultant first order equation a plot of T vs. u, first utilized by Becker [26], is very useful conceptually. The first integral must generally be obtained by numerical or graphical means, although analytic solutions are available in a few special cases. After the first integral is obtained connecting T and u, the second integral giving the actual shock profile must be obtained from either Eq. 5-5 or 5-6 (or a suitable combination).

A glance at Eq. 5-5 and 5-6 shows that in two special cases a first integral representable by a curve on the T, u plot is directly available, from which the complete solution may be obtained by a quadrature. These are the case $\mu'' = 0$ ($Pr'' = 0$), for which the right side of Eq. 5-5 is zero, and the case $k = 0$ ($Pr'' = \infty$), for which the right side of Eq. 5-6 is zero. These two integral curves on the T, u plot are termed "nonviscous" and "nonconducting," respectively. They play an important part in the geometry of the actual integral curves where neither μ'' nor k is zero. Another important special case appears if the enthalpy is a function of the temperature alone, i.e. if ρT is a function of pressure alone (this includes a perfect gas), and the longitudinal Prandtl number defined by

$$Pr'' = \frac{\mu'' c_p}{k} \tag{5-7}$$

is a constant and equal to 1. With the enthalpy a function of temperature alone, Eq. 5-4 may be expressed

$$\frac{\mu''}{m}\left(\frac{1}{Pr''}\frac{dh}{dx} + u\frac{du}{dx}\right) = h + \frac{u^2}{2} - h_0 \tag{5-8}$$

With $Pr'' = 1$ this has the integral

$$h + \frac{u^2}{2} = h_0 \tag{5-9}$$

which, since it satisfies conditions before and after the shock, is the one desired. This special solution was noted by Becker [*26*], and may be characterized by the phrase "constant total enthalpy." Since most gases are reasonably close to a perfect gas and have a value of Pr'' of the order of 1, the constant total enthalpy solution may be considered to be realistic for a gas, while the nonviscous and nonconducting solutions clearly are not.

The features of the solution may best be seen from a study of the T, u plot, here illustrated in Fig. D,5a. Integral curves are shown on this figure with an arrow indicating the direction of increasing x. The curve designated I from point 1 to point 2 is the solution desired. One other

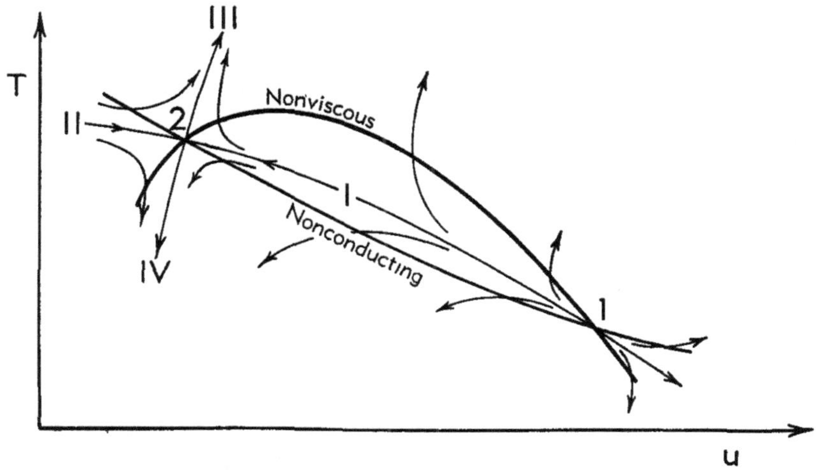

Fig. D,5a. Temperature-velocity plot.

integral curve reaches point 2 and is designated II; two solutions leave point 2 and are designated III and IV. All integral curves have infinite slope as they cross the nonviscous curve and have zero slope as they cross the nonconducting curve. If for a given case the longitudinal Prandtl number is changed then all the integral curves change, but the nonviscous and nonconducting curves remain unchanged. Ludford [*29*] has given a comprehensive presentation of the types of integral curves obtainable with a perfect gas.

This solution is, of course, only a first integral. The remainder of the solution involves the quadrature of Eq. 5-5 to obtain the relation between u and x, from which the solution for p or T may be obtained. Eq. 5-6 may be used as well and leads to the relation between T and x. The type of curve obtained is illustrated in Fig. D,5b. The velocity and temperature approach their end values for decreasing or increasing x, i.e. their values before and after the shock, in an asymptotic manner.

The nature of the solutions in the vicinity of points 1 or 2, which includes the asymptotic part of the actual solutions in terms of x, is here obtained with the simplifying assumption that the gas is a perfect one. Using the subscript $_1$ and letting

$$u = u_1 + \delta u \tag{5-10a}$$

$$T = T_1 + \delta T \tag{5-10b}$$

the equations corresponding to Eq. 5-5 and 5-6 may be expressed, dropping second order quantities,

$$\frac{\mu'' u_1}{m}\frac{d\delta u}{dx} = \mathcal{R}\delta T + 2u_1\delta u - \frac{p_0}{m}\delta u \tag{5-11}$$

$$\frac{\mu'' c_p}{mPr''}\frac{d\delta T}{dx} = c_v\delta T - u_1\delta u + \frac{p_0}{m}\delta u \tag{5-12}$$

The equations in this form are linear and admit a solution for δT and δu

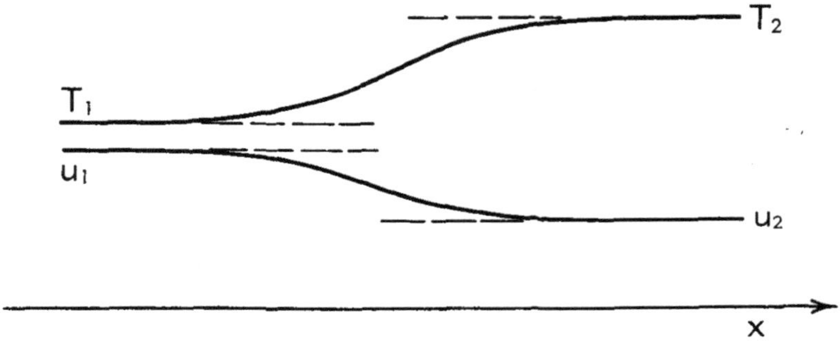

Fig. D,5b. Shock structure.

of the form $e^{\kappa m x/\mu''}$. This leads to the equation system

$$\left(c_v - \frac{\kappa c_p}{Pr''}\right)\delta T + \left(\frac{p_1}{\rho_1}\right)\frac{\delta u}{u_1} = 0 \tag{5-13}$$

$$-\mathcal{R}\delta T + \left(\frac{p_1}{\rho_1} - u_1^2 + \kappa u_1^2\right)\frac{\delta u}{u_1} = 0 \tag{5-14}$$

which has solutions if

$$(Pr'' - \gamma\kappa)(1 - \kappa)M_1^2 - (Pr'' - \kappa) = 0 \tag{5-15}$$

where

$$M_1^2 = \frac{u_1^2}{a_1^2} = \frac{u_1^2 \rho_1}{\gamma p_1} \tag{5-16}$$

Eq. 5-15 may be rewritten

$$\gamma M_1^2\kappa^2 - (\gamma M_1^2 + Pr''M_1^2 - 1)\kappa + Pr''(M_1^2 - 1) = 0 \tag{5-17}$$

of which the roots are

$$\kappa = \frac{(\gamma M_1^2 + Pr''M_1^2 - 1) \pm \sqrt{(\gamma M_1^2 - Pr''M_1^2 - 1)^2 + 4Pr''(\gamma - 1)M_1^2}}{2\gamma M_1^2} \tag{5-18}$$

This expression simplifies in various special cases:

$$Pr'' = 1 \qquad \kappa = 1, \frac{M_1^2 - 1}{\gamma M_1^2} \tag{5-19a}$$

$$Pr''M_1^2 \cong 0 \qquad \kappa = \frac{\gamma M_1^2 - 1}{\gamma M_1^2}, \frac{M_1^2 - 1}{\gamma M_1^2 - 1} Pr'' \tag{5-19b}$$

$$Pr''M_1^2 \cong \infty \qquad \kappa = \frac{Pr''}{\gamma}, \frac{M_1^2 - 1}{M_1^2} \tag{5-19c}$$

$$M_1^2 \cong 1 \qquad \kappa = \frac{\gamma + Pr'' - 1}{\gamma}, \frac{Pr''(M_1^2 - 1)}{\gamma + Pr'' - 1} \tag{5-19d}$$

$$\gamma \cong 1 \qquad \kappa = Pr'', \frac{M_1^2 - 1}{M_1^2} \tag{5-19e}$$

In Eq. 5-19b the results are not valid near $\gamma M_1^2 = 1$, and neither root changes sign as this point is traversed. The principal characteristics of these roots are that they are always real, that there is always one positive root, and that the other root is negative, zero, or positive as M_1^2 is less than, equal to, or greater than 1. With M_1^2 less than 1 the subscript $_2$ should properly have been used. At the point 2, with one positive and one negative root, the asymptotic results bear out the saddle-point nature of the solutions indicated by Fig. D,5a. At point 1, with two positive roots, the solutions on the T, u plot correspond to a node as indicated by Fig. D,5a.

The essential point here is that solutions can go toward point 2 in the sense of increasing x along the integral curves numbered I and II on Fig. D,5a, but all solutions touching point 1 must leave it. In other words, point 2 is "attainable" and point 1 is "unattainable." This shows again that the velocity behind a shock must be subsonic. The asymptotic parts of the actual solution correspond to the smaller positive root at point 1 and to the negative root at point 2.

As pointed out to the author by L. J. F. Broer,[3] the fact that the asymptotic solutions are small perturbations and linearizeable does not mean that the Navier-Stokes equations are valid there. The Navier-Stokes shear stress expression may be considered the first term in an expansion which contains other linear terms in higher order derivatives, as well as nonlinear ones. With shocks that are not weak the characteristic roots κ involved are not small and the other terms may come into play. This does not, however, affect the results that a supersonic point is not attainable while a subsonic point is.

A word of caution may be appropriate. Although the solutions near the points 1 and 2 on the T, u plot may be discussed in terms of mathematical stability, this is in no way connected with any question of physical stability. The perturbations which may increase exponentially

[3] Private communication.

do this not with increasing time but with increasing distance downstream under steady state assumptions. Downstream boundary conditions must be satisfied, and they control the solution.

The shock structure is now investigated using Eq. 5-5 and 5-6 for a perfect gas of constant ratio of specific heats in three cases, the constant enthalpy case ($Pr'' = 1$), the nonconducting case ($Pr'' = \infty$), and the nonviscous case ($Pr'' = 0$). The equations may be combined into a single one:

$$\frac{\mu''}{m}\left(u\frac{du}{dx} - \frac{\gamma - 1}{Pr''}\frac{dh}{dx}\right) = \frac{\gamma + 1}{2}u^2 - \gamma\frac{p_0 u}{m} + (\gamma - 1)h_0 \quad (5\text{-}20)$$

$$= -\Gamma(u_1 - u)(u - u_2)$$

where Γ is defined

$$\Gamma = \tfrac{1}{2}(\gamma + 1) \quad (5\text{-}21)$$

consistent with Eq. 1-27. The viscosity coefficient is assumed to be proportional to the temperature to the power s, and the cases $s = 0, \frac{1}{2}, 1$ are considered.

Case I ($Pr'' = 1$). The temperature is proportional to $u_{max}^2 - u^2$ where u_{max} is the limiting velocity of one-dimensional isoenergetic flow theory, equal to $\sqrt{2h_0}$. The coefficient of viscosity is chosen

$$\mu'' = \mu_I(u_{max}^2 - u^2)^s \quad (5\text{-}22)$$

in terms of which Eq. 5-20 becomes

$$\frac{\Gamma m}{\gamma\mu_I}dx = -\frac{(u_{max}^2 - u^2)^s u}{(u_1 - u)(u - u_2)}du \quad (5\text{-}23)$$

For $s = 0$ this has the integral

$$\frac{\Gamma m}{\gamma\mu_I}x = \frac{u_1}{u_1 - u_2}\ln(u_1 - u) - \frac{u_2}{u_1 - u_2}\ln(u - u_2) \quad (5\text{-}24a)$$

which is the solution obtained by Becker [*26*]. For $s = \frac{1}{2}$ the integral was found by Thomas [*30*], and is

$$\frac{\Gamma m}{\gamma\mu_I}x = \frac{u_1\sqrt{u_{max}^2 - u_1^2}}{u_1 - u_2}\ln\frac{u_1 - u}{u_{max}^2 - u_1 u + \sqrt{u_{max}^2 - u_1^2}\sqrt{u_{max}^2 - u^2}}$$

$$- \frac{u_2\sqrt{u_{max}^2 - u_2^2}}{u_1 - u_2}\ln\frac{u - u_2}{u_{max}^2 - u_2 u + \sqrt{u_{max}^2 - u_2^2}\sqrt{u_{max}^2 - u^2}}$$

$$+ \sqrt{u_{max}^2 - u^2} - (u_1 + u_2)\sin^{-1}\left(\frac{u}{u_{max}}\right) \quad (5\text{-}24b)$$

For $s = 1$ the integral is

$$\frac{\Gamma m}{\gamma\mu_I}x = \frac{u_1(u_{max}^2 - u_1^2)}{u_1 - u_2}\ln(u_1 - u)$$

$$- \frac{u_2(u_{max}^2 - u_2^2)}{u_1 - u_2}\ln(u - u_2)$$

$$- \tfrac{1}{2}u^2 - (u_1 + u_2)u \quad (5\text{-}24c)$$

Case II ($Pr'' = \infty$). The right-hand side of Eq. 5-6 is zero, and the temperature is proportional to $u_{\max}^2 - 2u_p u + u^2$ where $u_p = p_0/m$. The coefficient of viscosity is chosen

$$\mu'' = \mu_{II}(u_{\max}^2 - 2u_p u + u^2)^s \tag{5-25}$$

in terms of which Eq. 5-20 becomes

$$\frac{\Gamma m}{\mu_{II}}\, dx = -\frac{(u_{\max}^2 - 2u_p u + u^2)^s}{(u_1 - u)(u - u_2)}\, du \tag{5-26}$$

For $s = 0$ the solution is equivalent to Eq. 5-24a and was found by Rayleigh [*31*].

$$\frac{\Gamma m}{\mu_{II}}\, x = \frac{u_1}{u_1 - u_2} \ln (u_1 - u) - \frac{u_2}{u_1 - u_2} \ln (u - u_2) \tag{5-27a}$$

For $s = \frac{1}{2}$ the solution is

$$\begin{aligned}\frac{\Gamma m}{\mu_{II}}\, x = {} & \frac{u_1 \sqrt{u_{\max}^2 - 2u_p u_1 + u_1^2}}{u_1 - u_2} \times \\ & \ln \frac{u_1 - u}{u_{\max}^2 - u_p(u_1 + u) + u_1 u + \sqrt{u_{\max}^2 - 2u_p u_1 + u_1^2}\sqrt{u_{\max}^2 - 2u_p u + u^2}} \\ & - \frac{u_2 \sqrt{u_{\max}^2 - 2u_p u_2 + u_2^2}}{u_1 - u_2} \times \\ & \ln \frac{u - u_2}{u_{\max}^2 - u_p(u_2 + u) + u_2 u + \sqrt{u_{\max}^2 - 2u_p u_2 + u_2^2}\sqrt{u_{\max}^2 - 2u_p u + u^2}} \\ & + \sqrt{u_{\max}^2 - 2u_p u + u^2} - (u_1 + u_2 - u_p) \sinh^{-1}\left(\frac{u_p - u}{\sqrt{u_{\max}^2 - u_p^2}}\right)\end{aligned} \tag{5-27b}$$

For $s = 1$ the solution is

$$\begin{aligned}\frac{\Gamma m}{\mu_{II}}\, x = {} & \frac{u_1(u_{\max}^2 - 2u_p u_1 + u_1^2)}{u_1 - u_2} \ln (u_1 - u) \\ & - \frac{u_2(u_{\max}^2 - 2u_p u_2 + u_2^2)}{u_1 - u_2} \ln (u - u_2) \\ & + \tfrac{1}{2}u^2 + (u_1 + u_2 - 2u_p)u\end{aligned} \tag{5-27c}$$

Case III ($Pr'' = 0$). The right-hand side of Eq. 5-5 is zero, and the temperature is proportional to $u(u_p - u)$. The heat conduction coefficient is expressed

$$\frac{k}{c_p} = \mu_{III} u^s (u_p - u)^s \tag{5-28}$$

in terms of which Eq. 5-20 may be put in the form

$$\frac{\Gamma m}{\gamma \mu_{III}}\, dx = -\frac{u^s(u_p - u)^s(2u - u_p)}{(u_1 - u)(u - u_2)}\, du \tag{5-29}$$

For $s = 0$ the solution is

$$\frac{\Gamma m}{\gamma \mu_{\text{III}}} x = \frac{2u_1 - u_p}{u_1 - u_2} \ln (u_1 - u) - \frac{2u_2 - u_p}{u_1 - u_2} \ln (u - u_2) \tag{5-30a}$$

This solution was the earliest derived (actually in terms of the pressure), in the fundamental early paper of Rankine [*32*]. For $s = \frac{1}{2}$ the solution is

$$\frac{\Gamma m}{\gamma \mu_{\text{III}}} x = \frac{(2u_1 - u_p)\sqrt{u_1(u_p - u_1)}}{u_1 - u_2} \times \ln \frac{u_1 - u}{\frac{1}{2}u_p(u_1 + u) - u_1 u + \sqrt{u_1 u(u_p - u_1)(u_p - u)}} - \frac{(2u_2 - u_p)\sqrt{u_2(u_p - u_2)}}{u_1 - u_2} \times \ln \frac{u - u_2}{\frac{1}{2}u_p(u_2 + u) - u_2 u + \sqrt{u_2 u(u_p - u_2)(u_p - u)}} + 2\sqrt{u(u_p - u)} - 2(u_1 + u_2 - u_p) \sin^{-1}\left(\frac{2u - u_p}{u_p}\right) \tag{5-30b}$$

For $s = 1$ the solution is

$$\frac{\Gamma m}{\gamma \mu_{\text{III}}} x = \frac{(2u_1 - u_p)u_1(u_p - u_1)}{u_1 - u_2} \ln (u_1 - u) - \frac{(2u_2 - u_p)u_2(u_p - u_2)}{u_1 - u_2} \ln (u - u_2) + u(u_p - u) - 2(u_1 + u_2 - u_p)u \tag{5-30c}$$

In the nonviscous case, smooth solutions are only possible if $u_2 > \frac{1}{2}u_p$. If $u_2 < \frac{1}{2}u_p$ the solution proceeds from $u = u_1$ at $x = -\infty$ to the point for which $u = u_p - u_2$. At this point an isothermal discontinuity appears and u jumps to its final value u_2.

This difficulty may be clarified by a consideration of the T, u plot, repeated for $Pr'' \cong 0$ in Fig. D,5c. It is convenient to introduce the concept of the isothermal velocity of sound, defined by

$$a_{\text{isoth}}^2 = \left(\frac{\partial p}{\partial \rho}\right)_T \tag{5-31}$$

The point 3 at the maximum point of the nonviscous curve corresponds to $u = a_{\text{isoth}}$ and it lies to the left of, or to the right of, point 2 as the velocity u_2 is isothermally supersonic or subsonic. If u_2 is isothermally supersonic the solution merely follows the nonviscous curve. If u_2 is isothermally subsonic, the case depicted in Fig. D,5c, the solution follows the nonviscous curve from point 1 to point 4, and thence goes horizontally left to point 2. The velocity and temperature profiles for this case are shown in Fig. D,5d. The result is that there must be a "shock within the shock," an internal isothermal discontinuity in which viscosity or some other mechanism besides heat conduction must account for the dissipation. This was first noted by Prandtl [*26*].

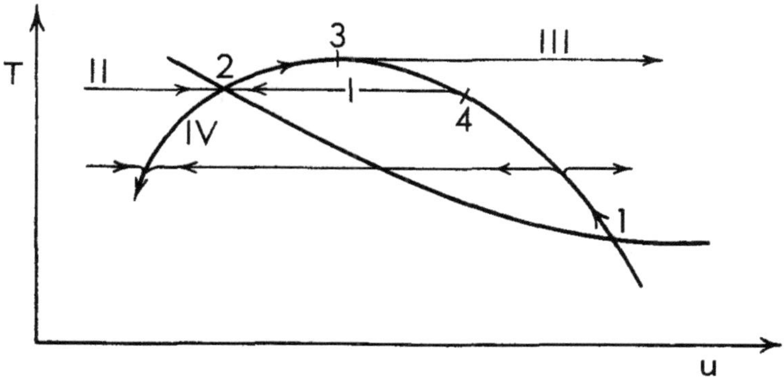

Fig. D,5c. Temperature-velocity plot for $Pr'' = 0$.

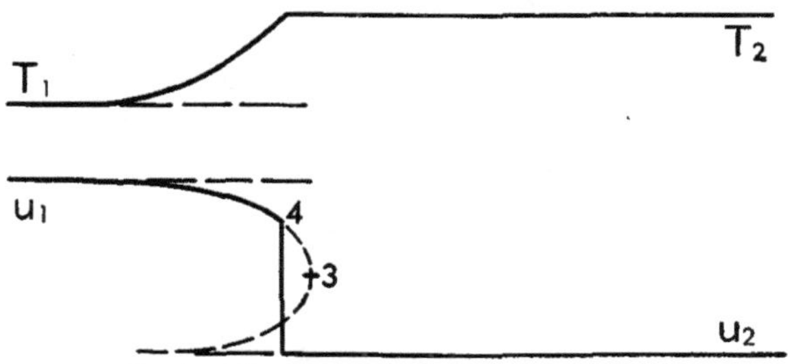

Fig. D,5d. Shock structure for $Pr'' = 0$.

In order to indicate the range which the longitudinal Prandtl number Pr'' may have, the following estimates taken from Table 3.1 of [*21*] are presented in Table D,5. The quantity Pr'' is the inverse of the quantity

Table D,5

Fluid	Pr''
Mercury	0.068
Acetone	20
Benzene	1,000
Glycerin	17,000
Argon, Helium	0.9
Neon	1.3
Air	1.4
Hydrogen	20
Carbon dioxide	800

$\mathcal{V}$ of [*21*]. The gases were taken at normal temperature and pressure, the liquids at 10 to 20°C.

Weak Shocks. In the analysis of weak shock structure, advantage may be taken of the fact that entropy changes are so small that they may be neglected except in a few places indicated by a separate order of magnitude estimate. Here an analysis is made which takes into account slow variations in time. The flow of interest is considered to be a perturbation on a uniform sonic flow of velocity a^*. The equations of motion are written

$$\frac{D\rho}{Dt} + \rho \frac{\partial u}{\partial x} = 0 \tag{5-32a}$$

$$\rho \frac{Du}{Dt} + \frac{\partial p}{\partial x} - \frac{\partial}{\partial x}\left(\mu'' \frac{\partial u}{\partial x}\right) = 0 \tag{5-32b}$$

$$\rho T \frac{Ds}{Dt} = \frac{\partial}{\partial x}\left(k \frac{\partial T}{\partial x}\right) + \mu'' \left(\frac{\partial u}{\partial x}\right)^2 \tag{5-32c}$$

in which the total or material derivative is defined

$$\frac{D}{Dt} = \frac{\partial}{\partial t} + u \frac{\partial}{\partial x} \tag{5-33}$$

and the energy equation has been replaced by an entropy equation.

In order to permit the treatment of a general fluid, several thermodynamic identities are needed. These are

$$dp = a^2 d\rho + \frac{\rho T \alpha}{c_p} ds \tag{5-34a}$$

$$c_p dT - T ds = \frac{\alpha T}{\rho} dp \tag{5-34b}$$

$$\frac{\alpha T}{\rho} = \frac{(\gamma - 1)c_p}{\rho a^2 \alpha} \tag{5-34c}$$

where α is defined in Eq. 1-28. These relations may be obtained from formulas given in standard thermodynamics texts, e.g. [*3*].

An altered continuity equation is obtained by substituting Eq. 5-34a and 5-32c in Eq. 5-32a, expressing the thermodynamic identity with material derivatives

$$\frac{Dp}{Dt} + a^2 \left[\rho \frac{\partial u}{\partial x} - \frac{\alpha}{c_p} \frac{\partial}{\partial x}\left(k \frac{\partial T}{\partial x}\right) - \frac{\alpha \mu''}{c_p}\left(\frac{\partial u}{\partial x}\right)^2\right] = 0 \tag{5-35}$$

At this point a number of assumptions are introduced. The last term in the bracket in Eq. 5-35 is small and is dropped. The material is assumed to be of uniform composition so that Eq. 5-34b may be applied to the temperature derivative term in Eq. 5-35. The contribution of the *ds* term in Eq. 5-34b is small and dropped here. Finally, the approximate relation

$$dp = -\rho u du \tag{5-36}$$

is used to change the dp in Eq. 5-34b and the $\partial p/\partial t$ part of the pressure derivative in Eq. 5-35 to derivatives of u. The result of these substitutions with Eq. 5-34c, is

$$-\rho u \frac{\partial u}{\partial t} + u \frac{\partial p}{\partial x} + \rho a^2 \frac{\partial u}{\partial x} + \frac{(\gamma - 1)ku}{c_p} \frac{\partial^2 u}{\partial x^2} = 0 \tag{5-37}$$

with products of derivatives dropped.

The momentum equation is multiplied by u, giving

$$\rho u \frac{\partial u}{\partial t} + \rho u^2 \frac{\partial u}{\partial x} + u \frac{\partial p}{\partial x} - \mu'' u \frac{\partial^2 u}{\partial x^2} = 0 \tag{5-38}$$

The pressure is eliminated between Eq. 5-37 and 5-38, giving

$$2\rho u \frac{\partial u}{\partial t} + \rho(u^2 - a^2) \frac{\partial u}{\partial x} - \mu'' u \left(1 + \frac{\gamma - 1}{Pr''}\right) \frac{\partial^2 u}{\partial x^2} = 0 \tag{5-39}$$

The quantity ν'' is defined

$$\nu'' = \frac{\mu''}{\rho}\left(1 + \frac{\gamma - 1}{Pr''}\right) \tag{5-40}$$

and it is noted that, approximately,

$$u - a = \Gamma(u - a^*) \tag{5-41}$$

where Γ is defined in Eq. 1-27. Further noting that $u + a$ is approximately equal to $2u$, Eq. 5-39 may be reexpressed:

$$\frac{\partial u}{\partial t} + \Gamma(u - a^*) \frac{\partial u}{\partial x} = \tfrac{1}{2}\nu'' \frac{\partial^2 u}{\partial x^2} \tag{5-42}$$

The classical solution of Taylor [*33*] for weak shock structure is the steady solution of Eq. 5-42

$$\Gamma(u - a^*) = -\Gamma A \tanh \frac{\Gamma A x}{\nu''} \tag{5-43}$$

with the time derivative term dropped. With the viscous term in ν'' dropped in Eq. 5-42, the equation governing first order wave structure appears (see, for example, [*34*]), while with the nonlinear term in $u - a^*$ dropped the equation governing low frequency acoustic dissipation appears, giving the classical theory of Kirchhoff [*35*]. In this approximation there is no distinction between forced and free waves.

A somewhat more general analysis leading to a basic equation equivalent to Eq. 5-42 has been made by Hayes [*36*]. It should also be noted that Eq. 5-42 is essentially the same as the equation extensively treated by Burgers [*37*]. Burgers' equation was first obtained for the case of a perfect gas (with viscosity and heat conduction) by Lighthill [*38*].

The weak shock structure solution was originally obtained by Taylor [*33*] for a perfect gas, but the present analysis shows his result to be valid for a general fluid. This generalization for the case of acoustic dissipation is attributed to Langevin by Truesdell [*21*, Sec. 7], who includes the generalization in his treatment of forced waves.

The proportionate contributions of viscosity and of heat conduction to the thickness of a weak shock are the same as to the absorption coefficient of low frequency sound waves, and this fact is a consequence of Eq. 5-42. The writer hypothesizes that the same will be true if other dissipative mechanisms such as relaxation and diffusion must be taken into account, provided that the weak shock is weak enough and the sound waves are at low-enough frequency.

The entropy change of Eq. 5-32c may be divided into two parts, into an entropy transport term

$$\rho T \left(\frac{Ds}{Dt}\right)_{\text{tr}} = T \frac{\partial}{\partial x}\left(\frac{k}{T}\frac{\partial T}{\partial x}\right) \tag{5-44}$$

which contributes no change to world entropy, and an entropy production term

$$\rho T \left(\frac{Ds}{Dt}\right)_{\text{pr}} = \frac{k}{T}\left(\frac{\partial T}{\partial x}\right)^2 + \mu''\left(\frac{\partial u}{\partial x}\right)^2 \tag{5-45}$$

which is always positive. This last may be reexpressed using the approximations made above, in particular Eq. 5-34b, 5-34c, and 5-36,

$$\rho T \left(\frac{Ds}{Dt}\right)_{\text{pr}} = \mu''\left(1 + \frac{\gamma - 1}{Pr''}\right)\left(\frac{\partial u}{\partial x}\right)^2 \tag{5-46a}$$

or

$$T \left(\frac{Ds}{Dt}\right)_{\text{pr}} = \nu''\left(\frac{\partial u}{\partial x}\right)^2 \tag{5-46b}$$

Thus the proportionate contributions of viscosity and of heat conduction to entropy production are the same as to weak shock thickness and to low frequency absorption. For the classical steady weak shock solution (Eq. 5-43), Eq. 5-46b may be expressed

$$\begin{aligned} Tuds &= \nu''\left(-\frac{\Gamma A}{\nu''}\operatorname{sech}^2\frac{\Gamma A x}{\nu''}\right)du \\ &= -\Gamma[A^2 - (u - a^*)^2]d(u - a^*) \end{aligned}$$

Replacing T by T_{av}, u by a^*, and Γ by Γ_{av}, and integrating from $u - a^* = A$ to $-A$, gives

$$T_{\text{av}}\Delta s = \frac{4\Gamma_{\text{av}}A^3}{3a^*} \tag{5-47}$$

Since $u_1 - u_2 = 2A$, this result is in agreement with Eq. 2-4 and 2-9.

The approximate relation between T and u given by Eq. 5-34b and 5-36 with $ds = 0$ may be used to eliminate the first derivative terms between Eq. 5-5 and 5-6 for the case of the steady shock wave. The result is

$$\left(\frac{p}{\rho} + u^2 - \frac{p_0 u}{m}\right) + \frac{Pr''}{\alpha T}\left(e - \frac{u^2}{2} + \frac{p_0 u}{m} - h_0\right) = 0 \qquad (5\text{-}48)$$

and may serve as an approximate first integral of Eq. 5-5 and 5-6 in cases in which the shock is only moderately weak. In Becker's case, $Pr'' = 1$ and $\alpha T = 1$, and his first integral follows immediately.

With the quantities Γ, a^*, and ν'' constant, general solutions to Burgers' equation may be obtained by a method due to Hopf [*39*] and Cole [*40*] which is related to the classical method for solving a Riccati differential equation.

In Eq. 5-42 the substitution

$$\Gamma(u - a^*) = -\nu'' \frac{\psi_x}{\psi} \qquad (5\text{-}49)$$

is made, leading directly to the relation

$$\frac{\psi_t - \frac{1}{2}\nu''\psi_{xx}}{\psi} = f(t) \qquad (5\text{-}50)$$

where $f(t)$ is an arbitrary function of t. The variable ψ times the exponential of minus the integral of $f(t)$ satisfies the same equation with f set equal to zero, or

$$\psi_t = \tfrac{1}{2}\nu''\psi_{xx} \qquad (5\text{-}51)$$

The multiplicative function of t does not enter into the expression (Eq. 5-49) for u, so that the ψ defined in Eq. 5-49 may be required to satisfy Eq. 5-51 with no loss of generality.

Eq. 5-51 is the classical heat conduction equation for which an extensive literature exists. The solution for a single steady shock of Eq. 5-43 is simply

$$\psi = \exp\frac{\Gamma^2 A^2 t}{2\nu''}\cosh\frac{\Gamma A x}{\nu''} \qquad (5\text{-}52)$$

Of the numerous solutions to Burgers' equation of interest in shock wave theory obtained by Cole [*40*], Howard [*41*], Lighthill [*38*], and others, four shall be cited here.

The coalescence of two shock waves. This solution was first found by Howard [*41*] and independently found and published by Lighthill [*38*]. The solution is given:

$$\psi = \exp\frac{(A + B)^2 t}{2\nu''}\cosh\frac{(A + B)x}{\nu''} + \tfrac{1}{2}\exp\frac{(A - B)^2 t}{2\nu''}\exp\frac{(A - B)x}{\nu''} \qquad (5\text{-}53)$$

It may be noted that for t large, ψ takes the form of Eq. 5-52 corresponding to a single shock of strength $\Gamma^{-1}(A + B)$. For $-t$ large the solution for u may be shown to approach

$$-\Gamma(u - a^*) \cong A \tanh \frac{A(x - Bt)}{\nu''} + B \tanh \frac{B(x + At)}{\nu''} \tag{5-54}$$

For the symmetric case $A = B$ the solution is

$$-\Gamma(u - a^*) = A \frac{\tanh \frac{A(x - At)}{\nu''} + \tanh \frac{A(x + At)}{\nu''}}{1 - \frac{1}{4} \exp \frac{2A^2t}{\nu''} \operatorname{sech} \frac{A(x - At)}{\nu''} \operatorname{sech} \frac{A(x + At)}{\nu''}} \tag{5-55}$$

This solution is plotted in Fig. D,5e.

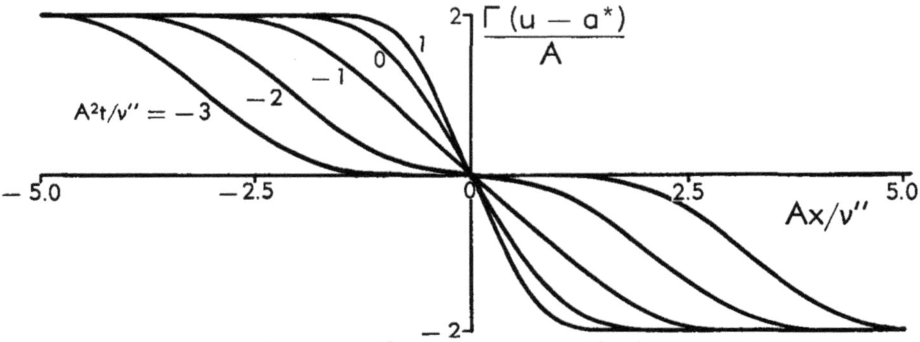

Fig. D,5e. Two equal shocks coalescing.

The asymptotic form of a compression pulse. Found by Lighthill [*38*], and independently by the writer, who considered it as a similar solution corresponding to a half-N wave, this solution is given by

$$\psi = (e^R + 1) - (e^R - 1) \operatorname{erf} \frac{x}{\sqrt{2\nu''t}} \tag{5-56}$$

and the resultant solution for u is shown in Fig. D,5f. The quantity R is a constant which is a Reynolds number, defined by Lighthill as

$$R = \ln \frac{\psi(-\infty)}{\psi(\infty)} \tag{5-57}$$

It should be noted that any solution to the heat equation (Eq. 5-51) is invariant under a change in the sign of x, which changes the sign of u. In the case here considered, the resulting solution is one for an expansion pulse, again corresponding to a half-N wave.

The whole symmetric N wave. This solution is due to Lighthill [*38*], and is

$$\psi = 1 + \left(\frac{t_0}{t}\right)^{\frac{1}{2}} \exp\left(-\frac{x^2}{2\nu''t}\right) \tag{5-58}$$

Lighthill here defines the Reynolds number of half the wave

$$R = \ln \frac{\psi(0)}{\psi(\infty)} = \ln \left[1 + \left(\frac{t_0}{t}\right)^{\frac{1}{2}}\right] \tag{5-59}$$

The resulting velocity profiles are shown in Fig. D,5g.

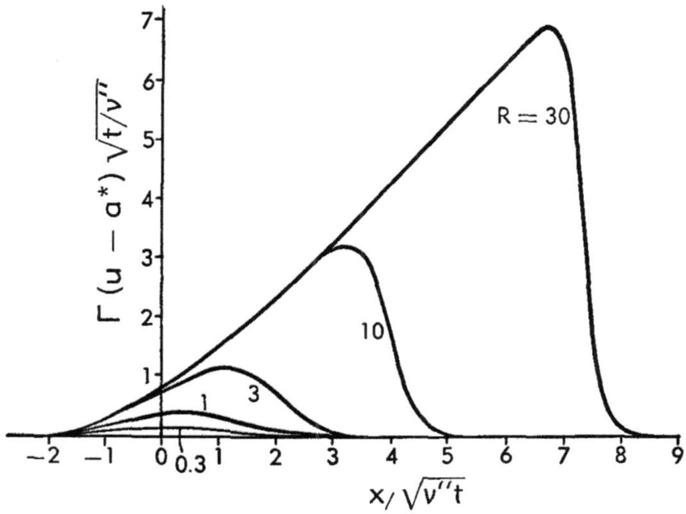

Fig. D,5f. Half-N wave or asymptotic compression pulse. After [*38*, Fig. 17, p. 335].

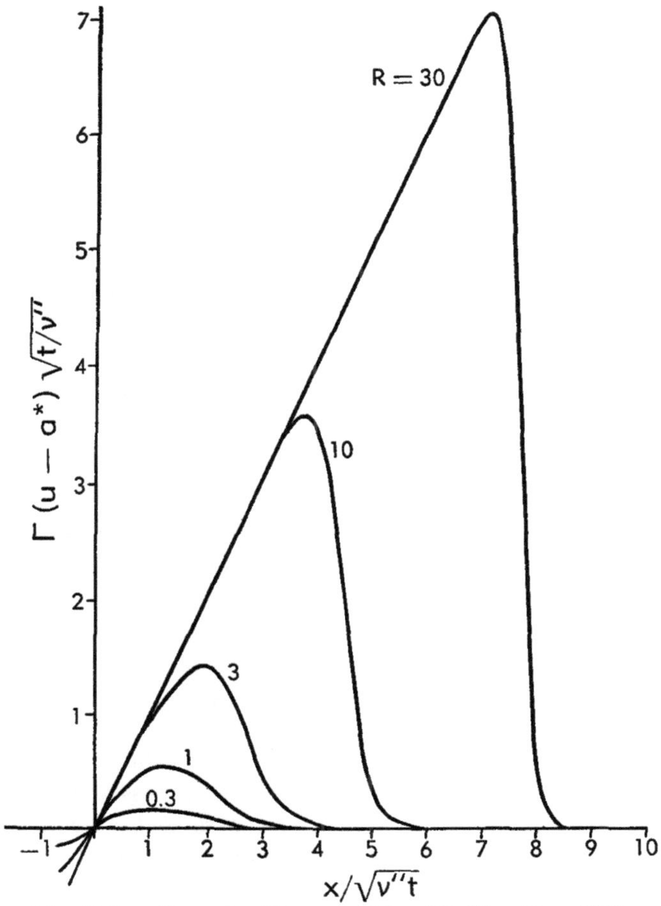

Fig. D,5g. Decaying N waves. After [*38*, Fig. 18, p. 338].

The decaying saw-tooth wave. This solution is also due to Lighthill [*38*]. The solution for ψ is

$$\psi = \frac{1}{\lambda}\sum_{-\infty}^{\infty} \exp\left[\frac{2\pi i n x}{\lambda} - \frac{2\pi^2 n^2 \nu'' t}{\lambda^2}\right] \tag{5-60a}$$

$$= \frac{1}{\sqrt{2\nu''\pi t}}\sum_{-\infty}^{\infty} \exp\left[-\frac{(x - n\lambda)^2}{2\nu'' t}\right] \tag{5-60b}$$

The solution for u, with the Reynolds number now defined for a single lobe or half period,

$$R = \ln\frac{\psi(0)}{\psi(\frac{1}{2}\lambda)} \tag{5-61a}$$

$$\cong \frac{\lambda^2}{8\nu'' t} - \ln 2 \qquad (t \text{ small}) \tag{5-61b}$$

$$\cong 4\exp\left(\frac{-2\pi^2\nu'' t}{\lambda^2}\right) \qquad (t \text{ large}) \tag{5-61c}$$

is shown in Fig. D,5h.

The analysis above, leading to Burgers' equation (Eq. 5-42), assumes that viscosity and heat conduction are the only dissipative effects which need to be taken into account. The author conjectures that Burgers' equation is still the appropriate equation for weak wave phenomena under certain conditions with other dissipative effects taken into account, with ν'' a sum of terms each arising from one or more of the effects. Thus, ν'' would have the form

$$\rho\nu'' = \tfrac{4}{3}\mu + \mu' + \mu_2 + \mu_3 + \mu_4 + \cdots \tag{5-62}$$

where μ and μ' are the quantities of Eq. 5-1, μ_2 is $k(\gamma - 1)/c_p$, μ_3 is a coefficient arising from diffusion if more than one molecular species is present, and μ_4 is a coefficient arising from relaxation phenomena; additional terms might account for radiation, etc. The conditions are that the characteristic times and distances involved in the wave phenomena be sufficiently large, e.g. that the transit time of a shock wave thickness or period of a sinusoidal sound wave be large compared with a characteristic relaxation time, or that the corresponding shock wave thickness or sound wavelength be large compared with the mean free path of a light-diffusing particle or of a photon. If Burgers' equation is valid, the equivalence between the coefficient controlling weak shock thickness and that controlling low frequency acoustic dissipation is generalized. It is also conjectured that Eq. 5-46b for entropy production would likewise maintain its form.

As a support for this conjecture may be cited the results of Dyakov [42], who gives a result for weak shock wave structure in a binary mixture which is of the form suggested. He indicates that the coefficient he calculates is the same as that obtained by Shaposhnikov and Goldberg [43] for sound absorption in a binary mixture with diffusion.

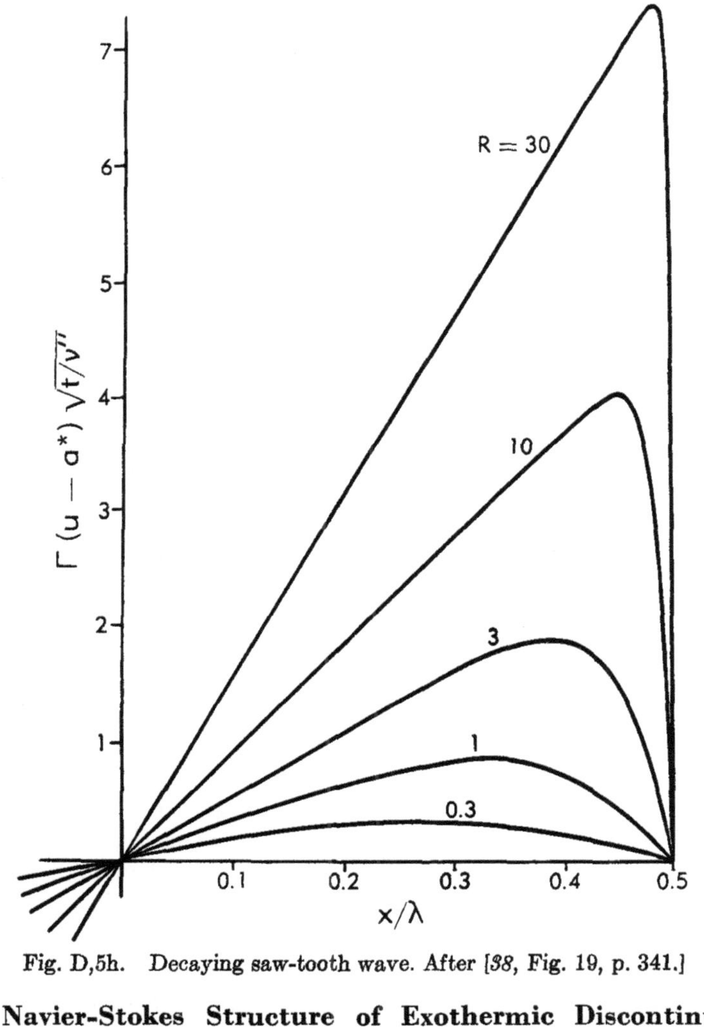

Fig. D,5h. Decaying saw-tooth wave. After [38, Fig. 19, p. 341.]

D,6. Navier-Stokes Structure of Exothermic Discontinuities. The structure of exothermic discontinuities is now investigated using the Navier-Stokes equations. In addition to the limitations discussed at the beginning of the previous section there are additional ones here, resulting from assumptions on the nature of the chemical processes involved. The principal assumption made, termed the "linear properties assumption," is that the quantities e, pv, and h vary within the discontinuity structure in a linear manner with a parameter which measures the extent to which the chemical reaction has proceeded. This generally implies that the chemical reaction may be considered a simple one, with negligible concentrations of intermediate products, that there is negligible relative diffusion of the various species occurring, that linear mixing laws hold for the mixture of unreacted and reacted material, and that thermal equilibrium exists locally. In particular, the assumptions of negligible diffusion and of negligible intermediate products may be expected to limit severely the applicability of this treatment. However, for the cases in which these assumptions are not too unreasonable it is hoped that a realistic picture of the hydrodynamics of one-dimensional combustion processes results. The use of the Navier-Stokes equations involves the assumption that neither the viscous stress nor the heat conduction has any explicit dependence upon the rate of reaction.

In the case of reactions for which the rate increases with increasing temperature it is convenient to use the concept of an "ignition temperature" below which the rate of reaction is so small as to be negligible. Such a concept is useful to forestall the question as to why the reaction involved has not partially taken place in the material before the discontinuity is reached. A reaction such as a condensation from a supersaturated vapor does not at all fit the ignition temperature concept and would require a separate treatment.

The parameter which measures the extent to which the reaction has proceeded is termed ϵ. The choice of the end values for ϵ is mostly a matter of convenience; this choice should generally be made in such a way as to simplify the equations as much as possible and is usually taken so that ϵ is the mass fraction of material which has reacted. The value $\epsilon = 0$ is then the choice for the state in front of the discontinuity. If the reaction can be considered to go to completion the choice $\epsilon = 1$ is made for the state behind the discontinuity. If the reaction does not go to completion but reaches an equilibrium state, the relation $\epsilon = 1$ may conveniently be used to denote total reaction and the final value of ϵ determined by the equilibrium conditions. In a case such as the condensation shock, account should be taken in the choice of ϵ of the fact that the Hugoniot curves for the vapor and for the saturated mixture intersect. In the development that follows, the reaction is assumed to go to completion and ϵ is chosen to be the mass fraction of reacted material, with $\epsilon = 1$ behind the front; altering this assumption would affect but slightly the main concepts of the treatment given, which thus may be considered representative.

As in the preceding article, the temperature-velocity plot is used, with the conceptual addition of ϵ as a third variable. It is convenient also to replace the variable u with the equivalent variable $v(=u/m)$; the dependence of the equations on the parameter m is thereby made somewhat clearer and the resultant plot involves only thermodynamic variables. The complete Hugoniot curves for all values of m and ϵ may be transcribed onto this space as a "Hugoniot surface," with the understanding, however, that only a single value of m is being considered at one time. The particular line on the Hugoniot surface corresponding to the chosen value of m is denoted the "equilibrium line." The concept of a space almost identical with the T, v, ϵ one was used by Friedrichs [44] in a treatment of a special case of the same problem, the case of a perfect gas whose gas constant does not change in the reaction.

The concepts of the "unreacted" and "reacted" surfaces, in this case the planes $\epsilon = 0$ and $\epsilon = 1$, are introduced, corresponding to the equations of state for the material before and after the reaction. The structure of the gasdynamic discontinuity is considered to be divided into three zones, the first occurring before the reaction has started and termed the "preheating zone" (on the unreacted surface), the second occurring during the reaction and termed the "reaction zone," and the third occurring after the reaction has finished and termed the "adjustment zone" (on the reacted surface).

On a constant ϵ plane of this T, v, ϵ space or on the unreacted or reacted surface, there may be drawn a figure similar to Fig. D,5a, with the integral curves of Art. 5 in general representing the projections of the actual integral curves. The nonviscous and nonconducting lines become surfaces in the extended space; these are not very significant except in cases of very small or very large Prandtl number. Of particular significance is the result of Art. 5 that a supersonic point is unapproachable. This result holds under the conditions of this section only for a reaction occurring in a plane ϵ = const, and so means that there can be no adjustment zone in a weak detonation or a strong deflagration. There must be two degrees of freedom in the parameters which govern the choice of the integral curve in order to ensure that the curve will go to such a supersonic point.

In the development which follows, the subscripts 1 and 2 are used as before to indicate conditions before and after the reaction. The subscripts u and b are used to indicate unreacted and fully reacted material, respectively; thus $\epsilon_u = 0$ and $\epsilon_b = 1$. The equations corresponding to Eq. 5-2, 5-5, and 5-6 may be expressed

$$u = mv \tag{6-1}$$

$$\frac{\mu'' m}{\rho}\frac{dv}{dx} = (1 - \epsilon)(pv)_u + \epsilon(pv)_b + m^2v^2 - p_0 v \tag{6-2}$$

$$\frac{k}{m}\frac{dT}{dx} = (1 - \epsilon)e_u + \epsilon e_b - \tfrac{1}{2}m^2v^2 + p_0 v - h_0 \tag{6-3}$$

with the linear properties assumption. To these equations may be added a fourth, for the time rate of change of ϵ,

$$mv\frac{d\epsilon}{dx} = f(T, v, \epsilon) \geqq 0 \tag{6-4}$$

which must come from chemical kinetic considerations, or in the case of a condensation, from nucleation considerations. The inequality of Eq. 6-4 states that the reaction is irreversible; this condition might need to be altered with an unusual definition of ϵ. The condition

$$0 \leqq \epsilon \leqq 1 \tag{6-5}$$

is a result of this inequality. The values of the constants m, p_0, and h_0 are presumed suitably chosen so that the conservation laws permit a solution. The constants are related to conditions in front of the discontinuity by

$$p_0 = p_1 + m^2 v_1 \tag{6-6a}$$

$$h_0 = h_1 + \tfrac{1}{2} m^2 v_1^2 \tag{6-6b}$$

as may be noted from Eq. 1-2 and 1-3.

Eq. 6-2 and 6-3 may be combined with the variable ϵ eliminated to give

$$\begin{aligned}(e_u - e_b)\mu'' mv \frac{dv}{dx} - [(pv)_u - (pv)_b] \frac{k}{m} \frac{dT}{dx} &= H(T, v) \\ &= (h_u + \tfrac{1}{2} m^2 v^2 - h_0)(pv)_b - (h_b + \tfrac{1}{2} m^2 v^2 - h_0)(pv)_u \\ &\quad + (m^2 v^2 - p_0 v)(h_u - h_b)\end{aligned} \tag{6-7}$$

If conditions in front of the discontinuity are fixed the function H may be expressed

$$H(T, v) = F + m^2 G \tag{6-8}$$

where

$$\begin{aligned}F(T, v) = {} & (h_1 - \tfrac{1}{2} h_u - \tfrac{1}{2} h_b)[(pv)_u - (pv)_b] \\ & - [p_1 v - \tfrac{1}{2}(pv)_u - \tfrac{1}{2}(pv)_b](h_u - h_b)\end{aligned} \tag{6-9a}$$

$$G(T, v) = \tfrac{1}{2}(v_1^2 - v^2)[(pv)_u - (pv)_b] - v(v_1 - v)(h_u - h_b) \tag{6-9b}$$

Of particular interest is the cylinder in the T, v, ϵ space determined by the equation

$$H(T, v) = 0 \tag{6-10}$$

This is termed the "invariant cylinder" and has the property that the T, v projection of any integral curve crossing it has a slope at the intersection which is independent of ϵ. The location of the invariant cylinder is itself independent of μ'' and k, hence also of Pr''.

If the right-hand sides of both Eq. 6-2 and 6-3 are set equal to zero, corresponding to hydrodynamic equilibrium, the result is the equilibrium line mentioned above. This line lies in the invariant cylinder and may be considered as the intersection of the Hugoniot surface and the invariant cylinder for the value of m of interest. The equilibrium line intersects the unreacted and reacted surfaces in the equilibrium points which correspond to conditions in front of or behind the discontinuity.

If the substances involved in the reaction are perfect gases with a gas constant that does not change in the reaction, then Eq. 6-2 is independent of ϵ and the nonviscous surface coincides with the invariant cylinder. With an exothermic reaction the unreacted nonviscous curve lies above (i.e. at higher temperatures) or below the invariant cylinder as the gas constant increases or decreases in the reaction.

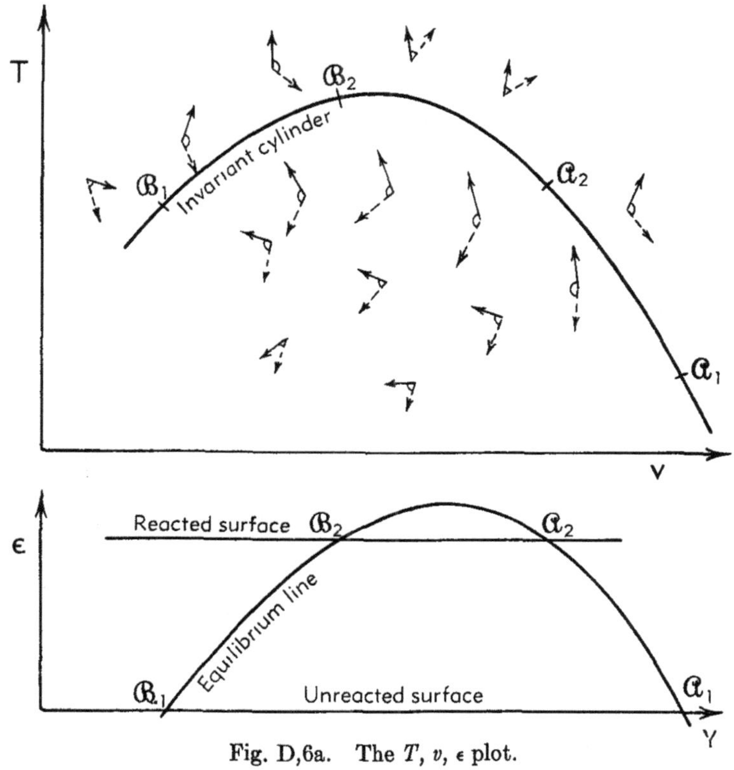

Fig. D,6a. The T, v, ϵ plot.

The results of previous articles may now be used to obtain a picture of the T, v, ϵ plot. There are generally two equilibrium points on the unreacted surface and two on the reacted surface. Of each of these pairs one point is supersonic and is designated $\mathcal{A}$, and the other is subsonic and is designated $\mathcal{B}$. The case in which the equilibrium line is tangent to the reacted surface corresponds to Ch-J conditions and may be considered as a limiting case. The case in which the equilibrium line does not intersect the reacted surface is not of interest. A solution of the type $\mathcal{A}_1$-$\mathcal{B}_2$ (from the unreacted point $\mathcal{A}$ to the reacted point $\mathcal{B}$) is a strong detonation, $\mathcal{A}_1$-$\mathcal{A}_2$ a weak detonation, $\mathcal{B}_1$-$\mathcal{B}_2$ a weak deflagration, and $\mathcal{B}_1$-$\mathcal{A}_2$ a strong deflagration.

Fig. D,6a illustrates what shall be called a normal configuration on the T, v, ϵ plot. On the T, v diagram may be drawn arrows (solid in Fig. D,6a), indicating the direction of integral curves on the unreacted surface. From the same points may be drawn arrows (dotted in Fig. D,6a), indicating the direction of integral curves on the reacted surface. The linear properties assumption requires that the direction of any integral curve (for which $0 \leqq \epsilon \leqq 1$) must lie in the angle between the two arrows which is less than 180°. On the invariant cylinder, and only there, the two arrows have the same angle. Between the points $\mathcal{B}_1$ and $\mathcal{B}_2$ and between $\mathcal{A}_2$ and $\mathcal{A}_1$ they have opposite directions. Elsewhere on the invariant cylinder they have the same direction. An integral curve crossing the invariant cylinder at a greater value of ϵ than that on the equilibrium curve at the same values of T and v does so in the direction of decreasing T, and conversely. These considerations and the condition

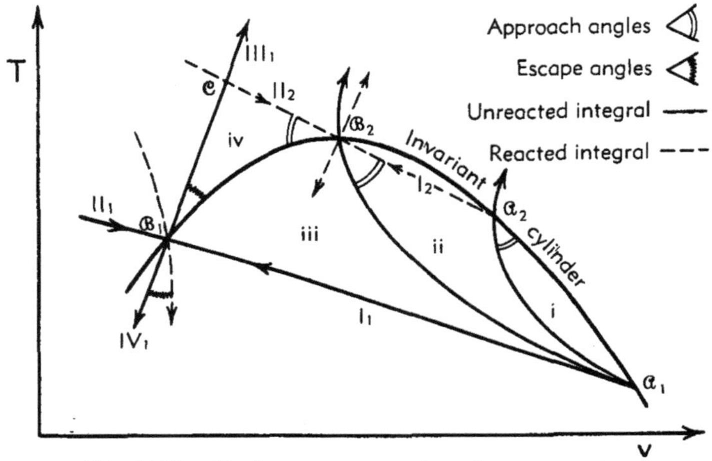

Fig. D,6b. Basic zones, approach and escape angles.

that ϵ cannot decrease provide limits for the location of an integral curve corresponding to a desired solution.

Fig. D,6b shows limiting boundaries on the T, v diagram. Four basic zones are defined: Zone *i* lies between the invariant cylinder and the unreacted integral curve from $\mathcal{A}_1$ to $\mathcal{A}_2$; zone *ii* lies between the unreacted integral curves from $\mathcal{A}_1$ to $\mathcal{A}_2$ and from $\mathcal{A}_1$ to $\mathcal{B}_2$, and below the reacted integral curve I_2 from $\mathcal{A}_2$ to $\mathcal{B}_2$; zone *iii* lies below the invariant cylinder and between the unreacted integral curves from $\mathcal{A}_1$ to $\mathcal{B}_2$ and from $\mathcal{A}_1$ to $\mathcal{B}_1(I_1)$; and zone *iv* lies in the space between the invariant cylinder, the unreacted integral curve III_1, and the reacted integral curve II_2. The intersection of curves III_1 and II_2 is designated $\mathcal{C}$, and corresponds to the highest temperature in the four zones. The order in which these zones are listed is significant, because an actual integral curve can only pass from one zone into the next and cannot pass into the one before. If an integral curve leaves a zone in any direction it cannot return. Any integral curve of interest must lie completely within the four zones. The points $\mathcal{A}_2$ and

$\mathcal{B}_2$ may only be approached from within certain angles, that in zone *i* in the case of $\mathcal{A}_2$, and those in zones *ii* and *iv* in the case of $\mathcal{B}_2$. Also, an integral curve may only escape $\mathcal{B}_1$ into two angles, with the one of physical interest in zone *iv*. In the Ch-J case, points $\mathcal{A}_2$ and $\mathcal{B}_2$ merge and zone *ii* disappears. However, the distinction between the two types of solutions remains because of the two distinct approach angles.

With the configuration shown in Fig. D,6a and D,6b the integral curve for a strong detonation could pass through any and all the zones in the proper order, though in most practical examples probably only zones *iii* and *iv* would be traversed. The curve for a weak detonation would have to lie completely in zone *i*, while that for a weak deflagration would lie completely in zone *iv*. A strong deflagration cannot exist at all with this configuration. This configuration, here termed normal, is characterized by the property that the unreacted integral curve III_1 lies entirely above the invariant cylinder. The case treated by Friedrichs [*44*] has a normal configuration.

The determinacy of the problem may now be discussed on the basis of the integral curves. One degree of freedom is available in the choice of an integral curve which leaves point $\mathcal{A}_1$, while none is available in the case of point $\mathcal{B}_1$. Considerations of the local geometry of the integral curves show that one degree of freedom is used up in requiring a curve to enter point $\mathcal{B}_2$, while two are used up in the case of point $\mathcal{A}_2$. An additional degree of freedom is available in letting the quantity m vary (the propagation velocity) with fixed initial conditions otherwise. The results are consistent with those of Art. 3 if there the reaction is considered able to furnish one quantity to determine the unknowns. A strong detonation may occur for any value of m above the minimum determined by Ch-J conditions. A weak detonation or a weak deflagration, if it exists, can only occur at a fixed value of m. In general, a strong deflagration cannot exist at all with given arbitrarily chosen initial conditions, though if permitted by other considerations a suitable choice of initial conditions might conceivably yield a solution. A detailed and more rigorous investigation of the determinacy question from a similar point of view was made by Friedrichs [*44*].

The actual integral curves may be determined for the limiting cases in which the reaction is relatively very slow or very sudden. In the case of a very slow reaction the integral curve must follow the equilibrium line except for a possible shock solution on an ϵ = const surface. In the case of a very sudden reaction, the integral curve would shift from the unreacted to the reacted surface with essentially no change in T or v. For example, the integral curve for a weak deflagration with a slow reaction follows the equilibrium line; with a sudden reaction it follows curve III_1 from $\mathcal{B}_1$ to $\mathcal{C}$, jumps to the reacted surface, and follows curve II_2 to $\mathcal{B}_2$. The T, v projections of these two limiting paths enclose zone *iv*. It may be noted that a strong deflagration is impossible with a slow reaction for any configuration, since this would require a negative shock [*2*, Sec. 93].

With a reaction for which the ignition temperature concept may be used, the integral curve must lie in the unreacted surface until the ignition temperature is reached. An illustrative solution for a weak deflagration is shown in Fig. D,6c.

Since chemical-kinetic expressions do not as a rule provide for a zero reaction rate at temperatures below the ignition temperature, in fact do not permit a strict definition of an ignition temperature, an integral curve

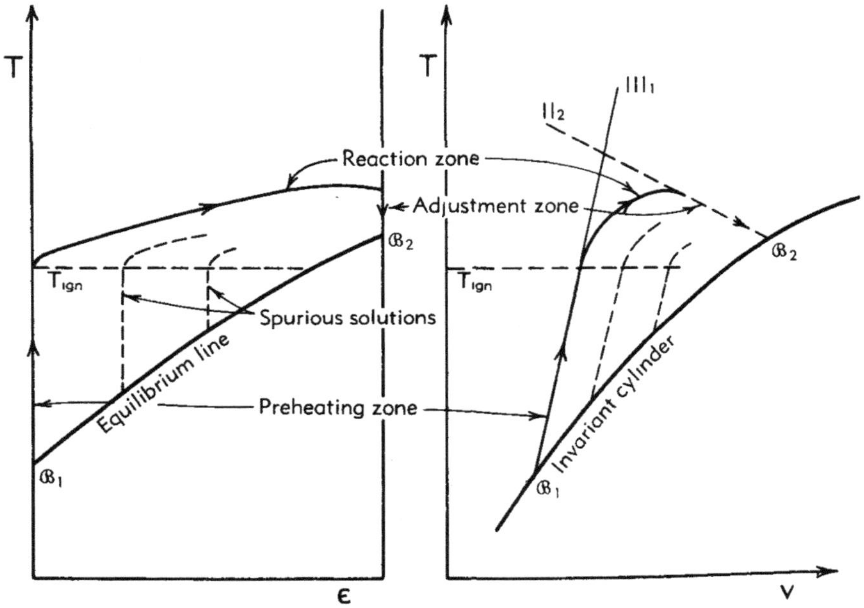

Fig. D,6c. Illustrative solution for weak deflagration.

may follow closely the equilibrium (slow reaction) line for a distance before leaving it. The reaction thickness or distance corresponding to the part of the integral curve lying along the equilibrium line is generally extremely great, so that the hydrodynamic discontinuity really starts with material which is already partially reacted. Hence, solutions given by such integral curves must be interpreted as spurious. The existence of these spurious solutions and the necessity of choosing the integral curves which lie closest to the unreacted surface near point 1 were pointed out by Burgers [*45*].

Some necessary conditions for the existence of the various kinds of discontinuities may now be stated. Such conditions may be obtained by considering the limiting case of the sudden reaction. For a Ch-J or strong detonation it is necessary for $T_{ign} < T_{\mathfrak{e}}$ for the case of interest. For a weak deflagration it is necessary for $T_{ign} < T_{\mathfrak{e}}$, where $T_{\mathfrak{e}}$ is the maximum within the range of possible solutions. For a weak detonation it is necessary for $T_{ign} < T_{\mathfrak{A}_2}$ under Ch-J conditions. With additional information about the actual integral curves or the reaction rates, better conditions may be found.

Sufficient conditions of like simplicity cannot be obtained, and additional information besides the ignition temperature must be given. The existence of the Ch-J detonation is sufficient for the existence of a strong detonation. The existence of a Ch-J detonation is probably generally ensured by the condition $T_{ign} < T_{\mathfrak{B}_1}$. It may be expected that a study of

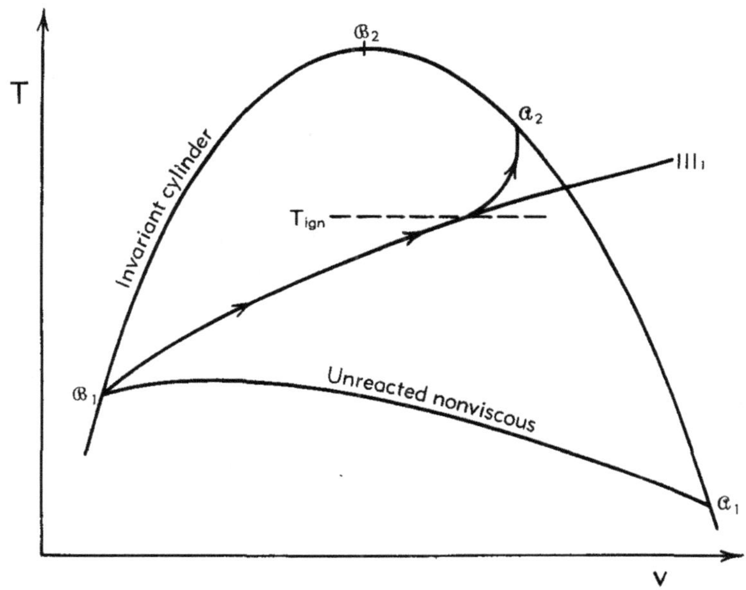

Fig. D,6d. Configuration permitting a strong deflagration.

the appropriate Ch-J diagrams and $m = 0$ diagrams could yield conclusions as to the existence of a weak detonation or a weak deflagration, apart from the question of stability.

A necessary condition for the existence of a strong deflagration may be obtained by demonstrating a configuration for which this solution is conceivably possible. This configuration is illustrated in Fig. D,6d. It is necessary that the unreacted integral curve III_1 pass under the point $\mathfrak{A}_2$ on the T, v plot for the case in question. Since the curve III_1 lies above the unreacted nonviscous curve it is necessary that the invariant cylinder also lie above the unreacted nonviscous curve. It may be seen that the proof of Friedrichs [*44*] of the nonexistence of strong deflagrations depends upon his assumption of perfect gases with a gas constant that does not change in the reaction, for which the unreacted nonviscous curve lies on the invariant cylinder. With perfect gases it is necessary that the gas constant decrease in the reaction. Conditions which would favor this configuration are a large decrease in pv, a small energy release in the reaction, and a small value of Pr''. A sharp ignition temperature with a very rapid reaction at higher temperatures would help the possibility of the existence of the integral curve. The initial conditions yielding a strong deflagration cannot be chosen arbitrarily because of the missing degree of freedom discussed above. Note that if a solution were to exist, a perturbation of the type discussed in Art. 4 would in general lead to a state for which no solution is available. This casts a shadow on the conclusion made in Art. 4 as to the instability of all strong deflagrations, and makes a stability investigation with finite perturbations yielding weak or Ch-J deflagrations advisable. In addition, the conditions which would favor the possible configuration are similar to those that would make the discontinuity endothermic in the sense of Art. 3 and would consequently make the term "strong deflagration" inapplicable.

To summarize the question of the existence of strong deflagrations, it may be said that although a complete unassailable rigorous proof of nonexistence has not yet been given, the evidence against existence is preponderant and may be taken as effective proof. This evidence may be listed:

1. The existence of finite perturbations which can change the strong deflagration to a family of weak or Ch-J deflagrations or to a Ch-J detonation.
2. The impossibility of finding a conceivably possible structure with the Navier-Stokes and linear properties assumptions, except for a reaction giving a very particular type of T, v configuration.
3. In case such a structure is conceivably possible the limitation on the possible original states because of the missing degree of freedom in choosing the integral curve.

To the writer's knowledge there has been no report of a true strong deflagration having been observed in practice.

With regard to the weak detonation, which has not been observed (again to the writer's knowledge) in explosion experiments, there is no theoretical evidence for nonexistence. It may only be said that chemical reaction rates are generally much smaller than would be necessary for a weak detonation. As mentioned above, the condensation shock which might be produced in a supersonic wind tunnel should be of the weak detonation type.

D,7. The Physics of Shock Waves. Although simple shock waves in a perfect gas can be treated adequately for most purposes from a purely hydrodynamic point of view, a much more general knowledge of physics is essential for an understanding of shock waves in more general materials. The relation is reciprocal, because experiments involving shock waves may often be used for the investigation of physical phenomena which would be much more difficult to study with other techniques. The fields of atomic physics, chemical physics, and statistical mechanics are particularly important here. The purpose of this section is to introduce qualitatively the physical processes which come into play in shock waves, particularly with very strong ones.

A basic distinction may be made between processes which affect only the structure of the shock wave and which do not affect the over-all shock relations, and processes which affect the equations of state of the material and the conservation relations, and which consequently do affect the over-all shock relations.

Weak shocks involve essentially isentropic changes in the equations of state. As indicated by the remark in Art. 5 on the classical sound absorption coefficient there is a close relation between the physics of the structure of weak shock waves and the physics of low frequency sound waves, even though the shocks and sound waves are basically distinct phenomena. As a general rule it may be said that any physical effect which is important in the propagation or attenuation of low frequency sound waves is also important in the structure of weak shock waves.

Physical processes which do affect the over-all shock relations are considered first, for shocks involving gases. Most processes of interest produce changes in the heat capacities and gas constant, and additional changes in the internal energy and enthalpy. The excitation of vibratory degrees of freedom at high temperature, governed by quantum mechanics laws, is a process which changes only the heat capacities. The extent to which certain other processes have proceeded depends upon equilibrium conditions governed by well-known thermodynamic laws. Such processes include dissociation, ionization, and various simple chemical reactions. In this category may also be put phase changes, such as the condensation of a condensation shock. With all of these processes the basic conservation laws given in Art. 1 do not require modification; the only complications introduced lie in the determination of the equations of state or of the details of the structure.

Several physical processes, however, do affect the basic conservation laws in a shock. Of these, undoubtedly the most important is radiation, by which a large amount of energy may be lost to the system if the material involved is suitably transparent. In a representative example a gas can become ionized in the shock and emit radiation through recombination of the ions and electrons. As a rule the influence of the momentum of the radiation may be neglected, and it is only the energy conservation law which must be modified or replaced. It may be noted that solutions of the deflagration type may occur with an external source of radiation, in particular for the case in which the source is downstream; the material downstream is transparent to the radiation in question; and the material upstream is strongly absorbent to at least part of the radiation.

If the shock in question is sufficiently strong or if the sound velocity (and temperature) in the original material is sufficiently high, the velocities involved might approach that of light. In such cases the conservation laws must be derived on the basis of relativistic mechanics (see Taub [*46*]). Straightforward laws like the ones of Art. 1., but with relativistic corrections, would likely have but little real significance because of the essential part that radiation effects would play. Relativistic effects may be considered generally negligible, even in the cosmical gas dynamics of astrophysics.

If the shock occurs in the presence of a magnetic field and the material has a sufficiently high conductivity, magneto-hydrodynamic effects appear. Such effects in the form of linear wave motion are now well known, and their influence in shock waves has been treated by De Hoffmann and Teller [*47*]. In cosmical shock waves, magneto-hydrodynamics is considered to play an important part.

In general, the conclusions reached in Art. 1, 2, and 3 are based upon a number of conventional assumptions as to physical conservation laws and thermodynamic behavior. In any new or unconventional situation the investigator must first make a critical assessment of the validity of these assumptions; only on the basis of such an assessment can he decide his subsequent course.

Physical processes which do not affect the over-all shock relations but only the shock structure are now considered. It is, of course, to be recognized that the processes which can affect the over-all relations play their own part in the structure and may do so even if they do not affect the over-all relations. For example, a diatomic gas might partly dissociate in the shock and then recombine. The major one of the effects, which only affect the structure, has already been discussed at the beginning of Art. 5, the failure of the Navier-Stokes viscous shear relations and the linear heat conduction relations in strong shocks. A simple physical interpretation of this failure for a monatomic gas is that, within a strong shock, the statistical velocity distributions are very different from the Maxwellian, different enough so that the Maxwellian distribution may not even be used as a first approximation. Methods for investigating this problem theoretically would have to take into account the fact that the mean free path is of the same order of magnitude as the shock thickness and would involve solutions to the differential-integral Boltzmann equation. This problem is more difficult than the related one-dimensional neutron diffusion problems, since in the neutron case the velocity characteristics of one party to each collision are known.

An important physical process in the structure of strong shocks in mixtures is that of diffusion. Under the influence of the strong pressure and temperature gradients in the shock, the various species making up the material diffuse relative to each other. This process can be very important in combustion waves. If there is ionization in the shock, diffusion can be expected to occur to a significant extent, because of the comparatively very high mobility or diffusivity of the free electrons. In this case there would also be local electric fields, resulting from the distribution of ions and electrons.

The structure of strong shocks may also be strongly affected by the phenomenon of relaxation. The term relaxation may be applied to the time delay of quite general processes in going to equilibrium, but is here applied specifically to the time delay in the excitation (or deexcitation) of the nontranslational degrees of freedom of a gas. The customary method of treating problems involving such relaxation is to consider the various degrees of freedom as comprising distinct thermodynamic systems, each of which may have its own distinct temperature. The interaction between two systems consists of a transfer of energy at a rate proportional to the temperature difference. This interaction law yields a definition for the relaxation time. If this relaxation time is large in comparison with the time between collisions in the gas behind a strong shock, there is a thin discontinuity corresponding to a shock in a monatomic gas followed by a relaxation zone in which the nontranslational degrees of freedom acquire energy and approach equilibrium. See, for example, [*48,49,50,51*].

The discussion in this entire section has been directed toward discontinuity phenomena in gases, although most of the results are applicable to liquids and, to a lesser extent, to solids. Special shock behavior may be expected in certain materials. One class of such materials includes those for which a phase change occurs; included would be substances near the critical point or near the triple point, and fogs or mists. An especially interesting case would be a liquid with vapor bubbles collapsing under the influence of a shock. Another class of materials would include anisotropic solids such as crystals and substances with various rheological properties. Materials with properties controlled by quantum mechanics effects may be mentioned, as the theoretical Bose-Einstein and Fermi-Dirac gases or liquid helium at a temperature low enough so that He II is present.

Finally, attention may be called to discontinuities which are not one-dimensional in the sense of this section but for which analogous conservation laws may be derived. An example of wide technical interest is the case of an incompressible fluid in a tube or pipe which distends under internal pressure; the discontinuities in this case give rise to the phenomenon known as water hammer, and are generally associated only with increases in the pressure.

D,8. Cited References.

1. Hugoniot, H. *J. école polytech.* *58*, 1–125 (1889).
2. Courant, R., and Friedrichs, K. O. *Supersonic Flow and Shock Waves*. Interscience, 1948.
3. Guggenheim, E. A. *Thermodynamics*. North Holland, Amsterdam, 1949.
4. Chapman, D. L. *Phil. Mag.* *47*, 90–104 (1899).
5. Jouguet, E. *Mécanique des Explosifs*. Doin et fils, Paris, 1917. Also *J. de Mathématiques, 6th series, 1*, 347–425 (1905); *2*, 5–86 (1906).
6. Bethe, H. A. The theory of shock waves for an arbitrary equation of state. *Office Sci. Research and Develop. Rept. 545*, 1942.
7. Kline, S. J., and Shapiro, A. H. Mémoires sur la mécanique des fluides. *Publs. sci. et tech. Ministère air, France*, 171–202 (1954).
8. Hayes, W. D. *Proc. Eighth Intern. Congress Theoret. and Appl. Mech., Istanbul*, 358 (1952). Abstract.
9. von Kármán, Th. *Proc. First U.S. Natl. Congress Appl. Mech.*, 673–685 (1951). Edwards, 1952. Also, von Kármán, Th., and Millan, G. *C. B. Biezeno Anniversary Volume on Applied Mechanics*, pp. 58–69. H. Stam, Haarlem, 1953.
10. Landau, L. *Acta Physicochim. U.R.S.S. 19*, 77–85 (1944).
11. Lewis, B., and von Elbe, G. *Combustion, Flames and Explosions of Gases*. Academic Press, 1951.
12. Truesdell, C. *J. math. pures et appl. 30*, 111–155 (1951).
13. Chapman, S., and Cowling, T. G. *The Mathematical Theory of Non-Uniform Gases*. Cambridge Univ. Press, 1939. Reissued with corrections, 1952.
14. Wang-Chang, C. S. On the theory of the thickness of weak shock waves. *Appl. Phys. Lab., Johns Hopkins Univ. Rept. CM-503*, 1948.
15. Zoller, K. *Z. Physik 130*, 1–38 (1951).
16. Herpin, A. *Mém. artillerie franç. 24*, 851–897 (1950).
17. Broer, L. J. F. *Appl. Sci. Research A3*, 349–360 (1953).
18. Travers, S. *Mém. artillerie franç. 24*, 421–624, 923–1006 (1950). Separately as thesis, Eyrolles, Paris, 1951.
19. Grad, H. *Commun. on Pure and Appl. Math. 5*, 257–300 (1952).
20. Mott-Smith, H. M. *Phys. Rev. 82*, 885–892 (1951).
21. Truesdell, C. *J. Rat. Mech. and Anal. 2*, 643–741 (1953).
22. Grad, H. *Commun. on Pure and Appl. Math. 2*, 331–407 (1949).
23. Jeffreys, H. *Proc. Roy. Soc. London A160*, 325–348 (1937).
24. Ikenberry, E., and Truesdell, C. *J. Rat. Mech. and Anal. 5*, 1–54 (1956). Also Truesdell, C. *J. Rat. Mech. and Anal. 5*, 55–128 (1956).
25. Sherman, F. S. A low-density wind-tunnel study of shock-wave structure and relaxation phenomena in gases. *NACA Tech. Note 3298*, 1955.
26. Becker, R. *Z. Physik 8*, 321–362 (1922). Also *NACA Tech. Mem. 505, 506*, 1929.
27. Gilbarg, D., and Paolucci, D. *J. Rat. Mech. and Anal. 2*, 617–642 (1953).
28. Truesdell, C. *J. Rat. Mech. and Anal. 1*, 125–300 (1952); *2*, 593–616 (1953).
29. Ludford, G. S. A. *J. Aeronaut. Sci. 18*, 830–834 (1951).
30. Thomas, L. H. *J. Chem. Phys. 12*, 449–452 (1944).
31. Rayleigh, Lord (J. W. Strutt). *Proc. Roy. Soc. London A84*, 247–284 (1910).
32. Rankine, W. J. M. *Phil. Trans. Roy. Soc. London 160*, 277–288 (1870).
33. Taylor, G. I. *Proc. Roy. Soc. London A84*, 371–377 (1910).
34. Hayes, W. D. *J. Aeronaut. Sci. 21*, 720–730 (1954).
35. Kirchhoff, G. *Ann. der Phys. und Chem.* (*Pogg. Ann.*) *134*, 177–193 (1868).
36. Hayes, W. D. Viscous first-order wave theory. Abstract to appear in *Proc. Ninth Intern. Congress Theoret. and Appl. Mech.*, Brussels, 1956.
37. Burgers, J. M. *Koningl. Ned. Akad. Wetenschap B57*, 45–72, 159–169, 403–433 (1954); *53*, 247–260, 393–406, 718–742 (1950). Also *Advances in Applied Mechanics I*, 171–199. Academic Press, 1948.
38. Lighthill, M. J. Viscosity effects in sound waves of finite amplitude. *Surveys in Mechanics, G. I. Taylor 70th Anniversary Volume*, 250–351. Cambridge, 1956.
39. Hopf, E. *Commun. on Pure and Appl. Math. 3*, 201–230 (1950).
40. Cole, J. D. *Quart. Appl. Math. 9*, 225–236 (1951).

41. Howard, L. N. Unpublished results, 1953.
42. Dyakov, S. P. *J. Exptl. Theoret. Phys. U.S.S.R. 27*, 283–287 (1954).
43. Shaposhnikov, I., and Goldberg, Z. *J. Exptl. Theoret. Phys. U.S.S.R. 23*, 425–429 (1952).
44. Friedrichs, K. O. On the mathematical theory of deflagrations and detonations. *Nav. Ord. Rept. 79-46*, 1946.
45. Burgers, J. M. *C. B. Biezeno Anniversary Volume on Applied Mechanics*, pp. 70–80. H. Stam, Haarlem, 1953.
46. Taub, A. H. *Phys. Rev. 74*, 328–334 (1948).
47. De Hoffmann, F., and Teller, E. *Phys. Rev. 80*, 692–703 (1950).
48. Bethe, H., and Teller, E. Deviations from thermal equilibrium in shock waves. *Aberdeen Proving Ground, BRL Rept. X-117*, c. 1941. Reprinted by *Eng. Research Inst., Univ. Mich.*, 1951.
49. Broer, L. J. F., and van den Bergen, A. C. *Appl. Sci. Research A4*, 157–170 (1954); *5*, 76–80 (1955).
50. Griffith, W. C., and Bleakney, W. *Am. J. Phys. 22*, 597–612 (1954).
51. Dyakov, S. P. *J. Exptl. Theoret. Phys. U.S.S.R. 27*, 728–734 (1954).

SECTION E

SHOCK WAVE INTERACTIONS

H. POLACHEK
R. J. SEEGER

E,1. Introduction. Shock waves are abrupt changes in the physical conditions (i.e. velocity, temperature, and entropy) of a moving fluid. Such abrupt changes can occur only when flow velocities exceed the acoustic velocity, and hence are characteristic of supersonic flows. The actual zones within which transition from one physical state to a second takes place are finite but extremely narrow (of the order of several mean free paths). The process or mechanism that results in this transition is very complex from a physical as well as mathematical point of view. Discussions of this transition zone and the mechanism by which it is brought about are contained in Sec. D, G, and H. For many purposes, however, a detailed knowledge of the physical conditions within the transition zone is not required. It is often possible to obtain an adequate description of this phenomenon by considering the shock wave as a mathematical discontinuity [*1,2,56*]. This section is based on this simplified point of view for the treatment of aerothermodynamic shock waves. The question of hydromagnetic shock waves [*57*] is not considered here.

E,2. Step-Shock Model. Consider the shock wave as a mathematical plane with cross section S, moving with constant velocity U within a fluid (Fig. E,2a). This plane acts as a boundary between two regions of different physical conditions. Let the physical state of the gas into which the shock wave is moving be specified by its pressure p_0, density ρ_0, and material velocity u_0; and let that behind the shock wave be specified by its pressure p, density ρ, and material velocity u.

The principles of conservation of mass, momentum, and energy are used to derive the relationship between the physical conditions ahead of and behind the shock front. The quantity of mass arriving at the shock front S during a small interval of time Δt is $\rho_0(U - u_0)\Delta t$ per unit area of S, whereas the amount of material leaving S is $\rho(U - u)\Delta t$. The principle of conservation of mass requires that

$$\rho_0(U - u_0)\Delta t = \rho(U - u)\Delta t$$

or

$$\rho_0(U - u_0) = \rho(U - u) \tag{2-1}$$

In a similar manner, by applying the principle of conservation of momentum, one obtains the relation

$$p - p_0 = \rho_0(U - u_0)(u - u_0) \tag{2-2}$$

Eliminating first u and then u_0 from these two equations, we obtain, respectively,

$$U - u_0 = \left[\frac{p - p_0}{(\rho_0/\rho)(\rho - \rho_0)}\right]^{\frac{1}{2}}$$

and

$$U - u = \left[\frac{p - p_0}{(\rho/\rho_0)(\rho - \rho_0)}\right]^{\frac{1}{2}} \tag{2-3}$$

On the other hand, the elimination of U from the same equations gives

$$u - u_0 = \left[\frac{(p - p_0)(\rho - \rho_0)}{\rho\rho_0}\right]^{\frac{1}{2}} \tag{2-4}$$

By applying the principle of conservation of energy, and keeping in mind that the work done on a thin layer of material in the neighborhood

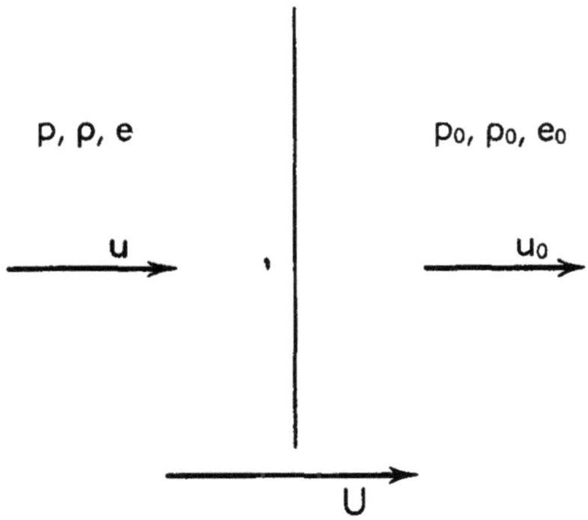

Fig. E,2a. Plane shock wave.

of the shock front is equal to the increase in internal energy, one arrives at the following relation:

$$pu - p_0u_0 = \rho(U - u)(e + \tfrac{1}{2}u_2) - \rho_0(U - u_0)(e_0 + \tfrac{1}{2}u_0^2)$$

where e and e_0 are the intrinsic energies per unit mass behind, and in front of, the shock wave, respectively. Upon substitution of Eq. 2-3 and 2-4 in the above, this reduces to

$$e - e_0 = \frac{(p + p_0)(\rho - \rho_0)}{2\rho\rho_0} \tag{2-5}$$

Eq. 2-3, 2-4, and 2-5 are known as the Rankine-Hugoniot [*3*,*4*] conditions across a shock wave. If these are combined with the thermodynamic properties of a fluid, including its equation of state, it is possible to compute the physical state behind the shock front when the state into which it moves and its strength (i.e. p_0/p) are known.

Other general properties of the fluid crossing a shock front are: (1) an increase in entropy of the third order expressed in terms of shock strength across the front, and (2) pressure, density, and temperature always greater behind the shock front than ahead of it.

Gases. For gases such as air at moderate temperatures it is possible to assume ideality, i.e. $\mathcal{R}T = p/\rho$, where $\mathcal{R}$ is the universal gas constant divided by the molecular weight of the gas. In addition, ideal gases possess the thermodynamic properties that the internal energy e is a function of the temperature T, and the specific heats are constant, i.e. $e = \text{const} \times T$. Specifically,

$$e = c_v T = \frac{c_v p}{\mathcal{R}\rho} \tag{2-6}$$

where $c_v = \mathcal{R}/(\gamma - 1)$ is the specific heat at constant volume and γ is the ratio of specific heat at constant pressure to that at constant volume

Substituting Eq. 2-6 in Eq. 2-5 we obtain

$$\frac{\rho_0}{\rho} = \frac{(\gamma + 1)p_0 + (\gamma - 1)p}{(\gamma - 1)p_0 + (\gamma + 1)p} \tag{2-7}$$

It is possible to express the conditions behind a shock front in terms of a single parameter. For this purpose it is convenient to use a measure of the intensity, the so-called shock strength ξ, defined as the ratio of pressures across the wave front, e.g. $\xi = p_0/p$. Furthermore, it is convenient to express all speeds as ratios with respect to the acoustic speed. Let $a_0 = (\gamma p_0/\rho_0)^{\frac{1}{2}}$ be the acoustic speed in front of the shock and $a = (\gamma p/\rho)^{\frac{1}{2}}$ be that behind the shock. Then Eq. 2-3 can be expressed by

$$M_0 \equiv \frac{U - u_0}{a_0} = \left[\frac{(\gamma + 1)/\xi + (\gamma - 1)}{2\gamma}\right]^{\frac{1}{2}} \tag{2-8}$$

where M_0 is the Mach number of the material ahead of the shock wave normalized with respect to the shock front. A useful plot $M_0(\xi)$ for air ($\gamma = 1.40$) is given in Fig. E,2b. Also,

$$\left.\begin{aligned} \tau &\equiv \frac{U - u_0}{a} = \frac{(\gamma - 1)\xi + (\gamma + 1)}{[2\gamma((\gamma + 1)\xi + (\gamma - 1))]^{\frac{1}{2}}} \\ \sigma &\equiv \frac{U - u}{a} = \left[\frac{(\gamma + 1)\xi + (\gamma - 1)}{2\gamma}\right]^{\frac{1}{2}} \end{aligned}\right\} \tag{2-9}$$

whereas Eq. 2-4 becomes

$$-\nu \equiv \frac{u - u_0}{a} = \tau - \sigma = \frac{2(1 - \xi)}{[2\gamma((\gamma + 1)\xi + (\gamma - 1))]^{\frac{1}{2}}} \tag{2-10}$$

Also, from Eq. 2-7 we obtain

$$\eta \equiv \frac{\rho_0}{\rho} = \frac{\sigma}{\tau} = \frac{(\gamma + 1)\xi + (\gamma - 1)}{(\gamma - 1)\xi + (\gamma + 1)} \tag{2-11}$$

It is noted that

$$\frac{a_0}{a} = \left(\frac{\xi}{\eta}\right)^{\frac{1}{2}} \tag{2-12}$$

It will be further noted that, in the derivation of the shock wave equations, the material in the region ahead of the shock front is assumed to be moving in a direction normal to it. This, however, places no restriction on the generality of the flow direction in this region, since any translational motion can be superimposed on the flow pattern without disturbing

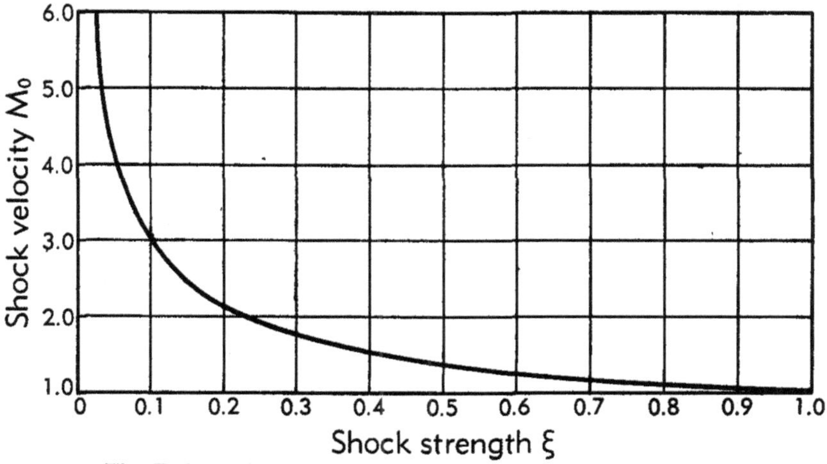

Fig. E,2b. Mach number M vs. shock strength ξ. $\gamma = 1.40$.

the validity of the derived relations. Thus we can superimpose a flow velocity component U_0 parallel to and ahead of the shock front as long as this is accompanied by an identical flow behind the shock front.

If we normalize with respect to the region ahead of the shock front, i.e. if we measure shock strength by the ratio $\xi' = p/p_0$, the basic shock wave equations (Eq. 2-9, 2-10, and 2-11) do not change in form. These may then be replaced by

$$\left.\begin{aligned} \tau' &\equiv \frac{U - u}{a_0} = \frac{(\gamma - 1)\xi' + (\gamma + 1)}{[2\gamma((\gamma + 1)\xi' + (\gamma - 1))]^{\frac{1}{2}}} \\ \sigma' &\equiv \frac{U - u_0}{a_0} = \left[\frac{(\gamma + 1)\xi' + (\gamma - 1)}{2\gamma}\right]^{\frac{1}{2}} \end{aligned}\right\} \tag{2-9'}$$

$$\nu' = \sigma' - \tau' \equiv \frac{u - u_0}{a_0} = \frac{2(1 - \xi')}{[2\gamma((\gamma + 1)\xi' + (\gamma - 1))]^{\frac{1}{2}}} \tag{2-10'}$$

$$\eta' = \frac{\rho}{\rho_0} = \frac{\sigma'}{\tau'} = \frac{(\gamma + 1)\xi' + (\gamma - 1)}{(\gamma - 1)\xi' + (\gamma + 1)} \tag{2-11'}$$

This invariance in the form of the shock wave equations proves to be highly useful in obtaining conditions for shock wave interaction patterns [*5,6,7*].

It is desirable at this point to call attention to several additional relationships which exist between the basic shock wave parameters σ, τ, ν and σ', τ', ν'. It follows directly from Eq. 2-9 and 2-10 that

$$\tau = \frac{(\gamma - 1)\sigma^2 + 2}{(\gamma + 1)\sigma} \tag{2-13}$$

and

$$\nu = \frac{2(\sigma^2 - 1)}{(\gamma + 1)\sigma} \tag{2-14}$$

Similarly from Eq. 2-9′ and 2-10′ it follows that

$$\tau' = \frac{(\gamma - 1)\sigma'^2 + 2}{(\gamma + 1)\sigma'} \tag{2-13′}$$

and

$$\nu' = \frac{2(\sigma'^2 - 1)}{(\gamma + 1)\sigma'} \tag{2-14′}$$

Careful measurements [8,9] with a shock tube show that the Rankine-Hugoniot relations derived above are experimentally verified for such shock waves over a wide range of shock strengths, viz. shock strengths $0.15 \leqq \xi \leqq 1.00$ in air and shock strengths $0.04 \leqq \xi \leqq 1.00$ in argon.

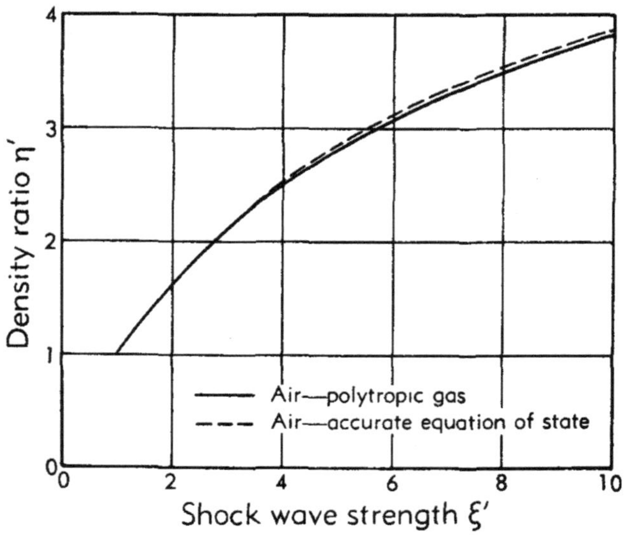

Fig. E,2c. Density ratio η' vs. shock strength ξ'. $\gamma = 1.40$.

For shock strengths outside the above ranges, involving higher temperatures and hence greater molecular excitation, the assumptions made above regarding thermodynamic behavior and ideality can no longer be used. The basic conservation equations (Eq. 2-1, 2-2, and 2-5) are, however, still valid, and provided that these are combined with accurate equation-of-state and thermodynamic data, valid results can be obtained for the physical states on both sides of a shock wave. The calculations to obtain such values then become extremely laborious. Tables giving the density, temperature, velocity, and entropy changes across shock fronts in air as a function of shock strength are contained in [10]. The results are tabulated over a range of temperatures to 15,000°K and of density to twenty times that of air at standard conditions.

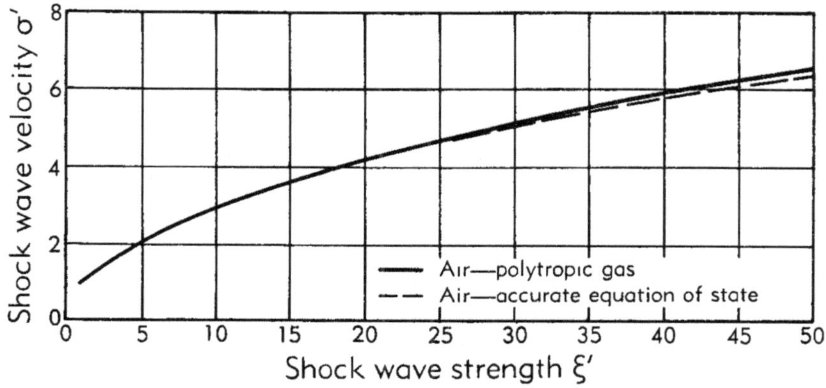

Fig. E,2d. Shock wave velocity $\sigma'(\xi')$ vs. shock strength ξ. $\gamma = 1.40$.

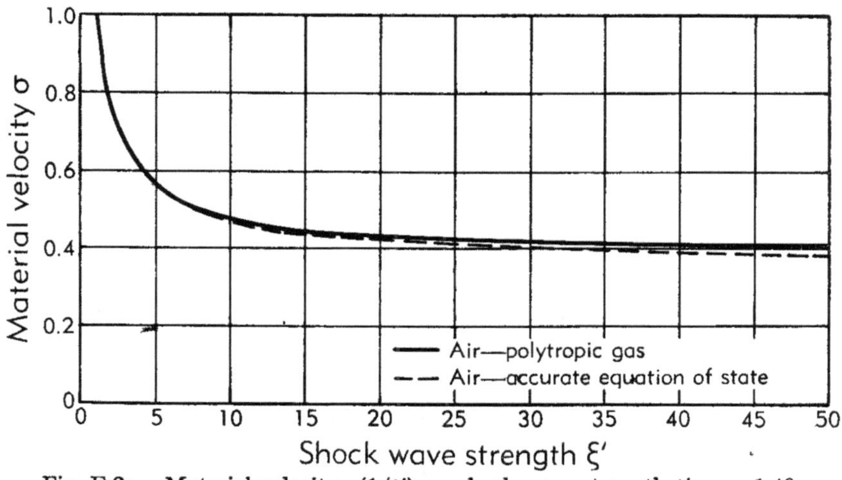

Fig. E,2e. Material velocity $\sigma(1/\xi')$ vs. shock wave strength ξ'. $\gamma = 1.40$.

In Fig. E,2c a plot of $\eta' = \rho/\rho_0$ as a function of $\xi' = p/p_0$ is given, which was computed by the use of equations derived in this section for an ideal gas and calculated for air in [10] on the basis of an accurate equation of state. It will be noted that substantially good agreement exists in the region $1.00 \leqq \xi' \leqq 10.0$. Fig. E,2d presents a similar plot for $\sigma'(\xi')$, the velocity of the shock front relative to the region ahead of the shock wave normalized with respect to the acoustic speed in that region; and Fig. E,2e presents $\sigma(\xi')$, the material velocity behind the shock front relative to the shock front (normalized with respect to the region behind the shock front). Note that the former is always supersonic whereas the latter is always subsonic. Plots for both an ideal gas and air are given. Good agreement is indicated here in the region $1.00 \leqq \xi' \leqq 50.0$.

Waterlike substances. A waterlike substance [*3*,*4*] is one for which the intrinsic energy is separable into one term that depends only upon its density and another term that depends only upon its specific entropy. In this case the pressure can be shown under all conditions to be a function of the density only. The phenomenon of the hydraulic jump for a liquid flowing in an open, shallow channel is described exactly by the waterlike relation $p = A\rho^2$ [*11*,*12*]. (Strictly speaking, we must describe a weak shock for an ideal gas with $\gamma = 2$ as a polytropic process, and not as an analogue of the hydraulic jump.) Using the same nomenclature as above, we have for all waterlike substances

$$\eta = \xi^{1/\gamma} \tag{2-11''}$$

where γ is the adiabatic exponent. (For a certain range of pressures, as in explosion phenomena, water itself can be regarded as a waterlike substance with $\gamma \cong 7$.) Two additional quantities can readily be expressed in terms of ξ, viz.

$$\tau = \frac{1}{\xi^{1/\gamma}}\left[\frac{1-\xi}{\gamma(\xi^{-1/\gamma}-1)}\right]^{\frac{1}{2}} \tag{2-9''}$$

and

$$-\nu = \pm\left[\frac{(1-\xi)(\xi^{-1/\gamma}-1)}{\gamma}\right]^{\frac{1}{2}} \tag{2-10''}$$

where the sign corresponds to that of $(1 - \xi)$. Here τ represents the speed of the shock relative to the medium ahead and ν signifies the change in the material speed across the shock front.

E,3. Normal Reflection of a Step-Shock at a Rigid Wall. As a first application of the Rankine-Hugoniot shock wave relations we shall now derive the physical conditions that exist following the collision of a shock wave of arbitrary strength with a rigid wall. In Fig. E,3a a shock wave of strength ξ is moving to the right in the direction of a rigid wall. The pressure, density, acoustic speed, and material velocity ahead of and behind the shock wave are designated by p_0, ρ_0, a_0, u_0, and p_1, ρ_1, a_1, u_1, respectively. Fig. E,3b shows the configuration after collision has occurred. The reflected shock wave moves away from the wall toward the left. The physical conditions adjacent to the wall have now changed as a result of the collision; these are designated by p_2, ρ_2, a_2, and u_2. It is our object to determine the changed physical conditions adjacent to the wall, and, as a consequence, the strength $\xi' = p_2/p_1$ of the resulting reflected wave. This can be accomplished simply by using the shock wave relations derived in the previous section.

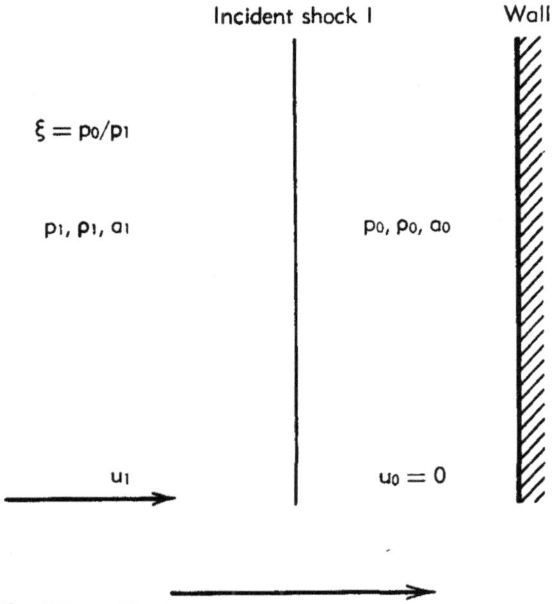

Fig. E,3a. Shock wave before collision with rigid wall.

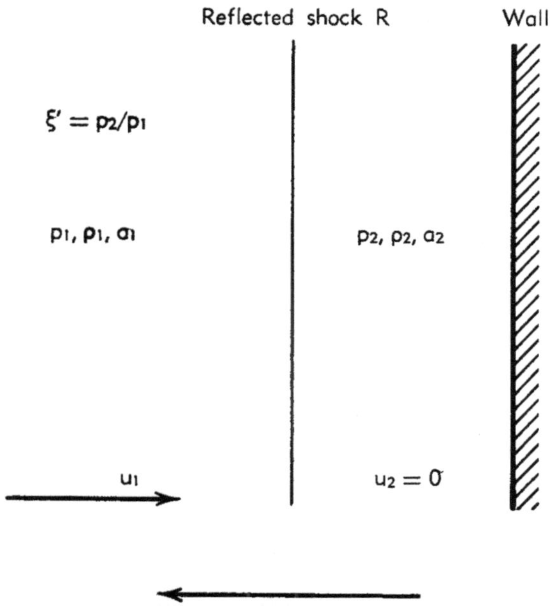

Fig. E,3b. Shock wave after collision with rigid wall.

Since the material at the wall remains stationary both before and after the collision, it follows from Eq. 2-10 and 2-10′ that $-\nu = (u_1 - u_0)/a_1 = u_1/a_1$ and that $\nu' = (u_1 + u_2)/a_1 = u_1/a_1$. Hence

$$-\nu = \nu' \tag{3-1}$$

is the basic condition to be satisfied for normal reflection of a shock wave at a rigid wall.

Using Eq. 2-14 and 2-14′ in connection with the result given in Eq. 3-1, one obtains the equivalent relation

$$\sigma - \frac{1}{\sigma} = \frac{1}{\sigma'} - \sigma' \tag{3-2}$$

This is satisfied by

$$\sigma' = \frac{1}{\sigma} \tag{3-3}$$

or, in terms of ξ, ξ', we obtain

$$\xi' = \frac{(3\gamma - 1) - (\gamma - 1)\xi}{(\gamma + 1)\xi + (\gamma - 1)} \tag{3-4}$$

For an acoustic wave it is well known that the excess pressure of the incident wave is equal to that of the reflected wave, i.e. $p_1 - p_0 = p_2 - p_1$. In the case of the normal reflection of a shock wave it may be

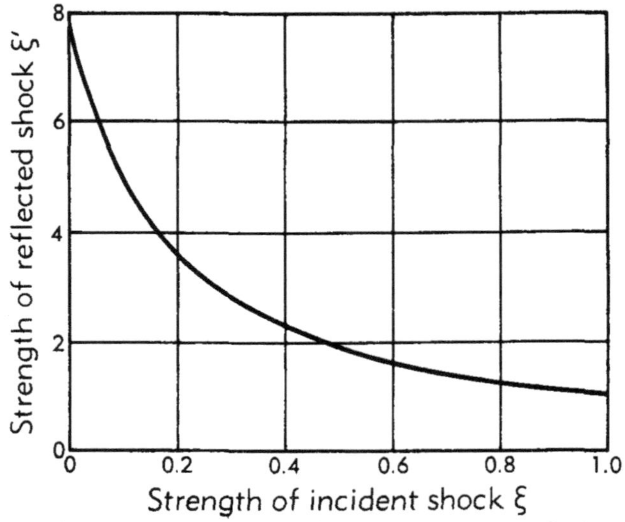

Fig. E,3c. Strength of reflected wave vs. strength of incident wave ξ—normal reflection. $\gamma = 1.40$.

deduced from Eq. 3-4 that the reflected shock wave is always "stronger" than the incident wave if strength is measured by the pressure difference (absolute compression), but it is "weaker" if strength is measured by the pressure ratio (relative compression). This statement is equivalent to the inequalities

$$p_1 - p_0 < p_2 - p_1, \quad \frac{p_1}{p_0} > \frac{p_2}{p_1} \tag{3-5}$$

In Fig. E,3c a plot is given of the strength ξ' of the reflected wave versus the strength ξ of the incident wave for normal reflection in air

($\gamma = 1.40$). It should be pointed out that in this case as well as in subsequent discussions, the results for very low values of ξ are only of qualitative and theoretical interest. It can be concluded on the basis of this simple theory, for example, that as the strength of the incident shock increases beyond bounds, the strength of the reflected shock is finite and approaches the limiting value 8 for air. (For a cylindrical wave, it becomes 17 and for a spherical wave 26.)

Experimental studies [*13*; *14*; *15*; *16*; *17*, pp. 358–377] of normal reflection of a plane shock at a rigid wall have been incidental to the broader program of oblique reflection (cf. Art. 5). In general, the simple step-shock model has been found adequate for describing this phenomenon.

E,4. One-Dimensional Interactions.

Gases. It can be shown [*1*,*2*] that the general problem of one-dimensional interaction of a shock wave with an interface separating two gaseous media or with another shock or rarefaction wave has a unique solution. The configuration produced in this manner contains a reflected wave that is either a shock wave or a rarefaction wave, depending upon the relative magnitude of a generalized acoustic impedance in the two media under consideration. This result may be deduced from an analysis of a material speed-pressure (u, p) plot of the various physical states before and after the interaction. Only the general procedure for the case involving the interaction of a shock wave with an interface is sketched here.

In Fig. E,4a, top, I_m represents the incident shock wave, traveling from the disturbed left l to an undisturbed middle m of the initial medium, while D is the interface front of a second medium on the right r. In Fig. E,4a, middle, R_l and T_r represent the two waves (shock or rarefaction) traveling to the left and right, respectively, after the collision of I_m with D has taken place. The states l and r remain the same, but a new middle state m^* has been produced by the interaction. In general, there exists in this region an interface D' across which there is no variation in pressure or velocity.

In Fig. E,4a, bottom, the pressure in each state is plotted against the material speed. Thus r and m are represented by the same point, while l is to the right (since $u_l > u_m$) and above m (or r). We draw S_m, which is the curve of all the possible states that can be connected to m by means of a shock wave (traveling toward the right). Naturally this curve must pass through l. We next draw T_r and R_l, which represent the totality of states that can be connected to r and l, respectively, either by means of a shock or a rarefaction wave. These are monotonically increasing or decreasing functions and can intersect only at one point, m^*. It is apparent that this intersection occurs on the upper part of T_r, which must represent a state that can be connected to r only by means of a shock wave (since the pressure at m^* is greater than that at r). Hence a shock wave is always transmitted. The reflected wave, however, may be either a shock wave or a rarefaction wave depending on whether l is below or above m^*. For the case $\gamma_m = \gamma_r$, it is shown that S_m lies either entirely above T_r or entirely

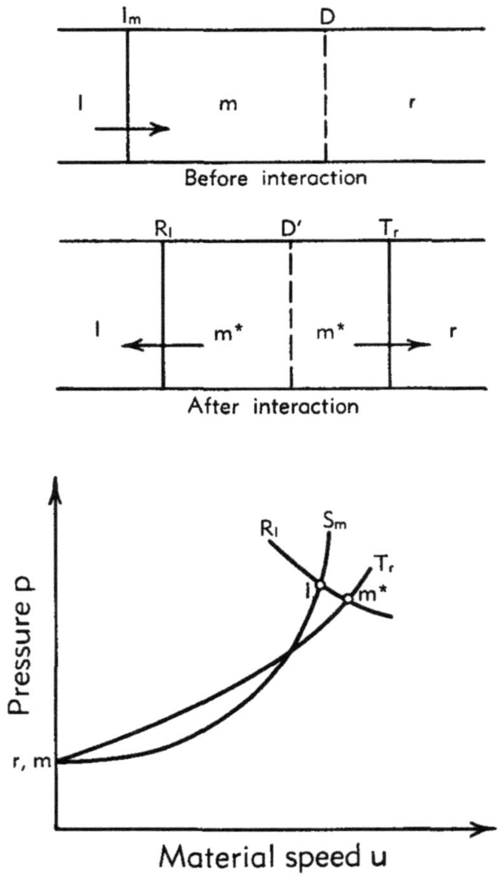

Fig. E,4a. Normal incidence of shock wave. Top, initial state. Middle, final state. Bottom, material speed-pressure diagram.

below it, depending on whether $\rho_r < \rho_m$ or $\rho_r > \rho_m$, respectively (ρ_r and ρ_m are the densities in regions r and m, respectively). In the general case, however, T_r and S_m intersect at the point

$$\tau(\xi) - \sigma(\xi) = \left(\frac{\xi}{\eta}\right)^{\frac{1}{2}}\left(\frac{a_r}{a_m}\right)\left[\tau''\left(\frac{1}{\xi}\right) - \sigma''\left(\frac{1}{\xi}\right)\right]$$

or

$$\left(\frac{1}{a_m}\right)[\gamma_m((\gamma_m + 1) + (\gamma_m - 1)\xi)]^{\frac{1}{2}} = \left(\frac{1}{a_r}\right)[\gamma_r((\gamma_r + 1) + (\gamma_r - 1)\xi)]^{\frac{1}{2}}$$

where ξ designates the strength of the incident shock wave, I_m.

Accordingly, if we designate the expressions on the left and right of the second equation by A_m and A_r, respectively, a shock or rarefaction wave will occur according to whether $A_m < A_r$ or $A_m > A_r$, respectively. For an acoustic wave ($\xi = 1$) the equality of these two quantities implies the equality of the acoustic impedances of the two media. Hence these quantities may be regarded as generalized expressions for the acoustic impedance.

Experimental studies [*18*] of one-dimensional interactions in gases in shock tubes have generally confirmed the theoretical predictions based upon the simple step-shock model. Typical cases, the head-on collision of two unequal shock waves and the head-on collision of a shock and a rarefaction wave, are shown in Fig. E,4b and E,4c, respectively.

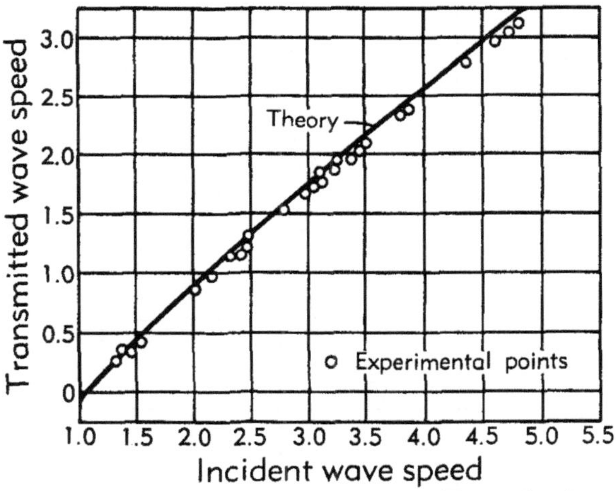

Fig. E,4b. Collision of two shocks. Transmitted wave speed vs. incident wave speed.

Waterlike substances. The one-dimensional interaction for waterlike substances and arbitrary fluids can be treated [*11*,*12*] in a manner similar to that for gases. In this case one has for u_n, the material speed of the normalized region, and for u, the material speed of the other one,

$$u_n = u - a_n\nu(\xi)$$

Put $\Phi_n(p) = a_n\nu(\xi)$; then

$$u = u_n + \Phi_n(p)$$

A shock traveling in the opposite direction will satisfy the relation

$$u = u_n - \Phi_n(p)$$

Likewise for a rarefaction wave,

$$u = u_n \pm \Psi_n(p)$$

where

$$\Psi_n(p) = -\frac{2a}{\gamma - 1}\left(\frac{a_n}{a} - 1\right)$$

The functions $\Phi_n(p)$ and $\Psi_n(p)$ represent physically the absolute change in material speed across a shock wave and a rarefaction wave of finite amplitude, respectively. By the use of these functions one can analyze the various types of interactions. It is significant that a waterlike substance does not always behave like an ideal gas (fictitious for $\gamma > \frac{5}{3}$) having the same value of γ. For instance, in the case of a waterlike substance, a rarefaction wave overtaking a shock wave will always produce a reflected shock wave. This is not the case for an ideal gas [*11,12*].

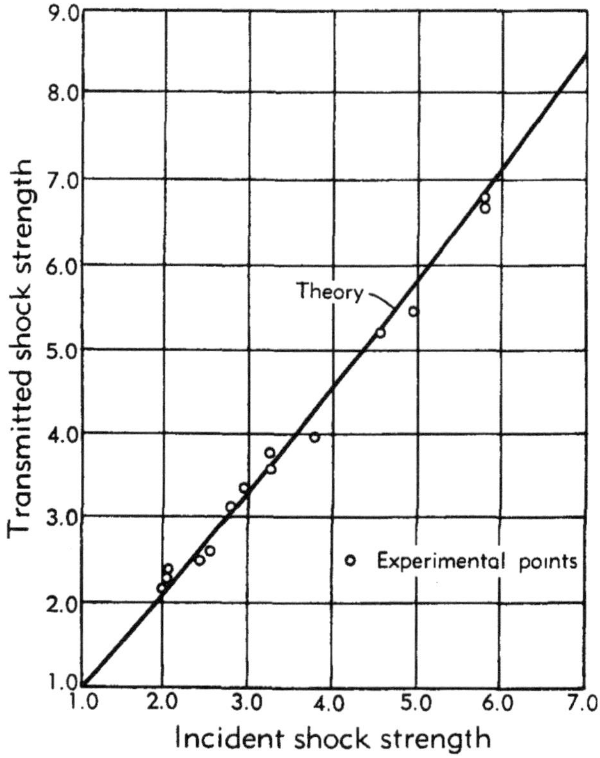

Fig. E,4c. **Collision of shock and rarefaction wave. Transmitted shock strength vs. incident shock strength.**

E,5. Oblique Reflection of a Step-Shock at a Rigid Wall. The more general properties of oblique reflection of shock waves at a rigid wall were studied carefully, both theoretically and experimentally, during and after World War II. These properties are highly significant in the understanding of damage due to air blast, and in a number of other military applications. For instance, it was found that head-on or normal reflection of a shock does not necessarily produce the strongest reflected shock wave. Oblique reflection results often in a stronger reflected shock wave and hence a higher peak pressure at the reflecting surface than does normal reflection. This air-burst effect was utilized in an attempt to optimize the blast damage radius due to the atomic bomb explosions at Hiroshima and Nagasaki.

One of the most interesting effects connected with oblique reflection of shock waves is the existence of the so-called Mach effect [*19*], named for E. Mach, who first observed the phenomenon. As the angle between the wall and the incident wave increases beyond a given limit (which is a function of the shock strength), an anomalous type of reflection pattern occurs, unlike that observed in acoustic reflection. Reflection which does not involve the Mach effect is called regular reflection.

The theory [*5,6,7*] for regular reflection of shock waves at a rigid boundary agrees well with experimental results [*10,13,14,15,16,17,20,61*] and provides the explanation for the occurrence of the Mach effect beyond a limiting or extreme angle of incidence. In the case of Mach reflection [*21,22*], however, the extent of agreement between theory and observed results is not well established. For this reason the two types of reflection are discussed separately here, with primary emphasis being given to regular reflection. These two phenomena are illustrated in the photographs in Plate E,5a for a wind tunnel and in Plate E,5b for a shock tube.

Regular Reflection.

Gases. A plane step-shock wave I, of strength ξ, is moving with constant velocity (in an ideal gas of negligible viscosity and thermal conductivity) and is incident at an angle ω upon an infinite, plane, rigid wall (as in Fig. E,5a, top). It is assumed that there is a plane reflected shock and that the pressure is uniform between it and the incident shock, as well as between it and the wall. Plate E,5b, bottom left, shows the observational validity of this assumption. It is required to find the strength ξ' (ratio of the pressure behind R to that in front of it) of the reflected shock R and the angle of reflection ω'. To facilitate derivation, consider the gas flow relative to an observer moving with the line of contact O (Fig. E,5a, bottom). The normal material speed τ, in front of the shock I, is expressed in terms of ξ in the first equation of Eq. 2-9, whereas the tangential material speed t along the shock I (both in front of and behind it) is determined by

$$t = \tau \cot \omega \tag{5-1}$$

On the other hand, the normal material speed σ behind the shock I is obtained from the second expression of Eq. 2-9, also in terms of ξ. Thus the material velocity q behind the shock I is completely determined. Resolving this velocity into components u and v along the axes x and y, respectively, we have

$$\left.\begin{aligned} u(\xi, \omega) &= t \cos \omega + \sigma \sin \omega \\ v(\xi, \omega) &= t \sin \omega - \sigma \cos \omega \end{aligned}\right\} \tag{5-2}$$

The normal material speed behind shock R is given in terms of ξ' by the first expression of Eq. 2-9′. It will be noted that τ', as well as t, τ, and σ, are normalized with respect to a, the acoustic speed in the region between I and R (in front of shock R). The tangential material speed t' along the shock R is given by

$$t' = \tau' \cot \omega' \tag{5-3}$$

The normal speed σ', in front of shock R, may be obtained in terms of ξ' from the second equation of Eq. 2-9′. Thus the x, y components (u', v',

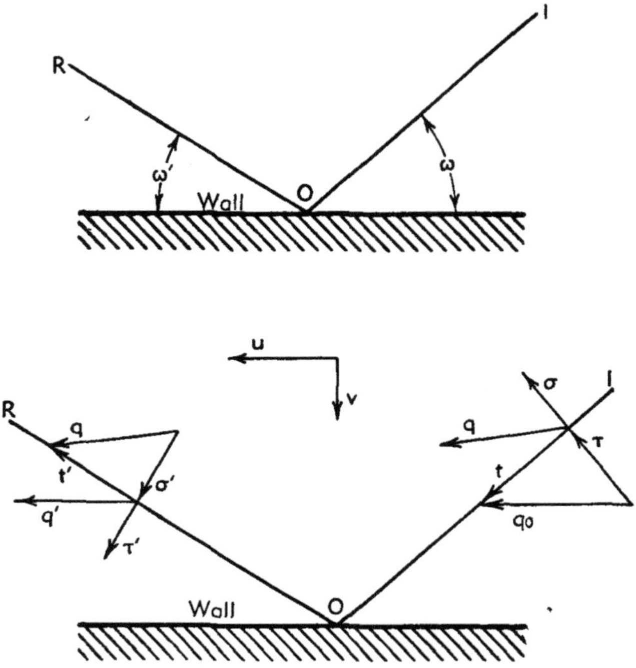

Fig. E,5a. Regular reflection of a shock wave at a rigid wall.

respectively) of the material velocity in front of shock R are given by

$$\left.\begin{aligned} u'(\xi', \omega') &= t' \cos \omega' + \sigma' \sin \omega' \\ v'(\xi', \omega') &= -t' \sin \omega' + \sigma' \cos \omega' \end{aligned}\right\} \tag{5-2′}$$

Equating these expressions for u' and v' to the expressions obtained above for u and v, we obtain a set of two equations which completely determine the conditions for regular reflection,

$$\left.\begin{aligned} u(\xi, \omega) &= u'(\xi', \omega') \\ v(\xi, \omega) &= -v'(\xi', \omega') \end{aligned}\right\} \tag{5-4}$$

Eliminating ω' from these two equations, we obtain, after considerable algebraic manipulation, the following basic quadratic expression for σ'^2

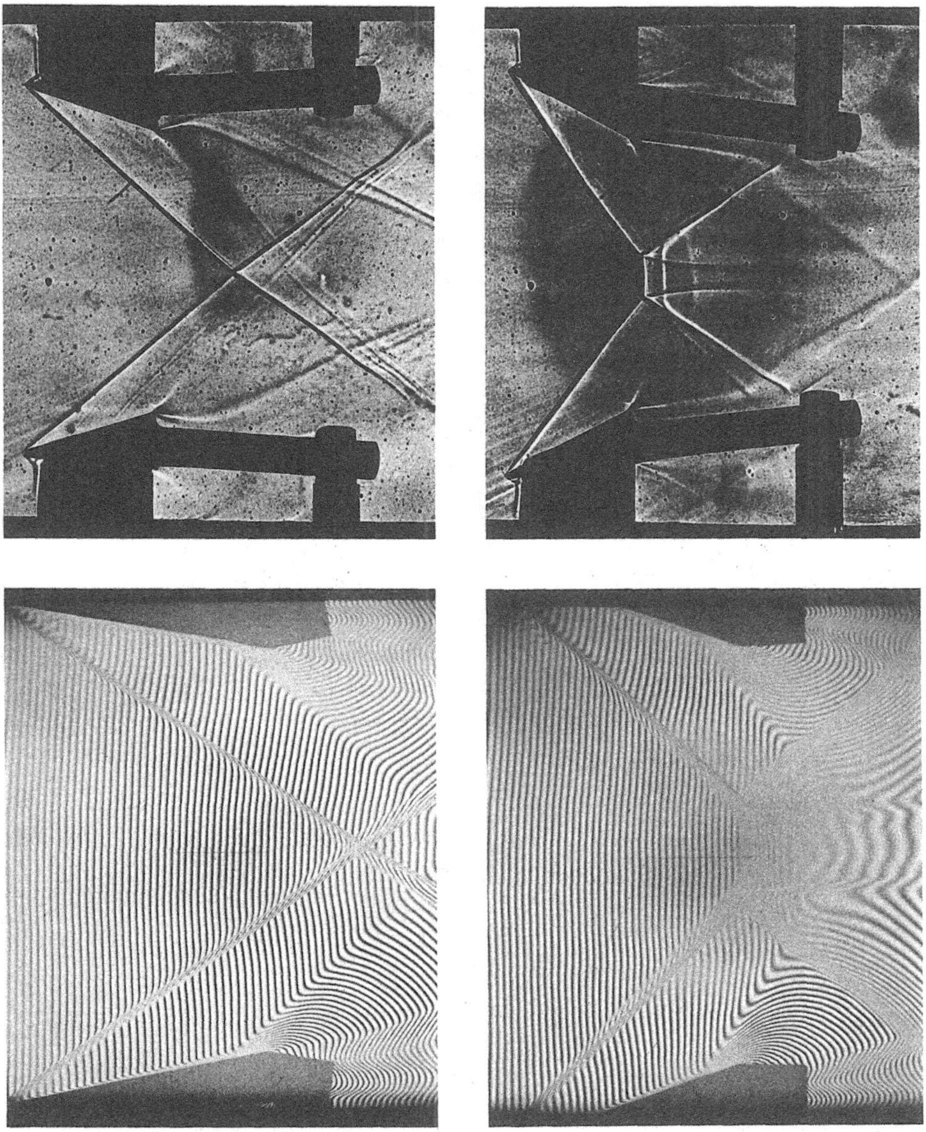

Plate E,5a. Top left, regular intersection, $\xi = 0.3$, $\omega = 40°$; top right, Mach intersection, $\xi = 0.2$, $\omega = 44°$ (shadowgrams, Great Britain National Physical Laboratory wind tunnel). Bottom left, regular intersection; bottom right, Mach intersection (interferograms, U.S. Naval Ordnance Laboratory wind tunnel).

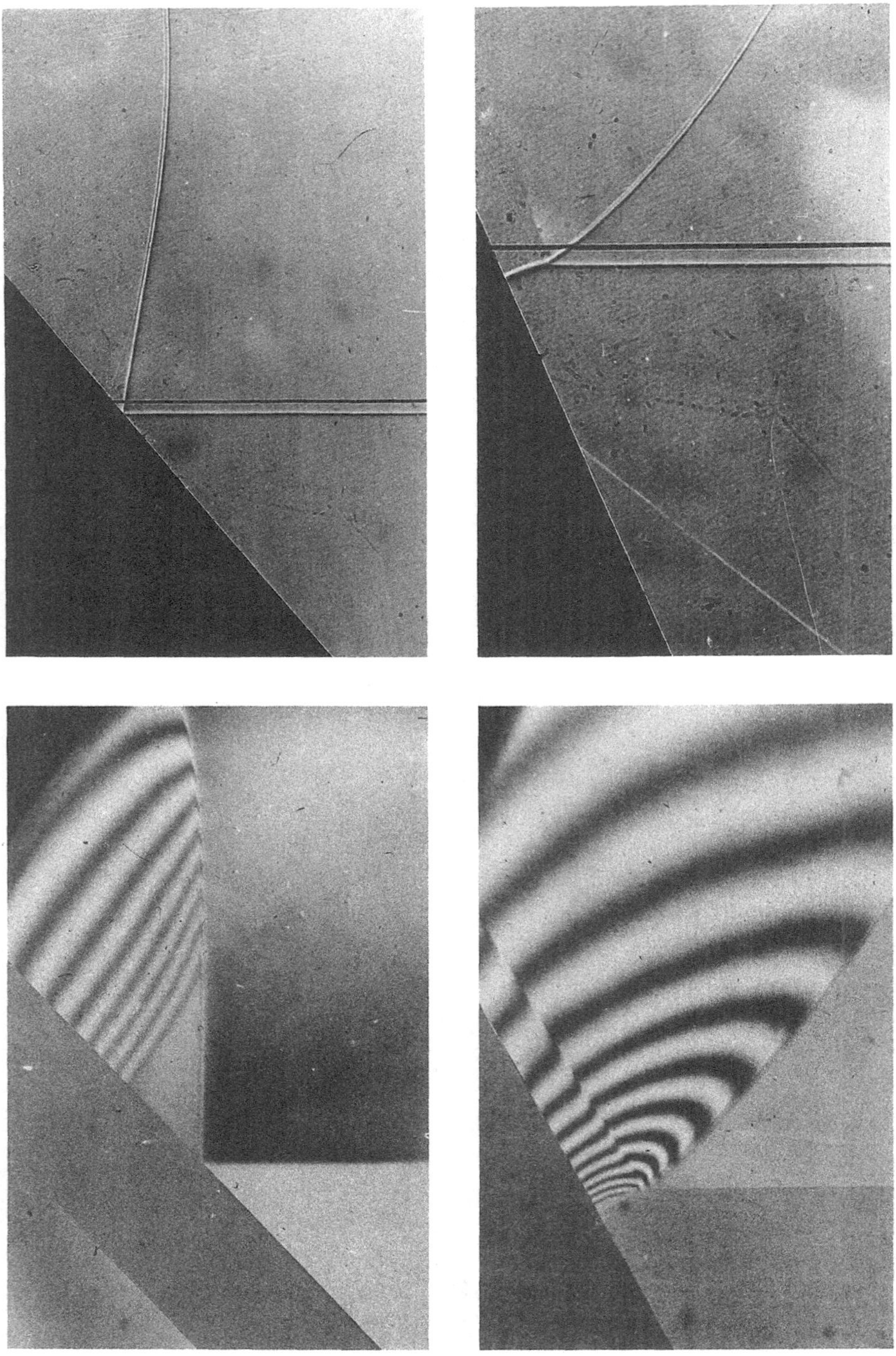

Plate E,5b. Top left, regular reflection, $\xi = 0.8$, $\omega = 48°$; top right, Mach reflection, $\xi = 0.8$, $\omega = 65°$ (shadowgrams, Princeton University shock tube). Bottom left, regular reflection, $\xi = 0.8$, $\omega = 45°$; bottom right, Mach reflection, $\xi = 0.8$, $\omega = 61°$ (interferograms, Princeton University shock tube).

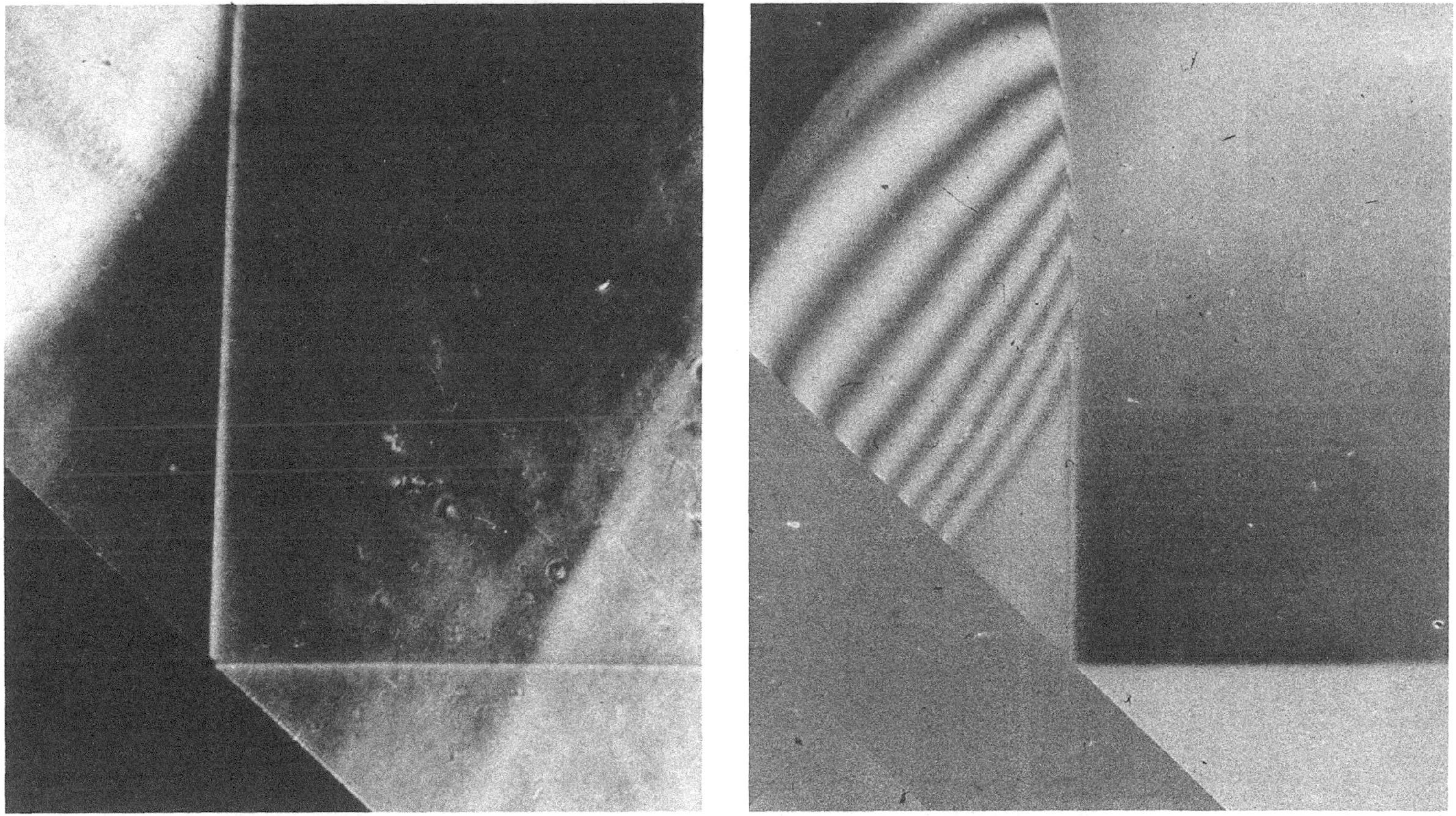

Plate E,5c. Left, catch-up phenomenon, $\xi = 0.8$, $\omega = 43°$, schlieren; right, catch-up phenomenon, $\xi = 0.8$, $\omega = 45°$, interferogram (Princeton University shock tube).

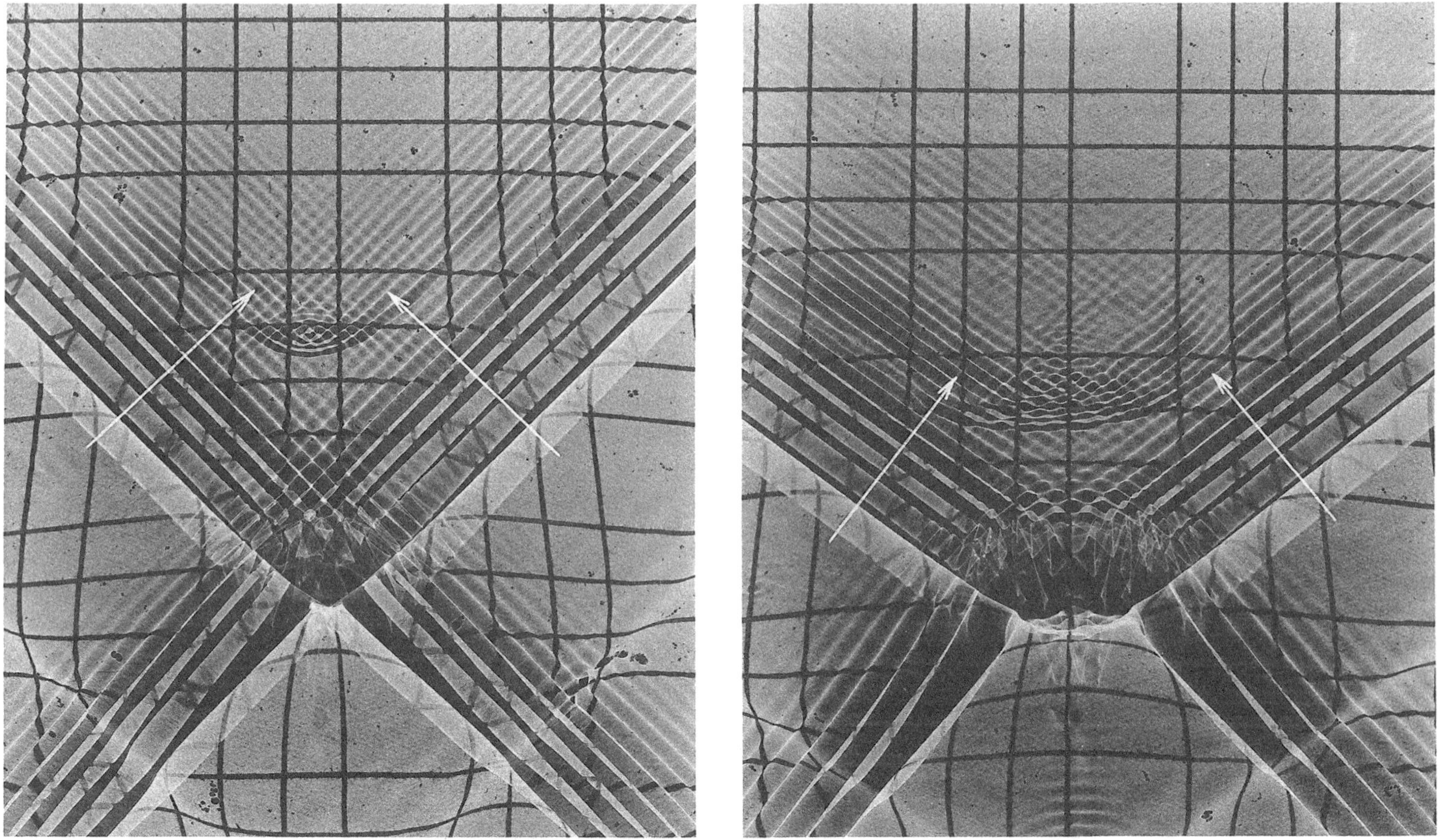

Plate E,5d. Left, regular intersection, $\xi = 0.50$; right, Mach intersection, $\xi = 0.50$ (shadowgrams, California Institute of Technology ripple tank, $\gamma = 2.00$).

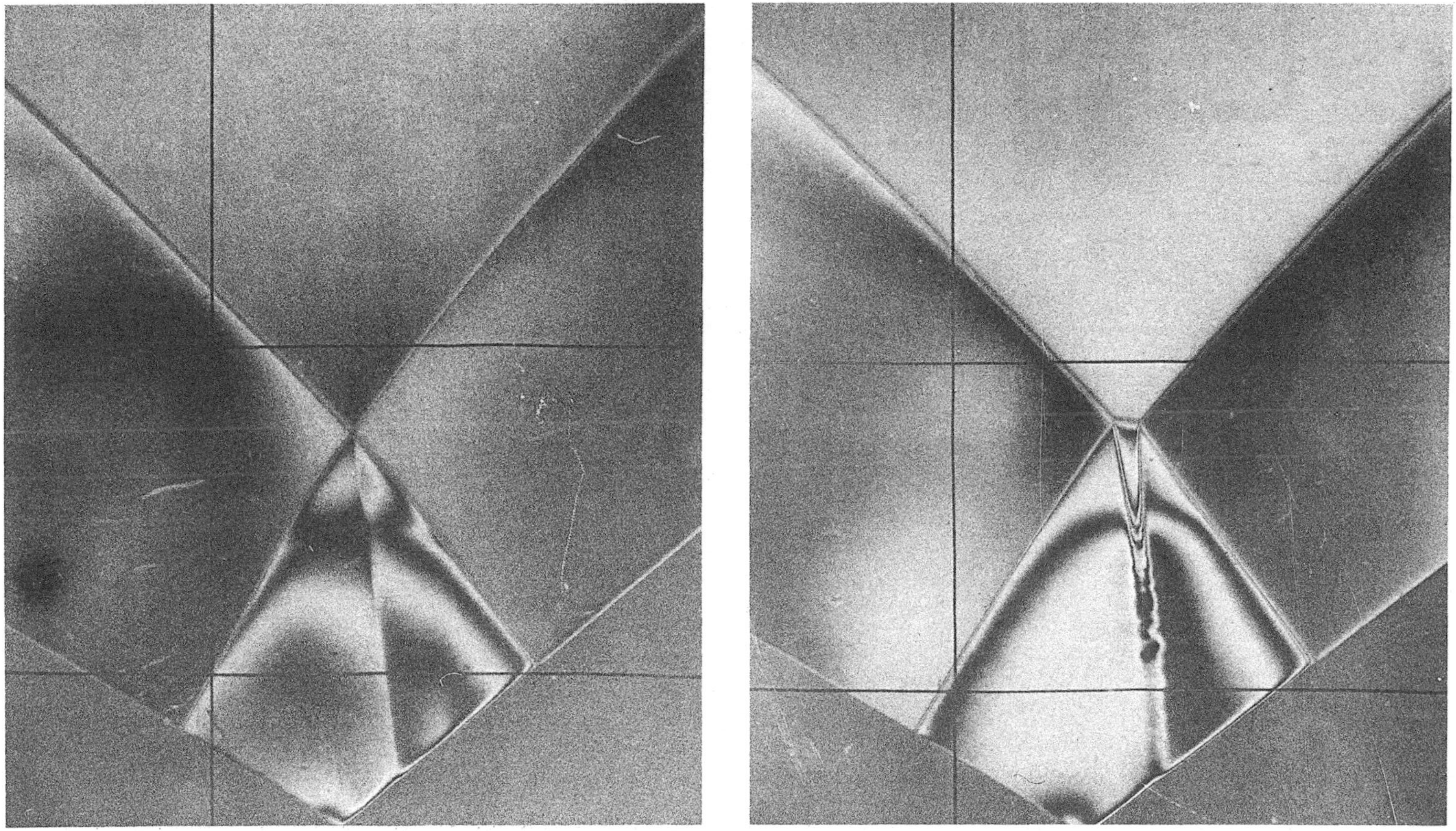

Plate E,5e. Left, regular intersection, $\xi_1 \cong 0.7$, $\xi_2 \cong 0.3$, $\omega_1 + \omega_2 \cong 81°$; right, Mach intersection, $\xi_1 \cong 0.7$, $\xi_2 \cong 0.3$, $\omega_1 + \omega_2 \cong 84°$ (interferograms, Princeton University shock tube).

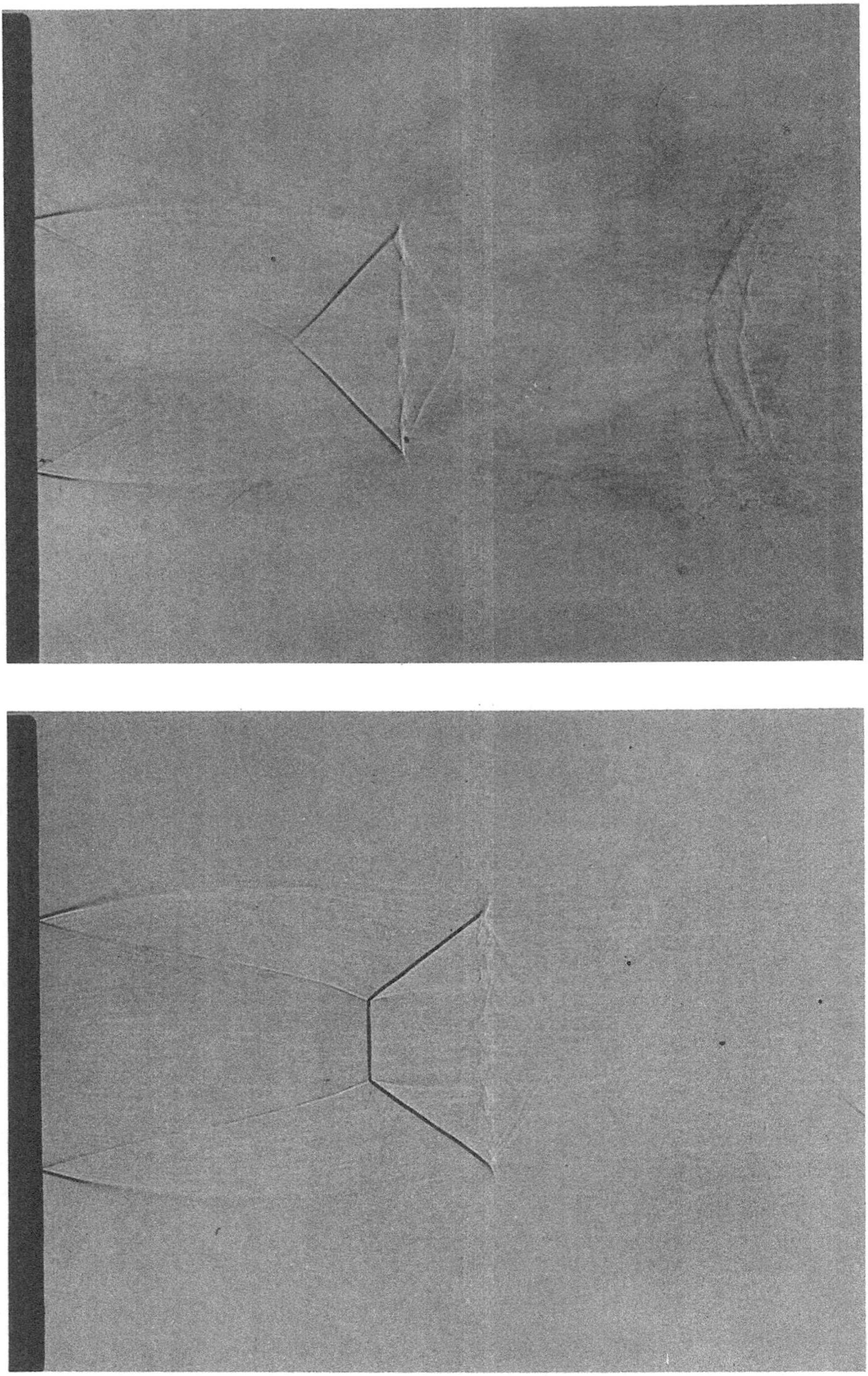

Plate E,5f. Top, regular intersection; bottom, Mach intersection (shadowgrams, Princeton University air jet).

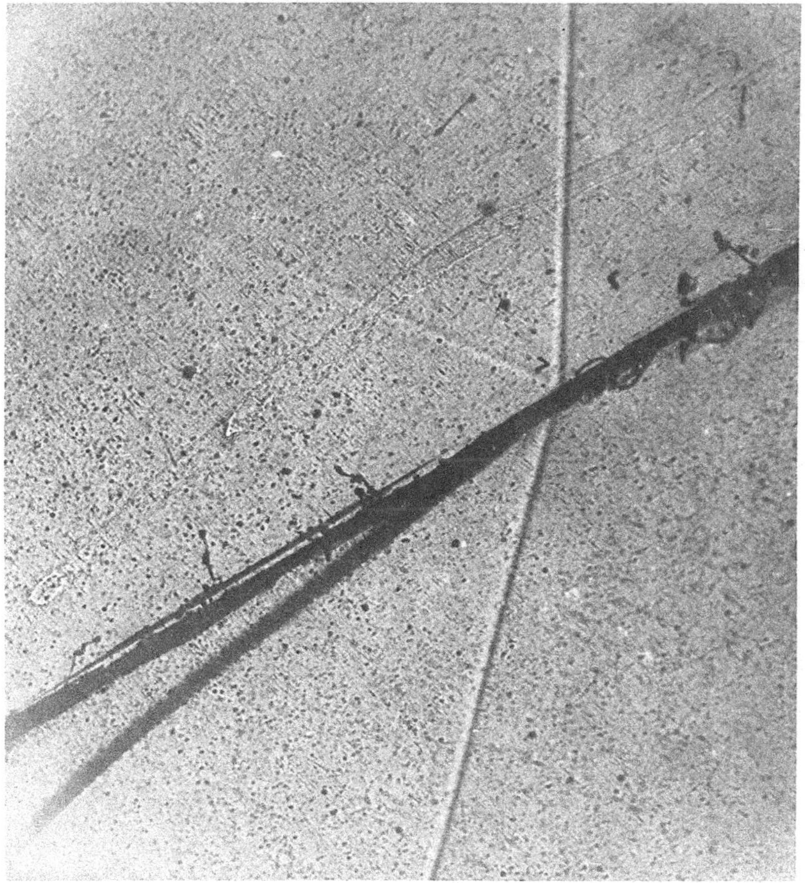

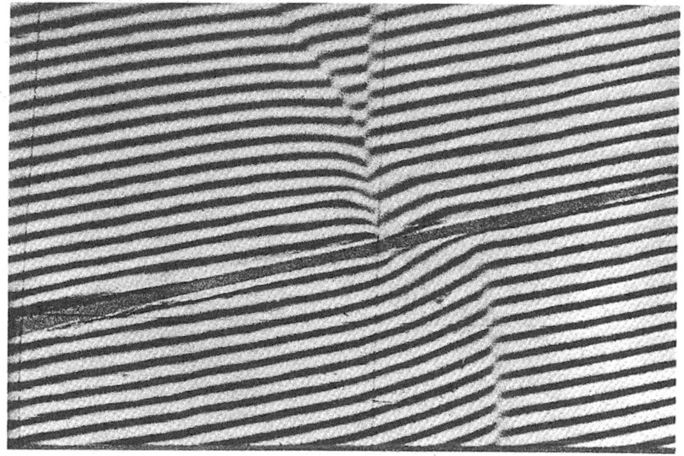

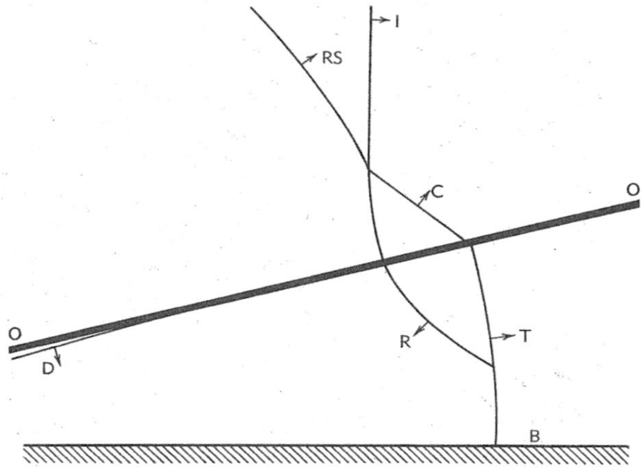

Plate E,6a. Left, regular refraction, air-CO_2 (shadowgram, Pennsylvania State College shock tube); right, irregular refraction, air-CH_4 (interferogram, Princeton University shock tube).

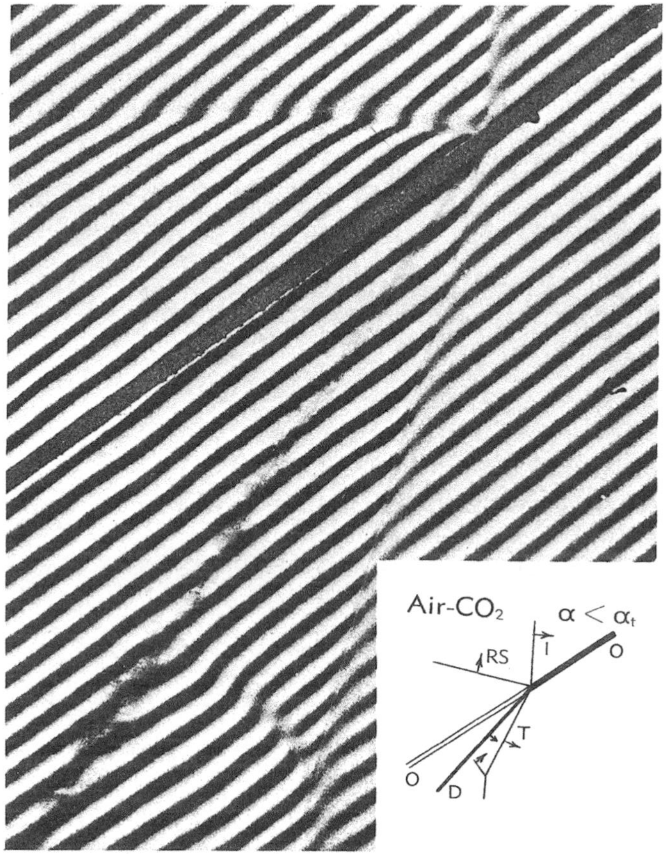

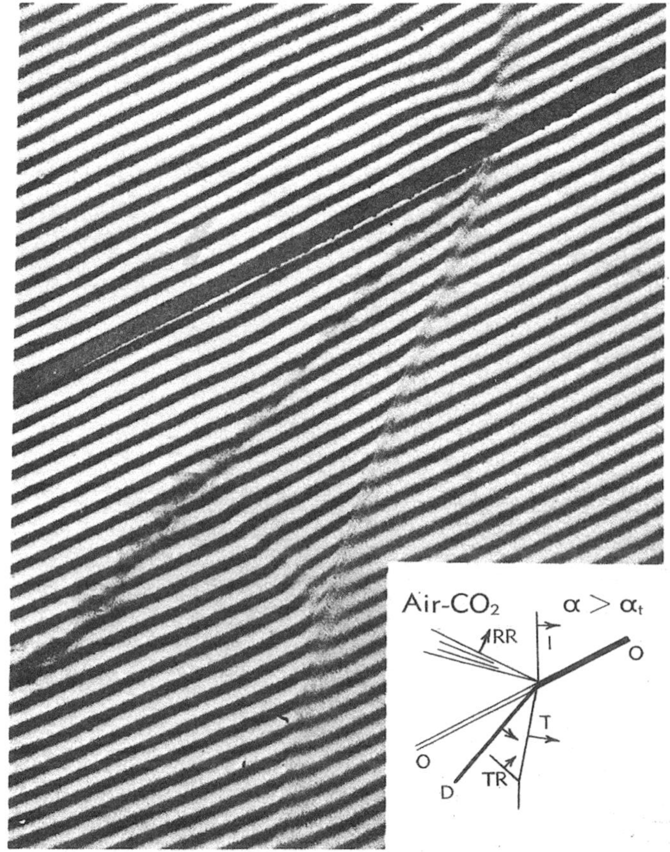

Plate E,6b. Regular refraction, air-CO_2, $\xi = 0.30$, $\alpha_t = 63°$. α is the angle between the incident shock and the interface; α_t is the angle at which the reflected wave changes phase. Left, $\alpha < \alpha_t$; right, $\alpha > \alpha_t$ (interferograms, Princeton University shock tube).

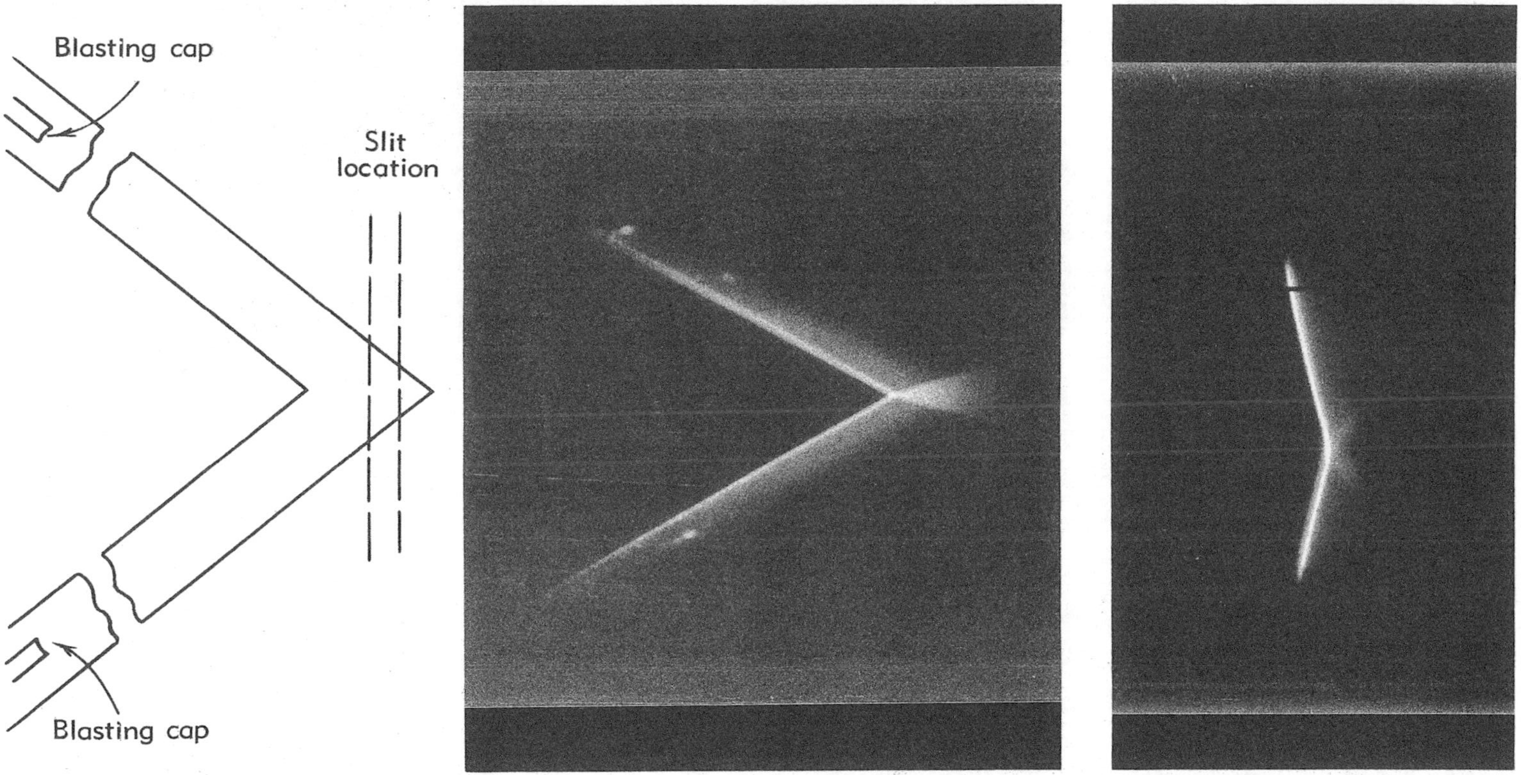

Plate E,7a. Left, geometric arrangement for plane detonation waves; center, regular interaction, $\omega = 30°$; right, Mach interaction, $\omega = 75°$ (Canada Defense Research Board streak camera records).

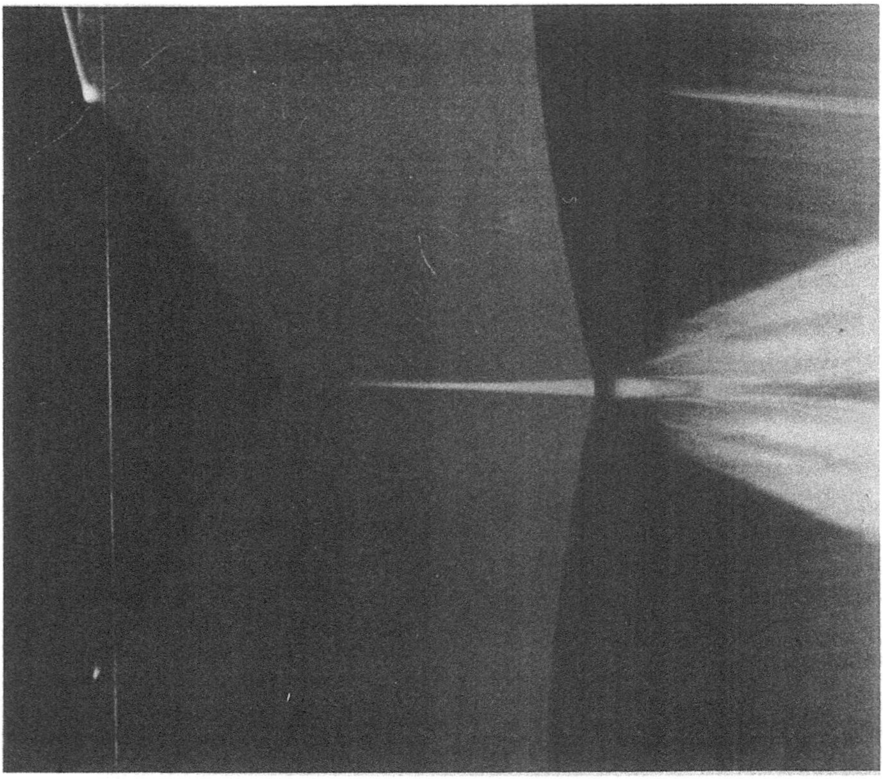

Plate E,7b. Mach interaction. Naval Ordnance Laboratory spherical detonations, $\omega \cong 42°$. Smear camera record.

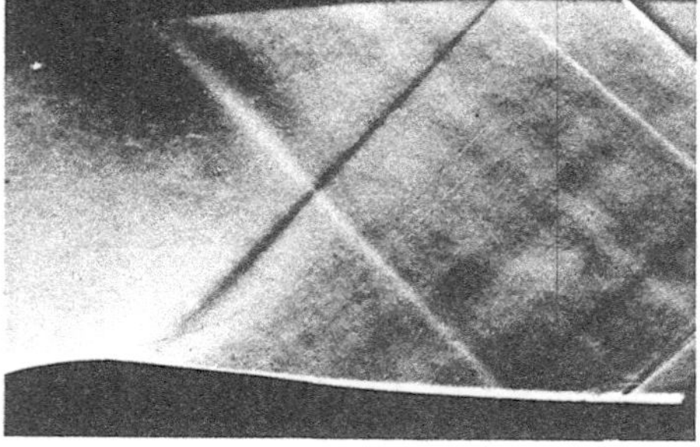

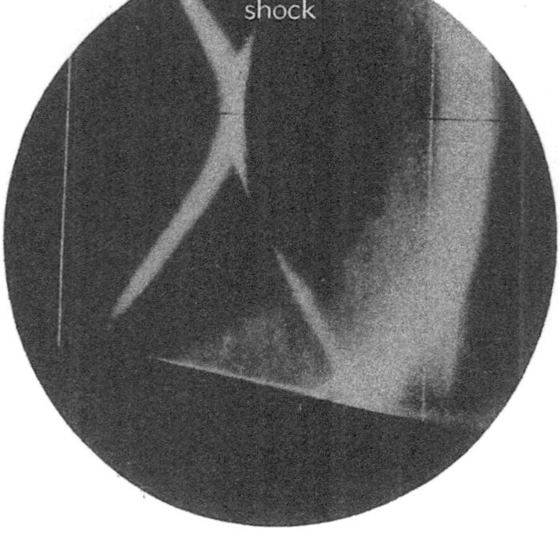

Plate E,7c. Left, regular interaction of condensation shocks; right, Mach interaction of condensation shocks (schlieren photographs, Great Britian Royal Aircraft Establishment wind tunnel).

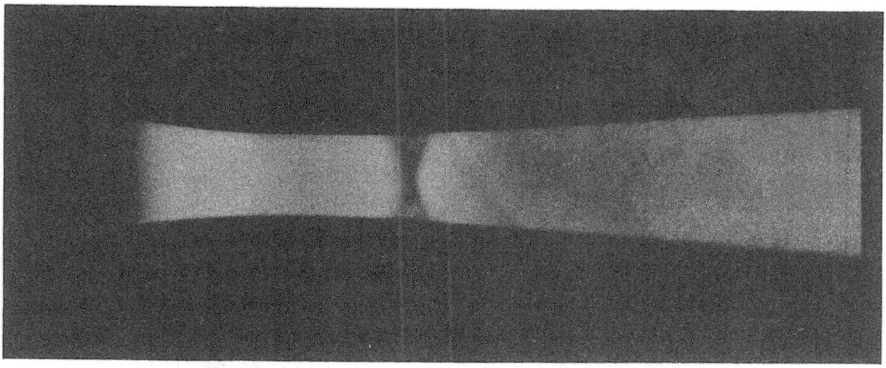

Plate F,1. Schlieren photograph of condensation shock.

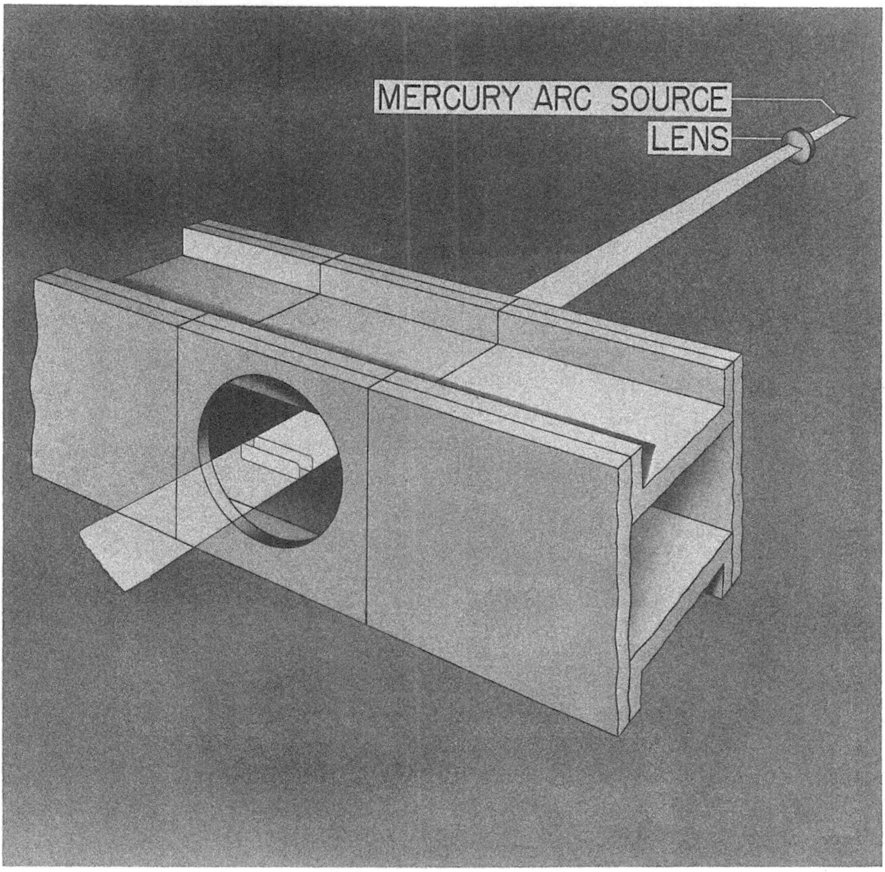

Plate F,5a. Schematic of typical light-scattering experiment.

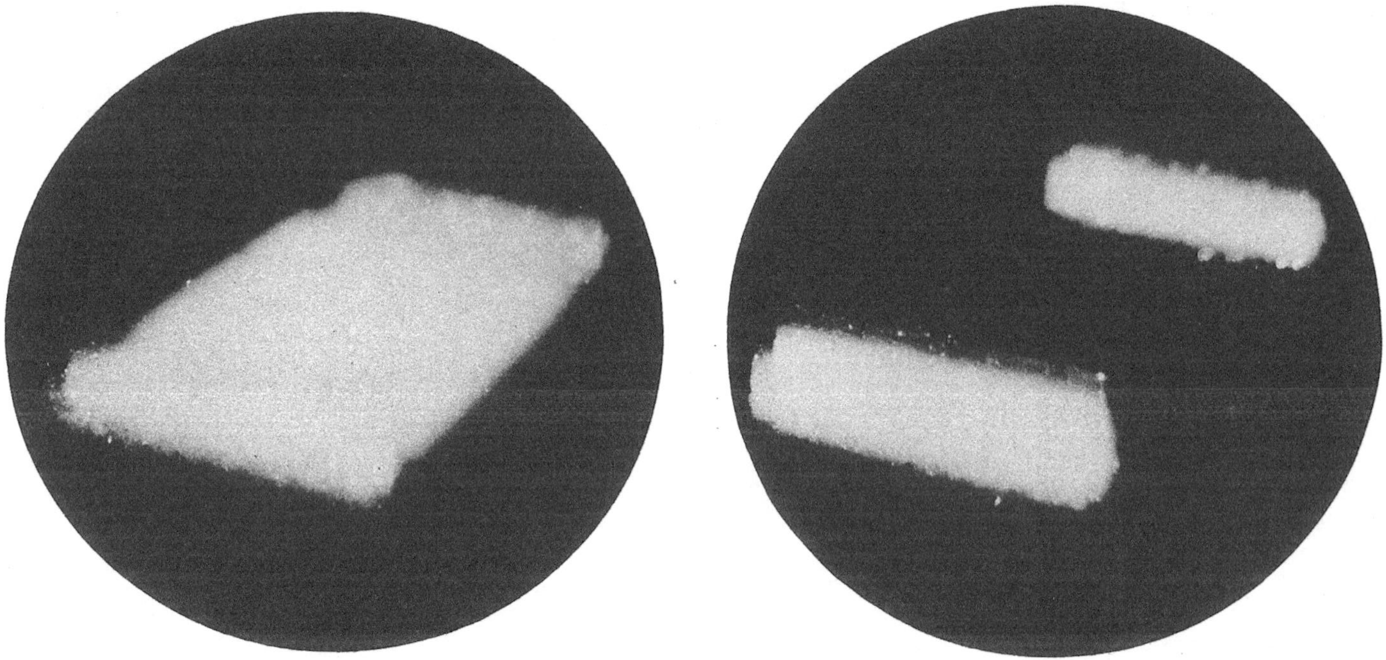

Plate F,5b. Light-scattering photographs with and without condensation.

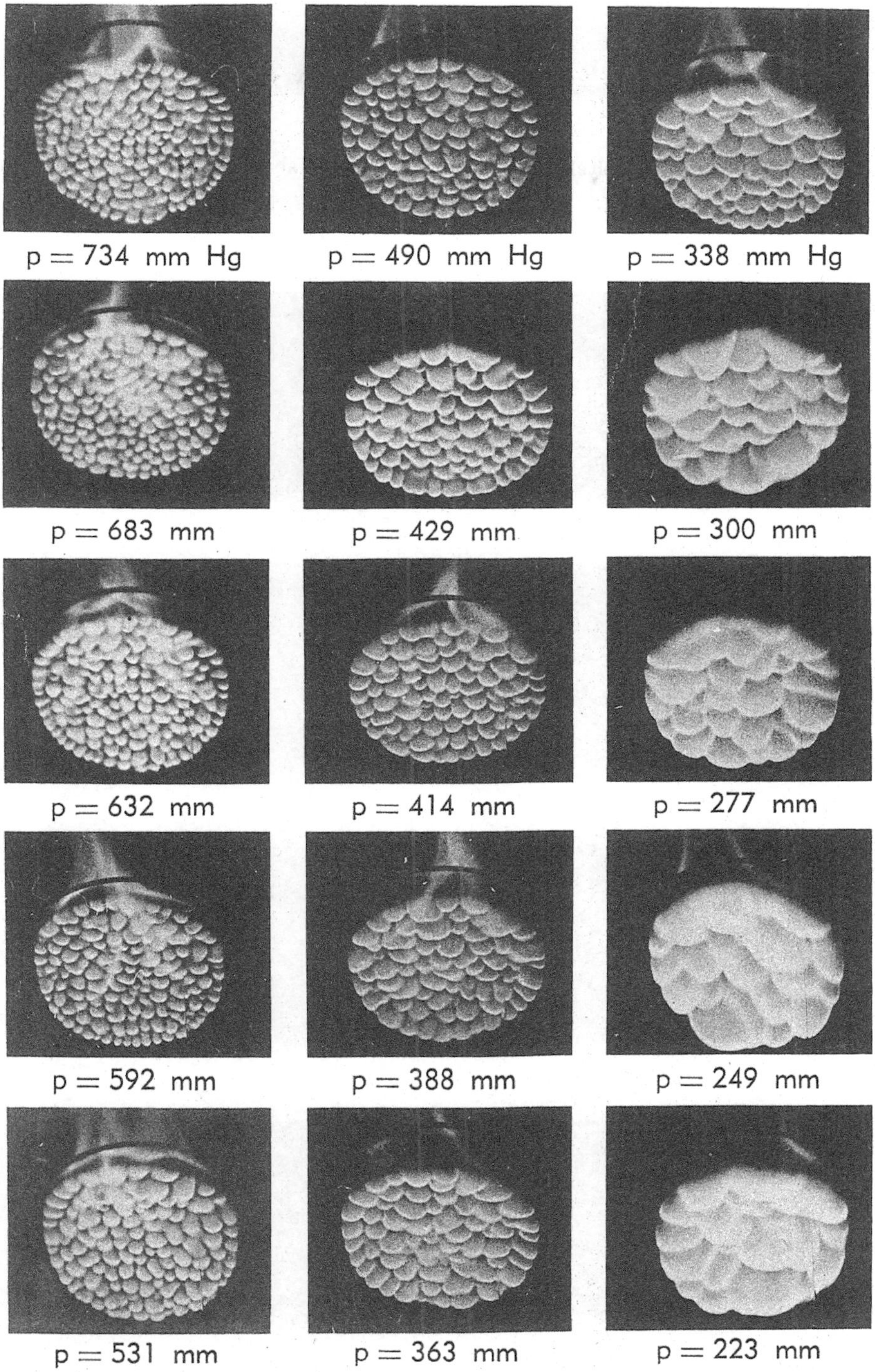

Plate G,9. Cellular flames in rich butane-air-nitrogen mixtures in 15.25-cm tube.

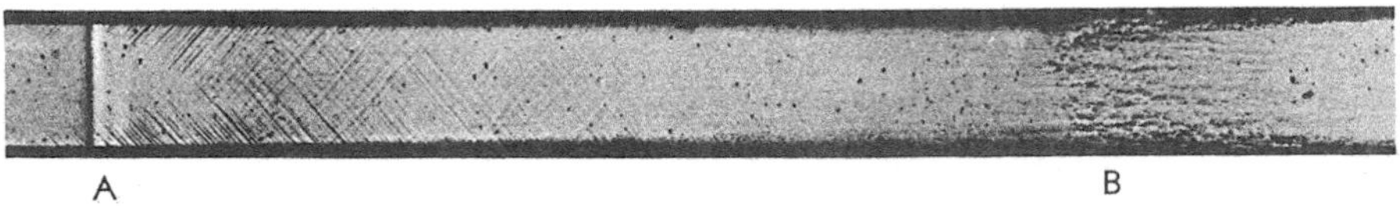

Plate G,11a. Shadow photograph showing compression wave *A* about 8 cm ahead of flame front *B*.

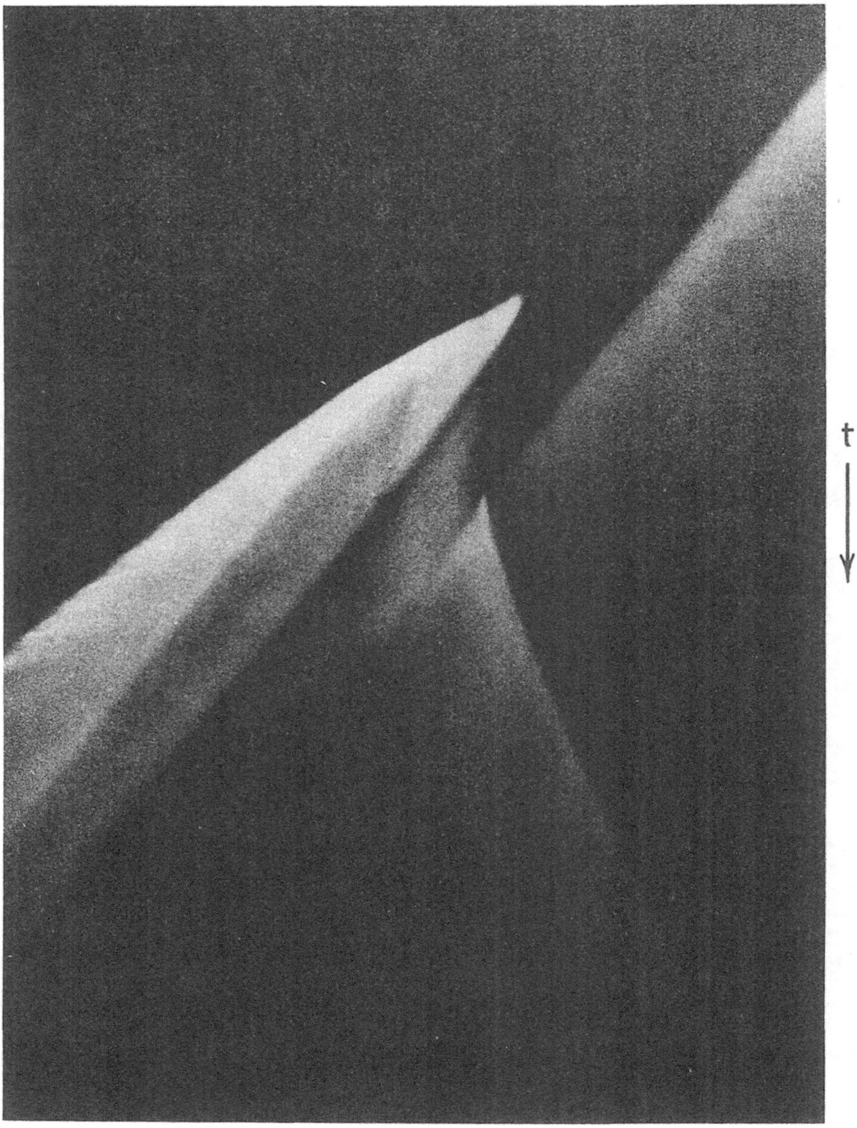

Plate G,11b. Detonation started ahead of flame front in mixture of $CO + \frac{1}{2}O_2$.

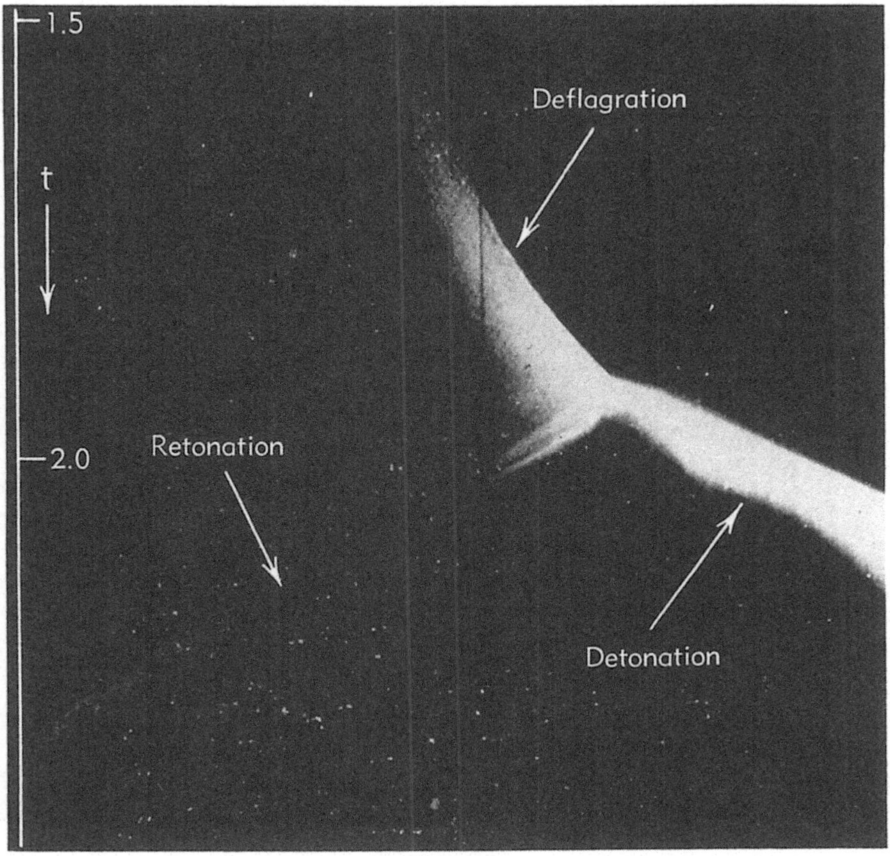

Plate G,11c. Deflagration transforming suddenly into detonation and retonation waves.

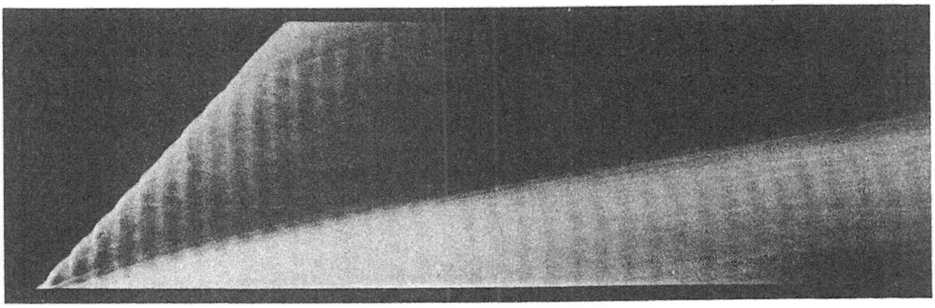

Plate G,13a. Spinning detonation.

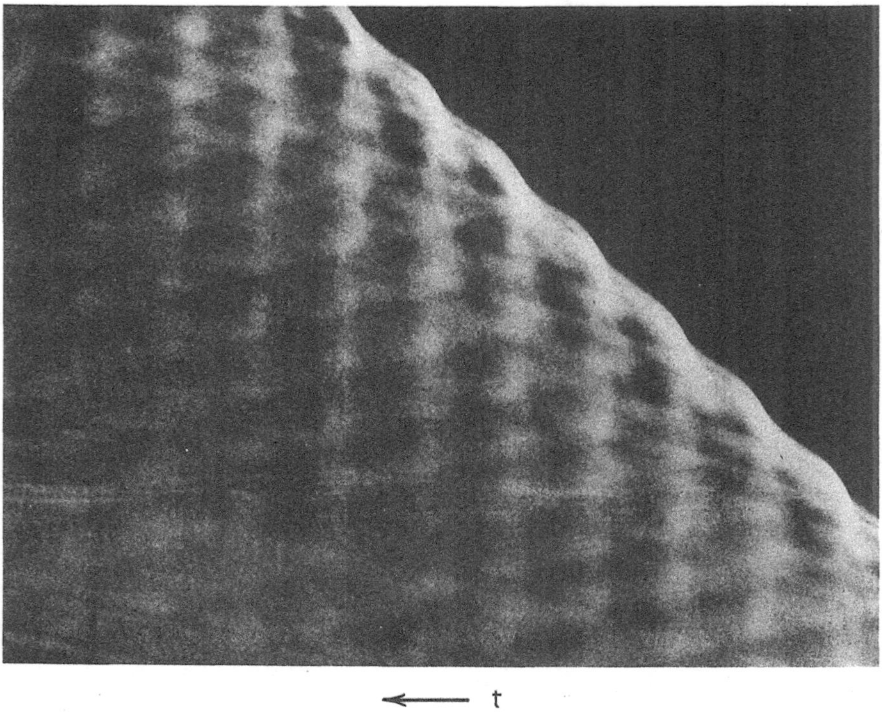

Plate G,13b. Spinning detonation.

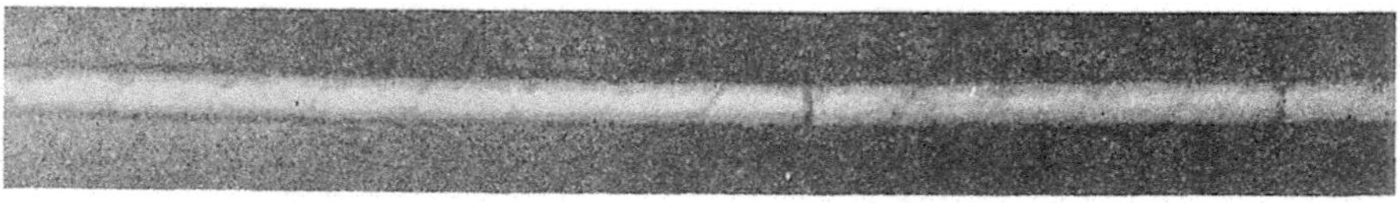

Plate G,13c. Helical path traced by detonation wave.

in terms of σ and ω,

$$(\gamma + 1)\sigma^4\sigma'^4 - \{2(\gamma + 1)^2\sigma^4 + 4\gamma\sigma^2(\sigma^2 - 1)\cos^2\omega + \sigma^2[(\gamma - 1)\sigma^2 + 2]^2\cot^2\omega\}\sigma'^2 + \{(\gamma + 1)^2\sigma^4 - 4(\sigma^2 - 1)^2\cos^2\omega + [(\gamma - 1)\sigma^2 + 2]^2\cot^2\omega\} = 0 \quad (5\text{-}5)$$

On the other hand, the condition $v = v'$ then gives ω' in terms of σ, ω, and σ',

$$\cos\omega' = -\frac{\sigma^{-1} - \sigma}{\sigma'^{-1} - \sigma'}\cos\omega \quad (5\text{-}6)$$

Numerical values for air ($\gamma = 1.40$) are shown here graphically. In Fig. E,5b, ω' is plotted as a function of ω for various values of ξ; in Fig.

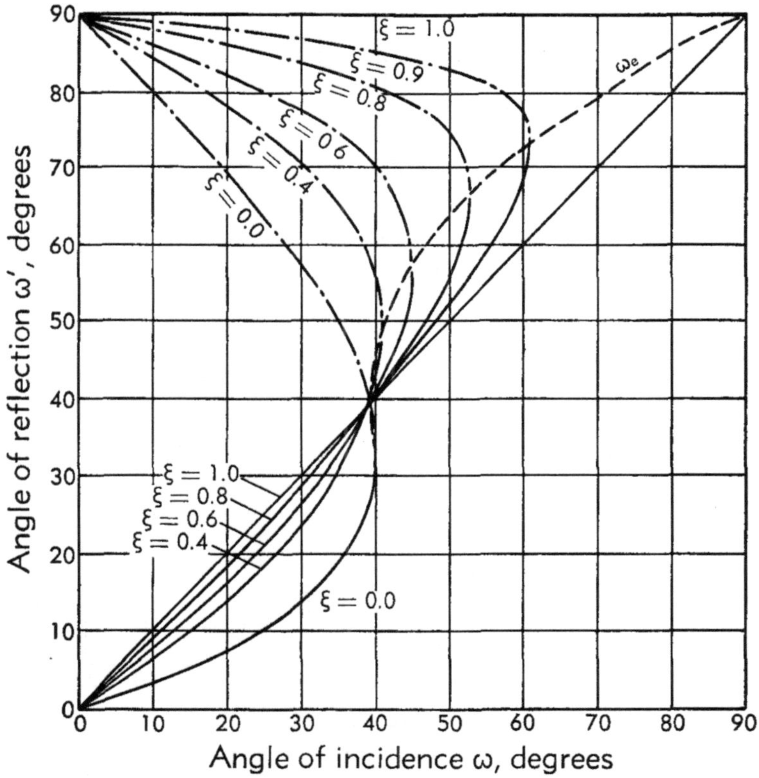

Fig. E,5b. Angle of reflection ω' vs. angle of incidence ω. $\gamma = 1.40$.

E,5c, ξ' is presented as a function of ω at various values of ξ. For each set of values ω and ξ, and whenever ω is below a limiting angle ω_e, two solutions are obtained. One solution is characterized by low values for ξ' and ω', and the other by high values for ξ' and ω'. On the other hand, when ω is greater than the extreme angle ω_e, no solutions exist. It is in this region that the anomalous Mach-type reflection configuration is observed.

From Eq. 5-5 it is possible to derive an expression for the extreme

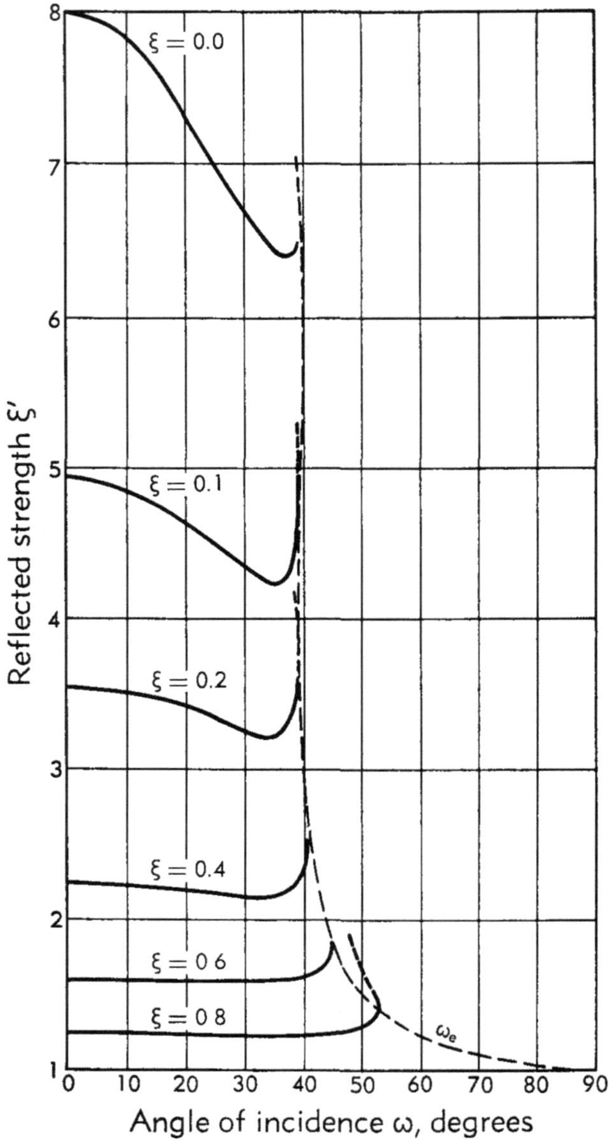

Fig. E,5c. Strength of reflected shock wave ξ' vs. angle of incidence.

angle ω_e, beyond which no regular reflection can exist. After some algebraic calculations the following cubic in $\sin^2 \omega_e$ is obtained, which determines ω_e for given $\sigma(\xi)$:

$$16\gamma^2(\sigma^2 - 1)^4 \sin^6 \omega_e - \{16(\gamma + 1)^2(\gamma\sigma^2 + 1)(\sigma^2 - 1)^2 + 16\gamma^2(\sigma^2 - 1)^4 - 8\gamma(\sigma^2 - 1)^2[(\gamma - 1)\sigma^2 + 2]^2\} \sin^4 \omega_e - \{4(\gamma + 1)^2(\sigma^2 - 1)[(\gamma - 1)\sigma^2 + 2]^2 - [(\gamma - 1)\sigma^2 + 2]^4 + 8\gamma(\sigma^2 - 1)^2[(\gamma - 1)\sigma^2 + 2]^2\} \sin^2 \omega_e - [(\gamma - 1)\sigma^2 + 2]^4 = 0 \quad (5\text{-}7)$$

Then ξ' and ω' can be obtained as before from Eq. 5-5 and 5-6, respectively (cf. Table E,5a).

Table E,5a

Minimum solutions $\gamma = 1.40$				Extreme solutions $\gamma = 1.40$		
ξ	ξ'_m	ω_m, deg.	ω'_m, deg.	ξ'_e	ω_e, deg.	ω'_e, deg.
0	6.379	37.29	22.04	6.983	39.970	32.97
0.05	5.095	36.17	22.64	5.654	39.517	35.29
0.1	4.256	35.24	23.17	4.780	39.288	37.47
0.2	3.210	33.77	24.05	3.677	39.326	41.55
0.3	2.574	32.63	24.81	2.991	39.893	45.44
0.4	2.141	31.71	25.46	2.509	40.945	49.26
0.5	1.825	30.94	26.05	2.145	42.534	53.18
0.6	1.584	30.29	26.59	1.852	44.781	57.28
0.7	1.392	29.72	27.08	1.607	47.973	61.75
0.8	1.237	29.21	27.53	1.392	52.736	66.90
0.9	1.109	28.75	27.95	1.194	60.822	73.54
0.95	1.052	28.54	28.15	1.098	68.025	78.28
1.0	1.000	28.34	28.34	1.000	90.000	90.00

Fig. E,5d shows the ξ, ω_e curve for $\gamma = 1.40$ dividing the region of regular reflection from that of Mach reflection. On this same graph observed shock tube data are plotted which indicate that regular reflection

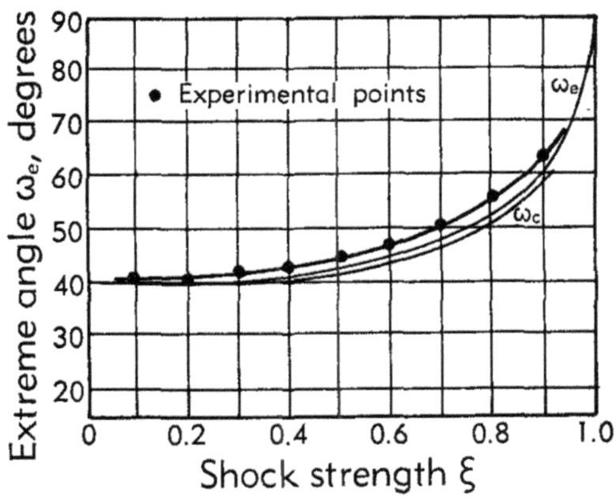

Fig. E,5d. Extreme angle ω_e vs. shock strength. Experimental data.

occurs for all shock strengths at angles slightly greater than the extreme angles. The phenomenon here is complicated by the experimental setup in which the reflecting surface has a corner from which an acoustic wave originates, catches up with the reflected shock, and affects its curvature (cf. Plate E,5c). The boundary condition at the wall can be met for a completely curved reflected shock [*23,24,25,26*] only if the flow behind it is sonic. The limiting angle of incidence ω_0 for this catch-up phenomenon occurs at angles slightly less than the extreme angles.

By the use of an argument based on the concept of continuity and because the angle of reflection for normal incidence is zero, it is possible to arrive at the conclusion that at small angles of incidence the solutions with small angles of reflection (and low values for ξ') are the physically realizable solutions. A systematic experimental study [*13,14,15,16,17*]

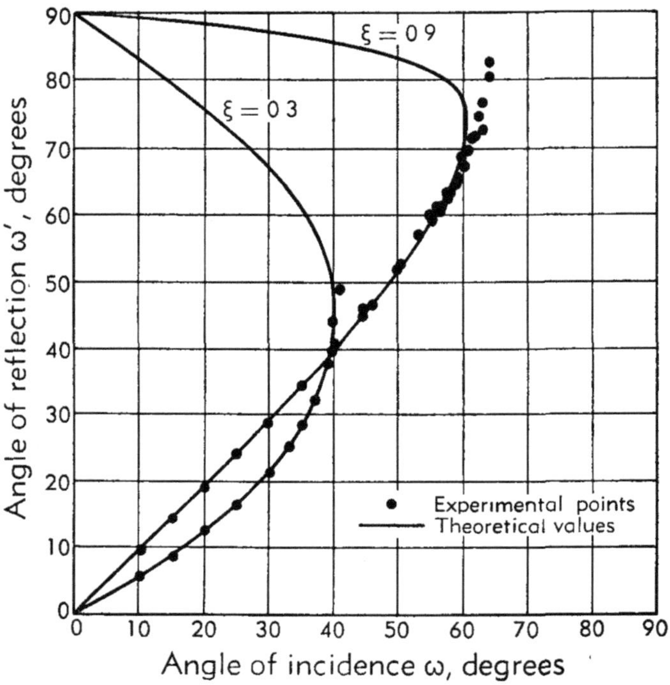

Fig. E,5e. Angle of reflection ω' vs. angle of incidence ω. Experimental data.

with a shock tube shows good agreement between observed values and those smaller angles predicted by the simple theory of regular reflection (cf. Fig. E,5e). Small deviations from the theoretical results for strong shocks ($\xi = 0.3$) may be expected from effects such as nonideality of the gas used and other small systematic discrepancies inherent in the theory and experiment.

Wind tunnel data [*21,22*] are too few to permit any adequate comparison with theory. What are available, however, for interaction of symmetric shocks (mathematically equivalent to reflection from a rigid wall) show satisfactory agreement. Wind tunnel phenomena are strictly stationary, whereas the shock tube produces pseudo-stationary flows [*5*] due to similitude in time, i.e. the physical variables are functions of ratios involving distance and time rather than distance and time separately.

Another family of solutions of particular interest for the regular reflection of shock waves is the set of solutions giving minimum reflected pressures. It can be shown that for $\gamma < 3$ there is always a particular angle ω_m for which the reflected pressure is a minimum. The value of ω_m may be obtained from the following fourth-degree equation in $\sin^2 \omega_m$ for given $\sigma(\xi)$:

$$4[(\gamma - 1)\sigma^2 + 2]^2(1 - \sin^2 \omega_m)^2\{[(\gamma - 1)\sigma^2 + 2]^2 + 4\gamma(\sigma^2 - 1)^2 \sin^4 \omega_m\}$$
$$- (\gamma + 1)\{[(\gamma - 1)\sigma^2 + 2]^2 + 4(\gamma\sigma^2 + 1)(\sigma^2 - 1) \sin^4 \omega_m\}^2 = 0 \quad (5\text{-}8)$$

Use is again made of Eq. 5-5 and 5-6 to obtain ξ' and ω'. The minimum solutions for $\gamma = 1.40$ are listed in Table E,5a.

It is to be noted that for a fixed value of ξ the angle of reflection ω' (lower branch, Fig. E,5b) is an increasing monotonic function of the angle of incidence ω. For $\gamma < 3$ there is one particular value of the angle of incidence ω_h for which the angle of reflection equals the angle of incidence. For angles of incidence smaller than this critical angle, the angle of reflection is less than the angle of incidence; whereas, for larger angles, the angle of reflection is greater than the angle of incidence. Inserting the condition for equality of these angles in Eq. 5-6, one obtains

$$\sigma'_h = \frac{1}{\sigma_h} \quad (5\text{-}9)$$

which gives a reflected pressure ξ'_h equal to that of head-on reflection (Eq. 3-3). The magnitude of ω_h is obtained by using the condition for equality of angles together with Eq. 5-9 in Eq. 5-5. Thus

$$\cos 2\omega_h = \frac{\gamma - 1}{2} \quad (5\text{-}10)$$

The value of ω_h for $\gamma = 1.40$ (air) is 39.23°. It will be noted by examining Fig. E,5c that $\omega = \omega'$ at $\omega = 39.23°$ for all values of ξ. At that angle, the strength of the reflected wave is always equal to the head-on value.

Waterlike substances. The basic conditions [*11,12*] governing the reflection of a shock wave at a rigid wall (Eq. 5-4) may be written in the form:

$$\left.\begin{aligned} \nu \cos \omega + \nu' \cos \omega' &= 0 \\ \nu \sin \omega - \nu' \sin \omega' &= \frac{\tau'}{\sin \omega'} - \frac{\tau}{\sin \omega} \end{aligned}\right\} \quad (5\text{-}4')$$

If the acoustic speed is supposed constant, then $\tau = \tau'$. Moreover, if we can neglect the change in material speed relative to the acoustic speed, the second equation states that the angle of incidence ω equals the angle of reflection ω'. In this case the first equation yields $|\nu| = |\nu'|$; or it can be shown that the change in pressure is doubled upon reflection. These relations hold only in the acoustic limit for which the above assumptions hold.

For the case of waterlike substances, it is convenient to combine the basic equations (Eq. 5-4′) into the following quadratic relation:

$$Lx^2 + Mx + N = 0 \tag{5-5′}$$

where

$$x \equiv \cos^2 \omega,$$
$$L \equiv \nu^2[(\tau'^2 - \sigma'^2) - (\tau^2 - \sigma^2)],$$
$$M \equiv \nu'^2[(\tau^2 - \sigma^2) + \sigma'^2] - \nu^2[(\tau'^2 - \sigma'^2) + \sigma^2],$$
$$N \equiv \nu'^2[\sigma^2 - \sigma'^2]$$

Table E,5b lists the "extreme solutions" of the above equation for the case of the hydraulic analogy ($\gamma = 2.00$).

Table E,5b

Minimum solutions $\gamma = 2.00$				Extreme solutions $\gamma = 2.00$		
ξ	ξ'_m	ω_m, deg	ω'_m, deg	ξ'_e	ω_e, deg	ω'_e, deg
0	—	—	—	∞	90.00	0
0.1	4.393	26.92	20.95	5.291	34.36	42.31
0.2	3.139	24.90	20.87	3.874	35.17	46.87
0.3	2.504	23.83	20.86	3.118	36.66	50.48
0.4	2.094	23.14	20.88	2.605	38.59	53.80
0.5	1.796	22.64	20.94	2.218	40.98	57.10
0.6	1.568	22.27	21.02	1.905	43.98	60.55
0.7	1.385	22.00	21.11	1.641	47.87	64.35
0.8	1.234	21.79	21.22	1.410	53.27	68.83
0.9	1.108	21.58	21.32	1.200	61.83	74.76
1.0	1.000	21.47	21.47	1.000	90.00	90.00

A systematic investigation [*27,28*] of such interactions has been made experimentally by the use of ripples on a water table (cf. Plate E,5d). Data for ξ varying between 0.70 and 0.28 indicate general agreement with theoretical values for the region where regular reflection takes place.

MACH REFLECTION.

Gases. It was pointed out in the previous subarticle that for angles of incidence greater than an extreme angle, ω_e, no regular reflection is theoretically possible. For such angles of incidence, Mach-type reflections are observed. This type of reflection pattern consists of three confluent shock waves meeting at a triple point, accompanied by a density discontinuity or slipstream originating at the same point.

Early theoretical investigations [*5,6,7,29*] of Mach or triple-shock configurations were carried out on assumptions used successfully for regular reflection. This simple theory assumes that the flow in the vicinity of the triple-shock pattern is steady, and that each region between adjacent shocks has uniform pressure. The predicted values are found to be in general agreement with the observed values obtained in the few wind tunnel experiments [*13,14,15,16,17*] and with shock tube data [*21,22*] for strong incident shocks. For weak shocks, however, this simple theory has been found to be inadequate.

The success of this so-called simple theory of shock wave interaction in the analysis of regular reflection patterns and its failure in the case of triple-shock wave configurations for weak shocks has caused much surprise. Considerable interest has been attracted by this apparent paradox [*30*].

Most of the attention has been given to the possibility of a peculiar type of mathematical singularity at the triple point, where the three shocks meet. The triple point for a Mach reflection in a shock tube is observed to move in a straight line from the reflecting corner in accord with the principle of similitude here. Its determination, of course, is outside the scope of the simple theory of stationary three-shock intersections. The consideration [*23,24,25,26*] of shock singularities leads to qualitative, but not quantitative, agreement between the theoretical value of this direction as a function of initial shock strength and of reflection angle. No singularities in shock curvature have actually been observed so that the treatment of the shocks as straight in the neighborhood of the triple point is reasonable. It is evident from Plate E,5b, bottom right, that the assumption of uniform pressure is not valid in the region between the reflected shock and the reflecting wall in a shock tube, except possibly very close to the triple point. To date, therefore, the problem of Mach reflection for weak shocks remains unsolved [*60*].

Some exploratory investigations have been made theoretically by the authors [*6,7*] and experimentally (cf. Plate E,5e) by Bleakney for the general case (no symmetry) of four-shock intersections, in an effort to find a clue to the three-shock enigma for weak shocks.

One would prefer complete solutions to the supersonic flow problems involved in shock interactions. Solutions have been obtained for various limiting conditions, viz. glancing incidence for weak shocks [*31*] and strong shocks [*32,33*], head-on incidence [*34*], and acoustic waves [*35*]. The observed data [*22,23,24,25*] agree generally with the theoretical predictions but they also present some new unsolved problems. Considerable empirical data have been systematically accumulated with respect to the so-called diffraction phenomena associated with the passage of shock waves over various bodies [*36*].

Special cases of regular and Mach interactions have been investigated in more complex phenomena such as the intersecting head-waves [*21,22*] of a projectile and supersonic air jets [*37,38*] (cf. Plate E,5f). Qualitative agreement with the simple theory for regular intersections is found. No systematic quantitative study has yet been made.

Waterlike substances. The simple theory [*11,12*] of three-shock intersections (uniform pressure between adjacent shocks) differs from that for ideal gases only in that the "slipstream" (separating the flow across two shocks from that through the third shock) for waterlike substances is not a density discontinuity. A waterlike substance has a range of weak shock strengths for which there are no simple three-shock solutions, whereas the same range for an ideal gas has a continuous set of such solutions. In this case, however, the minimum resultant pressure may be very large. For example, the minimum resultant pressure ξ' for $\gamma = 2.00$ is 3.3175 as $\xi \to 1$. Of particular interest also is the material flow when it is normal to the third shock. These so-called "stationary" solutions are given for $\gamma = 2.00$ in Table E,5c.

Table E,5c

"Stationary" solutions $\gamma = 2.00$			
ξ	ξ'_s	ξ	ξ'_s
0.0	∞	.6	3.509
0.1	4.441	.7	3.619
0.2	3.583	.8	3.740
0.3	3.379	.9	3.868
0.4	3.361	1.0	4.000
0.5	3.418		

The experimental investigation [*27,28*] of the shock interactions (cf. Plate E,5d) for waterlike substances, while less complete and precise than that for air, shows similar disagreement with the simple three-shock theory for weak shocks ($\xi \sim 0.70$), as well as significant discrepancies for strong shocks ($\xi \sim 0.28$). A few experimental investigations [*11,12*] of shock interactions in underwater explosion phenomena have been made for $\xi \sim 0.83$ and ω between 45° and 87°. There is qualitative agreement for the simple theory of regular intersections.

E,6. Refraction of Shocks in Gases.

Normal refraction. When a shock wave, traveling in one medium, enters into a second medium of different physical characteristics, refraction occurs. The problem of shock wave refraction at an interface separating two gaseous media was investigated theoretically for the case of normal incidence [*1,2*] and later experimentally [*18*], with generally good agreement between theory and experiment. A typical case, the head-on collision between a shock with a contact surface between air and argon, is shown in Fig. E,6a.

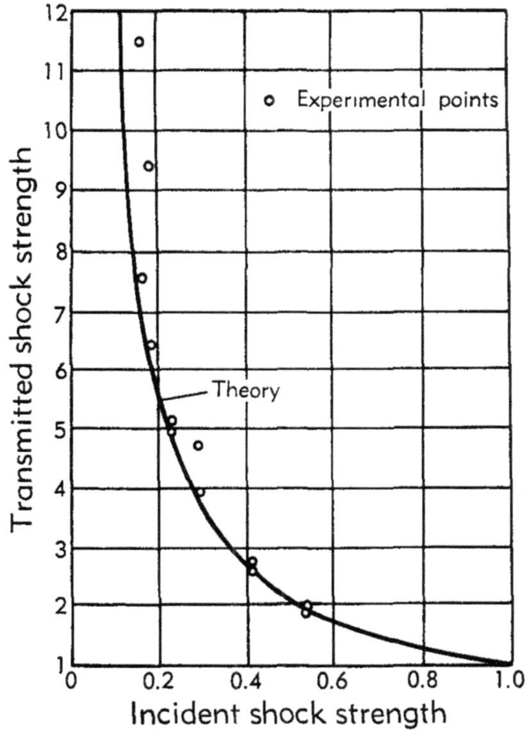

Fig. E,6a. Normal refraction of shock wave. Transmitted shock strength vs. incident shock strength.

Oblique refraction. Refraction for a shock wave at any oblique angle of incidence was developed in general [*6,7,39,40,41*] and in systematic detail [*40,41*] prior to the availability of any observational information [*42*]. As in the case of oblique reflection, both theory [*40,41*] and experment [*42*] show that, beyond certain limiting angles of incidence, so-called regular refraction (cf. Plate E,6a, left) ceases and irregular types (cf. Plate E,6a, right) of refraction patterns are observed. Existing theory applies only to regular refraction. Irregular refraction is still being investigated phenomenologically.

Consider a plane step-shock wave traveling in a uniform gaseous medium (Fig. E,6b) and impinging at an arbitrary angle ω upon a second medium with different physical characteristics. By analogy with the case of normal incidence (cf. Art. 4), it may be expected that a shock wave will be transmitted into the second medium, and that either a second shock wave or a rarefaction wave will be reflected back into the original medium. In the first instance, the resulting pattern will be a triple-shock configuration, in the second, a refraction configuration consisting of two confluent shocks and an angular rarefaction wave [*43*].

We shall specify the physical states in the two regions separated by an interface by the ratios of specific heats γ_0, γ_1, respectively; the acoustic speeds, a_0, a_1; and the pressures p_0, p_1. It is noted that for equilibrium $p_0 = p_1$. We postulate the absence of thermal conductivity and viscosity, and assume uniformity of pressure in the regions bounded by shock waves and/or rarefaction waves. Let the strength of the shock wave *I* (Fig. E,6b)

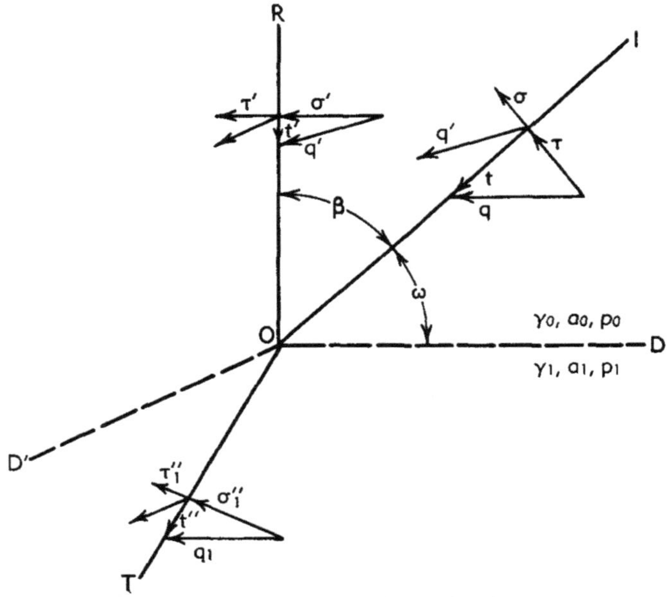

Fig. E,6b. Regular refraction of shock wave.

incident at the interface *D* of the undisturbed medium be measured by the ratio of the pressure in front of the shock wave to that behind it. Then we can immediately obtain the other physical quantities pertaining to the shock *I* in terms of the variables ξ and ω by the use of the Rankine-Hugoniot relations (Eq. 2-9, 2-10, and 2-11 and Eq. 2-9′, 2-10′, and 2-11′).

We formulate the problem from the stationary point of view of an observer traveling with the triple point *O*, which represents the intersection of the incident shock *I* and the interface *D*. As in the case of regular reflection, $\eta \equiv$ density ratio across shock *I* (normalized with respect to the region behind *I*), $\sigma \equiv$ material speed behind *I* and normal to it (normalized with respect to the acoustic speed behind *I*) and $\tau \equiv$ material speed in front of *I* and normal to it (normalized with respect to the acoustic speed behind *I*).

The tangential material speed t along the shock front I is unchanged across the shock and is given by

$$t = \tau \cot \omega \tag{6-1}$$

Thus the velocity of the gas flow in the region between the shocks I and R (reflected shock) is determined by the media characteristics (γ_0, γ_1, a_0, a_1, and p_0) and the shock wave characteristics (ξ and ω). Suppose β, the angle between the two shocks, is also known. Resolving the material velocity in that region into components σ' normal to the shock R and t' tangential to it, we obtain

$$\left.\begin{aligned} \sigma' &= t \sin \beta + \sigma \cos \beta \\ t' &= t \cos \beta - \sigma \sin \beta \end{aligned}\right\} \tag{6-2}$$

where a positive value of t' signifies the direction along R toward 0.

We now apply the Rankine-Hugoniot relations to the reflected shock R to obtain ξ' and τ' for the reflected shock wave (analogous to ξ and τ for the incident shock). We normalize with respect to the same region as that above, i.e. the region ahead of shock R. We thus obtain

$$\left.\begin{aligned} \xi' &= \frac{2\gamma_0 \sigma'^2 - (\gamma_0 - 1)}{\gamma_0 + 1} \\ \tau' &= \frac{(\gamma_0 - 1)\sigma'^2 + 2}{(\gamma_0 + 1)\sigma'} \end{aligned}\right\} \tag{6-3}$$

In this case τ' is the material speed behind the shock front R. The angle between the shock front R and the direction of flow behind it is given by

$$\delta_R = \tan^{-1}\left(\frac{\tau'}{-t'}\right) \tag{6-4}$$

In the case where a rarefaction wave rather than a shock is reflected, we must replace the Rankine-Hugoniot relations with the equivalent equations governing a Prandtl-Meyer angular rarefaction wave [*43*]. We include first the expression for the material speed behind the incident shock I (normalized with respect to the acoustic speed in the same region). Thus

$$q' = (\sigma^2 + t^2)^{\frac{1}{2}} \tag{6-5}$$

We denote by $\bar{q}$ the material speed behind the Prandtl-Meyer wave (normalized with respect to the acoustic speed in that region) and calculate the difference between the directions of material flow ahead of and behind this wave, which we denote by $\bar{\nu} - \nu'$. This is given by the expression [*43*]

$$\bar{\nu} - \nu' = \left(\frac{\gamma_0 + 1}{\gamma_0 - 1}\right)^{\frac{1}{2}} \left\{\tan^{-1}\left[\frac{(\gamma_0 - 1)(\bar{q}^2 - 1)}{\gamma_0 + 1}\right]^{\frac{1}{2}} - \tan^{-1}\left[\frac{(\gamma_0 - 1)(q'^2 - 1)}{\gamma_0 + 1}\right]^{\frac{1}{2}}\right\} + \left(\cos^{-1}\frac{1}{q'} - \cos^{-1}\frac{1}{\bar{q}}\right) \tag{6-6}$$

Next we replace Eq. 6-4 by an analogous equation (Eq. 6-4′) for the angle subtended by the incident shock I and the material speed behind the Prandtl-Meyer wave,

$$\lambda_R = 2\pi - \tan^{-1}\left(\frac{\sigma}{t}\right) + (\bar{\nu} - \nu') \tag{6-4′}$$

For the purpose of considering the gas flow across the transmitted shock T it is convenient to normalize with respect to the region ahead of T instead of that behind I, i.e. γ_1, a_1, p_1. Accordingly, let ξ'' represent the ratio of the pressure behind shock T to that ahead of it. It follows by virtue of the equality of pressures on both sides of the interface D' that

$$\xi'' = \frac{\xi'}{\xi} \tag{6-7}$$

for the case of a reflected shock wave, whereas

$$\xi'' = \frac{\xi'}{\xi} = \frac{1}{\xi}\left(\frac{\bar{p}}{p'}\right) = \frac{1}{\xi}\left[\frac{(\gamma_0 - 1)q'^2 + 2}{(\gamma_0 - 1)\bar{q}^2 + 2}\right]^{\frac{\gamma_0}{\gamma_0 - 1}} \tag{6-7′}$$

for a reflected rarefaction wave (where p' and $\bar{p}$ are the pressures before and after the Prandtl-Meyer wave, respectively).

Again we apply the Rankine-Hugoniot relations across the transmitted shock T, and obtain

$$\left.\begin{aligned} \sigma'' &= \left[\frac{(\gamma_1 + 1)\xi'' + (\gamma_1 - 1)}{2\gamma_1}\right]^{\frac{1}{2}} \\ \tau'' &= \frac{(\gamma_1 - 1)\sigma''^2 + 2}{(\gamma_1 + 1)\sigma''} \end{aligned}\right\} \tag{6-8}$$

The material speed in the region ahead of shock T (normalized with respect to the acoustic speed in that region) is given by

$$q_1 = \left(\frac{\tau}{\sin\omega}\right)\left(\frac{\eta}{\xi}\right)^{\frac{1}{2}}\left(\frac{a_0}{a_1}\right) \tag{6-9}$$

Also, the angle ϕ between shocks I and T, and t'' the material speed tangential to T are given by

$$\left.\begin{aligned} \phi - \omega &= \sin^{-1}\left(\frac{\sigma''}{q_1}\right) \\ t'' &= \sigma'' \cot(\phi - \omega) \end{aligned}\right\} \tag{6-10}$$

respectively, where a positive value of t'' signifies the direction along t toward O.

Finally, for the case of a reflected shock wave, the angle subtended by the reflected shock wave R and the direction of material velocity behind T is given by

$$\delta_T = 2\pi - \left[\phi + \beta + \tan^{-1}\left(\frac{\tau''}{-t''}\right)\right] \tag{6-11}$$

Thus the condition that the flow on both sides of the slipstream D' be parallel requires the equality

$$\delta_R = \delta_T \tag{6-12}$$

For the case of a reflected rarefaction wave, on the other hand, the two analogous equations are

$$\lambda_T = 2\pi - \phi - \tan^{-1}\left(\frac{\tau''}{-t''}\right) \tag{6-11$'$}$$

and

$$\lambda_R = \lambda_T \tag{6-12$'$}$$

The equations derived above constitute a complete mathematical solution of the simple theory for regular refraction. Interpretation of these results, however, presents a considerable number of difficulties due to the complexity of the governing equations. For instance, there are five independent parameters which must be considered. Also, for a fixed set of these parameters, it is possible to find a large number of mathematical solutions (elimination results in a single 12th-degree equation). Furthermore, in general, two types of refraction patterns exist; the first with a reflected shock wave, the second with a reflected rarefaction wave. One would like to predict which of the many solutions are physically realizable, and which of the two types of refraction configurations will occur under given conditions. In order to do this, one postulates that the physically likely solutions are those which can be continuously connected with the known solutions for the following two limiting cases: (1) infinitesimal (acoustic) waves at any angle of incidence and (2) finite shock waves at normal incidence.

For the acoustic limit case ($\xi \to 1$), Snell's law of refraction, viz.

$$\frac{\sin \omega}{\sin (\phi - \omega)} = \frac{a_0}{a_1} \tag{6-13}$$

is known to be physically satisfactory. The solution chosen for the general refraction problem, in order to be physically realizable, must thus converge to the above limit as $\xi \to 1$.

For the case of normal incidence ($\omega = 0$), a solution may be obtained simply from the equality of material speeds on both sides of the interface, both before and after the interaction in a manner similar to that used for the derivation of equations in the case of normal reflection of a shock wave (Art. 3). In the refraction problem, however, a rarefaction wave as well as a shock wave may be reflected. Using this type of argument one obtains, in the case of a reflected shock wave,

$$(\tau' - \sigma') + (\tau - \sigma) + \left(\frac{\xi}{\eta}\right)^{\frac{1}{2}} \left(\frac{a_1}{a_0}\right) (\tau'' - \sigma'') = 0 \qquad (6\text{-}14)$$

and for a reflected rarefaction wave,

$$\frac{2}{\gamma_0 - 1}\left[1 - \left(\frac{\bar{p}}{p'}\right)^{\frac{\gamma_0 - 1}{2\gamma_0}}\right] + (\tau - \sigma) + \left(\frac{\xi}{\eta}\right)^{\frac{1}{2}} \left(\frac{a_1}{a_0}\right) (\tau'' - \sigma'') = 0 \qquad (6\text{-}14')$$

For the refraction of a shock wave at normal incidence, one can use the method discussed for one-dimensional interactions (cf. Art. 4). In this manner we find a definite criterion to determine the type of refraction pattern at $\omega = 0$. We assume continuously varying solutions, so that the type of refraction pattern cannot change for larger angles of ω, until a certain transition angle ω_t is reached (discussed below). Therefore we can continue the solution of the problem from $\omega = 0$ to the angle of incidence ω_t with the same reflection pattern. Beyond ω_t a change will occur, in general, from one type of refraction pattern to the other.

If we stipulate further that no reflected wave (i.e. $\xi' = 1$) be produced at all, we obtain the following quadratic expression in terms of $S = \sin^2 \omega_t$, where ω_t is the so-called transition angle at which such a configuration occurs:

$$4\gamma_1(1 - \xi^2)\left(\frac{\gamma_0\alpha_1}{\gamma_1\alpha_0 b} - 1\right)S^2 + \left[\alpha_0\alpha_1\left(b - \frac{1}{b}\right) + 4(\gamma_1 - \gamma_0)(1 - \xi^2)\right]S + \alpha_0(\alpha_0 - \alpha_1 b) = 0 \qquad (6\text{-}15)$$

where the quantities α_0, α_1, and b are defined as follows:

$$\alpha_0 \equiv (\gamma_0 + 1) + (\gamma_0 - 1)\xi, \quad \alpha_1 \equiv (\gamma_1 + 1) + (\gamma_1 - 1)\xi$$

$$b \equiv \left(\frac{a_0}{a_1}\right)^2 \left(\frac{\gamma_1}{\gamma_0}\right)$$

Since no reflection wave exists at ω_t, any transition between a refraction pattern of the reflected shock variety to that with a reflected rarefaction, and vice versa, must take place at this angle. It can be shown that under the condition that b lies between α_0/α_1 and its reciprocal, i.e.

$$b < \frac{\alpha_0}{\alpha_1}, \quad b > \frac{\alpha_1}{\alpha_0} \quad (\text{if } \alpha_1 < \alpha_0) \qquad (6\text{-}16)$$

or

$$b > \frac{\alpha_0}{\alpha_1}, \quad b < \frac{\alpha_1}{\alpha_0} \quad (\text{if } \alpha_1 > \alpha_0)$$

no real solution for ω_t exists. On the other hand, if b does not lie between α_0/α_1 and its reciprocal, i.e. for the inequalities

$$b < \frac{\alpha_0}{\alpha_1}, \quad b < \frac{\alpha_1}{\alpha_0} \tag{6-16'}$$

or

$$b > \frac{\alpha_0}{\alpha_1}, \quad b > \frac{\alpha_1}{\alpha_0}$$

a unique solution must exist, in which case ω_t acts as a transition value between the two types of refraction configurations.

Table E,6a. List of refraction problems with numerical solutions.

Problem	γ_0	γ_1	$(a_0/a_1)^2$	Gas combinations
1	5/3	1.4	0.835	Argon-nitrogen
2	1.4	5/3	0.120	Air-helium
3	1.4	1.4	0.875	Oxygen-nitrogen
4	1.4	4/3	0.600	Air-methane
5	4/3	1.4	0.600	Carbon dioxide-air
6	1.1	1.4	0.190	Freon-air
7	1.1	5/3	0.020	Freon-helium
8	5/3	1.1	0.800	Krypton-propane
9	5/3	5/3	0.240	Krypton-neon
10	1.1	5/3	0.460	Freon-krypton
11	1.1	5/3	0.600	Propane-argon
12	1.1	1 4	0.013	Freon-hydrogen
13	1.4	5/3	1/0.835	Nitrogen-argon
14	5/3	1.4	1/0.120	Helium-air
15	1.4	1.4	1/0.875	Nitrogen-oxygen
16	4/3	1.4	1/0.600	Methane-air
17	1.4	4/3	1/0.600	Air-carbon dioxide
18	1.4	1.1	1/0.190	Air-freon
19	5/3	1.1	1/0.020	Helium-freon
20	1.1	5/3	1/0.800	Propane-krypton
21	5/3	5/3	1/0.240	Neon-krypton
22	5/3	1.1	1/0.460	Krypton-freon
23	5/3	1.1	1/0.600	Argon-propane
24	1.4	1.1	1/0.013	Hydrogen-freon

In addition to the special solutions discussed above, there are two a priori limitations on the magnitude of ω for which physical solutions of the refraction equations can exist: (1) For the reflected shock wave type the inequality $\omega \leqq \omega_L$, where

$$\sin^2 \omega_L = b \frac{\alpha_0}{\alpha_1} \tag{6-17}$$

must be satisfied in order that the angle $(\phi - \omega)$ be real; and (2) for both types of patterns $\omega \leqq \omega_1$ where

$$\tan \omega_1 = \tau(1 - \sigma^2)^{-\frac{1}{2}} \tag{6-18}$$

Table E,6b. Limiting angles of incidence.

Problem	ξ	ω_t, deg	ω_l, deg	ω_L, deg	Problem	ξ	ω_t, deg	ω_l, deg	ω_L, deg
1	1.0	55.257	90.000	66.034	7	1.0	7.9037	90.000	8.1301
	0.9	55.378	74.526	65.663		0.9	8.0308	73 554	8.1953
	0.7	55.800	65.566	64.899		0.7	8.2094	65.378	8.3317
	0.5	56.534	61.439	64.108		0.5	8.3936	62.929	8.4770
	0.3	57.698	60.007	63.284		0.3	8.5831	64.065	8.6322
	0.1	59.397	61.297	62.425		0.1	8.7770	70.061	8.7987
	0.0	60.382	63.435	61.982		0.0	8.8754	77.690	8.8866
2	1.0	18.594	90.000	20.268	8	1.0	58.227	90.000	63.435
	0.9	18.819	74.146	20.330		0.9	57.693	74.526	62.555
	0.7	19.285	65.457	20.460		0.7	56 815	65.566	60.826
	0 5	19.766	61.945	20.602		0.5	56.192	61.439	59.129
	0.3	20.242	61.439	20.755		0.3	55.781	60.007	57.454
	0.1	20.686	64.272	20.922		0.1	55.355	61.297	55.792
	0.0	20.884	67.792	21.012		0.0	54 910	63.435	54.964
3	1.0	43.089	90.000	69.295	9	1.0	26.100	90.000	29.334
	0.9	44.090	74.146	69.295		0.9	26.369	74.526	29.334
	0.7	46.442	65.457	69.295		0.7	26.944	65.566	29.334
	0.5	49.429	61.945	69.295		0.5	27.561	61.439	29.334
	0.3	53.392	61.439	69.295		0.3	28.193	60.007	29.334
	0.1	59.009	64.272	69.295		0.1	28.784	61.297	29 334
	0.0	62.833	67.792	69.295		0.0	29.029	63.435	29.334
4	1.0	39.594	90.000	50.769	10	1.0	unreal	90.000	42.706
	0.9	40.260	74.146	50.705		0.9	unreal	73.554	43.123
	0.7	41.785	65.457	50.572		0.7	unreal	65.378	44.009
	0.5	43.618	61.945	50.433		0.5	11.820	62.929	44.972
	0.3	45.803	61.439	50.286		0.3	20.739	64.065	46.025
	0.1	48.211	64.272	50.131		0.1	31.192	70 061	47.183
	0.0	49.268	67.792	50.051		0.0	37.558	77.690	47.807
5	1.0	35.459	90.000	50.769	11	1.0	unreal	90.000	50.769
	0.9	36.392	74.031	50.832		0.9	unreal	73.554	51.323
	0.7	38.526	65.433	50.965		0.7	unreal	65.378	52.512
	0.5	41.110	62.115	51.108		0.5	unreal	62.929	53.821
	0 3	44.268	61.905	51.259		0.3	unreal	64.065	55.275
	0.1	48.011	65.259	51.421		0.1	unreal	70.061	56.903
	0.0	49.887	69.303	51.505		0.0	unreal	77.690	57.795
6	1.0	21.947	90.000	25.842	12	1.0	6 4783	90.000	6.5470
	0.9	22.476	73.554	25.981		0.9	6 5203	73.554	6.5794
	0.7	23.577	65.378	26.269		0.7	6.6053	65.378	6.6466
	0.5	24.714	62.929	26.569		0.5	6.6914	62.929	6.7172
	0.3	25.830	64 065	26.884		0.3	6.7783	64.065	6.7916
	0.1	26.824	70.061	27.215		0.1	6.8652	70.061	6 8699
	0.0	27.230	77.690	27.386		0.0	6.9088	77.690	6.9107

Table E,6b (continued)

Problem	ξ	ω_t, deg	ω_1, deg	ω_L, deg	Problem	ξ	ω_t, deg	ω_1, deg	ω_L, deg
13	1.0	64.059	90.000	unreal	19	1.0	77.748	90.000	unreal
	0.9	64.563	74.146	unreal		0.9	78.552	74.526	unreal
	0.7	65.925	65.457	unreal		0.7	80.241	65.566	unreal
	0.5	67.952	61.945	unreal		0.5	82.049	61.439	unreal
	0.3	71.056	61.439	unreal		0.3	83.983	60 007	unreal
	0.1	76.122	64.272	unreal		0.1	86.050	61.297	unreal
	0.0	79.984	67.792	unreal		0.0	87.137	63.435	unreal
14	1.0	66.991	90.000	unreal	20	1.0	71.894	90.000	unreal
	0.9	68.195	74.526	unreal		0.9	72.247	73.554	unreal
	0.7	70.868	65.566	unreal		0.7	73.436	65.378	unreal
	0.5	73.947	61.439	unreal		0.5	75.478	62.929	unreal
	0.3	77.498	60.007	unreal		0.3	78.800	64.065	unreal
	0.1	81.574	61.297	unreal		0.1	84.147	70.061	unreal
	0.0	83.817	63.435	unreal		0.0	87.915	77.690	unreal
15	1.0	46.911	90.000	unreal	21	1.0	63.900	90.000	unreal
	0.9	48.059	74.146	unreal		0.9	65.043	74.526	unreal
	0.7	50.779	65.457	unreal		0.7	67.658	65.566	unreal
	0.5	54.295	61.945	unreal		0.5	70.814	61.439	unreal
	0.3	59.110	61.439	unreal		0.3	74.661	60.007	unreal
	0.1	66.410	64.272	unreal		0.1	79.373	61.297	unreal
	0.0	72.009	67.792	unreal		0.0	82.098	63.435	unreal
16	1.0	55.366	90.000	unreal	22	1.0	unreal	90.000	unreal
	0.9	56.623	74.031	unreal		0.9	unreal	74.526	unreal
	0.7	59.615	65.433	unreal		0.7	unreal	65.566	unreal
	0.5	63.493	62.115	unreal		0.5	16.848	61.439	unreal
	0.3	68.750	61.905	unreal		0.3	29.476	60.007	unreal
	0.1	76.277	65.259	unreal		0.1	44.916	61.297	unreal
	0.0	81.291	69.303	unreal		0.0	55.360	63.435	unreal
17	1.0	48.498	90.000	unreal	23	1.0	unreal	90.000	unreal
	0.9	49.930	74.146	unreal		0.9	unreal	74 526	unreal
	0.7	53.309	65.457	unreal		0.7	unreal	65.566	unreal
	0.5	57 647	61.945	unreal		0.5	unreal	61.439	unreal
	0.3	63.498	61.439	unreal		0.3	unreal	60.007	unreal
	0.1	71.949	64.272	unreal		0.1	unreal	61.297	unreal
	0.0	77.726	67.792	unreal		0.0	unreal	63.435	unreal
18	1.0	59.032	90.000	unreal	24	1.0	81.710	90.000	unreal
	0.9	60.792	74.146	unreal		0.9	82.338	74.146	unreal
	0.7	64.724	65.457	unreal		0.7	83.628	65.457	unreal
	0.5	69.308	61.945	unreal		0.5	84.985	61.945	unreal
	0.3	74.648	61.439	unreal		0.3	86.410	61.439	unreal
	0.1	80.784	64.272	unreal		0.1	87.902	64.272	unreal
	0.0	84.122	67.792	unreal		0.0	88.674	67.792	unreal

must hold in order that supersonic flow normal to the reflected wave can exist (i.e. $\sigma' \geqq 1$). These bounds have been used in determining the range of ω for which the numerical calculations have been carried out. Actually there are more stringent limitations on the range of ω for which solutions exist. These will be discussed subsequently in connection with the analysis of the computational results.

An extensive numerical survey of the regular refraction solutions has been carried out on a high speed calculator for twenty-four typical gas combinations in which the values of incident shock strength ξ and angle of incidence ω were permitted to vary over the entire physical range. Table E,6a lists the gas combinations used in this study.

It will be noted that problems 1–12 are related to problems 13–24. In the second group of problems the gases are inverted, so that the incident shock wave is located in the medium in which the transmitted shock is propagated in the initial group. The values of ξ are $\xi = 0, 0.1, 0.3, 0.5, 0.7, 0.9$, and 1.0; the values of ω have been varied from problem to problem in order to allow for the different limits in each case. The physically realizable solutions have been selected on the basis of the criteria discussed above. The values of the limiting angles ω_t, ω_1, and ω_L are given in Table E,6b.

Problems 3, 10, and 11 and the corresponding problems 15, 22, and 23, in which the gas combinations are inverted, represent a cross section of the types of solutions which may generally be expected. They yield solutions which begin at $\omega = 0$ with reflected shock waves as well as with rarefaction waves. Furthermore, some of the resulting curves pass through the transition stage at ω_t; in other cases, however, there is no solution of the transition equation (Eq. 6-15), so that the same refraction pattern is maintained throughout the range of ω.

The dependence of the strength ξ' of the reflected wave on the angle ω of the incident shock is shown for various values of the latter in Fig. E,6c and E,6d; E,6e and E,6f; E,6g and E,6h for problems 3, 15; 10, 22; and 11, 23; respectively.

For both problems 3 and 15, as may be seen from Table E,6b, the transition angle ω_t exists for all strengths ξ of the incident shock. In problem 15, however, the limiting angle ω_1, for which the material speed behind the incident shock becomes sonic, precedes the transition angle ω_t (i.e. $\omega_1 < \omega_t$ for strong shocks, viz $\xi = 0.1$ and 0.0), so that no transition can take place for these strengths. In problem 3, a (reflected) rarefaction wave solution exists for all values of ξ at normal incidence ($\omega = 0$); whereas for the gases inverted (problem 15) there is always a (reflected) shock wave for normal incidence. For all other incident shock strengths the curves pass through the transition stage at points on the physically plausible branch so that transition to the shock wave pattern may actually be expected in these cases. In problem 15, on the contrary,

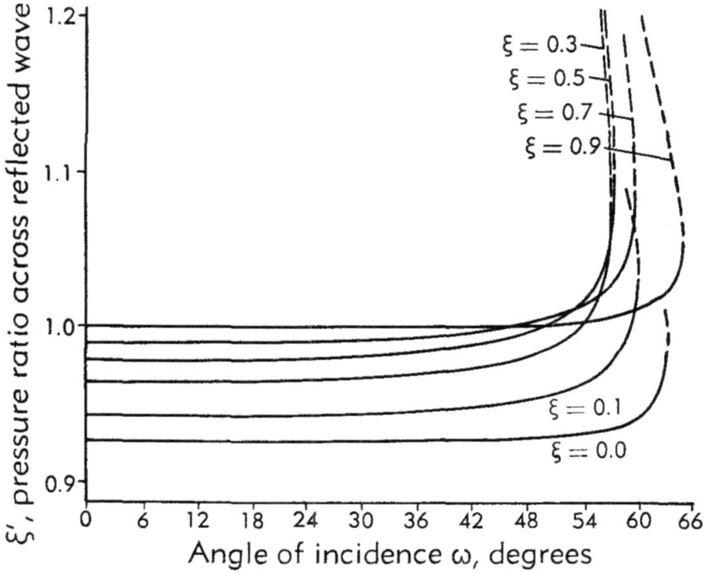

Fig. E,6c. Strength of reflected wave vs. angle of incidence. Problem 3: ω, ξ'.

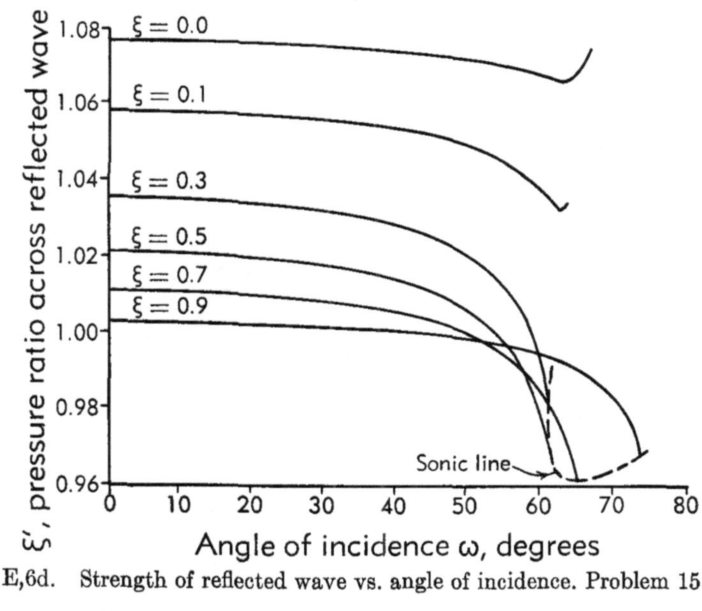

Fig. E,6d. Strength of reflected wave vs. angle of incidence. Problem 15: ω, ξ'.

for strong incident shocks ($\xi = 0.1$ and 0.0) which undergo no transition from the original shock wave pattern at normal incidence ($\omega = 0$), the curves also do not extend beyond an extreme angle ω_e. For weaker shocks, however, transition to a rarefaction pattern does occur and the curves now terminate at an angle of incidence ω_1, beyond which the material speed behind the incident shock is no longer sonic.

For problems 11 and 23, no solutions of the transition equation exist

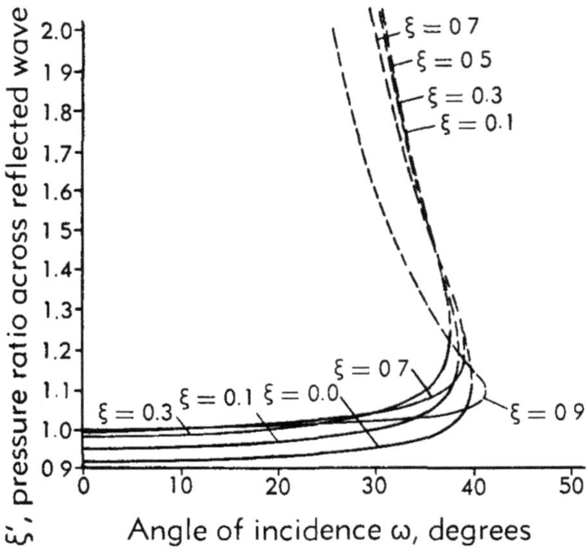

Fig. E,6e. Strength of reflected wave vs. angle of incidence. Problem 10: ω, ξ'.

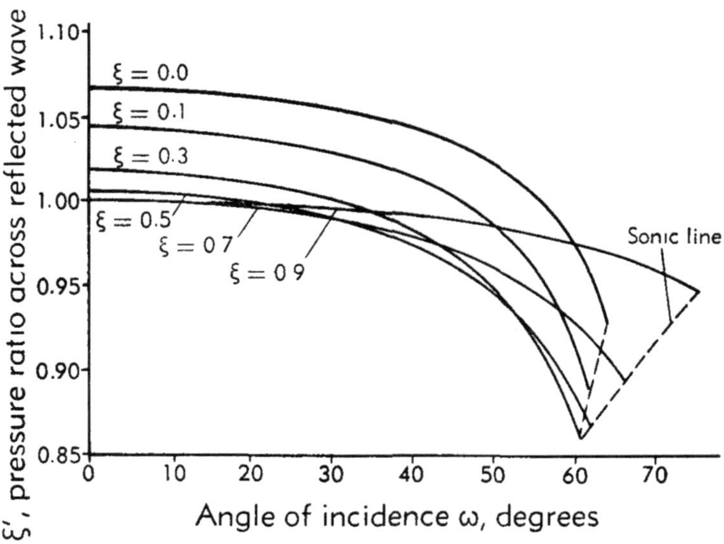

Fig. E,6f. Strength of reflected wave vs. angle of incidence. Problem 22: ω, ξ'.

at all; hence the type of pattern which occurs at normal incidence ($\omega = 0$) persists throughout the entire range of permissible values of angle of incidence ω. Thus a shock wave solution always occurs in problem 11, and a rarefaction wave solution when the gases are inverted (problem 23).

The production of shock wave refraction patterns under laboratory conditions presents practical difficulties. Plate E,6a shows an early shadowgram [*42*] of a regular refraction configuration produced in a shock

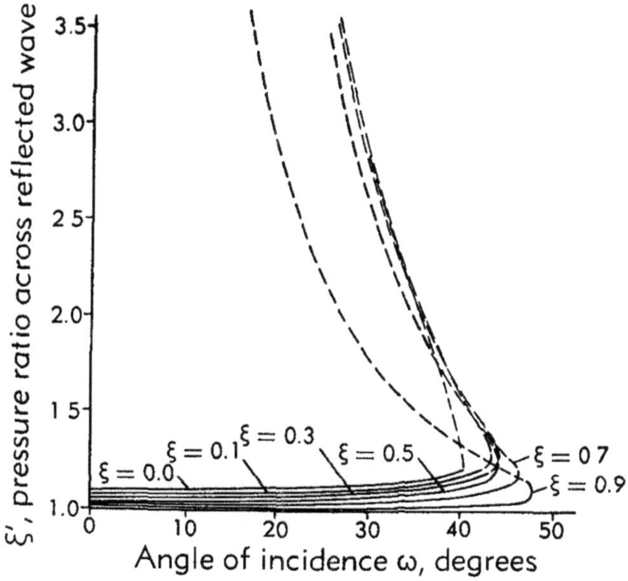

Fig. E,6g. Strength of reflected wave vs. angle of incidence. Problem 11: ω, ξ′.

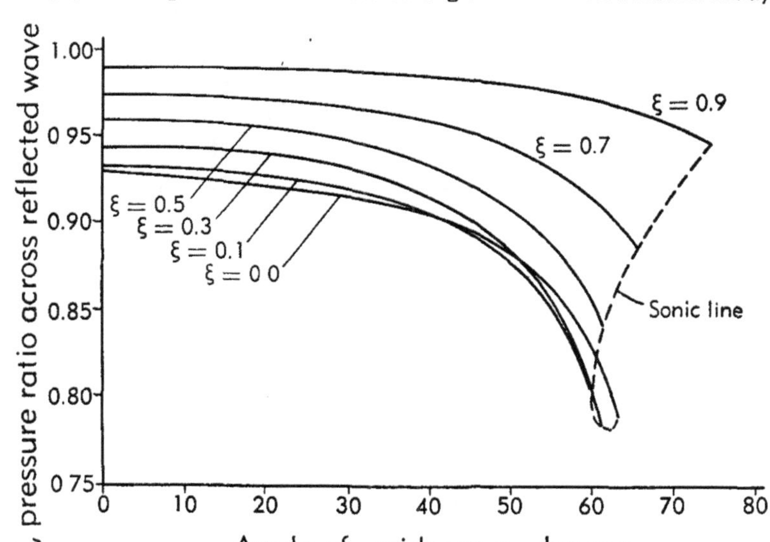

Fig. E,6h. Strength of reflected wave vs. angle of incidence. Problem 23: ω, ξ′.

tube by the separation of two gases (air and carbon dioxide) by means of a thin soap film.

A systematic experimental study [*43*] shows general agreement between the observed data and the theoretical values, where regular refraction is predicted, over a range of ξ values from 0.85 to 0.30. Typical data for the reflected wave for the air-CO_2 gas combination are shown in Fig. E,6i for the angle of reflection ω′ versus the angle of incidence ω, and in

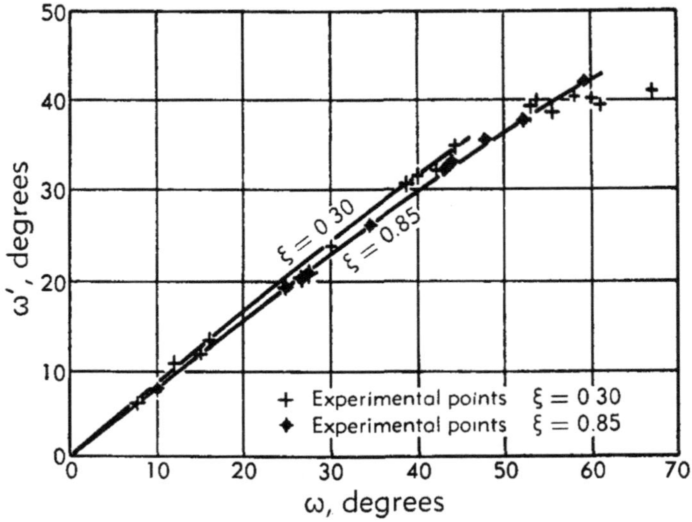

Fig. E,6i. Shock wave refraction. Angle of refraction vs. angle of incidence. Air-CO_2.

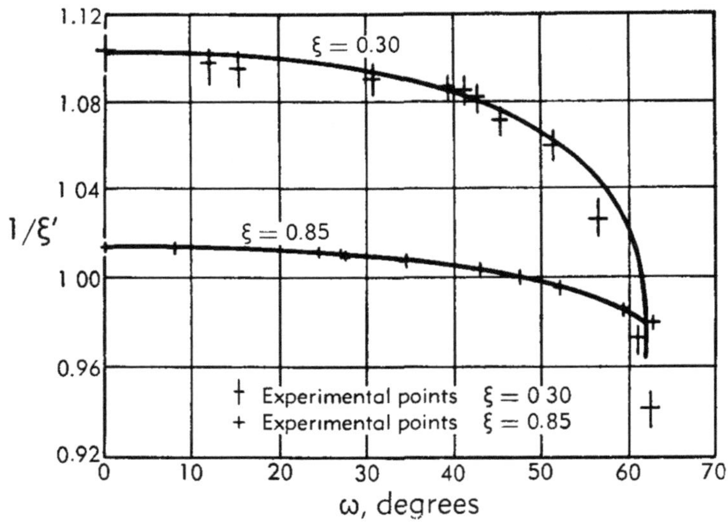

Fig. E,6j. Shock wave refraction. Strength of transmitted wave $1/\xi'$ vs. angle of incidence. Air-CO_2.

Fig. E,6j for the reciprocal strength $1/\xi'$ of the reflected wave vs. the angle of incidence ω.

Interferograms of typical configurations for regular refraction are given in Plate E,6b for the air-CO_2 combination. Plate E,6b, left, shows a reflected shock wave, and Plate E,6b, right, shows a reflected rarefaction wave. The predicted limiting angle of incidence for regular refraction to exist has been confirmed.

E,7. Aerothermodynamic Shock Waves in Gases.

Introduction. Shock waves frequently occur accompanied by an exchange of energy with the surrounding medium; for example, compression waves with conduction or combustion, including deflagration and detonation, and condensation shocks in supersonic wind tunnels. Considerable attention has been given to the mechanism of these processes [*44,45,46,47,48,49,50*]. One can profitably study the thermodynamic equilibrium across such shock fronts without detailed kinetic information. In this connection, one introduces the concept of an aerothermodynamic shock wave [*51*], i.e. a step-shock involving a gain or loss of heat Q at the discontinuity surface. Put

$$C_Q \equiv \frac{Q}{c_p T_0^0}$$

where $c_p T_0^0$ is the initial stagnation enthalpy of the ideal gas. Put

$$z \equiv \frac{M_0^2}{(M_0^2 - 1)^2}\left(M_0^2 + \frac{2}{\gamma - 1}\right)(\gamma^2 - 1)C_Q \tag{7-1}$$

where M_0 is the Mach number $(= (U - u_0)/a_0)$. Then from the conservation laws, one obtains

$$\begin{aligned} \frac{1}{\xi} &\equiv \frac{p}{p_0} = \frac{\gamma}{\gamma + 1}(M_0^2 - 1)[1 \pm \sqrt{1 - z}] + 1 \\ \eta &\equiv \frac{\rho_0}{\rho} = 1 - \frac{1}{\gamma + 1}\frac{M_0^2 - 1}{M_0^2}[1 \pm \sqrt{1 - z}] \\ M &= \frac{U - u}{a} = \left[(1 + C_Q)\left(\frac{1}{M_0^2} + \frac{\gamma - 1}{2}\right)\left(\frac{\rho}{\rho_0}\right) - \frac{\gamma - 1}{2}\right]^{-\frac{1}{2}} \end{aligned} \tag{7-2}$$

The energy exchange may be conveniently given in terms of a new parameter $K_0(M_0, C_Q)$, i.e.

$$K_0 \equiv \gamma\left(M_0^2 + \frac{2}{\gamma - 1}\right)C_Q$$

A given flow M_0 may have 0, 1, or 2 possible equilibrium states associated with it across a given aerothermodynamic shock, which is evident in the graphical relation between M and M_0 for typical values of K_0 in Fig. E,7a, where the two limiting curves represent a contact discontinuity and a Rankine-Hugoniot shock, respectively.

Interactions of Aerothermodynamic Shock Waves.

One-dimensional interactions. One-dimensional interactions of aerothermodynamic shock waves can be treated generally as in the previous cases (cf. Art. 4) by the introduction of an auxiliary monotonic function $\phi(K_0)$ of a slightly more complex form.

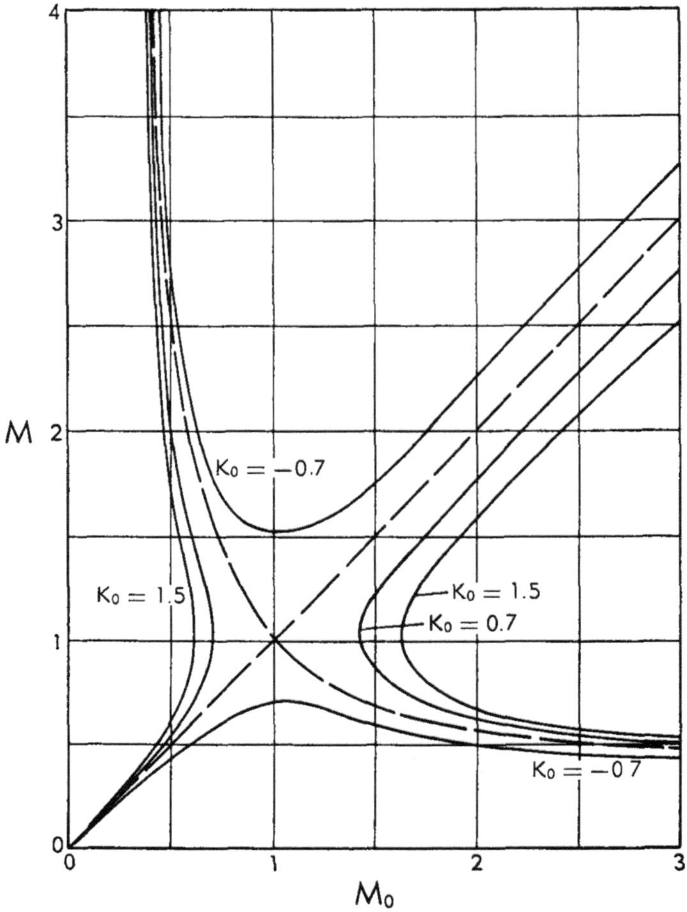

Fig. E,7a. Aerothermodynamic shock waves. $\gamma = 1.40$.

Oblique reflection of a step-shock from a rigid wall. An exploratory numerical analysis for $\gamma = 1.40$ has been made of the oblique reflection of an aerothermodynamic shock wave ($\xi = 0.3, 0.5, 0.7$) incident on a rigid wall. It is assumed that there is a reflected aerothermodynamic shock wave. The reflection curve has one branch which generally has the same characteristics as those for ordinary shocks. However, there exist new families of solutions. These heuristic calculations await experimental study. Some preliminary investigations have been made by G. Herzberg and G. R. Walker, and by T. D. Liddiard, Jr., on the interaction of detonation waves. Mach intersections for two detonation waves in NENO (undercooled) and in nitromethane are shown in the records of photographs of Plates E,7a and E,7b, respectively. Regular intersection is observed at more head-on incidence [*58*]. The effects of air humidity in producing condensation shocks have been investigated by many people (cf. summary review [*59*] by J. Lukasiewicz and J. K. Royle) and the usual

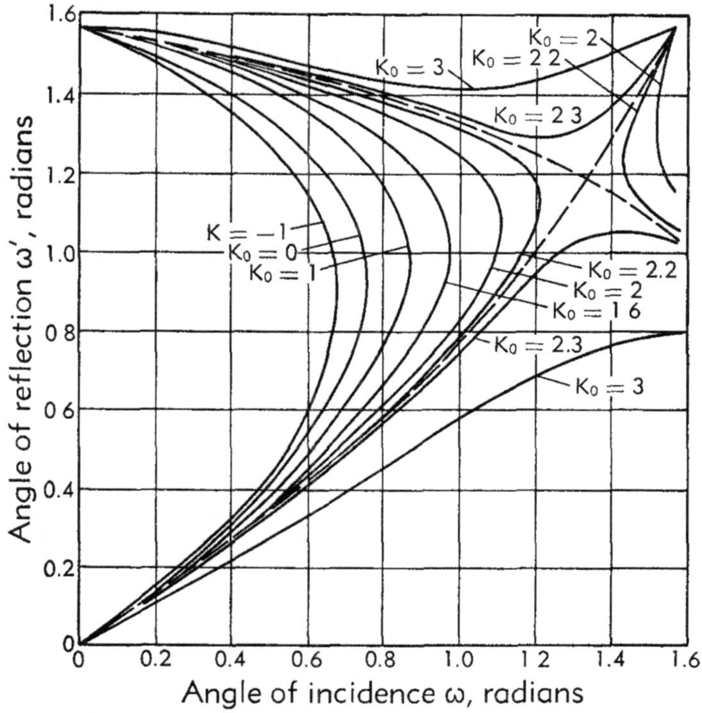

Fig. E,7b. Reflection of aerothermodynamic shock waves. $K = 0$, $\xi = 0.5$.

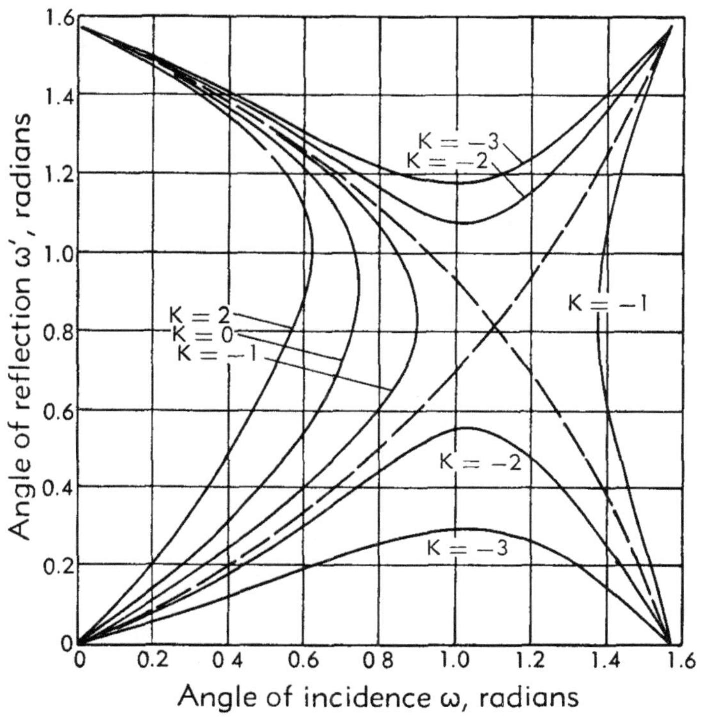

Fig. E,7c. Reflection of aerothermodynamic shock waves. $K_0 = 0$, $\xi = 0.5$.

pattern observed is regular interaction as in Plate E,7c, left. Mach interaction may be observed near the normal shock when starting supersonic flow, as indicated in Plate E,7c, right. If this simple theory is adequate, such curves might be employed to determine an unknown heat Q. Two sample curves for $\xi = 0.5$ and several values of K_0, K are given in Fig. E,7b and E,7c. K_0 is the energy exchange in the incident shock; and K is the energy exchange in the reflected shock.

E,8. Particle Models of the Continuum. In order to reduce the nonlinear equations of fluid dynamics to solvable cases, classical hydrodynamics introduced the concept of perfect incompressible fluids, viz. the perfect inviscid fluid and the perfect viscous fluid. Modern consideration of the added factor of compressibility has led to the above method of regarding shock waves as mathematical discontinuities and then investigating the basic equations for discontinuous solutions.

A different approach is to retain both the viscous and compressible terms and to look for large changes in solutions owing to small changes in the magnitudes of these terms. The consequent mathematical difficulties may be reduced by the use of a particle model of the continuum. In this case, the equations can be solved by means of a high speed digital calculator.

The particle model was first applied [*52,53*] to a compressible inviscid fluid, viz. the one-dimensional reflection of a shock wave in air from a rigid wall. It was then used for the propagation of an underwater explosion [*54*] from a spherical charge and from a cylindrical charge. Increasing the number of particles will increase the accuracy, provided the finite-difference representation is chosen to be numerically stable.

The particle model has also been used for the one-dimensional unsteady motion [*55*] of a compressible viscous fluid, viz. the sudden arrest of a gas-filled closed tube moving with uniform speed, and the sudden communication of two portions of gas at different pressures in a closed tube (the "shock-tube problem"). The success of these few applications hints hopefully of insights that may be gained through the heuristic use of "experimental mathematics," i.e. the use of numerical analysis for obtaining solutions with high speed digital calculators that may provide clues to mathematical understanding of the problem.

E,9. Cited References and Bibliography.

Cited References

1. Courant, R., and Friedrichs, K. O. *Supersonic Flow and Shock Waves.* Interscience, 1948.
2. Courant, R., and Friedrichs, K. O. Interaction of shock and rarefaction waves in one-dimensional motion. *Office Sci. Research and Develop. AMP Rept. 38.1R,* 1948.

3. Rankine, W. J. M. *Phil. Trans. Roy. Soc. London 160*, 277 (1870).
4. Hugoniot, H. *J. école polytech., 58*, 1–125 (1887–1889).
5. von Neumann, J. Oblique reflection of shocks. *Bur. Ord. Explosives Research Rept. 12*, 1943.
6. Polachek, H., and Seeger, R. J. Regular reflection of shocks in ideal gases. *Bur. Ord. Explosives Research Rept. 13*, 1944.
7. Polachek, H., and Seeger, R. J. On shock-wave phenomena: Interaction of shock waves in gases. *Proc. Symposia in Appl. Math. 1*, 1949.
8. Bleakney, W., Weimer, D. K., and Fletcher, C. H. The shock tube: A facility for investigations in fluid dynamics. *Rev. Sci. Instr. 20*, 807 (1949).
9. Griffith, W. C., and Bleakney, W. Shock waves in gases. *Am. J. Phys. 22*, 597 (1954).
10. Brinkley, S. R., Jr., Kirkwood, J. G., and Richardson, J. M. Tables of the properties of air along the Hugoniot curve and the adiabatics terminating in the Hugoniot curve. *Office Sci. Research and Develop. Rept. 3550*, 1944.
11. Polachek, H., and Seeger, R. J. Regular reflection in waterlike substances. *Bur. Ord. Explosives Research Rept. 14*, 1944.
12. Polachek, H., and Seeger, R. J. On shock-wave phenomena: Waterlike substances. *Nav. Ord. Lab. Rept. 1135*, 1949. Also *J. Appl. Phys. 22*, 640 (1951).
13. Smith, L. G. Photographic investigation of the reflection of plane shocks in air. *Office Sci. Research and Develop. Rept. 6271*, 1945.
14. Bleakney, W., and Taub, A. H. Interaction of shock waves. *Revs. Mod. Phys. 21*, 584 (1949).
15. Fletcher, C. H., Taub, A. H., and Bleakney, W. *Revs. Mod. Phys. 23*, 271 (1951).
16. White, D. R. An experimental survey of the Mach reflection of shock waves. *Proc. Midwestern Conf. on Fluid Mechanics*, 1952.
17. Ridenour, L. N., ed. *Modern Physics for the Engineer.* McGraw-Hill, 1954.
18. Glass, I. I., and Patterson, G. N. A theoretical and experimental study of shock-tube flows. *J. Aeronaut. Sci. 22*, 73 (1955).
19. Mach, E. *Vienna Acad. Sitzeingeberichte 77*, 819 (1878).
20. Wood, R. W. The interaction of shock waves. *Office Sci. Research and Develop. Progress Rept. 1996*, Nov. 1943.
21. Keenan, P. C., and Seeger, R. J. Analysis of data on shock intersections. *Progress Rept. I, Bur. Ord. Explosives Research Rept. 15*, 1944.
22. Polachek, H., and Seeger, R. J. Analysis of data on shock intersections. *Progress Rept. II, Nav. Ord. Rept. 417.*
23. Blackman, V. H., and Weimer, D. K. The reflection of shock waves from a blunt edge. *Phys. Rev. 94*, 508 (1954).
24. Taub, A. H. Curved shocks in pseudo-stationary flows. *Annals of Math. 58*, 501 (1953).
25. Taub, A. H. Singularities on shocks. *Am. Math. Monthly 61*, 11 (1954).
26. Taub, A. H. Determination of flows behind stationary and pseudo-stationary shocks. *Annals of Math. 62*, 300 (1955).
27. Einstein, H. A., and Baird, E. G. Progress reports on the analogy between surface shock waves on liquids and shocks in compressible gases. *Calif. Inst. Technol. Hydrodynamic Lab.*, Sept. 1946, July 1947.
28. Gilmore, F. R., Plesset, M. S., and Crossley, H. E., Jr. The analogy between hydraulic jumps in liquids and shock waves in gases. *J. Appl. Phys. 21*, 243 (1950).
29. Chandrasekhar, S. On the conditions for the existence of three shock waves. *Ballistic Research Lab. Rept. 367*, 1943.
30. Birkhoff, G. *Hydrodynamics: A Study in Logic, Fact and Similitude.* Princeton Univ. Press for Univ. Cincinnati, 1950.
31. Bargmann, V. On nearly glancing reflection of shocks. *Office Sci. Research and Develop Rept. 5171*, 1945.
32. Lighthill, M. J. *Proc. Roy. Soc. London A198*, 454 (1949).
33. Ting, L., and Ludloff, N. F. *J. Aeronaut. Sci. 18*, 143 (1951).
34. Lighthill, M. J. *Proc. Roy. Soc. London A200*, 554 (1950).
35. Keller, J. B., and Blank, A. *Commun. on Pure and Appl. Math. 4*, 75 (1951).
36. Bleakney, W., White, D. R., and Griffith, W. C. *J. Appl. Mech 17*, 439(1950).

37. Ladenburg, R. Interferometric study of supersonic phenomena. Part I. *Nav. Ord. Rept. 69-46*, 1946. Part. II. *Nav. Ord. Rept. 93-46*, 1946.
38. Ladenburg, R., van Voorhis, C. C., and Winckler, J. Analysis of supersonic air jets. Part II: Interferometric studies of faster than sound phenomena. *Phys. Rev. 76*, 662 (1949).
39. Taub, A. H. Refraction of plane shock waves. *Phys. Rev. 72*, 51 (1947).
40. Polachek, H., and Seeger, R. J. Refraction of shock waves at a free surface. *Proc. Sixth Intern. Congress Appl. Mech.*, 1948. *Nav. Ord. Lab. Mem. 9971*.
41. Polachek, H., and Seeger, R. J. On shock-wave phenomena: Refraction of shock waves at a gaseous interface. *Phys. Rev. 84*, 922 (1951).
42. Stoner, R. G., and Glauberman, M. H. *Phys. Rev. 76*, 882 (1949).
43. Jahn, R. G. The refraction of shock waves at a gaseous interface.
I. Regular refraction of weak shocks. *Princeton Tech. Rept. II-16*, 1954.
II. Regular refraction of strong shocks. *Princeton Tech. Rept. II-18*, 1955.
III. Irregular refraction. *Princeton Tech. Rept. II-19*, 1955.
44. Durand, W. F., ed. *Aerodynamic Theory*, Vol. 3. Springer, Berlin, 1934–1936.
45. von Neumann, J. Theory of detonation waves. *Office Sci. Research and Develop. Rept. 549*, 1942.
46. Friedrichs, K. O. On the mathematical theory of deflagrations and detonations. *Nav. Ord. Rept. 79-46*, 1946.
47. Lukasiewicz, J. Humidity effects in supersonic flow of air. *Roy. Aircraft Establishment Rept.*, Farnborough, 1947.
48. Lukasiewicz, J., and Royle, J. K. Effects of air humidity in supersonic wind tunnels. *Brit. Aeronaut. Research Council Tech. Rept. 2563*, 1953.
49. Reed, G. G., Jr. Some examples of weak detonations. *J. Chem. Phys. 20,* 1823 (1952).
50. Heybey, W. H., and Reed, G. G., Jr. Condensation shocks, weak detonations, and related phenomena. *Nav. Ord. Rept. 2779*, 1953.
51. Polachek, H., and Seeger, R. J. On shock-wave phenomena. Aerothermodynamics interactions. *Phys. Rev. 86*, 601 (1952).
52. von Neumann, J. Proposal and analysis of a new numerical method for the treatment of hydrodynamical shocks. *Office Sci. Research and Develop. Appl. Math. Rept. 108.1R*, 1944.
53. von Neumann, J., and Richtmyer, R. D. A method for the numerical calculation of hydrodynamic shocks. *J. Appl. Phys. 21*, 232 (1950).
54. Seeger, R. J. On computational techniques for certain problems in fluid dynamics. *Proc. of Symposium on Large-Scale Digital Calculating Machinery*, 1948.
55. Ludford, G., Polachek, H., and Seeger, R. J. On unsteady flow of compressible viscous fluids. *J. Appl. Phys. 24*, 490 (1953).
56. Patterson, G. N. *Molecular Flow of Gases*. Wiley, 1956.
57. de Hoffman, F., and Teller, E. *Phys. Rev. 80*, 692 (1950).
58. Herzberg, G., and Walker, G. R. *Natl. Research Council of Canada Rept. on Project XR-84*, 1945–1946.
Walker, G. R. *Canada Defence Research Board Rept. DR42*, Oct. 1951.
Eyring, H. *Office Sci. Research and Develop. Interim Rept.*, June–July 1944.
59. Lukasiewicz, J., and Royle, J. K. *Brit. Aeronaut. Research Council Repts. and Mem. 2563*, 1948.
60. Taub, A. H., and Clutterham, D. R. Numerical results on the shock configuration in Mach reflection. *Proc. Symposium in Appl. Math. 6*, 1955.
61. Lean, G. H. Experiments on the reflection of inclined shock waves. *Brit. Aeronaut. Research Council Rept. Aero. 7*, 495 (1943).
Lean, G. H. Report on further experiments on the reflection of inclined shock waves. *Brit. Aeronaut. Research Council Rept. Aero. 10*, 629 (1946).

Bibliography

Dailey, C. L., and Wood, F. C. *Computation Curves for Compressible Fluid Problems*. Wiley, 1949.

Emmons, H. W. *Gas Dynamics Tables for Air*. Dover, 1947.
Haworth, L. *Modern Developments in Fluid Dynamics—High Speed Flow*. Oxford Univ. Press, 1953.
Jones, D. M., Martin, P. M. E., and Thornhill, C. K. A note on the pseudo-stationary flow behind a strong shock diffracted or reflected at a corner. *Proc. Roy. Soc. London A209*, 238 (1951).
Ludloff, H. F. On the aerodynamics of blasts. *Advances in Appl. Mech. 3*, 109 (1953).
Lundquist, G. A. Shock wave formation in a shock tube. *J. Appl. Phys. 23*, 374 (1952).
Richardson, J. M. Hydrodynamic properties of sea water at the front of a shock wave. *J. Chem. Phys. 15*, 785–794 (1947).
Robinson, A. Wave reflexion near a wall. *Proc. Cambridge Phil. Soc. 3*, 528–544 (1951).
Sauer, R. *Theoretische Einfuhrung in die Gasdynamik*. Berlin, 1943.
Taub, A. H. Relativistic Rankine-Hugoniot equations. *Phys. Rev. 74*, 328–334 (1948).
Thomas, T. Y. The fundamental hydrodynamical equations and shock conditions for gases. *Math. Mag. 22*, 169–189 (1949).
Weyl, H. Shock waves in arbitrary fluids. *Commun. on Pure and Appl. Math. 2*, 103–122 (1949).

SECTION F

CONDENSATION PHENOMENA IN HIGH SPEED FLOWS

H. GUYFORD STEVER

F,1. Introduction. The acceleration of a compressible vapor to a high speed results in a change of state which, depending on the amount of expansion and the closeness of the original state to the saturation condition, may change the vapor from an unsaturated state to a supersaturated state. If condensation occurs, small liquid drops can be observed along with consequent changes in the properties of the flowing vapor, which are due primarily to the release of latent heat by the vapor as it condenses. Familiar examples of this condensation are the trails of fog from the tips of propellers and wings when aircraft are operated in regions of high relative humidity.

A case of more practical importance is the condensation of a vapor flowing in a convergent-divergent nozzle. Some of the condensation phenomena observed in this type of flow, when the vapor is steam, are illustrated in Fig. F,1 taken from Keenan [*1*], in which the static pressure of the vapor is plotted as a function of the distance along the nozzle. The pressure of the vapor decreases in close agreement with the well-known laws of isentropic expansion until it reaches p_{s_∞}, the saturation vapor pressure of the substance at the temperature of the stream at that point. If no condensation and no change of the ratio of specific heats were to occur, the drop in pressure of the vapor flowing through the nozzle would continue along an adiabatic line represented by the dashed line in the figure. Instead, however, the pressure follows the solid line, deviating somewhat from the normally expected adiabatic expansion. The exact point of deviation and the amount is often difficult to measure. The state of the vapor becomes supersaturated. At a point p_c there is an abrupt pressure rise followed again by a pressure fall. Following this point of pressure rise in the nozzle a fog can be observed downstream, easily in some cases or, if not, by the use of a carefully arranged light scattering experiment. In addition, if observation is made of the flow by schlieren or interferometric techniques, it is found at p_c that there is an abrupt change in the flow condition resembling that of a shock. In many cases, this shock has an X shape as illustrated in Plate F,1, which is a schlieren photograph of such a condensation shock, taken in the Massachusetts Institute of Technology Gas Turbine Laboratory.

The phenomena of condensation in flowing vapors are closely related to those encountered in cloud chamber expansions; C. T. R. Wilson, pioneer in the latter field, published, as early as 1897, an account of the behavior of water vapor mixed with air and other gases when subjected to rapid expansions.

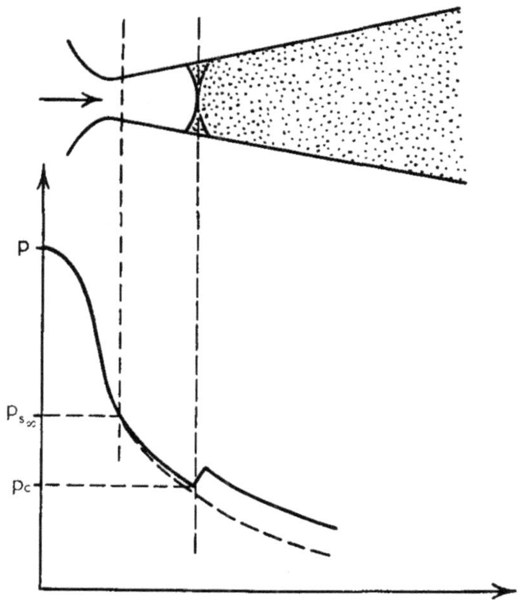

Fig. F,1. Static pressure in nozzle with condensation. After [*1*].

Stodola [*2*], in experiments on superheated steam, devised the experimental technique of light scattering to detect the occurrence of condensation, and also used static wall pressure measurements for the detection of the deviation in pressure from the adiabatic expansion. Since his early experiments there have been many investigations to determine the location of the point p_c at which the supersaturated state of the stream suddenly collapsed and to determine generally the nature of this flow. This point of abrupt change was detected by Yellott [*3,4*] and explained by Keenan [*1*] by considering that the fluid flow equations were satisfied on each side of a discontinuity at which a finite amount of heat was added to the stream. As a result of many investigations it was found that the amount of deviation from the adiabatic expansion line before the abrupt rise in pressure varies, and also that the abrupt rise in pressure accompanied by a condensation shock is found to occur at regions of fairly high supersaturation. Stodola first recognized the existence of this supersaturation.

F · *CONDENSATION PHENOMENA IN HIGH SPEED FLOWS*

In 1934 Hermann [5] and Wieselsberger began the investigation of the condensation of water vapor in supersonic wind tunnels. The experimental techniques which they used were similar to those used to study steam condensation. The phenomena of condensation of water vapor in air are similar to those of steam condensation, the small differences being attributable to the fact that the water vapor in the air represents only a small portion of the total gas flowing through the nozzle. Following these experiments, which were first reported in 1942, there have been many experimental investigations of the condensation of water vapor in air in subsonic high speed tunnels and in supersonic tunnels. Head [6] summarizes the results of many of these investigations for water vapor in air and also for steam.

In 1942, Wagner [7] recognized that the expansion of dry air at room temperature and atmospheric pressure in hypersonic wind tunnels would result in a supersaturated state for the nitrogen and oxygen in air when the Mach number became approximately five. From 1942 to date, a considerable amount of experimental and theoretical work has been done on the condensation of the principal components of air in hypersonic tunnels. In this case, most experimental observations have detected static pressure changes beginning near the saturation point p_{s_∞}, and condensation-fog light scattering, but the detection of a condensation shock or an abrupt collapse of the supersaturated state has not been definitely found, even though the expansion deviates considerably from the condensation-free isentrope.

The recognition of the deleterious effects of condensation in nozzles and wind tunnels has led designers to design apparatus to avoid the region of condensation. There are two techniques generally adopted for this purpose. The first is either to increase the stagnation temperature of the vapor by heating or to decrease the stagnation pressure in an attempt to prevent the vapor from reaching a saturated state in the expansion. The other method, applicable only to the case of the condensation of a minor component of a mixture of vapors and gases, such as the case of water vapor in air, is the partial elimination of the condensable component so that the condensation conditions of this minor component are avoided. For high speed wind tunnels, large and expensive drying apparatus has been employed to eliminate water vapor condensation. In the case of air condensing in hypersonic wind tunnels, it has been recognized that the first method becomes difficult for the cases of very high Mach numbers, eight or higher, and somewhat troublesome even for lower Mach numbers, from five to eight.

To eliminate the requirement for these costly techniques, considerable effort has been spent on an investigation of the condensation phenomena in high speed flow in order to determine the amount of supersaturation which can be achieved before the vapor condenses appreciably, and also to determine the deleterious effects of condensation on the desired experimental measurements.

The remainder of this section is devoted to an investigation of the theory of condensation and an application of that theory to experimental observations. The procedure will be to examine first the phenomena of condensing vapors when equilibrium conditions exist, then to investigate the kinetics of condensation, to apply the kinetic theory of condensation to the problem of condensation in high speed flow, to outline experimental techniques and results, and finally to compare theory and experiment.

F,2. Properties of Condensing Vapors. The dew point of a vapor is the state in which the vapor is in equilibrium with a flat surface of the liquid phase of the same substance. The locus of all dew points is called the saturation line. In this section the independent properties of state are taken as pressure and temperature; for those independent properties the saturation line is given by the equation,

$$p = p_{s_\infty}(T) \tag{2-1}$$

where the subscript s_∞ refers to the fact that the vapor pressure is taken over a liquid surface of infinite radius of curvature.

The slope of the saturation line can be found in terms of the temperature, the latent heat, and the volume change during the change of phase by applying one of the Maxwell relationships to a mixture of the liquid and vapor phase of a substance. As shown in many standard works on thermodynamics, this leads to the Clapeyron-Clausius relation,

$$\frac{dp_{s_\infty}}{dT} = \frac{1}{T}\frac{h_{lg}}{\Delta v} \tag{2-2}$$

where h_{lg} is the change in specific enthalpy h in changing state, or the latent heat, and Δv is the change in specific volume in the change of state. The slope of the saturation line at a given temperature is thus the ratio of the latent heat of vaporization to the product of the temperature and change in specific volume between the liquid and vapor state.

In the case of a vapor far from its critical point, where the perfect gas equation of state can be considered to hold, and where the specific volume of the vapor phase is considerably larger than the specific volume of the liquid phase, the Clapeyron-Clausius relation becomes

$$\frac{d}{dT}\ln p_{s_\infty} = \frac{h_{lg}}{\mathcal{R}T^2} \tag{2-3}$$

Here $\mathcal{R}$ is the specific gas constant. This slope of the saturation line can be compared with the slope of an isentropic expansion of a perfect gas with a ratio of specific heats γ, in which the pressure and temperature are related as follows:

$$p = p^0 \left(\frac{T}{T^0}\right)^{\frac{\gamma}{\gamma - 1}}$$

The slope of an isentropic expansion line is

$$\frac{d}{dT} \ln p = \frac{\gamma}{\gamma - 1} \frac{1}{T} = \frac{c_p}{\mathcal{R}T} = c_p T \frac{1}{\mathcal{R}T^2} \tag{2-4}$$

where c_p is the specific heat at constant pressure.

A comparison of c_pT and h_{lg} in Eq. 2-3 and 2-4 at the saturation line for vapors of interest shows that the saturation line has a greater slope than the isentropic expansion. Hence an isentropic expansion line for a vapor crosses the saturation line into the region of supersaturation as shown in Fig. F,2. Generally the same can be said for a mixture of vapors even though the slope of an isentrope for a pure vapor is not followed by that vapor when it is in a vapor mixture which expands isentropically [*6*].

Fig. F,2 shows the saturation lines for water, carbon dioxide, oxygen, and nitrogen plotted on logarithmic scales of p versus T. These saturation lines are plotted entirely from experimental data [*8*; *9*; *10*, Table 9.50; *11*; *12*]. On the same graph are two straight lines representing the isentropic expansion of vapors with ratios of specific heats of 1.4 and 1.3, expanding from a stagnation pressure of 1 atmosphere and a stagnation temperature of 293°K. On these straight lines are marked Mach numbers which result from these expansions. On this graph it is possible to move either straight line, representing an isentropic expansion parallel to itself, to any stagnation point p^0, T^0 in order to determine the Mach number, pressure, and temperature at which saturation is reached for an isentropic expansion from that stagnation point. It should be noted that each saturation line has an × marked at the triple point, below which the line represents the state of coexistence of the vapor and a flat solid surface. The derivatives of the saturation line are not continuous across the triple point.

In actuality, the condensation of interest in high speed flow often takes place at states which are well into the supersaturated region. If this supersaturation did not occur, and condensation took place at the saturation line, avoidance of condensation could be achieved by insuring that the stagnation point of the expansion was picked so that the final Mach number of the expansion was in the superheated region to the right of the saturation line. For some cases the amount of preheating to keep the entire isentropic expansion line to the right of the saturation line is large. Then it is of practical value to examine the amount of adiabatic undercooling, or overexpanding, that can be done before condensation effects become deleterious.

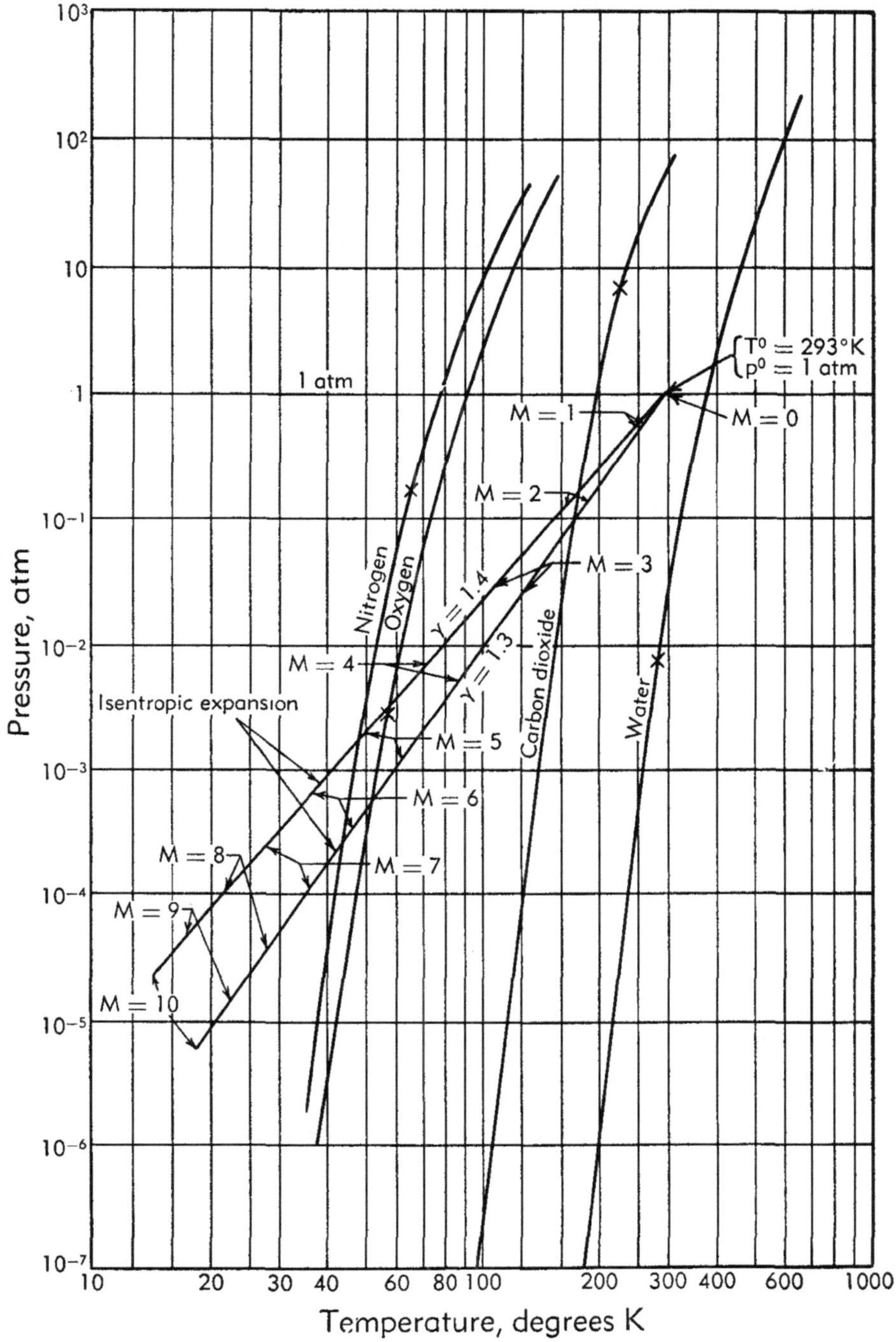

Fig. F,2. Saturation lines for water, carbon dioxide, oxygen, and nitrogen.

Nuclei of condensation, critical drop size. It is well known experimentally that condensation of a supersaturated vapor takes place, in the absence of a flat surface of the liquid phase, on nuclei of liquids or solids, or on the walls containing the vapor. In the problems of condensation which occur in high speed flow, the vapor which is in a supersaturated state is "insulated" from the solid surfaces by the thermal gradient in the boundary layer, in which the temperature increases from the free stream temperature in the flowing vapor almost to the stagnation temperature in the still vapor at the surface. This leaves as possible nuclei of condensation only foreign particles or spontaneously formed liquid or solid drops of the vapor. These nuclei are small and they must be extremely numerous to account for the observed condensation. In the case of a clean pure vapor, condensation must take place on spontaneously formed liquid or solid drops.

Because of the small size of the nuclei of condensation, it is necessary to study the conditions of equilibrium for a liquid drop of radius r and its surrounding vapor. This problem was investigated by Thomson [*13*], von Helmholtz [*14*], and Gibbs [*15*]. The equation relating the pressure of a vapor to the radius of a liquid drop in equilibrium with the vapor can be derived by considering a spherical drop of radius r, with g liquid molecules, in contact with its vapor at pressure p; the liquid and the vapor are at temperature T. Then the energy required to evaporate this drop is

$$E(g) = gE_\infty - 4\pi r^2\sigma \tag{2-5}$$

where E_∞ is the energy required to evaporate a single molecule from an infinite extent of liquid, and σ is the surface tension of the liquid, considered here to be a constant, independent of r. The subtraction of the surface energy $4\pi r^2\sigma$ of the drop is necessary because the surface molecules are closer energywise to the vapor state than the internal molecules, since the surface molecules have fewer molecular bonds holding them in the liquid phase. This surface energy becomes negligible compared to the volume energy gE_∞, when the radius increases without bound, because the volume energy is proportional to r^3 while the surface energy is proportional to r^2. It is seen that E_∞ is interpreted as the energy necessary to break all the molecular bonds of an internal molecule.

The energy required to evaporate one molecule from a drop of g molecules is, if a continuum is assumed,

$$\Delta E(g) = \left[\frac{dE(g)}{dg}\right] = E_\infty - 4\pi\sigma\frac{d(r^2)}{dg} = E_\infty - \frac{2\sigma V_{\text{liq}}}{r} \tag{2-6}$$

where V_{liq} is the volume of a liquid molecule. An application of Boltzmann's law shows that the vapor pressure about a drop of g molecules is proportional to the factor $e^{-\Delta E(g)/kT}$; hence

$$p = p_g = \text{const} \cdot e^{-\Delta E(g)/kT}$$

and

$$p_{s_\infty} = \text{const} \cdot e^{-E_\infty/kT}$$

where p_{s_∞} is the vapor pressure in equilibrium over a drop of infinite radius. Combining these, one finds Thomson's equation relating p, p_{s_∞}, r, and T in terms of the Boltzmann constant k, the surface tension σ of the liquid, and the volume V_{liq} of a liquid molecule

$$\frac{p}{p_{s_\infty}} = e^{2V_{\text{liq}}\sigma/rkT} \tag{2-7}$$

It can be shown that the equilibrium between vapor and liquid drop expressed by Eq. 2-7 is an unstable one by considering a single spherical drop in equilibrium with an infinite extent of vapor. If one molecule of the vapor condenses on the drop, the drop radius increases while the vapor pressure of the infinite extent of vapor remains unchanged. According to Eq. 2-7, the slightly enlarged drop has a lower equilibrium vapor pressure, so the existing vapor pressure is greater than the new equilibrium pressure. Condensation onto the drop will continue. On the other hand, if a single molecule of the drop evaporates to the gas, the smaller drop requires a higher equilibrium vapor pressure than that existing, and evaporation continues.

This unstable equilibrium expressed by Eq. 2-7 results in the concept of a critical size drop. If the pressure of a vapor at a given temperature is higher than the saturation pressure at that temperature, there exists a critical radius for a liquid drop such that the drop will grow larger if the radius of the drop is larger than the critical radius; and the drop will grow smaller if its radius is smaller than the critical radius.

The role of the nucleus of condensation becomes clear when the above critical size drop concept is considered. A similar physical interpretation must hold for a foreign particle acting as a nucleus of condensation, although the exact values of the volume and surface energy, which are important in the analysis of the critical drop size, are not known. That particles of foreign substances do act as nuclei has become extensively clear from many experiments in cloud chambers in which the number of condensed droplets decreases with the number of expansions, each expansion getting rid of some of the foreign particles since drops formed around the particles fall into the liquid, carrying the particles with them.

Spontaneously formed nuclei. In addition to foreign particles acting as nuclei, there are spontaneously formed drops of liquid always existing in a vapor. These result from fluctuations of the density giving rise to the formation of small drops of the liquid phase. The distribution of these small drops is given by the Gibbs formula

$$N_g = Ce^{-\Delta\Phi_g/kT} \tag{2-8}$$

where $\Delta\Phi_g$ is the energy of formation of a drop of g molecules, N_g is the number of drops of g molecules, k is Boltzmann's constant, and C is a constant equaling the total number of molecules, provided that the number of molecules contained in the liquid drops of all sizes is small compared to the number of molecules in the vapor. The energy of formation of a drop containing g molecules from the vapor is the difference of the thermodynamic (chemical) potential of the g molecules in the liquid drop state and the vapor state.

$$\Delta\Phi_g = g(\phi_{\text{liq}} - \phi_{\text{g}}) + 4\pi r^2\sigma \tag{2-9}$$

where ϕ_{liq} and ϕ_{g} are the thermodynamic potentials of single molecules in the liquid and vapor phases respectively, r is the radius of the drop of g molecules, and σ is the surface tension. The quantity g is related to the radius by

$$g = \frac{4\pi r^3}{3V_{\text{liq}}}$$

where V_{liq} is the volume of a liquid molecule. Both ϕ_{liq} and ϕ_{g} are functions of the state of the vapor. Substituting this relation for g into Eq. 2-9,

$$\Delta\Phi_g = \frac{4\pi}{3V_{\text{liq}}}(\phi_{\text{liq}} - \phi_{\text{g}})r^3 + 4\pi r^2\sigma \tag{2-10}$$

In an unsaturated state the vapor state is the stable one, a fact which can be expressed in terms of the thermodynamic potentials of a single molecule in the liquid and vapor states,

$$\phi_{\text{liq}}(p, T) > \phi_{\text{g}}(p, T)$$

(assuming the independent state variables to be pressure and temperature). With this relationship, it is seen that $\Delta\Phi_g$ has two positive terms, a volume energy term

$$\frac{4}{3}\frac{\pi}{V_{\text{liq}}}(\phi_{\text{liq}} - \phi_{\text{g}})r^3$$

and a surface energy term,

$$4\pi r^2\sigma$$

The distribution law (Eq. 2-8), with its negative exponent, denotes a stable distribution of liquid drops, the number of which decreases exponentially with increasing radius or increasing number of molecules in a drop.

As the state of the unsaturated vapor approaches the saturation line, where $\phi_{\text{liq}}(p_{s\infty}, T) = \phi_{\text{g}}(p_{s\infty}, T)$, the magnitude of the exponent of the distribution law decreases, so that the number of drops of given size becomes larger. The case of interest in condensation theory is the supersaturated state, where $\phi_{\text{liq}}(p, T) < \phi_{\text{g}}(p, T)$. Here the expression $\Delta\Phi_g$ has a positive and a negative term.

$$\Delta\Phi_g = -\frac{4\pi}{3V_{\text{liq}}}(\phi_{\text{g}} - \phi_{\text{liq}})r^3 + 4\pi r^2\sigma \tag{2-11}$$

thus exhibiting a maximum so that the distribution (Eq. 2-8) shows a minimum. To find this maximum of $\Delta\Phi_g$, consider the thermodynamic potential of a supersaturated vapor at this pressure and temperature and a liquid drop in equilibrium with it,

$$\Phi_g = N_{\text{g}}\phi_{\text{g}} + g\phi_{\text{liq}} + 4\pi r^2\sigma \tag{2-12}$$

where N_{g} is the number of molecules in the gaseous state. The liquid drop which is in equilibrium has the critical drop size r^*. Thermodynamic equilibrium is expressed by $\delta\Phi = 0$ or

$$\left(\frac{d\Phi_g}{dg}\right)_{r^*} = \left[-\phi_{\text{g}} + \phi_{\text{liq}} + 4\pi\sigma\frac{d(r^2)}{dg}\right]_{r^*} = 0$$

or (2-13)

$$(\phi_{\text{g}} - \phi_{\text{liq}})_{r^*} = \frac{2\sigma V_{\text{liq}}}{r^*}$$

This expresses $(\phi_{\text{g}} - \phi_{\text{liq}})_{r^*}$ in terms of r^*. Substituting this into the expression for $\Delta\Phi_g$ (Eq. 2-11), there results

$$\Delta\Phi_{g^*} = \frac{4\pi\sigma r^{*2}}{3} \tag{2-14}$$

Thus $\Delta\Phi_g$ increases as r increases up to this maximum value, and then decreases and becomes negative where the volume and surface terms are equal.

The distribution law (Eq. 2-8) thus describes for a supersaturated vapor a decreasing number of drops of given radius with increasing radius up to the critical size drop; above the critical radius the number of drops increases with radius. This is a distribution which obviously is unstable as, in fact, supersaturated states are.

To get a better evaluation of the constant C in the distribution law (Eq. 2-8), using a method suggested by Frenkel [*16*], it is possible to treat the equilibrium distribution of drops in an unsaturated state as a dilute solution of different substances in the vapor as a solvent. In this case, the different solutes are characterized by the number of molecules they have in a drop. His analysis leads to the expression

$$N_g = \frac{N_{\text{g}}^g}{(N_{\text{g}} + \sum N_{\text{g}})^g}(N_{\text{g}} + \sum N_g)e^{-\frac{1}{kT}[(\phi_{\text{liq}} - \phi_{\text{g}})g - 4\pi\sigma r^2]} \tag{2-15}$$

This gives the number of drops containing g molecules. When the number of liquid drops is small, the value is simplified to

$$N_g = N_g e^{-\frac{1}{kT}[(\phi_{\text{liq}} - \phi_g)g - 4\pi r^2\sigma]} \tag{2-16}$$

It is to be noted that this drop distribution is based on the assumption of equilibrium ($\phi_g < \phi_{\text{liq}}$) and that the resulting distribution is the stable distribution.

Summarizing, condensation can take place on nuclei provided that they are larger than the critical size. These nuclei may be foreign nuclei or spontaneously formed nuclei of the condensing vapor. Such spontaneously formed nuclei exist both in the unsaturated and supersaturated state, but in the supersaturated state the distribution of their numbers with respect to size is unstable. It now becomes necessary to investigate the rate at which these nuclei form and grow to the critical size.

F,3. The Kinetics of Condensation. In high speed flow in nozzles or wind tunnels, the state of a vapor may change extremely rapidly from an unsaturated state to a highly supersaturated state. The expressions for drop size distributions derived in Art. 2 apply only to equilibrium cases. For high speed flow, one of the problems is to determine the rate at which the distribution of drops changes from the stable distribution of the unsaturated state, in which the number of drops of given radius decreases exponentially with increasing radius, to the unstable distribution of the supersaturated state, in which the number of drops increases with increasing radius. Essentially this requires a determination of how rapidly drops of the critical size form.

Another aspect of the kinetics of condensation involves the growth of drops after the critical size has been reached. In some cases of condensation the particles grow to be visible, and in all cases they grow large enough to scatter visible light in detectable amounts (see IX,G).

When the nuclei formation rate and the drop growth rates are known, it is possible to employ the usual laws which determine channel flow to calculate the flow of the gas and the location of the condensation, as well as the pressure distribution and the other quantities of the flow. All of the above facets of the kinetics of condensation will be treated here.

Nuclei formation rate. In deriving expressions for the nuclei formation rate the assumption is made that the drop distribution (Eq. 2-8 and 2-9), derived for an unsaturated vapor ($\phi_g < \phi_{\text{liq}}$), can be used to describe conditions for a supersaturated vapor ($\phi_{\text{liq}} < \phi_g$). More particularly, it is assumed that this drop distribution holds for all drop sizes through the critical drop size up to some limiting drop size g_m, where the drop distribution is set equal to zero. By stopping the drop distribution at some maximum drop size, there is eliminated the difficulty, pointed out in Art.

2, that, for $\phi_{\text{liq}} < \phi_{\text{g}}$, the number of drops of given size increases with increasing size.

Volmer [*17*] who made important early contributions to nuclei formation rate theory assumed that, since all embryos which grow must first reach the critical size and then continue to grow as stable liquid drops, the critical size could be treated as a gate to condensation. The development of the nuclei formation rate then proceeds on a quasi-stationary point of view. In order to keep the state of the vapor the same it is assumed that all drops which had grown to the critical size were eliminated from the system and replaced by vapor molecules equal in number to those which the stable drops contained.

With these assumptions Volmer calculated the rate of condensation as the product of the number of critical size drops, the number of vapor molecules striking the surfaces of one critical size drop per unit time, and the number of molecules in a drop of critical size. The number of molecules striking unit area of a surface per unit time is a familiar factor in kinetic theory and is given by

$$b = \frac{p}{\sqrt{2\pi mkT}} \tag{3-1}$$

where m is the mass of a molecule of the vapor. Hence the number of molecules condensing per second is, according to Volmer,

$$\frac{dN_{\text{g}}}{dt} = N_{g^*} 4\pi r^{*2} b g^*$$

Using Eq. 2-8, 2-14, 2-16, and 3-1, the fractional condensation rate becomes

$$\frac{1}{N_{\text{g}}}\frac{dN_{\text{g}}}{dt} = \frac{16\pi^2 r^{*5} p}{3V_{\text{liq}}\sqrt{2\pi mkT}}\, e^{-4\pi\sigma r^{*2}/3kT} \tag{3-2}$$

This fractional condensation rate $1/N_{\text{g}} = dN_{\text{g}}/dt$ takes into account only the condensation which consists of drop nuclei growing to critical size and passing into the realm of stable drops. Actually the rate should include the condensation onto these drops as they grow larger than critical size. This latter factor is important, as is shown later, in high speed flow condensation. It is therefore more correct to use this theory of Volmer to express the nuclei formation rate $J(g^*)$ which is given by

$$J(g^*) = \frac{1}{g^*}\frac{dN_{\text{g}}}{dt} = \frac{4\pi r^{*2} p^2}{(kT)^{\frac{3}{2}}\sqrt{2\pi m}}\, e^{-4\pi\sigma r^{*2}/3kT} \tag{3-3}$$

This nuclei formation rate theory has been greatly modified by Farkas [*18*], by Kaischew and Stranski [*19*], particularly by Becker and Döring [*20*], and by Zeldovich [*21*] to take into account not only the growth of drops by the condensation of molecules but also the decay of drops by the evaporation of molecules. In the treatment of this theory of the kinetics of condensation, Frenkel's [*16*] description is again followed. Again the assumption is made that a drop which has grown to the upper limit size g_m is replaced by vapor molecules of number equal to those in the drop. This permits the assumption of an unchanging vapor state so that the distribution of drops in the subcritical size is unchanged.

Just as b represents the number of molecules striking a unit surface of a drop per unit time, a_g represents the number of molecules evaporating from a unit surface of a drop of g molecules per unit time. Unlike b, which comes from a consideration of the velocity distribution of gaseous molecules, a_g is dependent on the size of the drop. In a stable unsaturated vapor, where $\phi_g < \phi_{\text{liq}}$, the constancy of the drop distribution can be expressed by

$$N_g A_g a_g = N_{g-1} A_{g-1} b \tag{3-4}$$

where A_g is the surface area of a drop containing g molecules. If one assumes that the number of molecules in the embryonic drop under consideration is much greater than 1, the areas of the two neighboring drops can be assumed equal. Use of an approximate expression for N_g/N_{g-1} as determined from Eq. 2-16 leads to

$$a_g = b e^{\frac{1}{kT}\left[(\phi_{\text{liq}} - \phi_g) - \frac{2\sigma V_{\text{liq}}}{r}\right]} \tag{3-5}$$

This expression for a_g, the rate of evaporation of molecules per unit surface area for a drop of g molecules, has been derived for the case of a stable unsaturated vapor. However, there is no reason to believe that it would not hold as well for a supersaturated vapor where $\phi_{\text{liq}} < \phi_g$. In this case, using Eq. 2-13, the value of a_g for a critical size drop is

$$a_g = b \tag{3-6}$$

This is entirely reasonable since it was pointed out that there is a minimum in the drop distribution at $g = g^*$ so that $N_g^* \cong N_{g-1}^* \cong N_{g+1}^*$.

Also the expression for a_g (Eq. 3-5) is valid whether or not there is established an equilibrium in the drop distribution, provided that there is established a Maxwellian distribution of velocity in the vapor, a condition which has an extremely short relaxation time τ given, from standard texts in kinetic theory, as

$$\tau = p\lambda \sqrt{\frac{2m}{\pi (kT)^3}} \tag{3-7}$$

where λ is the mean free path. This time is very short, of the order of 10^{-10} sec for air at standard conditions, for example.

To calculate the rate of formation of drops of greater than critical size for a supersaturated vapor, it is necessary to use a nonequilibrium drop distribution which can be denoted by primed symbols. Thus N'_g is the number of drops of g molecules in the nonequilibrium distribution. For a short time the nonequilibrium drop distribution will be assumed to hold, and each drop as it grows to a stable drop size g_m will be replaced by g_m vapor molecules; thus again the process is quasi-stationary.

Let J_g be the increase per unit time in N_g, the number of drops of size g, which results from an unbalance between the evaporation from drops containing g molecules to form drops containing $g - 1$ molecules and the growth of $(g - 1)$-size drops to g-size drops.

$$J_g = N'_{g-1}A_{g-1}b - N'_gA_ga_g \tag{3-8}$$

Substituting Eq. 3-4

$$J_g = N_{g-1}A_{g-1}b\left(\frac{N'_{g-1}}{N_{g-1}} - \frac{N'_g}{N_g}\right) \tag{3-9}$$

Thus the rate of change of the number of drops of given size in the nonequilibrium distribution is

$$\frac{\partial N'_g}{\partial t} = J_g - J_{g+1} \tag{3-10}$$

This equation is the kinetic equation of condensation derived first by Becker and Döring [*20*] and solved by them and others with the assumption that the nonequilibrium drop distribution was similar to the equilibrium drop distribution for values for g less than the critical drop number, and that the distribution decreased to zero at a value of $g = g_m$ somewhat greater than the critical number.

For a small number of molecules in the embryonic drops, say 2, 3, 4, . . . , the quantities A_g, N_g, etc. in the expressions J_g change rapidly, but for large values, say $g > 10$, there will be only small changes, and the functions can be replaced by functions of a continuous variable. Furthermore the finite differences can be replaced by differential coefficients. This leads to an expression for $J(g)$

$$J(g) = -Ab\frac{\partial N'}{\partial g} - \frac{AbN'}{kT}\frac{\partial}{\partial t}[\Delta\Phi(g)] \tag{3-11}$$

This can be substituted into the kinetic equation (Eq. 3-10), again replacing $J_g - J_{g+1}$ by $-(\partial/\partial g)J(g)$,

$$\frac{\partial N'}{\partial t} = b\frac{\partial}{\partial g}\left(A\frac{\partial N'}{\partial g}\right) + \frac{b}{kT}\frac{\partial}{\partial g}\left\{AN'\frac{\partial}{\partial g}[\Delta\Phi(g)]\right\} \tag{3-12}$$

This differential equation should hold reasonably well for the kinetics of condensation, provided that the drops considered are made up of 10 or more molecules. It can be simplified by the approximation that the surface area is not a strong function of g compared to N'_g, so that the area A can be treated as a constant.

$$\frac{\partial N'}{\partial t} = bA\,\frac{\partial^2 N'}{\partial g^2} + \frac{bA}{kT}\frac{\partial}{\partial g}\left\{N'\,\frac{\partial}{\partial g}\,[\Delta\Phi(g)]\right\} \tag{3-13}$$

Integration of this differential equation by the method of Zeldovich [*21*] as outlined by Frenkel [*16*] leads to a critical drop formation rate

$$J(g^*) = \frac{p^2 V_{\text{liq}}}{(kT)^2}\sqrt{\frac{2\sigma}{\pi m}}\,e^{-4\pi\sigma r^{*2}/3kT} \tag{3-14}$$

or a fractional condensation rate

$$\frac{1}{N_g}\frac{dN_g}{dT} = \frac{g^* J(g^*)}{N_g} = \frac{p}{kT}\,\frac{4r^{*3}}{3}\sqrt{\frac{2\pi\sigma}{m}}\,e^{-4\pi\sigma r^{*2}/3kT} \tag{3-15}$$

These are the final expressions for the nuclei formation rate and the fractional condensation rate. They are markedly different from Eq. 3-2 and 3-3, the results of Volmer's calculation [*17*], but very nearly the same as those derived by Becker and Döring [*20*], who solved the kinetic equation by a different method. The ratio of these results and Becker and Döring's results is $g^{*1/3}$, a small factor in regions of appreciable supersaturation.

Eq. 3-14 and 3-15 for the rate of critical size nuclei generation and for the fractional condensation rate are based on a quasi-stationary assumption that the number of vapor molecules is kept constant by replacing all the molecules which are removed from the vapor system in critical size drops. In the real case the number of vapor molecules is decreasing. Moreover, instead of being removed from the system, all of those drops which have grown to the critical size continue to grow and, by virtue of their ever-increasing growth, continue to contribute markedly to the condensation rate. This growth rate is taken into account later.

Application of nuclei formation rate theory to pure vapors. Eq. 3-3 and 3-14 expressing the rate of nuclei formation, Eq. 3-2 and 3-15 expressing the fractional condensation rate, and Eq. 2-7 giving the value of the critical drop radius can be applied to a pure vapor for which the density, surface tension, and saturation vapor pressure are known as functions of temperature, and for which the molecular mass m is known. With these properties known, it is seen that the nuclei formation rate and the fractional condensation rate are determinable for every supersaturated state determined by p and T. The expressions could of course be modified to relate these quantities to any pair of independent state variables.

As an example of the use of these equations, Fig. F,3a, taken from Wegener and Smelt [*22*], shows a plot of Volmer's expression for the nuclei formation rate for water vapor plotted on a pressure versus temperature chart. Similarly Fig. F,3b, taken from Bogdonoff and Lees [*23*], shows a plot of Becker and Döring's expression for nuclei formation rate for nitrogen vapor plotted on a pressure versus temperature chart. In both cases there was necessarily some extrapolation of the experimental data on temperature dependence of density, surface tension, and saturation vapor pressure.

Also in Fig. F,3a and F,3b, lines of constant critical drop radius, expressed as a ratio of drop radius to molecule diameter D, are plotted. It is to be noted that the saturation line is a common member of the two

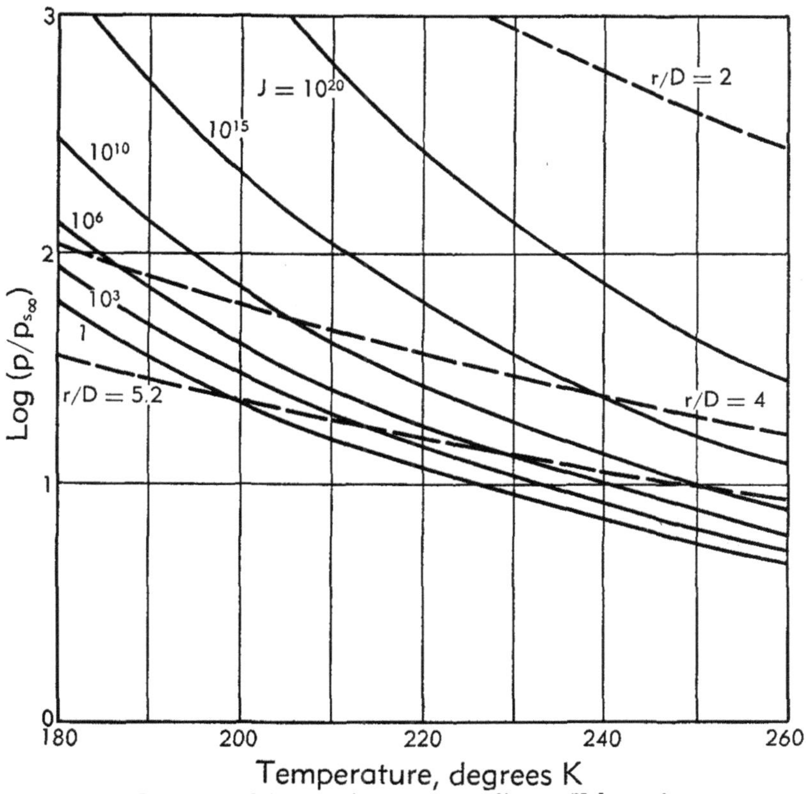

Fig. F,3a. Nuclei formation rate according to Volmer theory.

families of curves for the critical drop radius and for the nuclei formation rates, being the line of infinite critical drop radius and the line of zero nuclei formation rate. In the region of supersaturation, however, these two families of lines do not run parallel but cross at varying angles. It is important to note that high fractional condensation rates are not reached except for small values of critical drop radius. This means that, in the regions of supersaturation of interest in high speed flow condensation phenomena, small critical drop sizes are the rule.

Limitations of nuclei formation rate theory. The application of the nuclei formation rate theory to problems of condensation in high speed flow reveals certain weaknesses in our experimental and theoretical knowledge. For water vapor, oxygen, and nitrogen, which have had the theory applied, the surface tension σ as a function of temperature is not known at the low temperatures required. In the nuclei formation rate expressions the dominant term is the exponential

$$e^{-4\pi\sigma r^{*2}/3kT}$$

which can be written

$$e^{-16\pi\sigma^3 V_{\text{liq}}^2/3(kT)^3 \ln (p/p_{s_\infty})}$$

Because σ appears cubed, a small error in σ greatly alters the nuclei formation rate. Likewise an error in the density d of the liquid phase, which is

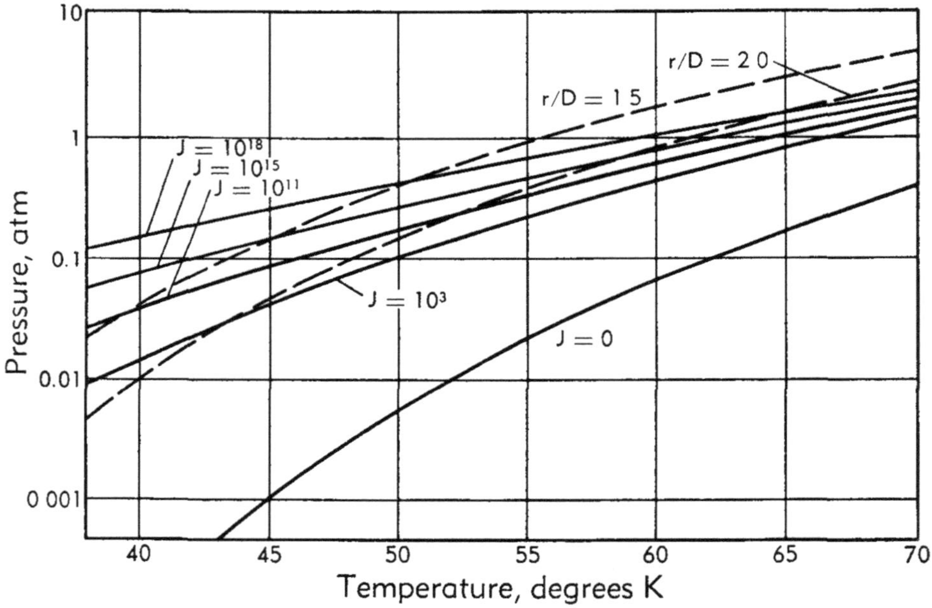

Fig. F,3b. Nuclei formation rate for nitrogen according to Becker-Döring theory.

related to the volume of a liquid molecule V_{liq} by $d = m/V_{\text{liq}}$, greatly affects the calculated result. Head [6] discusses the error introduced in the nuclei formation rate because of errors in σ.

The theory for nuclei formation rate applies to the formation of drops of the liquid phase, but actually there are many cases of interest where there must be solid nuclei formed directly from the condensing vapor. The values for the surface energy are then more difficult to assess, and even more difficult to employ in calculating nuclei formation rates.

As pointed out above, the regions of supersaturation which are of interest almost always involve critical drop radii which are small. The number of molecules contained in these small drops is often only a few. For these cases it becomes questionable to use the continuum theory which was introduced to transform the finite difference equations to differential equations, and the expressions used thus become approximate.

Another and perhaps more serious aspect of the small critical drop problem is involved in the estimation of the energy to form the drop from vapor molecules. As shown above, the nuclei formation rate at a given state is $J = Ce^{-\Delta\Phi/kT}$ where $\Delta\Phi$ is the energy of formation of the drop. Looking at this problem from the aspect of a continuum, $\Delta\Phi$ can be considered the difference between the work done against surface tension and the work done by the pressure forces

$$\Delta\Phi(r^*) = 4\pi r^{*2}\sigma - \tfrac{4}{3}\pi r^{*3}(p_{\text{liq}} - p_g) \tag{3-16}$$

Thompson showed from consideration of thermodynamic equilibrium

$$p_{\text{liq}} - p_g = \frac{2\sigma}{r^*} \tag{3-17}$$

Hence

$$\Delta\Phi(r^*) = \tfrac{4}{3}\pi r^{*2}\sigma \tag{3-18}$$

which checks Eq. 2-14.

These expressions hold only where the surface tension is a constant, which is certainly not the case for drops containing only a few molecules. For very small drops, the work of formation of the critical size drops must correctly be calculated by applications of statistical mechanics. Such attempts have led to excessive work when the drops have more than two or three molecules. Of course, for a large number of molecules the results are the same as the continuum theory. But in the region where the results are wanted, statistical mechanical approaches have failed to give the desired quantities readily and accurately. For these considerations see Mayer and Mayer [*24*].

The first attempt to correct for the effect of surface energy dependence on radius was made by Head [*6*]. He realized that the assumption that the surface tension was independent of the radius of the drop was in error. Physically it changes because the number of molecular bonds experienced by a surface molecule changes with radius of curvature of the surface. Head employed a radius dependency for surface tension which was calculated by Tolman [*25*] from a quasi-thermodynamical point of view and substantiated by Kirkwood and Buff [*26*] by statistical mechanical methods, which was

$$\sigma(r) = \sigma_\infty \frac{1}{1 + (2\delta/r)} \tag{3-19}$$

where σ_∞ is the surface tension for a flat surface, and δ is a length, almost constant, lying between 0.25 and 0.6 of the molecular radius in the liquid state. Head, investigating water vapor, used this to change the value of surface tension in Eq. 2-7 for critical size drop radius and Eq. 3-14 for nuclei formation rate.

Stever and Rathbun [27], investigating air condensation, extended Head's work by altering the expression for nuclei formation rate or fractional condensation rate in its derivation to take care of the radius dependency of the surface tension. For example, the nuclei formation rate as calculated by Stever after modifying the Zeldovich analysis is

$$J(g^*) = \frac{pV_{\text{liq}}}{kT}\sqrt{\frac{2}{\pi m}}\sqrt{\sigma(r^*) - r^*\left(\frac{\partial\sigma}{\partial r}\right)_{r^*}}\, e^{-\frac{8\pi}{kT}\left(\int_0^{r^*} r\sigma dr - \frac{1}{2}r^{*2}\right)} \tag{3-20}$$

Both of these corrections lead to much higher nuclei formation rates at a given supersaturated state. Head's predicted results agree quite well

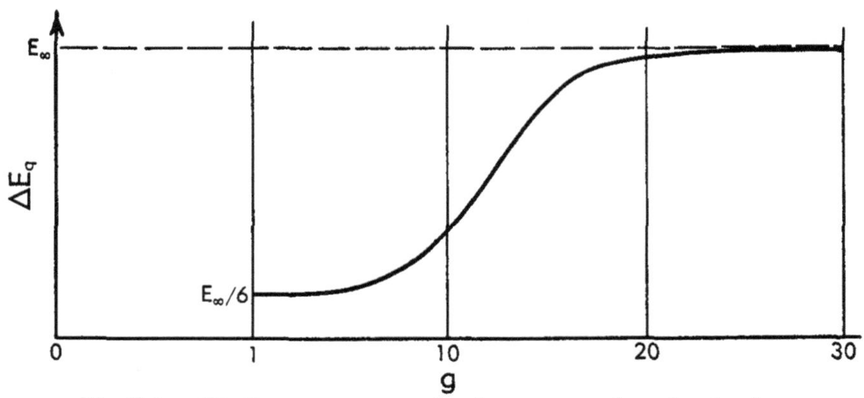

Fig. F,3c. Binding energy per molecule versus number of molecules.

with his experimental results on water vapor. A typical comparison of the nuclei formation rate curves with and without the correction can be taken from his results showing the line for a value of about 10^{16} nuclei per cubic centimeter per second for the corrected theory, coinciding with the line for one nucleus per cubic centimeter per second for the uncorrected theory.

Bogdonoff and Lees [23] have approached this problem somewhat differently. They have examined data from Taylor, Eyring, and Sherman [28] on binding energies of crystals of sodium atoms and some calculation on nitrogen drop binding energies by Reed [29]; they conclude that the binding energy per molecule ΔE_g stays almost constant up to a reasonable number of molecules in the drop, say 10, and then increases as shown schematically in Fig. F,3c. For the energy of formation of a drop, $\Delta\Phi$, one uses

$$\Delta\Phi = -gkT \ln \frac{p}{p_{s_\infty}} + (gE_\infty - E_g) \tag{3-21}$$

where again the first term is a volume energy equal to $g(\phi_g - \phi_{\text{liq}})$ and the second term is a "surface" energy. This can be compared to the previous expression as calculated before,

$$\Delta\Phi = -gkT \ln \frac{p}{p_{s\infty}} + 4\pi r^2 \sigma \tag{3-22}$$

Using the estimated curve of E_g in Fig. F,3c one can plot schematically the energy of formation of the drop. This is plotted in Fig. F,3d and compared to $\Delta\Phi$ as calculated using a constant value of surface tension. There are two things to note. First, the maximum of $\Delta\Phi$ occurs at a higher value of g; thus the size of the critical drop is larger for given conditions of state. Second, the value of the maximum $\Delta\Phi$ is larger using Lees' curve. The energy of formation of the critical size drops is larger, thus the rate of formation of critical size drops is smaller. Thus Lees comes to a conclusion which is opposite from that of Head and Stever. This difference of opinion

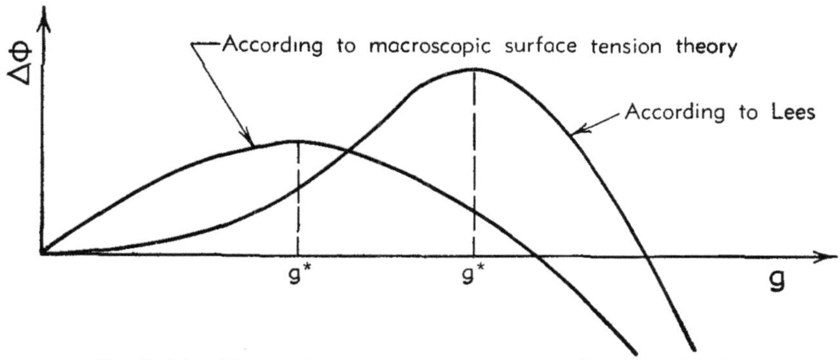

Fig. F,3d. Energy drop formation versus number of molecules.

can be settled finally only when the energy of formation of small drops is accurately calculated or measured. The approximate calculations by Reed [*29*] is a major step in this direction. However, from his results on nitrogen, it is still not possible to determine the nature of the correction.

Condensation of vapor onto small drops. The rate at which a supersaturated vapor condenses depends not only on the rate of formation of critical size drops but also on the rate of condensation onto these stable drops. The growth of such small drops depends upon the molecular transport of molecules onto the surface of the drop, and the conduction away from the drop of the latent heat of the condensing molecules. If the latent heat of vaporization were not conducted away, the drop would reach an equilibrium state because its temperature would become higher than that of the surrounding vapor and would cease to grow. For drops that are small compared to the mean free path of the vapor molecules the processes which govern the condensation are molecular, involving the condensation of a fraction of the impinging molecules and the evaporation of other molecules which conduct the latent heat away; these latter molecules leave with higher energy on the average, but in fewer numbers, than the impinging molecules. If, on the other hand, the drop is large compared to the mean free path of the vapor, the transport of the latent heat away from the drop will depend upon gaseous heat conduction through a well-defined temperature gradient in the gas around the drop.

First the case of a very small drop is considered in the fashion suggested by Oswatitsch [30]. Let the drop be of radius r, temperature T_{liq}, and saturation pressure $(p_{s\infty})_{liq}$. The vapor surrounding this drop is at a pressure and temperature p and T. From textbooks on kinetic theory the heat transfer per unit time is given as

$$Q = \sqrt{\frac{2}{\pi}}\, p \sqrt{\frac{k}{mT}}\,(T_{liq} - T)a \tag{3-23}$$

where a is an accommodation coefficient which varies from 0 to 1, depending on the vapor and the liquid surface. We can rewrite this equation

$$Q = Cp\sqrt{\frac{k}{mT}}\,(T_{liq} - T) \tag{3-24}$$

where C is a constant usually of the order of 1.

Let f be the fraction of the molecules which, impinging on a surface, condense; then the amount of water condensing on a unit surface area per unit time is again found from kinetic theory to be $fp\sqrt{kT/2\pi m}$. If h_{lg} is the latent heat, the heat transported per unit time per unit area is $h_{lg}fp\sqrt{kT/2\pi m}$. We can equate the heat arriving due to condensation and leaving due to molecular transport to be

$$h_{lg}fp\sqrt{\frac{kT}{2\pi m}} = Cp\sqrt{\frac{k}{mT}}\,(T_{liq} - T) \tag{3-25}$$

or in other words

$$\frac{fh_{lg}T}{\sqrt{2\pi}} = C(T_{liq} - T)$$

The increase in mass M of the drop is

$$\frac{dM}{dt} = \frac{1}{h_{lg}}\,4\pi r^2 Cp\sqrt{\frac{k}{mT}}\,(T_{liq} - T) \tag{3-26}$$

since the amount of heat arriving equals the amount conducted away.

$$M = \frac{4\pi r^3}{3}\rho, \qquad dM = 4\pi r^2 \rho\, dr$$

$$\frac{dr}{dt} = \frac{1}{h_{lg}}\frac{Cp}{\rho}\sqrt{\frac{k}{mT}}\,(T_{liq} - T) \tag{3-27}$$

Integrating

$$r = \frac{1}{h_{lg}} \frac{Cp}{\rho} \sqrt{\frac{k}{mT}} (T_{liq} - T)t + r_0 \tag{3-28}$$

Hence a linear law results for the radius as a function of time.

On the other hand, if we treat the drop as a large one, so that the normal gaseous heat conduction determines the heat removal, the droplet growth is, according to Oswatitsch [*30*],

$$r = \sqrt{\frac{2k(T_{liq} - T)}{h_{lg}\rho_{liq}} t} + r_0 \tag{3-29}$$

Thus the formula (Eq. 3-28) derived for the growth of the drops, assuming molecular processes to be the governing ones, yields a higher rate than the one derived assuming that normal gaseous heat conduction is the governing process. For water vapor in air, and for steam, Oswatitsch uses typical figures of the supersaturation and concludes that the process changes nature at a radius of about $r = 10^{-4}$ cm. This value seems to be large when one considers that the length of the mean free path is less than this by a factor of at least 10.

For the initial part of the growth of the spontaneously formed drops the radius can be calculated on the basis of Eq. 3-28, derived assuming molecular processes to be governing.

The role of nuclei of foreign materials. It is generally agreed that atmospheric condensation which occurs in the formation of fog, clouds, or precipitation takes place on nuclei of foreign particles or "dust" particles at slight degrees of supersaturation. The number of these particles per cubic centimeter has been variously estimated from 10^3 to 10^6 in air. Stodola [*2*] has compared the slow and rapid expansion of steam, showing that for the slow case (0.05 sec duration) there is practically no supersaturation. He concluded that the foreign nuclei act as condensation nuclei in the slow case, but that they do not affect appreciably the condensation which takes place very rapidly at high degrees of supersaturation.

It has been pointed out previously that the amount of condensation per unit length of nozzle can become fairly large, resulting in condensation shocks. That these are not caused by condensation onto foreign particles can be shown by a calculation of the amount of water vapor which condenses on these in a given unit of distance. Taking an example from Oswatitsch [*30*], consider 10^5 foreign nuclei per cubic centimeter traveling in a gas for a distance of 10 centimeters where there is a supercooling of 30°C. This 10 centimeters will be traversed in 3×10^{-4} seconds if the velocity is 3.3×10^4 centimeters per second, a slightly supersonic velocity for the lower temperatures involved in getting 30° supercooling. For this case the formula for drop growth, using the faster growth formula, shows a radius of growth of about 3×10^{-5} centimeters in this distance. The amount of water condensing then becomes

$$\frac{4\pi r^3}{3} \rho\eta = 10^{-8} \qquad \text{g/cm}^3$$

The heat released by this amount of water condensing is very slight and would not appreciably affect the stream quantities. It is concluded therefore that the number of nuclei must be much greater, too great in fact to be accounted for by foreign particles. We then fall back on the spontaneous formation of drops from the vapor as described earlier, and whose rate of formation is given by Eq. 3-14. In many instances the state of affairs may be better represented by condensation on both foreign nuclei and spontaneously formed nuclei, with different growth rates for each. It is possible that the above calculation is in error because the condensation rate on foreign nuclei may be governed by quite different laws from the rates on spontaneously formed nuclei.

There is one high speed flow condensation phenomenon in which condensation on nuclei of foreign material plays the dominant role. This is the case where there is a mixture of two vapors, one of which constitutes a minor component which is present in only small amounts. If the saturation conditions of the minor component are reached at higher temperature and pressure, hence earlier in an expansion, than those of the major component, a large amount of nucleation of the minor component may take place. This nucleation may not condense sufficient vapor to affect the stream properties to a measurable extent, yet the small drops of the minor component may act as nuclei of condensation for the major component, once the stream conditions have reached the saturation, or better, slightly supersaturated conditions of the major component. Thus condensation of the major component will take place at only slight degrees of supersaturation. An example of this phenomenon is discussed later.

F,4. Condensation in Flowing Vapor. Having developed the theories of the kinetics of condensation of a vapor in a supersaturated state, it is now possible to apply these theories to the condensation which takes place in a high speed gas stream. There are three separate problems to be treated here. First the effects of heat addition to a one-dimensional stream are considered. Next, the equations governing the condensation, that is the nuclei formation rate and drop growth rate equation, are incorporated into the equations which govern flow. Finally, a thermodynamic equilibrium theory of condensation is developed.

The effects of heat addition in compressible flow. The most important effect of condensation on the flowing vapor is due to the release of the latent heat of the condensing portions of the vapor. The change in density of the vapor due to removal of a portion in the condensed phase is usually a relatively minor effect. An analysis of the flow before and after the addition of a given amount of heat for a one-dimensional flow is easily presented and is quite illuminating concerning the effect of condensation on flow. This analysis will be carried out assuming that friction, and changes in cross-sectional area, as well as the small changes in mass flow of the vapor due to a small fraction of vapor condensing can be neglected. For the case of supersonic flow in which the duct cross section generally varies, this simple analysis could be considered to hold only for the abrupt condensation shock. As pointed out, generally these shocks are somewhat extended. However, one-dimensional analysis gives approximate information concerning the effects.

This so-called diabatic flow has been developed by numerous authors including Hawthorne and Cohen [*31*], Heybey [*32*], and Hicks [*33*]. Lukasiewicz [*34*] has a treatment of this subject applied to condensation, which will be followed here. The theoretical treatment is the same as that for an adiabatic normal shock except for an additional term in the energy equation giving the heat input to the gas. Using subscripts $_1$ and $_2$ to denote the vapor before and after the heat is added (or removed), and superscript ^{0} for stagnation conditions, the equations of state, continuity, momentum, and energy which define the problem are as follows:

State
$$\frac{p_1}{\rho_1 T_1} = \frac{p_2}{\rho_2 T_2} \tag{41-}$$

Continuity
$$\rho_1 q_1 = \rho_2 q_2 \tag{4-2}$$

Momentum
$$\rho_1 q_1^2 + p_1 = \rho_2 q_2^2 + p_2 \tag{4-3}$$

Energy
$$Q + \frac{\gamma}{\gamma - 1}\frac{p_1}{\rho_1} + \frac{q_1^2}{2} = \frac{\gamma}{\gamma - 1}\frac{p_2}{\rho_2} + \frac{q_2^2}{2} \tag{4-4}$$

Here Q is the heat added to the vapor (in energy units), p, ρ, T, q are the pressure, density, temperature, and velocity respectively, and γ is the ratio of specific heats. The stagnation temperatures for state 1 and state 2 are T_1^0 and T_2^0 where

$$c_p T_1^0 = c_p T_1 + \frac{q_1^2}{2} \tag{4-5}$$

and

$$c_p T_2^0 = c_p T_2 + \frac{q_2^2}{2} = c_p T_1^0 + Q \tag{4-6}$$

Let

$$Q' = \frac{Q}{c_p T_1^0} = \frac{T_2^0}{T_1^0} - 1 \tag{4-7}$$

The following solutions result for the pressure, temperature, density, and Mach number of the final flow in terms of the initial flow

$$\frac{p_2}{p_1} = \frac{\gamma}{\gamma + 1}(M_1^2 - 1)[1 \pm \sqrt{1 - A}] + 1 \tag{4-8}$$

$$\frac{\rho_1}{\rho_2} = \frac{q_2}{q_1} = 1 - \frac{1}{\gamma + 1}\frac{M_1^2 - 1}{M_1^2}[1 \pm \sqrt{1 - A}] \tag{4-9}$$

$$\frac{T_2}{T_1} = \frac{p_2}{p_1}\frac{\rho_1}{\rho_2} \tag{4-10}$$

$$M_2 = \left[(1 + Q')\left(\frac{1}{M_1^2} + \frac{\gamma - 1}{2}\right)\left(\frac{q_1}{q_2}\right)^2 - \frac{\gamma - 1}{2}\right]^{-\frac{1}{2}} \tag{4-11}$$

where

$$A = \frac{M_1^2}{(M_1^2 - 1)^2}\left(M_1^2 + \frac{2}{\gamma - 1}\right)(\gamma^2 - 1)Q' \tag{4-12}$$

Fig. F,4a, from Hawthorne and Cohen [*31*], shows graphically the effects of heat additions in compressible flows. The equation relating the Mach numbers before and after the heat addition, M_1 and M_2, is plotted

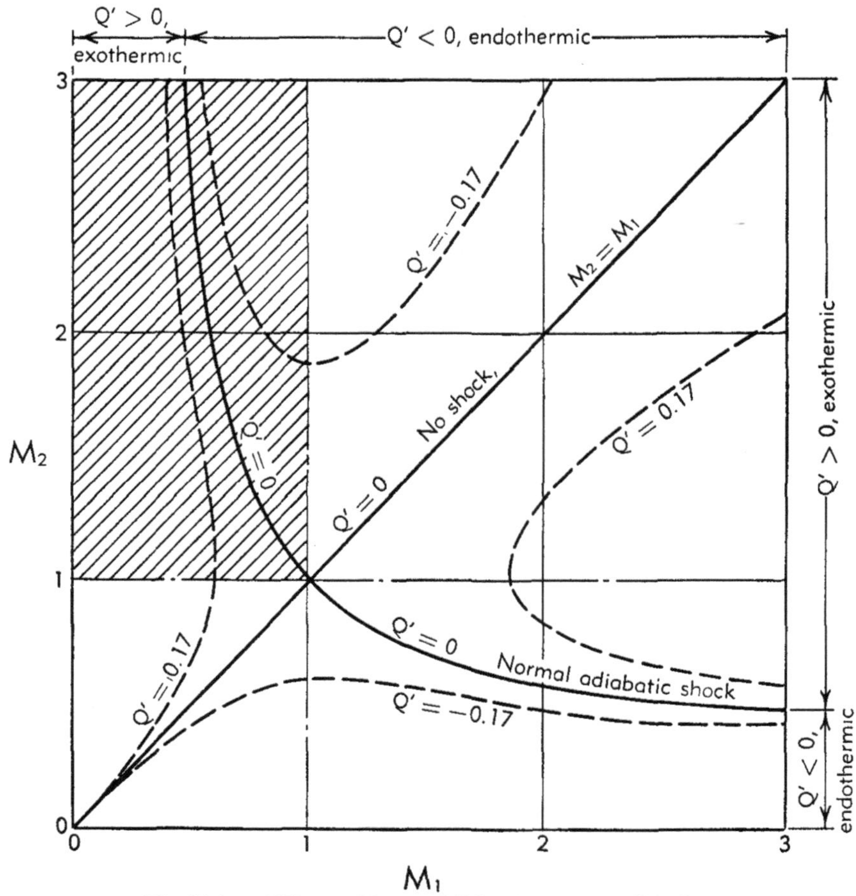

Fig. F,4a. Effects of heat addition in compressible flow

for three values of the quantity Q', which is a measure of the heat added. These values are 0, 0.17, and -0.17. The position of the solutions for other values of Q' is obvious.

The region of subsonic initial Mach numbers ($M_1 < 1$) and supersonic final Mach numbers ($M_2 > 1$) is excluded on physical grounds because it is a region of entropy decrease. The line $Q' = 0$ and $M_2 = M_1$ is shown to represent unchanged flow; another line $Q' = 0$ is shown representing the adiabatic normal shock in which the Mach number is supersonic before the shock ($M_1 > 1$) and subsonic after the shock ($M_2 < 1$).

There are two regions of positive heat additions which correspond to the case of condensation. The first of these is in the subsonic region. It is seen that condensation with its attendant release of latent heat to the stream results in an increase in the Mach number in this subsonic region. Furthermore, for every subsonic Mach number there is a maximum heat release, or amount of condensation, which will just make the Mach number sonic ($M_2 = 1$). If the stream conditions are such as to result in more condensation than this, it is evident that the flow cannot be steady, and, in fact, the flow will "choke." This flow becomes particularly sensitive when the initial subsonic Mach number is nearly sonic, for there no heat can be added, hence condensation cannot take place in a steady sonic flow.

It can also be shown from the above relationships that for the subsonic case under discussion, the heat addition due to condensation results in a decrease in the static and stagnation pressures.

For condensation occurring in the supersonic region, that is, the region for which $M_1 > 1$ and $Q' > 0$, the situation is more complex. Here again, for a given initial Mach number, there is a maximum amount of heat addition due to condensation which can take place. This maximum is given by

$$Q'_{\max} = \frac{Q_{\max}}{c_p T_1^0} = \frac{(M_1^2 - 1)^2}{M_1^2\left(M_1^2 + \dfrac{2}{\gamma - 1}\right)(\gamma^2 - 1)} \tag{4-13}$$

and it occurs when $M_2 = 1$. If, for a given value of M_1, the heat added is less than this maximum there are two possible solutions, one with the final Mach number subsonic and one with it supersonic. There is no solution for the case of $Q' > Q'_{\max}$ for a given M_1.

It can be shown that there is always a stagnation pressure decrease for supersonic condensation. Also, with regard to the static pressure change, there is always an increase, one which is greater for the case of change to the subsonic flow than for the change to supersonic flow.

Without analyzing the time duration or the mechanism of condensation, it is seen that certain conclusions can be drawn about the effects of this condensation on supersonic and subsonic flow.

Application of nuclei formation rate and drop growth rate theory to high speed flow. The application of nuclei formation rate and drop growth theory to high speed flow in convergent-divergent nozzles has been made by Oswatitsch [30]. He developed the basic equations for the compressible flow of a mixture of an inert gas, a condensible vapor, and drops of its liquid phase in a channel of varying area. He included in these equations terms for the rate of formation of liquid drops from the supersaturated vapor and for the rate of growth of the existing and newly formed drops. For the rate of formation of drops, Oswatitsch used the work of Becker and Döring, et al., developed in Art. 3. He used his own work for the drop growth expressions. The resulting equations determining the flow cannot be solved analytically but permit numerical stepwise calculations. The results of such a calculation agree with the pressure distribution measurements for steam condensation, which were made by other workers. This is described later. An account of Oswatitsch's methods and results is given below, although it is simplified to cover only the case of a vapor and drops of its liquid phase in one-dimensional flow.

The physical problem to be considered is the flow of a vapor in the supersaturated state, together with some of its liquid phase in the form of small drops in a channel of slowly varying cross section $A(x)$ where x is the distance along the channel. For the sake of simplicity of analysis the liquid is considered to occupy a stream tube of area A_{liq} and the vapor an area A_{g}. Thus

$$A = A_{\text{liq}} + A_{\text{g}} \tag{4-14}$$

Quantities referring to the gas and liquid are henceforth denoted by subscripts $_{\text{g}}$ and $_{\text{liq}}$.

In general both the temperature and the stream velocity of the liquid and gas phases are different, and must be denoted by separate terms.

The mass flow rate M is given by

$$M = M_{\text{g}} + M_{\text{liq}} \tag{4-15}$$

where

$$\begin{aligned} M_{\text{g}} &= A_{\text{g}}\rho_{\text{g}}q_{\text{g}} \\ M_{\text{liq}} &= A_{\text{liq}}\rho_{\text{liq}}q_{\text{liq}} \end{aligned} \tag{4-16}$$

where ρ and q are density and velocity of flow respectively. The assumption is made above that the velocity of all the liquid drops is the same.

The *condition of continuity* is obtained readily as

$$M = A_{\text{g}}\rho_{\text{g}}q_{\text{g}} + A_{\text{liq}}\rho_{\text{liq}}q_{\text{liq}} \tag{4-17}$$

It is not possible to assume that the number of drops is constant because the vapor is in a supersaturated state and, as shown in the last section, it is generating critical size drops at a rate of J drops per unit volume per unit time. Let n denote the number of drops per unit volume. The number of drops passing a given point along the nozzle per unit time is $nq_{liq}A$. The increase in this number per unit length of the nozzle is

$$\frac{d}{dx}(nq_{liq}A) = JA \tag{4-18}$$

$$q_{liq}A\frac{dn}{dx} + n\frac{d}{dx}(q_{liq}A) = JA \tag{4-19}$$

$$\frac{dn}{dx} = \frac{J}{q_{liq}} - \frac{n}{q_{liq}A}\frac{d}{dx}(q_{liq}A) \tag{4-20}$$

Thus there are two causes of the change in number of drops per unit volume per unit path; new drops form and the volume of the gas containing a given number of drops changes.

An *equation of motion* must be derived for both the vapor and for the liquid. In addition to the usual pressure forces acting on the vapor, there is the drag force of the drops. Let D be the drag of a drop. Then the equation of motion for the vapor is

$$A_g\rho_g dx\frac{dq_g}{dx}q_g = -A_g dp - A dx n D \tag{4-21}$$

This states that the acceleration of the vapor mass $A_g\rho_g dx$ at point x is due to the sum of the reduction in the pressure forces $-A_g dp$ and the resistance of the $Adxn$ drops in this air mass.

The acceleration of the liquid mass in this element of length dx at point x is a result of three causes: a drop in pressure acts on the liquid; the vapor exerts a force on the drops; and vapor condenses onto the drops already formed and into newly formed drops, which may have a lower velocity than the vapor. With these three accelerating terms, the equation of motion for the liquid phase is

$$A_{liq}\rho_{liq}dx\frac{dq_{liq}}{dx}q_{liq} = -A_{liq}dp + AdxnD + (q_g - q_{liq})dM_{liq} \tag{4-22}$$

Rewriting these two equations with the aid of the continuity equation, the dynamic equations become

$$q_g\frac{dq_g}{dx} = -\frac{1}{\rho_g}\frac{dp}{dx} - \frac{n}{\rho_g}\frac{A}{A_g}D$$

$$M_{liq}q_{liq}\frac{dq_{liq}}{dx} = -M_{liq}\frac{1}{\rho_{liq}}\frac{dp}{dx} + \frac{q_{liq}}{q_g}(M - M_{liq})\frac{n}{\rho_g}\frac{A}{A_g}D$$
$$+ \rho_{liq}(q_g - q_{liq})\frac{dM_{liq}}{dx} \tag{4-23}$$

The *energy equation* is also required. One source of heat is the latent heat of vaporization h_{lg} released to the stream by the condensing vapor;

a quantity $h_{lg}dM_{liq}$ is released to the vapor in the element of stream dx in a unit time. Heat is also produced by the frictional drag of the drops; this can be calculated since it is equal to the energy given the drops per unit time or, for one drop, $D(q_g - q_{liq})$. Finally the vapor which condenses is slowed down upon condensing; the energy release to the stream in this action for the length dx is $\frac{1}{2}(q_g - q_{liq})^2 dM_{liq}$. Letting u be the internal energy per unit mass, the first law of thermodynamics becomes

$$A_g\rho_g dx\left(\frac{du_g}{dx} + p_g\frac{dv_g}{dx}\right) + A_{liq}\rho_{liq}q_{liq}\frac{du_{liq}}{dx}$$
$$= h_{lg}\frac{dM_{liq}}{dx} + ADq_g\left(1 - \frac{q_{liq}}{q_g}\right) + \frac{1}{2}(q_g - q_{liq})^2\frac{dM_{liq}}{dx} \quad (4\text{-}24)$$

or

$$M_g\left(\frac{du_g}{dx} + p_g\frac{dv_g}{dx}\right) + M_{liq}\frac{du_{liq}}{dx}$$
$$= h_{lg}\frac{dM_{liq}}{dx} + M_g\frac{A}{A_g}\frac{n}{\rho_g}D\left(1 - \frac{q_{liq}}{q_g}\right) + \frac{1}{2}(q_g - q_{liq})^2\frac{dM_{liq}}{dx} \quad (4\text{-}25)$$

The latent heat of vaporization is the difference in the enthalpy $u + pv$ of the vapor and liquid states or

$$h_{lg} = u_g + p_g v_g - u_{liq} - p_{liq}v_{liq} \quad (4\text{-}26)$$

Multiplying the first equation of motion by M_g and adding the two equations of motion to the energy equation and using

$$\frac{dM_{liq}}{dx} = -\frac{dM_g}{dx}$$

there results an energy equation

$$\frac{d}{dx}\left[M_g\frac{q_{liq}^2}{2} + M_{liq}\frac{q_{liq}^2}{2} + M_g(u_g + p_g v_g) + M_{liq}(u_{liq} + p_{liq}v_{liq})\right] = 0 \quad (4\text{-}27)$$

This energy equation appears in a form not containing any frictional quantities.

If the quantity dM_{liq}/dx were explicitly stated, the system of equations would be complete. To find this relation the expression for both the rate of formation of stable drops and for the rate of growth of these drops must be employed. It is now assumed that the vapor is reasonably free from foreign nuclei so the use of Eq. 3-14 for the critical size drop formation rate and Eq. 3-28 for the drop growth rate permits evaluation of the quantity dM_{liq}/dx. At any position x in the stream downstream from the point x_0 at which condensation begins (strictly speaking, the saturation line), there are drops of liquid of varying sizes depending on the distance upstream at which they were formed. Taking a particular value of x, say

$x = x'$, the contribution to the liquid phase at the position x due to the formation of drops in the region x' to $x' + dx'$ is $m(x', x)J(x')A(x')dx'$ where $m(x', x)$ is the mass of a drop formed at x' upon arrival at x and is determined by the drop growth rate (Eq. 3-28).

The total mass $M_{\text{liq}}(x)$ of the liquid phase passing the point x per unit time can be found by integration of this expression from the position $x' = x_0$ to $x' = x$. Thus

$$M_{\text{liq}}(x) = \int_{x'=x_0}^{x'=x} m(x', x)J(x')A(x')dx' \tag{4-28}$$

At the point x, the mass of a drop formed is zero, $m(x, x) = 0$; at the point x_0, the rate of formation is zero $J(x_0) = 0$. Differentiation of the integral leads to the quantity $dM_{\text{liq}}(x)/dx$.

$$\frac{dM_{\text{liq}}(x)}{dx} = \int_{x'=x_0}^{x'=x} \frac{dm(x', x)}{dx} J(x')A(x')dx' \tag{4-29}$$

This *equation of condensation* completes the set of equations describing the flow.

There are simplifications which can be made. Calculations show that the drops do not grow to very large size because of the short time they are present in the nozzle. Then it is possible to assume that they are moving with the speed of the vapor. Hence, $q_{\text{liq}} = q_g$. This also means that the drag of the drops is zero, $D = 0$.

Another simplification results from the assumption that A_g is equal to A, justified because the amount of the vapor which condenses is small; even if it were large, the difference in density between the vapor and liquid states is so great that this approximation is good. It is further assumed that the vapor follows the perfect gas equation of state. Then the continuity equation becomes, if it is also assumed that the drops have all grown to an equal size,

$$M = A\rho_g q_g + A_{\text{liq}}\rho_{\text{liq}}q_{\text{liq}} = A\rho_g q_g + Aq_{\text{liq}}n\frac{4\pi}{3}r^3\rho_{\text{liq}} \tag{4-30}$$

It can also be expressed

$$M = A\rho_g q_g + \int_{x'=x_0}^{x'=x} m(x', x)J(x')A(x')dx \tag{4-31}$$

Taking the logarithm of both sides and differentiating, the *equation of continuity* is given in differential form as

$$\frac{1}{A}\frac{dA}{dx} + \frac{1}{\rho_g}\frac{d\rho_g}{dx} + \frac{1}{q}\frac{dq}{dx} = \frac{1}{M_{\text{liq}} - M}\frac{dM_{\text{liq}}}{dx} \tag{4-32}$$

The *momentum equation* becomes simply

$$q_g\frac{dq_g}{dx} = -\frac{1}{\rho_{\text{liq}}}\frac{dp}{dx} \tag{4-33}$$

The *energy equation* becomes

$$\frac{d}{dx}\left[M\frac{q^2}{2} + M_g c_p T - M_{liq} h_{lg}\right] = 0 \tag{4-34}$$

Finally, the *equation of condensation* becomes

$$\frac{dM_{liq}}{dx} = \int_{x'=x_0}^{x'=x} \frac{dm(x', x)}{dx} J(x')A(x')dx' \tag{4-35}$$

These four equations govern the condensing flow of any vapor in a channel. It is not possible to integrate them analytically, but it is possible to carry out the computations numerically for any case where the nozzle characteristics and the properties of the vapor are known. The results of such a computation are discussed later.

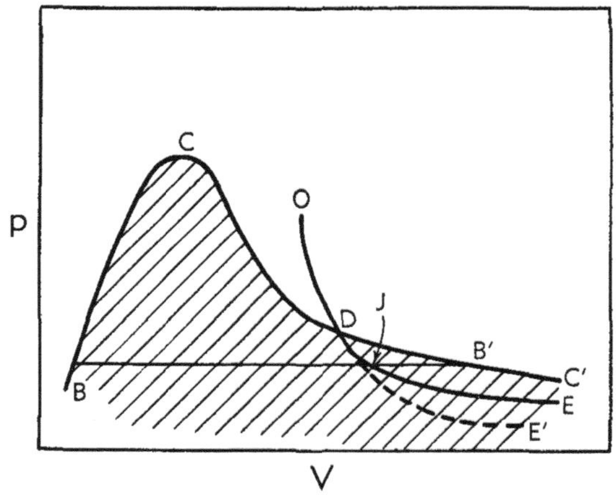

Fig. F,4b. Schematic pressure volume diagram.

Isentropic expansion of a saturated vapor and its condensed phase. The condensation of air in hypersonic wind tunnels takes place apparently at relatively low supersaturation ratios. This is probably condensation taking place onto nuclei of water and/or carbon dioxide, which exist in the air as minor components, and which condense well before saturation of nitrogen or oxygen is reached. The condensation of the water and/or carbon dioxide does not seriously affect the stream properties because of the small percentage in which they exist. However, a large number of nuclei are formed in the condensation of these minor components, which then act as nuclei for oxygen and nitrogen condensation at low degrees of supersaturation of these principal components.

Condensation close to the dew point suggested to Wegener, et al. [*35*] a somewhat simplified model in which the expansion process with condensation present is carried out reversibly. The expansion of the vapor follows an isentrope to the dew point, and from there follows the isentrope of a saturated vapor and its coexistent liquid phase. Fig. F,4b taken from Wegener [35] shows such an expansion line ODE, on a pressure volume diagram and compares it to a perfect gas isentrope ODE' that the vapor would have followed if supersaturation without condensation occurred. The slope of these isentropes gives the quantity $dp/d\rho$ from which the speed of sound a can be computed at any point in the expansion by

$$a^2 = \frac{dp}{d\rho} \tag{4-36}$$

This is seen to have an abrupt change at the dew point where the slope of the expansion line ODE changes abruptly.

If the point D where the expansion line meets the dew point line is known, it is possible to approximate the fraction f of the vapor which is condensed as follows. For the vapor the perfect gas law holds,

$$p = \rho_g \mathcal{R} T \tag{4-37}$$

The average density of the vapor and condensed phase is given by

$$\frac{1}{\rho} = \frac{1-f}{\rho_g} + \frac{f}{\rho_{liq}} \tag{4-38}$$

where ρ_{liq}, the density of the liquid phase, is much larger than ρ_g. Generally $1 - f$ is also much larger than f. Hence

$$\rho = \frac{\rho_g}{1-f} \tag{4-39}$$

and

$$p = \rho(1 - f)\mathcal{R}T \tag{4-40}$$

The temperature of the expansion is determined by the saturation line from the pressure alone.

$$T = T_{s\infty}(p) \tag{4-41}$$

The process is considered isentropic so that the entropy per unit mass s is a constant where

$$s = s_g - f\frac{h_{lg}}{T} \tag{4-42}$$

Here, h_{lg} is the latent heat and is dependent on the temperature. The entropy is known from the initial conditions of the expansion; $h_{lg}(T)$ and $T_{s\infty}(p)$ are known properties of the substance.

Applying this set of equations to a flow in a channel whose cross section is known as a function of the distance along the channel, it is seen that the measurement of the static pressure p along the nozzle is sufficient to determine the other quantities. First, the determinations of p fix T and also h_{lg}. Second, Eq. 4-42 and 4-40 determine ρ and f. Then the speed of sound (Eq. 4-36) can be calculated from a plot of the measured pressure versus the calculated density.

Integrating the equation of motion, a relation for the velocity of the flow at any point x in terms of the velocity $q(x_1)$ at a particular point x_1 and the calculated ρ and the measured p is given by

$$q(x) = \sqrt{[q(x_1)]^2 - 2\int_{x_1}^{x}\left(\frac{dp}{dx}\right)\frac{dx}{\rho}} \tag{4-43}$$

The velocity can then be computed. Finally, the Mach number of the flow can be computed.

$$M = \frac{q}{a}$$

Summarizing, if the assumption is made that a condensation process in a flowing vapor takes place isentropically so that the vapor is in equilibrium with its condensed phase at the dew point, the measurement of static pressure along a channel of known cross section is sufficient to determine the entire flow. This of course assumes that the properties of the vapor, the saturation line and latent heat, are known.

F,5. Experimental Techniques. There are several techniques which permit the detection and/or measurement of condensation in high speed flow. The detection of condensation fog can be made directly by light scattering experiments. The effect of the condensation on the static and dynamic pressure of the flowing gas can be measured. Changes in density, including the formation of condensation shocks, can be measured by schlieren and interferometric photography. The influence of condensation on the Mach number with the consequent change in the shock waves formed on wedges or other bodies in the stream can also be detected and measured by interferometry and schlieren photography.

Light scattering experiments. The detection of condensation fog by light scattering requires a very simple experimental setup. Plate F,5a from Becker [*36*] shows a setup of a typical light scattering experiment. A light source, in this case a mercury arc, and a lens are arranged to direct a nearly parallel light beam through the observation windows of a wind tunnel. Plate F,5b from the same reference shows typical results of a photograph taken at an angle to the axis of the light beam. If no condensation is present in the flow, there is light scattering from the windows only; if condensation is present, the drops in the gas between the two windows also scatter light. These photographs were taken in the 11-inch square hypersonic tunnel of the Langley Aeronautical Laboratory, in which the Mach number of the flow without condensation is about 6.9. The condensation detected is that of oxygen and/or nitrogen in the air. The two photographs were taken for two different stagnation temperatures, one 1160°R to insure that the oxygen and nitrogen were not saturated in the expansion to $M = 6.9$, and the other at 540°R where the oxygen and nitrogen definitely had been expanded into the saturation region.

This technique has been used by many investigators. Sometimes, if the experimental setup permits it, the beam of light is directed along the flow axis instead of across the flow. Such a technique permits photography of the position where the light scattering is first detectable in the flow. Stever and Rathbun [*27*] have used the light scattering technique with a very small beam, about $\frac{1}{8}$ in. in cross-sectional diameter, to detect the condensation of the principal components of air within a quarter of an inch of the throat in a two-dimensional hypersonic tunnel, indicating that time lags in the formation of condensation are not appreciable. Wegener and Lundquist [*37*] have used an extension of this technique to detect condensation of water vapor in a shock tube at extremely low values of temperature, thus showing that the concept of condensation cutoff mentioned earlier is probably not a valid one. They employed a sensitive photomultiplier tube to detect light scattering beyond the sensitivity of the eye or normal photography.

To make light scattering experiments more quantitative is quite difficult. A discussion of the light scattering technique is given in more detail in IX,G. The measurement of light intensity and wavelength shift as a function of the scattering angle will yield some information about the size of the drops and the number (exact information about one if the other is known). So far, this has not been used extensively, although Stodola [*2*] and Yellott and Holland [*4*] have tried to estimate particle size from a change in color of the light and from the polarization of scattered light. More recently, Durbin [*38*] has succeeded in applying the technique quantitatively to condensation in hypersonic tunnels. His results are quoted later.

Dynamic and static pressure measurement. The measurement of dynamic and static pressures in flow in which there is condensation is very illuminating. As an example, Fig. F,5a from Wegener, et al. [*35*] shows the effect, as the stagnation temperature is changed at constant stagnation pressure, on the static pressure on the wall in a hypersonic tunnel of nominal Mach number 7.6. For reference, dashed lines are plotted for the nozzle, illustrating the potential flow pressure ratios and Mach numbers as a function of distance along the nozzle for two conditions, one for a cold throat at the beginning of a run, and the other for a hot throat later during the run. The solid circles were measured when the air was preheated to a stagnation temperature of 290°C so that the air was unsaturated in the expansion. It is interesting to note the deviation of these measured points from the potential flow line. The open circles are experimental points for a case in which there was no preheating (10°C). It is noted that the pressure ratios do not exactly coincide with the preheated runs even at points in the nozzle before the air reached saturation. This is probably due to a slight difference in the nozzle throat contours due to the thermal expansion. After the air reaches saturation, the pressure measurements deviate from the unsaturated expansion measurements more rapidly. It is seen that a final Mach number of about 7.3 is reached for the preheated air. The Mach number of the nonheated air is unassignable. One point of interest is that there is no abrupt static pressure rise along the curves with condensation present. It is somewhat difficult to pick the point at which condensation begins to deviate this flow.

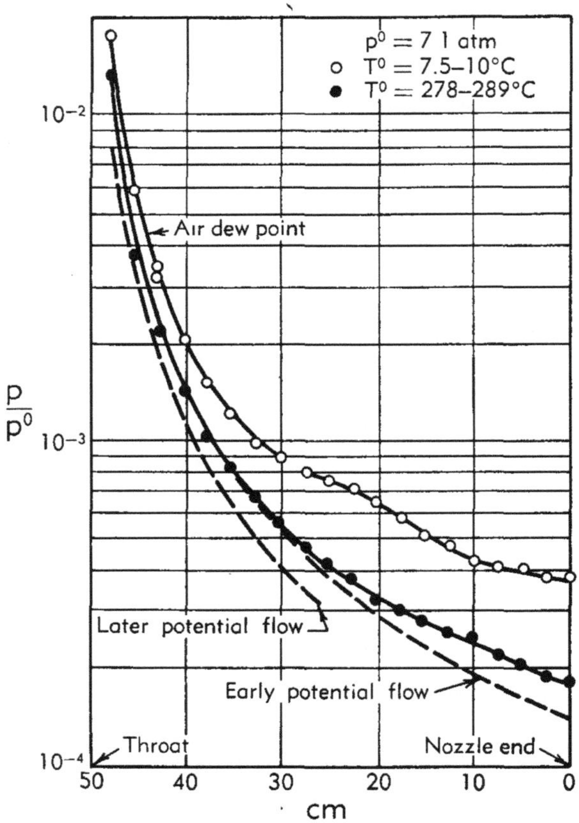

FIG. F,5a. Static pressure measurements in a hypersonic wind tunnel with condensation present.

For the case of water vapor in air, Fig. F,5b taken from Head [*6*] shows typical results of the pressure ratio p/p^0 plotted against distance along the nozzle. For reference, a dashed curve showing the isentropic pressure ratio as a function of distance along the nozzle is given in the figure. Three cases, for relative humidities of 43.8 per cent, 62.2 per cent, and 86.5 per cent, are plotted, and in each case the stagnation pressures are somewhat different. It is seen that the static pressure ratio begins to deviate from the isentropic curve, the deviation increasing with relative humidity. Somewhat downstream from the saturation point, there is a rather abrupt rise in the static pressure ratio curve, and at this point one normally finds by schlieren measurements a condensation shock. It is interesting to note that the position of the condensation shock or the

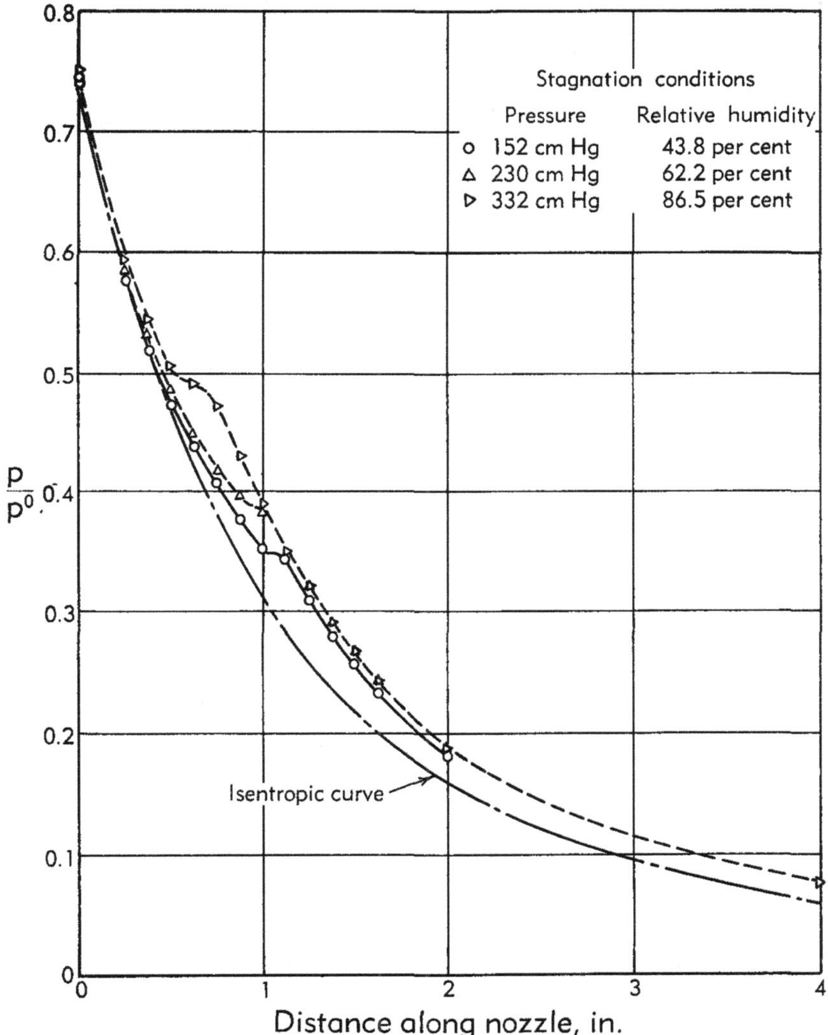

Fig. F,5b. Static pressure measurements for water vapor.

abrupt rise in pressure ratio shifts to lower values of the pressure ratio as the relative humidity is decreased.

Another type of experiment is one involving the ratio of static pressure to stagnation pressure in which this ratio is measured at a particular station in the nozzle for different conditions of stagnation temperature. If no condensation were present this pressure ratio would remain constant at a given point in the nozzle, regardless of the value of the stagna-

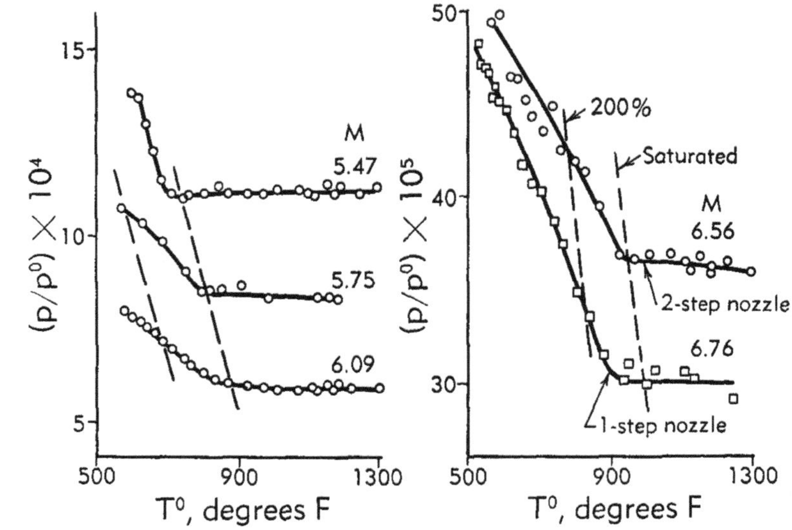

Fig. F,5c. Effect of condensation on static-to-stagnation pressure ratio versus stagnation temperature.

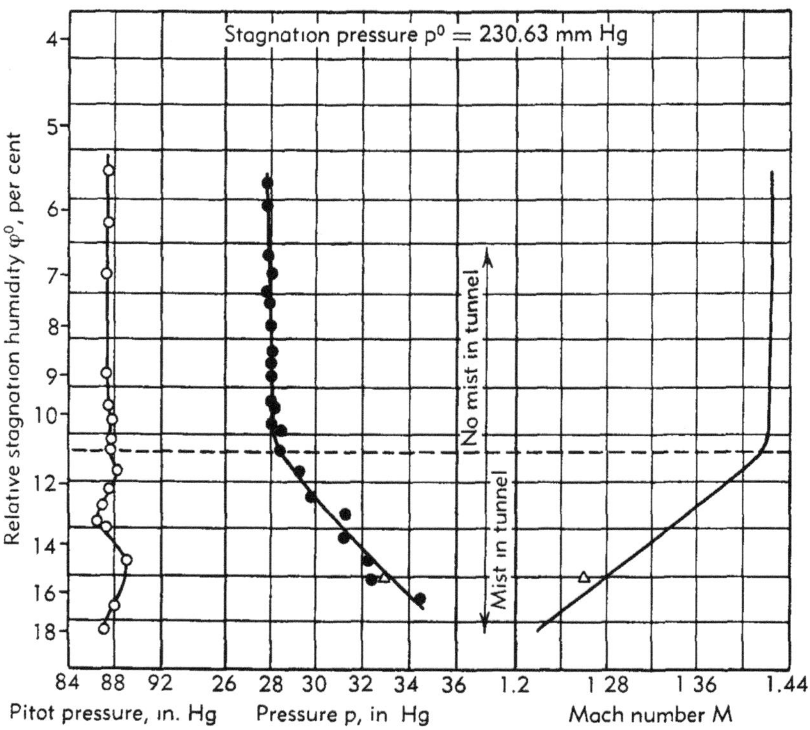

Fig. F,5d. Total head pressure, static pressure, and Mach number variation with relative stagnation humidity.

tion temperature. Fig. F,5c taken from the results recorded by Becker [*36*] shows such measurements for five stations in a hypersonic wind tunnel of nominal Mach number of 7. Stations at which the condensation-free Mach number are 5.47, 5.57, 6.09, 6.56, and 6.76 are shown. The pressure

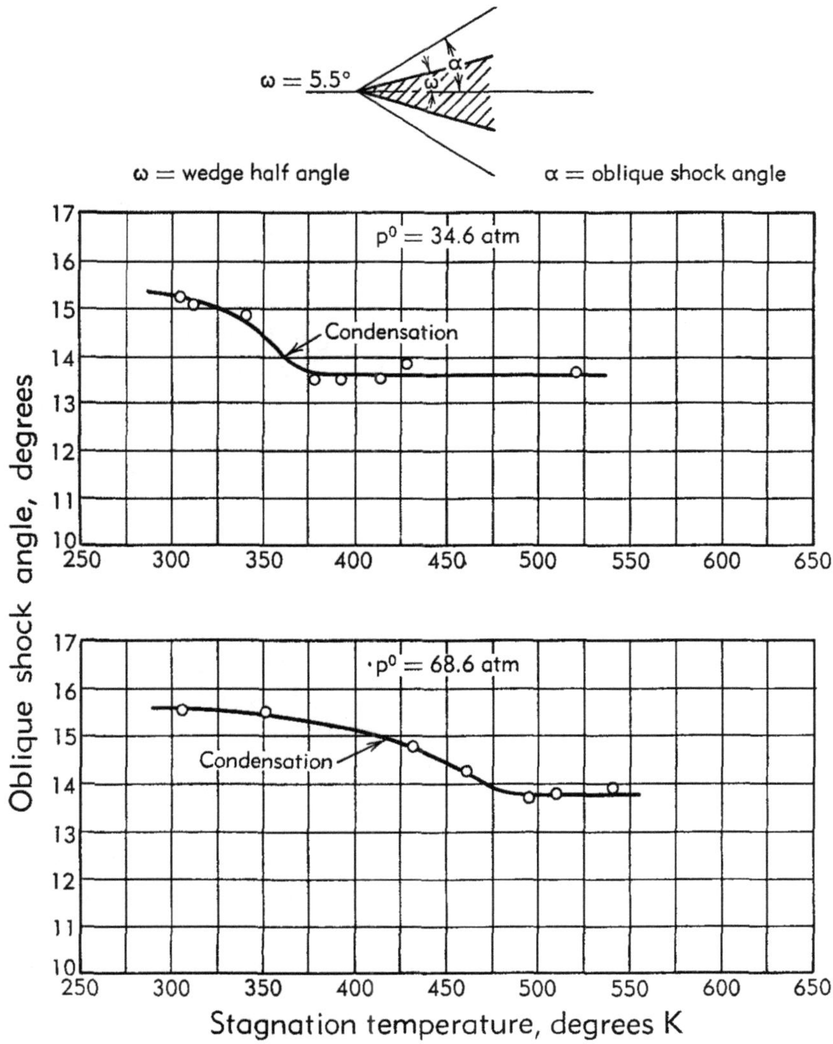

Fig. F,5e. Effect of stagnation temperatures on oblique shock angle from 11° wedge in hypersonic tunnel.

ratio begins to increase shortly after the saturation line for air is reached by lowering the stagnation temperature. The break in the curves indicating condensation occurs between the air saturation line and the 200 per cent saturated line.

A similar experiment, reported by Lukasiewicz [*34*], is shown in Fig. F,5d in which the stagnation pressure of humid air in several runs was kept at a constant value and the stagnation temperature was varied. Here the total head pressure, the static pressure, and also the Mach number at which mist was observed are plotted versus stagnation temperature. The regions of these curves where mist was observed in the tunnel are indicated. This gives a very good indication of how these three quantities are affected by the condensation occurring in the flow.

Oblique shock wave schlieren photographs. The change of Mach number due to condensation can be utilized to detect the point at which condensation occurs. Fig. F,5e from Stever and Rathbun [*27*] shows the shock wave angle, measured on a given wedge from schlieren photographs taken of that shock wave for different values of the stagnation temperature in a hypersonic wind tunnel with a nominal Mach number of 7. The shock wave angle remains constant for high values of the stagnation temperature where condensation does not exist. As the stagnation temperature is lowered it is found that the shock angle begins to increase, indicating a decrease in the Mach number of the flow. The point where the curve breaks coincides with the point at which condensation is detected by other means.

Schlieren and interferometric photography. The quantitative use of schlieren and interferometer photography to detect the abrupt changes in flow caused by sudden condensation is not extensive. It does illustrate, however, the condensation shock phenomenon. Plate F,1 is a schlieren photograph of the condensation shock in a nozzle.

F,6. Experimental Results. As pointed out before, the importance of steam nozzles, subsonic and supersonic wind tunnels, and also hypersonic wind tunnels, has resulted in a concentration of experimental and theoretical efforts on the problems of condensation of water vapor, water vapor in air, and nitrogen and oxygen in air. In the previous section, in which the experimental technique used to detect and measure condensation are described, some experimental results on these substances are given. From that data it is clear that, in some cases, highly supersaturated states are reached before sudden condensation takes place. In other cases, however, the condensation begins gradually at only slight degrees of supersaturation and continues as the expansion continues; in this type of condensation, found in hypersonic wind tunnels for example, the vapor does not reach a point at which the sudden rise in pressure, characteristic of a shock wave, occurs.

There are many plots on which experimental results can be portrayed. For example, the experimental points of pressure and temperature, at which condensation is detected for an isentropic expansion, can be plotted on a pressure-temperature diagram. It is found that the locus of experimental points at which condensation is detected is roughly parallel to the saturation line. Another common method of showing experimental results is to plot the saturation ratio versus the temperature. Here the saturation ratio is the ratio of the partial pressure of the vapor and the normal saturation pressure corresponding to the temperature.

If one compares the experimental results obtained by different experimenters using different apparatus, a rather wide scatter of experimental points is noted. Head [*6*] has compared a large number of experimental results from different sources for water vapor in air and for water vapor in a steam nozzle. They show a marked scatter for many reasons. In the first place, the experimental method used to detect condensation differs from experiment to experiment. There is rough agreement among the methods of detecting condensation, but all methods have a certain degree of uncertainty. Apparently there is not precise agreement among the methods.

But over and beyond the uncertainty or the scatter in the experimental points as detected by different methods, there is another systematic variation of results which was pointed out by Oswatitsch [*30*] and others and experimentally investigated by Head [*6*] and especially by Wegener and Smelt [*22*]. This systematic variation in the results is due to the difference in the temperature gradient in the expansion used in the different experiments. The kinetic effects of condensation are such that the rapidity with which the temperature varies with distance along the nozzle is an important factor in delaying the onset of condensation. This effect is shown clearly in Fig. F,6a, taken from Wegener and Smelt [*22*]. In this figure they have plotted many results taken in their own apparatus together with results of other workers. Although the experimental points spread rather widely, probably due to the fact that the initial relative humidities vary markedly, there is a definite trend toward an increase in the amount of supercooling measured by ΔT with an increase in the temperature gradient in the nozzle in which the measurements were made. The nozzle used by Wegener and Smelt permitted a rather wide variation in this temperature gradient. It is seen, as the temperature gradient varies from about 10°/cm to 90°/cm, that there is an increase in the supercooling from about 50° to 90°. Such changes in the values of supercooling correspond to widely different values of the supercooling ratio.

Without taking account of the variation in the temperature gradient in the expansion, Lukasiewicz [*34*] has plotted the experimental results of Binnie and Woods [*39*], Yellott [*3*], Oswatitsch [*40*], and the National Physical Laboratories, measurements in the 11 × 11-in. supersonic tunnel, showing the variation in the supercooling versus the relative humidity of the stagnation conditions for both water vapor and water vapor in air. These roughly agree as seen in Fig. F,6b.

For the case of condensation of water vapor in air, many experimenters have found that the point at which the spontaneous collapse of the supersaturated state occurs, or the point at which the condensation shock is obtained in a given apparatus, depends upon the initial relative humidity, that is, the condensation shock point moves downstream as the initial relative humidity is decreased. The absolute humidity, in a series of experiments by Head [6], was kept constant and the relative humidity was varied by altering the stagnation pressure. The results of a series of runs taken this way are shown in Fig. F,5b. The isentropic curve calculated from the area ratio for the nozzle is plotted. When dry nitrogen

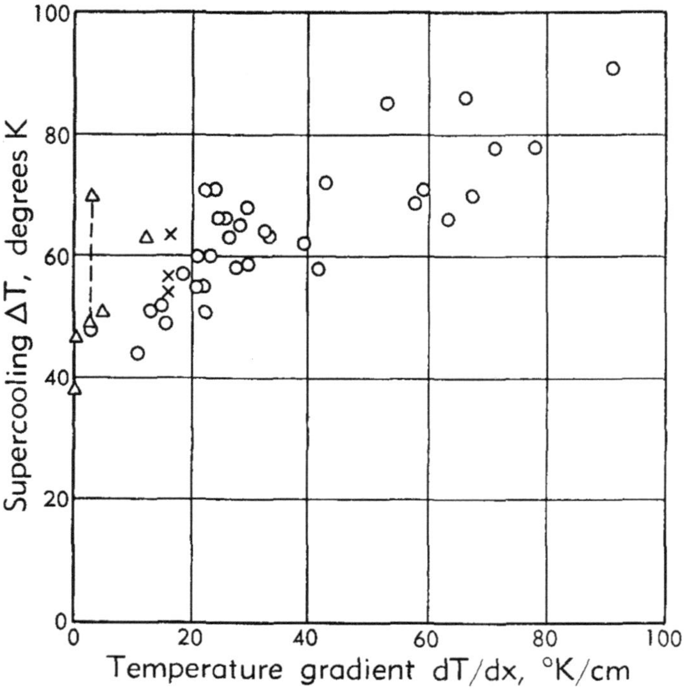

Fig. F,6a. Supercooling ΔT versus temperature gradient dT/dx for all initial humidities.

was used in the nozzle, the expansion followed closely the calculated isentropic curve. There are several things to note in the experimental results on condensation for the three runs which are listed in this figure. First of all, the deviation of the pressure curves begins well before the point at which the spontaneous collapse of the supersaturated state occurs. Secondly, the amount of deviation depends upon the stagnation relative humidity, increasing with increasing relative humidity. It is generally found that the intensity of the spontaneous collapse or the pressure jump decreases with decreasing relative humidity. However, the total amount of water vapor which is condensed, including that which condenses before the condensation shock, in the condensation shock, and after the condensation shock, is almost constant if the initial absolute humidity is fixed.

This can be seen in Fig. F,5b where all of the curves tend to converge to a single curve after the condensation shock.

In Head's experiment on water vapor in air he was unable to detect any condensation of the water vapor for those cases in which the expansion took the water vapor to a temperature below approximately 150°K. This result seemed to agree with a result obtained in cloud chamber experiments by Cwilong [*41*], which indicated that there was a minimum temperature below which no condensation would start. However, a recent

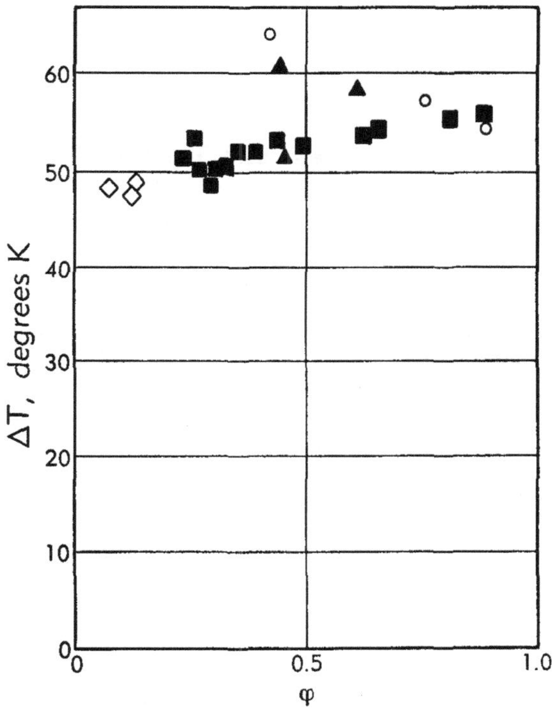

Fig. F,6b. Supercooling of steam and humid air versus relative humidity.

experiment by Wegener and Lundquist [*37*], in which more sensitive instrumentation was used, namely, a photomultiplier tube to detect the scattering of light by the condensation fog, showed that condensation could be detected down to a temperature of 129° for water vapor in air. Wegener and Lundquist's experiments were carried out in a shock tube. There did not appear to be a cutoff temperature below which condensation could not take place.

The experimental results obtained by several observers on the condensation of nitrogen and oxygen in water vapor seem to agree that the degree of supersaturation obtained for this condensation of air in hypersonic wind tunnels is much less than the supersaturation obtained with water. The most plausible explanation for this experimental result is that the oxygen and nitrogen are condensing on nuclei of previously condensed water vapor and carbon dioxide. By many experimenters, the water vapor was reduced to a very low level, perhaps one part in ten thousand to one part in a million. Recent experiments by Nagamatsu, Willmarth, and Arthur [*42,43*] have shown high degrees of supersaturation in nitrogen in which the water vapor and carbon dioxide content is greatly reduced. The addition of small amounts of carbon dioxide and water vapor greatly decreases the degree of supersaturation obtainable. From Fig. F,2a it is seen that both water vapor and carbon dioxide would be expected to condense before the temperatures of saturation of oxygen and nitrogen are reached. Fig. F,5a and F,5c show the amount of supersaturation reached in air before condensation is detected.

F,7. Correlation of Experiment and Theory. The most thorough investigation of the correlation of experiment and theory was carried out by Oswatitsch [*30*]. He used the experimental data of Yellott [*3*] and

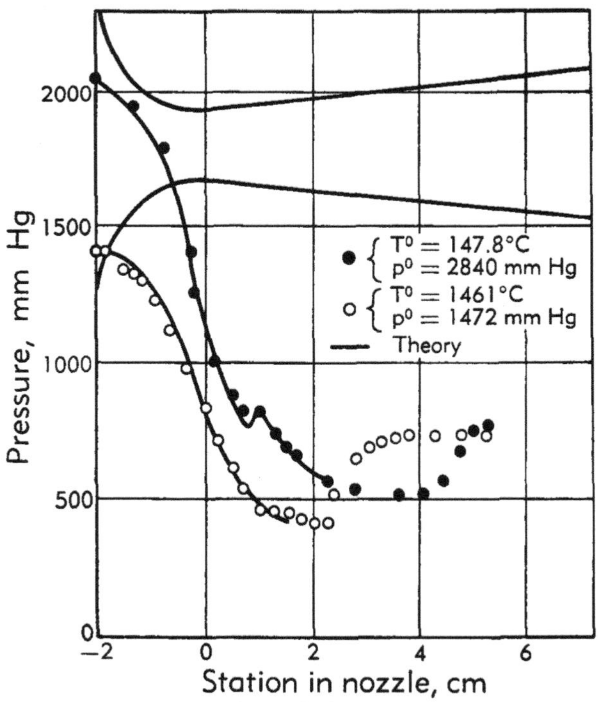

Fig. F,7a. Comparison of theoretical calculations and measured static pressures.

Binnie and Woods [*39*] on the condensation of steam in nozzles, together with some of his own experimental results. Eq. 4-32, 4-33, 4-34, and 4-35 were employed in a numerical calculation. The nozzle areas as a function of the distance along the nozzle were known. In Fig. F,7a, the experimental points of Yellott for two different runs are compared with the results of the numerical integration of the equations represented by the solid line. As shown, there is excellent agreement. The rise of pressure at the right in the figure is due to a compression shock in the nozzle and has

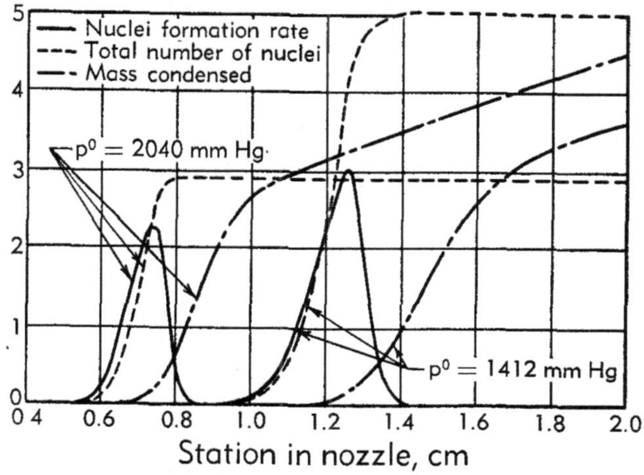

Fig. F,7b. Nuclei formation rate, total drop number, and total condensation versus position in nozzle. The nuclei formation rate is given in (nuclei per cm³ per cm) $\times 10^{-14}$, the total number of nuclei in (nuclei per cm³) $\times 10^{-13}$, and the mass condensed in (fraction of mass in liquid state) $\times 10^{-2}$.

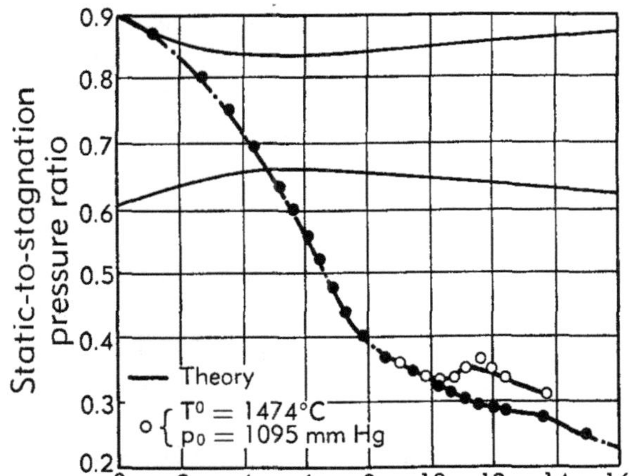

Fig. F,7c. Comparison of theoretical calculations and measured static pressures.

nothing to do with condensation. Fig. F,7b shows some interesting plots of the results of the calculations for these two cases. For each run there is a plot of the rate of formation of nuclei, the total number of drops, and the total amount of condensed water. The last quantity is calculated using Eq. 3-28.

In Fig. F,7c the results of calculations by Oswatitsch for the experiments of Binnie and Wood are plotted. Here the solid circles represent flow in the nozzle as measured with stagnation conditions such that there was no condensation. The open circles represent the pressure measurements with condensation conditions present. Again there is good agreement of these experimental points with prediction.

Another interesting factor which can be gained from the experimental results of Yellott and the theoretical results of Oswatitsch relates to the drop size. From the equation it is possible to calculate the mean size of the drops, which turned out to be about 0.5×10^{-6} cm. Yellott had estimated from scattering measurements that the drops were much smaller than the wavelength of blue light or 0.4×10^{-4} cm. Keenan, commenting on Yellott's work, estimated from the pressure jump a size of 0.3×10^{-6} cm. These results show that the equation for drop size as a function of time is best taken here as that determined by molecular processes (Eq. 3-28). The size of drop calculated in this experiment is considerably smaller than the 5.0×10^{-6} cm measured by Durbin [*38*] in light scattering experiments on hypersonic condensation.

Examination of Oswatitsch's curves in Fig. F,7b gives very interesting insight into the process of condensation in nozzles governed by spontaneous nucleation and drop growth. It can be seen that the process of drop formation is pretty well completed before the most rapid rise of total condensation occurs. The steep portions of the curves representing the total amount of water condensed occur after the nucleation is completed. The points of abrupt pressure rise coincide with the points of greatest slope on the total condensation curve. This is to be expected since the total condensation rather than the nuclei formation is the process which releases heat to the stream.

Another point worth noting is that the distance between the points of maximum drop formation and maximum condensation rate is different in the two runs, being larger for the run in which the pressure rise occurs farther downstream. This seems to explain why the point of the condensation shock measured in terms of supersaturation ratio depends so strongly on the temperature gradient in the nozzle as described in the previous section.

Because the nuclei formation rate (Eq. 3-14) varies so strongly with pressure and temperature, it is generally found that at some point in the flow there is a rather abrupt change in these rates from extremely low values to extremely high values, depending naturally on the rate of expansion in the channel. Many workers in the field of condensation have thus taken this to indicate that a selection of a given condensation rate or a given nuclei formation rate can be used as a criterion for condensation. This is true only in a qualitative sense, as can be seen from the above-quoted calculations of Oswatitsch and the experimental results of Head and Wegener. At present, criteria based on this assumption vary over wide limits. It is believed that the only accurate way to determine the condensation characteristics for a given nozzle is by the method developed by Oswatitsch. It can be hoped that other calculations of this nature will be carried out in the future.

This discussion brings up another point about condensation theory. Oswatitsch's calculations were carried out using a value of surface tension which did not vary with the radius of the surface. In Art. 4 it is pointed out that this was probably not the case. Several authors [*6,23,27,29*] have tried to correct the condensation theory for this weakness. Lees' results

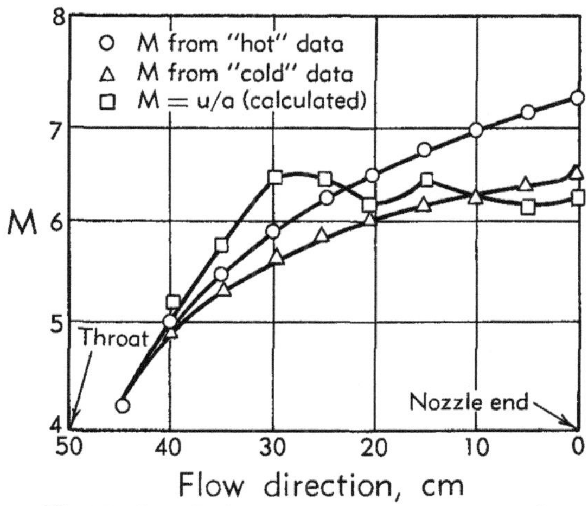

Fig. F,7d. Mach number measurements and calculation according to Wegener theory.

indicate that a "cutoff" limit of condensation at extremely high supersaturations might exist. Head and Stever's results, based on surface tension calculations by Tolman, Kirkwood, and Buff, indicate that the condensation rates and the nuclei formation rates should be higher at high supersaturation ratios instead of lower as Lees' results indicate. Unfortunately there is no good proof of any of these theories, and there probably will not be until much more thorough applications of the complete theory are made along the lines started by Oswatitsch.

The experimental test of the approximate theory developed by Wegener, et al. [*35*] and described in Art. 4 is illustrated in Fig. F,7d showing a plot of the Mach number versus the position along the nozzle in their hypersonic tunnel. The Mach number is plotted for a heated run in which no condensation is detected, for a cold run in which condensation is definitely present, and as calculated for the cold run using the approximate theory. It is seen that the theoretical and cold-run experimental points are in poor agreement when condensation first starts, probably due to the rapid drop in speed of sound used in the calculation. However, the experimental and calculated curves agree much better as condensation proceeds.

In conclusion it can be said that the phenomena associated with condensation in high speed flow are generally quite well understood. There does remain a considerable amount of work in making certain aspects of the theoretical calculations more accurate and more readily usable. Also there is a great need for more accurate measurement of the fundamental properties of common vapors over a wider range of the state variables so that more accurate calculations can be carried out.

F,8. Cited References.

1. Keenan, J. H. *Thermodynamics*. Wiley, 1941.
2. Stodola, A. *Steam and Gas Turbines*. McGraw-Hill, 1927.
3. Yellott, J. I. *Engineering 137*, 303 (1934).
4. Yellott, J. I., and Holland, C. K. *Engineering 143*, 647 (1937).
5. Hermann, R. *Luftfahrtforschung 19*, 201 (1942).
6. Head, R. M. *Investigations of Spontaneous Condensation Phenomena. Ph.D. Thesis*, Calif. Inst. Technol., 1949.
7. Wagner, C. Serechnungen über die Möglichkeiten zur Erzielung hoher Machscher Zanlen in Windkanalversuchen. *WVA Arch. A-385/1*, 1942.
8. *Landolt-Bornstein Physikalisch-Chemische Tabellen*, 5th ed., Vol. I. Springer, Berlin, 1923.
9. Hodgman, C. D., ed. *Handbook of Chemistry and Physics*, 28th ed. Chemical Rubber Publ. Co., 1944.
10. *NBS-NACA Tables of Thermal Properties of Gases*. U.S. Government Printing Office, 1949.
11. Aoyama, S., and Kanda, E. *Science Repts. Tohoku Imp. Univ. 24*, 107–115 (1935).
12. Perry, J. H. *Chemical Engineers' Handbook*, 3rd ed. McGraw-Hill, 1950.
13. Thomson, Sir W. (Lord Kelvin). *Proc. Roy. Soc. Edinburgh VII*, 63–68 (1870); *Phil. Mag. 4*, 448–452 (1871).
14. von Helmholtz, R. *Ann. Phys. 27*, 508–543 (1886).
15. Gibbs, J. W. *Collected Works*. Longmans, Green & Co., 1931.
16. Frenkel, J. *Kinetic Theory of Liquids*. Clarendon Press, Oxford, 1946.
17. Volmer, M. *Kinetik der Phasenbilding*. Steinkopff, Dresden, 1939.
18. Farkas, L. *Z. physik. Chem. A125*, 236–242 (1927).
19. Kaischew, R., and Stranski, I. *Z. physik. Chem. 826*, 317–326 (1934).
20. Becker, R., and Döring, W. *Ann. Phys. 5*, 719–752 (1936).
21. Zeldovich, Y. B. *J. Exptl. Theoret. Phys. U.S.S.R. 12*, 525–538 (1942).
22. Wegener, P., and Smelt, R. Summary of Naval Ordance Laboratory research on liquefaction phenomena in hypersonic wind tunnels. *Nav. Ord. Lab. Mem. 10772*, 1950.
23. Bogdonoff, S. M., and Lees, L. Study of the condensation of the components of air in supersonic wind tunnels. Part I. Absence of condensation and tentative explanation. *Princeton Univ. Aeronaut. Eng. Lab. Rept. 146*, 1949.
24. Mayer, J. E., and Mayer, M. G. *Statistical Mechanics*. Wiley, 1940.
25. Tolman. R. C. *J. Chem. Phys. 17*, 333–337 (1949).
26. Kirkwood, J. G., and Buff, F. P. *J. Chem. Phys. 17*, 338–343 (1949).
27. Stever, H. G., and Rathbun, K. C. Theoretical and experimental investigation of condensation of air in hypersonic wind tunnels. *NACA Tech. Note 2559*, 1951.
28. Taylor, H. S., Eyring, H., and Sherman, A. *J. Chem. Phys. 1*, 68 (1933).

29. Reed, S. G., Jr. *On the Early Stages of Condensation. Ph.D. Thesis,* Catholic Univ. of America, 1951.
30. Oswatitsch, K. *Z. angew. Math. u. Mech. 22,* 1–14 (1942).
31. Hawthorne, W. R., and Cohen, H. Pressure losses and velocity changes due to heat release and mixing in frictionless, compressible flow. *Roy. Aircraft Establishment Rept. E3997,* 1944.
32. Heybey, W. Analytische Behandlung de Geraden Kondensationsstosses. *WVA Arch. 66/72,* 1942.
33. Hicks, B. L. Addition of heat to a compressible fluid in motion. *NACA Advanced Confid. Rept. E5A29,* 1945.
34. Lukasiewicz, J. Humidity effects in supersonic flow of air. *Roy. Aircraft Establishment Rept. Aero. 2211-SD 20,* 1947.
35. Wegener, P., Reed, S., Jr., Stollenwerk, E., and Lundquist, G. *J. Appl. Phys. 22,* 1077–1083 (1951).
36. Becker, J. V. *J. Appl. Phys. 21,* 619–628 (1950).
37. Wegener, P., and Lundquist, G. *J. Appl. Phys. 22,* 233 (1951).
38. Durbin, E. J. Optical methods involving light scattering for measuring size and concentration of condensation particles in supercooled hypersonic flow. *NACA Tech. Note 2441,* 1951.
39. Binnie, A. M., and Woods, M. W. *Proc. Inst. Mech. Engrs. London 138,* 229 (1938).
40. Oswatitsch, K. Die Nebelbildung in Windkanalen und Ihr Einfluss auf Modelversuche. *Luftfahrtforschung,* 1941.
41. Cwilong, B. M. *Proc. Roy. Soc. London A190,* 137–143 (1947).
42. Nagamatsu, H. T., and Willmarth, W. W. Condensation of nitrogen in a hypersonic nozzle. *GALCIT Hypersonic Wind Tunnel Mem. 6,* Jan. 1952. Also *J. Appl. Phys. 23,* 1089–1095 (1952).
43. Arthur, P. D., and Nagamatsu, H. T. Effects of impurities on the supersaturation of nitrogen in a hypersonic nozzle. *GALCIT Hypersonic Wind Tunnel Mem. 7,* Mar. 1952.

SECTION G

GAS DYNAMICS OF COMBUSTION AND DETONATION

CHAPTER 1. AEROTHERMODYNAMIC PROBLEMS OF COMBUSTION

TH. VON KÁRMÁN

G,1. General Aspects of the Combustion Problem. With the advent of jet propulsion it became necessary to broaden the field of aerodynamics to include problems which before were treated mostly by physical chemists; for example, the problems related to combustion, especially combustion in high speed flow. The propagation of flames and of detonations is treated in detail in the second volume of this series, especially from the viewpoint of chemical kinetics. In this section the general problem of propagation will be considered from the viewpoint of aerothermodynamics.

In order to formulate the concept of velocity of combustion or of detonation, one considers the case of a unidimensional flow of gas, or mixture of gases, in which a chemical reaction takes place. Instead of a progressive wave, a steady motion is considered. Then the problem is to determine the magnitude of the velocity ahead of the reaction zone for which no displacement of the zone occurs so that the entire process can be described as a steady motion.

As in the case of a one-dimensional flow of a gas through a constant cross section with heat addition treated in Sec. B, we have three conditions to fulfill: the conservation of matter, momentum, and energy. It is necessary, however, to take into account viscous forces acting in the flow direction and heat conduction in the same direction. We denote the velocity by u; the temperature, pressure, and density of the gas by T, p, and ρ, respectively; the coefficient of viscosity by μ; the coefficient of heat conduction by k; the specific heat referred to unit mass at constant pressure by c_p; and by q the heat addition per unit mass due to the chemical reaction occurring between $x = -\infty$ and the section considered. We then have the following three equations for the conservation of mass, momentum, and energy:

$$\rho u = m \tag{1-1}$$

$$-\tfrac{4}{3}\mu \frac{du}{dx} + p + \rho u^2 = i \tag{1-2}$$

$$-k \frac{dT}{dx} + m(c_p T + \tfrac{1}{2}u^2) - \tfrac{4}{3}\mu u \frac{du}{dx} - mq = E \tag{1-3}$$

The quantities m, i, and E are constant for the entire process. In particular, m is the mass flow, which we want to determine in such a way that the process is stationary.

Because of the presence of different chemical components the following changes in the usual interpretation of these equations in aerodynamics are necessary:

1. The specific heat c_p is a mean value for the mixture; it is actually variable with the composition and the temperature.
2. The so-called average transport coefficients μ and k are also functions of the composition of the mixture and of the temperature.
3. The mean velocity u is defined as the mean mass velocity.

Due to the presence of diffusion of the components of the mixture the velocity of an arbitrary component is defined as $u_j = u + u_{jd}$ where u_{jd} is the velocity of diffusion of the jth component into the mixture; since u is defined as the mass velocity, $\sum_j \rho_j u_{jd} = 0$, where ρ_j is the density of the jth component, and $\sum_j \rho_j = \rho$. The product $\rho_j u_j = \rho_j u + \rho_j u_{jd}$ we call the mass flux of the jth component.

In order to solve the complete problem, we must also know the following:

1. The equation of state for the mixture.
2. The laws of diffusion between the components.
3. The laws of production and disappearance of components by means of chemical reactions.
4. The amount of heat release due to chemical reactions.

In our discussion one essential restriction is introduced. We assume that the amount of heat added is determined by a single parameter, which we choose as a measure of the progress of the reaction. We want to distinguish between the reactants, i.e. the original partners in the reaction, and the products. If there is only one possible way for a chemical change to take place, we may choose any one of the products for characterizing the progress of the reaction; such a measure is, for example, the amount of the chosen component produced between the initial and an arbitrary cross section of the flow pattern. Then the amount of all the other components produced or destroyed by the reaction, and also the total heat added, are determined by this single parameter.

If we denote the parameter chosen for characterizing the progress of the reaction by ϵ, where $\epsilon = 0$ corresponds to the initial cross section, then the heat added q can be written in the form $q = Q\epsilon$. The parameter ϵ refers to unit mass flow, so that Q is the heat liberated by chemical reaction per unit mass of the mixture and can be computed from the enthalpies of formation of the components.

It appears convenient to introduce the quantity w defined by

$$w = m \frac{d\epsilon}{dx} \tag{1-4}$$

The quantity w is determined by the chemical kinetics of the reaction as a function of the local temperature, pressure, and chemical composition of the mixture. It is proportional to the quantity known as "the rate of reaction" which is defined as the number of moles of the selected species produced per unit volume in unit time. It is evident that the amounts of all other species produced per unit volume in unit time are proportional to w. The corresponding values for other species we may denote by indices such as w_j.

It is easily seen that the quantity $m\epsilon$ is equal to the mass flux of the chosen product component through an arbitrary cross section. In fact if we consider the conservation of mass for this component, which was not existent in the initial composition, it is seen that its mass flux through an arbitrary cross section has to be equal to

$$m\epsilon = \int_{-\infty}^{x} w dx \tag{1-4a}$$

We now introduce analogous quantities for the other components, denoted by ϵ_j.

According to our definition of the mass flux, $m\epsilon_j = \rho_j u_j$. Introducing the value of $u_j = u + u_{jd}$ as defined above and the mass fraction $Y_j = \rho_j/\rho$, one obtains

$$m\epsilon_j = mY_j + \rho_j u_{jd} \tag{1-5}$$

If the diffusion is neglected, $\epsilon_j = Y_j$.

In general the initial value of ϵ_j will be different from zero. Denoting the initial value by ϵ_{j0}, the difference $\epsilon_j - \epsilon_{j0}$ will be equal, in analogy with Eq. 1-4a, to

$$m(\epsilon_j - \epsilon_{j0}) = \int_{-\infty}^{x} w_j dx \tag{1-6}$$

The diffusion velocities are determined by the space gradients of the mass fractions Y, according to the diffusion laws. These relations, however, are rather complicated in the general case of n components. They are only simple in the case of binary diffusion when only two components are present. Therefore we want to restrict ourselves to the case of two components. Our conclusions remain valid, in principle, for the general case. The difficulties are merely arithmetical.

For the binary case we have

$$\rho \mathfrak{D} \frac{dY}{dx} = m(Y - \epsilon) \tag{1-7}$$

where $\mathfrak{D}$ is the binary diffusion coefficient between the two components and Y is the mass fraction of the product, which we have chosen for characterizing the progress of the reaction. Evidently for $\mathfrak{D} = 0$, $Y = \epsilon$.

Finally we have the equation of state for the mixture, viz.

$$\frac{p}{\rho} = \mathfrak{R}_m T \tag{1-8}$$

where the gas constant per gram $\mathfrak{R}_m$ of the mixture, in general, changes with the chemical composition. If the reaction does not alter the mean molecular weight of the mixture, $\mathfrak{R}_m$ is a constant.

Eq. 1-1, 1-2, 1-3, 1-4, 1-7, and 1-8 determine the six unknown quantities p, ρ, T, u, ϵ, and Y, provided the boundary conditions are given. However, the solution of this system of differential equations, without placing restrictions on the relative order of magnitude of the various parameters, has not been obtained.

The theories of detonation and deflagration, which we shall discuss here, are based on the assumption that the time scale of the chemical processes which depend on the reaction rate and the time scale of the mechanical processes which manifest themselves in the transport coefficients (viscosity, heat conduction, and diffusion) are of different order. This assumption corresponds to our picture of these two classes of processes: chemical changes are produced by collisions between activated particles, i.e. at a relatively slow rate compared with the number of collisions which lead to the transfer of mass, momentum, and energy.

The time scale of the chemical change is determined by the reaction rate w. In fact the ratio ρ/w has the dimension of time; evidently the change of ϵ per unit time in a stationary flow is equal to (cf. Eq. 1-4)

$$u \frac{d\epsilon}{dx} = \frac{w}{\rho} \tag{1-9}$$

We may call the reciprocal of this quantity the chemical time t_{chem}.

On the other hand the time scales of the mechanical collisions responsible for the transport phenomena are measured by the expressions μ/p, k/pc_p and $\rho\mathfrak{D}/p$.

If we choose μ/p as the characteristic mechanical time t_{mech}, it is evident that

$$\frac{k}{pc_p} = \frac{1}{Pr} t_{\text{mech}} \tag{1-10}$$

and

$$\frac{\rho\mathfrak{D}}{p} = \frac{1}{Sc} t_{\text{mech}} \tag{1-11}$$

where $Pr = \mu c_p/k$ denotes the Prandtl number and $Sc = \mu/\rho\mathfrak{D}$ the Schmidt number.

We assume now that the nondimensional ratio

$$\alpha = \frac{t_{\text{mech}}}{t_{\text{chem}}} = \frac{\mu w}{p\rho} \tag{1-12}$$

is small compared with unity. In order to investigate the bearing of this assumption on our fundamental equations, we change the independent variable from x to the nondimensional variable ϵ. Then, using the parameter α, we obtain the following differential equations

$$p\left[-\tfrac{4}{3}\alpha\left(\frac{1}{u}\frac{du}{d\epsilon}\right) + 1 + \gamma M^2\right] = i \tag{1-13}$$

$$mc_pT\left[-\frac{1}{Pr}\frac{\alpha}{\gamma M^2}\left(\frac{1}{T}\frac{dT}{d\epsilon}\right) - \frac{4}{3}\frac{\gamma-1}{\gamma}\alpha\left(\frac{1}{u}\frac{du}{d\epsilon}\right) + 1 + \frac{\gamma-1}{2}M^2 - \frac{Q}{c_pT}\epsilon\right] = E \tag{1-14}$$

$$\frac{1}{Sc}\frac{\alpha}{\gamma M^2}\frac{dY}{d\epsilon} = Y - \epsilon \tag{1-15}$$

In these equations the expression $u^2\rho/p$ is replaced by γM^2, where M is the local Mach number and γ is the ratio of the specific heats $\gamma = c_p/c_v$; it is also assumed that $c_p = \mathfrak{R}_m\gamma/(\gamma - 1)$ is constant through the reaction zone. It should be noted that the ratio Q/c_pT is one of the similarity groups introduced by Damköhler in his paper of 1936 [*1*].

It is seen that the viscosity terms are always small in comparison with unity, so that we may write

$$p(1 + \gamma M^2) = i \tag{1-16}$$

$$mc_pT\left[-\frac{1}{Pr}\frac{\alpha}{\gamma M^2}\left(\frac{1}{T}\frac{dT}{d\epsilon}\right) + 1 + \frac{\gamma-1}{2}M^2 - \frac{Q}{c_pT}\epsilon\right] = E \tag{1-17}$$

$$\frac{1}{Sc}\frac{\alpha}{\gamma M^2}\frac{dY}{d\epsilon} = Y - \epsilon \tag{1-18}$$

It is seen that the character of the equations is different in the two cases when γM^2 is of the order of unity and when it is so small that $\alpha/\gamma M^2$ may become of the order of unity.

In the first case all the terms containing α can be neglected. Hence one falls back on the theory of ideal compressible fluids with heat addition, as it was discussed in VI,A,5. Since the time required for a thermomechanical shock is small in comparison with the time required for the thermochemical process, it is justifiable to assume that first a shock occurs, and that the combustion (i.e. heat addition) follows the shock. This process was considered in detail by von Neumann [*2*], Döring [*3*], and Zeldovich [4]; see also G,11. It always starts with a supersonic velocity in front of the shock. In other words, such a wave, which is called a "detonation wave," propagates with a supersonic velocity in an unburned gas mixture. The diagram in Fig. G,1 shows that the relative velocity between the wave and the burned gas must be subsonic or in the limiting case sonic, provided that the chemical reaction is exothermic, i.e. involves continuous heat addition. Hence, if $\alpha \ll 1$, a detonation, starting and ending with supersonic velocity, is impossible.

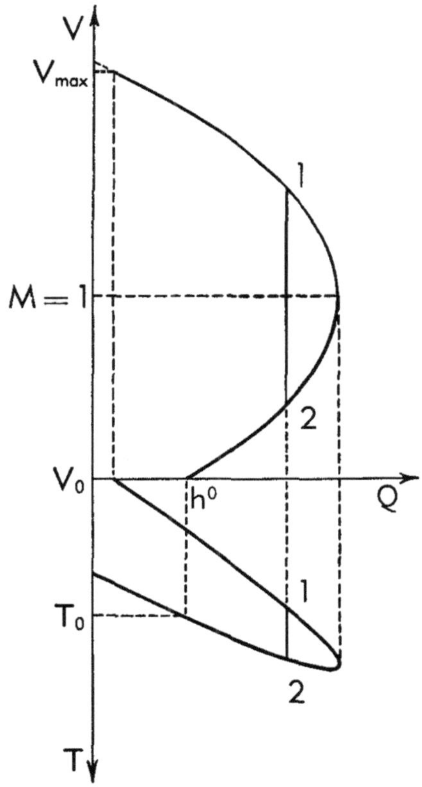

Fig. G,1. Velocity and temperature produced in a gas flowing in a cylindrical pipe by addition of heat.

The value of the relative velocity between wave and burned gas depends on the boundary conditions imposed on the entire process. In the case of a plane detonation wave in a one-dimensional flow with constant cross-sectional area, the two possible continuations of the flow are either constant velocity and constant pressure or an isentropic expansion in which the velocity in every cross section is equal to the local velocity of sound. Consequently the velocity has to be equal to the velocity of sound at the end of the combustion. Such a detonation is called a Jouguet-Chapman detonation; it is the rule in most practical cases and for the flow from $-\infty$ to $+\infty$ which we considered, it is the only stationary solution.

In other cases, for example when a detonation wave proceeds in a tube with one closed end, it can be shown that a Jouguet-Chapman wave also occurs. However, it is possible to construct cases in which the relative velocity between the detonation wave and the burned gas is subsonic. Imagine, for example, a mixture of two gases able to detonate beyond a certain ignition temperature, held under pressure in a container; we allow the mixture to expand at high velocity and assume that its temperature is inferior to the ignition temperature. Then we let the mixture pass through a converging-diverging Laval nozzle into a space, the pressure of which we can regulate. If we increase the pressure at the exit, we obtain a shock in the nozzle. Let us assume that the temperature behind the shock is superior to the ignition temperature; then combustion will follow the shock. However, assume that the amount of released heat is not sufficient to accelerate the gas to sound velocity; then the flow remains subsonic.

In the case of the Jouguet-Chapman detonation the velocity of propagation of the wave, which for the stationary case corresponds to the velocity ahead of the shock, can be easily calculated, provided the total amount of released heat is known. The process starts at the upper supersonic branch of a diagram similar to that given in Fig. G,1. For example at point 1 the shock leads to the subsonic branch point 2 keeping the value of the enthalpy constant. To determine the location of the starting point (1), i.e. the detonation velocity, the enthalpy difference between the start and the sonic point must be made equal to the heat released by the reaction.

It has to be noted, however, that the theory of detonation indicated above applies only if $\alpha \ll 1$, i.e. if one is justified in assuming that the combustion essentially follows the shock, in other words, the chemical reaction during the passage of the shock is negligible. If this condition is not fulfilled, for example if the rate of reaction w is very large, one has to solve the simultaneous equations (Eq. 1-2, 1-3, and 1-7), including the terms containing viscosity and heat conduction. Friedrichs [*5*] made an attempt to discuss the mathematical aspects of this general problem (to be sure, with no diffusion), aiming to determine the possible modes of transition, including the simultaneous processes of shock and combustion.

According to a classification due to Jouguet, there are four cases based on the supersonic or subsonic character of the velocities ahead of and beyond the transition:

from supersonic to subsonic	strong detonation
from supersonic to supersonic	weak detonation
from subsonic to subsonic	weak deflagration
from subsonic to supersonic	strong deflagration

Evidently the case we have discussed above is a strong detonation (including the Jouguet-Chapman detonation as limiting case). We will treat the case of weak deflagration in the subsequent lines of this section. Friedrichs seems to have reached the conclusion that weak detonation may be possible in the case of very high reaction velocities, whereas strong deflagrations are impossible.

For the case that γM^2 is of the order of α or smaller, we obtain from Eq. 1-16 the simple conclusion p = const—in other words the variations of the pressure through the combustion zone are negligible. Eq. 1-17 takes the form

$$\frac{1}{Pr}\frac{\alpha}{\gamma M^2}\frac{1}{T}\frac{dT}{d\epsilon} = 1 - \frac{Q}{c_pT}\epsilon + \frac{E}{mc_pT} \tag{1-19}$$

or

$$\frac{\lambda w}{m^2c_p}\frac{dT}{d\epsilon} = T - \frac{Q}{c_p}\epsilon - \frac{E}{mc_p} \tag{1-20}$$

We defined the parameter ϵ in such a way that in the initial state of the mixture $\epsilon = 0$; the corresponding temperature is $T = T_0$. Now we denote the value of ϵ for the final state by ϵ_f: the corresponding value of the temperature is T_f, which corresponds to the constant pressure adiabatic combustion temperature of the mixture. Then

$$E = mc_pT_0 \tag{1-21}$$

and

$$q = \frac{c_p(T_f - T_0)}{\epsilon_f} \tag{1-22}$$

Thus we obtain for Eq. 1-20 the form

$$\frac{kw}{m^2c_p}\frac{dT}{d\epsilon} = (T - T_f) + (T_f - T_0)\left(1 - \frac{\epsilon}{\epsilon_f}\right) \tag{1-23}$$

This equation, together with the diffusion equation (Eq. 1-18), determines the problem. Substituting in Eq. 1-15 the values of α and γM^2, we obtain

$$\frac{\rho\mathfrak{D}w}{m^2}\frac{dY}{d\epsilon} = Y - \epsilon \tag{1-24}$$

The two equations are interconnected since w is a given function of Y and the temperature.

The boundary conditions for $\epsilon = \epsilon_f$ are easily defined. Obviously T must be equal to T_f and $Y = \epsilon_f$. However, there is some difficulty in formulating the boundary conditions for $\epsilon = 0$. It can be easily seen that a solution for which $T = T_0$ when $\epsilon = 0$ cannot exist. This is also physically evident since T_0 is the temperature of the unburned gas mixture and therefore burning cannot start at $T = T_0$. Most theories assume that there exists an ignition temperature T_i at which the reaction starts.

Since both of the equations (Eq. 1-23 and 1-24) are of the first order, no solution can satisfy more than two boundary conditions for arbitrary values of m^2. Evidently we have to deal with an eigenvalue problem for m^2, where m is the mass flow which makes the stationary flow problem possible. In other words, if we put $m = \rho_0 u_0$ and if ρ_0 is the density of the mixture at $x = -\infty$, the u_0 which corresponds to the eigenvalue of m^2 furnishes the value of the laminar flame velocity or the velocity of propagation of the deflagration wave at constant pressure.

If we determine the eigenvalue of m^2, we have a solution of Eq. 1-23 and 1-24 which satisfies the condition $T = T_i$ at the cold boundary and the conditions $T = T_f$ and $Y = \epsilon_f$ at the hot boundary. The value of Y resulting from this solution for $\epsilon = 0$ we will denote by Y_i.

The heating of the mixture from the initial value T_0 to T_i takes place by heat conduction. The composition of the mixture has been changed during this heating period by diffusion. In other words, in order to complete the picture we have to solve the system of equations

$$k \frac{dT}{dx} = mc_p(T - T_0) \tag{1-25}$$

$$\rho \mathfrak{D} \frac{dY}{dx} = mY \tag{1-26}$$

for the domain $-\infty < x < 0$ with the conditions $T = T_i$ and $Y = Y_i$ for $x = 0$. For $x = -\infty$ one obtains, due to the nature of the equations, $T = T_0$ and $Y = 0$.

From the physical point of view, it is a weakness of such a theory that—as experimental investigations have shown—the mechanism of ignition is too complicated to be taken into account by a fixed value of an ignition temperature T_i, i.e. by the simple assumption that the reaction rate is equal to zero for $T < T_i$ and is equal to the value given by chemical kinetics for $T \geqq T_i$. The flame velocity appears in the theory as a function of a fictitious ignition temperature, whereas from experimental facts it appears that there exists a definite value of the laminar flame velocity which is independent of the specific process which produced the ignition of the flame. Semenov, Zeldovich, and their followers [*6*] have assumed that the ignition temperature always lies very close to the adiabatic flame temperature. Then, provided the reaction rate decreases exponentially with the reciprocal of the temperature, as it does for example in the case which follows Arrhenius' law, the flame velocity becomes practically independent of the specific choice of the ignition temperature. Millán and the present author [*7*] investigated the question of the dependence of the eigenvalue in the flame equation on the choice of the ignition temperature. They found that the calculated flame velocity becomes infinite when $T_i = T_0$ and zero for $T_i = T_f$. However, the increase to infinity and the fall to zero take place when T_i is very close to T_0 and T_f respectively.

Over a great portion of the interval $T_0 < T_i < T_f$ the eigenvalue is practically constant.

Using an analytical approach to the solution of the flame equations it is easy to take care of this situation and to obtain a fair approximation for the essentially constant magnitude of the eigenvalue and therefore the flame velocity. On the other hand, when a point-to-point numerical integration method is used, the determination of the eigenvalue is difficult. Therefore Hirschfelder, Curtiss, and Campbell [*8*, pp. 190–211] introduced the artifice of assuming the existence of a small heat sink and a semipermeable diffusion filter for $T = T_0$, i.e. for the "cold boundary." This model also leads to the same conclusion concerning a definite value of the flame velocity.

In the above considerations we made two restrictions: First, the number of components was restricted to two so that the law of binary diffusion could be used. This restriction is not essential; when more than two components are present, Eq. 1-7 will be replaced by a system of equations valid for n components. The conclusion concerning the existence of an eigenvalue remains the same.

Second, and more essential, is the assumption of a single chemical reaction. As a matter of fact, such a case is rather rare in practice. A rather large number of practical problems involve chain reactions, which without some simplification do not fit into the scheme given above. However, in recent years much work has been done to simplify the system of such reactions, especially by extending the "steady state condition," widely applied in classical kinetics for the determination of radical concentrations in chain reactions, to the theory of flames. This assumption allows us to find semianalytical methods, which lead to results in fair accordance with numerical point-to-point integration of the original complete system of equations. The reader will find a review of the state of the art and ample references in the author's survey paper presented at the Sixth Symposium (International) on Combustion [*9*].

It must be noted that the flame theory sketched in this section deals only with laminar flames. The question of combustion in turbulent flows is reviewed in Vol. II of this series. Also, some important questions of combustion of laminar and turbulent character, e.g. the mechanism of ignition, the influence of cold walls on ignition and combustion, and the limits of inflammability, are discussed in the same volume.

Although the general mechanics of reacting continua is of great theoretical interest for aeronautical engineers, in many cases it is sufficient to assume the existence of a laminar or turbulent flame velocity and to treat the aerothermodynamic problems connected with combustion in the sense indicated in Chap. 2 of this section. The data derived from the combustion theory furnish the transition conditions for a discontinuity in the flow. It is then possible to describe and compute the flow phenomena occurring in flames, combustion chambers, and industrial combustion devices by known aerothermodynamic methods. This branch of aerothermodynamics, which we may call "aerothermochemistry," is still in its infancy but promises to become one of the most interesting and fascinating branches of fluid mechanics.

CHAPTER 2. FLOW DISCONTINUITIES ASSOCIATED WITH COMBUSTION

H. W. EMMONS

G,2. Theory of the Flame Front. In a general case of combustion a mixture of substances moves about in space while various chemical reactions occur. The composition of the mixture will vary from place to place because of the changes brought about by the reactions and sometimes also because the fluids entering the combustion region are not the same everywhere.

If all fluid entering the combustion region is identical, the reactants (fuel and oxidizer) are premixed. For premixed fluids the reaction in the combustion region maintains itself by propagation of the reaction according to some mechanism into the newly introduced and as yet nonreacting fluid.

If fluids of differing composition enter the combustion region in different places, some or all of the reactions that are to take place will be delayed in their propagation by the necessity that the reactants must first come into contact before they can react.

If we define a flame as the flow and reaction process which converts reactants to products of combustion, a certain portion of space can (at least in principle) be picked out of a combustion-aerodynamic field which contains the flame. The flame thus defined has a volume which extends more or less completely across the combustion region and has a thickness in the direction of the streamlines which depends upon the rate of reaction and the stream velocity.

The chemical reactions occurring in a premixed reactive mixture proceed at rates which depend upon the composition of the gases and their state. In general, reactions proceed faster with increased pressure and temperature. However, since the actual reactions occurring in a given combustible mixture are very complex,[1] no simple universal relation between the reaction rate and the composition, temperature, and pressure exists.

[1] The over-all reaction, say $2H_2 + O_2 \rightarrow 2H_2O$, almost never indicates the detailed process. The actual reactions involve atomic and ionized hydrogen, oxygen, radicals OH^-, and perhaps more obscure combinations like HO_2.

A flame is never stationary relative to the combustible medium. It may, however, be stationary in space if the fluid moves at just the correct velocity at every point to maintain conditions constant. The velocity of the unburned premixed reactants normal to a constant temperature surface in the flame is called the flame speed. A flame propagates because some mechanism ignites the fresh combustible mixture which enters the flame. Reactions are accelerated by raising the temperature or density, or by introducing highly energetic atoms or radicals from the flame itself into the entering fluid.

Thus heat from the high temperature products of combustion is conducted into and accelerates the reactions in the reactants. This is called "thermal flame propagation." Various chemical species—especially atomic hydrogen—may diffuse from the products of combustion (chemical equilibrium at a high temperature implies an appreciable concentration of hydrogen atoms) or from intermediate products into the reactants and initiate reactions there. This will be called diffusive flame propagation. Finally, the motions of the products of reaction may be such that the reactants experience a large pressure rise in the flame and thus the reactions may be propagated. This will be called pressure flame propagation.[2]

The flame in premixed reactants can often be divided for consideration into two separate parts: (1) a heating (and diffusion) zone in which the reactions are proceeding at a very low rate but where the mixture temperature is being raised by heat conduction from the products of combustion into the reactant stream, and (2) a reaction zone in which reactions occur vigorously and produce the products of combustion very rapidly.

Of course the heating zone is preceded by a region in which the reactants flow without appreciable reaction or heat transfer from the flame. Also, the reaction zone is followed by a region in which the products of combustion flow away from the flame without significant change.

The first attempt to derive a quantitative relation between flame speed and the properties of the reacting materials was that of Mallard and Le Chatelier [*10*], who assumed that for a given gas mixture an ignition temperature T_i exists below which no reaction proceeds and above which the reaction proceeds rapidly. If we imagine that reactants flow from $x = -\infty$ along the x axis (see Fig. G,2a), the temperature rises as a result of conduction until the ignition temperature T_i is reached at x_i. Then the reaction proceeds, raising the temperature further to its final value T_f.

Mallard and Le Chatelier [*10*] noted that the energy conducted from

[2] The pressure changes are only important in explosive reactions which sometimes are called detonations as distinguished from flames. Detonations will be discussed in chapter 3 of this section. No further consideration of pressure propagation will be given in this part.

the reaction zone into the reactants at x_i must equal that required to heat the gas from the initial temperature T_0 to T_i. Thus

$$k\left(\frac{dT}{dx}\right)_{x=x_i} = \rho uc(T_i - T_0) \tag{2-1}$$

Furthermore, the temperature gradient at x_i must be related in some manner to the temperature rise produced by the reaction. Mallard and Le Chatelier assumed a linear relation. Numerous later studies have replaced this linear relation by better assumptions involving some reaction mechanism. Many of the references may be found in [*11*; *12*, Chap. 3, Sec. 6].

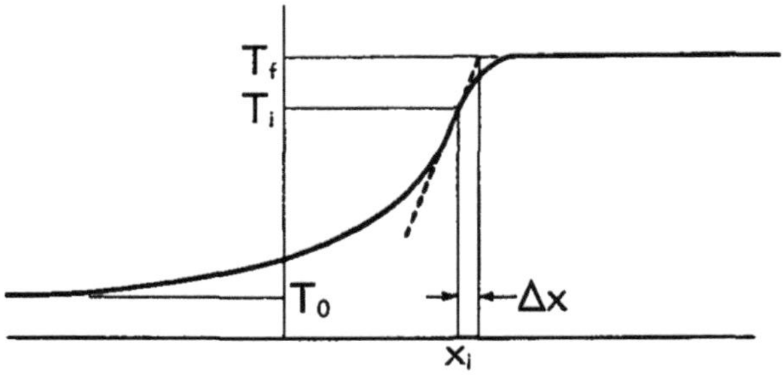

Fig. G,2a. Rise of temperature through a flame.

If it is assumed that the reaction proceeds at a rate α (fraction of the total mass flowing which becomes the product of combustion per unit time, $\sec^{-1} = g/\sec$ per g) and that in the reaction zone this reaction rate is constant, then the time required to complete the reaction in a given element of fluid will be $\Delta t = 1/\alpha$. The fluid flows a distance $\Delta x = u/\alpha$ during this reaction period and the temperature rises from the ignition temperature T_i to the final temperature T_f. Therefore Δx is an estimate of the thickness of the reaction zone. Finally, the temperature gradient at x_i may be approximated as

$$\left(\frac{dT}{dx}\right)_{x=x_i} = \frac{T_f - T_i}{\Delta x} = \frac{\alpha(T_f - T_i)}{u} \tag{2-2}$$

Now, eliminating the temperature gradient between Eq. 2-1 and 2-2 gives

$$S^2 = u^2 = \alpha\kappa\left(\frac{T_f - T_i}{T_i - T_0}\right) \tag{2-3}$$

where $\kappa = k/c\rho$ = thermal diffusivity. The quadratic relationship between the property $\kappa = k/c\rho$ and the flame speed was verified experimentally by Bartholomé [*13*].

The appearance of an ignition temperature in Eq. 2-3 greatly decreases its value. Experiments on ignition temperatures lead to results which depend considerably upon the apparatus used. At best an ignition temperature is illy defined and cannot be predicted from basic physical and chemical constants of the combustible mixture. More recent effort has therefore been directed toward a basic formulation of the problem in mathematical form followed by attempts at solution by some form of approximation. Hirschfelder, et al. [*14*] have set up the general one-dimensional equations of motion of a multicomponent mixture of reacting ideal gases starting from kinetic theory. These equations are not reproduced here because, while they are undoubtedly the most exact formulation of the problem so far attempted, no useful solutions have yet been published. The very complexity of the equations makes solutions practically obtainable only by numerical (computing machine) computations and tends to obscure the phenomena. Also, the detailed knowledge of the chemistry required is usually unobtainable at present.

Therefore, it has been profitable to examine special cases by approximate methods in order to discover further features of flame propagation. In every case the assumption of a constant reaction rate α is replaced by an assumed concentration, density, and (or) temperature dependence. The chemical complexity of the usual flame combustion processes is so great that the assumed reaction rates can only be considered as suggestive of the type of phenomena occurring. A typical reaction rate assumption is the first order rate equation

$$\alpha = \alpha_0(1 - Y)e^{-E/RT} \tag{2-4}$$

where α_0 is a constant, Y is the fraction of a unit mass of mixture which has already become a product, and α is the change of composition produced by chemical reaction only (not by diffusion or convection) and is equal to $u(d\epsilon/dx)$ where ϵ is the "flowing mass fraction," i.e. the mass fraction of products of combustion which moves by convection and diffusion through a section at x.

If we now consider a one-dimensional flow of a combustible mixture from $-\infty$ along the x axis with a stationary flame in the neighborhood of the origin and the products of combustion flowing to $x = +\infty$, the motion is described by the continuity equation for the mixture,

$$m = \rho u = \text{const} \tag{2-5}$$

the continuity equation for the products of combustion [*15*, pp. 5–41],

$$\frac{d}{dx}\left(\rho\mathfrak{D}\frac{dY}{dx}\right) - m\frac{dY}{dx} + \rho\alpha = 0 \tag{2-6}$$

and the energy equation,

$$\frac{d}{dx}k\frac{dT}{dx} - m\frac{d}{dx}[c_p^1(1 - Y) + c_pY]T + \rho Q\alpha = 0 \tag{2-7}$$

where $\mathfrak{D}$ is the coefficient of the diffusion of the products of combustion into the reactants, c_p^1 is the specific heat at constant pressure of the reactants (assumed constant), c_p is the specific heat at constant pressure of the products (assumed constant), and Q is the heat of reaction. The momentum equation is usually not written because the pressure is assumed constant throughout the system. This is a reasonable approximation since the flame speeds are usually low—of the order of 2 ft/sec—and hence velocity and pressure changes across the flame are also generally small. See Eq. 1-16 (Chap. 1). Hence the equation of state supplies a relation between the properties Y, ρ, and T.

Eq. 2-6 and 2-7 are to be compared with Eq. 1-7 and 1-3 (of Chap. 1). Eq. 2-6 and 2-7 differ from the previous ones in the following respects.

1. They are written for a volume element and hence the previous equations are a first integral of the present ones.
2. The viscous energy transfer is neglected, which is always possible in flames.
3. The kinetic energy is neglected since flame speeds are low.
4. Account is taken of the differing specific heats of the products and reactants.
5. The notation is somewhat different but terms are identified thus by the appropriate definitions

$$\rho\alpha = m\frac{d\epsilon}{dx} \quad \text{and} \quad \rho Q\alpha = \frac{d}{dx}(mq)$$

Eq. 2-5, 2-6, and 2-7 together with some reaction rate relation such as Eq. 2-4, an equation of state, and values of the diffusion coefficient, thermal conductivity, and other properties are now solvable for $Y(x)$ and $T(x)$ if suitable boundary conditions are selected.

The customary boundary conditions assumed are that at $x = -\infty$, $T = T_0$, $Y = 0$, and that the reaction is assumed to be stopped because of the low temperature, i.e. $\alpha = 0$. These conditions imply that $dT/dx = 0$. If a steady flame exists, the products of combustion ($Y = 1$) pass to $x = \infty$ at a temperature T_f given by

$$T_f = \frac{Q}{c_p} + \frac{c_p^1}{c_p}T_0 \tag{2-8}$$

which follows either from the differential equations (Eq. 2-6 and 2-7) or directly from the fact that the heat of reaction and enthalpy of the entering fluid must be carried out as enthalpy by the products of combustion. These boundary conditions are examined more critically in a later part of this article.

The system of equations is of second order in Y and T and hence only four boundary conditions are required for a solution if m is given initially.

However, since there are five conditions, m must be chosen so that a solution is possible. The velocity of the reactants given by Eq. 2-5 is then the flame speed.

This equation system has been solved analytically in various approximate ways, numerically, and by machine computation. The solutions are of the type expected, namely, the temperature rises very slowly at first from T_0 at $x = -\infty$, then more rapidly as the reaction zone is approached, and finally the temperature asymptotically approaches T_f at $x = +\infty$. The composition simultaneously changes from $Y = 0$ to $Y = 1$.

To illustrate a typical approximation, consider a flame in a mixture of constant thermal conductivity and specific heats. Also suppose that the products of combustion and the reactants have the same specific heat. Eq. 2-7 reduces to

$$k\frac{d^2T}{dx^2} - mc_p\frac{dT}{dx} + \rho Q\alpha = 0 \tag{2-9}$$

For the heating zone the reaction rate α is assumed negligible whereupon an integration of Eq. 2-9 reproduces Eq. 2-1, and if again integrated shows an exponential temperature rise from T_0 to the ignition temperature T_i. For the reaction zone it is assumed that because of the strong dependence of α upon the temperature, the entire reaction proceeds essentially at the maximum temperature T_f, i.e. the ignition temperature differs but little from T_f. Under these conditions little energy is removed from the reaction zone by convection because the fluids enter and leave at only slightly different temperatures. The heat of reaction is therefore removed by conduction into the reactants. Thus the middle term in Eq. 2-9 is neglected and, on noting that

$$\frac{d^2T}{dx^2} = \frac{1}{2}\frac{d}{dT}\left(\frac{dT}{dx}\right)^2$$

an integration yields

$$\left(\frac{dT}{dx}\right)^2_{x=x_i} = \int_{T_i}^{T_m} \frac{2\rho Q\alpha}{k}\,dT \tag{2-10}$$

Now by equating $(dT/dx)_{x=x_i}$ in the heating and reaction zones at their common boundary, Eq. 2-1 and 2-10 give for the flame speed

$$S = \frac{\kappa}{T_i - T_0}\sqrt{\int_{T_0}^{T_f} \frac{2\rho Q\alpha}{k}\,dT} \tag{2-11}$$

This equation reduces to Eq. 2-3 if α is assumed constant except for a constant factor of $\sqrt{2}$. The ignition temperature can be eliminated from Eq. 2-11 by using the assumption that $T_i = T_f$ in the denominator and $T_i = 0$ in the integral. Such seeming contradiction is reasonable since α is so strongly temperature dependent that nearly all of the value of the integral is obtained very near $T_i = T_f$ regardless of the lower limit assumed. Thus this approximation to the flame speed becomes (using Eq. 2-8)

$$S = \kappa \sqrt{\frac{2}{Q} \int_0^{T_f} \frac{\alpha c_p}{\kappa} \, dT} \tag{2-12}$$

For the special case $\kappa = \text{const}$ (essentially $\rho = \text{const}$ since k and c_p have already been assumed constant), α given by Eq. 2-4 for a first order reaction, and $\mathfrak{D} = \kappa$ (approximately correct by kinetic theory of gases), then Eq. 2-6 and 2-9 give

$$c_pT - YQ = c_pT_0 = c_pT_f - Q = \text{const} \tag{2-13}$$

as the relation between the composition Y and the temperature T required in the evaluation of α in Eq. 2-4. Note that if $\mathfrak{D} \neq \kappa$ then the simple relation (Eq. 2-13) is not true; only in this case does the diffusion exactly balance the energy conducted as heat.

Now with Eq. 2-13 and 2-4, the flame speed for a first order reaction equation (Eq. 2-12) becomes

$$S = \frac{RT_f^2}{E(T_f - T_0)} \, (2\alpha_0 \kappa e^{-E/RT_f})^{\frac{1}{2}} \tag{2-14}$$

To compare this or any other flame speed equation with experimental results requires evaluation of the various physical and chemical constants involved in the formula. All the constants except α_0 and E are obtainable at least approximately in the literature. Since only an over-all reaction was assumed, a reaction that does not actually occur in the manner assumed, the values of α_0 and E are best found from the flame speed formula itself. With the two constants adjusted in this way, all of the approximate flame speed equations yield results with generally correct trends, e.g. [*13,16,17*]. The works of Hirschfelder [*14*] and von Kármán and Penner [*15*, pp. 5–41] have dealt with reactions whose chemical simplicity appears sufficient to make a somewhat more accurate treatment possible. Thus for cases where the assumed simplicity was real, rather exact predictions were possible.

The failure to predict the flame speed precisely might be ascribed to any one of several causes:

1. The approximate solution is inadequate.
2. The physical constants are not accurately known.
3. The chemistry of the problem is not adequately described by the equations used.
4. The equations used do not properly account for all important phenomena.

These will be discussed in turn.

The test of the adequacy of the approximations can be made by an iterative improvement or by a direct numerical solution of the differential equations. This has been done for a methane air mixture [*17*] by use of a differential analyzer using the best information available for the reactants, products, and over-all rate of reaction. The differential analyzer solution gives $S = 207$ cm/sec. By an approximate method $S = 105$ cm/sec, while an experimental value for this mixture is 49 cm/sec. It is clear that the method of solving the differential equations is not alone responsible for remaining discrepancies.

The lack of knowledge of physical constants is certainly the source of some of the discrepancies since, especially for gas mixtures at high temperatures, there are numerous gaps in the data and since some of what is available is of low precision. Much additional work is needed to supply the data necessary for all aspects of the aerodynamics of combustion.

A most important set of properties on which little information is available is that of the proper reaction rate relations for various mixtures. This problem is much more fundamental than mere property measurement since it is connected with the chemistry of the problem. It is probably not possible to write an over-all reaction rate relation of general validity and reasonable precision. For precision, it will probably be necessary to know the details of the chemical processes and a reaction rate relation for each. This fact brings us to the third cause for discrepancies between the present predicted and experimental values. If the details of the chemical process are important, then a separate continuity equation for each species is required. The attack in this direction has been started by Hirschfelder, et al. [*14*] and von Kármán and Penner [*15*, pp. 5–41].

The inadequacy of the boundary conditions used to compute flame speeds has been pointed out by Emmons, Harr, and Strong [*18*] and can be seen as follows: In Eq. 2-6 and 2-7 write

$$\frac{d}{dx}\rho\mathfrak{D}\frac{dY}{dx} = \frac{dY}{dx}\left(\frac{d}{dY}\rho\mathfrak{D}\frac{dY}{dx}\right) \quad \text{and} \quad \frac{d}{dx}k\frac{dT}{dx} = \frac{dY}{dx}\left(\frac{d}{dY}k\frac{dT}{dx}\right)$$

which results in

$$\frac{d}{dY}\frac{\rho\mathfrak{D}}{m}\frac{dY}{dx} - 1 + \frac{\rho\alpha}{m(dY/dx)} = 0 \tag{2-15}$$

$$\frac{d}{dY}\frac{k}{m}\frac{dT}{dx} - T(c_p - c_p') - [c_p'(1 - Y) + c_pY]\frac{dT}{dY} + \frac{\rho Q\alpha}{m(dY/dx)} = 0 \tag{2-16}$$

In the heating zone where the reaction rate is assumed to be negligible, the term containing α in these equations might be dropped. If so, the solution curves near the origin are dY/dx proportional to Y and dT/dx nearly proportional to $T - T_0 = Y^K$ (where K is a positive constant dependent upon the fluid properties). We now note, however, that if α is merely small—not actually zero—then the term neglected is not negligible for $dY/dx \to 0$. In fact, dY/dx can be zero only for $\alpha = 0$. For a first order reaction with a rate given by Eq. 2-4, the only possible conditions for zero dY/dx are $Y = 1$ (reaction completed as assumed at $x = \infty$) and $T = 0$ (not assumed as a starting condition). Similar conclusions hold for dT/dx. The solution curves look as shown in Fig. G,2b. Starting at $Y = 1$ with any value of m (hence any value of flame speed) a solution to Eq. 2-15 and 2-16 can be constructed step by step. Thus unless the reaction rate actually *is* zero at a finite temperature, the assumptions of $T = T_0$, $Y = 0$, $dY/dx = 0$, and $dT/dx = 0$ are impossible as boundary conditions.

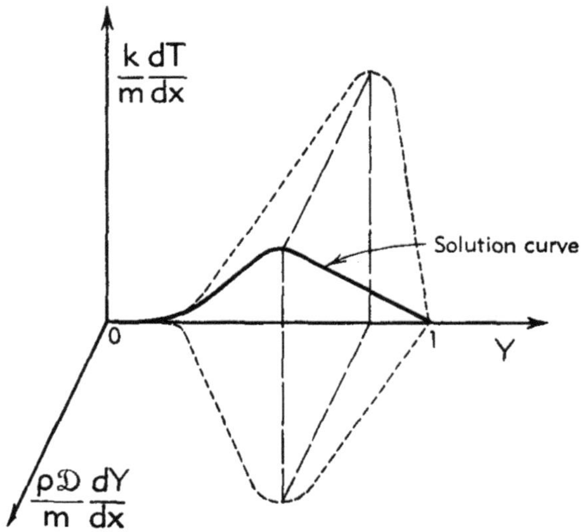

Fig. G,2b. Nature of solution of flame speed equations (Eq. 2-15 and 2-16).

Fig. G,2c shows the variation of flame speed with initial temperature gradients for several fuel heating values for a thermally propagated flame [*18*]. For high heating values there is a considerable range of temperature gradient with an essentially fixed value of the flame speed. This suggests that flames often propagate at well-established speeds, not because a unique eigenvalue exists but because the high propagation rates appropriate to very low temperature gradients are made impossible by some phenomena not included in Eq. 2-6 and 2-7. The phenomena probably important are the conduction of heat to pipe walls and other objects and the catalytic destruction of active diffusing particles by walls. In fact, these phenomena have been invoked many times to explain in a semi-quantitative way the stability of a flame on a burner. Future work will probably show them to be essential for the existence of a unique flame speed.

To summarize, then, the general mechanisms of flame propagation are known to be heat conduction and diffusion of active chemical species from the reaction zone into the heating zone. The knowledge of chemical reaction mechanisms and physical-chemical properties is not sufficient to permit high (or sometimes even reasonable) accuracy of flame speed prediction. Approximate methods of solution of appropriate differential equations have already been worked out adequately to supply the flame speed of readily combustible mixtures. However, near the composition

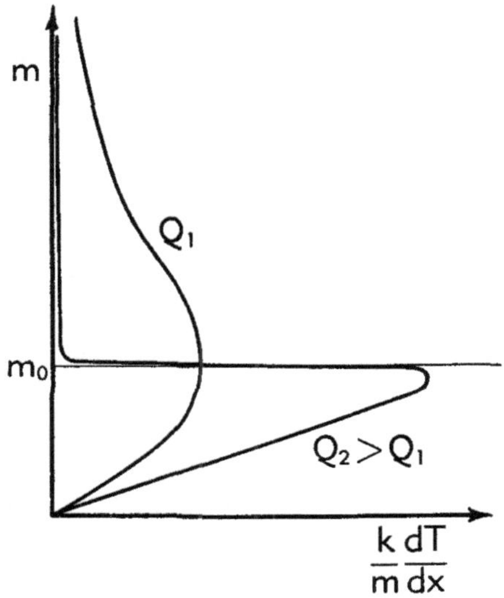

Fig. G,2c. Variation of flame speed with initial temperature gradient in combustible mixture.

limits of combustion and for problems of transition from flame to detonation, terms involving heat conduction and diffusion normal to streamlines will probably be required to get sufficient precision of description of the phenomena.

G,3. The Flame Front as a Discontinuity. In the heating zone of a flame the temperature rises exponentially from T_0 to near T_f. The characteristic length for this process is $\Delta x_{\text{heating}} = \kappa/S$ (Eq. 2-1). The characteristic length for the reaction zone $\Delta x_{\text{reaction}} = S/\alpha$ (Eq. 2-2). Thus the total streamline length of the flame is approximately

$$\Delta x = \frac{\kappa}{S} + \frac{S}{\alpha} = \frac{\kappa}{S}\frac{T_f - T_0}{T_i - T_0} \cong \frac{\kappa}{S} \tag{3-1}$$

Note that the assumption that the ignition temperature T_i differs but little from the final temperature T_f is equivalent to supposing the reaction zone to be very small compared to the heating zone. Since the thermal diffusivity of a combustible gas mixture at atmospheric pressure is about $\kappa = 0.4$ cm^2/sec while the flame speed is approximately 50 cm/sec, the order of the thickness of the entire flame region is $\Delta x = 0.008$ cm. When dealing with combustion processes in devices of customary laboratory and industrial size, the flame can be considered to be a discontinuity in fluid properties, separating the region occupied by the reactants from the region in which the products of combustion flow.

We are thus led to the question: Cannot the aerodynamics of combustion be studied as the flow of a (compressible) fluid with discontinuities embedded within it? For many aerodynamic purposes such an approach is not new. Shock waves are treated in this manner. The shock conditions must be known, i.e. the relations between the change of properties across an arbitrary shock wave must be known so that the fluid on the two sides of the discontinuity can be properly related. In the next three articles the discontinuity properties of flame fronts and the general conditions that these impose on the flow of reactants and products are considered.

Before doing so, however, the limitations of this process should be noted. In the first place it is clear that only when the size of the phenomena of interest is large compared to the flame thickness is it permissible to treat the flame as a discontinuity. From Eq. 3-1 it is seen that either a small flame speed or a high thermal diffusivity will lead to a thick flame front. The flame speed depends in a complex way on the many factors discussed in Art. 2 while the thermal diffusivity varies almost inversely with the pressure and depends in a complex way on the mixture composition and temperature. At low pressures (a few tenths of an atmosphere), flame fronts become very thick. Their thickness may exceed by a factor of 5 or so the dimensions of the burner which supplies the combustible. Under such conditions it is necessary to consider all the details of the chemical processes to get a proper description of the combustion aerodynamics processes. When the flame front is sufficiently thin, the effects of the chemistry, heat conduction, and diffusion are confined to a narrow flame zone and serve to control its spread (flame speed) and its thickness without affecting the over-all flow. Under these conditions the situation is similar to the effect of thermal conductivity and viscosity on a shock wave. The thickness is affected, but the flow before and after the shock depends only upon the over-all shock conditions.

A second restriction on the use of a flame as a discontinuity is imposed by the radius of curvature of the flame. If the radius of curvature is of the same order as the flame thickness—as perhaps at the tip of a Bunsen flame—the flame speed may be markedly affected by the curvature. This is because the thermal and diffusive mechanisms of flame propagation are greatly affected by the curvature. (The fact that the stability of flames seems to depend upon this radius-of-curvature effect is discussed in Art. 9.)

G,4. The Normal Flame Front.

General conditions for flow with a source of thermal energy. We assume now a stationary flame front in a one-dimensional flow. The flame will convert reactants with properties denoted by subscript $_1$ to products of combustion denoted by subscript $_2$. Between sections 1 and 2 of Fig.

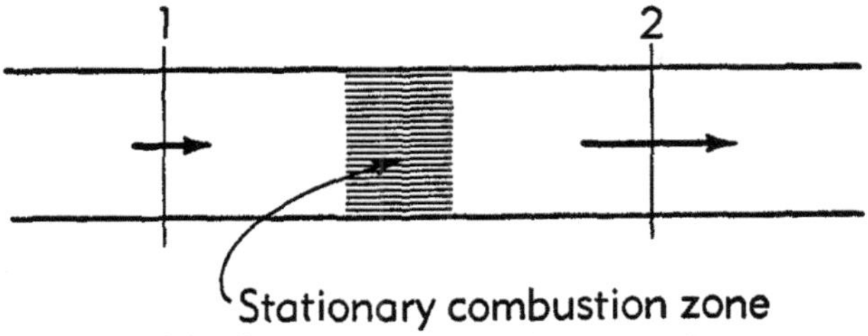

Fig. G,4a. Combustion in a channel.

G,4a, the reacting fluid will be assumed to be isolated from the surroundings, i.e. there will be no leaks (no gain or loss of mass), no friction, and no heat transfer. This method of analysis has been applied to more general problems by Shapiro, Hawthorne, and Edelman [*19*].

Conditions at 1 and 2 are therefore related by

(i) Conservation of mass

$$m = \frac{w}{A} = \rho_1 u_1 = \rho_2 u_2 \tag{4-1}$$

(ii) Conservation of momentum

$$p_1 + mu_1 = p_2 + mu_2 \tag{4-2}$$

(iii) Conservation of energy

$$h_1 + \frac{u_1^2}{2} = h_2 + \frac{u_2^2}{2} \tag{4-3}$$

For solution, these equations must be supplemented by appropriate relations between the properties indicated. We assume ideal gases

$$p_1 = \rho_1 \mathcal{R}_1 T_1, \qquad p_2 = \rho_2 \mathcal{R}_2 T_2 \tag{4-4}$$

where $\mathcal{R}_1$ and $\mathcal{R}_2$ are the gas constants for the reactants and products of combustion, respectively, and are related to the universal gas constant R by

$$\mathcal{R}_1 = \frac{R}{\mathfrak{M}_1}, \qquad \mathcal{R}_2 = \frac{R}{\mathfrak{M}_2} \tag{4-5}$$

where $\mathfrak{M}_1$ and $\mathfrak{M}_2$ are the corresponding mean molecular weights.

The enthalpy is dependent upon the specific heat and the composition through the temperature integral of the constant pressure-specific heat and the constant pressure heat of reaction. For simplicity we will assume constant specific heats and write

$$h_1 = c_{p_1}T_1 + Q \tag{4-6}$$

$$h_2 = c_{p_2}T_2 \tag{4-7}$$

where Q is related to the constant pressure heat of reaction at a temperature T_0 by

$$Q_p(T_0) = h_1(T_0) - h_2(T_0) = Q + (c_{p_1} - c_{p_2})T_0 \tag{4-8}$$

Substituting Eq. 4-6 and 4-7 into Eq. 4-3 gives the energy equation in the form

$$\frac{u_1^2}{2} + c_{p_1}T_1 + Q = \frac{u_2^2}{2} - c_{p_2}T_2 \tag{4-9}$$

Consider now Eq. 4-1, 4-2, 4-4, and 4-9, the continuity, momentum, gas law, and energy equations for steady one-dimensional flow. If all conditions at section 1 of Fig. G,4a are considered known and the energy release by reaction Q is known, then the conditions at section 2 (u_2, p_2, ρ_2, and T_2) can be computed.

To study the nature of the solution, start with Eq. 4-2 and eliminate m by Eq. 4-1 and p_1, p_2 by Eq. 4-4, this result is obtained:

$$u_2\mathcal{R}_1T_1 + u_2u_1^2 = u_1\mathcal{R}_2T_2 + u_2^2u_1 \tag{4-10}$$

Now by elimination of T_2 between Eq. 4-9 and 4-10 and the substitution of

$$\gamma_2 = \frac{c_{p_2}}{c_{v_2}}, \quad \gamma_1 = \frac{c_{p_1}}{c_{v_1}}, \quad \mathcal{R} = c_p - c_v, \quad a_1^2 = \gamma_1\mathcal{R}_1T_1 \tag{4-11}$$

we obtain the equation for u_2 in the form

$$(u_2u_1)^2 - \frac{2\gamma_2}{\gamma_2+1}\left(\frac{a_1^2}{\gamma_1} + u_1^2\right)(u_2u_1) + \frac{2(\gamma_2-1)}{\gamma_2+1}\left(\frac{a_1^2}{\gamma_1-1} + \frac{u_1^2}{2} + Q\right)u_1^2 = 0 \tag{4-12}$$

As a check on this equation we note that for $Q = 0$ the two solutions represent: (1) nothing happens at all ($u_2 = u_1$), and (2) a shock wave occurs

$$u_1u_2 = (a_1^*)^2 \tag{4-13}$$

where

$$(a_1^*)^2 = \frac{2}{\gamma+1}(a_1^0)^2 = \frac{2}{\gamma+1}\left(a_1^2 - \frac{\gamma-1}{2}u_1^2\right)$$

= square of the critical speed of sound

If $Q \neq 0$, i.e. a reaction does occur with a heat release, we again get two solutions, this time given by

$$u_1u_2 = \frac{\gamma_2}{\gamma_2+1}\left(\frac{a_1^2}{\gamma_1}+u_1^2\right) \pm \left\{\left[\frac{\gamma_2 a_1^2-\gamma_1 u_1^2}{\gamma_1(\gamma_2+1)}\right]^2 - \frac{2u_1^2}{\gamma_2+1}\left[\frac{(\gamma_2-\gamma_1)a_1^2}{\gamma_1(\gamma_1-1)}+(\gamma_2-1)Q\right]\right\}^{\frac{1}{2}} \tag{4-14}$$

If combustion occurs in a stream of reactants with given properties, the stream of products will leave with one of two possible sets of properties.

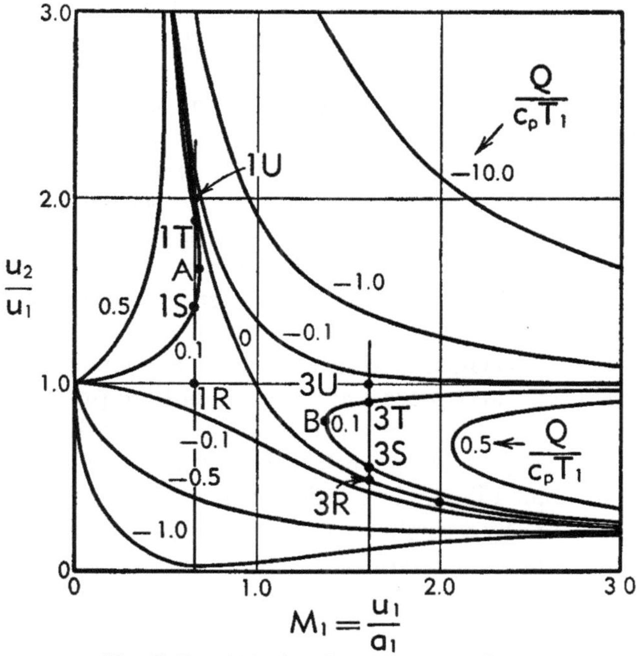

Fig. G,4b. Velocity change across a flame.

In the special case that the radical is zero, there is only one possible set of conditions which the exit stream may have. This special condition is

$$u_2 = a_2^* \tag{4-15}$$

that is, the combustion products are flowing at a Mach number $M_2 = u_2/a_2 = 1$ and the passage is blocked.

The possible velocities for the products of combustion for the case $\gamma_2 = \gamma_1 = 1.40$ are shown in Fig. G,4b. In this figure it is seen that for fluid entering the passage with a sufficiently low subsonic velocity corresponding to point 1 there appear to be four possible velocities at which the fluid may leave the passage in steady flow. Similarly, four exit conditions are possible if the fluid enters with a sufficiently high supersonic velocity. Consider this last case first. The fluid entering with velocity at point 3 may leave, with no change whatever, point 3U; with a simple compression shock without any combustion, point 3R; with a simple combustion wave, point 3T; or with a combined combustion and compression shock, point 3S. The two cases without combustion are of course well known. They are easily produced experimentally. With combustion each case has been observed, although the details are less clear. A condensation shock in a supersonic wind tunnel (this is combustion in the sense that heat is released) seems to be of type 3T of Fig. G,4b. A detonation wave may be of type 3S, although in most cases the wave changes to point B in the steady state.

Next consider fluid entering the passage with a velocity corresponding to point 1. This fluid may leave, without any change, point 1R; with a simple combustion, point 1S; with an "expansion shock," point 1U; or with an "expansion shock" plus a combustion, point 1T. Only the flows corresponding to points 1R and 1S are experimentally observed. The "expansion shock" (point 1U) is impossible since the entropy would decrease from 1 to 2 (Fig. G,4a) contrary to the second law of thermodynamics. The simple second law application is not enough to indicate the impossibility of a combined expansion and combustion. On the other hand, the second law may be extended, as suggested by its statistical interpretation, to state that the entropy must increase or in the limit remain constant for each isolated mechanism in a complex process. Thus both the expansion shock process and the combustion process take place from a molecular point of view without interaction with the other and without affecting any external work or heat transfer. Thus the process represented by point 1T may be considered as consisting of an expansion (1R to 1U) followed by a combustion (1U to 1T); since the first is impossible the whole process is impossible.

It is important to note at this point that this analysis satisfies continuity of mass, momentum, and energy for steady flow, i.e. a stationary wave between sections 1 and 2 of the passage, and leaves out of consideration all questions of stability and mechanism, either or both of which may make the indicated transformation impossible. For example, it was noted in Art. 2 that mixtures have fairly definite propagation rates controlled by the effect of pressure rise, thermal conductivity, composition, and other properties. The propagation rate required for the mixture at point 1 of Fig. G,4b is u_1. If this is not equal to or greater than the rate permitted by the internal mechanisms, no products can be produced in the steady process assumed. (If u_1 is less than S, the flame travels toward the inlet.) If the rates are equal, a normal flame front occurs in the passage. If the internal mechanism gives a flame speed of S less than the fluid velocity, a stationary combustion process can result if the flame front is stable in a position oblique to the passage. This last case does not strictly fall within the requirements of the over-all equations (Eq. 4-1, 4-2, and 4-3), since generally the products of combustion are of neither uniform velocity nor properties across the passage. In spite of this, however, the overall treatment is useful for some practical purposes.

Point 1S then can be attained only if $u_1 \geqq S$, since otherwise the flame will flash back. Finally, there is a unique point, $u_1 = S$, at which this analysis applies with strict precision because only then is there a truly normal flame. Even in this case a plane flame may be unstable, so that experimentally an irregular nonsteady flame is obtained. In the remainder of this section it is this unique deflagration condition that will be under consideration.

There is a maximum subsonic velocity (point A) and a minimum supersonic velocity (point B) between which no possible steady flow with combustion exists. The existence of point A (at an inlet Mach number of 0.14 for a stoichiometric mixture of hydrocarbon and air) imposes a serious limitation on the maximum flow into a flame, regardless of the propagation mechanisms available. Furthermore, the transition from deflagration to detonation cannot occur slowly in a succession of steady states. The change must necessarily occur as a transient so rapidly that the nonsteady terms omitted from Eq. 4-1, 4-2, and 4-3 can make transient inlet flame speeds between A and B of Fig. G,4b possible.

The diagram of possible steady flows in a combustion wave, Fig. G,4b, has not been very extensively used to date. From the point of view that the very important transient processes, such as passage from deflagration to detonation, stability of flames, etc., can only be precisely understood after the possible steady conditions are established, the more extensive use of the steady flow diagram seems desirable.

There are several difficulties connected with the construction of steady flow diagrams for a specific combustible mixture because such a diagram presupposes a knowledge of the properties of the combustible and the resulting products of combustion. In particular the heat of reaction Q is required. The principal difficulty is associated with the composition of the combustion products. The assumption of equilibrium chemical composition under the conditions (temperature and pressure) at the flame exit is good in many cases; however, the required computations are very extensive—in most cases prohibitive. This is one aspect of combustion aerodynamic problems to which the large scale computing machine would be expected to contribute in a decisive way.

G,5. Properties of a Normal Flame Treated as a Discontinuity. The most important property of the flame front is its rate of propagation into the unburned mixture, the flame speed. There are a number of experimental methods in common use for determining this value, all of which suffer from a lack of knowledge of the complete aerodynamics of the process.

1. Constant pressure or constant volume bomb (a soap bubble or a metal container). The combustion process causes gas movements which can only be corrected for in an approximate way.
2. The Bunsen burner method. In this method the velocity distribution entering the flame must be known, since it is the component of this velocity normal to the flame which is the flame speed. Usually the velocity distribution is assumed to be unaltered by the flame. Occasionally some correction is made by observing streamlines by illuminating entrained particles of solid. In neither case is high precision obtained. Another approach is based upon the fact that a constant flame speed implies a constant flow per unit area of flame, and hence a flame area dependent upon the total flow

$$w = A\rho S \tag{5-1}$$

In this method the difficulties are associated with (1) the difficulty of measuring A, (2) the difficulty of determining w (some of the flow near a wall or a flame apex may not be burned), and (3) the S probably is not exactly constant over the whole flame surface.

All of these methods, when carefully used, give moderate precision and some typical values of flame speed are shown in Fig. G,5a. These results were obtained by the Bunsen burner flame angle measurements by Johnson [*20*]. We first note the very small values of the flame speed relative to the speed of sound. Also the kinetic energy of the fluid is small compared to the heat of reaction of the normal fuel-air mixtures. Thus the steady flow energy equation (Eq. 4-3) shows that the final temperature (hence temperature jump) depends primarily upon the energy released by the chemical reactions. The energy equation written for the flame gives

$$h_2^0 = h_1^0 + Q \tag{5-2}$$

Thus if we limit our attention for the present to perfect gases of constant specific heat and define the stagnation temperature ratio n (see Eq. 4-9),

$$n = \frac{T_2^0}{T_1^0} = 1 + \frac{Q}{c_p T_1^0} \tag{5-3}$$

Then n is constant for a given combustible mixture at a given *stagnation* temperature. Since kinetic energies are generally very small in comparison, we have approximately

$$\frac{T_2}{T_1} = n \tag{5-4}$$

The pressure jump by Eq. 4-2,

$$\Delta p = -\rho_1 S \Delta u \tag{5-5}$$

is generally very small compared to the absolute pressure.

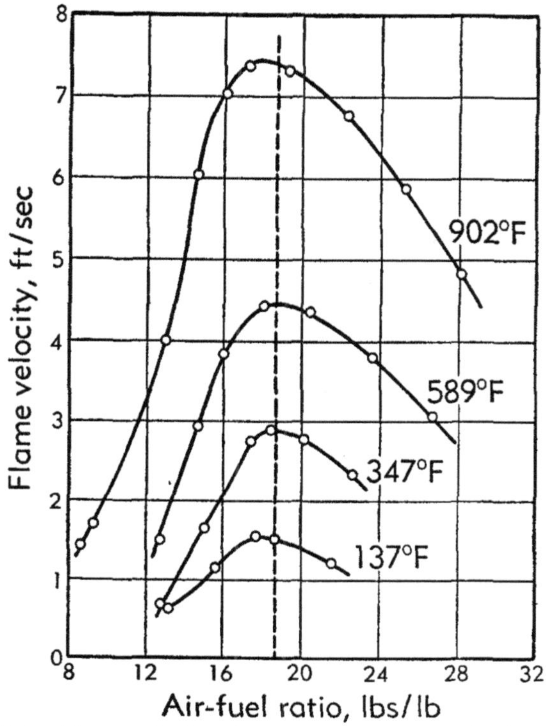

Fig. G,5a. Flame velocity of natural gas at atmospheric pressure.

For gases, then, the density jump depends primarily upon the temperature, with a small additional effect from the change of composition. Finally the velocity change follows from the continuity equation (Eq. 4-1). Thus, closely

$$\frac{\rho_1}{\rho_2} = \frac{u_2}{u_1} = n \tag{5-6}$$

the pressure jump is

$$\Delta p = -(n - 1)\rho_1 S^2 \tag{5-7}$$

and the change in stagnation pressure is

$$\Delta p^0 = -\frac{(n - 1)\rho_1 S^2}{2} \tag{5-8}$$

Typical values of the density ratio are given in Fig. G,5b as computed by Scurlock [*21*].

For a density ratio of $n = 8$ and a flame speed of $S = 2$ ft/sec the products of combustion leave the flame front at $u_2 = 16$ ft/sec, the pressure jump is $\Delta p = 0.065$ lb/ft², and the change in stagnation pressure is $\Delta p^0 = 0.033$ lb/ft². It will be noted that the pressure variations are indeed very small and for many purposes the process may be considered one of constant pressure. It is not permissible, however, to assume that the whole flow process (as on a Bunsen burner) is a constant pressure process and that therefore the divergence of streamlines is caused by temperature-induced density changes, etc. Since the dynamic pressure $\frac{1}{2}\rho S^2 = 0.0046$ lb/ft², the available pressure differences are large enough to cause very drastic flow redistributions.

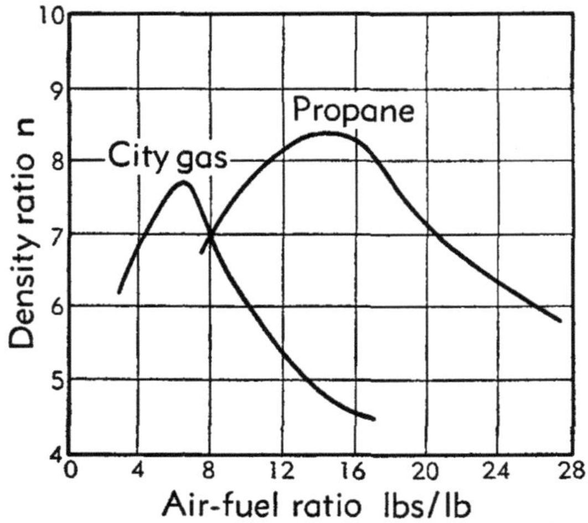

Fig. G,5b. Density ratio for several combustible mixtures.

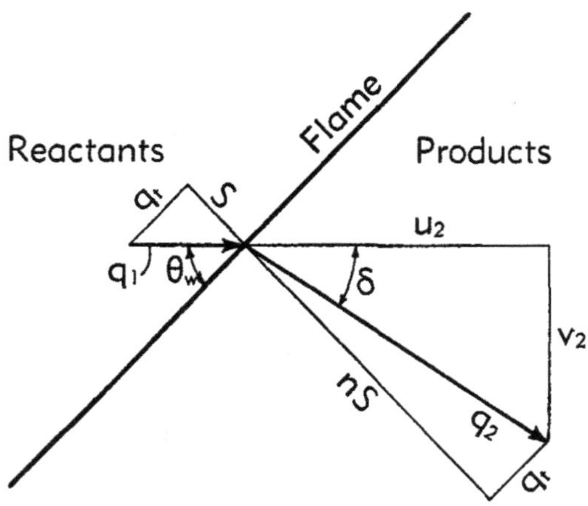

Fig. G,6a. Oblique flame velocity diagrams.

G,6. The Oblique Plane Flame Front. The discontinuity properties of a normal flame front discussed in the previous article are easily extended to oblique plane flame fronts. The simplest way to approach the study of oblique flames is to start with a normal flame, as in Fig. G,6a, which travels at velocity S and from which the products of combustion move at rate λS. (We use λ instead of n since the stagnation temperatures are altered by this experiment.)[3] We now imagine ourselves on a vehicle traveling down the page at velocity q_t parallel to the flame front. The flame front is stationary relative to the vehicle (this would not be true if we did not travel parallel to the front) and the inlet and exit velocities have changed to

$$q_1 = \sqrt{S^2 + q_t^2} = \frac{S}{\sin \theta_w} \qquad (6\text{-}1)$$
$$q_2 = \sqrt{\lambda^2 S^2 + q_t^2}$$

It will be observed that the component of velocity normal to the flame increases by the factor λ while the tangential component is the same on both sides of the flame. As a consequence, the flow angle changes abruptly across the oblique flame. If the change of direction is δ, it is related to the flame angle θ_w and density ratio λ by

$$\frac{\tan(\theta_w + \delta)}{\tan \theta_w} = \lambda$$

or $\qquad$ (6-2)

$$\tan \delta = \frac{(\lambda - 1)\tan \theta_w}{1 + \lambda \tan^2 \theta_w}$$

Since the flame speed has a value independent of the angle of obliquity (it probably depends upon the flame curvature because of the thermal-diffusive propagation mechanisms), Eq. 6-1 shows θ_w as a function of q_1. To see what a flame does to a gas stream, it is convenient to plot a flame polar (in analogy to a shock polar), as in Fig. G,6b. This figure shows all possible velocities that can exist immediately behind a flame for the given density ratio λ (assumed const $= n$) and for all possible flame angles if the combustible mixture enters the flame along the x axis. The line marked "q_1 at $\theta = 0$" shows the magnitude of the combustible mixture velocity drawn in the direction of the velocity of the products of combustion. The actual direction is of course $\theta = 0$ as noted.

[3] Since the tangential velocity component q may be large, even if S is not, we should reexamine Eq. 4-1, 4-2, and 4-3 to check the legitimacy of the assumption that density ratio $= n$. The energy equation is now $c_pT_1 + \frac{1}{2}S_1^2 + \frac{1}{2}q_t^2 + Q = c_pT_2 + \frac{1}{2}q_{n_2}^2 + \frac{1}{2}q_t^2$. Thus $n = T_2^0/T_1^0 = 1 + (Q/c_pT_1^0)$, a constant for all angles of obliquity. The density ratio now is (to second order in the velocity) $\rho_1/\rho_2 = n + (n-1)[(\gamma - 1)/2]q_t^2/(a_1^0)^2 + n(\gamma + 1)S^2/(a_1^0)^2 + \cdots$. For a normal flame front, $q_t = 0$. For $n = 8$, $\gamma = 1.4$, $S = 2$, $q_t =$ sonic velocity, the terms have a relative magnitude $\rho_1/\rho_2 = 8 + 1.16 + 0.0054$. Thus the third term is negligible but the second is not. Tsien [*22*] effectively includes this term by using $\lambda = (c_pT_2 + \frac{1}{2}q_{n_2}^2)/(c_pT_1 + \frac{1}{2}S^2)$, the ratio of normal stagnation temperatures instead of the actual stagnation temperature ratio n. The relation of λ to n is $\lambda = [(n - \frac{1}{2}q_t^2)/c_pT_1^0]/[(1 = \frac{1}{2}q_t^2)/c_pT_1^0]$.

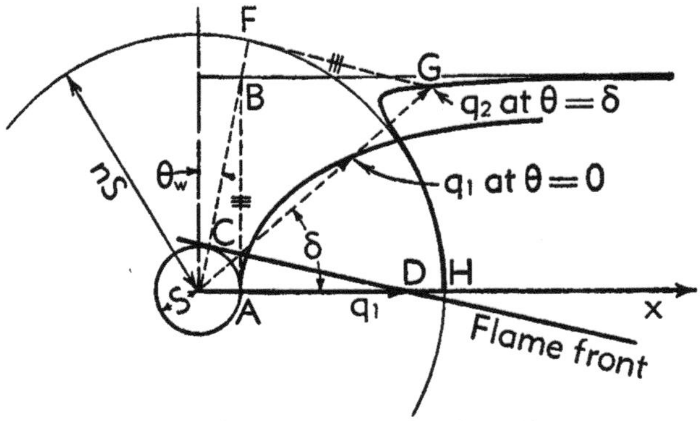

Fig. G,6b. Flame polar ($n = 6$).

The flame polar is most easily constructed as follows, referring to Fig. G,6b:

1. Draw the flame velocity circles with radii S and nS.
2. Draw AB normal to the mixture velocity direction x.
3. Draw radius OF at the flame angle θ_w.
4. Draw tangent FG equal in length to AB.

Then it follows:

1. OG is the velocity (magnitude and direction) of the products of combustion immediately after the flame.
2. OB is the magnitude of the velocity of the reactants immediately before the flame—the direction is OX.
3. Angle HOG is the velocity deviation δ.
4. CD is the flame front direction.
5. Angle ODC = angle OBA = the flame angle θ_w.

Flame polars for various values of n are shown in Fig. G,6c.

It is seen that for flame angles sufficiently near 90° the magnitude of the velocity of the products does not change much, but the direction changes considerably. Approximately (by Eq. 6-2)

$$\delta = \frac{\lambda - 1}{\lambda}(90 - \theta_w) \tag{6-3}$$

The deviations are shown in Fig. G,6d. As the flame angle decreases from 90°, the component u_2 of the velocity of products, in the direction of the reactant velocity, at first decreases. It reaches a minimum of

$$u_{2_{\min}} = 2\sqrt{\lambda - 1}\,S = 2q_1 \tag{6-4}$$

at a flame angle given by

$$\sin\theta_w = \frac{1}{\sqrt{\lambda - 1}} \tag{6-5}$$

Thereafter u_2 increases indefinitely as the flame angle θ_w approaches zero.

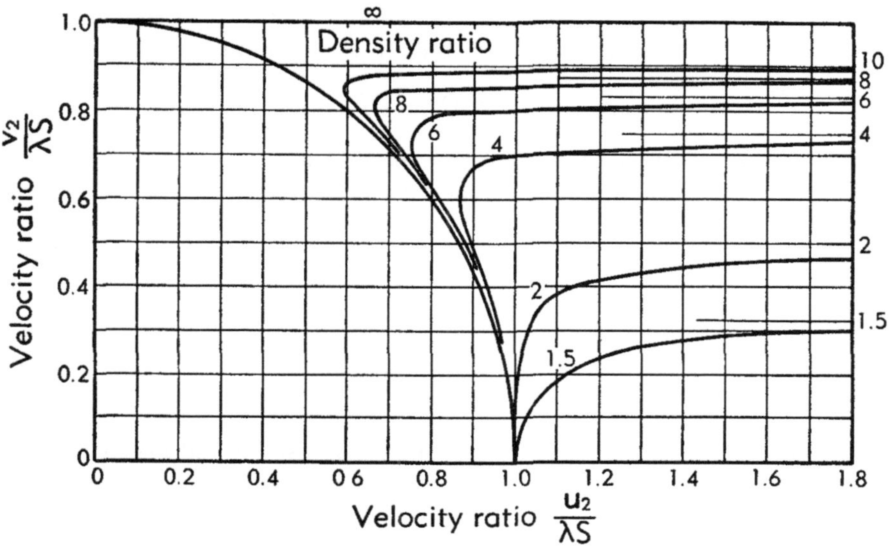

Fig. G,6c. Flame polars for various density ratios showing velocity (direction and magnitude) of combustion products resulting from horizontal reactant velocity.

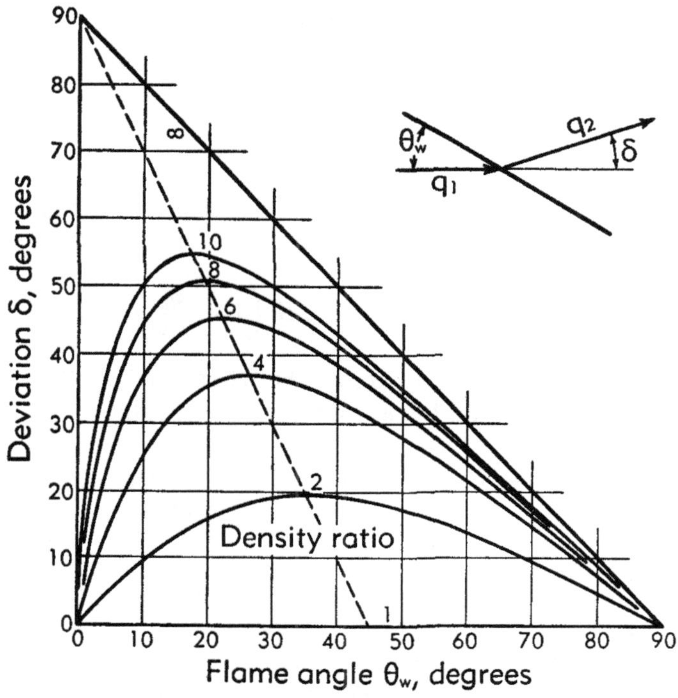

Fig. G,6d. Velocity deviations produced by a flame for various density ratios.

The component v_2 of the velocity of products normal to the reactant velocity increases monotonically with θ_w to a maximum value of

$$v_{2_{\max}} = (\lambda - 1)S \tag{6-6}$$

The deviation of the stream first increases and then decreases as the flame becomes increasingly oblique. It reaches a maximum of

$$\tan \delta_{\max} = \frac{\lambda - 1}{2\sqrt{\lambda}} \tag{6-7}$$

at a flame angle of

$$\tan \theta_w = \frac{1}{\sqrt{\lambda}} \tag{6-8}$$

A simple relation which is independent of flame speed and density ratio exists between the maximum deviation and the flame angle at which it occurs, i.e.

$$2\theta_w + \delta_{\max} = \frac{\pi}{2} \tag{6-9}$$

Fig. G,6d shows the stream deflection by a flame as a function of the flame angle θ_w and the density ratio λ.

The change of static pressure across the flame required to change the normal velocity in keeping with the density ratio is the same for all angles of obliquity

$$p_2 - p_1 = -(\lambda - 1)\rho_1 S^2 \tag{6-10}$$

Since the tangential velocity component is not changed by the flame, the total pressure change does depend upon the angle of obliquity of the flame

$$p_2^0 - p_1^0 = -\frac{\lambda - 1}{2}\rho_1 S^2 \left(1 + \frac{q_t^2}{\lambda S^2}\right) \tag{6-11}$$

The entropy increases abruptly across the flame front, both because of the combustion process and the abrupt dynamic effects.

$$e^{-\frac{\Delta s}{\mathfrak{R}}} = \frac{p_2^0}{p_1^0}\left(\frac{T_2^0}{T_1^0}\right)^{-\frac{\gamma}{\gamma-1}} = n^{-\frac{\gamma}{\gamma-1}}\left[1 - \frac{\lambda - 1}{2}\frac{\rho S^2}{p_1^0}\left(1 + \frac{q_t^2}{\lambda S^2}\right)\right] \tag{6-12}$$

G,7. General Flame Fronts. Only recently has effort been concentrated on the aerodynamic phenomena associated with combustion processes. As a consequence, only a few aspects of the problem have been investigated, which leaves most of the field yet to be studied.

One of the aspects receiving some attention is the vorticity produced by the flame front. The following simple development is taken from Emmons, et al. [*23*]. The equation of motion for a frictionless incompressible fluid can be put in the form (see Lamb [*24*]).

$$\mathbf{q} \times \mathbf{\Omega} = \frac{1}{\rho} \nabla p^0 \tag{7-1}$$

If we confine our attention to plane flow, ζ is normal to the plane and hence

$$\zeta = \frac{1}{\rho q} \frac{\partial p^0}{\partial \nu} \tag{7-2}$$

where ν is the distance in the plane normal to the streamline, since p^0—the stagnation pressure—is constant along a streamline. Now, introducing the stream function

$$\rho q = \frac{\partial \psi}{\partial \nu} \tag{7-3}$$

gives

$$\zeta = \frac{dp^0}{d\psi} \tag{7-4}$$

Thus the vorticity on each streamline is constant in the unburned gas and in the burned gas but may change abruptly at the flame front where p^0 changes discontinuously. By differentiation of Eq. 6-11 with respect to ψ, (assuming λ constant), we obtain

$$\zeta_2 - \zeta_1 = -\frac{\lambda - 1}{\lambda} \tfrac{1}{2}\rho_1 S^2 \frac{d \cot^2 \theta_w}{d\psi} \tag{7-5}$$

Since

$$\rho_1 S = \frac{\partial \psi}{\partial t} \tag{7-6}$$

where t is the distance along the flame front

$$\zeta_2 - \zeta_1 = -\frac{\lambda - 1}{\lambda} \frac{S}{2} \frac{d \cot^2 \theta_w}{dt} \tag{7-7}$$

which shows that the change of vorticity in the fluid is related to the rate of change of the flame angle along the flame. Note that θ_w is the angle between the flame and a streamline in the unburned fluid so that a straight flame front does not necessarily mean constant θ_w.

Eq. 7-5 can be put into another form. Substitute for cot θ_w the ratio of velocity components, and write

$$\frac{d \cot^2 \theta_w}{d\psi} = \frac{1}{S^2} \left(\frac{dq_t^2}{d\psi}\right)_{\text{along the flame}} \tag{7-8}$$

Now by Bernoulli's equation

$$\tfrac{1}{2}\rho_1 q_t^2 + \tfrac{1}{2}\rho_1 S^2 = p_1^0 - p_1 \tag{7-9}$$

Substitute Eq. 7-9 into Eq. 7-8

$$\tfrac{1}{2}\rho_1 S^2 \frac{d \cot^2 \theta_w}{d\psi} = \zeta_1 - \left(\frac{dp_1}{d\psi}\right)_{\text{along the flame}} \tag{7-10}$$

Now p_1 is a function of position in the plane of the flow and not a function of the stream function only. Hence, if s is the coordinate along, and ν the coordinate normal to a streamline

$$\frac{dp_1}{d\psi} = \frac{1}{\rho_1 q_1}\left(\frac{\partial p_1}{\partial \nu} + \frac{\partial p_1}{\partial s}\frac{ds}{d\nu}\right) \tag{7-11}$$

Since

$$\frac{\partial p_1}{\partial \nu} = -\frac{\rho_1 q_1^2}{R_1} \quad \text{and} \quad \frac{\partial p_1}{\partial s} = -p_1 q_1 \frac{\partial q_1}{\partial s}$$

where R_1 is the radius of curvature of the streamlines in the unburned fluid, we get finally,

$$\zeta_2 = \frac{\zeta_1}{\lambda} - \frac{\lambda - 1}{\lambda}\left(\frac{q_1}{R_1} + \frac{\partial q_1}{\partial s}\right) \tag{7-12}$$

The vorticity in the burned fluid is equal to that in the unburned fluid decreased by a factor λ, because each fluid element is stretched by this factor in passing through the flame; and also by an amount dependent upon the curvature of the streamlines in the unburned fluid and the rate of expansion of the stream tubes.

The corresponding equations for a steady flame in the three-dimensional case are derived as follows: Vazsonyi [*25*] has shown that for the steady flow of a frictionless nonheat-conducting fluid, that the equation of motion can be put in the form

$$\mathbf{q} \times \mathbf{\Omega} = \nabla h^0 - T\nabla s \tag{7-13}$$

where the stagnation enthalpy and the entropy are constant on each streamline, but may differ from one streamline to the next. Since both h^0 and s change discontinuously across the flame front, the vorticity ζ will in general also change discontinuously.

For an ideal gas this equation takes the form

$$\mathbf{q} \times \mathbf{\Omega} = (T^0 - T)\nabla s + \frac{\nabla p^0}{\rho^0} \tag{7-14}$$

while, for an incompressible fluid, $T_0 = T$ and

$$\mathbf{q} \times \mathbf{\Omega} = \frac{\nabla p^0}{\rho} \tag{7-15}$$

in agreement with Eq. 7-1.

Now consider a section of flame front with the two streamlines meeting at a point A (Fig. G,7). Place a rectangular coordinate system with origin at A, with the n axis normal to the flame, and t and ν axes in the flame. The momentum equation written for a volume element at A in any direction tangent to the flame shows that the tangential velocity of the burned fluid is the same as the tangential velocity of the unburned fluid. The normal velocity component will change as previously discussed (see Fig. G,4b). Thus the velocity vectors before ($\mathbf{q}_1$) and after ($\mathbf{q}_2$) the flame lie in one plane which is perpendicular to the flame. If we now rotate the axes about n so that the t axis lies along the intersection of the flame and the velocity plane in the direction of the tangential velocity while the ν axis is normal to the velocity plane, the velocity vectors have components,

$$\begin{aligned} &\mathbf{q}_1(S, q_t, 0) \\ &\mathbf{q}_2(q_{2n}, q_t, 0) \end{aligned} \tag{7-16}$$

Next we observe that the unburned gases come from a region which determines the distribution between streamlines of h_1^0 and s_1. If the gases

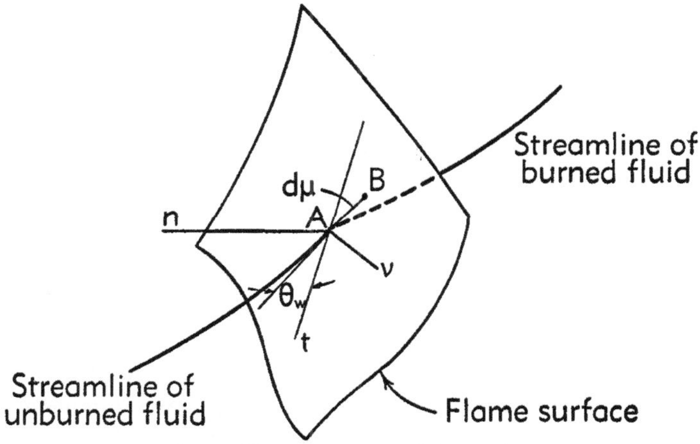

Fig. G,7. Three-dimensional flame element.

flow with negligible friction and heat transfer from a large reservoir at a constant temperature and pressure, h_1^0 and s_1 will be constants. On the other hand, if the unburned gases are supplied to the flame from a long insulated tube in laminar flow, the gas will come from a region of uniform pressure p_1 and temperature T_1—hence of uniform entropy s—but of nonuniform velocity—hence of nonuniform stagnation enthalpy h_1^0. In any case, a value of h_1^0 and s_1 is set at each point of the flame by the source of the streamline in the unburned fluid.

The flame itself causes discontinuous changes of fluid properties determined by the local conditions in the unburned fluid. Thus also the fluid properties h_2^0 and s_2 at each point behind the flame will be given by

$$\begin{aligned} h_2^0 - h_1^0 &= \Delta h^0 \\ s_2 - s_1 &= \Delta s \end{aligned} \tag{7-17}$$

where Δh^0 and Δs are the changes occurring across the flame.

By Eq. 4-9

$$\Delta h^0 = Q \tag{7-18}$$

the heat of reaction of the mixture. The entropy change Δs is given by the solution to Eq. 4-1, 4-2, and 4-3, which may be given in closed form but is more conveniently computed from intermediate results.

Now to determine the change in vorticity across the flame, consider a second point B, in the flame a distance $d\mu$ away from point A.

Eq. 7-13 applied to the burned and unburned fluids gives,

$$\begin{aligned} \boldsymbol{\mu} \cdot \mathbf{q}_1 \times \boldsymbol{\Omega}_1 &= \frac{\partial h_1^0}{\partial \mu} - T_1 \frac{\partial s_1}{\partial \mu} \\ \boldsymbol{\mu} \cdot \mathbf{q}_2 \times \boldsymbol{\Omega}_2 &= \frac{\partial h_2^0}{\partial \mu} - T_2 \frac{\partial s_2}{\partial \mu} \end{aligned} \tag{7-19}$$

where $\boldsymbol{\mu}$ is the unit vector along $d\mu$.

Now if we make the reasonable assumption that Q is constant, Eq. 7-17 give,

$$\begin{aligned} \frac{\partial h_1^0}{\partial \mu} &= \frac{\partial h_2^0}{\partial \mu} \\ \frac{\partial s_2}{\partial \mu} &= \frac{\partial s_1}{\partial \mu} + \frac{\partial \Delta s}{\partial \mu} \end{aligned} \tag{7-20}$$

Combining Eq. 7-19 and 7-20

$$\boldsymbol{\mu} \cdot \mathbf{q}_2 \times \boldsymbol{\Omega}_2 = \boldsymbol{\mu} \cdot \mathbf{q}_1 \times \boldsymbol{\Omega}_1 + (T_1 - T_2) \frac{\partial s_1}{\partial \mu} - T_2 \frac{\partial \Delta s}{\partial \mu} \tag{7-21}$$

This equation relates the vorticity before and after the flame. Since $d\mu$ is any direction in the flame front, there are only two independent relations here. Let the vorticity vectors before and after the flame be resolved into components along the flame normal and along the two tangential directions t, ν as used in Fig. G,7:

$$\begin{aligned} &\boldsymbol{\Omega}_1(\zeta_{1n}, \zeta_{1t}, \zeta_{1\nu}) \\ &\boldsymbol{\Omega}_2(\zeta_{2n}, \zeta_{2t}, \zeta_{2\nu}) \end{aligned} \tag{7-22}$$

By choosing $\boldsymbol{\mu}$ along the t direction, Eq. 7-21 gives

$$\zeta_{2\nu} = \frac{S}{q_{2n}} \zeta_{1\nu} - \frac{T_1 - T_2}{q_{2n}} \frac{\partial s_1}{\partial t} + \frac{T_2}{q_{2n}} \frac{\partial \Delta s}{\partial t} \tag{7-23}$$

With $\boldsymbol{\mu}$ along ν,

$$\zeta_{2t} = \frac{S}{q_{2n}} \zeta_{1t} + \frac{T_1 - T_2}{q_{2n}} \frac{\partial s_1}{\partial \nu} - \frac{T_2}{q_{2n}} \frac{\partial \Delta s}{\partial \nu} \tag{7-24}$$

Finally

$$\zeta_{1n} = \zeta_{2n} \tag{7-25}$$

which follows from the definition of ζ since velocity components parallel to the flame are preserved.

We can now see the general picture. Unburned fluid moves toward the flame in accordance with the equations of motion. Thus viscosity can be neglected except within boundary layer regions. The fluid outside of the boundary layers may, however, carry vorticity produced in some previous boundary layer (or shock wave). If so, this vorticity is carried along in accordance with Eq. 7-13 wherein h^0 and s are constant on streamlines. Then across the flame the normal component of vorticity is preserved while the tangential components are decreased by the ratio of normal velocities because of the volume expansion of the fluid and are further changed by amounts dependent upon the initial entropy distribution and the entropy change across the flame.

Irrotational motion of the unburned gases with constant h^0 (and therefore s) does not imply irrotational motion of the burned gases. Such is only likely to be the case for a plane flame front, since only in this case is the change of entropy across the flame likely to be the same everywhere. In general, curved flames mean rotational motion of the burned gas.

Notice that while vorticity is altered by a flame, there is no necessary connection between the flame and the production of turbulence. If the unburned gases are turbulent, the nature of the turbulence will be altered, usually made more violent, by the flame. If the initial flame carried isotropic turbulence, the burned gases would carry initially anisotropic turbulence. However, flame sheets, steady or nonsteady, produce shear flows, not turbulent flows.

G,8. Steady Flames. The considerations of the preceding chapter make the computation of steady flame possible, at least in principle.

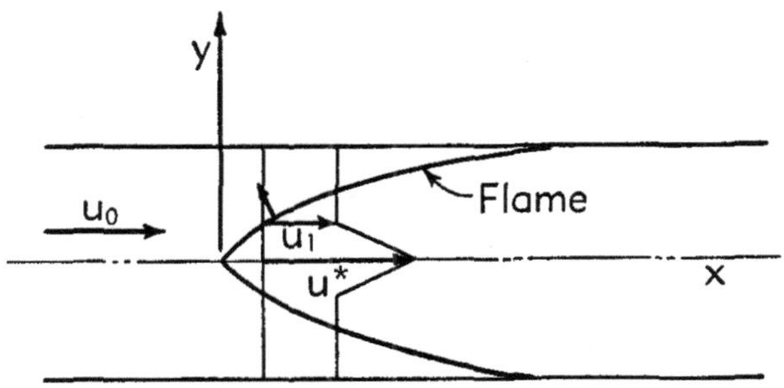

Fig. G,8a. A flame in a channel.

Scurlock [*21*] solved for the flame front in a channel (Fig. G,8a) by a one-dimensional treatment in which the pressure was assumed constant at each channel section. The momentum and continuity equations written in the direction of the channel axis and expressed in terms of the fraction burned as the independent variable gave equations permitting a numerical solution of this problem. On the basis of Scurlock's work, Tsien [*22*] obtained analytical solutions of the same problem by assuming a linear velocity profile in the products of combustion. Finally, this same problem was solved by Ball [26] using a relaxation method and the general equations of motion of a fluid in two dimensions. Velocity profiles of these three treatments are compared in Fig. G,8b for the case $n = 4.5$, $S = 0.4u_0$ (see Fig. G,8a for notation). The appearance of the streamlines and flame is shown in Fig. G,8c.

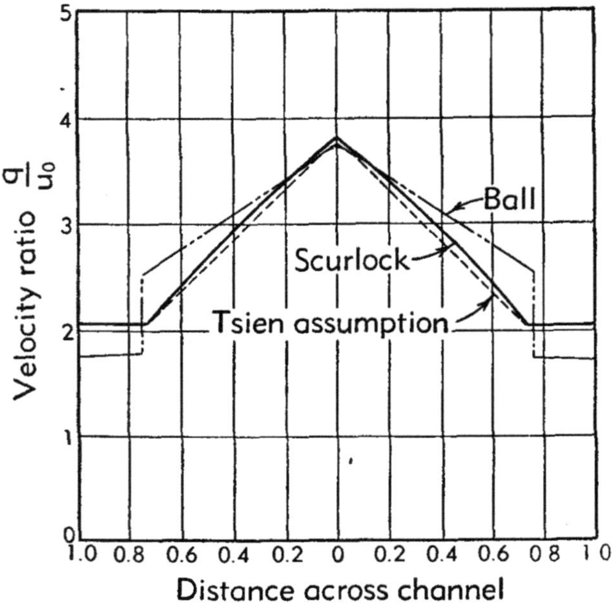

Fig. G,8b. Velocity distribution across channel for 50 per cent of initial mixture burned. Density ratio $\lambda = 4.5$.

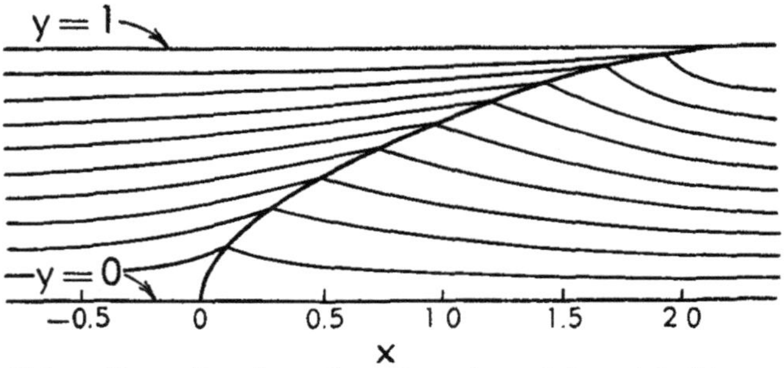

Fig. G,8c. Streamlines for a flame in a channel ($\lambda = 4.5$, $S/u_0 = .4$).

The simplified analysis of the problem by Tsien [22] is as follows: For incompressible fluids the Bernoulli equations for the unburned and burned fluids are

$$\begin{aligned} p_0 - p &= \tfrac{1}{2}\rho_1(u_1^2 - u_0^2) \\ p_0 - p &= \tfrac{1}{2}\rho_2(u^{*2} - u_0^2) \end{aligned} \tag{8-1}$$

where p is the static pressure at an arbitrary section of the channel while p_0 is the static pressure of the unburned gases at the inlet. The velocities are defined in Fig. G,8a. These equations neglect both the static and total pressure drop across the flame. By elimination of the pressures

$$\frac{u^*}{u_0} = \left[1 + n\left(\frac{u_1^2}{u_0^2} - 1\right)\right]^{\frac{1}{2}} \tag{8-2}$$

where

$$n = \frac{\rho_1}{\rho_2}$$

Assuming a linear velocity distribution in the burned gases, we can write the velocity as a function of the unburned gas velocity ratio u_1/u_0.

$$\frac{u_2}{u_0} = \left\{1 + n\left[\left(\frac{u_1}{u_0}\right)^2 - 1\right]\right\}^{\frac{1}{2}} - \left\{\left[1 + n\left(\left(\frac{u_1}{u_0}\right)^2 - 1\right)\right]^{\frac{1}{2}} - \frac{u_1}{u_0}\right\}\frac{y}{y_1} \tag{8-3}$$

where y_1 is the width of the flame at the channel section where the unburned fluid has the velocity u_1.

Now by the continuity relation, we have

$$\rho\int_0^{y_1} u_2 dy + \rho_1(b - y_1)u_1 = \rho_1 b u_0 \tag{8-4}$$

where b is the channel width. By substituting Eq. 8-3 for u_2 into Eq. 8-4 the ratio of the flame width to the channel width (y_1/b) can be expressed in terms of the velocity ratio (u_1/u_0)

$$\frac{y_1}{b} = \frac{2n\left(\frac{u_1}{u_0} - 1\right)}{(2n - 1)\frac{u_1}{u_0} - \left\{1 + n\left[\left(\frac{u_1}{u_0}\right)^2 - 1\right]\right\}^{\frac{1}{2}}} \tag{8-5}$$

The fraction f of gas burned is

$$f = 1 - \frac{\rho_1(b - y_1)u_1}{\rho_1 b u_0} = 1 - \frac{u_1}{u_0}\left(1 - \frac{y_1}{b}\right) \tag{8-6}$$

The relation between the flame width and fraction burned is easily computed directly from Eq. 8-5 and 8-6 by considering u_1/u_0 as a parameter. The results are shown in Fig. G,8d. The flame shape and position in the channel can now be constructed graphically by starting at the flame holder and drawing the flame step by step, using the flame speed S and the unburned gas velocity u_1.

The flame is seen to cause the unburned fluid to be crowded from the center of the channel toward the wall. As the flame polar figure (Fig. G,6b) shows, there are only two ways for the flame to reach a wall along which there is to be no separation: either normally or tangentially. Ball [26], who has made the most detailed analysis of this point to date, was unable to find a solution with a flame normal to the wall at the 100-per cent burned point. Thus the unburned fluid velocity becomes very large (theoretically ∞) in this neighborhood. For real (compressible) fluids there is a limit velocity, and hence the flame cannot reach the wall tangentially.

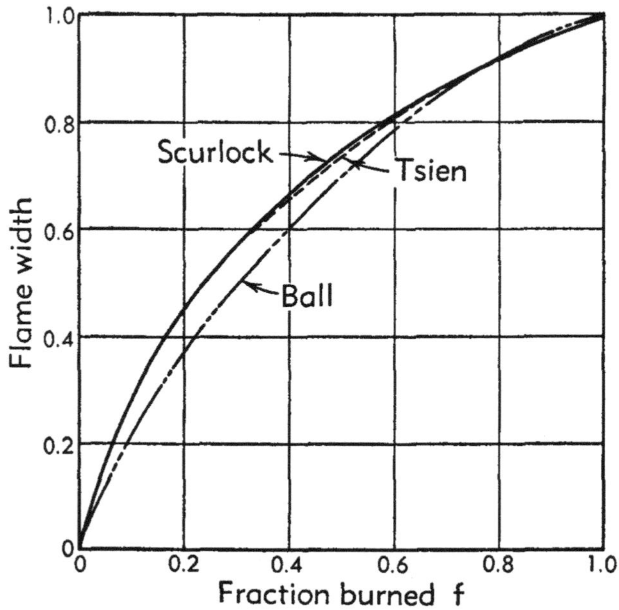

Fig. G,8d. Width of the flame in a channel (incompressible fluid). $\lambda = 4.5$.

In view of the accuracy of the simple analysis, Tsien [*22*] applied the same assumptions (pressure constant on channel sections and linear velocity distribution in the burned gases) to the combustion of ideal gases. Without giving the derivation which follows the previous one, step by step, the result is shown in Fig. G,8e. We note that there is a maximum possible fraction burned. The reason for this effect is that any additional burning would produce additional volume flow in the channel. This can be accommodated only by an increase of velocity which in turn means a pressure drop. When the unburned fluid reaches sonic velocity, a further velocity increase is possible only by an increase in cross section of the unburned fluid stream. Thus the flow in the channel blocks. Note that the unburned fluid causes the blocking rather than the burned fluid (see discussion of Fig. G,4b).

The critical inlet Mach numbers at which the combustion blocks the passage are shown in Fig. G,8f. No experimental evidence is yet available with which to check these results, and indeed some modification may be required by the failure of some of the assumptions. However, such experimental deviations would be expected to be small.

So far only the flame in a channel of uniform area has been considered.

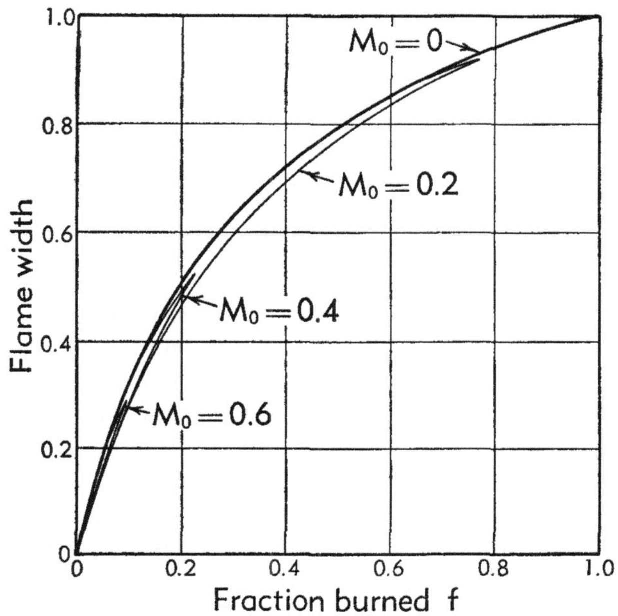

Fig. G,8e. Width of the flame in a channel (compressible fluid). $\lambda = 6$.

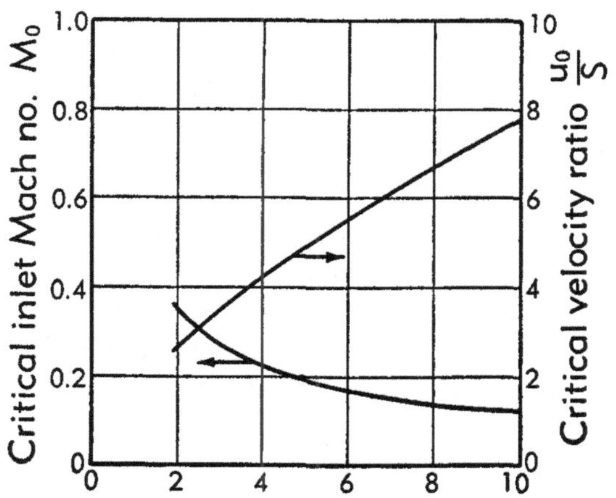

Fig. G,8f. Critical inlet combustion blocking Mach numbers.

The important case of a Bunsen burner flame has not yet been solved. The free jet boundaries are of course a major difficulty in this problem. Combustion problems can be set up as integral equations by considering the free streamlines as vortex sheets and the flame front as a source sheet. Such techniques have not been studied to the point of yielding useful solutions.

The absence of complete combustion for aerodynamic reasons has practical implications in combustion chamber design. The absence of complete combustion has generally been supposed to be associated with premature quenching of the flame by walls. A wall can quench a flame by removal of heat or catalytic effects, or both. The tip of a Bunsen burner flame may be open or closed, depending upon circumstances. The explanation has again usually been one of increase or decrease of flame speed by the flame tip curvature. The channel solutions suggest that aerodynamic effects may also be very important. However, at present these questions are among the unsolved problems.

G,9. Stability of Flames. Anyone who has looked at flames at all critically is struck by the infinite variety of effects to be observed. While the simple theory shows that a plane flame front should always be possible, it is rarely observed. Even if we observe cases where a burner cone or other simple flame figure should be possible, we find that the cone is ribbed with or without rotation, is irregular and steady, or is simply unsteady. Under some conditions it is unstable and either blows back or blows off. Very few of these effects have been clarified in basic principle. In fact, few have a clear indication of whether they depend upon physical, chemical, or aerodynamic causes, or some combination of them.

A start has been made on the problem of the instability of flames as a combination of aerodynamic and physical factors. Landau [*27*] has analyzed a plane flame and found it to be unstable with respect to disturbances of all wavelengths. Recently, Markstein [*28*], in an experimental and theoretical study, considered the breakup of plane flame fronts into cells. Some of the experimental flames viewed from the unburned gas are shown in Plate G,9. The cell size was found to vary inversely as the one-third power of the molecular weight of the fuel and the three-fourths power of the mixture pressure. The experimental results did not disclose the basic reason for the cell structure, but an extension of Landau's analysis is very suggestive. The basic reason for the instability of a plane flame of constant flame speed is indicated by Fig. G,9a. The unburned gases flowing toward a disturbed flame have streamlines as indicated, with flow diverted from the forward flame front toward the downstream portion. If the flame speed is constant, the amplitude of the disturbance will grow since the approach speed of the unburned gases will be low at point A and high at point B. Markstein's extension of this theory consists in supposing that the flame speed is not constant but depends upon the radius of curvature of the flame.

$$S = S_0\left(1 + \mu\frac{L}{R} + \cdots\right) \tag{9-1}$$

where S, the flame speed, depends upon the flame speed S_0 of the plane flame front as well as the radius of curvature R of the actual flame front. The characteristic length L is taken as the flame thickness, of the order of α/S_0 according to Eq. 3-1.

If the flame is concave upstream, R is taken as positive (as at B of Fig. G,9a), then the flame speed would be greater according to the thermal flame propagation view, because of the fact that heat is conducted into unburned gas from the sides as well as from downstream. Hence μ would

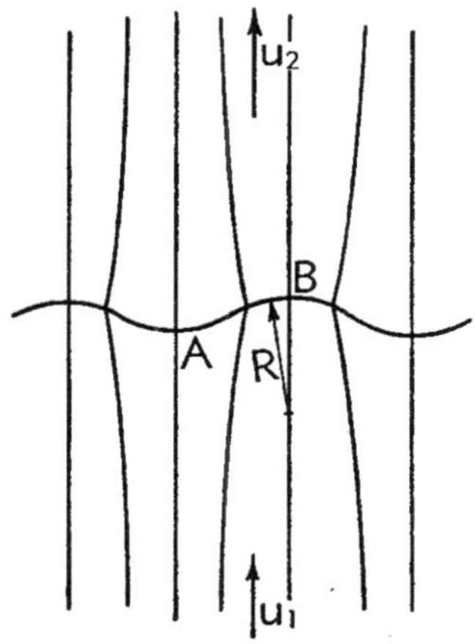

Fig. G,9a. Aerodynamic effects of oscillating flame.

be expected to be positive in Eq. 8-1. The effect of this increase of flame speed (at B, Fig. G,9a) is to increase the stability of the flame because the flame propagates fastest where the fluid speed is greatest.

To analyze this effect in greater detail, we start with a combustible mixture in a channel flowing at the flame speed $u = S_0$, burning at a plane flame, and leaving at a uniform speed $u = \lambda S_0$. We will assume incompressible fluids throughout the analysis but cannot neglect the pressure drop across the flame, which for the undisturbed flame is given by Eq. 5-7:

$$p_{0_2} - p_{0_1} = -(\lambda - 1)\rho_1 S_0^2 \tag{9-2}$$

We now perturb this solution by writing

$$\begin{aligned} u_1 &= S_0 + u_1' \\ v_1 &= v_1' \\ p_1 &= p_{0_1} + p_1' \end{aligned} \tag{9-3}$$

for the unburned fluid $(\quad)_1$. Similar equations are written for the products of combustion $(\quad)_2$.

$$\begin{aligned} u_2 &= \lambda S_0 + u_2' \\ v_2 &= v_2' \\ p_2 &= p_{0_2} + p_2' \end{aligned} \tag{9-4}$$

By inserting Eq. 9-3 into the continuity equation and the two equations of motion for a two-dimensional flame, and rejecting second order terms, we get for the unburned gas:

$$\frac{\partial u_1'}{\partial x} + \frac{\partial v_1'}{\partial y} = 0$$

$$\frac{\partial u_1'}{\partial t} + S_0 \frac{\partial u_1'}{\partial x} = -\frac{1}{\rho_1}\frac{\partial p_1'}{\partial x} \qquad (9\text{-}5)$$

$$\frac{\partial v_1'}{\partial t} + S_0 \frac{\partial v_1'}{\partial x} = -\frac{1}{\rho_1}\frac{\partial p_1'}{\partial y}$$

Three similar equations exist for the burned gases.

The six differential equations for the burned and unburned gases must be solved for the four perturbation velocities and two perturbation pressures, subject to the boundary conditions that the perturbations all be

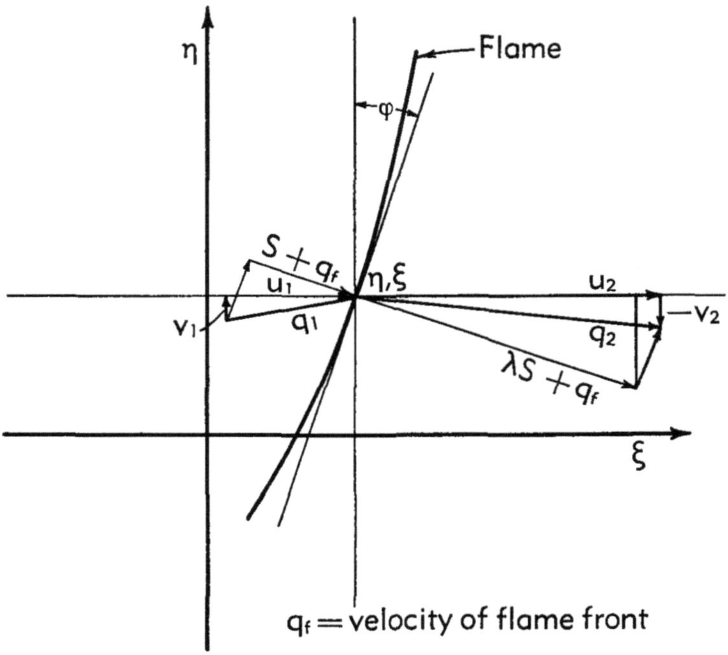

Fig. G,9b. Velocity triangles for a nonstationary flame.

zero far upstream ($x = -\infty$) and that the flow at the flame front satisfy the discontinuity conditions there. These latter conditions are derived from Fig. G,9b where $\xi(y, t)$ is the flame displacement from its original position. The flame velocity along the x axis is

$$\frac{\partial \xi}{\partial t} = u_1 - S \cos \varphi = u_2 - \lambda S \cos \varphi$$

The velocity along the y axis is

$$v_1 + S \sin \varphi = v_2 + \lambda S \sin \varphi \qquad (9\text{-}6)$$

and the pressure difference is

$$p_2 - p_1 = -(\lambda - 1)\rho_1 S^2$$

The solution of this system is carried out for a harmonically disturbed flame, i.e.

$$\xi = \frac{A_4}{h} g(y, t) \tag{9-7}$$

where $g(y, t) = e^{ihy+\delta t}$

A_4/h is the amplitude factor of the perturbation

$h = 2\pi/\lambda'$ is the wave number, with λ' the wavelength, of the perturbation

δ is the damping factor.

The sign of δ is to be found from the solution, since for $\delta > 0$ the flame displacement amplitude grows with time.

The solution of the differential equations (Eq. 9-5) which satisfies the conditions at $x = -\infty$ is

$$\begin{aligned}
\frac{u_1'}{S_0} &= A_1 e^{hx} g(y, t) \\
\frac{v_1'}{S_0} &= iA_1 e^{hx} g(y, t) \\
\frac{p_1'}{\rho_1 S_0^2} &= -(\alpha\lambda + 1) A_1 e^{hx} g(y, t) \\
\frac{u_2'}{S_0} &= (A_2 e^{-hx} + A_3 e^{-\alpha hx}) g(y, t) \\
\frac{v_2'}{S_0} &= i(-A_2 e^{-hx} - \alpha A_3 e^{-\alpha hx}) g(y, t) \\
\frac{p_2'}{\rho_1 S_0^2} &= (\alpha - 1) A_2 e^{-hx} g(y, t)
\end{aligned} \tag{9-8}$$

where A_i are constants to be determined by the flame front conditions, and where

$$\alpha = \frac{\delta}{\lambda h S_0}$$

Before using the flame front conditions we note that the perturbation is to be small, and hence we may approximate

$$\cos\varphi \cong 1, \quad \sin\varphi \cong \frac{\partial \xi}{\partial y}, \quad \frac{1}{R} \cong -\frac{\partial^2 \xi}{\partial y^2} \tag{9-9}$$

Hence the flame speed itself is given by Eq. 9-1 as

$$S = S_0(1 + \sigma A_4 g(y, t)) \tag{9-10}$$

where $\sigma = \mu h L$.

Now the substitution of the solution equations (Eq. 9-7 and 9-8) into the flame equations (Eq. 9-6) gives, on again neglecting second order terms,

$$\begin{aligned} A_1 - (\alpha\lambda + \sigma)A_4 &= 0 \\ A_2 + A_3 - \lambda(\alpha + \sigma)A_4 &= 0 \\ A_1 + A_2 + \alpha A_3 - (\lambda - 1)A_4 &= 0 \\ (\alpha\lambda + 1)A_1 + (\alpha - 1)A_2 + 2\sigma(\lambda - 1)A_4 &= 0 \end{aligned} \tag{9-11}$$

This system of linear equations has a nontrivial solution for the A_i only if the coefficient determinant is zero. Hence

$$\begin{vmatrix} 1 & 0 & 0 & -(\alpha\lambda + \sigma) \\ 0 & 1 & 1 & -\lambda(\alpha + \sigma) \\ 1 & 1 & \alpha & -(\lambda - 1) \\ \alpha\lambda + 1 & \alpha - 1 & 0 & 2\sigma(\lambda - 1) \end{vmatrix} = 0 \tag{9-12}$$

By getting zeros in place (1, 4) and (3, 3), this determinant reduces to

$$(\alpha - 1)[\lambda(\lambda + 1)\alpha^2 + 2\lambda(\sigma + 1)\alpha + 2\sigma\lambda - \lambda + 1] = 0 \tag{9-13}$$

which then yields the three roots

$$\begin{aligned} \alpha_{1,2} &= \frac{1}{\lambda + 1}\left\{-(\sigma + 1) \pm \left[1 + \lambda - \frac{1}{\lambda} + \sigma^2 - 2\sigma\lambda\right]^{\frac{1}{2}}\right\} \\ \alpha_3 &= 1 \end{aligned} \tag{9-14}$$

The root α_3 has no physical significance since it leads to $A_i = 0$, while the negative radical sign leads only to stable solutions. Thus the unstable solutions are to be found only from the positive sign of the radical. Hence

$$\frac{\delta L}{S_0} = \frac{\lambda}{\lambda - 1}\chi\left\{\left[1 + \lambda - \frac{1}{\lambda} + (\mu\chi)^2 - 2\lambda\mu\chi\right]^{\frac{1}{2}} - (1 + \mu\chi)\right\} \tag{9-15}$$

where

$$\chi = \frac{2\pi L}{\lambda'}$$

yields the conditions for stability.

This solution is shown in Fig. G,9c for $\lambda = 5$. The line $\mu = 0$ is Landau's solution [*27*] and corresponds to unstable solutions for all wavelength disturbances. As qualitatively noted before, the effect on the flame disturbances is stabilizing if μ is positive, i.e. if the flame speed increases when the flame is concave toward the unburned gas. We note, however, that regardless of how rapidly the flame speed increases with flame radius of curvature the flame is unstable for sufficiently long wavelengths (i.e. sufficiently small $2\pi L/\lambda'$). This, too, is physically plausible since for a fixed amplitude of disturbance (ξ assumed small) an increase of wavelength causes a decrease of radius of curvature and hence a decrease of its beneficial effects.

The occurrence of a range of instability checks well with the experimental results since the resulting maximum suggests that a specific size of unstable cell would be observed, which was indeed the case in Plate G,9.

Markstein [28] has shown that the thermal theory of flame propagation leads to a flame speed-radius of curvature dependence of about $\mu = 1$ with $L = \alpha/S_0$. Thus reasonable values (1 cm) for the flame cell size are estimated. The occurrence of completely stable flames for lean hydrocarbon-air mixtures has two suggestive explanations. Such mixtures may,

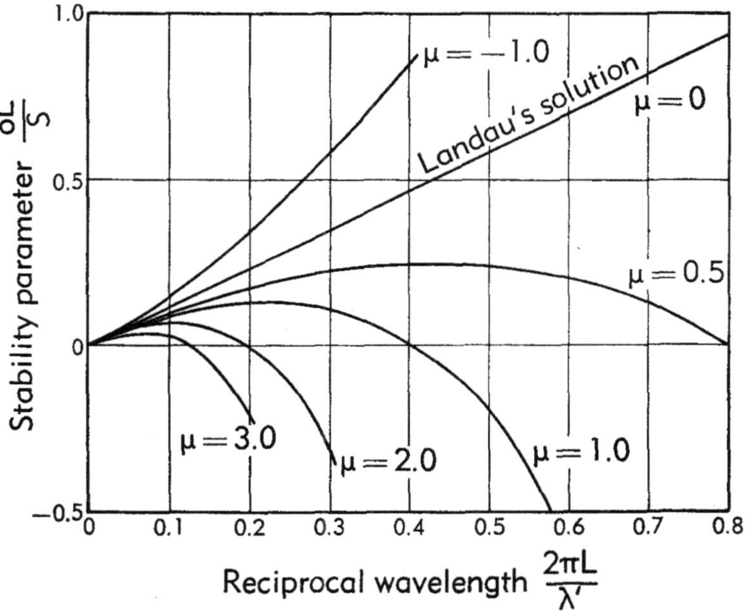

Fig. G,9c. Stability parameter for a flame.

for some reason not yet known, have μ values that are very large. In this case the range of unstable wavelength may be restricted to such large values so as to be unobservable on the largest size of apparatus used so far. Another possible explanation to be coupled with this is the stabilizing effect of gravity. A flame front (so far as gravity effects are concerned) is like the interface between two fluids of different density. With the unburned fluid below the flame, any flame perturbation is counteracted by gravity since the high point of the wave (downstream portion of perturbed flame) tends to fall back toward a plane surface. A mathematical analysis of this effect by Markstein [28] gives the expected stabilizing influence of gravity. It would be interesting to repeat the flame stability experiments with the unburned fluid admitted at the top, so as to test the destabilizing effect of gravity to be expected in this case.

CHAPTER 3. GAS DYNAMICAL ASPECTS OF DETONATION

G. I. TAYLOR
R. S. TANKIN

G,10. The Chapman-Jouguet Theory of Detonation. The idea that detonation is propagated by means of a shock wave at which very rapid chemical action takes place led Chapman and Jouguet to study the dynamical consequences of such an hypothesis.

If a detonation is advancing at a uniform speed U into an unburned stationary explosive of density ρ_1, the equations of continuity, momentum, and energy are

$$\rho_2(U - u_2) = \rho_1 U \tag{10-1}$$

$$p_2 + \rho_2(U - u_2)^2 = p_1 + \rho_1 U^2 \tag{10-2}$$

$$\frac{p_2}{\rho_2} + \tfrac{1}{2}(U - u_2)^2 + e_2 = \frac{p_1}{\rho_1} + \tfrac{1}{2}U^2 + e_1 \tag{10-3}$$

p_1, ρ_1, and e_1 are the initial pressure, density, and energy per unit mass of the explosive and p_2, ρ_2, and e_2 are the corresponding quantities in the products of combustion, while u_2 is their velocity. u_2 is positive when the products of combustion move in the same direction as the shock wave relative to the unburned explosive. From Eq. 10-1 it is seen that u_2 is positive or negative according to whether ρ_2 is greater or less than ρ_1. Eliminating u_2 from Eq. 10-1 and 10-2

$$\frac{p_2 - p_1}{(1/\rho_1) - (1/\rho_2)} = \rho_1^2 U^2 \tag{10-4}$$

and eliminating U and u_2 from Eq. 10-1, 10-2, and 10-3

$$\tfrac{1}{2}(p_1 + p_2)\left(\frac{1}{\rho_1} - \frac{1}{\rho_2}\right) = e_2 - e_1 \tag{10-5}$$

This is usually described as the Hugoniot equation. The change in energy $e_2 - e_1$ which occurs on passing through the zone of reaction is due partly to the release of chemical energy and partly to the change in temperature.

If the composition of the reaction products is known, the difference in chemical energy between the explosive at some standard temperature and the products at the same temperature can be found from tables which give the experimentally determined heats of formation of the various products or their differences. As an example, the reaction between two volumes of hydrogen and one of oxygen may be considered. The energy released is usually given in calories per gram molecule. Thus the equation,

$$H_2 + \tfrac{1}{2}O_2 = H_2O + 57{,}113 \text{ cal} \tag{10-6}$$

means that, if 2 grams of hydrogen could react completely with 16 grams of oxygen at 0°K, energy amounting to 57,113 calories would be given out if the products of combustion were cooled to 0°K. In considering reactions as they actually occur, the equation which represents the conservation of energy must take account of the total energy of the material before and after the reaction. Thus if the temperatures on the two sides of the reaction zone are T_1 and T_2, the total heat energy per gram molecule[1] of the unburned mixture $H_2 + \frac{1}{2}O_2$ may be written

$$(E_{T_1} - E_0)_{H_2} + \tfrac{1}{2}(E_{T_1} - E_0)_{\frac{1}{2}O_2} \tag{10-7}$$

where $(E_{T_1} - E_0)_{H_2}$ is the difference per mole of hydrogen in energy content at temperature T_1 and that at 0°K. The mean specific heat in the temperature range T_1 to T_2 is evidently

$$\frac{(E_{T_2} - E_{T_1})}{\mathfrak{M}(T_2 - T_1)}$$

where $\mathfrak{M}$ is written for the molecular weight of the mixture. The heat energy of the products of decomposition, which in the case represented by Eq. 10-6 consist only of H_2O, is $(E_{T_2} - E_0)_{H_2O}$ and the energy change which occurs when 2 grams of H_2 react with 8 grams of O_2 is therefore

$$E_2 - E_1 = (E_{T_2} - E_0)_{H_2O} - (E_{T_1} - E_0)_{H_2} \\ - \tfrac{1}{2}(E_{T_1} - E_0)_{O_2} - 57{,}113 \text{ cal} \tag{10-8}$$

Values of $(E_T - E_0)$ have been tabulated for a large variety of elements and compounds so that $e_2 - e_1$ in Eq. 10-5 can be determined when the nature of the reaction is known, and also the temperature on the two sides of the reaction zone. The equation of state which determines the relationship among pressure, density, and temperature must also be regarded as known. In the case of gaseous reactions at pressures which are not too high, the equation of state relates the partial pressure of each component to its partial density, and the equation of state of the whole mixture relates the sum of the partial pressures to the pressure and density by the equation

$$p = RT\rho\frac{n}{M} \tag{10-9}$$

where n is the number of moles of reacting mixture and M their total mass. In the reaction of Eq. 10-6, $H_2 + \frac{1}{2}O_2 = H_2O$, the value of M is

[1] Here E is used to denote energy per mole in distinction to the use of e which is energy per unit mass of the mixture.

$2 + \frac{1}{2}(32) = 18 = \mathfrak{M}$ the molecular weight of the products. Thus the equations of state before and after the reaction in this case are

$$p_1 = RT_1\rho_1\left(\frac{\frac{3}{2}}{18}\right)$$

$$p_2 = RT_2\rho_2\left(\frac{1}{18}\right) \tag{10-10}$$

The equations of state in gaseous reactions will be taken as

$$p_1 = RT_1\rho_1\frac{n_1}{M} \tag{10-11}$$

$$p_2 = RT_2\rho_2\frac{n_2}{M}$$

The Hugoniot equation, Eq. 10-5, may be written in the form

$$\tfrac{1}{2}RT_1n_1(y+1)\left(1-\frac{1}{\mu}\right) = \mathfrak{M}(e_2 - e_1) \tag{10-12}$$

where $y = p_2/p_1$ and $\mu = \rho_2/\rho_1$. It has already been pointed out that the chemical and heat energies of gases are usually expressed in terms of calories per mole. In the case of the reaction, $H_2 + \frac{1}{2}O_2 = H_2O$, $\mathfrak{M}(e_2 - e_1) = E_2 - E_1$, the expression given in Eq. 10-8.

In gaseous reactions involving high temperatures the products of decomposition are in a state of partial ionization so that n_2 depends on the temperature T_2, but before considering such complications it is of interest to study the Hugoniot curve for ideal gases which possess a constant specific heat. In that case, if c_{v_1} and c_{v_2} are the specific heats of the mixture before and after reaction,

$$\mathfrak{M}(e_2 - e_1) = (c_{v_2}T_2 - c_{v_1}T_1)\mathfrak{M} - Q \tag{10-13}$$

where Q is the heat given out per gram molecule of the reaction. Using Eq. 10-12 and 10-13, this becomes

$$\tfrac{1}{2}(y+1)\left(1-\frac{1}{\mu}\right) = \left(\frac{c_{v_2}}{n_2}\frac{y}{\mu} - \frac{c_{v_1}}{n_1}\right)\frac{\mathfrak{M}}{R} - \frac{Q}{RT_1n_1} \tag{10-14}$$

writing

$$x = \frac{1}{\mu} \qquad b = \frac{2c_{v_1}}{n_1}\frac{\mathfrak{M}}{R} + \frac{2Q}{RT_1n_1}$$

$$a = \frac{2c_{v_2}}{n_2}\left(\frac{\mathfrak{M}}{R}\right), \qquad c = \frac{ab + a + b}{(a+1)^2}$$

Eq. 10-14 may be written in the form

$$\left(x - \frac{1}{a+1}\right)\left(y + \frac{1}{a+1}\right) = c \tag{10-15}$$

It will be noticed that the assumption that c_{v_2} and n_2 are constant is equivalent to assuming that the ratio of the specific heats in the burned gases (γ_2) is constant, in fact $(\mathfrak{M}/n)(c_p - c_v) = R$; so that

$$a = \frac{2c_{v_2}}{n_2}\left(\frac{\mathfrak{M}}{R}\right) = \frac{2}{\gamma_2 - 1}; \quad \frac{1}{a+1} = \frac{\gamma_2 - 1}{\gamma_2 + 1} \tag{10-16}$$

It will be seen that in this ideal case the Hugoniot curve is a rectangular hyperbola. For very high values of y

$$x = \frac{1}{a+1} \quad \text{or} \quad \mu = a + 1 = \frac{\gamma_2 + 1}{\gamma_2 - 1} \tag{10-17}$$

The density therefore has the limiting value which it would have at very high pressures if the shock wave were not the seat of release of chemical energy. The details of the calculation of this curve are given below.

Calculation of Hugoniot curve in a particular case. (No dissociation; chemical reaction assumed to be known.) The simplest way to understand the method used for calculating solutions to the Hugoniot equation is to follow the details in a particular case. For this reason the mixture $H_2 + \frac{1}{2}O_2 + 2.5N_2$ will be discussed. In the first instance, it is assumed that the reaction is complete and that no dissociation occurs in the products of combustion. The reaction may then be written

$$H_2 + \tfrac{1}{2}O_2 + 2.5N_2 = H_2O + 2.5N_2 + 57{,}113 \text{ calories} \tag{10-18}$$

and $$n_1 = 4.0,\ n_2 = 3.5,\ M = 2 + 16 + 70 = 88$$

The unburned mixture is assumed to be at 300°K and at 1 atmosphere pressure. From the equations of state for gaseous reactions (Eq. 10-11)

$$y = \mu \frac{T_2}{T_1}\frac{n_2}{n_1} = 0.002917T_2\mu \tag{10-19}$$

so that Eq. 10-12 is written

$$\left(1 - \frac{1}{\mu}\right)(0.002917T_2\mu + 1) = 2\mathfrak{M}\frac{(e_2 - e_1)}{n_1RT_1} \tag{10-20}$$

and using $R = 2.0$ calories, then

$$\frac{2}{n_1RT_1} = \frac{1}{1200}$$

Writing X for $\mathfrak{M}(e_2 - e_1)$, Eq. 10-20 becomes

$$\mu^2 - \mu\left(1 - \frac{342.9}{T_2} + \frac{X}{3.5T_2}\right) - \frac{342.9}{T_2} = 0 \tag{10-21}$$

Table G,10a. Energy content of gases, $E_T - E_0$, in calories. From [*31*].

Temp., °K	H_2	O_2	N_2	CO	NO	OH	CO_2	H_2O	CH_4	O_3	HCN	Br_2	HBr	N_2O
200	965	987	992	992	1095	—	—	1192	—	1229	1310	1134	994	—
250	1197	—	—	—	—	—	—	—	1500	—	—	1454	1243	1362
300	1440	1486	1489	1489	1616	1523	1660	1791	1815	1950	1629	1783	1493	1712
400	1936	1998	1987	1989	2132	2034	2403	2409	2530	2786	2333	2456	1991	2492
500	—	—	—	—	—	—	—	—	3375	3660	3106	3140	2493	3357
600	2936	3088	3006	3017	3196	3048	4135	3687	4360	4627	3933	3830	3003	4296
700	—	—	—	—	—	—	—	—	5480	5644	4809	4524	3525	5293
800	3947	4265	4078	4110	4332	4069	6107	5073	6730	6702	5726	5222	4061	6339
900	—	—	—	—	—	—	—	—	—	7793	6682	5923	4612	7428
1000	4978	5511	5216	5270	5534	5118	8247	6577	—	8909	7672	6626	5179	8553
1200	6044	6802	6410	6485	6786	6200	10503	8200	—	—	9737	8037	6355	10889
1400	7151	8123	7646	7741	8074	7340	12844	9920	—	—	11909	9455	7579	13312
1600	8293	9476	8912	9025	9389	8525	15246	11740	—	—	14162	10880	8843	—
1800	9478	10852	10207	10334	10724	9740	17698	13655	—	—	16485	—	—	—
2000	10700	12248	11528	11665	12075	10985	20187	15650	—	—	18858	—	—	—
2200	11954	13667	12857	13011	13439	12255	22703	17700	—	—	—	—	—	—
2400	13234	15110	14200	14365	14814	13565	25248	19800	—	—	—	—	—	—
2600	14545	16570	15550	15725	16197	14890	27819	21945	—	—	—	—	—	—
2800	15881	18049	16914	17096	17588	16235	30406	24125	—	—	—	—	—	—
3000	17231	19544	18287	18476	18985	17607	33012	26330	—	—	—	—	—	—
3200	18593	21061	19668	19860	20387	19000	—	—	—	—	—	—	—	—
3500	20650	23366	21743	21947	22500	21105	—	—	—	—	—	—	—	—

In gases, X is a function of T_2 only; so that

$$X = (E_{T_2} - E_0)_{H_2O} + 2.5(E_{T_2} - E_0)_{N_2} - (E_{300} - E_0)_{H_2} - \tfrac{1}{2}(E_{300} - E_0)_{O_2} - 2.5(E_{300} - E_0)_{N_2} - 57{,}113$$

The values of all these quantities have been tabulated. They are given in Table G,10a. As an example, the value of X for $T_2 = 2600°$ is taken from the table.

$$(E_{2600} - E_0)_{H_2O} = 21{,}945$$

$$2.5(E_{2600} - E_0)_{N_2} = (2.5)(15550) = 38{,}875$$

$$(E_{300} - E_0)_{H_2} + \tfrac{1}{2}(E_{300} - E_0)_{O_2} + 2.5(E_{300} - E_0)_{N_2} = 1440 + 743 + 3722 = 5905$$

$$X = 21{,}945 + 38{,}875 - 5905 - 57{,}113 = -2198 \text{ calories}$$

The value of μ calculated from Eq. 10-21, using these values of T_2 and X, is $\mu = 0.761$ and $x = 1.32$. The value of y is: $y = 0.002917 \times 2600 \times 0.761) = 5.77$. The corresponding values for other values of T_2 are given in Table G,10b.

Table G,10b. Calculation of density ratios, pressure ratios, and velocities for a mixture of $H_2 + \frac{1}{2}O_2 + 2.5N_2$ using the Hugoniot equations and assuming no dissociation.

Temp., °K	μ	y	$x = \frac{1}{\mu}$	$\frac{y-1}{1-x}$	U, m/sec	u_2, m/sec
Calculated using data from [*29*, 1938 ed.].						
3200	2.24	21.3	0.447	36.2	1900	1050
3000	1.811	15.85	0.55	33.7	1870	840
2800	1.317	10.76	0.76	40.6	2060	625
2600	0.761	5.77	1.32	−14.9		
2200	0.1950	1.25	5.13	−0.06		
1800	0.0627	0.33	15.9		68	−1010
1600	0.0483	0.225	20.7	0.039	63	−1240
Recalculated using data from [*29*, 1951 ed.].						
2800	1.2406	10.133	0.8060	47.07	2109	

It is difficult to cover the whole range of possible conditions in one diagram; accordingly, the high pressure part down to $y = 1.5$ (if $p_1 = 1$ atm this corresponds to $p_2 = 1.5$ atm) is shown in Fig. G,10a. Here the scale of y is reduced relative to x in the ratio 10:1, while in Fig. G,10b the part of the curve representing pressures less than 1.25 atmospheres is shown with the scale of x relative to y reduced in the ratio 10:1. The point 0 in both figures represents the initial state of the mixture, namely, $x = 1$, $y = 1$.

Calculation of Hugoniot curve assuming equilibrium of products of decomposition and dissociation for the same mixture. At the high temperature attained in detonation some of the products of the reaction will decompose and may be dissociated. To find the equilibrium composition at any given temperature it is necessary to think of all possible constituents; thus in the example given there could be conceivably N_2, N, O_2, O, H_2, H, OH, NO. The relative proportions of the constituents are determined by using equilibrium constants which have already been determined for a large number of reactions.

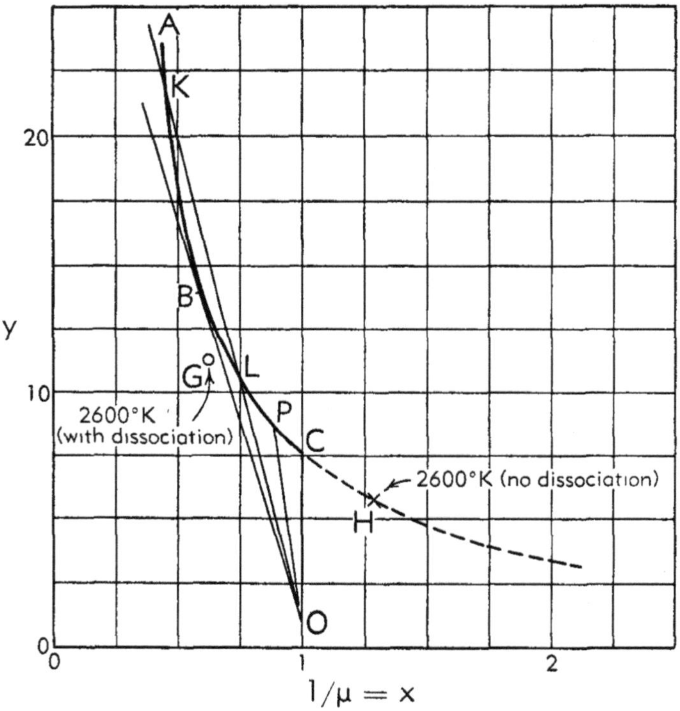

Fig. G,10a. Hugoniot curve for mixture of $H_2 + \frac{1}{2}O_2 + 2.5N_2$ without allowing for dissociation except at point *G*.

A table of these constants is given in Lewis and von Elbe's book [*29*, p. 382] and is reproduced here in Table G,10c. The calculation is somewhat complicated and may be most easily understood by following it through in a particular case. The case chosen is that of the mixture $H_2 + \frac{1}{2}O_2 + 2.5N_2$ for which the Hugoniot curve has already been discussed, assuming that the reaction was complete and that no decomposition of the products occurs. The equilibrium constant K_1, which must be used to determine the ratio of the constituents H_2O, H_2, and O_2 in the reaction $H_2 + \frac{1}{2}O_2 = H_2O$ is $K_1 = p_{H_2}\sqrt{p_{O_2}}/p_{H_2O}$ where p_{H_2}, p_{O_2}, and

p_{H_2O} are the partial pressures expressed in atmospheres of H_2, O_2, and H_2O in the products of combustion. If there are α moles of H_2, β of O_2, and γ of H_2O and the total number of moles in the products is n_2 and n_1 is the initial number, so that

$$p_{H_2} = \alpha \frac{T_2}{T_1} \frac{p_1}{n_1} \frac{\rho_2}{\rho_1}, \quad K_1 = \left(\frac{T_2 p_1}{T_1 n_1}\right)^{\frac{1}{2}} \mu^{\frac{1}{2}} \frac{\alpha \sqrt{\beta}}{\gamma}$$

Similar equilibrium equations can be written for all the possible reversible reactions between molecules or atoms in the products of decompositions at any particular temperature; however, some of these equations correspond with such small amounts of reacting material that they may be

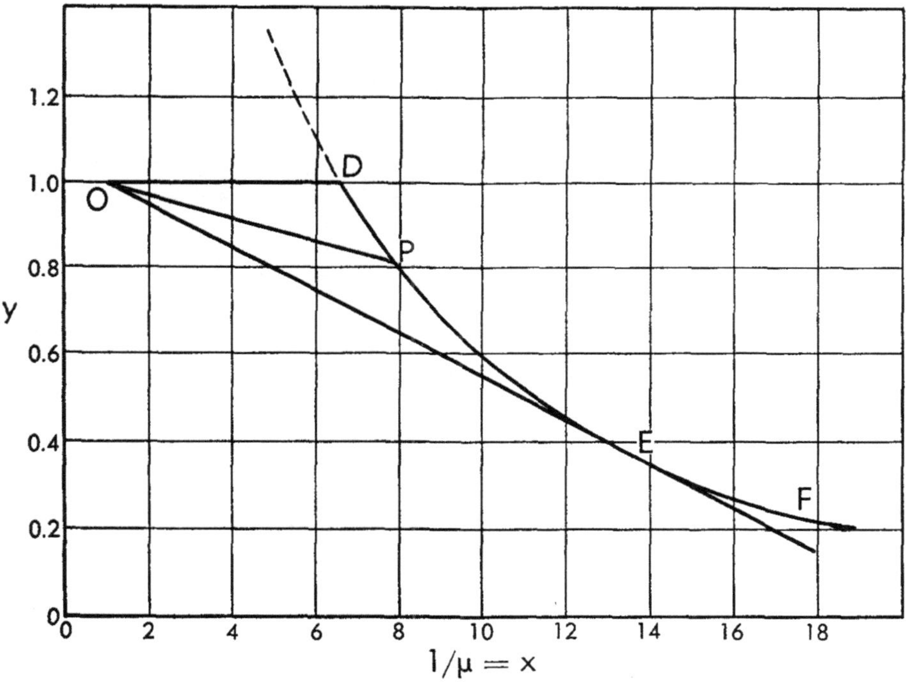

Fig. G,10b. Hugoniot curve for mixture of $H_2 + \frac{1}{2}O_2 + 2.5N_2$ without allowing for dissociation.

left out in considering the equilibrium composition. At any assumed temperature, therefore, the values of K corresponding to all the possible reversible reactions must be considered and any that are very small may be left out, unless for any reason it is required to know how much of a partial product is present even though it be small in amount.

Since the equilibrium constants may be taken as depending only on temperature, it is convenient to calculate the pressure and density for a point on the Hugoniot curve corresponding to an arbitrary assumed tem-

Table G,10c. Equilibrium constants of various reactions and their energies of reaction at absolute zero. From [31].

Reaction	$2H \rightleftharpoons H_2 + 102{,}480 \pm 20$	$2O \rightleftharpoons O_2 + 117{,}350 \pm 700$	$2N \rightleftharpoons N_2 + 169{,}220 \pm 500$	$2Cl \rightleftharpoons Cl_2 + 56{,}900 \pm <100?$	$2Br \rightleftharpoons Br_2 + 45{,}230 \pm <18$	$2I \rightleftharpoons I_2 + 35{,}795 \pm 100$	$CO + \frac{1}{2}O_2 \rightleftharpoons CO_2 + 66{,}760 \pm 30$	$H_2 + \frac{1}{2}O_2 \rightleftharpoons H_2O + 57{,}113 \pm 10$	$OH + \frac{1}{2}H_2 \rightleftharpoons H_2O + 63{,}000 \pm 1000$	$CO + H_2O \rightleftharpoons CO_2 + H_2 + 9{,}646 \pm 40$	$C(\beta\text{-graphite}) + \frac{1}{2}O_2 \rightleftharpoons CO + 27{,}404 \pm 100$	$O_3 \rightleftharpoons 1.5O_2 + 34{,}513 \pm 240$
Temp., °K	1	2	3	4	5	6	7	8	9	10	11	12
300	−70.23	−80.2	−118.1	−36.45	−28.00	−21.3	−44.72	−39.77	−43.3	−4.947	−24.08	−28.29
400	−51.35	−58.6	−86.9	−25.92	−19.61	−14.6	−32.43	−29.26	−31.7	−3.167	−19.23	−22.15
600	−32.41	−36.9	−55.8	−15 33	−11.21	− 7.93	−20.07	−18.64	−20.0	−1.433	−14.41	−15.98
800	−22.88	−26.1	−40.2	− 9.994	− 6.99	− 4.59	−13.89	−13.28	−14.07	−0.610	−11.98	−12.89
1000	−17.13	−19.48	−30.9	− 6.770	− 4.45	− 2.57	−10.20	−10.05	−10.53	−0.147	−10.52	−11.03
1200	−13.28	−15.10	−24.6	− 4.610	− 2.74	− 1.21	− 7.755	− 7.90	− 8.17	+0.145	− 9.530	− 9.79
1400	−10.51	−11.97	−20.1	− 3.060	− 1.52	− 0.23	− 5.999	− 6.34	− 6.47	+0.341	− 8.817	− 8.89
1600	− 8.429	− 9.61	−16.8	− 1.891	− 0.593	+ 0.48	− 4.715	− 5.20	− 5.20	+0.485	− 8.277	− 8.22
1800	− 6.803	− 7.772	−14.0	− 0.977	—	—	− 3.690	− 4.27	− 4.19	+0.580	− 7.850	− 7.70
2000	− 5.496	− 6.298	−12.0	− 0.245	—	—	− 2.862	− 3.52	− 3.40	+0.658	− 7.504	− 7.29
2200	− 4.424	− 5.091	−10.3	+ 0.356	—	—	− 2.193	− 2.91	− 2.74	+0.717	− 7.221	− 6.95
2400	− 3.529	− 4.078	− 8.83	+ 0.858	—	—	− 1.648	− 2.41	− 2.19	+0.762	− 6.980	− 6.66
2600	− 2.769	− 3.228	− 7.60	+ 1.284	—	—	− 1.206	− 2.00	− 1.74	+0.794	− 6.777	− 6.42
2800	− 2.115	− 2.495	− 6.56	+ 1.650	—	—	− 0.811	− 1.63	− 1.34	+0.819	− 6.595	− 6.21
3000	− 1.548	− 1.858	− 5.65	+ 1.966	—	—	− 0.470	− 1.31	− 0.999	+0.840	− 6.440	− 6 03
3200	− 1.051	− 1.290	− 4.85	—	—	—	—	—	—	—	—	− 5.87
3500	− 0.409	− 0.577	− 3.83	—	—	—	—	—	—	—	—	− 5.68
4000	+ 0.449	+ 0.379	− 2.46	—	—	—	—	—	—	—	—	− 5.40
5000	+ 1.654	+ 1.715	− 0.524	—	—	—	—	—	—	—	—	− 5.02
6000	—	—	—	—	—	—	—	—	—	—	—	—

Table G,10c (continued)

Reaction	$O_2 + O \rightleftharpoons O_3 + 24{,}137 \pm 600$	$NO + \frac{1}{2}O_2 \rightleftharpoons NO_2 + 12{,}775 \pm 150$	$NO \rightleftharpoons \frac{1}{2}O_2 + \frac{1}{2}N_2 + 21{,}400 \pm 100?$	$\frac{1}{2}H_2 + \frac{1}{2}Cl_2 \rightleftharpoons HCl + 21{,}984 \pm 10$	$\frac{1}{2}H_2 + \frac{1}{2}Br_2 \rightleftharpoons HBr + 11{,}990 \pm 100?$	$\frac{1}{2}H_2 + \frac{1}{2}I_2(g) \rightleftharpoons HI + 972 \pm 14$	$\frac{1}{2}S_2(g) + O_2 \rightleftharpoons SO_2 + 85{,}860 \pm 200$	$SO + \frac{1}{2}O_2 \rightleftharpoons SO_2 + 79{,}300 \pm <500$	$2S(g) \rightleftharpoons S_2(g) + 102{,}590 \pm 230$	$HCN \rightleftharpoons \frac{1}{2}H_2 + \frac{1}{2}N_2 + C(\beta\text{-graphite}) + 31{,}090 \pm ?$	$C(\beta\text{-graphite}) + 2H_2 \rightleftharpoons CH_4 + 16{,}174 \pm <200$	$C(g) \rightleftharpoons C(\beta\text{-graphite}) + 123{,}600 \pm 550$	$C(g) \rightleftharpoons \frac{1}{2}C_2(g) + 62{,}250 \pm ?$
Temp., °K	13	14	15	16	17	18	19	20	21	22	23	24	25
300	−11.83	−6.42	−15.04	−16.58	−9.37	−1.42	−59.33	−54.24	−69.92	−20.76	−8.935	—	—
400	−7.16	−3.94	−11.13	−12.56	−7.16	−1.23	−43.56	−39.68	−51.00	−15.12	−5.580	—	—
600	−2.48	−1.44	−7.194	−8.52	−4.93	−0.94	−27.78	−25.09	−32.01	−9.49	−2.047	—	—
800	−0.13	−0.18	−5.231	−6.49	−3.80	−0.81	−19.88	−17.80	−22.48	−6.68	−0.180	—	—
1000	+1.29	+0.57	−4.052	−5.26	−3.11	−0.72	−15.14	−13.42	−16.74	−5.01	+0.985	—	—
1200	+2.24	+1.08	−3.267	−4.436	−2.65	−0.66	−11.99	−10.50	−12.89	−3.90	+1.779	—	—
1400	+2.92	+1.43	−2.706	−3.847	−2.32	−0.62	−9.73	−8.41	−10.14	−3.10	+2.354	—	—
1600	+3.42	+1.69	−2.285	−3.403	−2.07	−0.59	−8.04	−6.85	−8.08	−2.51	+2.790	—	—
1800	+3.82	+1.89	−1.959	−3.057	—	—	—	—	−6.49	−2.06	+3.129	—	—
2000	+4.15	+2.07	−1.695	−2.780	—	—	−5.68	−4.67	−5.18	−1.69	+3.397	−5.455	−3.76
2200	+4.41	+2.18	−1.479	−2.554	—	—	—	—	−4.11	—	+3.616	−4.212	—
2400	+4.63	+2.29	−1.300	−2.362	—	—	−4.10	−3.21	−3.25	—	+3.801	−3.187	—
2600	+4.81	+2.39	−1.150	−2.201	—	—	—	—	−2.48	—	+3.960	−2.315	—
2800	+4.97	+2.48	−1.019	−2.062	—	—	−2.98	−2.17	−1.85	—	+4.098	−1.573	—
3000	+5.11	+2.56	−0.907	−1.941	—	—	—	—	−1.29	—	+4.218	−0.926	−1.40
3200	+5.23	—	−0.807	—	—	—	—	—	—	—	—	—	—
3500	+5.39	—	−0.680	—	—	—	—	—	—	—	—	—	—
4000	+5.60	—	−0.513	—	—	—	—	—	+0.669	—	—	—	−0.21
5000	+5.89	—	−0.279	—	—	—	—	—	+1.854	—	—	—	+0.51
6000	—	—	—	—	—	—	—	—	—	—	—	—	+1.00

perature. Assuming this to be $T_2 = 2600°K$, the equilibrium values are found for the possible reactions:

$$H_2 + \tfrac{1}{2}O_2 = H_2O + 57{,}113 \text{ cal},$$
$$\log K_1 = -2.002, \quad K_1 = 0.0100 \qquad (10\text{-}22)$$

$$OH + \tfrac{1}{2}H_2 = H_2O + 63{,}000 \text{ cal},$$
$$\log K_2 = -1.74, \quad K_2 = 0.0182 \qquad (10\text{-}23)$$

$$2H = H_2 + 102{,}480 \text{ cal},$$
$$\log K_3 = -2.769, \quad K_3 = 0.00170 \qquad (10\text{-}24)$$

$$2O = O_2 + 117{,}350 \text{ cal},$$
$$\log K_4 = -3.228, \quad K_4 = 0.00059 \qquad (10\text{-}25)$$

$$NO = \tfrac{1}{2}O_2 + \tfrac{1}{2}N_2 + 21{,}400 \text{ cal},$$
$$\log K_5 = -1.1503, \quad K_5 = 0.0708 \qquad (10\text{-}26)$$

The constant K_4 is so small that the dissociation of oxygen may be left out of consideration. The equations representing the combustion may therefore be taken as $H_2 + \tfrac{1}{2}O_2 + 2.5N_2 = \alpha H_2 + \beta O_2 + \gamma H_2O + wH + rOH + sNO + tN_2$.

When $T_2 = 2600°K$, $T_1 = 300°K$, $p_1 = 1$ atm, $n_1 = 4$

$$\left(\frac{T_1 n_1}{T_2 p_1}\right)^{\frac{1}{2}} = (0.4615)^{\frac{1}{2}} = 0.6793$$

The equations of equilibrium are:

$$\frac{\alpha\sqrt{\beta}}{\gamma} = (0.0100)(0.679)\mu^{-\frac{1}{2}} = 0.00679\mu^{-\frac{1}{2}} \qquad (10\text{-}27)$$

$$\frac{r\sqrt{\alpha}}{\gamma} = (0.0182)(0.679)\mu^{-\frac{1}{2}} = 0.01236\mu^{-\frac{1}{2}} \qquad (10\text{-}28)$$

$$\frac{w^2}{\alpha} = (0.0017)(0.4615)\mu^{-1} = 0.000784\mu^{-1} \qquad (10\text{-}29)$$

$$\frac{s}{\sqrt{\beta t}} = (0.0708) \qquad (10\text{-}30)$$

Beside these equations there are three which represent the fact that none of the three elements involved is lost in the reaction. These are for hydrogen

$$2 = 2\alpha + 2\gamma + w + r \qquad (10\text{-}31)$$

for oxygen

$$1 = 2\beta + \gamma + r + s \qquad (10\text{-}32)$$

for nitrogen

$$5.0 = 2t + s \qquad (10\text{-}33)$$

To calculate the point on the Hugoniot curve corresponding with $T_2 = 2600°K$ using the form (Eq. 10-12) it is first necessary to calculate $e_2 - e_1$ for the assumed composition at this temperature. Using the figures from Table G,10a,

$$\begin{aligned}\mathfrak{M}(e_2 - e_1) = {} & \overset{\mathrm{H_2}}{14{,}545\alpha} + \overset{\mathrm{O_2}}{16{,}570\beta} + \overset{\mathrm{H_2O}}{21{,}945\gamma} + \overset{\mathrm{O}}{14{,}890r} + \overset{\mathrm{NO}}{16{,}197S} \\ & + \overset{\mathrm{N_2}}{15{,}550t} + \overset{\mathrm{H}}{\tfrac{3}{2}R(2600)w} - \overset{\text{Eq. 10-22}}{57{,}113\gamma} + \overset{\text{Eq. 10-23}}{63{,}000r} + \overset{\text{Eq. 10-24}}{102{,}480w} \\ & + \overset{\text{Eq. 10-26}}{21{,}400S} - \overset{\text{(energy at 300°K)}}{5905}\end{aligned} \quad (10\text{-}34)$$

Here the energies of the products at 2600°K are given under the appropriate chemical symbols and the energies of the reactions are marked by appropriate references to the reversible reactions considered. Using Eq. 10-12, 10-19, and 10-34, the Hugoniot equation can be expressed in terms α, β, r, s, t, w, and μ. The eight equations (Eq. 10-27 to 10-34) suffice to determine their eight unknowns, but the numerical work is very laborious because the equations are not linear. One method for obtaining a solution is to regard the amounts of dissociated products as small. The problem can then be treated as one of small perturbations of the problem in which the composition is supposed known. In the case under consideration this can be achieved by writing Eq. 10-27, 10-28, 10-29, and 10-30 in the form

$$\beta = (0.00670)^2\gamma^2\mu^{-1}\alpha^{-2} \quad (10\text{-}35)$$

$$r = 0.01236\gamma\mu^{-\frac{1}{2}}\alpha^{-\frac{1}{2}} \quad (10\text{-}36)$$

$$w = (0.000784)^{\frac{1}{2}}\mu^{-\frac{1}{2}}\alpha^{\frac{1}{2}} \quad (10\text{-}37)$$

$$s = 0.0708t^{\frac{1}{2}}\beta^{\frac{1}{2}} \quad (10\text{-}38)$$

In the first approximation in which dissociation is neglected, that is $t = 2.5$, $\gamma = 1$, it was found that $\mu = 0.761$. If these values are inserted into Eq. 10-35, 10-36, 10-37, and 10-38 it is found that β, r, w, and s are small if α be within a certain range of small numbers. Assuming that this requirement is met, an equation for α can be obtained by substituting from Eq. 10-35, 10-36, 10-37, and 10-38 in Eq. 10-39, which is obtained from Eq. 10-31 and 10-32,

$$2\alpha + w - r - 2s - 4\beta = 0 \quad (10\text{-}39)$$

This equation can be solved approximately by inserting trial values of α. Corrected values of γ and t can then be found by inserting the values of β, r, and s in Eq. 10-32 and 10-33.

In this way values can be obtained for α, β, γ, r, s, t, and w, and n_2 can be found by summing them. These can then be substituted in Eq. 1-34 to put in $(e_2 - e_1)$ and the value so obtained inserted in the Hugoniot equation (Eq. 10-12), which can be treated as an equation for μ. This second approximation to μ turned out to be 1.76, which is very far from the first approximation 0.761. Using the values of γ and t obtained in the second approximation together with $\mu = 1.76$, new values can be inserted on the right-hand sides of Eq. 10-35, 10-36, 10-37, and 10-38 and the process repeated. In this way the following sequence was found for μ.

Approximation number	1	2	3	4	Estimated value
μ	0.76	1.76	1.47	1.54	1.52

The estimated values which would satisfy all the equations were found to be

μ	α	β	γ	r	s	t	w	n_2
1.52	0.051 H_2	0.010 O_2	0.929 H_2O	0.041 OH	0.011 NO	2.494 N_2	0.005 H	3.560

The corresponding value of y is 11.6 and the point on the Hugoniot curve corresponding with $T_2 = 2600$ is therefore $y = 11.6$, $1/\mu = 0.66$. This point is shown at G in Fig. G,10a. It will be seen that it is rather close to the Hugoniot curve obtained without considering dissociation, but far from the point H (Fig. G,10a) which corresponds to the same temperature.

Properties of the Hugoniot curve. The most important geometrical property of the Hugoniot curve is that it provides a simple method for studying Eq. 10-15. Eq. 10-4, which may be written

$$\frac{y-1}{1-x} = \frac{\rho_1 U^2}{p_1} \tag{10-40}$$

connects the velocity U of propagation of the reaction zone into the unburned gas with pressure and density changes. Similarly, if U_2 is the velocity of the combustion wave relative to the burned gas, the equation shows that $U_2 = U - u_2 = U/\mu$. The line OP connecting the point P on the curve with O therefore makes with the x axis an angle α such that

$$\tan\alpha = \frac{\rho_1 U^2}{p_1} \tag{10-41}$$

α is taken positive when the line OP slopes towards the positive y axis as shown in Fig. G,10a.

The values of U calculated by Eq. 10-40 for the mixture $H_2 + \frac{1}{2}O_2 + 2.5N_2$ without allowing for dissociation are given in column 6 of Table G,10b. It will be noticed that some spaces in this column are blank. This is because there is a range of points on the Hugoniot curve for which $(y-1)/(1-x)$ is negative. The value of U calculated for these points using Eq. 10-40 would be imaginary. That part of the curve in Fig. G,10a and G,10b therefore does not correspond with any physically possible wave of combustion and is shown dashed.

Eq. 10-40 provides a criterion in the light of which the physical properties of the wave of combustion corresponding with any point on the Hugoniot curve may be classified. As the point moves down the curve from K (Fig. G,10a), y decreases so that U decreases to a minimum value at the point B where the tangent from the point $x = 1$, $y = 1$ touches the curve. As the representative point moves farther down the curve, U increases until at C, where $x = 1$, U becomes infinite. The next point which corresponds with real physically possible conditions is D (Fig. G,10b) where $y = 1$. There $U = 0$. D, therefore, corresponds with a possible deflagration which travels so slowly that the pressure difference necessary to accelerate the gas, when its density decreases at the flame front, is negligibly small. As the representative point moves down from D, U increases until the point E is reached where the tangent from O touches the curve. This corresponds with the maximum velocity with which a deflagration could travel into unburned gas. The two points B and E are both critical points and have the property that a small variation in the pressure change at the shock point produces no variation in U. A small change in pressure in the region behind the shock front might be caused by a small disturbance or sound wave propagated in the burned gas, and the small increase in pressure would then be related to the small change in density by

$$\frac{dp_2}{d\rho_2} = \left(\frac{dp_2}{d\rho_2}\right)_{\text{isentropic}} \tag{10-42}$$

where $(dp_2/d\rho_2)_{\text{isentropic}}$ refers to isentropic changes in pressure and density in gas behind the shock front. Waves are propagated through the burned gas with velocity

$$a_2 = \sqrt{\left(\frac{dp_2}{d\rho_2}\right)_{\text{isentropic}}}$$

Waves propagated with velocity a_2 in gas moving with velocity u_2 are propagated relative to the unburned gas with velocity $a_2 + u_2$. At the point on the Hugoniot curve for which $u_2 + a_2 = U$, a small disturbance would be propagated through the products of combustion in such a way that it would remain fixed relative to the zone of reaction. Under this condition, Eq. 10-42 applies to the pressure and density in the disturbance as well as to the condition immediately behind the zone of reaction. This may be expressed by the equation

$$\left(\frac{dp_2}{d\rho_2}\right)^{\text{crit}}_{\text{isentropic}} = \left(\frac{dp_2}{d\rho_2}\right)^{\text{crit}}_{\text{Hugoniot}} \tag{10-43}$$

which represents the condition that the gas be in the critical condition represented in Fig. G,10a and G,10b by B and E and for which

$$\frac{dU}{d\rho_2} = 0$$

Thus

$$a_2^2 = \left[\frac{dp_2}{d\rho_2}\right]_{\text{Hugoniot}}^{\text{crit}} = -\frac{1}{\rho_2^2}\left[\frac{dp_2}{d(1/\rho_2)}\right]_{\text{Hugoniot}}^{\text{crit}} = -\frac{p_1x^2}{\rho_1}\left[\frac{dy}{dx}\right]_{\text{Hugoniot}}^{\text{crit}} \tag{10-44}$$

and from Fig. G,10a and G,10b it will be seen that

$$-\left[\frac{dy}{dx}\right]_{\text{Hugoniot}}^{\text{crit}} = \frac{y-1}{1-x} = \frac{\rho_1U^2}{p_1} \tag{10-45}$$

And therefore $a_2 = xU_x$ and $u_2 = U(1 - x)$. It will be noticed that the velocity of the zone of reaction relative to the products of combustion is $U_2 = U - u_2$ so that at the critical points $U_2 = a_2$.

The critical points divide the Hugoniot curve into the following regions:

AB, $U < u_2 + a_2$
B, U is minimum for detonation into gas (p_1, ρ_1)
BC, $U > u_2 + a_2$
CD no possible state
DE, $U < u_2 + a_2$
E, U is maximum for deflagration into gas (p_1, ρ_1)
EF, $U > u_2 + a_2$

The point F is marked in Fig. G,10b at the end of the curve shown. In fact the region $U > u_2 + a_2$ extends to the point at which the Hugoniot curve cuts $y = 0$, at a definite value of x.

In considering the physical meaning of points on the Hugoniot curve it is useful to conceive ideal experiments which might produce the corresponding conditions. Referring to Fig. G,10a, it will be seen that at very high pressures the velocity is also very high and in the limit when U and p tend to infinity, ρ tends to the same finite limit (see Eq. 10-17) that would be attained if the products of combustion were compressed by a piston forcing them along a tube at very high speed. Values of u_2 corresponding to the calculated points on the Hugoniot curve for the mixture $H_2 + \frac{1}{2}O_2 + 2.5N_2$ are given in column 7 of Table G,10b. As the representative point moves down the Hugoniot curve (Fig. G,10a) from K to C, the values of u_2 decrease from infinity to zero. It might be thought therefore that the whole range of possible states of detonation might be produced by firing gas in a tube through which a piston was moving at the speed u_2. This, however, is not necessarily the case. If a detonation wave represented by a point in the section KB of the Hugoniot curve is set up in the way described, a variation in the speed of the piston will cause a wave to be propagated in the gas which will reach the detonation front because in that region $U < u_2 + a_2$. Thus a new steady condition can be set up corresponding with the changed value of u_2. If the detonation wave corresponding with a point in the section where $U > u_2 + a_2$ could be set up in this way, a change in the speed of the piston could not give rise to a change in U.

It seems unlikely that detonations corresponding to points where $U > a_2 + u_2$ could be produced by moving a piston and simultaneously igniting the gas, but another method can be imagined which would materialize the conditions corresponding to any point in the region where $U > u_2 + a_2$. Suppose that in a tube of large cross section, a small detonating cord were laid parallel to the axis. The velocity U with which the flame was propagated in the cord would control the rate of propagation of the zone of reaction. The conditions appropriate to that value of U would then be set up behind the flame front. If such an experiment were tried it would probably be more convenient to set up a large number of spark plugs in the tube and fire them in succession. The velocity of propagation could then be controlled at will. It is true that an experiment of this kind would not produce a sharp-front shock wave in which the whole combustion was completed almost instantaneously, but the equations do not require this. All that is needed is that the zone of reaction shall move at a constant speed. The length of the zone of combustion in the tube is immaterial. It is worth noting that when the mixture $CO + \frac{1}{2}O_2$ detonates, the zone of reaction extends over a considerable length of the tube in which it is contained.

At the critical points B and E in Fig. G,10a and G,10b, the Hugoniot curve is tangential to the lines representing isentropic states of the burned gas. The entropy is therefore a maximum or a minimum. It has been shown (Sec. D) that at the point B it is a minimum and at the point E it is maximum. This consideration has made some writers conclude that the Chapman-Jouguet hypothesis could be predicted by a purely thermodynamic principle. That this is a mistake can be appreciated from the consideration that states corresponding with points above B could certainly be materialized by moving a piston in a tube behind a detonating gas. To understand the physical significance of the hypothesis, it is necessary to find out whether it is possible for the flow condition which the Hugoniot equation requires behind the zone of combustion to be consistent with a possible field of flow in the products of combustion.

If the velocity of detonation U is constant, the pressure and density, and therefore the entropy are constant immediately behind the zone of reaction, and since the reaction is complete there, the products of combustion must expand isentropically. Among the various kinds of isentropic expansions which have been studied is that which occurs when one end of a long tube containing high pressure gas is suddenly opened. A wave of rarefaction travels down the tube, its front moving into the stationary gas with the velocity of sound. Waves of this type are called progressive waves of finite amplitude. They can be regarded as being composed of a number of small wavelets, each of which propagates a small change of pressure δp and velocity δu into the air which has already been moved by previous wavelets (Sec. C). Since each of these wavelets can be regarded as a sound wave, δu and δp are related by the equation

$$\delta p = \rho a \delta u \tag{10-46}$$

where a is the velocity of sound, ρ the density, and

$$a^2 = \frac{dp}{d\rho} \tag{10-47}$$

Combining Eq. 10-46 and 10-47

$$dp = \rho \frac{du}{a}$$

so that

$$u - u_0 = \int_{a_0}^{a} a \frac{d\rho}{\rho} \tag{10-48}$$

where u_0 is the velocity at a point where $a = a_0$, $p = p_0$.

From the manner in which this solution has been constructed it will be seen that any given value of p, ρ, or u is propagated through the gas with relative velocity a. The absolute velocity of any constant condition is therefore $u + a$.

Imagine a plane surface moving perpendicular to itself with velocity U and emitting gas in which the sound velocity is a_1 at relative velocity $U - u_1$. Such a gas could move as a progressive wave of rarefaction provided $U = a_1 + u_1$. If the Hugoniot wave satisfies the Chapman-Jouguet hypothesis the plane behind the zone of reaction in detonation is such a surface so that the detonation front would be followed by a single progressive wave provided that the boundary conditions were such as to permit its existence.

In the case of a gas for which $p\rho^{-\gamma}$ is constant in isentropic motion,

$$\frac{a}{a_0} = \left(\frac{\rho}{\rho_0}\right)^{\frac{\gamma-1}{2}} = \left(\frac{p}{p_0}\right)^{\frac{\gamma-1}{2\gamma}} \tag{10-49}$$

Integrating Eq. 10-48

$$u - u_0 = \frac{2}{\gamma - 1}(a - a_0) \tag{10-50}$$

$u + a$ is therefore linearly related to u. If the density increases at the detonation front in the ratio $\mu:1$ these formulas can be written:

$$u = \frac{2a}{\gamma - 1} + U\left[1 - \frac{1}{\mu} - \frac{2}{\mu(\gamma - 1)}\right]$$

$$u + a = \frac{\gamma + 1}{2}u + U\left[1 - \frac{\gamma + 1}{2}\left(1 - \frac{1}{\mu}\right)\right]$$

If it is assumed that detonation suddenly starts at a point 0 in a tube, say at one end of it, since the unburned gas is at rest, the whole field of flow can be calculated. At time R/U after the initiation, the detonation front has moved down the tube through a distance R. It is followed by a progressive wave so that the point where the velocity of the burned gas

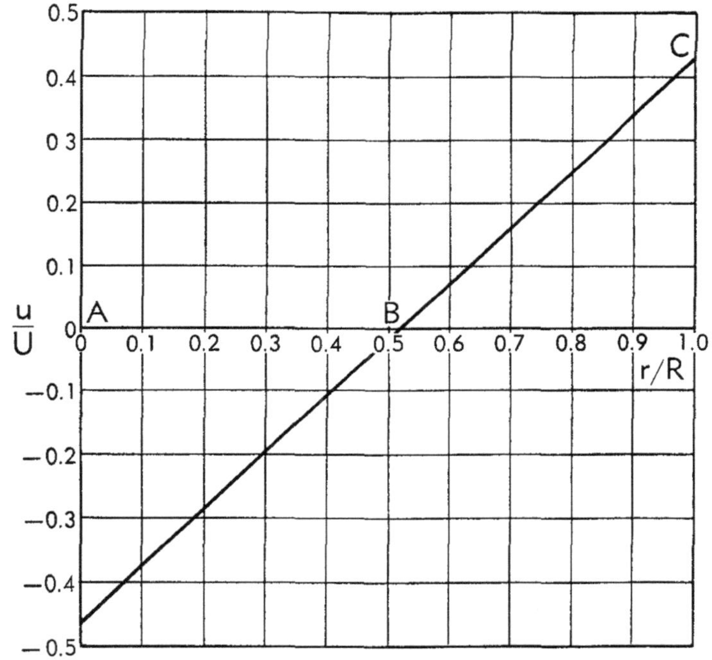

Fig. G,10c. Velocity distribution for detonation of $H_2 + \frac{1}{2}O_2$ at 1 atm pressure. r = distance from point of initiation; R = distance detonation front has traveled.

is u has traveled a distance $r = (u + a)(R/U)$. Since $u + a$ is linearly related to u, a diagram showing the distribution of velocity along the tube is a straight line. The particular case when $\mu = 1.78$, $\gamma = 1.215$, $y = 18.05$, corresponding to detonation of $H_2 + \frac{1}{2}O_2$ at 1 atmosphere pressure, is shown in Fig. G,10c. In this figure the abscissas represent $r/R = (u + a)/U$, where r is the distance from the point of initiation, and R is the distance through which the detonation front has traveled. It will be seen that the line representing u as a function of r/R cuts the axis $u = 0$ at the point

$$\frac{r}{R} = \frac{a}{U} = \left[1 - \left(1 - \frac{1}{\mu}\right)\frac{\gamma - 1}{2}\right] = 0.515$$

At $r = 0$, $u + a = 0$, so that

$$\frac{u}{U} = \left(1 - \frac{1}{\mu} - \frac{2}{\gamma - 1}\right) = 0.438 - 0.902 = -0.464$$

If the tube is open at the end, the products of combustion will therefore flow out through the open end with velocity $= 0.464U$. If p_a is the pressure of the atmosphere, the pressure of the gas at the exit is given by

$$\frac{p}{p_0} = \left(\frac{a}{a_0}\right)^{\frac{2\gamma}{\gamma-1}} \text{ and } p_0 = 18.05p_a, \quad a_0 = \frac{U}{1.78} = 0.5618U, \quad a = 0.464U$$

so that

$$\frac{p}{p_a} = 18.05\left(\frac{0.464}{0.5618}\right)^{\frac{2(1.215)}{0.215}} = 18.05(0.8259)^{11.302} = 2.073 \text{ atm}$$

It will be seen therefore that the pressure inside the tube has fallen from 18 atmospheres behind the detonation front to 2 atmospheres at the open end of the tube. If the firing end of the tube is closed it is possible for the whole of the gas between the firing end and the point on the curve where $u = 0$ to be at rest and at the pressure corresponding to this point. The gas between this point and the detonation front moves forward just as though the firing end were open. The column of gas which is at rest grows at the rate $0.515U$ and the pressure at the closed end is $18.05(0.515/0.5618)^{11.302} = 6.73$ atm. This condition is represented by the bent line ABC in Fig. G,10c.

Detonation front followed by piston. It has already been seen that if a piston is moved behind a reaction front at a speed greater than the particle velocity in the critical Chapman-Jouguet condition, the speed of this reaction front will be greater than the critical detonation velocity.

Consider the case when the velocity of the zone of reaction U is controlled by some external agency to a value which is greater than the critical. The points K and L in Fig. G,10a represent the two possible conditions at the detonation front which could occur. Corresponding with these points the reaction front velocities are U_K and U_L and the corresponding gas velocities

$$u_K = U_K\left(1 - \frac{1}{\mu_K}\right) \quad \text{and} \quad u_L = U_L\left(1 - \frac{1}{\mu_L}\right)$$

Since $\mu_K > \mu_L$, $u_K > u_L$. If U_p, the piston velocity, is exactly equal to u_K or u_L conditions corresponding with the points K and L on the Hugoniot curve will be set up and the space between the reaction front and the piston will contain gas moving uniformly at the speed of the piston. If U_p is not equal either to u_K or u_L the simple conditions envisaged in the discussion of the Hugoniot equation do not describe the whole field of flow between the reaction front and the piston. There are two possible ways in which gas velocities can change from u_K or u_L at the front to U_p at the piston, namely through a wave of expansion or through a shock wave. The sequence of flow fields corresponding with piston velocities ranging from 0 to U_K is shown in Fig. G,10d. In this figure gas velocity

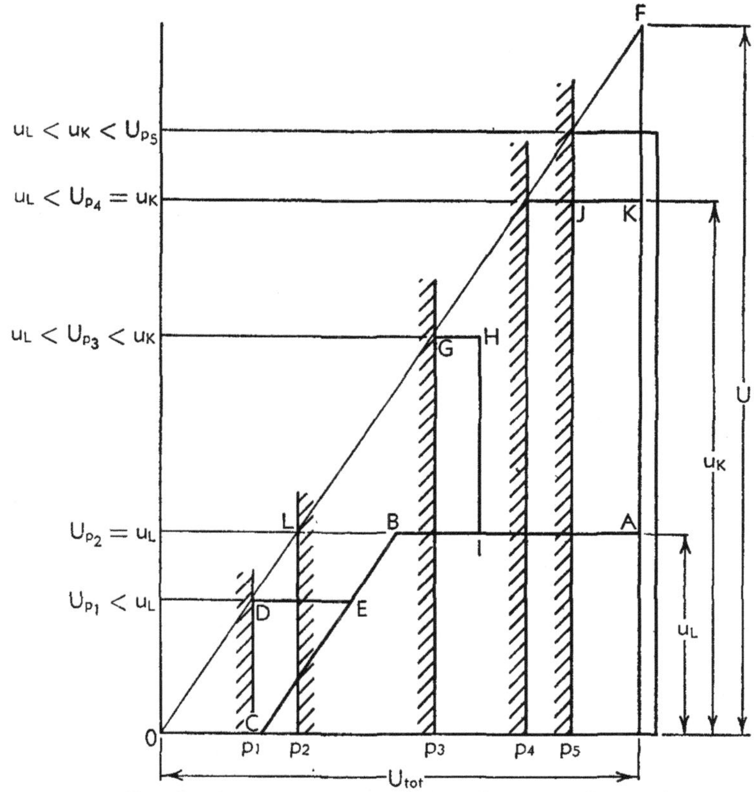

Fig. G,10d. Sequence of velocity distribution in detonation with piston velocity varying from 0 to U_K.

is represented by the ordinates, and distance from the point of initiation of the reaction (assumed coincident with the point where the piston starts moving with velocity U_p) is displayed by the abscissas. One may distinguish the following possibilities:

When $U_p = 0$, e.g. when the reaction starts from the closed end of a cylinder, the gas velocity u_L is uniform from A in the reaction front to the point B (Fig. G,10d) to which the wave of rarefaction has penetrated. Then the velocity falls off to the point C where it becomes zero. The gas is at rest between 0 and C. The distribution in the range $0CB$ is of the same character that is shown in Fig. G,10c.

When $U_{p_1} < u_L$, then the velocity is constant and equal to U_{p_1} from D to E (Fig. G,10d) and the point D must be on the line connecting 0 to F which represents the velocity of the reaction front. The distribution of velocity is shown by $DEBA$.

When $U_{p_2} = u_L$, the velocity is constant and represented by LBA.

When $u_L < U_{p_3} < u_K$, a shock wave is produced and the distribution is represented by $GHIA$.

If $U_{p_4} = u_K$, then as the velocity of the shock wave through the products of combustion, which themselves are moving with velocity u_L, approaches u_K the flow behind it must become identical with u_K since the Hugoniot equation concerns itself only with conditions in front of and behind a zone, the detailed structure of which is immaterial. This condition is represented by JK in Fig. G,10d.

If $U_{p_5} > u_K$, then a shock wave must penetrate into the unburned gas ahead of the zone of reaction when the velocity of that zone is controlled. In fact, no doubt, the detonation wave would move ahead faster than the igniting flame and the conditions would be those represented by a point on the Hugoniot curve higher up than K.

Distribution of pressure and temperature in the zone of combustion. The Hugoniot equations have so far been applied only to cases where combustion is assumed to be complete. However, this is not necessary. To apply the Hugoniot equation, all that is necessary is that at a section which moves at the uniform speed of detonation the conditions shall be uniform over the section. If combustion is uniform at all points in each section the Hugoniot equation can be applied to every section between the shock wave in the original mixture and the region where the reaction is complete. In spinning detonation (see Art. 13) it is not possible to apply the Hugoniot equation to any sections except the initial shock wave which runs in advance of the zone of chemical action and the final sections where combustion is complete. There are, however, cases where spinning detonation does not occur and yet the reaction is not instantaneous. In such cases it is possible to describe the Hugoniot curve appropriate to the case when any given fraction of the total chemical energy has been released. Fig. G,10e shows such a family of curves corresponding to release of 0, 30, 60, 90, and 100 per cent of the total energy. To find the condition of the gas at any point in the reaction zone, which may be assumed to progress at a uniform speed u, a line may be drawn through the point $p/p_0 = 1$, $\mu = \rho_0/\rho = 1$, so that

$$\frac{p}{p_0} - 1 = \left(1 - \frac{\rho_0}{\rho}\right)\left(\frac{\rho_0 U^2}{p_0}\right)$$

Three such lines are drawn in Fig. G,10e. One of them $A_0A_1A_2A_3A_4$-$A_5A_6A_7A_8O$ cuts all the curves twice. The upper part of $A_0A_1A_2A_3A_4$

represents a physically possible detonation wave which can only be maintained by means of a piston moving at the speed of the gas corresponding with the point A_4, i.e.

$$U\left[1 - \left(\frac{1}{\mu}\right)_{A_4}\right]$$

If such a wave were established and the piston were removed, a wave of rarefaction would move forward and the pressure would drop until the condition represented by the line $B_0B_1B_2B_3B_4$ was attained. At that

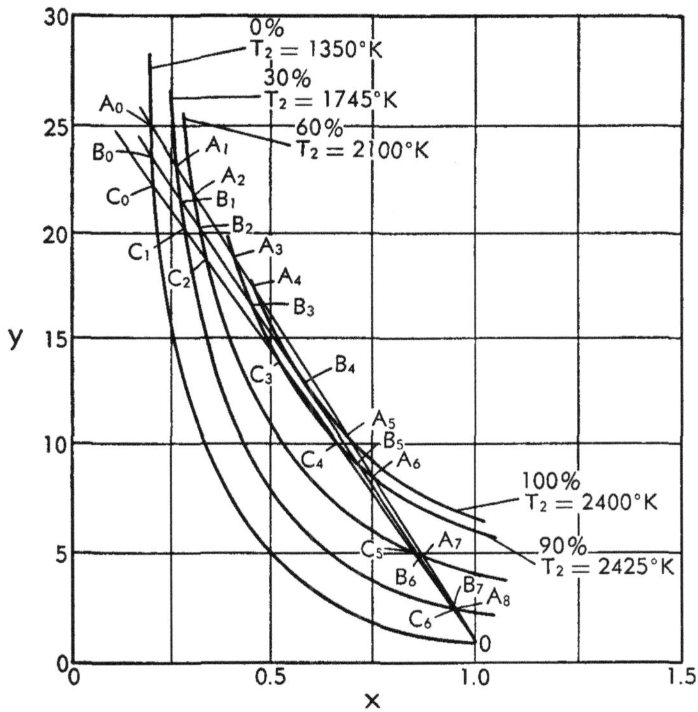

Fig. G,10e. Family of Hugoniot curves corresponding to fraction reacted in 20 per cent H_2 air.

critical state the reaction of the burned gas moving in a Riemann wave of rarefaction would supply the place of the piston and the wave would remain stable. The zone of reaction would travel at uniform speed, retain uniform thickness, and be followed by a Riemann wave of ever-increasing length.

The values of p and ρ corresponding with the parts of the zone of reaction where 0, 30, 60, 90, and 100 per cent of the reaction has taken place are represented by the points $B_0B_1B_2B_3B_4$. The distribution of pressure behind the initial shock wave depends on the rate of reaction. If the rate of reaction were uniform and complete after time t_1 the dis-

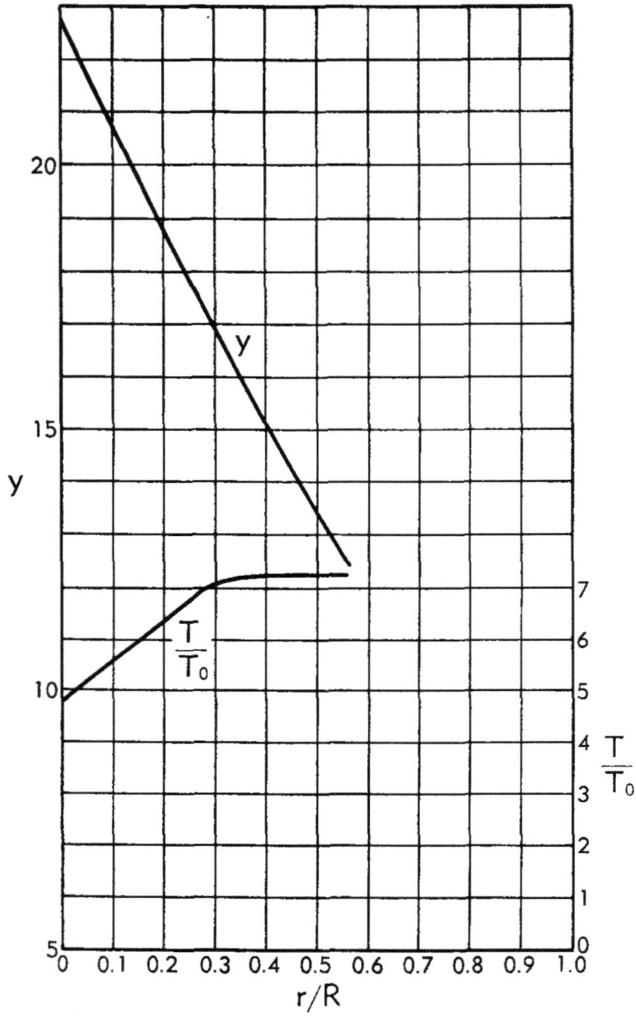

Fig. G,10f. Distribution of temperature and pressure for a Chapman-Jouguet wave with uniform rate of reaction.

tance r moved by the gas relative to the shock wave while a fraction $\delta q/q$ of the reaction took place would be

$$r = (U - u)\delta t = \frac{U\delta t}{\mu} = \frac{Ut_1}{\mu}\left(\frac{\delta q}{q}\right) \tag{10-51}$$

For the Chapman-Jouguet wave $B_0B_1B_2B_3B_4$ and the uniform rate of reaction, the distribution of pressure is shown in Fig. G,10f. If the number of molecules is constant through the reaction, the temperature is given by

$$\frac{T}{T_0} = \left(\frac{p}{p_0}\right)\left(\frac{\rho_0}{\rho}\right)$$

This is also shown in Fig. G,10f. It will be noticed that the pressure falls and the temperature rises through the reaction zone.

It is worth noticing that the line $A_5A_6A_7A_8O$ in Fig. G,10e represents a dynamically possible wave in which an initiator of chemical action travels faster than the Chapman-Jouguet speed. A zone of reaction follows in which the pressure rises gradually from p_0, its value at O, to the value corresponding with complete reaction and represented by A_5. At this point the burned gas is released from the zone of reaction with the condition that $u + a < U$. The zone of reaction is therefore followed by a column of gas moving at constant speed and at constant pressure. Into this column a Riemann wave of rarefaction follows the zone of reaction but at a smaller speed.

It will be noticed also that the line $C_0C_1C_2C_3C_4C_5C_6O$ cannot represent any real condition since no point on it corresponds with complete reaction. In fact no line through O can cut all the Hugoniot curves which has a slope less than the critical line OB_4 but greater than the slope of the second tangent curve (see line OE of Fig. G,10b) to the Hugoniot curve from the point O.

G,11. Transformation from Deflagration to Detonation. Gaseous mixtures which are capable of being detonated do not usually detonate immediately when they are ignited. Wave camera photographs (see Art. 13) show that when such mixtures are ignited a flame starts from the ignition point and moves away from it with increasing speed. The unburned mixture is forced away by the expansion of the burning gas and the pressure and temperature of the unburned gas are therefore raised. This increases the speed of the flame and thus increases the pressure in front of it still further.

Schlieren or shadow photographs taken with a wave camera show that a shock wave forms in front of the flame. Plate G,11a is a shadow photograph showing the shock wave at A and the zone of combustion at B. In some cases the new source of ignition appears in front of the flame and develops immediately into detonation waves. Plate G,11b shows a reaction front moving through the mixture $CO + \frac{1}{2}O_2$. It enters from the left with velocity 1275 m/sec. The bright point in the center is where a detonation wave starts in front of the flame. The detonation front begins to move forward at 3260 m/sec and later slows down to 1980 m/sec. The normal detonation speed in this mixture is 1740 m/sec. It is difficult to deduce from such photographs the velocity of the detonation wave relative to the explosive mixture which is already moving forward behind the initial shock wave. Plate G,11c shows a case in which a flame front increases in speed until suddenly it becomes a detonation; at the same time a shock wave, called by Dixon [*30*] a retonation wave, moves backwards from the point where the change occurs.

The physical conditions which control the velocity of propagation of a flame are not fully understood, and even if they were, it would be difficult to give a complete dynamical description of the phenomenon occurring when detonation starts at a point inside a gas. It seems worthwhile, however, to study the dynamics of some ideal cases in which the relationship between deflagration where density decreases at the flame front, and detonation, where it increases, can be described mathematically.

Table G,11a. Hugoniot curve for $C_2H_2 + O_2$ mixture at 288°K and 1 atmosphere pressure.

$\frac{1}{\mu} = x$	y	xy	$(x - 0.08)(y + 0.08)$
1.2	20.3	24.3	22.8
1.15	21.2	24.3	22.7
1.10	22.3	24.5	22.8
1.05	23.3	24.5	22.8
1.00	24.8	24.8	22.9
0.95	26.1	24.8	22.8
0.90	27.8	25.0	22.9
0.85	29.6	25.2	22.8
0.80	31.7	25.4	22.9
0.75	34.1	25.6	22.9
0.70	36.9	25.8	22.9
0.65	39.8	25.9	22.8
0.55	48.0	26.4	22.6
0.50	54.0	27.0	22.7

It has been seen that often the Hugoniot equation together with those of the progressive wave make it possible to describe the flow associated with flame fronts that move faster than the critical speed D, which is associated with the upper critical point B of Fig. G,10a.

The Hugoniot equation also describes deflagrations propagating into still gas at speeds less than the second critical speed (point E) associated with the front shown in Fig. G,10b. It is possible for flames to be propagated at speeds intermediate between the two critical speeds, but such flames evidently cannot be propagated into stationary gas. It is clear, in fact, that a region of forward-moving unburned gas must move ahead of the flame. If the flame moves at a constant speed s relative to the stationary gas it will force the unburned gas to move forward at speed u_1, and this will produce a shock wave moving at speed U into the stationary gas. The Hugoniot equation can be applied to the flame front moving in the gas which has already been compressed, but it is not the same equation that would be applied if the flame were moving into the still gas. The gas behind the flame front must then be considered in relation to the conditions at the rear end of the tube in which the gas is contained.

The complete description of such a system in any particular case using accurately calculated Hugoniot relationships would be very complicated. Fortunately, it is found that actual Hugoniot curves are very close to the ideal ones represented by Eq. 10-15. For example, the Hugoniot curve for mixture $C_2H_2 + O_2$ was calculated by Manson [*31*] and values for $x = \rho_1/\rho_2$, $y = p_2/p_1$ taken from his published curve for gas mixture at 288°K are given in columns 1 and 2 of Table G,11a. Column 3 gives xy. It will be seen that xy steadily increases as y increases. Column 4 gives the values of $(x - 0.08)(y + 0.08)$. It will be seen that this product is nearly constant. With very little error therefore the Hugoniot curve for $C_2H_2 + O_2$ burned at atmospheric pressure may be taken in the ideal form

$$(x - 0.08)(y + 0.08) = 22.8 \tag{11-1}$$

In order that Eq. 11-1 (Manson's calculations which include dissociation) may be made identical with Eq. 10-15 and 10-16 (with no dissociation), it is necessary that $(\gamma_2 - 1)/(\gamma_2 + 1) = 0.08$ $(=k_2)$ so that $\gamma_2 = 1.174$. Equating the constant term, it is found that

$$22.8 = \frac{k_2}{k_1} + k_2^2 + \frac{2k_2Q}{RT_1n_1} \tag{11-2}$$

where $k_1 = (\gamma_1 - 1)/(\gamma_1 + 1)$ and γ_1 the ratio of specific heats for the unburned mixture may be taken for the purpose of this calculation as 1.4. In the cases to which the Hugoniot equation is to be applied, the unburned gas will be heated to temperature T_1 by the first shock wave. The Hugoniot equation must therefore be written

$$(x - 0.08)(y + 0.08) = \frac{k_2}{k_1} - k_2^2 - \left(22.8 - \frac{k_2}{k_1} + k_2^2\right)\left(\frac{288}{T_1}\right) \tag{11-3}$$

or

$$(x - 0.08)(y + 0.08) = 0.474 - 22.326\left(\frac{288}{T_1}\right)$$

The expression (Eq. 11-3) is a good representation of the results of Manson's calculation at values of T_2 other than 288°K. At $x = 0.6$, for instance, the comparison is shown below.

T_1 °K	288	350	500	700
Manson's value	22.7	18.8	13.1	9.4
Formula (Eq. 11-3)	22.8	18.9	13.3	9.4

The following symbols will be used to represent the conditions of the gas in the various regions into which it is divided by the shock wave and the flame.

	Pressure	Density	Ratio of specific heats	Gas velocity	T	Velocity of sound
Undisturbed mixture	p_0	ρ_0	γ_0	0	T_0	a_0
Between shock wave and flame	p_1	ρ_1	$\gamma_1 = \gamma_0$	u_1	T_1	a_1
Close behind flame	p_2	ρ_2	γ_2	$u_1 + u_2$		a_2
Waves of rarefaction in products (if any)	p_3	ρ_3	$\gamma_3 = \gamma_2$	u_3		

In addition let:

U = the velocity of the shock wave into the undisturbed mixture
S = the velocity of the flame relative to the gas in region 1
U_1 = the absolute velocity of the flame front
$= u_1 + S$
$p_1/p_0 = y_1$, $p_2/p_1 = y$, and $\rho_1/\rho_2 = x$.

The equations to be satisfied at the shock wave are

$$\frac{U^2}{a_0^2} = \frac{1}{2\gamma_0}[\gamma_0 - 1 + (\gamma_0 + 1)y_1] \tag{11-4}$$

$$\frac{u_1}{a_0} = (y_1 - 1)\sqrt{\frac{2}{\gamma_0(\gamma_0 - 1) + \gamma_0(\gamma_0 + 1)y_1}} \tag{11-5}$$

$$\frac{\rho_1}{\rho_0} = \frac{\gamma_0 - 1 + (\gamma_0 + 1)y_1}{\gamma_0 + 1 + (\gamma_0 - 1)y_1} \tag{11-6}$$

$$\frac{T_1}{T_0} = \frac{y_1\rho_0}{\rho_1} \tag{11-7}$$

The equations to be satisfied at the flame front are the Hugoniot equation (Eq. 11-3) with the corresponding equation for S, namely

$$S^2 = \frac{p_1}{\rho_1}\left(\frac{y - 1}{1 - x}\right) \tag{11-8}$$

using $a_0^2 = \gamma_0(p_0/\rho_0)$ and the gas equation (Eq. 11-7). Eq. 11-8 may be written

$$\frac{S^2}{a_0^2} = \frac{T_1}{\gamma_0 T_0}\left(\frac{y - 1}{1 - x}\right) \tag{11-9}$$

The remaining equation is

$$\frac{u_2}{S} = 1 - x \tag{11-10}$$

One more equation is required to represent the condition of confinement of the gas behind the flame. In the case when the containing tube is closed at the rear end the necessary conditions would be satisfied by

$$u_1 + u_2 = 0 \tag{11-11}$$

but it will be seen later that under certain circumstances this condition can be replaced by another.

There are now eight equations relating the quantities U, S, u_1, u_2, T_1/T_0, ρ_1/ρ_0, x, y, y_1. To complete the solution it would be necessary to know how S depends on the physical condition of the gas in front of the flame. In the absence of such knowledge we may assume a series of values for S and determine a corresponding set of values for the remaining eight unknowns.

In solving the equations, it is more convenient to assume arbitrarily the value of y_1. U, u_1, ρ_1/ρ_0, and T_1/T_0 are then found. A value can then be assumed for y; x can then be found from Eq. 11-1. S and u_2 can then be found from Eq. 11-9 and 11-10. It will in general be found that the condition imposed by the (closed or open) rear end of the tube, e.g. Eq. 11-11, is not satisfied, but by taking a series of values for y one can be found which satifies the required condition. The results of this calculation are set forth in Table G,11b. Assuming first that the tube is closed at the rear end and that the burned gas is at rest, up to the flame front, the values of the variables which satisfy Eq. 11-11 together with Eq. 11-3, 11-9, and 11-10 are given for a series of values of y_1 in Table G,11b. It will be seen that U is always greater than U_1 so that it is impossible for the flame front to catch the shock wave if the gas behind the flame front is at rest. It is indeed obvious that this must be the case if the pressure in this burned gas is greater than that of the undisturbed gaseous mixture, for in that case the rate of increase of momentum in the gas must be positive and the only place in the field where there is any momentum is in the space between the flame front and the shock wave. The distance between these two must therefore continually increase.

When the velocity of the flame front is very low it moves more slowly than sound waves in the stationary burned gas. The effect of any change in the confinement at the rear end of the tube will be propagated forward to the flame front. At a certain critical velocity, however, the velocity of the flame front becomes greater than a_2, the velocity of sound in the burned gas. At this stage it is possible to alter the conditions anywhere behind the flame without alteration of the conditions immediately behind it. When the flame front moves at a greater speed than this critical value

Table G,11b. Results of calculations for closed-end tube where shock wave precedes flame.

y_1	$\frac{u_1}{a_0}$	$\frac{U}{a_0}$	$\frac{\rho_1}{\rho_0}$	y	$\frac{S}{a_0}$	$\frac{U_1}{a_0}$	x	$x'y'$	$\frac{T_1}{T_0}$	y_1y	$\frac{U}{a_0} - \frac{U_1}{a_0}$	$\frac{a_2}{a_0}$	$\frac{S - u_2}{a_0}$
2.0	0.524	1.362	1.625	0.9793	0.03465	0.5586	16.15	18.63	1.231	1.96	0.80	4.04	0.56
10	2.18	2.95	3.81	0.7465	0.217	2.40	11.08	9.08	2.626	7.46	0.55	4.27	2.40
20	3.264	4.158	4.654	0.56	0.413	3.677	8.93	5.67	4.297	11.26	0.480	4.26	3.69
30	4.074	5.085	5.028	0.45	0.576	4.650	8.03	4.216	5.967	13.5	0.43	4.25	4.63
50	5.34	6.56	5.38	0.32	0.848	6.18	7.26	2.874	9.302	16.25	0.43	4.25	6.19
70	6.35	7.75	5.540	0.246	1.070	7.42	6.94	2 274	12.65	17.95			
100	7.6316	9.2675	5 6698	0.183	1.3335	8.965	8.69		17.637	19.4			
150	9.381	11.275	5.7756	0.125	1.712	1.093	6.50	1.3334	25.97	21.1			
27	3.85	4.93	4.93	4.78	0.530	4.25	8.26	4.558	5.47				

it can be followed by a progression (expansion) wave of finite amplitude of exactly the same nature as that which follows a detonation. For a flame speed equal to or greater than the critical it is therefore *possible* to replace the equation $u_1 + u_2 = 0$ by

$$U_1 = u_1 + u_2 + a_2 \tag{11-12}$$

In fact the point on the Hugoniot curve which corresponds with this condition is the lower critical point E in Fig. G,10b. It is also possible to satisfy all the conditions by other combinations of the variables for which U_1 is greater than $u_1 + u_2 + a_2$. For reasons which will be mentioned later, however, attention will be confined to cases which satisfy Eq. 11-12.

Critical flame speed. To find the critical flame speed beyond which the condition of confinement has no effect, it is necessary first to calculate a_2. This is given by

$$a_2^2 = \gamma_1 \frac{p_2}{\rho_2} = \gamma_2 y p_1 \frac{x}{\rho_1} = \left(\frac{\gamma_2 y y_1 x}{\rho_1/\rho_0}\right)\left(\frac{p_0}{\rho_0}\right)$$

so that

$$\frac{a_2^2}{a_0^2} = \frac{\gamma_2}{\gamma_0}\frac{y_1 y x}{\rho_1/\rho_0} = \frac{\gamma_1 T_1 x y}{\gamma_0 T_0} \tag{11-13}$$

The values of a_2/a_0 calculated by this method are given in column 13 of Table G,11b. In the case when the burned gas is at rest, so that $u_1 + u_2 = 0$, the critical condition is given by $U_1 = a_2$. Comparing columns 7 and 13 it will be seen that when $y_1 < 20$, $U_1 < a_2$, while when $y_1 > 30$, $U_1 > a_2$. By graphical interpolation the critical is found to occur when $y_1 = 27$, $x = 8.26$, $S/a_0 = 0.530$, $U_1/a_0 = a_2/a_0 = 4.25$.

If the flame moves into the unburned gas faster than $S = 0.530a_0$ it is followed by a progressive wave which insulates it from any effects due to varying the confinement at the rear end of the explosion tube.

Effect of further increase in flame speed, second critical speed. For flame speeds greater than the critical value calculated above, the condition (Eq. 11-12) can be applied and the corresponding value of U_1 determined. The values of the variables obtained using this condition are given in the lower section of Table G,11c. Comparing the values of U and U_1 (columns 3 and 7 of Table G,11c) it will be seen that between $y_1 = 70$ and $y_1 = 100$ a second critical speed exists at which $U = U_1$. Graphical interpolation shows that this second critical occurs at $y_1 = 88$ and the corresponding values of the other variables are given at the foot of Table G,11c. It will be noticed that the second critical flame velocity is $8.69a_0$.

For flame speeds greater than $8.69a_0$, $U_1 > U$, the flame catches up to the shock wave. Such conditions can exist but cannot persist. If a solution of the equations for a given U which does not satisfy Eq. 11-12 is considered, the critical point for which $U_1 = U$ would have involved a higher value of S. Thus, if all the states corresponding with increasing burning velocity were considered, the first for which $U_1 = U$ would be that for which Eq. 11-12 is satisfied. This is the reason why only those solutions have been considered here.

Table G,11c. Results of calculations where flame speed $U_1 = u_1 + u_2 + a_2$.

y_1	$\frac{u_1}{a_0}$	$\frac{U}{a_0}$	$\frac{\rho_1}{\rho_0}$	y	$\frac{S}{a_0}$	$\frac{U_1}{a_0}$	x	$\frac{U - U_1}{a_0}$	$\frac{u_1 + u_2}{a_0}$	$\frac{a_2}{a_0}$	$\frac{\rho_2}{\rho_0}$
30	6.08	5.08	5.02	0.495	.579	4.652	7.41	0.433	0.364	4.288	0.666
50	5.34	6.56	5.38	0.52	.9105	6.24	4.85	0.33	1.83	4.44	1.06
70	6.35	7.75	5.64	0.538	1.24	7.59	3.70	0.16	3.00	4.59	1.496
100	7.6316	9.2675	5.6698	0.572	1.755	9.386	2.75	−0.119	4.54	4.81	1.95
150	9.381	11.275	5.775	0.64	2.68	12.06	1.93	−0.79	6.59	5.30	2.74
88	7.15	8.69	5.63	0.555	1.546	8.69	3.08	−0.01	3.94	4.76	

Critical

$$y_1 y = 48.805$$

$$x\frac{\rho_0}{\rho_1} = 0.5464$$

$$\frac{\rho_2}{\rho_0} = 1.831$$

Based on Hugoniot curve representing change from p_0, ρ_0 to p_2, ρ_2.

$$\frac{U_1^2}{a_0^2} = \frac{1}{1.4}\,\frac{(47.805)}{0.4536} = 75.2$$

$$\frac{U_1}{a_0} = 8.69$$

It is clear that in the second critical condition when $U_1 = U$ the zone of high pressure air between the flame and shock wave remains of constant thickness. The Hugoniot equation must therefore apply between the undisturbed gas p_0, ρ_0 and the gas p_2, ρ_2 behind the flame. Also, since the condition (Eq. 11-12) is satisfied, the flow corresponds with the *upper* critical point. Thus in the second critical condition the *lower* critical point in the Hugoniot curve for the change at the flame front is identical with the *upper* critical point in the Hugoniot curve representing the whole change from p_0, ρ_0, to p_2, ρ_2. It is a simple matter to prove this in the present case. The condition that this point (p_2, ρ_2) on the Hugoniot curve representing the change from p_0, ρ_0 to p_2, ρ_2 is the upper critical is

$$-\frac{dy}{dx} = \frac{y - 1}{1 - x}$$

and y is connected with x by the Hugoniot equation $(x - 0.08)(y + 0.08) = 22.8$. The solution of these two equations is

$$\begin{aligned} y &= 48.95 \\ x &= 0.545 \end{aligned} \tag{11-14}$$

The corresponding value of the velocity of propagation U is given by

$$\frac{U^2}{a_0^2} = \frac{p_0(y-1)}{a_0^2/\rho_0(1-x)} = \frac{1(y-1)}{\gamma_0(1-x)} = \frac{47.95}{(1.4)(0.455)} = 75.4$$

so that

$$\frac{U}{a_0} = 8.68 \tag{11-15}$$

Comparing this with the figures for $y_1 = 88$, it will be seen that

$$\frac{U}{a_0} = \frac{U_1}{a_0} = 8.69$$

Also

$$\frac{p_2}{p_0} = y_1 y = (88)(0.555) = 48.9$$

and

$$\frac{\rho_0}{\rho_2} = \frac{\rho_0}{\rho_1} x = \frac{3.08}{5.63} = 0.547 \text{ compared with Eq. 11-14}$$

Since the density of the gaseous mixture $C_2H_2 + O_2$ is nearly the same as that of air, a_0 will be nearly the same as the velocity of sound in air, which may be taken as 340 m/sec. Thus in this mixture the condition of confinement of the gas behind the flame affects the flow when the velocity of burning is less than $S = (0.54)(a) = 185$ m/sec. When S is greater than 185 m/sec the rate at which it propagates into the gas mixture does not depend on the confinement behind it, but a compression wave is propagated ahead of it. As the speed with which the flame penetrates the compressed gas increases up to $S = (1.54)(a_0) = 520$ m/sec, the shock wave always moves faster than the flame. When S exceeds 520 m/sec, the flame catches up with the shock wave and no steady condition is possible until the steady detonation wave is established.

These calculations do not give a correct description of what actually happens when a flame "burns to detonation." They are designed to illustrate the existence of critical rates of flame propagation and to show in a qualitative way how a deflagration can be transformed into a detonation. In many actual cases, detonation starts at a point ahead of the flame front in the gas which has been heated by compression. It seems that when the gas has been heated or otherwise activated to such an extent that the flame is propagated at a speed greater than the higher critical speed, 520 m/sec in the case of $C_2H_2 + O_2$, any point at which a flame starts will become a center of detonation rather than deflagration, and the flame will move at $8.69a_0 = 2960$ m/sec.

Chapman-Jouguet hypothesis. It has been seen that to materialize the conditions which give rise to points in the part KB (Fig. G,10a) of the Hugoniot curve it is necessary to control u, the velocity of combustion products, while for points in the section BC, U, the velocity of the reaction zone into the unburned gas, must be controlled. In experiments on detonation, neither is controlled, yet experiments show that a very definite value D of U is attained. Chapman and Jouguet made the assumption that the conditions behind a detonation wave are those represented by the critical point B. It will be shown later that the reaction of the expanding gases behind the detonation front is such that the pressure, density, and velocity of the gas are equal to those calculated using this assumption.

To find the minimum value of U from calculations in which a series of values are assumed for T_1, it is convenient to plot U against x and to draw a smooth curve through the resulting points. In the case of the mixture $H_2 + \frac{1}{2}O_2 + 2.5N_2$ at atmospheric pressure and $T_1 = 300°K$, the minimum value of U calculated without taking account of dissociation and using the data from which Table G,10b was constructed is $U = 1860$ m/sec. The minimum value of U taking account of dissociation has been calculated by Lewis and Von Elbe for this and some similar mixtures at $T_1 = 291°K$. Their results are given in Table G,11d. It will be seen that

Table G,11d. Comparison of calculated and experimental detonation velocities in mixtures of hydrogen, oxygen, and nitrogen. p_1 = atm, $T_1 = 291°K$.

Explosive mixture	p_2, atm	T_2, °K	Detonation velocity, m/sec		Concentration, % of burned gas	
			Calc	Exptl*	OH	H
$(2H_2 + O_2)$	18.05	3583	2806	2819	25.3	6.9
" $+ 1O_2$	17.4	3390	2302	2314	28.5	1.8
" $+ 3O_2$	15.3	2970	1925	1922	13.5	0.2
" $+ 5O_2$	14.13	2620	1730	1700	6.3	0.07
" $+ 1N_2$	17.37	3367	2378	2407	14.7	3.3
" $+ 3N_2$	15.63	3003	2033	2055	5.5	0.9
" $+ 5N_2$	14.39	2685	1850	1822	2.1	0.2
" $+ 2H_2$	17.25	3314	3354	3273	5.9	6.5
" $+ 4H_2$	15.97	2976	3627	3527	1.2	3.0
" $+ 6H_2$	14.18	2650	3749	3532	0.3	1.1

* From [*30*].

the calculated detonation velocity for the mixture $H_2 + \frac{1}{2}O_2 + 2.5N_2$ is 1850 m/sec.

The values of U originally calculated by Jouguet for a number of mixtures are given in Table G,11e, together with a number of experimental values. These were calculated without considering dissociation using specific heats calculated from measurements made by exploding gases in closed vessels. In such experiments dissociation certainly occurred, but in analyzing the results of such experiments no dissociation was allowed for. The result is that the numerical agreement obtained by Jouguet between his calculations and observation was better than it would have been if he had had more accurate data about specific heats.

Calculations of detonation velocity in gaseous mixtures at atmospheric pressure and temperature of 291°K have been made by Lewis and Von Elbe using the equations of chemical equilibrium to determine the

Table G,11e. Calculated and observed detonation velocities for various gas mixtures.

Gas mixture	T_1, °K	T_2	$\frac{\rho_2}{\rho_1}$	$\frac{p_2}{p_1}$	Velocity, m/sec	Velocity observed, m/sec
$H_2 + O$	10	3956	1.879	17.5	2629	{2810, 2821
$H_2 + O$	100	3981	1.864	12.9	2615	2790
$H_2 + O + 5H$	10	2596	1.79	14.4	3526	3530
$H_2 + O + 5N$	10	2596	1.79	14.4	1798	1822
$H_2 + O + 5O$	10	2596	1.79	14.4	1692	1707
$CO + O$ + humidity	10	3852	1.887	17.2	1664	1676
$CO + O$ + humidity	35	3748	1.88	15.6	1669	1738
$CO + H_2 + O_2$	10	3900	1.881	17.3	1984	{2008, 2143
$C_2H_2 + 3O_2$	10	4890	1.91	28.8	2120	2220
$C_2H_2 + 10O_2$	10	3560	1.84	22.0	1858	1850
$C_2H_2 + O_2$	10	5570	1.84	54.5	3091	2961
$C_2N_2 + O_2$	10	5960	1.837	58.2	2645	2728
$C_2N_2 + O_2 + 2N_2$	10	4244	1.8	33.7	2214	2166
$CH_4 + O_2$	10	3050	1.835	29.8	2477	2528
$C_2N_2 + 2O_2$	10	5150	1.914	34.8	2075	{2195, 2321
$CH_4 + 2O_2$	10	4080	1.904	27.4	2220	{2287, 2322
$CH_4 + 4O_2$	10	3570	1.86	23.4	2139	2166
$H + Cl$	10	3880	1.787	24.5	1851	1729
$H + Cl + H_2$	10	2400	1.73	14.7	2000	1855
$N_2O + H_2$	10	3933	1.865	25.9	2350	{2284, 2305

composition of the products of combustion and reliable data for their energy content. Their results for a number of gaseous mixtures are given in Table G,11d together with experimental determinations. It will be seen that the agreement is very good except in cases where there is a great excess of a neutral gas or of one of the constituents over the stoichiometric mixture. In such cases the calculated velocity is higher than the observed velocity. This indicates that in these cases the mixture has not attained the equilibrium composition. It will be seen later that other evidence confirms this view.

Detonation when the gaseous products are perfect gases with constant specific heat and fixed composition. In this case the Hugoniot equation may be written in the form

$$(x - k_2)(y_2 + k_2) = c \tag{11-16}$$

and

$$c = \frac{k_2}{k_1} - k_2^2 + \frac{2k_2Q}{RT_1n_1}$$

where

$$k_1 = \frac{\gamma_1 - 1}{\gamma_1 + 1}, \qquad k_2 = \frac{\gamma_2 - 1}{\gamma_2 + 1}$$

The values x, y where the tangent from $x = y = 1$ falls on the curve are

$$x = \frac{c}{1 + k_2} + k_2 \pm \sqrt{\frac{c^2}{(1 + k_2)^2} - \frac{c(1 - k_2)}{(+ k_2)}}$$

$$y = \frac{c}{1 - k_2} - k_2 \pm \sqrt{\frac{c^2}{(1 - k_2)^2} - \frac{c(1 + k_2)}{(1 + k_2)}}$$

G,12. Spherical Detonation. The possibility that a steady plane detonation wave can exist depends on two physical conditions. The first is that the chemical action can proceed at a sufficiently high speed and the other is that constant values of pressure and temperature can be maintained behind the reaction zone by the dynamical reaction of the burned gas. This second condition is satisfied in plane detonation, and experimentally by detonation in a tube, because in one dimension a progressive wave of finite amplitude can be propagated so as to maintain the required physical conditions at the back of the reaction zone. If an explosive is ignited at a point it may be expected that a spherical flame will spread out from it. If it is possible for the reaction of the burned gas moving spherically to maintain the same constant conditions behind a very thin reaction zone that are maintained behind a plane front by a plane progressive wave, spherical detonation would be dynamically possible.

Spherical waves of finite amplitude are not dynamically possible in the sense that plane ones are. It will be shown, however, that it is possible for an expanding spherical motion to exist in the burned gas which by its dynamical reaction can maintain the constant physical conditions necessary for the propagation of spherical detonation.

Assume that a region can exist in which u, the radial velocity, p, ρ, and a are functions of r/t only, r being the radial coordinate of a point and t the time from the initiation of the motion.

The equation of motion is:

$$\frac{\partial u}{\partial t} + u\frac{\partial u}{\partial r} = -\frac{1}{\rho}\frac{\partial p}{\partial r} \tag{12-1}$$

The condition that u, p, and ρ depend on r/t only is

$$\left(\frac{\partial}{\partial t} + w\frac{\partial}{\partial r}\right)(u, p, \rho) = 0 \tag{12-2}$$

where $w = r/t$. Using Eq. 12-2, Eq. 12-1 becomes

$$(u - w)\frac{du}{dw} = -\frac{1}{\rho}\frac{dp}{dw} \tag{12-3}$$

The equation of continuity is

$$\frac{\partial \rho}{\partial t} + u\frac{\partial \rho}{\partial r} + \rho\left(\frac{\partial u}{\partial r} + \frac{2u}{r}\right) = 0 \tag{12-4}$$

In view of $w = r/t$, Eq. 12-4 may be written

$$\left(\frac{u - w}{\rho}\right)\frac{d\rho}{dw} + \frac{du}{dw} + \frac{2u}{w} = 0 \tag{12-5}$$

Writing $a^2 = dp/d\rho$, so that

$$\frac{dp}{dw} = a^2\frac{d\rho}{dw}$$

hence with Eq. 12-3 and 12-5,

$$\frac{du}{dw}\left[1 - \left(\frac{w - u}{a}\right)^2\right] = -\frac{2u}{w} \tag{12-6}$$

It will be noticed that the equivalent equation for one-dimensional motion is

$$\frac{du}{dw}\left[1 - \left(\frac{w - u}{a}\right)^2\right] = 0$$

so that if a one-dimensional motion has the characteristic that u, p, and ρ are functions of w only, either

$$\frac{du}{dw} = 0 \quad \text{or} \quad w = a + u$$

The isentropic equation of state in the expanding burned gas gives a as a function of ρ or conversely ρ as a function of a^2, so that in Eq. 12-3 $(1/\rho)(dp/dw)$ may be written

$$\frac{1}{\rho}\frac{dp}{da^2}\frac{da^2}{dw}$$

Since $(1/\rho)(dp/da^2)$ may be regarded as a function of a^2. Eq. 12-5 and 12-6 determine u and a as functions of r when the appropriate boundary conditions are satisfied. Eq. 12-3 and 12-5 may be written in a nondimensional form using the variables

$$\xi = \frac{u}{w}, \quad \eta = \frac{a^2}{w^2}, \quad z = \ln w + \text{const} \tag{12-7}$$

$$\frac{d\eta}{d\xi} = \frac{2\eta}{\xi}\left[\frac{\eta - (1-\xi)^2 + Z\xi(1-\xi)}{3\eta - (1-\xi)^2}\right] \tag{12-8}$$

$$\frac{dz}{d\xi} = -\frac{1}{\xi}\left[\frac{\eta - (1-\xi)^2}{3\eta - (1-\xi)^2}\right] \tag{12-9}$$

where

$$Z = \frac{\rho}{a^2}\frac{da^2}{d\rho}$$

For numerical work it is convenient to change the variables, putting

$$\psi = \xi\eta^{-\frac{1}{2}} = \frac{u}{a}$$

In the case of a perfect gas,

$$Z = \gamma_2 - 1 \tag{12-10}$$

γ_2 being the ratio of specific heats for the burned gas.

Boundary conditions. If D is the velocity of detonation and R the radius of the wave at any time, $w/D = r/R$, then

$$\frac{u}{D} = \frac{u}{w}\frac{r}{R} = \xi\frac{r}{R} \tag{12-11}$$

$$\frac{a}{D} = \left(\frac{r}{R}\right)\sqrt{\eta}$$

If μ is the density ratio at the detonation front, the equation of continuity is

$$\frac{u}{U} = \left(1 - \frac{1}{\mu}\right) = \xi_1 \tag{12-12}$$

where ξ_1 is the value of ξ at the detonation front. The Chapman-Jouguet condition is $u + a = D$ or, from Eq. 12-11, $\xi_1 + \sqrt{\eta_1} = 1$ where η_1 is the value of η at the detonation front.

Using these boundary values of ξ and η, Eq. 12-8 and 12-9 may be used to calculate z and η as functions of ξ. The constant in Eq. 12-7 may conveniently be chosen so that $z = \ln(r/R)$; at $r = R$, $\eta_1 = (1 - \xi_1)^2 - 1/\mu^2$ so that $dz/d\xi = 0$. The rate of change in u and a with radius therefore becomes infinite when $r = R$. In physical problems infinite rates of change cannot occur. This does not prevent the motion described by Eq. 12-8 and 12-9 from being a good representation of a possible motion in a real gas. The error is likely to be of the order

$$\frac{\text{(thickness of reaction zone)}}{\text{(radius of detonation front)}}$$

To apply the equation it is necessary to expand the solution of Eq. 12-9 near $\xi = \xi_1$ in terms of

$$\left(1 - \frac{1}{\mu} - \xi\right)$$

Thus, near $\xi = \xi_1$

$$z = -\frac{\mu^2}{\mu - 1}\left(\frac{1}{2} + \frac{1}{4S}\right)\left(1 - \frac{1}{\mu} - \xi^2\right) \tag{12-13}$$

Example. As an example the distribution of velocity and pressure in the spherical detonation of the mixture $C_2H_2 + O_2$ initially at atmospheric pressure has been calculated. Manson [*31*] finds for this mixture $D = 2.96 \times 10^5$ cm/sec; $\mu = 1.87$, $Z = 0.13$.[2] The values of u/D calculated from Eq. 12-8 and 12-11 are given in Table G,12a and are shown in Fig. G,12a. To obtain the pressure it is first necessary to calculate the pressure rise at the shock wave. This is obtained by the Hugoniot equation:

$$y_1 = 1 + \frac{\rho_0 U^2}{p_0}\left(1 - \frac{1}{\mu}\right)$$

The density of $C_2H_2 + O_2$ at 288°K is $\rho_0 = 0.00124$ g/cm³ and, using Manson's values for D and μ, this gives $y_1 = 51.0$.

Since
$$\frac{p}{p_1} = \left(\frac{a}{a_1}\right)^{\frac{2\gamma}{1-\gamma}} = \left[\frac{\eta}{\eta_1}\left(\frac{r}{R}\right)^2\right]^{\frac{\gamma}{1-\gamma}}$$

$$\frac{p}{p_0} = y_1\left(\frac{\eta}{\eta_1}\right)^{\frac{\gamma}{\gamma-1}}\left(\frac{r}{R}\right)^{\frac{2\gamma}{\gamma-1}}$$

Values of p/p_0 calculated using this formula are given in Table G,12a and are shown in Fig. G,12a. It will be seen that the radial velocity of the gas is reduced to zero at $r/R = 0.50$ and that the pressure is constant and equal to 16 atmospheres from the center to $r/R = 0.50$.

Spherical deflagration behind a shock wave. The analysis of spherical detonation involves the conception of an infinitely thin layer in which

[2] This corresponds with $\gamma_2 = 1.13$. The figure is slightly different from the value $\gamma_2 = 1.175$ used in calculating the figures of Table G,11b. This difference is due to the fact that Manson's value is that actually calculated by him, taking into account the dissociation and variation in specific heat at the temperature of the detonation front. The value 1.175 which was used in the calculations of Table G,11b was obtained as a mean when the calculated points on the Hugoniot curve were approximated by the nearest hyperbola in the form which applies to a perfect gas.

the gas is suddenly compressed and then immediately expanded at a rate which is initially infinite. It has long been recognized that the assumption that the zone of compression is infinitely thin is a justifiable approximation in discussing macroscopic motions in a gas. It has in fact been shown that shock waves in which the pressure change is an atmosphere, or more, have a thickness of a few mean free paths. Similar considerations show that a shock wave of expansion associated with chemical action

Table G,12a. Distribution of u, a_0, and p in spherical detonation of mixture $C_2H_2 + O_2$. $D = 2960$ m/sec, $p_0 = 1$ atm.

ξ	$\psi = \frac{u}{a_0}$	$\frac{r}{R}$	$\frac{u}{D} = \xi\frac{r}{R}$	$\frac{a_0}{D} = \frac{\xi}{\psi}\frac{r}{R}$	$\frac{p}{p_0}$
0.465	0.870	1.000	0.465	0.535	50.2
0.450	0.843	0.999	0.450	0.533	48.1
0.440	0.825	0.998	0.439	0.533	46.9
0.420	0.786	0.995	0.418	0.532	45.5
0.400	0.746	0.989	0.396	0.530	43.5
0.380	0.705	0.981	0.373	0.529	41.1
0.360	0.662	0.970	0.349	0.527	39.6
0.340	0.618	0.956	0.325	0.526	37.5
0.320	0.574	0.938	0.300	0.523	34.5
0.300	0.528	0.918	0.275	0.521	32.3
0.280	0.483	0.895	0.251	0.519	30.3
0.260	0.437	0.871	0.226	0.517	28.3
0.240	0.393	0.845	0.203	0.516	26.7
0.220	0.350	0.817	0.180	0.514	25.1
0.200	0.308	0.790	0.158	0.512	23.3
0.180	0.269	0.762	0.137	0.510	22.4
0.160	0.231	0.734	0.117	0.509	21.2
0.140	0.195	0.706	0.099	0.507	20.2
0.120	0.161	0.678	0.081	0.506	19.3
0.100	0.129	0.651	0.065	0.505	18.6
0.080	0.099	0.623	0.050	0.503	17.9
0.060	0.071	0.596	0.036	0.503	17.4
0.040	0.045	0.568	0.023	0.502	16.9
0.020	0.021	0.540	0.011	0.502	16.4
0	0	0.502	0	0.502	15.9

might also be only a few mean free paths in thickness if every collision between the reacting constituents was effective. Since only a fraction of the collisions are in fact effective, the thickness of the zone of reaction must be greater than that of an equivalent compression shock wave, but even so its thickness is in general small enough to make the assumption that it is infinitely thin valid in discussing macroscopic gas flow.

The peculiar feature of the spherical detonation is that the compression is immediately followed by a rarefaction in which the change in pressure is initially infinite. This situation may be understood more readily if it is conceived as a limiting case when the velocity of a wave of spherical deflagration is equal to that of the initial shock wave. For this reason the hydrodynamical problems presented by a spherical sheet of flame with an assigned rate of flame propagation into the unburned gas will be discussed. The problem is strictly analogous to the plane problem discussed in Art. 11 and for that reason the deflagration in the same mixture, $C_2H_2 + O_2$, has been calculated.

The equations which represent a uniformly expanding spherical flow can be applied both to burned gas and to the flow between the flame and

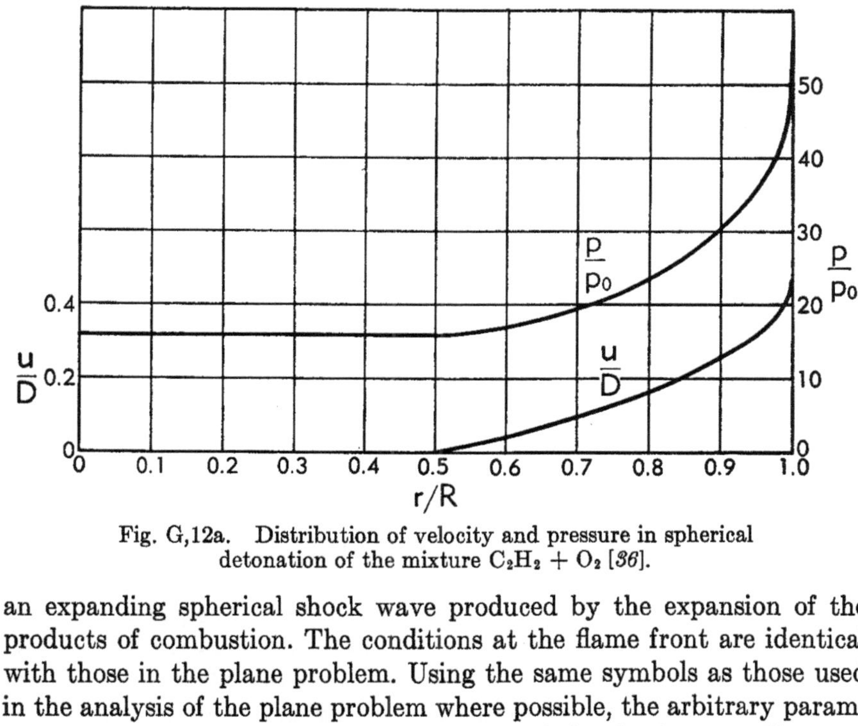

Fig. G,12a. Distribution of velocity and pressure in spherical detonation of the mixture $C_2H_2 + O_2$ [*36*].

an expanding spherical shock wave produced by the expansion of the products of combustion. The conditions at the flame front are identical with those in the plane problem. Using the same symbols as those used in the analysis of the plane problem where possible, the arbitrary parameter may be taken as y_1, the pressure ratio of the shock wave. If ξ_1 and η_1 are the values of ξ and η in the unburned gas immediately behind the shock wave it is found that

$$\xi_1 = \frac{2(y_1 - 1)}{(\gamma_0 - 1) + (\gamma_0 + 1)y_1}$$

$$\eta_1 = \frac{2\gamma_0 y_1[(\gamma_0 + 1) + (\gamma_0 - 1)y_1]}{[(\gamma_0 - 1) + (\gamma_0 + 1)y_1]^2}$$

$$\frac{\rho_1}{\rho_0} = \frac{(\gamma_0 - 1) + (\gamma_0 + 1)y_1}{(\gamma_0 + 1) + (\gamma_0 - 1)y_1}$$

If $a_0^2 = \gamma_0(p_0/\rho_0)$ the velocity U of the shock wave is generally

$$\frac{U^2}{a_0^2} = \frac{(\gamma_0 - 1) + (\gamma_0 + 1)y_1}{2\gamma_0} \quad \text{and} \quad \frac{T_1}{T_0} = \frac{\rho_0 y_1}{\rho_1}$$

Using these values of ξ_1 and η_1, Eq. 12-8 can be used to calculate η as a function of ξ and Eq. 12-9 to calculate r/R_1, R_1 being the radius of the shock wave. In this way a set of corresponding values of ξ, η, and r/R_1 is found for any given value of y_1. The figures when $y_1 = 10$, $\gamma_0 = 1.40$ are given in columns 1, 2, and 3 of Table G,12b. To find what rate of burning corresponds with any particular value of y_1 it is necessary to assume a set of possible radii for the flame front and to calculate radial velocity and other properties of the burned gas by applying the Hugoniot equation at each radius. As in the plane case, a value is chosen which can be associated with dynamically possible spherical expansion in the burned gas. The velocity of a point at which ξ, η, and r/R are constant is $r/t = w$. This is conveniently expressed in the nondimensional form r/a_0t, where

$$\frac{r}{a_0 t} = \left(\frac{U}{a_0}\right)\left(\frac{r}{R_1}\right) \tag{12-14}$$

For the case $y_1 = 10$ values of r/a_0t are given in column 4, Table G,12b. The radial velocity of the gas, u, is found from

$$\frac{u}{a_0} = \frac{r}{a_0 t} \tag{12-15}$$

and the rate of propagation S of the flame into the unburned gas is given by

$$\frac{S}{a_0} = \frac{r}{a_0 t} - \frac{u}{a_0} \tag{12-16}$$

Values for $y_1 = 10$, S/a_0 found by subtracting column 5 from column 4 are given in column 6.

It is necessary to apply the Hugoniot equation to the gas in the state represented by each row in Table G,12b. This equation (Eq. 11-3) requires a knowledge of T/T_0. The quantity T_1/T_0 has already been given. The change in temperature due to the adiabatic compression of the unburned gas after passing through the shock wave is represented by T/T_1. From

$$\frac{T}{T_1} = \frac{a^2}{a_1^2} = \frac{\eta}{\eta_1}\left(\frac{r}{R_1}\right)^2$$

so that

$$\frac{T}{T_0} = \frac{T_1}{T_0} \cdot \frac{\eta}{\eta_1}\left(\frac{r}{R_1}\right)^2 \tag{12-17}$$

This may be substituted in the Hugoniot equation

$$(x - 0.08)(y + 0.08) = 0.474 + 22.326\,\frac{T_0}{T} = A \tag{12-18}$$

Values of T/T_0 are given in column 8 in Table G,12b.

Table G,12b. Calculation of spherical flame front where $y_1 = 10$.

$\frac{r}{R_1}$	ξ	η	$\frac{r}{a_0t}$	$\frac{u}{a_0}$	$\frac{S}{a_0}$	$\left(\frac{a}{a_0}\right)^2$	$\frac{T}{T_0}$	$\frac{1-y}{x-1}$	A
1.00	0.73700	0.3010	2.9518	2.17754	0.7743	1.000	2.623	0.320	8.97
0.9953	0.7500	0.3047	2.9379	2.20342	0.7345	1.00	2.623	0.287	8.97
0.9773	0.800	0.3198	2.8848	2.30784	0.5770	1.015	2.662	0.1750	8.85
0.9584	0.850	0.3346	2.8290	2.43294	0.3961	1.021	2.678	0.0818	8.80
0.9408	0.900	0.3490	2.7770	2.49930	0.2777	1.026	2.691	0.0400	8.77
0.9242	0.950	0.3625	2.7280	2.59160	0.1364	1.030	2.701	0.0096	8.73
0.9086	1.000	0.3746	2.6820	2.6820	0	1.030	2.701	0	8.73
$\frac{r}{R_1}$	$\frac{1-y}{x-1}$	x_2	x_1	$\frac{T}{T_0}$	$\frac{S}{a_0}$	$\frac{S}{a_0}(x_2-1)$	$\frac{S}{a_0}(1-x_1)$	$\frac{u}{a_0}$	
0.9242	0.0096	105.0	8.75	2.701	0.136	14.1	1.054	2.592	
0.932	0.020	45.34	9.74	2.70	0.196	8.68	1.712	2.548	
0.937	0.030	25.55	11.53	2.70	0.240	5.89	2.525	2.520	
0.939	0.035	17.50	14.38	2.70	0.259	4.27	3.47	2.509	

Critical:

$r/R_1 = 0.9369$
$u/a_0 = 2.520$
$S/a_0 = 0.240$
$U_1/a_0 = (S/a_0) + (u/a_0) = 2.76$

r is the radius of the flame front
R_1 is that of the shock wave
$A = 0.474 + 22.326T_0/T$

To find x and y, the quantities which define the changes in density and pressure at the flame front, it is necessary to find another equation between them. This is provided by the fact that the velocity of flame propagation into the unburned gas is S. Thus

$$S^2 = \left(\frac{1 - y}{x - 1}\right)\frac{p_1}{\rho_1} \tag{12-19}$$

p_1 and ρ_1 being the pressure and density of the gas in front of the flame. Since

$$\begin{aligned}\frac{p_1}{\rho_1} &= \frac{p_0}{\rho_0}\frac{T_1}{T_0} = \frac{a_0^2}{\gamma_0}\frac{T_1}{T_0}\\ \frac{1 - y}{x - 1} &= \frac{S^2\gamma_0 T_0}{a_0^2 T_1}\end{aligned} \tag{12-20}$$

Values of $(1 - y)/(x - 1)$ are given in column 9 of Table G,12b. Eliminating y from Eq. 12-18 and 12-20 a quadratic is obtained for x which has two real or two imaginary roots. In the case of $y_1 = 10$ it is found that except for the two lowest rows the roots are imaginary. When r/R_1 is less than 0.939 the two roots are real.

To attain the conditions described by the numbers in Table G,12b it is in general necessary to have a spherically expanding inner boundary surface. The case corresponding with $S/a_0 = 0$, for instance, is that of the air wave surrounding a sphere which is expanding at a rate $2.682a_0$. Such an expansion can be obtained in effect by an expanding spherical flame front. A physically possible state for the gas inside the flame front is one in which the gas is at rest. The condition that this shall be so is that

$$\frac{u}{a_0} + (1 - x)\frac{S}{a_0} = 0 \tag{12-21}$$

To find the point where this condition is satisfied, both $(S/a_0)(x - 1)$ and u/a_0 are plotted in a diagram like Fig. G,12b. The curve representing $(S/a_0)(x - 1)$ starts at $r/R_1 = 0.9086$ when $S/a_0 = 0$. It increases with r/R_1 up to 0.939 when $(S/a_0)(x - 1) = 3.9$. The part of the curve above this point corresponds with the larger of the two values of x. The values of u/a_0 are plotted and the crossing point $r/R_1 = 0.937$, $u/a_0 = 2.520$ gives $S/a_0 = 0.240$, $x = 11.5$, $y = 0.686$.

When the flame speed is $0.240a_0$, it is situated on a sphere of radius $0.937R_1$. The rise in pressure at the shock wave is 10 atm.

To find the pressure behind the flame front it is necessary to calculate the pressure p immediately in front of it. Using the adiabatic relationship

$$\frac{p}{p_1} = \left(\frac{T}{T_1}\right)^{\frac{\gamma_0}{\gamma_0 - 1}} = \left(\frac{T}{T_1}\right)^{3.5} \quad \text{where} \quad \gamma_0 = 1.4$$

In the case $y_1 = 10$ this is $(1.026)^{3.5} = 1.095$ so that the pressure is

$$\frac{yp_1}{p_0} = (1.095)(0.886)(10) = 7.3 \text{ atm}$$

Critical burning velocities. As S increases, point P of Fig. G,12b where $(u/a_0) + (1 - x)S/a_0 = 0$ moves up the curve representing $(x - 1)S/a_0$ against r/R_1. At a certain value of y_1 it reaches the point

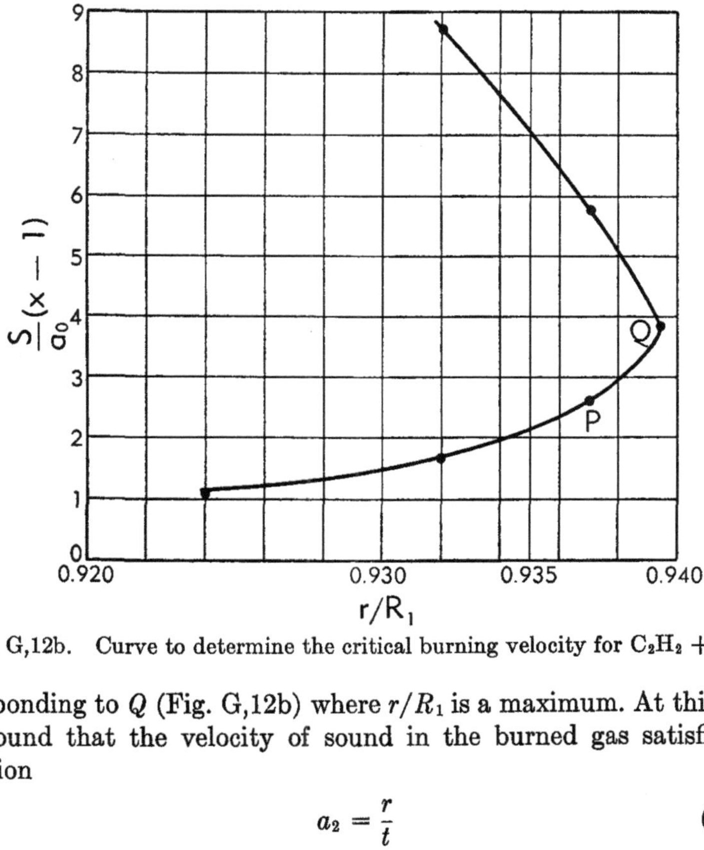

Fig. G,12b. Curve to determine the critical burning velocity for $C_2H_2 + O_2$.

corresponding to Q (Fig. G,12b) where r/R_1 is a maximum. At this point it is found that the velocity of sound in the burned gas satisfies the condition

$$a_2 = \frac{r}{t} \tag{12-22}$$

(i.e. small disturbances propagated in the burned gas move at the same speed as the flame front). The value of a_2 is found by a formula analogous to Eq. 11-13. Using

$$\left(\frac{a_2}{a_0}\right)^2 = \frac{\gamma_1}{\gamma_0}\frac{T}{T_0}xy \tag{12-23}$$

The calculations described for $y_1 = 10$ were repeated for $y_1 = 2, 20,$ 30, and 70 (Tables G,12c, G,12d, G,12e, and G,12f). The results are given in Table G,12g.

Table G,12c. Calculation of spherical flame front where $y_1 = 2.0$.

$\frac{r}{R_1}$	ξ	η	$\frac{r}{a_0t}$	$\frac{u}{a_0} = \xi \frac{r}{at}$	$\frac{S}{a_0}$	$\left(\frac{a}{a_1}\right)^2$	$\frac{T}{T_0}$	$\frac{1-y}{x-1}$	A
1.00	0.3846	0.6672	1.3632	0.5243	0.8389	1.000	1.2308	0.803	18.62
0.9925	0.400	0.6752	1.3530	0.5412	0.8188	1.000	1.2308	0.765	18.62
0.9687	0.450	0.7189	1.3205	0.5942	0.7263	1.0111	1.243	0.596	18.50
0.9448	0.50	0.7641	1.2880	0.6440	0.6440	1.0223	1.256	0.463	18.30
0.9215	0.55	0.8108	1.2562	0.6909	0.5653	1.0319	1.269	0.353	18.10
0.8993	0.60	0.8578	1.2259	0.7355	0.4904	1.0398	1.279	0.264	18.0
0.8782	0.65	0.9038	1.1972	0.7782	0.4190	1.0447	1.285	0.192	17.8
0.8543	0.70	0.9503	1.1646	0.8152	0.3494	1.0395	1.279	0.133	17.8
0.8360	0.75	0.9963	1.1396	0.8547	0.2849	1.0436	1.284	0.0885	17.8
0.8186	0.80	1.0416	1.1159	0.8927	0.2232	1.0461	1.287	0.0540	17.8
0.8025	0.85	1.0866	1.0940	0.9299	0.1641	1.0488	1.291	0.0292	17.8
0.7877	0.90	1.1301	1.0738	0.9664	0.1074	1.0509	1.293	0.0125	17.7
0.7736	0.95	1.1725	1.0546	1.0019	0.0527	1.0517	1.294	0.0030	17.7
0.7606	1.00	1.2138	1.0368	1.0368	0.00	1.0524	1.295	0	17.7

$\frac{r}{R_1}$	$\frac{1-y}{x-1}$	A	x_1	x_2	$\frac{S}{a_0}$	$(1 - x_1)\frac{S}{a_0}$
0.7606	0	17 7			0	
0.770	0.017	17.7	18.3	569.7	0.038	0.668
0.775	0.037	17.7	19.1	250.9	0.058	1.05
0.780	0.067	17.7	20.4	129.9	0.078	1.515
0.785	0.104	17.7			0.097	
0.7877	0.125	17.7	21.6	65.8	0.1074	22.2
0.790	0.014	17.7			0.116	

Critical:
$r/R_1 = 0.774$
$S/a_0 = 0.054$
$U_1/a_0 = r/a_0t = 1.055$

$A = 0.474 + 22.326T_0/T$
r is radius of flame front
R_1 is radius of shock wave

Table G,12d. Calculation of spherical flame front where $y_1 = 20$.

$\frac{r}{R_1}$	ξ	η	$\frac{r}{a_0 t}$	$\frac{u}{a_0}$	$\frac{S}{a_0}$	$\left(\frac{a}{a_1}\right)^2$	$\left(\frac{T}{T_0}\right)$	$\frac{1-y}{x-1}$
1.000	0.7851	0.2495	4.1580	3.2644	0.8936	1.0	4.2974	0.261
0.9763	0.850	0.2645	4.05946	3.45054	0.60892	1.01	4.351	0.1192
0.9586	0.900	0.2761	3.98586	3.58727	0.39859	1.019	4.37	0.0509
0.9417	0.950	0.2871	3.91559	3.71981	0.19578	1.023	4.39	0.0122
0.9258	1.00	0.2974	3.84948	3.84948	0	1.027	4.41	0
$\frac{r}{R_1}$	A	x_1	x_2					
1.000	5.67							
0.9763	5.60							
0.9586	5.58	7.53	19.73					
0.9417	5.50	6.00	77.00					
0.9258	5.53	5.23	0					
0.960		8.35	12.35					

Critical:

$S/a_0 = 0.430$

$r/R_1 = 0.9617$

$U_1/a_0 = (S/a_0) + (u/a_0) = 3.99$

$u/a_0 = 3.56$

r is radius of flame front

R_1 is radius of shock wave

$A = 0.474 + 22.326 T_0/T$

Table G,12e. Calculation of spherical flame front where $y_1 = 30$.

$\frac{r}{R_1}$	ξ	η	$\frac{r}{a_0 t}$	$\frac{u}{a_0}$	$\frac{S}{a_0}$	$\left(\frac{a}{a_1}\right)^2$	$\frac{T}{T_0}$	$\frac{1-y}{x-1}$	A
1.000	0.80110	0.23076	5.0851	4.0737	1.0114	1.000	5.967	0.240	4.21
0.9931	0.820	0.23501	5.0500	4.1410	0.9090	1.0044	5.979	0.193	4.20
0.9858	0.84	0.23945	5.0129	4.2108	0.8021	1.0084	5.955	0.151	4.20
0.9785	0.86	0.24181	4.9758	4.2792	0.6966	1.0030	5.991	0.1135	4.20
0.9715	0.88	0.24606	4.9402	4.3474	0.5928	1.0064	6.003	0.0820	4.19
0.9645	0.90	0.25020	4.9046	4.4141	0.4905	1.0086	6.015	0.0560	4.18
0.9577	0.92	0.25430	4.8700	4.4804	0.3896	1.0107	6.027	0.0352	4.17
0.9509	0.94	0.25836	4.8354	4.5453	0.2901	1.0123	6.038	0.0195	4.16
0.9442	0.96	0.26227	4.8013	4.6092	0.1921	1.0132	6.044	0.0085	4.16
0.9378	0.98	0 26606	4.7688	4.6734	0.0954	1.0140	6.050	0.0021	4.16
0.9315	1.00	0.26972	4.7368	4.7368	0	1.0142	6.050	0	

Critical:

$r/R_1 = 0.9703$

$S/a_0 = 0.574$

$U_1/a_0 = (S/a_0) + (u/a_0) = 4.934$

r is radius of flame front

R_1 is radius of shock front

$A = 0.474 + 22.326T_0/T$

Table G,12e (continued)

$\frac{r}{R_1}$	$\frac{S}{a_0}$	$\frac{1-y}{x-1}$	A	x_1	x_2	$(1-x_1)\frac{S}{a_0}$	$(1-x_2)\frac{S}{a_0}$	$\left(\frac{a_2}{a_0}\right)_1$	
0.9315	0.000								
0.935	0.050	0.00058	4.16						
0.94	0.125	0.00361	4.16	3.97	296.0	0.371	36.9		
0.945	0.200	0.00928	4.16	4.01	113.6	0.603	22.5		
0 95	0.275	0.01753	4.16	4.171	58.56	0.872	15.56		
0.955	0.349	0.0283	4.17	4.29	35.01	1.149	11.89		
0.960	0.420	0.0410	4.17	4.52	22.96	1.148	9.22	4.433	
0.965	0.495	0.0571	4.18	4.83	15.41	1.90	7.12	4.420	
0.97	0.570	0.0758	4.19	6.167	9.161	2.94	4.65	4.347	
0.968	0.540	0.06793	4.19	5.183	11.495	2.42	5.67		
0.967	0 555	0.07175	4.19	5.611	10.511	2.56	5.28		
0.9706	0.580	0.07843	4.19	6.699	8.151	3.31	4.15	4.319	
0.9707	0.580								

$\frac{r}{R_1}$	$\left(\frac{a_2}{a_0}\right)_2$	$x_1\frac{S}{a_0}-\left(\frac{a_2}{a_0}\right)_1$	$x_2\frac{S}{a_0}-\left(\frac{a_2}{a_0}\right)_2$
0.96	3.450	−2.53	+6.19
0.965	3.873	−2.02	+3.76
0.97	4.201	−0.832	+1.02
0.976	4.247	−0.44	+0.47

Table G,12f. Calculation of spherical flame front where $y_1 = 70$.

$\frac{r}{R_1}$	ξ	η	$\frac{r}{a_0 t}$	$\frac{u}{a_0}$	$\frac{S}{a_0}$	$\left(\frac{a}{a_1}\right)^2$	$\frac{T}{T_0}$	$\frac{1-y}{x-1}$	A
1.000	0.8195	0.21011	7.7552	6.3554	1.3998	1.000	12.635	0.2171	2.240
0.9998	0.8200	0.21022	7.7536	6.3579	1.3957	1.000	12.636	0.2158	2.241
0.9925	0.8400	0.21426	7.6970	6.4655	1.2315	1.0045	12.693	0.1673	2.233
0.9854	0.86	0.21824	7.6420	6.5721	1.0699	1.0086	12.745	0.1257	2.226
0.9782	0.88	0.22215	7.5861	6.6758	0.9103	1.0117	12.784	0.0907	2.220
0.9712	0.90	0.22598	7.5318	6.7786	0.7532	1.0145	12.820	0.0615	2.216
0.9643	0.92	0.22973	7.4783	6.8800	0.5983	1.0167	12.847	0.03901	2.212
0.9576	0.94	0.23338	7.4264	6.9808	0.4456	1.0186	12.871	0.02160	2.209
0.9509	0.96	0.23692	7.3744	7.0794	0.2950	1.0196	12.884	0.00946	2.207
0.9444	0.98	0.24037	7.3240	7.1775	0.1405	1.0203	12.893	0.00233	2.205
0.9380	1.00	0.24368	7.2744	7.2744	0	1.0204	12.894	0	2.204
$\frac{r}{R_1}$	x_1	x_2							
0.9782	2.4123	10.569							
0.9712	2.290	16.262							
0.9854	2.620	7.052							
0.9925	3.270	4.264							

Critical for $(u/a_0) + (S/a_0)(1 - x) = 0$
$r/R = 0.9851$
$U_1/a_0 = 7.64$
Critical for $U_1/a_0 = (a_2/a_0) + (u/a_0) + (S/a_0)(x - 1)$
$u/a_0 = 6.3688$
$U_1/a_0 = 7.6205$
$r/R = 0.9933$

r is radius of flame front
R_1 is radius of shock wave
$A = 0.474 + 22.326T_0/T$

Table G,12g. Results of calculations where products are at rest inside spherical burning surface.

y_1	$\frac{r}{R_1}$	x	y	$\frac{U_1}{a_0}$	$\frac{S}{a_0}$	$\frac{a_2}{a_0}$	$\frac{a_2}{a_0} - \frac{U_1}{a_0}$	
2.0	0.774	19.54	0.82	1.055	0.054	4.169	3.113	
10.0	0.9369	11.50	0.766	2.765	0.240	4.225	1.46	
20.0	0.9617	9.28	0.526	3.999	0.430	4.231	0.24	
30.0	0.9703	8.67	0.407	4.934	0.574	4.213	−0.712	
70.0	0.9851	7.185	0.233	7.640	1.0633	4.280	−3.36	

Table G,12h. Results of calculations where spherical burning surface is followed by a wave of expansion.

y_1	$\frac{r}{R_1}$	$\frac{U_1}{a_0}$	x	$\frac{S}{a_0}$	y	xy	$\frac{a_2}{a_0}$	$x\frac{S}{a_0} - \frac{a_2}{a_0}$
30	0.9707	4.942	7.4	0.5797	0.6924	3.644	4.285	4.29 − 4.2850 = −0.005
70	0.9933	7.6205	3.717	1.2517	0.534	1.987	4.60	4.65 − 4.60 = +0.050

The values of r/R_1, x, and y which correspond to those of the burned gas at rest are given in columns 2, 3, and 4. U_1/a_0, S/a_0, and a_2/a_0 are given in columns 5, 6, and 7. When the burned gas is at rest, values of $(U_1/a_0) - (a_2/a_0)$, which is the relative velocity of sound waves in the burned gas and the flame front, are given in column 8. Plotting $(U_1/a_0) - (a_2/a_0)$ against y_1 it is found that the critical value of y_1 for which $U_1 = a_2$ is $y_1 = 22.5$ and the critical rate of burning is given by $S/a_0 = 0.46$, which is slightly less than the critical values $S/a_0 = 0.530$ and $y_1 = 27$ (see Table G,11b) for the plane case. As soon as the flame speed exceeds $0.46a_0$, the velocity of sound is less than that of the flame front. It is then possible for a new regime to occur in which the burned gas is expanding and the condition $a_2 + S = U_1$ is satisfied at the flame front.

This condition may be written

$$x\frac{S}{a_0} - \frac{a_2}{a_0} = 0 \tag{12-24}$$

and it corresponds with the point Q of Fig. G,12b, where the greatest possible value of r/R_1 for the particular value of y occurs. The values of r/R_1, S/a_0, a_2/a_0, and U_1/a_0 for these waves are given in Table G,12h. For $y_1 = 30$, which is only slightly above the critical value $y_1 = 22.5$, the maximum value of r/R_1 is 0.9707 which is only slightly greater than the value 0.9703, the value corresponding with that of stationary burned gas. For $y_1 = 70$, the flame front is at a radius within 0.7 per cent of the shock wave. Comparing the calculated values of S/a_0 and U_1/a_0 with those for the plane case at the same value of y_1 in Tables G,11g and G,12h, it will be seen that they are nearly identical. It might be expected that when the radius of the wave front is 0.9933 of that of the shock wave the curvature of the shock wave would have little effect on the motion. In fact the flame front overtakes the shock wave when $S/a_0 = 1.546$ as it does in the plane case (see Table G,11b) and the spherical detonation discussed in Fig. G,12a results.

Comparison between spherical and plane detonation. The analysis shows that there is very little difference between spherical and plane deflagration and that the flame velocity which corresponds with the change from deflagration to detonation, preceded by a compressed gas, is the same in both cases, namely 524 meters per second in the case of $C_2H_2 + O_2$.

Little experimental evidence has been published about spherical detonation. Jouguet [*32*] considered that spherical detonation waves moving with constant speed could not exist. It seems that his objection can be reduced to an objection to the existence of a pressure distribution in which the rate of change in pressure approaches infinity as the detonation front is approached. In the author's opinion this objection has exactly the same kind of validity as an objection to the use of shock waves in theoretical discussion, on the ground that the region in which the properties change is not in fact infinitely thin.

There is little experimental evidence about spherical shock waves. Lafitte [*33*] found that ignition by spark at the center of a spherical glass bulb of 24-cm diameter containing $CS_2 + 3O_2$ did not produce detonation. When the same mixture was fired by a detonator containing 1 gram of fulminate of mercury, a spherical wave was produced which traveled at the same speed as the detonation wave produced when the same mixture was exploded in a tube. Similar results were obtained with $2H_2 + O_2$.

It seems that there is evidence that spherical detonation waves can be produced but there is no evidence that a spherical deflagration flame can increase in speed until detonation occurs at or near the flame. The analysis suggests no method by which either plane or spherical flames can "burn to detonation" for it assumes the existence of a constant rate of burning. If the rate of burning were simply a function of the temperature or of some activations produced at the shock wave it would in fact remain constant and the whole system would expand uniformly. The increase in flame velocity which is observed when gaseous mixtures burn to detonation in an open-ended tube seems to depend in some way on the existence of its wall. Deflagration started by a spark in a gaseous mixture takes longer to burn to detonation in a tube of large cross section than it does in a small one. It is of course possible that the temperature of the unburned gas in front of a flame traveling down a tube increases with the distance it has traveled because the length of the column of compressed, unburned gas increases. The friction of the wall must cause a rise in temperature between the shock wave and the flame, and this rise must increase as the length of the column of compressed unburned gas increases.

G,13. Spinning Detonation in Gaseous Mixtures.

Apparatus for the study of detonation. The propagation of flame and detonation in gaseous mixtures can best be investigated by means of a rotating drum camera. The gas is contained in a glass tube; the image of this tube is projected by the camera lens onto the generator of a rotating cylinder on which a photographic film is wound. The image of a luminous portion of the tube is at any instant a line on the film. If m represents the magnification of the camera, that is, the ratio of the size of the image to the size of the object, a flame front moving with velocity U will produce an image which travels with velocity mU parallel to the axis of the drum. If the drum rotates with circumferential speed V, the flame front sweeps out a line on the drum which makes, with the generator, an angle $\tan^{-1}(V/mU)$. If the peripheral speed of the drum is known, the velocity U can be determined by measuring the photograph. Plate G,13a is a rotating mirror photograph of the detonation of $2CO + O_2$ in a tube of

1.3-cm diameter. Plate G,13b is another photograph of the same mixture. It has been shown that the zone of intense reaction represented by the wavy sloping line in Plates G,13a and G,13b is preceded by a shock wave (not visable in the photographs) in the unburned gas which probably travels at a uniform speed. A detonation head of very intense activity seems to travel around the inner surface of the tube while remaining in, or close to, the shock wave. The result is that this head moves in a spiral of about 3 diameters pitch. The waviness of the line representing the detonation front in Plate G,13a shows that this head does not progress uniformly along its spiral path. It glows very brightly once during each revolution. On the other hand, the path of the head was shown to be a spiral because it made a spiral mark on the inside of a specially prepared tube (see Plate G,13c). In most photographs the streaks appear to be continuous as in Plate G,13a, but some show the spotted appearance of Plate G,13b. This spotted appearance was attributed by Bone, Fraser, and Wheeler [*34,35*] to interference between forward-moving gas locally heated at one of the points of intense activity and backward-moving sound waves. It can be proved that if, and only if, the Chapman-Jouguet condition is satisfied, $D = u + a$, the forward-moving particles will meet the backward-moving waves simultaneously. Then the illuminated spots appearing at these crossing places will be on a line parallel to the tube. Bone, Fraser, and Wheeler, in fact, attributed the whole of the streaky appearance in their photographs to this effect. If this were the complete explanation, it could be shown that the streaks would cease to be parallel to the tube in the gas some distance behind the detonation point where D is greater than $u + a$. It has been found that this is not the case.

The fact that the streaks are nearly always parallel to the tube shows that instants of intense activity occur at all points along the tube at the same time. This could happen if the products of combustion were in a state of violent lateral oscillation, which would cause a region of increased pressure and increased temperature to appear simultaneously at all sections of the tube. Strong confirmation of this explanation is found when the frequencies of the simpler modes of transverse oscillation are calculated and compared with observation.

Theory of spinning detonation. Oscillations in which the motion is confined to transverse planes are governed by the equations:

$$p' = \phi e^{i2\pi nt} \tag{13-1}$$

$$\nabla^2\phi + \frac{4\pi^2 n^2 \phi}{a^2} = 0 \tag{13-2}$$

where p' is the fluctuation in pressure, n is the frequency, and a the velocity of sound. A solution of Eq. 13-2 is

$$\phi = J_s\left(\frac{2n\pi r}{a}\right)\cos s\theta \tag{13-3}$$

where r and θ are polar coordinates of a point in a transverse plane and s is a whole number or zero. $J_s(z)$ is the Bessel function of order s and argument z. If the gas is contained in a tube of radius R, the boundary condition is

$$\left(\frac{\partial \phi}{\partial r}\right)_{r=R} = 0$$

or

$$\frac{\partial}{\partial n}\left[J_s\left(\frac{2\pi n r}{a}\right)\right]_{r=R} = 0 \tag{13-4}$$

If the gas oscillates transversely in the mode for which $s = 1$, the smallest root of the Bessel function is

$$\frac{2n\pi r}{a} = 1.84 \tag{13-5}$$

so that

$$n. = \frac{1.84a}{\pi d} \tag{13-6}$$

where d is the tube diameter.

If the gas oscillates in the mode for which $s = 2$, i.e. a mode in which maximum pressure is attained simultaneously at opposite ends of a diameter, then

$$n = \frac{3.05a}{\pi d} \tag{13-7}$$

If the gas is contained in a tube whose section is a rectangle of sides l_1 and l_2, the principal modes of oscillation are parallel to the sides and the fundamental frequencies are

$$n_1 = \frac{a}{2l_1}, \qquad n_2 = \frac{a}{2l_2} \tag{13-8}$$

Higher frequencies are also possible; for instance,

$$n_3 = \frac{a}{l_1}, \qquad n_4 = \frac{a}{l_2} \tag{13-9}$$

For a tube in the form of an equilateral triangle with side l, the fundamental frequency is

$$n = \frac{2}{3}\frac{a}{l} \tag{13-10}$$

The main difficulty in applying these formulas is to determine a. In some cases this has been possible because streaks on the photographs represent the bodily motion of the gas behind the front so that u is known. Near the detonation point, $D = a + u$, so that a can be found knowing D. In the case shown in Plate G,13b, two sets of streaks can be seen.

Those sloping in the same direction as the detonation front give u, which is found to be 780 m/sec. The streaks sloping in the reverse direction are taken to be the traces of sound waves emitted at each increase in activity of the detonation head. They therefore determine $a - u$; measurement of the photograph gave $a - u = 320$ m/sec. The value of a is therefore $a = 780 + 320 = 1100$ m/sec and $u + a = 780 + 1100 = 1880$ m/sec. The measured value of D was 1980 m/sec so that the Chapman-Jouguet relationship is verified, so far as the accuracy of this kind of measurement permits.

Measurements of the spin frequency n in $2CO + O_2$ mixtures contained in tubes varying from 0.362 to 2.54 cm in diameter are given in column 3 of Table G,13a. Values calculated from Eq. 13-5 are given in

Table G,13a. Detonation velocity D, spin frequency n, in mixture $2CO + O_2$ contained in circular tubes of diameter $2R$.

$2R$, cm	D, cm/sec	n, cps	n, cps calculated assuming $a = 1100$ m/sec	a, m/sec agreement between theory and observed	$\frac{P}{2R}$, pitch–diameter ratio
0.362	1760	148,000	177,000	926	3.28
0.415	1760	132,000	155,000	940	3.23
1.28	1720	44,800	50,800	965	2.00
1.30	1750	44,300	49,500	985	3.04
1.50	1750	39,800	42,400	1035	2.90
2.54	1795	23,900	25,300	1040	2.95

column 4, assuming $a = 1100$ m/sec. It will be seen that for the larger tube the calculated value of n is very close to that observed. For the smaller tubes the calculated value of n is larger than the observed value. It seems likely that the value of a is smaller in the smaller tubes, possibly owing to cooling by the wall of the tube. For this reason the value of a which would correspond with exact agreement between theory and experiment was calculated and is given in column 5 of Table G,13a. The results of experiments made with rectangular tubes are given in Table G,13b. In this table the first 4 columns record observed quantities. Column 5 gives the frequency obtained from columns 3 and 4. To calculate the frequency of transverse oscillations, values of a are taken from column 5 of Table G,13a, assuming that a depends on the perimeter of the tube. A rectangular tube of section 2.2×1.35 cm, for instance, is taken as equivalent (as far as a is concerned) to a circular tube with a diameter of 2.26 cm.

The values appropriate to each rectangular tube are given in column 6. The two lowest frequencies calculated from Eq. 13-8 for oscillations parallel to each of the sides are given in column 7 to 10. The calculated frequency which is nearest to that observed is set in bold type. It will be seen that the theory is well verified. Detonation in a tube of triangular section with 1.7-cm sides gave a spin frequency of 34,000. Using $a = 985$ m/sec gives $n = 38{,}500$, which again is in reasonable agreement with the theory.

Spinning detonation occurs in other mixtures besides $2CO + O_2$. In general it appears only in a short length of tube close behind the detonation point, but near the limits of detonation of $H_2 + O_2$ mixtures, i.e. with the largest possible excess of O_2 or H_2 spin phenomena are very well marked. The way in which bursts of activity appear periodically over a considerable length of tube behind the shock front is very striking. It seems that the shock front can play the part which is similar to that of a

Table G,13b. Detonation velocity D, pitch P, and n for detonation of $2CO + O_2$ contained in tubes of rectangular section $l_1 \times l_2$. n_1, n_2, n_3, n_4 are frequencies calculated for oscillations parallel to the sides of the tube.

l_1, cm	l_2, cm	D, cm/sec	P, cm	$n = \frac{D}{P}$ observed	a, cm/sec	n_1	n_2	n_3	n_4
1.2	0.98	1695	4.20	40,300	950	**39,500**	48,300	79,000	96,600
2.2	1.35	1780	7.28	23,800	1030	**23,300**	60,500	46,600	121,000
2.2	1.1	1770	3.75	47,200	1030	23,300	**46,600**	**46,600**	93,200
3.0	2.4	1780	3.23	55,000	1040	17,300	21,600	34,600	**43,200**
		to	to	to					
		1820	4.5	40,300					
Mean values		1800	3.86	46,500					

fixed wall so far as the transverse oscillations are concerned. In all cases of detonation in which marked transverse oscillations occur over a long length of the tube there is independent evidence that combustion is complete in the region close to the shock front. To maintain a violent transverse oscillation over a long length of tube it is necessary that some feature of the combustion shall be capable of forcing it. If a considerable amount of gas has escaped combustion as it passes the detonation front, combustion will occur at greater intensity in the portions of the tube where the pressure and temperature attain their greatest value during an oscillation. This action would act as a strong agent for forcing the oscillation.

G,14. Solid and Liquid Explosives. In discussing gaseous explosions, the physical laws which have been assumed are (1) that the equation of state is that of a perfect gas, (2) that the energy is entirely in the form of heat motions (translational, vibrational, and rotational), and

(3) that the composition of the products of decomposition is determined by the equations of chemical equilibrium of the gas mixtures.

In solid or liquid explosives some of these assumptions are untrue: (1) is untrue because the volume occupied by the molecules is an appreciable fraction of the total volume; (2) is untrue because the energy of the mutual interactions between the molecules, which is analogous to the energy of elastic compression, is comparable with that due to heat motions. Many formulas have been developed to represent the equation of state for condensed gases, and the earliest attempts to describe detonation in solid explosives made use of one or the other of these, retaining the assumption that the energy of each constituent depends only on its temperature. The simplest is a modification of van der Waals' equation due to Abel, in which specific volume $(1/\rho)$ is reduced by a constant covolume b. The equation of state is taken as

$$p\left(\frac{1}{\rho} - b\right) = nRT \tag{14-1}$$

where n is the number of gram molecules per gram of the products of decomposition which are in the form of gas. If it is assumed that the equilibrium composition does not change with temperature near the critical point on the Hugoniot curve, the pressure, density, and velocity of the detonation wave can be expressed in finite terms. For this purpose a new variable K is introduced:

$$K = \frac{\dfrac{1}{\rho} - b}{\dfrac{1}{\rho_0} - \dfrac{1}{\rho}} \tag{14-2}$$

Using Eq. 10-5, the Hugoniot equation is found to be

$$e_T - e_0 - q = \frac{1}{2}\frac{nRT}{K} \tag{14-3}$$

provided that the pressure p_0 is so small compared with p as to be negligible. The equation of state is

$$p\left(\frac{K}{1+K}\right)\left(\frac{1}{\rho_0} - b\right) = nRT \tag{14-4}$$

The Chapman-Jouguet equation is

$$\frac{p}{\dfrac{1}{\rho_0} - \dfrac{1}{\rho}} = -\left[\frac{dp}{d(1/\rho)}\right]_{\text{Hugoniot}} = -\frac{(1+K)^2}{\left(\dfrac{1}{\rho_0} - b\right)}\left(\frac{dp}{dK}\right)_{\text{Hugoniot}} \tag{14-5}$$

or

$$p = -(1+K)\left(\frac{dp}{dK}\right)_{\text{Hugoniot}} \tag{14-6}$$

To find $(dp/dK)_{\text{Hugoniot}}$ it is necessary to differentiate Eq. 14-4 and 14-3 and eliminate $(dT/dK)_{\text{Hugoniot}}$. Thus

$$\left(\frac{dp}{dK}\right)_{\text{Hugoniot}} = \frac{nRT}{K\left(\frac{1}{\rho_0} - b\right)}\left[\frac{1 + \frac{2c}{nR}}{1 - \frac{2cK}{nR}}\right] \tag{14-7}$$

where $c = de/dT$ = specific heat of formed products. Combining Eq. 14-4, 14-6, and 14-7 and writing K_{cr}, p_{cr}, and T_{cr} for the values of K, p, and T at the critical point on the Hugoniot curve, it is found that

$$1 + \frac{2c}{nR} = -1 + \frac{2cK_{cr}}{nR}$$

so that

$$K_{cr} = 1 + \frac{nR}{c} \tag{14-8}$$

and

$$\mu = \left[\frac{\rho}{\rho_0}\right]_{cr} = \frac{K_{cr} + 1}{K_{cr} + b\rho_0} \tag{14-9}$$

The detonation pressure is:

$$p_{cr} = \left(\frac{1 + K_{cr}}{K_{cr}}\right)\left(\frac{\rho_0 nRT_{cr}}{1 - b\rho_0}\right) \tag{14-10}$$

The detonation velocity equation is:

$$D^2 = \frac{p_{cr}}{\rho_0^2\left(\frac{1}{\rho_0} - \frac{1}{\rho}\right)} = \left(\frac{1 + K_{cr}}{1 - b\rho_0}\right)^2 \frac{nRT_{cr}}{K_{cr}} \tag{14-11}$$

To determine the critical values it is first necessary to assume a composition for the products of combustion. From tables such as those by Lewis and von Elbe the values of $e_T - e_0$ can be found for a range of temperatures and $c_v = de_T/dT$ is known for any assumed temperature. q can also be obtained from tables corresponding with the assumed reaction, and n is known when the reaction is known. Taking any assumed temperature, both sides of Eq. 14-3 can be evaluated and by numerical interpolation the particular value of T which satisfies Eq. 14-8 can be found.

Having found the critical values of K and T it is possible to find the other critical values, such as μ and D, if b is known. It is not possible to measure b at the high pressures which occur during detonation in solids. On the other hand it is a simple matter to measure the velocity D with which a detonation front travels down a stick of explosive. For this reason the best use that can be made of Eq. 14-11 is to insert the measured values of D and the calculated values of K_{cr} and T_{cr} and to calculate b.

The values calculated in this way are very much smaller than those found in more direct experiments at lower pressures. Analysis of experiments in which the detonation velocity has been measured with the same explosive packed to several initial densities does not give the same value of b. The simple theory involving constant covolume must therefore be rejected.

A number of more complicated equations of state have been tried. That used by Jones and Miller [37] is

$$p\left(\frac{v'}{N'}\right) = RT + bp + cp^2 + dp^3 \tag{14-12}$$

where v' is the volume per mole and N' the number of moles involved in the reaction. Now, b, c, and d depend on the temperature and composition

Table G,14a. Values of p^* at 3000°K. N' = total number of moles of gaseous products in volume v'.

p, atm	p^*/p	v'/N'
1×10^4	2.88	51.1
5×10^4	98.1	27.5
1×10^5	2.23×10^3	21.6
2×10^5	2.31×10^5	16.5

of the products of combustion, but in the absence of any knowledge from other sources, Jones and Miller assumed them to be constant. Later they determined these constants in the case of the detonation of TNT ($C_7H_5N_3O_6$) by finding the values which most nearly predicted the observed values of D for a range of initial loading densities of the TNT. Jones and Miller's analysis of the detonation of TNT is too complicated to reproduce, but the method can be described briefly. First they define E_1 as the energy per mole of gas arising from the mutual interactions of the gas molecules and show that

$$E_1 = -\tfrac{1}{2}cp^2 - \tfrac{2}{3}dp^3 \tag{14-13}$$

Next they define a pressure p^* such that

$$\ln\frac{p^*}{p} = \frac{bp + \frac{1}{2}cp^2 + \frac{1}{3}dp^3}{RT}$$

and show that p^* plays the same part as p in the equations for the equilibrium composition. The values of p^* at 3000°K and v'/N' are given in Table G,14a. Assuming that the products of detonation of TNT can be represented by

$$C_7H_5N_3O_6 = 1.5N_2 + x_1H_2O + x_2H_2 + x_3CH_4 + x_4CO_2 + x_5CO + x_6C$$

The reactions considered are

$$CO + H_2O = CO_2 + H_2$$
$$C(\beta \text{ graphite}) + 2H_2 = CH_4$$
$$CO + \tfrac{1}{2}O_2 = CO_2$$
$$C(\beta \text{ graphite}) + \tfrac{1}{2}O_2 = CO$$

if $K_1(T)$, $K_2(T)$, $K_3(T)$, $K_4(T)$ are the equilibrium constants tabulated in Lewis and von Elbe's tables (Table G,10c). The equations for $x_1, \ldots,$ x_6 are

$$\begin{aligned} x_3 + x_4 + x_5 + x_6 &= 7 \\ 2x_1 + 2x_2 + 4x_3 &= 5 \\ x_1 + 2x_4 + x_5 &= 6 \\ x_5x_1 &= K_1(T)x_2x_4 \\ x_2^2 &= K_2(T)\left(\frac{N'}{p^*}\right)x_3 \\ x_5^2 &= \frac{K_3(T)}{K_4(T)}\left(\frac{N'}{p^*}\right)x_4 \end{aligned} \tag{14-14}$$

Writing

$$a_0 = \frac{5K_3}{TK_1^2K_4}, \quad a_1 = \frac{K_3(K_1 - 2)}{TK_1^2K_4}, \quad a_2 = 1 - \frac{4K_3}{K_1^2K_2K_4}$$

$$a_4 = \frac{2T}{K_2}, \quad \lambda = \frac{N'T}{p^*}$$

x_1, x_3, x_4, x_5, x_6 can be eliminated. The equation for x_2 is

$$(\lambda a_0 + \lambda a_1 x_2 + a_2 x_2^2)\left(2.5 - x_2 - \frac{a_3 x_2^2}{\lambda}\right) - 6x_2^2 = 0$$

It will be seen from Table G,14a that λ is extremely small at the high pressures attained in detonation of TNT. For this reason $x_2 = x_5 = 0$, and it is found that

$$\begin{aligned} x_1 &= 2.5 - \phi(T) \\ x_3 &= \phi(T) \\ x_4 &= 1.75 + \phi(T) \end{aligned}$$

where

$$\phi(T) = \frac{\left[\left(\frac{49}{4} + \frac{240K_3}{K_1^2K_2K_4}\right)^{\frac{1}{2}} - \left(\frac{7}{2} + \frac{20K_3}{K_1^2K_2K_4}\right)\right]}{4\left(1 - \frac{4K_3}{K_1^2K_2K_4}\right)}$$

Values of $\phi(T)$ calculated from values of K given in the tables are

T°K	1000	2000	3000
$\phi(T)$	0.470	0.603	0.666

The composition of the gas in the detonation front is therefore determined and the energy in kilocalories per mole can be found for each of the constituents. In this way it was found that the heat energy of the products in equilibrium at temperature T is quite accurately given by

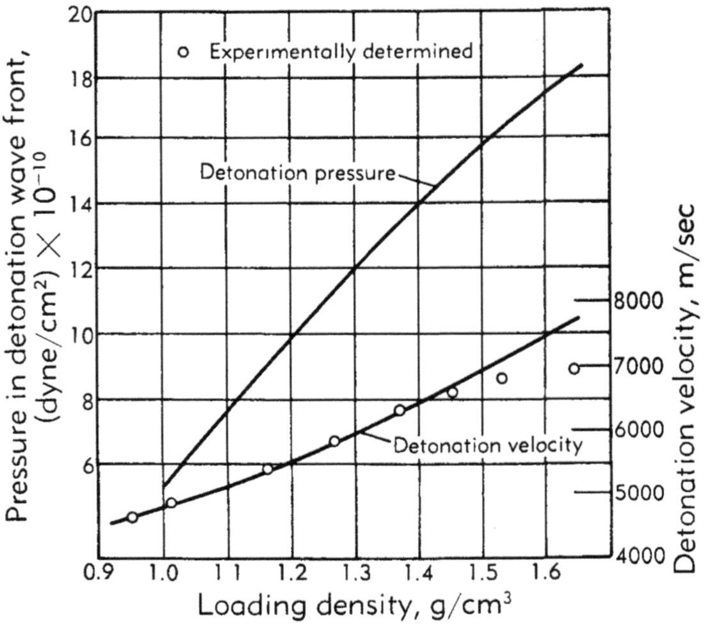

Fig. G,14a. Detonation pressure and velocity for various loading densities. Experimentally determined (a) detonation pressure, (b) detonation velocity.

$0.0875T - 38$ in the range of temperatures which may occur in detonation. The total energy change in the gas is therefore

$$E = 0.0875T - 38 + N'E_1$$

where E_1 is defined by Eq. 14-13. The heat liberated can be found from thermochemical tables giving the heats of formation of each of the products. It will be seen from Eq. 14-15 that Q is a linear function of $\phi(T)$. In fact it was found that in kilocalories per mole

$$Q = 292.7 - 3.888\phi(T) \tag{14-15}$$

so that, in the range from 1000 to 3000°K, Q varies by less than 0.25 per cent. Q was therefore taken as constant and equal to 290 kilocalories per mole or 1.27 kilocalories per gram. It will be noticed that this is not the same as the heat given out by the explosion after the products have expanded and cooled to NTP.

The determination of D from the equation of state, the Chapman-Jouguet condition, and the Hugoniot equation is complicated. Jones and Miller used measurements of D by Friedrichs (1933–1936) for a range of loading densities from 0.94 to 1.65 g/cm³. These are shown graphically in Fig. G,14a. The velocities obtained by choosing the constants in the equations of state (Eq. 14-12) as

$$b = 25.4, \quad c = -0.104, \quad d = 2.33 \times 10^{-4}$$

are also shown in Fig. G,14a. These constants correspond with a unit of pressure of 10^9 dynes/cm². It will be seen that this choice of constants gives good agreement for loading densities less than 1.5. The calculated pressure is also shown in Fig. G,14a.

Table G,14b. Temperature, pressure, and volume during adiabatic expansion of detonation wave front in TNT at loading density of 1.5 g/cm³.

Temperature T, °K	Pressure p, dyne/cm²	Specific volume v, cm³/g	γ
3400	15.88×10^{10}	0.5136	3.36
3200	10.77	0.5911	2.63
3000	7.809	0.6676	2.76
2800	5.833	0.7421	2.93
2600	4.371	0.8189	3.10
2400	3.238	0.9022	3.27
2200	2.299	1.002	3.30
2000	1.517	1.136	3.05
1800	8.898×10^9	1.355	2.76
1700	6.245	1.540	2.39
1600	4.318	1.797	2.09
1500	2.931	2.162	1.86
1400	1.947	2.693	1.68
1300	1.264	3.484	1.54
1200	7.980×10^8	4.697	1.44
1100	4.882	6.607	1.37
1000	2.880	9.718	1.32
900	1.626	14.97	1.29
800	8.707×10^7	24.29	1.27
700	4.360	41.79	1.27
600	1.999	77.35	1.27
500	8.159×10^6	157.0	1.27
400	2.818	362.8	1.27

Expansion of the products of combustion for TNT. Jones and Miller applied their equations of state to determine the pressure-volume relationship during adiabatic expansion of the products of decomposition behind the detonation front. Their calculations are complicated. The results for loading densities of 1.5 and 1.0 g/cm³ are given in Table G,14b and G,14c. The quantity

$$\gamma = \frac{\rho}{p}\frac{dp}{d\rho}$$

is given in each table because it is needed in calculations such as that of the spherical detonation wave in TNT.

One of the most striking results is that as the temperature and pressure fall, the amount of CO rises rapidly. In the detonation wave it is very small but by the time the products occupy 230 cm^3, in the case where the initial loading density is 1.5 g/cm^3 or about 190 cm^3 when it is

Table G,14c. Temperature, pressure, and volume during adiabatic expansion of detonation wave front for TNT at loading density of 1.0 g/cm^3.

Temperature T, °K	Pressure p, $dyne/cm^2$	Specific volume v, cm^3/g	γ
3800	8.224×10^{10}	0.7035	2.43
3600	6.905	0.7563	2.54
3400	5.759	0.8123	2.69
3200	4.754	0.8727	2.77
3000	3.851	0.9414	2.91
2800	3.045	1.021	2.98
2600	2.334	1.116	2.97
2400	1.726	1.235	2.88
2200	1.191	1.406	2.62
2000	7.420×10^{9}	1.684	2.22
1800	3.929	2.242	1.90
1700	2.783	2.688	1.73
1600	1.930	3.322	1.59
1500	1.307	4.245	1.48
1400	8.620×10^{8}	5.624	1.41
1300	5.511	7.752	1.32
1200	3.400	11.13	1.28
1100	2.011	16.76	1.25
1000	1.131	26.50	1.23
900	5.948×10^{7}	44.69	1.22
800	2.942	79.57	1.21
700	1.314	155.0	1.20
600	5.184×10^{6}	335.7	1.20
500	1.725	839.3	1.20
400	4.486×10^{5}	2580	1.20

1.0 g/cm^3, the amount of CO is equal to the amount of CO_2. Fig. G,14b shows the calculated compositions during expansion from detonation at 1.5 g/cm^3 density. As the specific volume of the products increases, the composition ceases to change much. As the temperature decreases, the rate of reaction rapidly decreases, so that in a rapid expansion a time must come when the chemical equilibrium cannot even be attained approximately. At a certain temperature the products of combustion may be regarded as having attained a fixed composition and further expansion may be calculated on the assumption that the composition is constant.

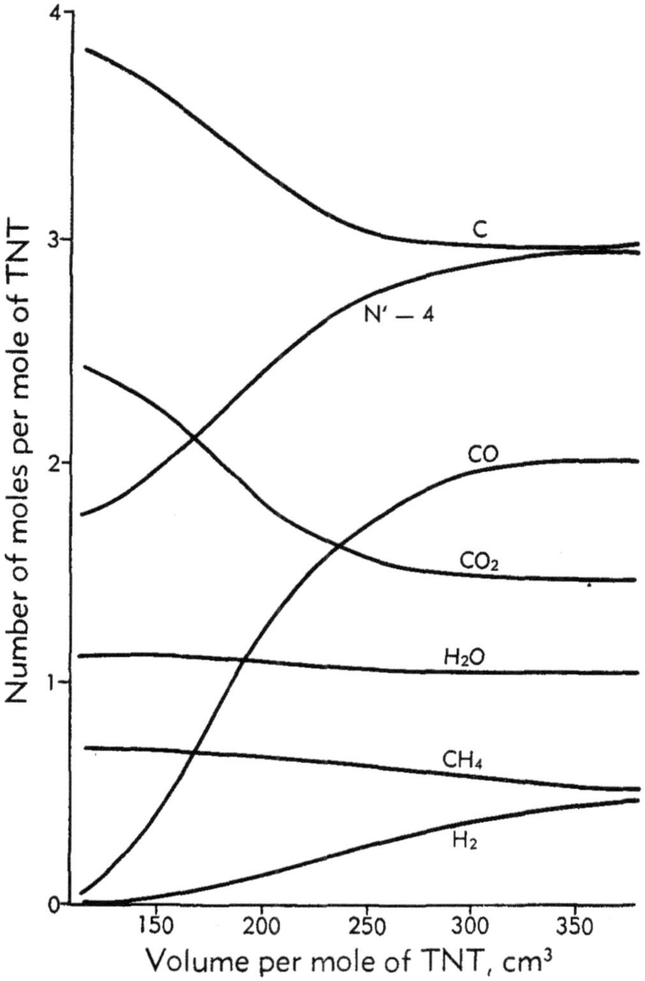

Fig. G,14b. Chemical composition of products of detonation for a loading density of 1.5 g/cm³.

G,15. Cited References.

1. Damköhler, G. *Z. Electrochem. 42,* 846 (1936).
2. von Neumann, J. Theory of detonation waves. *Natl. Defense Research Comm. Div. B, Office Sci. Research and Develop. Progress Rept. 549,* 1942.
3. Döring, W. *Ann. Physik 43,* 421 (1943).
4. Zeldovich, Y. B. On the theory of propagation of detonation. *J. Exptl. Theoret. Phys. U. S. S. R. 10,* 542 (1940). Transl. as *NACA Tech. Mem. 1261,* 1950.
5. Friedrichs, K. O. On the mathematical theory of deflagrations and detonations. *New York Univ. Inst. for Math. and Mech. Rept. 79-46,* 1946.
6. Semenov, N. N., and Zeldovich, Y. B. *NACA Tech. Mem. 1026,* 1942 (transl. from *Progr. Phys. Sci. U. S. S. R. 24, 4,* 1940); *1084,* 1946 (transl. from *J. Exptl. and Theoret. Phys. U. S. S. R. 10,* 1940).
7. *Biezeno's Anniversary Volume.* Delft, Netherlands, 1953.
8. *Fourth Symposium on Combustion.* Williams & Wilkins, 1953.
9. *Sixth Symposium on Combustion.* Reinhold, 1957.

10. Mallard, E., and Le Chatelier, H. Combustion des melanges gazeux explosifs. *Ann. mines 8*, 274–568 (1883).
11. Markstein, G. H., and Polanyi, M. Flame propagation—a critical review of existing theories. *Cornell Aeronaut. Lab. Bumblebee Series Rept. 61*, 1947.
12. Jost, W. *Explosion and Combustion Processes in Gases*. McGraw-Hill, 1946.
13. Bartholomé, E. The flame velocity in stationary flames. *Naturwissenschaften 36*, 171–175, 206–213 (1949).
14. Hirschfelder, J. O., et al. Theory of propagation of one-dimensional flames, I, II. *Univ. Wisconsin Nav. Ord. Rept. 9938*, 1948.
15. *Agard Combustion Colloquium*, Cambridge Univ., 1953. Butterworth's, London, 1954.
16. Corner, J. The effect of diffusion of the main reactants on flame speeds in gases. *Proc. Roy. Soc. London A198*, 388–405 (1949).
17. Bartholomé, E., Dreyer, H. J., and Lesemann, K. J. Bestimmung der Flammengeschwindigkeit einer Wärmeflamme durch losen eines eigenwertproblems mit der integrieranlage I.P.M.-Ott. *Z. Elektrochem. 54*, 246–252 (1950).
18. Emmons, H. W., Harr, J. A., and Strong, P. Thermal flame propagation. *Harvard Univ. Computation Lab. Rept. HUX-9*, 1950.
19. Shapiro, A. H., Hawthorne, W. R., and Edelman, G. M. The mechanics and thermodynamics of one-dimensional gas flow. *J. Appl. Mech. 14*, 317–336 (1947).
20. Johnson, W. C. Flame velocities of gases and vapors by the bunsen burner method. *Westinghouse Research Lab. Rept.*, 1947.
21. Scurlock, A. C. Flame stabilization and propagation in high velocity gas streams. *Mass. Inst. Technol. Fuels Research Lab. Meteor Rept. 19*, 1948.
22. Tsien, H. S. Influence of flame front on the flow field. *J. Appl. Mech. 18*, 188–194 (1951).
23. Emmons, H. W., Ball, G. A., and Maier, A. D. Development of a combustion tunnel. *Harvard Univ. Combustion Tunnel Lab. Rept.*, Feb. 1950.
24. Lamb, H. *Hydrodynamics*, 6th ed. Dover, 1945.
25. Vazsonyi, A. On rotational gas flows. *Quart. Appl. Math. 3*, 29–37 (1945).
26. Ball, G. Two dimensional flame in a laminar flow channel. *Harvard Univ. Combustion Tunnel Lab. Rept.*, Sept. 1951.
27. Landau, L. On the theory of slow combustion. *Acta Physiocochim. U.R.S.S.19*, 77–85 (1944).
28. Markstein, G. H. Experimental and theoretical studies of flame-front stability. *J. Aeronaut. Sci. 18*, 199–209 (1951).
29. Lewis, B., and von Elbe, G. *Combustion, Flames and Explosions of Gases*. Academic Press, 1951.
30. Dixon, H. B. The rate of expansion in gases. *Phil. Trans. Roy. Soc. London A184*, 97-188 (1893).
31. Manson, N. Propagation des détonations et des deflagrations dans les mélanges gazeux. *Office Natl. d'Etudes et Recherches Aéronaut. et Inst. Franç des Pétroles*, Paris, 1947. Also *Compt. rend. 222*, 46 (1946).
32. Jouguet, E. Sur la propagation des réactions chimiques dans les gaz. *J. de Math. Pures et Appliquées*, 1905–1906.
33. Lafitte, M. P. *Recherches Experimentales sur L'onde Explosive et L'onde de Choc. Thesis*, Univ. Paris, 1925.
34. Bone, W. A., Fraser, R. P., and Wheeler, W. H. *Trans. Roy. Soc. London A235*, 29 (1936).
35. Bone, W. A., and Fraser, R. P. *Trans. Roy. Soc. London A230*, 363 (1932).
36. Taylor, G. I. The dynamics of the combustion products behind plane and spherical detonation fronts in explosives. *Proc. Roy. Soc. London A200*, 235–247 (1950).
37. Jones, H., and Miller, A. R. Detonation of solid explosives. *Proc. Roy. Soc. London A194*, 480–507 (1948).

SECTION H

FLOW OF RAREFIED GASES

S. A. SCHAAF
P. L. CHAMBRÉ

CHAPTER 1. INTRODUCTION

H,1. Introduction. The mechanics of rarefied gases has been the subject of many investigations since the time of Maxwell [*1*, pp. 26, 681]. Until comparatively recently, these studies were confined to the case of very slow speeds and in general to "internal" flow geometries associated typically with vacuum installations. Summaries are given in I,I and in standard texts on the kinetic theory of gases, e.g. Kennard [*2*] or Loeb [*3*]. Since about 1946, there has been a considerable revival of interest and activity in the field, largely due to the possibility of flight at very high altitudes and at very high speeds. A growing body of both experimental and theoretical results has been obtained, with emphasis on the aerodynamic problem of very high speed flow of a rarefied gas past a body which is wholly "submerged." This activity was given impetus in pioneering articles by Tsien [*4,5*], which formulated many of the significant problems as well as the general lines of research which have subsequently proved successful.

A rarefied gas flow is a flow in which the length of the molecular mean free path $\bar{l}$ is comparable to some significant dimension L of the flow field. The gas then does not behave entirely as a continuous fluid but rather exhibits some characteristics of its coarse molecular structure. The dimensionless ratio $\bar{l}/L$ is denoted by K and is called the Knudsen number. A rarefied gas flow is thus one for which K is not negligibly small. For some considerations, L may be a characteristic dimension of the body itself or the diameter of an internal flow conduit. For other considerations, L may be the boundary layer thickness, the diameter of a wind tunnel probe, or the thickness of the shock transition zone. In particular, one may expect to encounter rarefied gas effects in those regions of the flow possessing very sharp gradients, i.e. regions in which the velocity, pressure, or temperature change appreciably in the space of a few mean free paths regardless of whether or not the absolute density of the gas flow is especially low.

The Knudsen number K is related to the more familiar parameters of fluid mechanics, the Mach number M and the Reynolds number Re. From kinetic theory, one defines $\bar{l}$ by the relation (Eq. 1-1)[1]

$$\nu = \tfrac{1}{2}\bar{l}\bar{v}_m \tag{1-1}$$

where ν is the kinematic viscosity and $\bar{v}_m$ is the mean molecular speed. The mean speed $\bar{v}_m$ is related to the sound speed a as follows:

$$a = \bar{v}_m \sqrt{\frac{\pi\gamma}{8}} \tag{1-2}$$

where γ is the isentropic exponent; hence one obtains

$$\bar{l} = 1.26\sqrt{\gamma}\,\frac{\nu}{a} \tag{1-3}$$

Combining these results yields the fundamental relation

$$K = 1.26\sqrt{\gamma}\,\frac{M}{Re} \tag{1-4}$$

where K and Re are both based on the same characteristic length L.

H,2. Flow Regimes. The division of gas dynamics into various regimes, based on characteristic ranges of values of an appropriate Knudsen number, has been proposed by Tsien [*4*], Roberts [*6*], Donaldson [*7*], and Siegel [*8*], among others. The various criteria which have been suggested are in some disagreement. In the present treatment the terms "continuum flow," "slip flow," "transition flow," and "free molecule flow" are defined so that they correspond to flows in which, roughly speaking, the density levels are, respectively, ordinary, slightly rarefied, moderately rarefied, and highly rarefied.

For flows of high Reynolds number, i.e. $Re \gg 1$, the significant characteristic dimension of the flow field, which is of importance in the determination of viscous effects, is the boundary layer thickness δ rather than a dimension L typical of the body itself. Since

$$\frac{\delta}{L} \sim \frac{1}{\sqrt{Re}} \tag{2-1}$$

the corresponding Knudsen number K is given by

$$K \sim \frac{M}{\sqrt{Re}} \tag{2-2}$$

Ordinary gas dynamics hence prevails for $M/\sqrt{Re} \ll 1$ and $Re \gg 1$. On

[1] The factor $\frac{1}{2}$ is sometimes replaced by 0.499; the present application does not seem to warrant such a refinement.

the other hand, for very small Reynolds numbers, the Stokes type "slow flow" occurs and the characteristic dimension itself is the significant parameter. Also, for any internal flow, only the diameter L of the conduit is of significance. Here the appropriate Knudsen number is simply K based on the body dimension and ordinary low speed continuum dynamics prevails for $M/Re \ll 1$. For flows in which the value of the appropriate Knudsen number is small but not negligible, some departure from

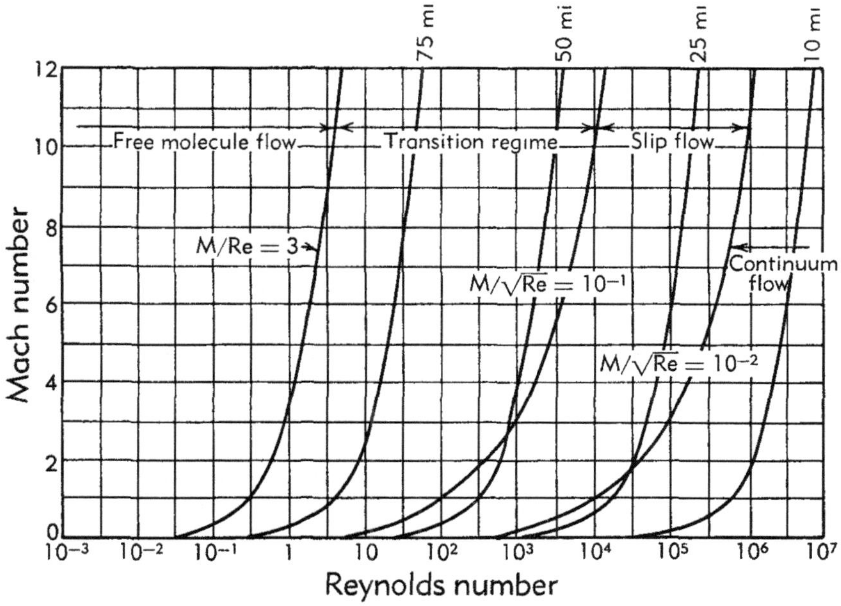

Fig. H,2a. The regimes of gas dynamics.

continuum gas dynamics phenomena may be expected to occur. As shown in more detail below, one of the more striking of these effects is the phenomenon of "slip"; the layer of gas immediately adjacent to a solid surface is no longer at rest but has a finite tangential velocity. The term "slip flow" is thus appropriate for flows of small but not negligible Knudsen number. The change from ordinary continuum gas dynamics to this regime is of course gradual; but to fix ideas, the slip flow regime on the basis of our present experimental evidence is defined by the following limits:

$$
\begin{aligned}
&0.01 < \frac{M}{\sqrt{Re}} < 0.1, \quad Re > 1 \\
&0.01 < \frac{M}{Re} < 0.1, \quad Re < 1
\end{aligned}
\tag{2-3}
$$

These boundaries are depicted in Fig. H,2a.

In the slip flow regime, so defined, the mean free path is of the order of 1 to 10 per cent of the boundary layer thickness or other characteristic dimension of the flow field. Slip flow effects may thus be expected to be approximately of this same order. A typical example, for the slip correction in the slow speed flow, can be seen in Eq. 17-3. For the boundary layer range, consult Eq. 19-5. It may be observed that both these examples are in agreement with the above criterion, Eq. 2-3. It should be noted that in the slip flow regime, either the Reynolds number must be very small or the Mach number large. Thus true rarefaction effects such as slip occur only in coincidence with either strong viscous or compressibility effects. In the slip flow regime, as defined above, these latter phenomena very often dominate rarefaction effects associated with the coarse molecular structure of the gas, and really large scale deviations from continuum behavior are then not apparent until the "transition" regime, defined below, is reached. It might also be observed that in the hypersonic range, the boundary layer thickness is no longer given by Eq. 2-1, so that appropriate modifications in the parameter $M/\sqrt{Re}$ should be made here.

For extremely rarefied flows, the mean free path $\bar{l}$ is much greater than a characteristic body dimension L. Under these circumstances no boundary layer is formed. Molecules reemitted from a surface do not collide with free stream molecules until far away from the body. One may consequently neglect any distortion of the free stream velocity distribution due to the presence of the body. Here the flow phenomena are mostly governed by molecule-surface interaction. This regime of fluid mechanics is termed "free molecule flow" and may be defined on the basis of present experimental evidence by

$$\frac{M}{Re} > 3 \tag{2-4}$$

In the transition regime between the slip flow and free molecule regimes, the mean free path is of the same order as a typical body dimension. Surface collisions and free stream intermolecular collisions are of more or less equal importance, and the analysis becomes extremely complicated. Present knowledge about this transition regime is very much more limited than that in the free molecule or the slip regime. Except for a few special theoretical results which are presented in subsequent articles, the information which is available is mostly empirical. For some theoretical considerations, that portion of the transition regime immediately adjacent to the free molecule flow regime becomes of special importance. It corresponds physically to the region in which the effects of a few intermolecular collisions begin to distort the free stream velocity distribution. This regime has not been given a separate designation, however, and in the present section it is not considered separately. To summarize: the terms "slip flow," "transition flow," and "free molecule flow" as defined above are considered sufficient to designate flows of slight, moderate, and great rarefaction, respectively.

These various flow regimes are shown in Fig. H,2a, together with the corresponding altitudes in miles above sea level, provided that the characteristic body dimension is taken as 1 foot. Present knowledge as to the constitution and density of the upper atmosphere, and consequently of the variation of the mean free path with altitude, is somewhat limited [*9,10*]. The calculations are based on the NACA "proposed" standard atmosphere tables [*10*]. The results are given in Fig. H,2b.

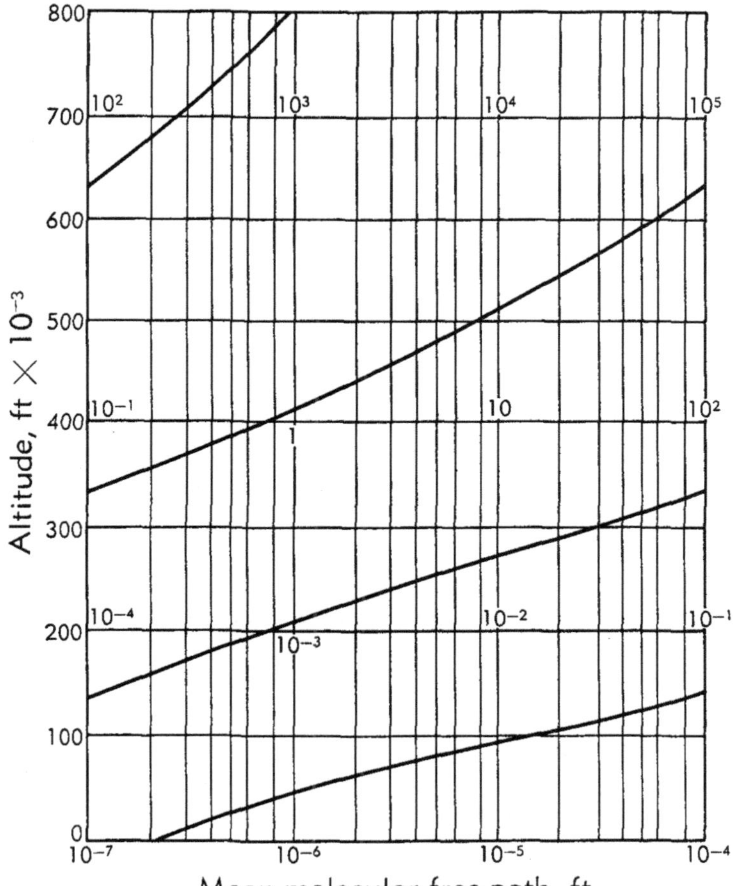

Fig. H,2b. Mean molecular free path of the adopted atmosphere.

A few numerical examples drawn from Fig. H,2a and H,2b may serve to relate these phenomena to typical conditions. Using 10 feet as a characteristic dimension for a high altitude missile, slip effects are expected to begin to be of importance at an altitude of 30 miles for $M = 1$ and at 20 miles for $M = 4$. Free molecule flow is established at about 90 miles for any Mach number. Compressible laminar boundary layer studies with a free stream Mach number greater than about 3 may be expected to be accompanied by some slip flow effects if the Reynolds number is of the order of 10^5 or less. Likewise, hypersonic wind tunnel design, operation, and instrumentation may involve slip flow phenomena unless very high stagnation pressures are used. For example, if the stagnation pressure is 100 atmospheres at normal temperature, the Reynolds number per inch in a test section at $M = 10$ is of the order 300,000. This is well within the slip regime. Very small hot wire probes used in boundary layer measurements or in turbulence studies may exhibit rarefaction effects at wind tunnel densities only slightly less than normal [*11*]. It might also be appropriate to recall that the mean free path is not the distance between molecules. At an altitude of 80 miles, for example, the mean free path is about one foot, but there are still 10^{13} molecules per cm^3. Most meteors are entirely vaporized by air friction at altitudes of this order, entirely in free molecule flow.

CHAPTER 2. FREE MOLECULE FLOW

H,3. Introduction. The free molecule flow regime is the regime of extreme rarefaction. The molecular mean free path is by definition many times the characteristic dimension of the body which is assumed to be located in a gas flow of infinite extent. The molecules which hit the surface of the body and are then reemitted on the average travel very far before colliding with other molecules. It is consequently valid to neglect the effect of the reemitted particles on the incident stream, at least so far as effects on the body itself are concerned. The incident flow is therefore assumed to be entirely undisturbed by the presence of the body. This is the basic assumption of free molecule flow theory. It is a consequence of this basic assumption that no shock waves are expected to form in the vicinity of the object. The "boundary layer" will be very diffuse and has no effect on the flow incident on the body.

Theoretical calculations of the external heat transfer and aerodynamic characteristics of bodies submerged in a free molecule flow field may be carried out by treating the flows of incident and reflected or reemitted molecules separately. In calculating the flux of momentum or energy incident on the surface it is assumed that the approaching gas is in local Maxwellian equilibrium. The results should hence be applied to very high altitude considerations with some care, since present knowledge as to the state and composition of the upper atmosphere is limited.

Early external free molecule flow calculations by Epstein [*12*] were confined to the case of small mean flow velocities while those of Zahm [*13*] treated the other extreme, neglecting the thermal motions of the molecules as compared to their mean flow. There is a close connection between this limiting case of free molecule flow and the "Newtonian flow" approximation of hypersonic theory [*14*]. More recently, calculations which take proper account of both the thermal and macroscopic motions have been carried out by a number of investigators. The treatment presented in this chapter follows, to a considerable extent, that of Stalder, et al. [*15,16*], Tsien [*4*], and Sänger [*17*].

H,4. Reflected Molecules. The determination of the flux of momentum or energy carried by molecules reflected or reemitted from the body surface requires a specification of the interaction between the impinging particles and the surface. The complete specification of this interaction would necessitate the theoretical or experimental determination of the velocity distribution function of the reemitted molecules, given their incident velocities. Fortunately, it is not necessary to have such detailed information in the free molecule flow regime. It is sufficient to know only certain average parameters which characterize the interaction phenomena and which are known as the accommodation and reflection coefficients.

For energy considerations one utilizes the "thermal accommodation coefficient" α, introduced by Smoluchowski [*18*] and Knudsen [*19*], which is defined by

$$\alpha = \frac{dE_i - dE_r}{dE_i - dE_w} \tag{4-1}$$

dE_i and dE_r are respectively the energy fluxes incident on and reemitted from a differential surface element per unit time. The quantity dE_w is the energy flux that would be carried away if all incident molecules were reemitted with a Maxwellian distribution corresponding to the surface temperature T_w. The term α is thus a measure of the degree to which the molecules have their mean energy "accommodated" to what it would be if the returning molecules were issuing with energy dE_w. For perfect accommodation $\alpha = 1$, while for the case of vanishing energy exchange with the surface, $dE_i = dE_r$ and hence $\alpha = 0$. It is implicitly assumed in the definition of α that all the energies associated with those molecular degrees of freedom which enter into an energy exchange with the surface are accommodated to the same degree. Available experimental evidence seems to indicate that this is approximately true for the translational and rotational energy components, while the vibrational energy is affected to a much less extent in a surface collision (see Herzfeld [*20*, pp. 228–251]). If necessary, it is possible to introduce separate accommodation coefficients for each energy component. Such refinements are, however, not considered in the present treatment. Experimentally determined values of α for various typical air-surface combinations as obtained by Wiedmann [*21*] are given in Table H,4a. See also Devienne [*22,23*].

The traditional treatment of momentum transfer originally due to Maxwell [*1*, p. 708] and subsequently followed in many free molecule flow calculations will be slightly modified. It has been customary to suppose that a fraction $(1 - \sigma)$ of the incident molecules is reflected "specularly." This reflection consists of a reversal of the normal velocity component which produces a normal momentum transfer to the surface and leaves the tangential velocity component unchanged. The remaining incident fraction σ is assumed to be reflected "diffusely," i.e. the molecules issue with a Maxwellian velocity distribution at a temperature not necessarily equal to that of the surface. It is this fraction σ which contributes the tangential momentum transfer to the surface.

Table H,4a. Thermal accommodation coefficient α for air [*21*].

Surface	α
Flat lacquer on bronze	0.88–0.89
Polished bronze	0.91–0.94
Machined bronze	0.89–0.93
Etched bronze	0.93–0.95
Polished cast iron	0.87–0.93
Machined cast iron	0.87–0.88
Etched cast iron	0.89–0.96
Polished aluminum	0.87–0.95
Machined aluminum	0.95–0.97
Etched aluminum	0.89–0.97

There is, however, preliminary experimental evidence that the interaction for air on typical engineering surfaces is not of such a simple type [*24*]. Consequently the single parameter σ may not suffice to describe the reflection phenomenon completely. In order to specify both the tangential and normal force components of the reflected flux the following two coefficients are introduced in analogy to Eq. 4-1

$$\sigma = \frac{\tau_i - \tau_r}{\tau_i - \tau_w}, \quad (\tau_w = 0) \tag{4-2}$$

$$\sigma' = \frac{p_i - p_r}{p_i - p_w} \tag{4-3}$$

where τ and p are respectively the tangential and normal momentum components. The subscripts $_i$ and $_r$ refer to the incident and reflected flux while p_w and τ_w denote respectively the normal and tangential momentum components of the molecules which are reemitted with a Maxwellian distribution at the surface temperature T_w. (Obviously τ_w must be zero.)

For the hypothetical cases of (i), entirely specular reflection with vanishing energy exchange, one would have $\alpha = \sigma = \sigma' = 0$, while for (ii), entirely diffuse reflection, which has been completely accommodated to the surface temperature T_w, one would have $\alpha = \sigma = \sigma' = 1$. In general, it is to be expected that the three parameters will be independent.

No information is available at the present time with respect to the quantity σ', but experimental values of σ determined at low speeds for some gas-surface combinations are given in Table H,4b. Inspection of Tables H,4a and H,4b reveals that, for many combinations, α and σ are close to unity, which indicates that the molecules are reemitted nearly diffusely. It is thus to be expected that σ' will also be close to unity. In the calculations which follow, the two limiting cases of completely diffuse and completely specular interaction are presented.

Table H,4b. Values of reflection coefficient σ [*25*].

Gas and surface combination	σ
Air or CO_2 on machined brass or old shellac	1.00
Air on oil	0.895
CO_2 on oil	0.92
H_2 on oil	0.93
Air on glass	0.89
He on oil	0.87
Air on fresh shellac	0.79

It should be emphasized that the quantities α, σ, and σ' are over-all phenomenological averages. Additional interaction parameters could be defined. There is indication that α and possibly σ may depend on the surface temperature, gas temperature, and pressure; absorbed gas film characteristics [*20*, Pt. 2, Sec. 4]; and possibly the velocity and direction of the mean flow with respect to the surface [*24*]. Hence the tabulated values should be used with caution.

H,5. Free Molecule Flow Heat Transfer. Convective heat transfer calculations for a surface element in a steady, uniform, free molecule flow have been made by Sänger and Bredt [*26*], Luntz [*27*], and Stalder and Jukoff [*28*]. An energy balance for a differential surface element dA is written in the form

$$dQ = dE_i - dE_r \tag{5-1}$$

where dQ is the total amount of heat removed from dA per unit time. If $\alpha \neq 0$, one obtains on elimination of the reflected energy flux with the help of Eq. 4-1,

$$dQ = \alpha(dE_i - dE_w) \tag{5-2}$$

For the incident molecules, the energy flux dE_i is broken up into two components, the first $dE_{i,tr}$, which is due to the translational motion of the molecules, while the second $dE_{i,int}$ is due to the internal degrees of freedom such as rotation or vibration. For the determination of $dE_{i,tr}$ one utilizes the assumption that the distribution of velocities for the incident molecules relative to an observer moving with the steady gas velocity U is Maxwellian. Then the number density of molecules per unit volume of phase space with total absolute velocity components v_{m_1}, v_{m_2}, and v_{m_3}, relative to the surface element shown in Fig. H,5, is given by [2]

$$f = \frac{\rho_\infty}{m(2\pi \mathcal{R} T_\infty)^{\frac{3}{2}}} \exp\left[-\frac{(v_{m_1} - U\sin\theta)^2 + (v_{m_2} + U\cos\theta)^2 + v_{m_3}^2}{2\mathcal{R}T}\right] \tag{5-3}$$

where m is the molecular mass, $\mathcal{R}$ the specific gas constant, T_∞ and ρ_∞ the free stream temperature and density of the gas, and θ the angle of

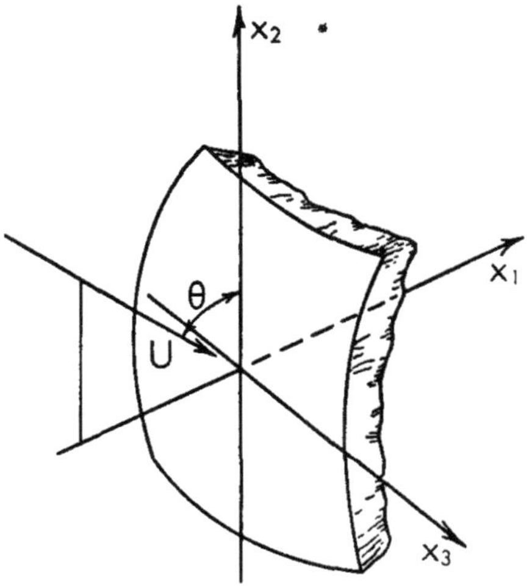

Fig. H,5. Coordinate system for free molecule flow.

attack between the surface element dA and the flow direction. The flux of translational energy onto dA is then

$$dE_{i,tr} = \int_{-\infty}^{\infty}\int_{-\infty}^{\infty}\int_0^{\infty} \tfrac{1}{2}mv_m^2 v_{m_1} f dv_{m_1} dv_{m_2} dv_{m_3} dA \tag{5-4}$$

Integrating, one obtains

$$dE_{i,tr} = \rho\mathcal{R}T\sqrt{\frac{\mathcal{R}T}{2\pi}}\{(S^2+2)e^{-(S\sin\theta)^2} + \sqrt{\pi}\,(S^2 + \tfrac{5}{2})(S\sin\theta)[1 + \operatorname{erf}(S\sin\theta)]\}dA \tag{5-5}$$

where S denotes the "molecular speed ratio" which can also be expressed in terms of the Mach number M and the isentropic exponent γ by

$$S = \frac{U}{\sqrt{2\mathcal{R}T}} = \sqrt{\frac{\gamma}{2}}M \tag{5-6}$$

The number of molecules N_i that are incident on dA per unit time is given by

$$N_i = \int_{-\infty}^{\infty} \int_{-\infty}^{\infty} \int_{0}^{\infty} v_{m_1} f dv_{m_1} dv_{m_2} dv_{m_3} dA \tag{5-7}$$

which on integration yields

$$N_i = \frac{\rho}{m} \sqrt{\frac{\mathfrak{R}T}{2\pi}} \{e^{-(S \sin \theta)^2} + \sqrt{\pi}\,(S \sin \theta)(1 + \operatorname{erf}[S \sin \theta])\} dA \tag{5-8}$$

Each of these molecules carries by the principle of equipartition on the average $j(m\mathfrak{R}T/2)$ units of internal energy, where j denotes the number of internal degrees of freedom which partake in energy exchange with the surface; j is related to the appropriate γ by

$$j = \frac{5 - 3\gamma}{\gamma - 1} \tag{5-9}$$

The flux of internal energy incident on dA per unit time is hence

$$dE_{i,\text{int}} = \frac{5 - 3\gamma}{\gamma - 1} \frac{m\mathfrak{R}T}{2} N_i \tag{5-10}$$

The total flux of incident energy is then obtained by combining Eq. 5-5, 5-8, and 5-10.

By a similar line of reasoning one computes the energy flux dE_w in Eq. 5-2. This flux is due to particles which issue from the surface with no macroscopic velocity and in Maxwellian equilibrium at the temperature T_w. A comparison of Eq. 5-5 and 5-8 reveals that, for a gas at rest (i.e., $S = 0$) with a temperature T_w, the average translational energy carried by molecules across a unit area per unit time is $2m\mathfrak{R}T_w$. The reflected molecules may be thought of as having issued from such a hypothetical gas. Together with the appropriate internal energy contribution one then obtains

$$dE_w = (4 + j) \frac{m\mathfrak{R}T_w}{2} N_w = \frac{\gamma + 1}{2(\gamma - 1)} m\mathfrak{R}T_w N_w \tag{5-11}$$

where N_w is the number of particles reemitted from dA per unit time. Since for a steady state N_w must equal N_i, one obtains finally from Eq. 5-2, 5-5, 5-8, 5-10, and 5-11:

$$dQ = \alpha\rho\mathfrak{R}T \sqrt{\frac{\mathfrak{R}T}{2\pi}} \left(\left[S^2 + \frac{\gamma}{\gamma - 1} - \frac{\gamma + 1}{2(\gamma - 1)} \frac{T_w}{T} \right] \{e^{-(S \sin \theta)^2} + \sqrt{\pi}\,(S \sin \theta)[1 + \operatorname{erf}(S \sin \theta)]\} - \tfrac{1}{2} e^{-(S \sin \theta)^2} \right) dA \tag{5-12}$$

H,6. Heat Transfer Characteristics of Typical Bodies in Free Molecule Flow. Since dA is in general a function of θ, the basic relation (Eq. 5-12) may be integrated for any specific body shape in order to determine the total heat transfer Q. It is assumed in the following that the surface is everywhere convex towards the gas stream, which excludes the possibility of molecular inter-reflection.

In general the temperature T_w varies over the surface. However, if the thermal conductivity of the body is sufficiently large and the magnitude of heat convected to or from the surface is sufficiently small, T_w is very nearly constant. In the following it is assumed that this condition is always met.

The following two problems are of greatest interest, namely the determination of the recovery or equilibrium surface temperature T_r corresponding to the case of $Q = 0$ and the evaluation of the heat transfer Q to or from the body. In order to compare the results with continuum theory, the equilibrium and heat transfer characteristics are most conveniently expressed in terms of a modified thermal recovery factor r' and a modified Stanton number St' introduced by Oppenheim [*29*], both of which depend only on the speed ratio

$$\begin{aligned} r' &= \frac{T_r - T_\infty}{T_\infty^0 - T_\infty} \frac{\gamma + 1}{\gamma} \\ St' &= \frac{Q}{A\rho U c_p (T_r - T_w)} \frac{\gamma}{\alpha(\gamma + 1)} \end{aligned} \tag{6-1}$$

T_∞^0 is the stagnation temperature of the flow which is given by

$$T_\infty^0 = T_\infty \left(1 + \frac{\gamma - 1}{\gamma} S^2\right)$$

A the total heat transfer surface area, and c_p the specific heat of the gas at constant pressure.

By direct integration of Eq. 5-12 one obtains the following results: For the flat plate, at angle of attack θ, with front and rear surface in perfect thermal contact and A the total area of both sides of the plate, one has

$$\begin{aligned} r' &= \frac{1}{S^2} \left[2S^2 + 1 - \frac{1}{1 + \sqrt{\pi}\,(S \sin \theta) \operatorname{erf}(S \sin \theta) e^{(S \sin \theta)^2}}\right] \\ St' &= \frac{1}{4\sqrt{\pi}\,S} \left[e^{-(S \sin \theta)^2} + \sqrt{\pi}\,(S \sin \theta) \operatorname{erf}(S \sin \theta)\right] \end{aligned} \tag{6-2}$$

For a flat plate, at angle of attack θ, with front and rear surfaces insulated from one another and A, the area of one side of the plate, one has for the front side

$$r' = \frac{1}{S^2}\left\{2S^2 + 1 - \frac{1}{1 + \sqrt{\pi}\,(S\sin\theta)[1 + \operatorname{erf}(S\sin\theta)]e^{(S\sin\theta)^2}}\right\} \tag{6-3}$$

$$St' = \frac{1}{4\sqrt{\pi}\,S}\left\{e^{-(S\sin\theta)^2} + \sqrt{\pi}\,(S\sin\theta)[1 + \operatorname{erf}(S\sin\theta)]\right\}$$

For the rear side one need only replace θ by $(-\theta)$ in these equations.

For the right circular cylinder with axis normal to the direction of flow and A the surface area without end contributions, one obtains from the calculations of Stalder, Goodwin, and Creager [*15,16*],

$$r' = \frac{(2S^2+3)I_0\left(\frac{S^2}{2}\right) + (2S^2+1)I_1\left(\frac{S^2}{2}\right)}{(S^2+1)I_0\left(\frac{S^2}{2}\right) + S^2 I_1\left(\frac{S^2}{2}\right)}$$

$$St' = \frac{e^{-S^2/2}}{4\sqrt{\pi}}\left[\frac{S^2+1}{S}I_0\left(\frac{S^2}{2}\right) + SI_1\left(\frac{S^2}{2}\right)\right] \tag{6-4}$$

where I_0 and I_1 are the modified Bessel functions. For a sphere with A the surface area one obtains from the calculations of Sauer [*30*]

$$r' = \frac{(2S^2+1)\left[1 + \frac{1}{S}\operatorname{ierfc}(S)\right] + \frac{2S^2-1}{2S^2}\operatorname{erf}(S)}{S^2\left[1 + \frac{1}{S}\operatorname{ierfc}(S)\right] + \frac{1}{2S^2}\operatorname{erf}(S)} \tag{6-5}$$

$$St' = \frac{1}{8S^2}\left[S^2 + S\operatorname{ierfc}(S) + \tfrac{1}{2}\operatorname{erf}(S)\right]$$

where ierfc (S) is the integrated complementary error function [*31*]. These results are exhibited in Fig. H,6a and H,6b.

A detailed investigation of the surface temperature of a flat plate at different angles of attack in the presence of radiation effects has been carried out by Stalder and Jukoff [*28*]. The calculations cover flight speeds up to 6.8 mi/sec and an altitude range of 75 to 190 miles.

It may be observed from Fig. H,6a that for all three bodies, and for the frontal side of the insulated flat plate, the recovery factors are always greater than unity. Hence, in contrast to a continuum flow, the equilibrium temperature in free molecule flow is greater than the local stagnation temperature of the gas. The anomaly can be explained by a consideration of the magnitudes of the incident and reemitted molecular energy fluxes. This striking effect has been verified experimentally by Stalder, Goodwin, and Creager for a cylinder in both a monatomic and diatomic gas [*15,16*].

It should be noted that the recovery factor, and hence also the equilibrium temperature T_r, are independent of the thermal accommodation

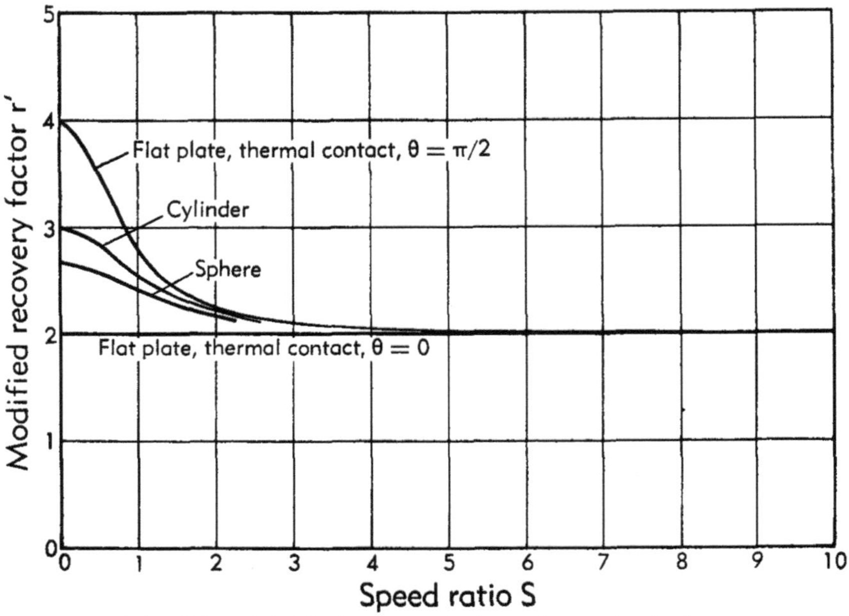

Fig. H,6a. Modified recovery factor in free molecule flow.

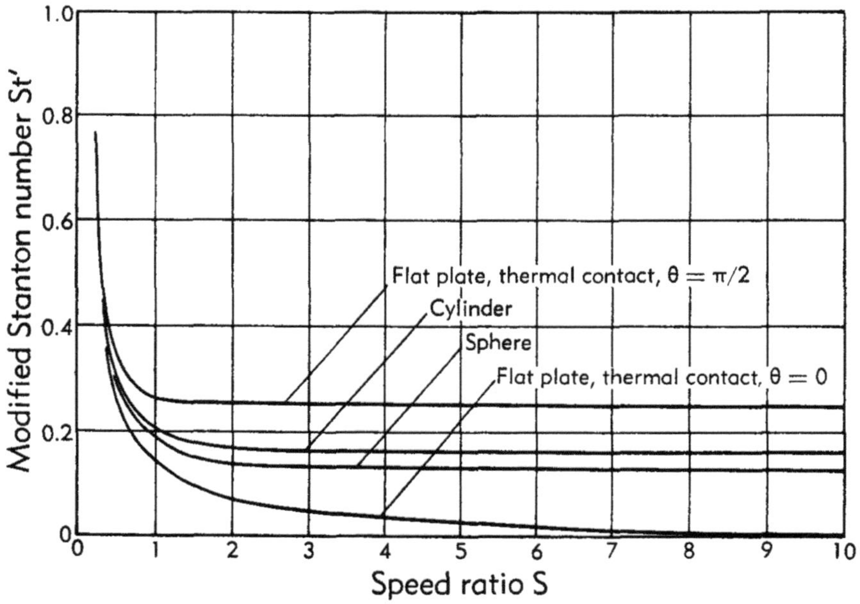

Fig. H,6b. Modified Stanton number in free molecule flow.

coefficient. This is true if convection is the only mechanism of heat transfer present. The equilibrium surface temperature of a body obtained in the presence of convection, radiation, and internal conduction usually involves a balance between the three modes of energy transfer and is therefore dependent on α. On the other hand, if $\alpha = 0$ the energy exchange by convection ceases.

H,7. Aerodynamic Forces in Free Molecule Flow. Aerodynamic forces on typical body geometries have been calculated by many authors. The general method of attack has been nearly the same in all cases. A summary of the references is given in Table H,7. The force $d\mathbf{F}$ on the

Table H,7. Summary of references treating aerodynamic characteristics of bodies in free molecule flow.

Plate	Cylinder	Sphere	Cone	Ogive	Ellipsoid	Composite bodies
[4]	[27]	[27]	[27]	[33]	[32]	[33]
[17]	[32]	[32]	[32]	[34]	[37]	[34]
[27]	[34]	[33]	[33]	[36]		[37]
[32]	[35]	[34]	[34]	[37]		
[33]	[36]	[35]	[35]			
[34]	[37]		[37]			
[35]						

surface element dA consists of a component $d\mathbf{F}_i$ due to the incident molecules and a component $d\mathbf{F}_r$ due to reemitted molecules. It is convenient to resolve the stress $d\mathbf{F}_i/dA$ into a component p_i normal to the surface (pressure) and a component τ_i tangential to the surface (shear). With the help of the coordinate system shown in Fig. H,5 one obtains

$$\begin{aligned} p_i &= \int_{-\infty}^{\infty}\int_{-\infty}^{\infty}\int_0^{\infty} mv_{m_1}^2 f dv_{m_1} dv_{m_2} dv_{m_3} \\ &= \frac{\rho_\infty U^2}{2\sqrt{\pi}S^2}\{(S\sin\theta)e^{-(S\sin\theta)^2} \\ &\qquad + \sqrt{\pi}\,[\tfrac{1}{2} + (S\sin\theta)^2][1 + \operatorname{erf}(S\sin\theta)]\} \end{aligned} \tag{7-1}$$

$$\begin{aligned} \tau_i &= \int_{-\infty}^{\infty}\int_{-\infty}^{\infty}\int_0^{\infty} mv_{m_1}v_{m_2} f dv_{m_1} dv_{m_2} dv_{m_3} \\ &= -\frac{\rho_\infty U^2\cos\theta}{2\sqrt{\pi}S}\{e^{-(S\sin\theta)^2} + \sqrt{\pi}\,(S\sin\theta)[1 + \operatorname{erf}(S\sin\theta)]\} \end{aligned} \tag{7-2}$$

where f is the molecular distribution function given by Eq. 5-3. From the definition of the reflection coefficients σ and σ' in Eq. 4-2 and 4-3, one has

$$\begin{aligned} p_r &= (1 - \sigma')p_i + \sigma' p_w \\ \tau_r &= (1 - \sigma)\tau_i \end{aligned} \tag{7-3}$$

The net pressure p and shear τ are then given by

$$\begin{aligned} p &= p_i + p_r = (2 - \sigma')p_i + \sigma' p_w \\ \tau &= \tau_i - \tau_r = \sigma\tau_i \end{aligned} \tag{7-4}$$

The pressure p_w exerted by molecules leaving the surface with no macroscopic velocity component and in Maxwellian equilibrium at temperature T_w is obtained with the help of Eq. 5-8 and 7-1. A comparison reveals that for a gas at rest (i.e. $S = 0$) with a temperature T_w the average normal momentum component carried by molecules across a unit area per unit time is $\frac{1}{2}m\sqrt{2\pi \mathcal{R} T_w}$. The reflected particles may be thought of as having issued from such a hypothetical gas. Since for a steady state the number of incident molecules must equal the reflected number where N_i is given by Eq. 5-8, one has

$$p_w = \tfrac{1}{2}m\sqrt{2\pi \mathcal{R} T_w}\, N_i \tag{7-5}$$

Combining the results of Eq. 7-1, 7-2, 7-3, 7-4, and 7-5 one obtains

$$\begin{aligned} p = \frac{\rho_\infty U^2}{2S^2}\Bigg[\left(\frac{2-\sigma'}{\sqrt{\pi}}S\sin\theta + \frac{\sigma'}{2}\sqrt{\frac{T_w}{T}}\right)e^{-(S\sin\theta)^2} \\ + \Bigg\{(2-\sigma')[(S\sin\theta)^2 + \tfrac{1}{2}] \\ + \frac{\sigma'}{2}\sqrt{\frac{\pi T_w}{T}}(S\sin\theta)\Bigg\}[1 + \operatorname{erf}(S\sin\theta)]\Bigg] \end{aligned} \tag{7-6}$$

$$\tau = -\frac{\sigma\rho_\infty U^2\cos\theta}{2\sqrt{\pi}}\{e^{-(S\sin\theta)^2} + \sqrt{\pi}(S\sin\theta)[1 + \operatorname{erf}(S\sin\theta)]\} \tag{7-7}$$

The total force acting on a body in a particular direction may be obtained by integrating the appropriate components of p and τ over the surface. Inspection of Eq. 7-6 and 7-7 reveals that the net forces in general depend on the surface temperature T_w (and hence α) as well as on both σ and σ'. In the following specific applications it is, for the sake of simplicity, assumed that the body is at a uniform temperature T_w equal to the free stream temperature and that either the reflection is completely diffuse, i.e. $\sigma = \sigma' = 1$ or that it is completely specular, i.e. $\sigma = \sigma' = 0$. In a more realistic calculation the surface temperature distribution must be determined from Eq. 5-12 taking the internal conduction and external radiation characteristics of the body into account. The force components should then be integrated over the surface with appropriate local values for α, σ, σ', and T_w. A number of investigations listed in Table H,7 have been concerned with the determination of the aerodynamic forces under the assumption, originally due to Maxwell, that a fraction σ of the incoming molecules is reflected diffusely and the remainder $(1 - \sigma)$ specularly. In this case Eq. 7-3 is replaced by

$$\begin{aligned} p &= (2-\sigma)p_i + \sigma p_r \\ \tau &= \sigma\tau_i \end{aligned} \tag{7-8}$$

where p_r is the pressure exerted by molecules leaving the surface in Maxwellian equilibrium at a temperature T_r which is not necessarily equal to the surface temperature T_w. T_r can be found with the help of the accommodation coefficient from Eq. 4-1.

H,8. Aerodynamic Force Characteristics of Typical Bodies in Free Molecule Flow. The basic relations given in Eq. 7-6 and 7-7 may be applied to a particular geometry in order to obtain by integration the lift and drag components acting on the body. The results are most conveniently expressed in terms of a drag coefficient C_D and a lift coefficient C_L. These are given by

$$C_D = \frac{D}{\frac{1}{2}\rho U^2 A} \qquad C_L = \frac{L}{\frac{1}{2}\rho U^2 A} \tag{8-1}$$

where L and D are the total lift and drag forces respectively, and A is the reference area of the body which must be specifically defined in each case. By direct integration, the details of which may be found in the references listed in Table H,7, one obtains with the assumption of uniform surface temperature the following:

For a flat plate at an angle of attack θ and A the area of one side of a plate, one has for diffuse reflection

$$C_{D,\text{diff}} = \frac{2}{\sqrt{\pi}\, S}\left[e^{-(S\sin\theta)^2} + \sqrt{\pi}\, S\sin\theta\left(1+\frac{1}{2S^2}\right)\operatorname{erf}(S\sin\theta) + \frac{\pi S}{S_w}\sin^2\theta\right] \tag{8-2}$$

$$C_{L,\text{diff}} = \frac{\cos\theta}{S^2}\left[\operatorname{erf}(S\sin\theta) + \sqrt{\pi}\,\frac{S^2\sin\theta}{S_w}\right]$$

and for specular reflection,

$$C_{D,\text{spec}} = \frac{4\sin\theta}{\sqrt{\pi}\, S^2}\left\{(S\sin\theta)e^{-(S\sin\theta)^2} + \sqrt{\pi}\left[\tfrac{1}{2} + (S\sin\theta)^2\right]\operatorname{erf}(S\sin\theta)\right\} \tag{8-3}$$

$$C_{L,\text{spec}} = \cot\theta\, C_{D,\text{spec}}$$

For a right circular cylinder with its axis normal to the flow and A the projected area (no end contributions), one has for diffuse reflection:

$$C_{D,\text{diff}} = \frac{\sqrt{\pi}\, e^{-S^2/2}}{S}\left[\left(S^2+\frac{3}{2}\right)I_0\left(\frac{S^2}{2}\right) + \left(S^2+\frac{1}{2}\right)I_1\left(\frac{S^2}{2}\right)\right] + \frac{\pi^{\frac{3}{2}}}{4S_w} \tag{8-4}$$

and for specular reflection,

$$C_{D,\text{spec}} = \frac{4}{3}\left(C_{D,\text{diff}} - \frac{\pi^{\frac{3}{2}}}{4S_w}\right) \tag{8-5}$$

For a sphere with A the projected area, one has for diffuse reflection

$$C_{D,\text{diff}} = \frac{e^{-S^2/2}}{\sqrt{\pi}\, S^3}(1 + 2S^2) + \frac{4S^4 + 4S^2 - 1}{2S^4}\,\text{erf}\,(S) + \frac{2\sqrt{\pi}}{3S_w} \qquad (8\text{-}6)$$

and for specular reflection,

$$C_{D,\text{spec}} = C_{D,\text{diff}} - \frac{2\sqrt{\pi}}{3S_w} \qquad (8\text{-}7)$$

The quantity S_w which appears in these results is defined by

$$S_w = \frac{U}{\sqrt{2\mathcal{R}T_w}} \qquad (8\text{-}8)$$

and thus requires knowledge of the as-yet-unspecified surface temperature. The above results are shown in Fig. H,8a, H,8b, H,8c, and H,8d for the special case when the surface and free stream temperatures are

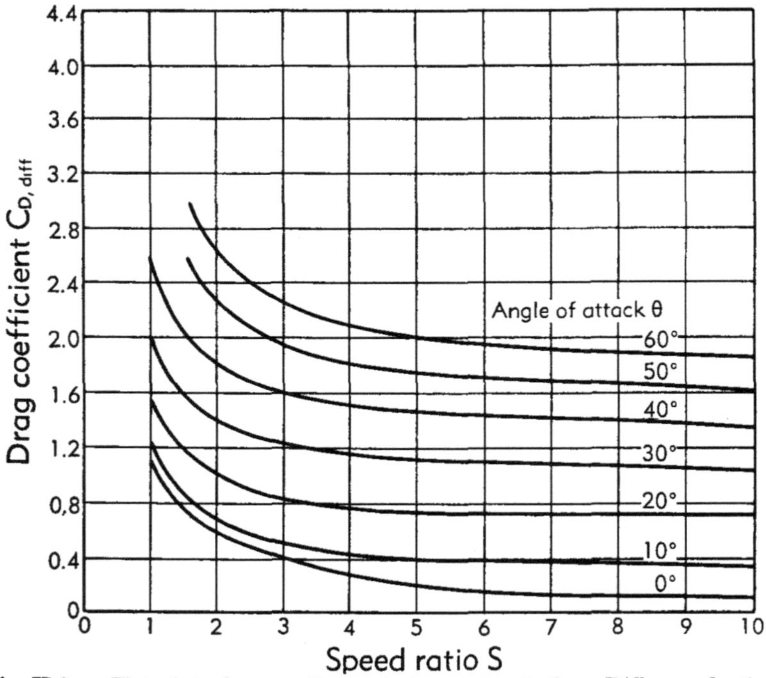

Fig. H,8a. Flat plate drag coefficient in free molecule flow. Diffuse reflection.

assumed equal, i.e. $S_w = S$. In this instance the equations do not contain γ and therefore apply to both monatomic and diatomic gases. A more realistic assumption can be made in the special case in which radiation and internal conduction effects are absent. Then the surface temperature equals the equilibrium temperature which can be obtained from the appropriate recovery factor in Art. 6. In this case the aerodynamic coefficients depend on γ and are somewhat larger than those for $S_w = S$, since the recovery factor is larger than unity.

The drag on a cylinder in free molecule flow has been determined experimentally by Stalder, Goodwin, and Creager [*15*] in satisfactory

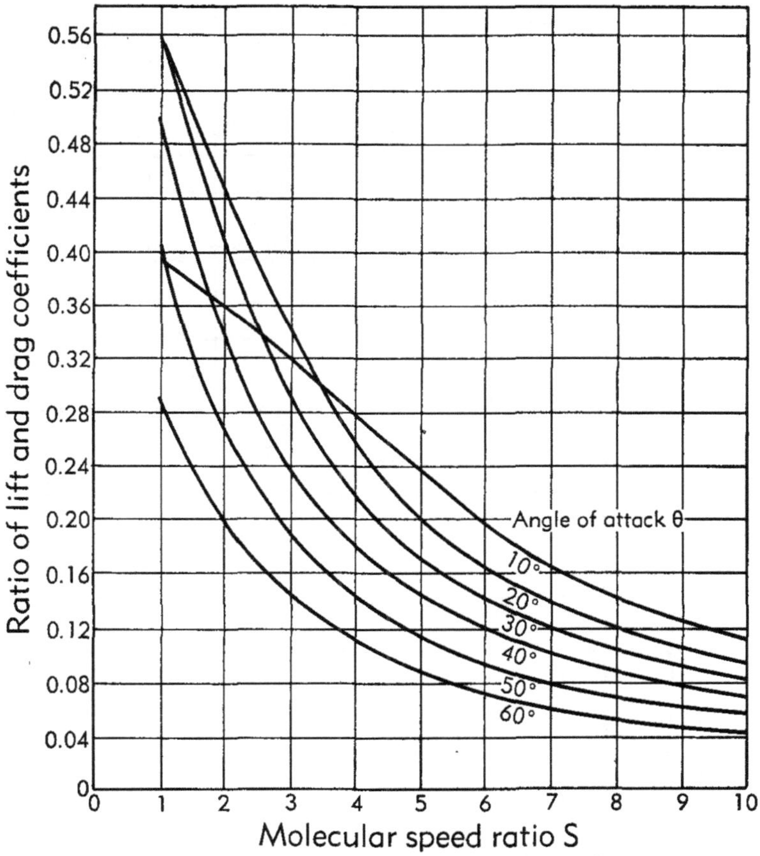

Fig. H,8b. Flat plate lift-drag ratio. Diffuse reflection.

agreement with the theoretical results. The aerodynamic characteristics of flat plate airfoils has been the subject of a number of investigations (see Table H,7). Polar lift and drag diagrams indicate in general a poor flight efficiency under diffuse molecular reflection in contrast to the very high efficiency for specular reflecting surfaces.

H,9. Nonuniform, Unsteady, and Surface-Interacting Free Molecule Flows. The foregoing discussion was confined to uniform external flows which can be described by the Maxwellian equilibrium distribution given by Eq. 5-3. If the flow is no longer uniform due to the presence of

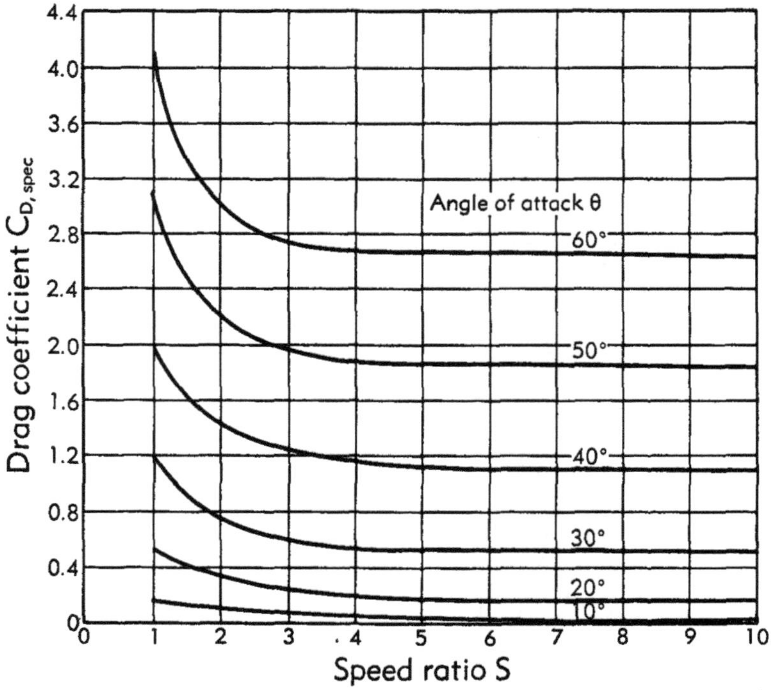

Fig. H,8c. Flat plate drag coefficient in free molecular flow. Specular reflection.

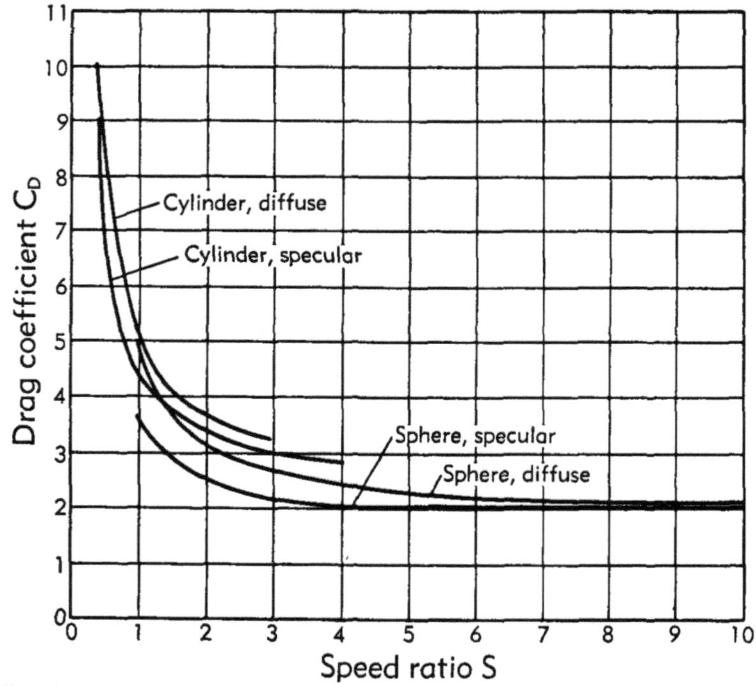

Fig. H,8d. Drag coefficients for sphere and cylinder in free molecule flow.

shear stresses and heat flux components one may use as a first approximation a distribution function proposed by Maxwell [*1*, p. 681], and given in Eq. 12-9. The methods of the previous articles have been applied to the calculation of the heat transfer and the aerodynamic forces on a cylinder [*38,39*] for this case. It is found that the effect on the pertinent parameters due to the nonuniformity is in general small compared to a uniform flow, except at low speeds or for flows with very large gradients.

Tsien [*5*] first suggested that a cylinder of sufficiently small diameter could serve as a temperature probe in a free molecule flow. From a corrected reading of the equilibrium temperature and the knowledge of the local stagnation temperature one can then determine the Mach number of the stream. Such an instrument has recently been used by Sherman [*40*] in a study of the structure of a shock wave where the nonuniformity of the flow affects the reading of the device.

A case of unsteady motion in a free molecule flow has been treated by Burgers [*41*]. The relevant feature of the problem is that gas is in contact with a surface which has a large variable molecular speed ratio. The state of the gas at any point, at any instant, is dependent in part on those molecules which left the surface at an earlier time. Their number and energy is determined by the surface velocity at the moment of their reflection. Consequently, even with completely diffuse reemission an appreciable deviation from the Maxwellian distribution appears in the reflected gas.

For very large molecular speed ratios there arises the possibility that the flow may begin to cause ablation from the surface of the body. This is in part due to the fact that the heat transfer to a rapidly moving object becomes so great that surface melting or vaporization may occur, and in part due to the mechanical action of the aerodynamic forces. Under these extreme conditions one may consider the molecules essentially at rest and the transfer of momentum and energy due to those particles which are intercepted by the body in its travel. If the reflected molecular contributions are considered negligible compared to the incoming molecular flux, Eq. 5-12, 7-1, and 7-2 reduce to

$$Q = \alpha \tfrac{1}{2}\rho U^3 A \sin\theta \tag{9-1}$$

and [*13*]

$$\begin{aligned} p_i &= \tfrac{1}{2}\rho U^2 2 \sin^2\theta \\ \tau_i &= \tfrac{1}{2}\rho U^2 2 \sin\theta\cos\theta \end{aligned} \tag{9-2}$$

where $A \sin\theta$ is the projected cross section of the body perpendicular to the direction of flight. Thomas and Whipple [*42*] have summarized and correlated the available experimental heat transfer data in terms of the parameter $\Lambda = Q/\tfrac{1}{2}\rho U^3 S'$ where S' is the effective heat transfer area of the body. From meteor observations up to speeds of 40 mi/sec, where considerable surface ablation occurs, they conclude that $\Lambda \leqq 0.1$ and that it decreases with decreasing speed. If the velocity becomes sufficiently high to melt the surface material and if radiation effects are negligible one can estimate the mass loss per unit time per unit area $= \Lambda\frac{1}{2}\rho U/L$ where L is the heat of vaporization or fusion according to which process the mass loss is due. Eq. 9-2 is the well-known Newton resistance law for inelastic impact. Meteor observations [*42*] indicate that the drag coefficient based on the projected area perpendicular to the direction of flight is of the order of one half.

Sänger [*43*] has made some estimates of the effects on the energy transfer to the surface due to excitation, dissociation, and ionization of molecules which might be produced by extremely high impact speeds between particles and the body. In general these effects decrease the energy transfer to the surface due to a partial transfer of the energy into higher molecular energy states. For air, Sänger estimates that only elastic reflection begins to be affected at speeds larger than 1.24 mi/sec.

H,10. Transition Regime Calculations. The foregoing discussion has been confined to the case of free molecule flow in which the effects of molecular encounters have been entirely neglected. With increasing density the effect of these collisions begins to be of importance and the low density part of the transition regime is entered. The general formulation of the problem in this regime is due to Jaffé [*44*] and is based on a perturbation expansion of the Maxwell-Boltzmann equation in inverse powers of the molecular mean free path. The distribution function f, which satisfies this equation, is assumed to be of the form

$$f = f_0\left[1 + \frac{L}{l}\varphi_1 + \left(\frac{L}{l}\right)^2\varphi_2 + \cdots\right] \tag{10-1}$$

where f_0 is the Maxwellian equilibrium distribution and L a typical macroscopic dimension.

It is no longer sufficient to specify certain over-all boundary conditions on the energy and momentum fluxes as presented in Eq. 4-1, 4-2, and 4-3. One must in addition specify conditions on f itself. A number of special calculations have been carried out in this regime by Heineman [*32*], Keller [*45*], Wang-Chang [*46*], Wang-Chang and Uhlenbeck [*47*], Szymanski [*48*], Lunc [*49*], Lunc and Lubonski [*50*], and Kryzwoblocki [*51*]. The calculations are quite formidable and no results of direct aerodynamic interest are available. Experimental data obtained by Bowyer [*52*] provide confirmation of the theoretical predictions [*46*] for the torque transmitted between concentric rotating cylinders in this part of the transition regime.

CHAPTER 3. SLIP FLOW

H,11. Introduction. The slip flow regime is the flow regime of slight rarefaction. The gas density is just slightly less than that characteristic of a completely continuum flow. In the slip flow regime there are really three separately important, but interrelated, parameters, the Mach number M, the Reynolds number Re, and the appropriate Knudsen number, either M/Re or $M/\sqrt{Re}$. These parameters serve to indicate the importance of compressibility, viscosity, and rarefaction effects, respectively. For the Knudsen number to be in the range indicated in Eq. 2-3, i.e. of the order of 0.01 to 0.1, it is clear that either M must be large, or Re small, (or both). Hence rarefaction effects in the slip flow regime are associated with, and in fact often dominated by, very strong compressibility or viscosity effects. In general, it is to be expected that boundary layers will be laminar; they will often be quite thick and in fact the Reynolds number may be so low that boundary layer theory is not strictly applicable. It is also to be expected that interaction effects between these thick viscous layers and the supersonic inviscid flow field [*53,54,55*] will be of considerable importance. Because of the complexity of these interrelations in the slip flow regime there are few, if any, solutions yet available for specific flow situations which take proper account of viscosity, compressibility, and rarefaction effects. The general effect of the slip flow and temperature jump boundary conditions of Eq. 14-3 is to reduce skin friction and heat transfer, other factors being similar.

There is a considerable and rapidly growing body of experimental data of aerodynamic interest covering portions of the slip flow as well as the transition flow regimes. Some of this information can be discussed in terms of comparatively approximate theories while some of it is only empirical. This material is summarized in Art. 15 to 21. However, the basic and fundamental problem of the correct formulation of the proper differential equations and boundary conditions for describing slip flows in general is not yet resolved in a satisfactory way. Until recently, it had been generally considered that some modification of the Navier-Stokes equations, such as the Burnett equations [*4*] or Thirteen Moment equations [*56*], or those proposed by Koga [*57*] together with slip velocity and temperature jump boundary conditions (see Art. 14) could be regarded as applicable to the slip flow regime. This view has been held in spite of several well-recognized difficulties associated with the Burnett and Thirteen Moment equations, and also in spite of the lack of any experimental verification of their validity. Recent experimental and theoretical findings, although not yet completely conclusive, now make the validity of these equations appear less likely and in fact seem to indicate that the Navier-Stokes equations, together with slip velocity and temperature jump boundary conditions, are not only adequate but probably superior. In the next article, a very brief derivation of the Thirteen Moment and Burnett equations is given, after which some of the difficulties met are discussed.

H,12. The Thirteen Moment and Burnett Equations. Both of these sets of equations are intended as approximations to the basic Maxwell-Boltzmann equation (Eq. 12-4) of the kinetic theory of monatomic gases. Each represents the third step in a successive approximation scheme. The Thirteen Moment equations were implicitly indicated by Maxwell [*1*, p. 681] but the method seemed applicable only to "Maxwell molecules," i.e. molecules which repelled each other according to the inverse fifth power of their distance of separation. A method of solution applicable to any molecular model was suggested by Hilbert, and calculations were carried out by Enskog [*58*], Burnett [*59*], Chapman and Cowling [*60*], and others. More recently, Grad [*56*] has completed the calculations as formulated by Maxwell and obtained the Thirteen Moment equations for the Maxwell molecule. It has also been shown [*61*, p. 237] that the Thirteen Moment equations contain the Burnett equations as a special case for Maxwell molecules. The Burnett equations for other molecular models differ only slightly from those for Maxwell molecules. These results are presented in complete details in [*56*; *60*; *62*, Chap. 7; *63*], but the essential elements are briefly recalled here.

The gas is supposed to consist of molecules of mass m, having only three degrees of freedom, so that the state of the gas may be given in terms of a distribution function,

$$f(x_i, v_{m_i}, t), \qquad i = 1, 2, 3 \tag{12-1}$$

which is the number density of molecules at point x_i with velocity v_{m_i} at time t, within the phase space volume element,

$$d\mathbf{x}d\mathbf{v}_m = dx_1dx_2dx_3dv_{m_1}dv_{m_2}dv_{m_3} \tag{12-2}$$

The ordinary macrosopic variables of gas dynamics—the density ρ, the velocity u_i, the stress τ_{ij} the temperature T, and the heat flux q_i—are given in terms of f by

$$\rho = m \int_{-\infty}^{\infty} \int_{-\infty}^{\infty} \int_{-\infty}^{\infty} f dv_{m_1}dv_{m_2}dv_{m_3} \equiv m \int f d\mathbf{v}_m \equiv mn$$

$$u_i = \frac{1}{n} \int v_{m_i} f d\mathbf{v}_m$$

$$P_{ij} = m \int (v_{m_i} - u_i)(v_{m_j} - u_j) f d\mathbf{v}_m \equiv m \int V_i V_j f d\mathbf{v}_m \equiv p\delta_{ij} - \tau_{ij} \tag{12-3}$$

$$T = \frac{1}{n\mathcal{R}} \int V^2 f d\mathbf{v}_m = \frac{p}{\mathcal{R}\rho}$$

$$q_i = \frac{1}{2} m \int V_i V^2 f d\mathbf{v}_m$$

where $\mathcal{R}$ is the specific gas constant.

The distribution function obeys a conservation equation, i.e. the Maxwell-Boltzmann equation,

$$\frac{\partial f}{\partial t} + \sum_{i=1}^{3} \left(v_{m_i} \frac{\partial f}{\partial x_i} + X_i \frac{\partial f}{\partial v_{m_i}} \right) = \Delta_c f \tag{12-4}$$

where X_i is an external force field and $\Delta_c f$ is the time rate of change of f due to intermolecular collisions. The term $\Delta_c f$ is given by

$$\Delta_c f = \iiint d\mathbf{v}_{m_1} \iint d\Omega[gI(g, \theta)(f'f_1' - ff_1)] \tag{12-5}$$

where $g = |\mathbf{v}_{m_1} - \mathbf{v}_m|$ is the relative velocity of the given molecule and a colliding molecule whose velocity is $\mathbf{v}_{m_1}$, the quantity $gI\ (g, \theta)\ d\Omega$ is the probability that the colliding molecule will be scattered into $d\Omega$ at θ, and f' and f_1' denote the distribution functions for molecules after collision, with velocities $\mathbf{v}_m'$ and $\mathbf{v}_{m_1}'$. Multiplying Eq. 12-4 by m, v_{m_i}, and V^2, respectively, and integrating over all velocities furnishes the usual macroscopic flow equations (assuming $X_i \equiv 0$):

$$\frac{\partial \rho}{\partial t} + \sum_{i=1}^{3} \frac{\partial(\rho u_i)}{\partial x_i} = 0$$

$$\frac{\partial u_i}{\partial t} + \sum_{j=1}^{3} \left(u_j \frac{\partial u_i}{\partial x_j} + \frac{1}{\rho} \frac{\partial P_{ij}}{\partial x_j} \right) = 0 \tag{12-6}$$

$$\frac{\partial p}{\partial t} + \sum_{i=1}^{3} \left(\frac{\partial(pu_i)}{\partial x_i} + \frac{2}{3} \frac{\partial q_i}{\partial x_i} + \frac{2}{3} \sum_{j=1}^{3} P_{ij} \frac{\partial u_i}{\partial x_j} \right) = 0$$

The collision term disappears identically. This system of equations is incomplete since there are more unknowns ρ, p, u_i, τ_{ij} (13 in all) than equations (5 in all). The Hilbert-Enskog-Chapman-Burnett method obtains the necessary additional relations in terms of a "solution" of Eq. 12-4. It is assumed that the collision term is dominant and that f may be determined by successive iterations on the collision term. This is roughly equivalent to supposing that f is given in terms of a power series in the mean free path $\bar{l}$. Eq. 12-4 is reduced to a sequence of linear integral equations, which are in turn solved approximately by replacing them by systems of simultaneous linear equations. The first approximation to f is the equilibrium distribution

$$f^{(0)} = \frac{n}{(2\pi \Re T)^{\frac{3}{2}}} e^{-\frac{V^2}{2\Re T}} \tag{12-7}$$

The substitution of Eq. 12-7 gives the Euler equations of ideal-fluid theory. Similarly, the second approximation $f^{(1)}$ gives the Navier-Stokes equations, while the third approximation $f^{(2)}$ gives the Burnett equations. It is a consequence of the iteration method of solution that f, and hence τ_{ij} and q_i, are given, apparently exactly if the calculation were continued, in terms of space gradients of ρ, u_i, and T only. This is a form of the "Hilbert paradox" and indicates the special nature of the solutions of Eq. 12-4, to which the iteration method is restricted. Distribution functions which are not of this form are simple to formulate. For example, a distribution function for which there is a time-dependent viscous shear stress, $\tau(t)$, with *no* space variation of the velocity, is given by

$$f = f^{(0)} \left[1 + \frac{V_1 V_2}{2p\Re T} \tau(t) \right] \tag{12-8}$$

To obtain the Thirteen Moment equations, one starts with the substitution of

$$f = f^{(0)} \left[1 - \sum_{i=1}^{3} \sum_{j=1}^{3} \frac{\tau_{ij}}{2p\Re T} V_i V_j - \sum_{i=1}^{3} \frac{q_i}{p\Re T} V_i \left(1 - \frac{V^2}{5\Re T} \right) \right] \tag{12-9}$$

into Eq. 12-4. Multiplying successively by each of the fourteen quantities m, v_{m_i}, $\frac{1}{2}mV^2$, V_iV_j, $\frac{1}{2}V_iV^2$, and integrating over all molecular velocities, yields Eq. 12-6 and eight additional independent equations in the form ($\bar{A}_{ij}$ denotes $A_{ij} + A_{ji} - \frac{2}{3}\delta_{ij} \sum_{k=1}^{3} A_{kk}$):

$$\frac{\partial \tau_{ij}}{\partial t} + \sum_{k=1}^{3} \frac{\partial}{\partial x_k}(u_k \tau_{ij}) - \frac{2}{5}\overline{\frac{\partial q_i}{\partial x_j}} + \sum_{k=1}^{3} \overline{\tau_{ik}\frac{\partial u_j}{\partial x_k}} - p\overline{\frac{\partial u_i}{\partial x_j}} = -\frac{p}{\mu}\tau_{ij}$$

$$\begin{aligned} \frac{\partial q_i}{\partial t} &+ \sum_{k=1}^{3} \left[\frac{\partial (u_k q_i)}{\partial x_k} + \tfrac{7}{5} q_k \frac{\partial u_i}{\partial x_k} + \tfrac{2}{5} q_k \frac{\partial u_k}{\partial x_i} + \tfrac{2}{5} q_i \frac{\partial u_k}{\partial x_k} \right. \\ &\left. - \Re T \frac{\partial \tau_{ik}}{\partial x_k} - \tfrac{7}{2}\tau_{ik}\Re \frac{\partial T}{\partial x_k} + \sum_{j=1}^{3} \frac{\tau_{ij}}{\rho}\frac{\partial P_{jk}}{\partial x_k} \right] + \tfrac{5}{2} p \Re \frac{\partial T}{\partial x_i} = -\frac{2}{3}\frac{p}{\mu} q_i \end{aligned} \tag{12-10}$$

These equations are exact for "Maxwell" molecules, i.e. molecules which repel each other with a force proportional to the inverse fifth power of the distance of separation. For other models, certain very small additional terms are present which have been ignored above. The terms,

$$-\frac{p}{\mu}\tau_{ij} \quad \text{and} \quad -\frac{2}{3}\frac{p}{\mu}q_i \tag{12-11}$$

are obtained from the collision term. All other terms arise in the convection terms. The quantity μ is identified as the coefficient of viscosity and is given by

$$\mu = 0.243\left(\frac{2m}{A}\right)^{\frac{1}{2}} m\mathcal{R}T \qquad \text{Maxwell molecule } (F \sim r^{-5}) \tag{12-12}$$

and

$$\mu = \tfrac{5}{8}m\sqrt{\pi\mathcal{R}T}\,\frac{1}{2\pi d^2} \qquad \text{hard sphere, diameter } d$$

The system of 13 equations, i.e. Eq. 12-6 and 12-10, constitutes a sufficient system for the determination of the 13 unknowns listed previously. It is observed that τ_{ij} and q_i do not depend in any simple way on ρ, u_i, and T, or their gradients of any order. For a system which is uniform in space, Eq. 12-10 reduces, for example, to

$$\frac{\partial\tau_{ij}}{\partial t} + \frac{p}{\mu}\tau_{ij} = 0 \tag{12-13}$$

This has the solution

$$\tau_{ij} \sim e^{-pt/\mu} \tag{12-14}$$

The quantity μ/p is then a "relaxation" time (of the order of the time between successive collisions) for the disappearance of the stress.

The Euler, Navier-Stokes, and Burnett equations are all contained in the system of Eq. 12-6 and 12-10 if it is assumed that τ_{ij} and q_i can be given in terms of successive approximations by iteration on the collision term in Eq. 12-10. One has, for the nth approximation to the stress tensor and heat flux vector,

$$\tau_{ij}^{(n)} = \mu\,\overline{\frac{\partial u_i}{\partial x_j}} - \frac{\mu}{p}\left[\frac{D\tau_{ij}^{(n-1)}}{Dt} + \tau_{ij}^{(n-1)}\sum_{k=1}^{3}\frac{\partial u_k}{\partial x_k} - \frac{2}{5}\,\overline{\frac{\partial q_i^{(n-1)}}{\partial x_j}} + \overline{\sum_{k=1}^{3}\tau_{ik}^{(n-1)}\frac{\partial u_j}{\partial x_k}}\right]$$

$$\begin{aligned} q_i^{(n)} = -k_{\text{th}}\frac{\partial T}{\partial x_i} - \frac{3\mu}{2p}\Bigg(\frac{Dq_i^{(n-1)}}{Dt} + \sum_{j=1}^{3}\Bigg\{&\tfrac{7}{5}q_i^{(n-1)}\frac{\partial u_j}{\partial x_j} \\ &- \mathcal{R}T\frac{\partial\tau_{ij}^{(n-1)}}{\partial x_j} + q_j^{(n-1)}\left[\frac{\partial u_i}{\partial x_j} + \frac{2}{5}\left(\frac{\partial u_i}{\partial x_j} + \frac{\partial u_j}{\partial x_i}\right)\right] \\ &- \tfrac{7}{2}\tau_{ij}^{(n-1)}\mathcal{R}\frac{\partial T}{\partial x_j} + \sum_{k=1}^{3}\frac{\tau_{ij}^{(n-1)}}{\rho}\frac{\partial P_{jk}^{(n-1)}}{\partial x_k}\Bigg\}\Bigg) \end{aligned} \tag{12-15}$$

where $k_{th} = 15\mathcal{R}\mu/4$ is identified as the coefficient of thermal conductivity, and D/Dt is the material derivative.

Beginning with $p_{ij}^{(0)} = q_i^{(0)} = 0$, i.e. the Euler equations, one obtains as the next approximation

$$\begin{aligned} \tau_{ij}^{(1)} &= \mu \overline{\frac{\partial u_i}{\partial x_j}} \\ q_i^{(1)} &= -k_{th} \frac{\partial T}{\partial x_i} \end{aligned} \tag{12-16}$$

These yield the Navier-Stokes equations (for a monatomic gas) when substituted into Eq. 12-6. As the next step, one obtains

$$\begin{aligned} \tau_{ij}^{(2)} &= \mu \overline{\frac{\partial u_i}{\partial x_j}} - \frac{\mu}{p}\left[\frac{D}{Dt}\left(\mu \overline{\frac{\partial u_i}{\partial x_j}}\right) + \mu \overline{\frac{\partial u_i}{\partial x_j}} \sum_{k=1}^{3} \frac{\partial u_k}{\partial x_k} \right. \\ &\qquad \left. + \frac{2}{5}\overline{\frac{\partial}{\partial x_j}\left(k_{th}\frac{\partial T}{\partial x_i}\right)} + \mu \sum_{k=1}^{3} \overline{\frac{\partial u_i}{\partial x_k}\frac{\partial u_j}{\partial x_k}}\right] \\ q_i^{(2)} &= -k_{th}\frac{\partial T}{\partial x_i} + \frac{3\mu}{2p}\left\{\frac{D}{Dt}\left(k_{th}\frac{\partial T}{\partial x_i}\right) + \sum_{j=1}^{3}\left[\tfrac{7}{5}k_{th}\frac{\partial T}{\partial x_i}\frac{\partial u_j}{\partial x_j}\right.\right. \\ &\qquad + \mathcal{R}T\frac{\partial}{\partial x_j}\left(\mu \overline{\frac{\partial u_i}{\partial x_j}}\right) + k_{th}\frac{\partial T}{\partial x_j}\left[\frac{\partial u_i}{\partial x_j} + \frac{2}{5}\left(\frac{\partial u_i}{\partial x_j} + \frac{\partial u_j}{\partial x_i}\right)\right] \\ &\qquad \left.\left. + \tfrac{7}{2}\mu \overline{\frac{\partial u_i}{\partial x_j}}\mathcal{R}\frac{\partial T}{\partial x_j} - \frac{\mu}{p}\sum_{k=1}^{3}\overline{\frac{\partial u_i}{\partial x_j}}\frac{\partial p}{\partial x_k}\delta_{jk}\right]\right\} \end{aligned} \tag{12-17}$$

These yield the Burnett equations (for Maxwell molecules) when substituted into Eq. 12-6. For non-Maxwell molecules, some of the numerical coefficients in Eq. 12-17 are slightly different, but that is the only difference. It can be shown [*4*] that the ratios $(\tau_{ij}^{(2)} - \tau_{ij}^{(1)})/\tau_{ij}^{(1)}$ and $(q_i^{(2)} - q_i^{(1)})/q_i^{(1)}$ are in general proportional to M^2/Re so that the higher order approximations would become of importance in the slip flow regime, provided, of course, that they are correct.

It is of historical interest to note that those parts of the expression for $\tau_{ij}^{(2)}$ given in Eq. 12-17, which depend on temperature gradients, were obtained by Maxwell [*1*, p. 681] in order to explain the radiometer force in the slip flow regime. This effect is now thought to be due rather to the "thermal creep" boundary effect [*61*, p. 333] and explainable in terms of the Navier-Stokes stresses only.

H,13. Difficulties Associated with the Thirteen Moment and Burnett Equations. Some of the basic difficulties associated with these equations are as follows:

1. They are confined to monatomic gases. Since air is composed, for the most part, of diatomic nitrogen and oxygen, this is a serious restriction for aerodynamic applications. At normal temperatures, the internal energy in air is distributed equally over the translational and rotational degrees of freedom of the diatomic molecules, while the vibrational degrees of freedom are almost completely unexcited. Following any abrupt change in the state of the gas, e.g. upon passage through a shock wave, the internal energy is redistributed over the translational and internal degrees of freedom of the molecules, and a new equilibrium is achieved only after a sufficient number of molecular collisions, as indicated in Eq. 12-14. It seems reasonably well established experimentally [*40,64,65,66*] that the rotational degrees of freedom for air adjust in a relaxation time which corresponds to only a very few additional collisions. Under these circumstances, Wang-Chang and Uhlenbeck [*67*] have shown that the only macroscopically apparent effect—at least to the Navier-Stokes level of approximation—is the phenomenon of "bulk viscosity," i.e. the stress is given by

$$\tau_{ij} = \mu \overline{\frac{\partial u_i}{\partial x_j}} + \lambda \delta_{ij} \sum_{k=1}^{3} \frac{\partial u_k}{\partial x_k} \tag{13-1}$$

The experiments seem to indicate that, for air, the magnitude of the second coefficient of viscosity λ is about $\frac{2}{3}$ the ordinary shear modulus μ. At higher temperatures, in air, the vibrational degrees of freedom begin to be excited and the readjustments of the internal energy associated with these modes of motion have very much longer relaxation times. Then, no simple formulas, as given by Eq. 13-1, are valid and phenomena become very complicated.

Almost nothing is known of such relaxation phenomena to the next order of approximation. Truesdell [*68,69*] has obtained general formulas for the stress tensor and heat flux vector, subject only to the necessary restrictions of dimensional and geometrical invariance, but which follow from the basic—and crucial—assumption that these quantities are given in terms of derivatives of the velocity, pressure, and temperature. Truesdell's formulas contain many more terms, at the Burnett level, than are given in Eq. 12-17. Possibly the missing terms can be associated with such relaxation effects. The kinetic theory approach to the problem has been formulated by Wang-Chang and Uhlenbeck [*67*], who have shown the relation of the second viscosity coefficient to the relaxation effect and have indicated the necessary calculations for determining higher order terms. A "Seventeen Moment" set of equations, including a second "temperature" associated with the internal degrees of freedom and the three components of the corresponding heat flux vector, has been formulated by Mostov, et al. [*70*] for a molecular model consisting of a rough sphere. This model has also been considered by Kohler [*71*]. These investigations also exhibit bulk viscosity phenomena.

In summarizing, it is clear that Eq. 12-17 are not complete for air, even to the Navier-Stokes level of approximation.

2. Present experimental evidence seems contrary to either the Burnett or Thirteen Moment equations. The crucial experiment to test the validity of either of these sets of equations must be carried out for a flow system having large gradients far away from any surface so as to avoid any ambiguities connected with boundary conditions or wall interference. Two such experiments are the determination of the internal structure of weak normal shock waves, and the determination of the dispersion and absorption of ultrasonic sound waves. The former has been studied by Greene and Andersen [*64,72*] and Sherman and Talbot [*40,73*], while the latter has been investigated most extensively by Greenspan [*65*]. Solutions are obtainable for the Navier-Stokes, Burnett, and Thirteen Moment equations for both cases.

The initial shock structure studies [*40*] were carried out in air and in helium at shock strengths above $M = 1.8$. Here they showed complete agreement with the Navier-Stokes prediction (including a bulk viscosity term for the case of air). At this shock strength the Burnett profile is quite different from the Navier-Stokes profile and the Thirteen Moment equations cannot be solved. The more recent shock structure studies [*73*] were carried out in argon at lower Mach numbers. For these conditions, all three sets of equations may be solved; the experimental results were in best agreement with the Navier-Stokes relations, particularly in the initial part of the shock wave which was undisturbed by downstream expansions. The sound dispersion and absorption experiments using an acoustic interferometer have not been completely free from wall interference and the results are not conclusive; they seem in better agreement with the Navier-Stokes equations, however, than with the other sets.

3. Recent theoretical results also seem contrary to the Burnett or Thirteen Moment equations. Several "exact" solutions of the Boltzmann equation itself have been obtained, which do not seem to be in accord with the basic assumptions underlying the Burnett or Thirteen Moment formulations. The linearized Boltzmann equation has been solved by Wang-Chang and Uhlenbeck for the case of heat transport between parallel plates [*47*] and for the case of flow near a surface [*74*]. The molecular distribution function f has the property for both these cases that $f \rightarrow f^0$ as $\bar{l} \rightarrow 0$, but f is not expandable in a power series in $\bar{l}$. Functions of the type $e^{-x/\bar{l}}$ occur. The case of simple shear in an infinite medium has

been solved by Ikenberry and Truesdell [*75,76*] in terms of the complete Boltzmann equation. Here again the distribution function cannot be adequately approximated by either the Burnett or Thirteen Moment representations. For all of these theoretical cases it is noteworthy that the Navier-Stokes results are in general a better approximation than either the Burnett or Thirteen Moment results.

4. Even if the Burnett equations were valid, there is an additional and unresolved problem connected with the formulation of proper boundary conditions. The order of the Burnett equations is higher than the order of the Navier-Stokes equations, so that additional boundary conditions seem necessary. At the Euler level of approximation the equations are of first order, and only one boundary condition, usually that the normal velocity vanish at a surface, can be prescribed. At the Navier-Stokes level of approximation, the equations are of second order and two boundary conditions, usually that the normal and the tangential velocity vanish at the surface, must be prescribed. Solutions are thus singular in the viscosity (or equivalently, the mean free path), and the boundary layer solution (for $\nu \to 0$) cannot be obtained by perturbation schemes starting with the inviscid solution. At the Burnett level of approximation, the equations are of third order and it is to be expected that an additional boundary condition must be prescribed, although there is no clear guide as to which one. Schamberg [*77*] has suggested that solutions which are accurate to the Burnett level may be obtained by perturbation series techniques starting with the Navier-Stokes solution. If this is the case no additional conditions are necessary, or in fact, can be used and the question would be only one of improving the form of the boundary conditions (to include slip and temperature jump effects). Although the Euler-to-Navier-Stokes equivalent of this suggestion definitely does not work, and in fact, completely loses the whole boundary layer phenomenon, only a solution obtained directly from the Maxwell-Boltzmann equation could provide confirmation, or the reverse, of this hypothesis. The solution obtained by Wang-Chang and Uhlenbeck [*47*] may be reduced to macroscopic terms to give the heat flux in terms of the temperature drop and the mean free path $\bar{l}$. For small $\bar{l}$, the heat flux agrees with that predicted by the Navier-Stokes equations and the simple temperature jump boundary condition of Eq. 14-3, below. The correction term, however, is nonanalytic in $\bar{l}$ and so a perturbation calculation will not work. Although these results thus strongly indicate that Schamberg's suggestion will not work, they do not definitely prove it because, among other things, the Burnett terms disappear identically for this case. For the Thirteen Moment equations, such boundary condition uncertainties do not occur. Grad [*56*] has shown that the characteristic curves for these equations are such as to require the same number of boundary conditions as for the Navier-Stokes equations. Hence approximating the Thir-

teen Moment equation by the Navier-Stokes equations can be done unambiguously.

In conclusion, it seems best at present to rely on the Navier-Stokes equations, including a second viscosity coefficient, together with the slip velocity and temperature jump boundary conditions given in Eq. 14-3, to provide the basic system for the slip flow regime. At lower densities, in the beginning of the transition regime, the theoretical results suggest that such solutions may remain fairly good, but that any corrections must be of a very complicated nature, possibly involving complete solutions of the Maxwell-Boltzmann equation. The Navier-Stokes equations of course do not remain valid at very low densities, well into the transition regime; however, the Burnett and Thirteen Moment equations do not seem able even to predict when the Navier-Stokes relations break down.

H,14. Slip Velocity and Temperature Jump Boundary Conditions. It has been recognized since the time of Smoluchowski [*18*,*78*] and Knudsen [*19*] that there is a slip velocity and a temperature jump at the interface between a solid and a gas at low pressures. An approximate analysis in a monatomic gas adjacent to an isothermal surface serves to relate the slip velocity and the gradient of the tangential velocity. Near the wall the gas consists of molecules, one half of which have just come off the wall, the other half of which have come, on the average, from a layer of gas a mean free path away. If y is the normal coordinate, one has

$$u(0) = \tfrac{1}{2}\left(\left[u(0) + \bar{l}\left(\frac{\partial u}{\partial y}\right)_0\right] + \left\{(1-\sigma)\left[u(0) + \bar{l}\left(\frac{\partial u}{\partial y}\right)_0\right] + \sigma \cdot 0\right\}\right) \tag{14-1}$$

where σ is the fraction of diffusely reflected molecules (whose average tangential velocity is thus 0) and $1 - \sigma$ is the fraction of specularly reflected molecules (whose average tangential velocity is the same as that of the molecules incident from the layer a distance $\bar{l}$ above the wall). Hence

$$u(0) = \frac{2-\sigma}{\sigma}\,\bar{l}\left(\frac{\partial u}{\partial y}\right)_0 \tag{14-2}$$

It is observed that, even for completely diffuse reflection, i.e. $\sigma = 1$, there is a definite slip velocity. At normal densities $\bar{l}$ is so small that the effect is completely negligible. Very much more elegant calculations, involving approximate solutions of the Maxwell-Boltzmann equation in the vicinity of a wall have been carried out by many investigators, e.g.

[*22,23,56,74,77,79,80,81,82*]. The results given by Kennard [*2*] are (within a numerical factor close to unity):

$$u(0) = \frac{2-\sigma}{\sigma}\,\bar{l}\left(\frac{\partial u}{\partial y}\right)_0 + \frac{3}{4}\frac{\mu}{\rho T}\left(\frac{\partial T}{\partial x}\right)_0$$

$$T(0) - T_w = \frac{2-\alpha}{\alpha}\,\frac{2\gamma}{\gamma+1}\,\frac{\bar{l}}{Pr}\left(\frac{\partial T}{\partial y}\right)_0 \qquad (14\text{-}3)$$

where α is accommodation coefficient defined in Eq. 4-1.

There is some disagreement with the temperature jump condition for polyatomic gases, and Welander [*81*] and Wang-Chang and Uhlenbeck [*74*] also disagree with the factor $(2-\alpha)/\alpha$. For macroscopic considerations these differences are not serious since they only correspond to different macroscopically determined values for the coefficients σ and α. The values tabulated in Table H,4a and H,4b for these quantities are in agreement with the forms of these relations indicated in Eq. 14-3. Schamberg [*77*] has obtained correction terms to the order $\bar{l}^2$ for similar equations for monatomic gases, using the Burnett distribution function. These results are reproduced by Grad's boundary conditions upon iteration in the same manner as the Burnett equations were shown to be obtainable from the Thirteen Moment equations.

Special mention should be made of the term

$$\frac{3}{4}\frac{\mu}{\rho T}\left(\frac{\partial T}{\partial x}\right)_0$$

in Eq. 14-3, according to which a temperature gradient along a surface induces a flow in the direction of increasing temperature. The currents set up by such an induced flow are called the "thermal creep" [*61*, p. 333]. This phenomenon leads to a pressure gradient along a tube which has a temperature gradient, even when there is no flow, provided that the average pressure is low enough. This can become of some importance in connection with wind tunnel pressure lines and is discussed in I,I.

CHAPTER 4. EXPERIMENTAL RESULTS IN SLIP FLOW AND TRANSITION REGIMES

H,15. Introduction. In the following sections a brief review is presented of the experimental data which are available in the slip and transition regimes for various geometries of aerodynamic interest. A similar review of results for internal flow geometries which are primarily of vacuum engineering interest is given in I,I.

There are also quite a few theoretical analyses available, mostly based upon simplified forms of the Navier-Stokes equations, e.g. the boundary layer equations or the Oseen equations, which include slip and temperature jump boundary condition effects. Often it is assumed that the flow is incompressible, but the results are still presented in terms of the Mach number as well as the Reynolds number. Here the Mach number arises by virtue of the boundary conditions of Eq. 14-3 and the relation $\bar{l}/L \sim M/Re$, so that M is thus not an indication of compressibility at all, but rather of the effect of slip or temperature jump. These analyses are included but should be used with this limitation in mind.

Some of the recent hypersonic experiments have approached the range of slip flow effects, both because of the high Mach number and because of the comparatively low test section densities often used in the hypersonic wind tunnels. It is felt that these experiments are more properly discussed in another place and the present treatment considers only the Mach number range, $0 < M < 6$. For this range the presently known experimental results, particularly for supersonic flow, are due mostly to two laboratories, the Ames Laboratory of the National Advisory Committee for Aeronautics and the Low Density Laboratory of the University of California at Berkeley.

H,16. Couette Flow. The transport of momentum by a rarefied gas between parallel plates or concentric cylinders in relative tangential motion was first studied by Millikan and Van Dyke [*25,83*] for the case of very low speeds. Using the Navier-Stokes equations for the parallel Couette flow, and the slip velocity boundary condition of Eq. 14-3, the drag force per unit area of plate is

$$\tau_w = \tfrac{1}{2}\rho U^2 \frac{2}{Re\left(1 + 2\dfrac{2-\sigma}{\sigma}\dfrac{\bar{l}}{h}\right)} \tag{16-1}$$

U is the relative velocity, h the distance between the plates, $Re = Uh\rho/\mu$, and where it has been assumed that the viscosity μ and the density ρ are constant. This result may be put in the form

$$C_D M = \frac{2}{\dfrac{Re}{M}\left(1 + 2.51\sqrt{\gamma}\,\dfrac{2-\sigma}{\sigma}\dfrac{M}{Re}\right)} \tag{16-2}$$

by using Eq. 1-4, where C_D is the appropriate drag coefficient.

For concentric rotating cylinders with very small annular gap, the result is very closely the same. Millikan and Van Dyke used these results to determine σ by measuring the torque transmitted by a rarefied gas across the annular gap between concentric rotating cylinders (see Table H,4b). The calculation has been extended to include the effects of the Burnett terms and of the "second order" slip boundary condition terms by Schamberg [77], and the additional effect of variable viscosity with temperature by Lin and Street [84]. These results are given in the form of power series in $\bar{l}$, however, and it has been difficult to determine the number of terms in the series that are consistent with the assumptions made. In view of Wang-Chang and Uhlenbeck's more recent result [47] for the heat transport between parallel plates, the existence of a solution of the fundamental Maxwell-Boltzmann equation in the assumed power series form now seems rather unlikely. A solution of the Thirteen Moment

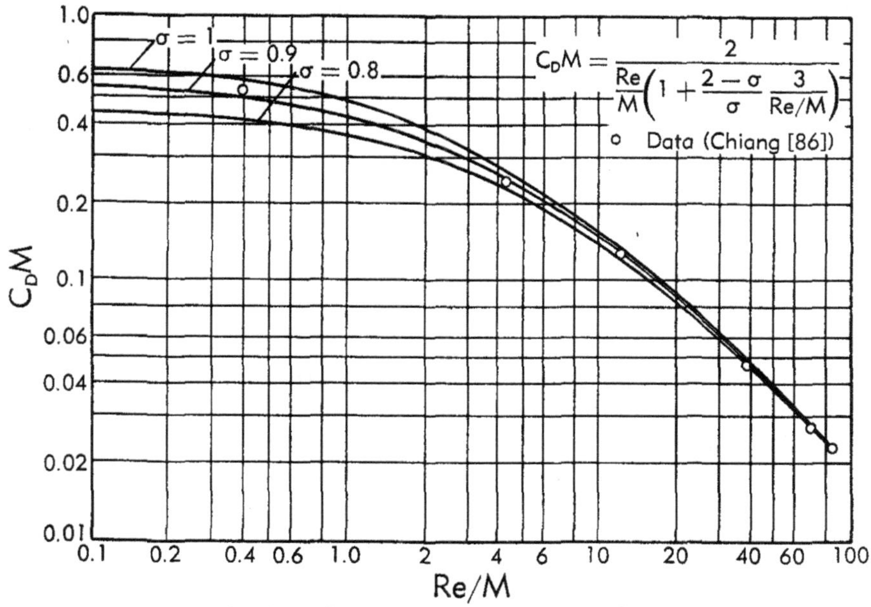

Fig. H,16. Drag coefficient for Couette flow.

equations has also been obtained by Rose [85]. However, to investigate possible deviations from Eq. 16-2, and also to determine how independent of speed the parameter σ is, several investigators have repeated Millikan's experiments recently, at higher speeds. Chiang [86], using air, covered the entire slip and transition range. No change in σ was detectable even for peripheral velocities approaching the sonic range. His results are indicated in Fig. H,16, where the remarkable extent of the validity of Eq. 16-2 is apparent. Similar experimental results have also been obtained by Beams, et al. [87] and Kuhlthau [88].

Incidentally, the limiting value of Eq. 16-2 for $M/Re \to \infty$, is

$$C_DM \to \frac{0.798}{\sqrt{\gamma}\,\frac{2-\sigma}{\sigma}} \tag{16-3}$$

which is the correct value for free molecule flow. This is of course only a coincidence, by which the proper solution of the Boltzmann equation in the low density portion of the transition regime [89] has the same limiting value as Eq. 16-3, as $\bar{l} \to \infty$. The stability of the laminar slip flow between concentric rotating cylinders has been investigated theoretically [90] and experimentally by Chiang [86], with the result that slip has the effect of rendering the flow slightly more stable.

Heat transfer between parallel plates in slip flow may be analyzed in the same way as momentum transfer. If the plates are at different temperatures, T_1 and T_2, and also in relative motion, then the heat flow q per unit area arises from two sources, the viscous dissipation in the flow and the heat conduction due to the temperature difference. From the Navier-Stokes equation and the temperature jump boundary condition, one obtains

$$\frac{hq}{\mu U^2} = \frac{1}{\left(1 + 2\frac{2-\sigma}{\sigma}\frac{\bar{l}}{h}\right)^2} + \frac{T_1 - T_2}{(\gamma - 1)PrM^2T_0\left(1 + \frac{2-\alpha}{\alpha}\frac{4\gamma}{\gamma+1}\frac{\bar{l}}{Prh}\right)} \tag{16-4}$$

where T_0 is the average temperature and Pr is the Prandtl number. It is observed that the temperature jump boundary condition reduces the heat transport. The Mach number in all these expressions is a measure of the slip and temperature jump and not of the compressibility.

H,17. Spheres. Slip flow over a sphere at very low Reynolds and Mach numbers has been investigated by Basset [91, p. 270]. Using the Navier-Stokes equations for incompressible flow and the Stokes "slow-flow" simplification of ignoring the inertia terms in comparison to viscous and pressure terms, the effects of slip were incorporated in a boundary condition of the form

$$\tau_t = \beta u_t \tag{17-1}$$

where u_t is the tangential slip velocity, τ_t the tangential shear stress, and β the slip coefficient. In terms of Eq. 14-3 one has

$$\beta = \frac{\mu}{\frac{2-\sigma}{\sigma}\bar{l}} \tag{17-2}$$

Basset's result for the drag coefficient C_D for a sphere of radius a in a stream of velocity U, density ρ and viscosity μ can be written in the form

$$C_D = \frac{12}{Re}\left(\frac{1 + 2\frac{2-\sigma}{\sigma}\frac{\bar{l}}{a}}{1 + 3\frac{2-\sigma}{\sigma}\frac{\bar{l}}{a}}\right) \tag{17-3}$$

where Re is the Reynolds number $Re = U\rho a/\mu$. A solution of the Thirteen Moment equations for the flow past a sphere, including the effects of compressibility and heat flow, has been obtained by Goldberg [92] which leads to a drag coefficient in the form

$$C_D = \frac{12}{Re}\left[\frac{\left(1 + \frac{15}{2}\frac{2-\alpha}{\alpha}\frac{\bar{l}}{a}\right)\left(1 + 2\frac{2-\sigma}{\sigma}\frac{\bar{l}}{a}\right) + \frac{6}{\pi}\frac{\bar{l}^2}{a^2}}{\left(1 + \frac{15}{2}\frac{2-\alpha}{\alpha}\frac{\bar{l}}{a}\right)\left(1 + 3\frac{2-\sigma}{\sigma}\frac{\bar{l}}{a}\right) + \frac{9}{5\pi}\left(4 + 9\frac{2-\sigma}{\sigma}\frac{\bar{l}}{a}\right)\frac{\bar{l}^2}{a^2}}\right] \tag{17-4}$$

As with Couette flow, Eq. 16-3, this result has the same limiting value for $\bar{l}/a \to \infty$ as the free molecule flow result, Eq. 8-6, for $S \to 0$.

Millikan [93] used an approximation to Basset's result in the form

$$C_D = \frac{12}{Re}\frac{1}{1 + \text{const}\cdot\frac{M}{Re}} \tag{17-5}$$

valid for small M/Re, and an approximation due to Epstein [12] for the free molecule range (i.e. Eq. 8-6 with $S \to 0$) in the form

$$C_D \sim \frac{\text{const}}{M} \tag{17-6}$$

Combining Eq. 17-5 and 17-6, Millikan suggested an empirical equation of the form

$$C_D = \frac{12}{Re\left[1 + \frac{\bar{l}}{a}\left(A + Be^{-C\bar{l}/a}\right)\right]} \tag{17-7}$$

valid over the entire range of M/Re, but restricted to small values of both M and Re. For liquid oil drops in air, Millikan obtained the empirical values of $A = 1.22$, $B = 0.41$, and $C = 8.75$. Eq. 17-7 is the solid curve in Fig. H,17a.

At higher Mach numbers the foregoing analysis breaks down and the situation becomes very complicated due to compressibility effects, including a detached shock wave at supersonic velocities. Experimental values for C_D have been obtained by Kane [94] over the range $2.05 < M < 2.81$; $15 < Re < 768$. No Mach number effect was detected over this range and Kane proposed an empirical formula for C_D in the form

$$C_D = \left(0.97 + \frac{1.32}{\sqrt{Re_1}}\right)\left(1 + \frac{1.0}{\sqrt{Re_1}}\right)^2 \tag{17-8}$$

where Re_1 is the Reynolds number based on the flow conditions behind the normal shock and the sphere diameter. This formula is the dashed curve in Fig. H,17a, which also indicates the experimental points for the supersonic experiments. The difference between the dashed and the solid curves is presumably due to the difference in the Mach number. At high Reynolds numbers the skin friction component of the drag is relatively less important than that of the wave drag and the value of the drag coefficient is higher at high Mach numbers. At low Reynolds numbers, however, the skin friction component is of greater relative importance and at the same Reynolds numbers is less at high Mach numbers

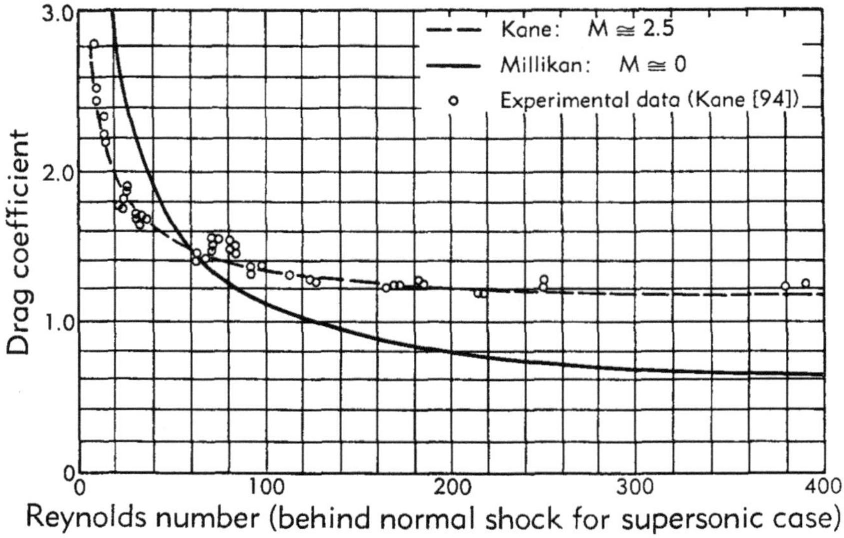

Fig. H,17a. Sphere drag coefficient.

due to the correspondingly increased effect of slip. The drag coefficient curves for two different Mach numbers thus cross each other, the higher Mach number curve lying below the low Mach number at low Reynolds number and above it at high Reynolds number.

Heat transfer to a sphere in slip flow has been investigated experimentally for supersonic flow by Drake and Backer [*95*] and Eberley [*96*], and for subsonic flow by Kavanau [*97*]. A theoretical estimate of the effect of the temperature jump at the surface on the over-all heat transfer was given in [*95*] for incompressible flow, using a modified form of Oseen's equation. A somewhat simpler semiempirical analysis for the over-all heat transfer has been suggested by Kavanau, who argued from Eq. 14-3 that the heat transfer with a temperature jump for a sphere of radius a should be the same as the heat transfer without a temperature jump but for a sphere of radius $(a - \text{const}\,\bar{l})$. Defining the Nusselt number Nu by

$$Nu = \frac{Q}{\pi D (k_{\text{th}})_{\text{r}} (T_{\text{w}} - T_{\text{r}})} \tag{17-9}$$

in the usual notation, with *Re* based on the diameter, Kavanau obtained by this assumption

$$Nu = \frac{Nu^{(0)}}{1 + 3.42 \dfrac{M}{RePr} Nu^{(0)}} \tag{17-10}$$

where the constant 3.42 has been determined empirically. The quantity $Nu^{(0)}$ is the value the Nusselt number would have if there were no temperature jump, an expression for which has been given by Drake and Backer [*95*]

$$Nu^{(0)} = 2 + \frac{2}{\pi^2} \int_0^\infty \frac{(1 - e^{-\pi V^2})(1 + V^4)^{-1} dV}{J_1^2\left(V\sqrt{\dfrac{2RePrU_{av}}{U}}\right) + Y_1^2\left(V\sqrt{\dfrac{2RePrU_{av}}{U}}\right)} \tag{17-11}$$

Here U_{av} is the "average" velocity over the sphere, which has been taken to be $0.77U$. This equation is the solid curve of Fig. H,17b and H,17c. The supersonic data at $M \sim 2.5$ and $M \sim 3.2$ are shown in Fig. H,17b

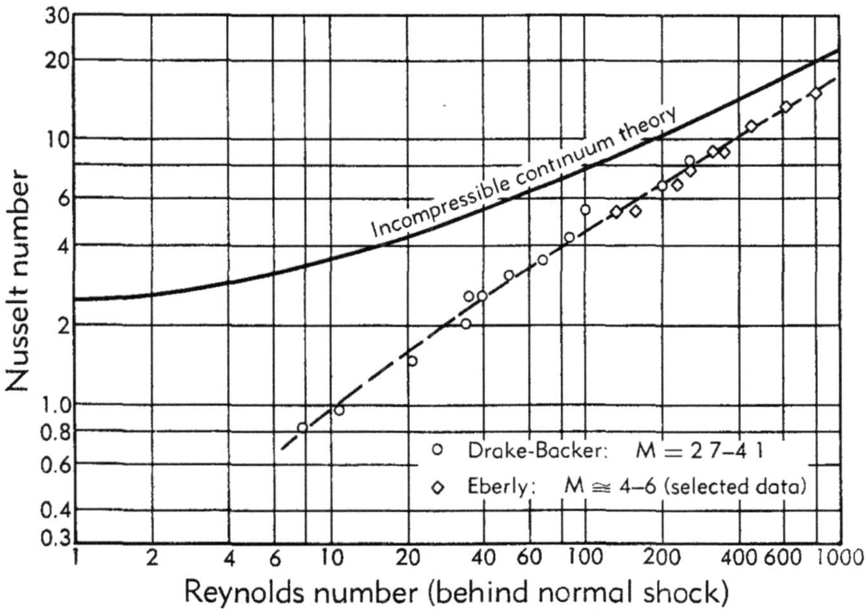

Fig. H,17b. Convective heat transfer coefficient for spheres in supersonic flow.

The subsonic data are shown in Fig. H,17c, together with the corresponding theoretical slip flow and free molecule flow results. The decrease in heat transfer, due primarily to the effective thermal contact resistance associated with the temperature jump, is apparent for both cases.

A somewhat more surprising result than this decrease in the heat

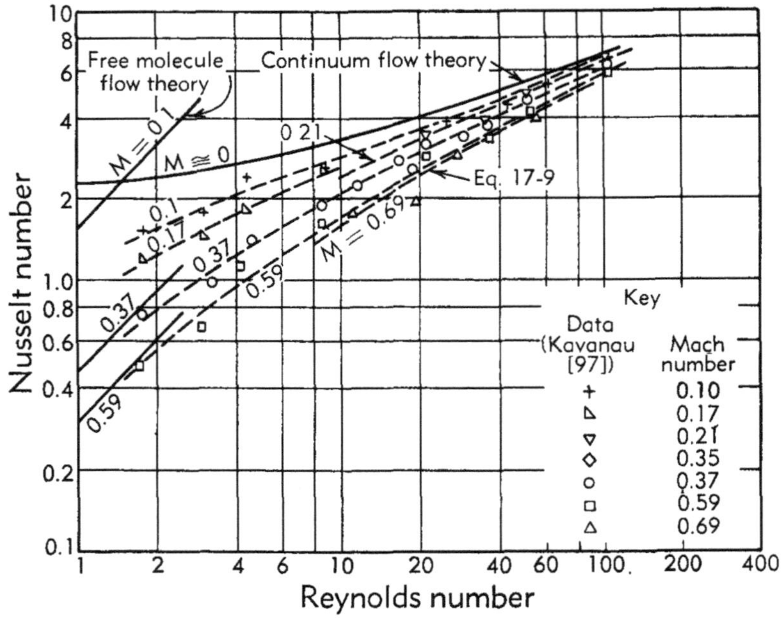

Fig. H,17c. Heat transfer coefficients for spheres in subsonic flow.

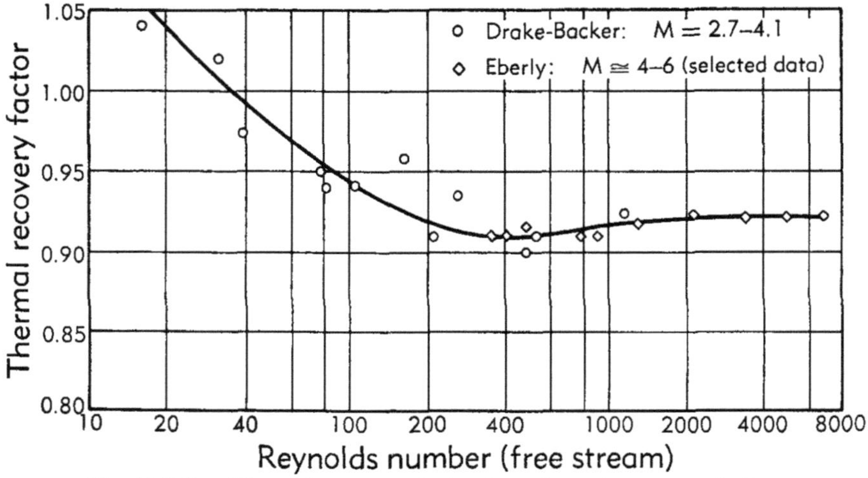

Fig. H,17d. Thermal recovery factor for spheres in supersonic flow.

transfer rate is the trend to values greater than unity of the thermal recovery factor r defined by

$$r = \frac{T_r - T_e}{T_e^0 - T_e} \tag{17-12}$$

where T_r, T_e and T_e^0 are the recovery or adiabatic sphere temperature, the local stream gas temperature, and the local stagnation temperature respectively. This increase is seen in Fig. H,17d. It has already been noted that the recovery factor in free molecule flow for any convex body is greater than unity. It is apparent from the experimental values indicated in Fig. H,17d that this trend starts in the slip flow regime. An analysis has been made [98] of this phenomenon by integrating the energy equation along the forward stagnation stream line, utilizing Homann's velocity distribution [99]. For an adiabatic sphere this leads to an expression for the recovery factor for the stagnation point of the form

$$r = 1 + \frac{f(Re)}{Re}, \qquad f(Re) > 0 \tag{17-13}$$

where $f(Re)$ is a slowly varying function. This result, together with the experimental data indicated in Fig. H,17d, suggests that the viscous layer surrounding the sphere is so thick in this Reynolds number range that heat is generated by dissipation more rapidly than it can be conducted away; this is of course quite different from the laminar boundary layer case where the recovery factor is approximately equal to $\sqrt{Pr}$. The result is somewhat similar to the increase in the impact pressure (IX,B,1,) which is also observed at low Reynolds numbers. The experimental accuracy of the subsonic measurements was not sufficient to determine the recovery factor satisfactorily.

H,18. Cylinders. The incompressible flow at low Reynolds number past a right circular cylinder with the axis perpendicular to the free stream velocity was investigated theoretically for the no-slip case by Lamb [100] using the Oseen equations, which partially account for the inertia terms. Tsien [4] extended Lamb's results to the case of slip and determined a drag coefficient in the form

$$C_D = \frac{4\pi}{Re\left[\ln\frac{4}{Re} - 1.28 + 1.26\sqrt{\gamma}\,\frac{2-\sigma}{\sigma}\,\frac{M}{Re}\right]} \tag{18-1}$$

where Re is the Reynolds number based on the radius a of the cylinder and C_D is the drag per unit length divided by $a\rho U^2$. There is as yet no experimental data on cylinder drag in the slip flow regime, the results of Finn [101] being just barely in the continuum range, while those of Stalder, Goodwin, and Creager [15] are in free molecule flow.

The heat transfer to cylinders in the slip flow and transition regime has been determined experimentally by Stalder, Goodwin, and Creager [16], Kovásznay and Tormarck [102], and Laufer and McClellan [103]. Their results cover the range $2.0 < M < 3.3$, $0.28 < Re < 203$, and are presented in Fig. H,18a. The Knudsen and Reynolds numbers are based on free stream conditions and the cylinder diameter. The thermal recovery factor, as seen in Fig. H,18b, exhibits the same sort of increase at

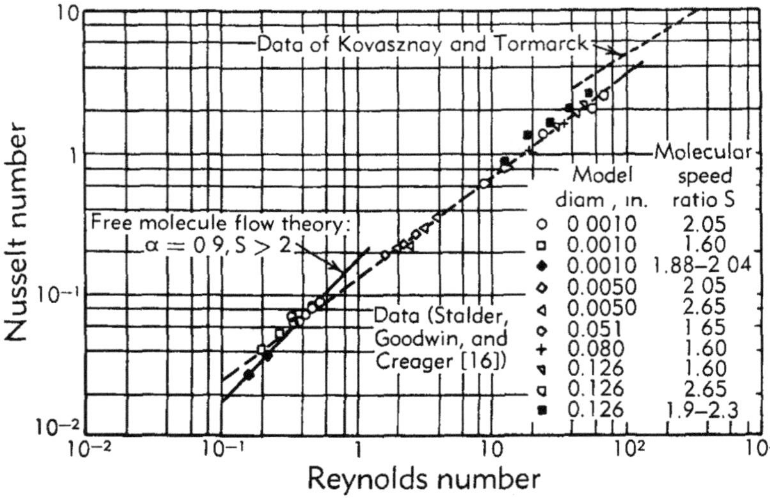

Fig. H,18a. Convective heat transfer coefficients for transverse cylinders in supersonic flow. (Courtesy National Advisory Committee for Aeronautics.)

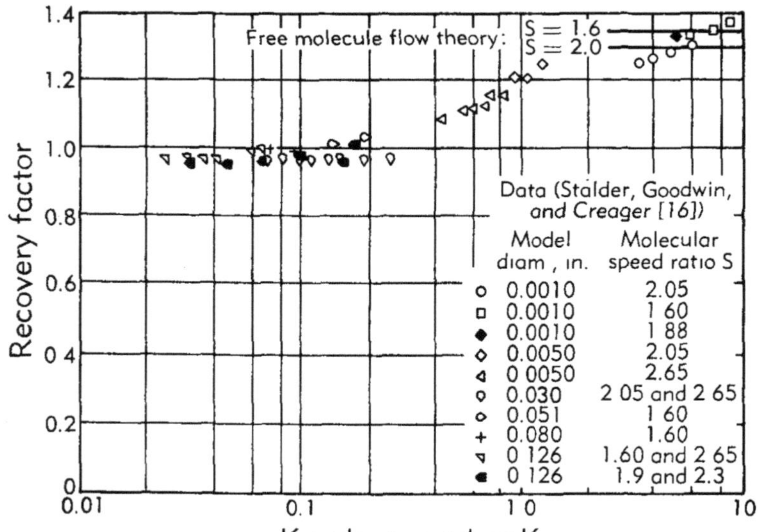

Fig. H,18b. Thermal recovery factor for transverse cylinders in supersonic flow. (Courtesy National Advisory Committee for Aeronautics.)

low densities as that already noted for spheres. A theoretical analysis of the effect of the temperature jump boundary condition on the heat transfer rate for cylinders has been presented by Sauer and Drake [*104*]. An analysis of heat transfer to cylinders in the Oseen range, neglecting the effect of temperature jump, and thus outside the slip flow regime has also been given by Cole and Roshko [*11*].

H,19. Flat Plates. Slip flow past a flat plate at zero angle of attack has been studied quite extensively. It was hoped to isolate skin friction phenomena by careful consideration of this geometry. A number of theoretical analyses [*7,105,106,107,108,109*] have been presented for determining the effect of slip on the boundary layer flow for an isothermal, semi-infinite flat plate. The stream function ψ for the flow, which satisfies the usual boundary layer equations of motion but which obeys the slip boundary condition of Eq. 14-3, can be obtained readily for small $\bar{l}$ by perturbation series methods in the form,

$$\psi = \psi_0 + \frac{2 - \sigma}{\sigma} \bar{l} \frac{\partial \psi_0}{\partial y} + \cdots \tag{19-1}$$

where y is the normal direction, and where ψ_0 is the stream function for the case of no slip. The local skin friction coefficient c_f is given by

$$c_f = \frac{2\mu}{\rho U^2} \left(\frac{\partial^2 \psi_0}{\partial y^2} + \frac{2 - \sigma}{\sigma} \bar{l} \frac{\partial^3 \psi_0}{\partial y^3} + \cdots \right)_{y=0} \tag{19-2}$$

However, since

$$\left(\frac{\partial^3 \psi_0}{\partial y^3} \right)_0 \sim \left(\frac{dp}{dx} \right)_\infty = 0$$

for the flat plate boundary layer solution without pressure gradient, there is no effect, to order $\bar{l}$, of slip on the local skin friction. To order $\bar{l}^2$, moreover, the boundary layer equations are no longer valid due, for example, to interaction effects between the viscous layer and the inviscid outer flow [*53,54,55*]. These interaction effects arise because the boundary layer displaces the inviscid flow outward so that the resulting "free stream" pressure has a negative gradient along the surface; this, in turn, affects the boundary layer so as to increase the skin friction.

A reliable method for extending boundary layer theory to lower Reynolds numbers is not yet available [*109,110,111,112,113,114,115,116*]. There is thus no adequate theory at present for predicting the effect of slip on the local skin friction of a flat plate at zero angle of attack, the basic difficulty stemming from the fact that the pressure gradient for the ideal flow, and therefore the slip correction term of order $\bar{l}$, vanishes. The flat plate thus seems to be an exceptional case. The incompressible boundary layer flow over a wedge of semi-vertex angle θ satisfies the Falkner-Skan equation [*117*].

$$\begin{aligned} \psi &= \sqrt{u_e \nu x}\, f(\eta), \quad \eta = \sqrt{u_e/\nu x} \cdot y \\ m f'^2 - \tfrac{1}{2}(m + 1) f f'' &= m + f''' \end{aligned} \tag{19-3}$$

where the velocity at the edge of the boundary layer is given by $u_e = cx^m$ with $m = \theta/(\pi - \theta)$. This can easily be solved for the case of slip in the form of the perturbation series indicated in Eq. 19-1 to yield an expression for the local skin friction coefficient

$$c_f = c_{f,0}\left[1 - \frac{2-\sigma}{\sigma}\frac{\theta}{\pi-\theta}\frac{\sqrt{Re}}{f''(0)}\frac{\bar{l}}{x} + \cdots\right] \tag{19-4}$$

where $c_{f,0}$ is the coefficient for the no-slip case, and $Re = u_e x/\nu_1$, x being the distance from the vertex. The quantity $f''(0)$ depends on θ, and can be found in [*117*]. There is thus a nonzero correction term of order $\bar{l}$ except for the flat plate case of $\theta = 0$. The case $\theta = \pi/2$ is closely related to the stagnation point flow case considered in [*105*]. Eq. 19-4 may also be put in the form

$$c_f = c_{f,0}\left[1 - \frac{2-\sigma}{\sigma}\frac{\theta}{\pi-\theta}\frac{1.26\sqrt{\gamma}}{f''(0)}\frac{M}{\sqrt{Re}} + \cdots\right] \tag{19-5}$$

by the use of Eq. 1-4; here M is an indication of slip, not of compressibility. It is observed that slip has the expected effect of reducing the local skin friction coefficient.

Experimental values of skin friction in the slip flow range have been determined in both supersonic and subsonic flow [*109*]. The total drag coefficient based on total drag and the area of one side, for a series of finite flat plates at zero angle of attack, was determined. The increase in the skin friction due to the trailing edge, which has been determined theoretically by Kuo [*112*] for incompressible no-slip flow, was thus necessarily present. This increase is probably of secondary importance in the supersonic case but explains the increase in the total drag coefficient down to $Re = 15$ observed by Janour [*118*] in incompressible flow. The experimental data for the slip flow case covered the range $3 < Re < 500$, for $M \cong 0.2$ and 0.6, and $34 < Re < 2020$ for $2.5 < M < 3.8$. This data is presented in Fig. H,19, together with a curve representing the data of Janour (and the theory of Kuo) covering the same Reynolds number range, but for incompressible no-slip flow. It is observed that the measured skin friction for the subsonic case is generally lower than the corresponding continuum results of Janour, in general agreement with Eq. 19-5. The supersonic data, on the other hand, are generally higher than the continuum values. For $Re \cong 1000$ the skin friction increases with increasing Mach number, in agreement with the interaction effect. For $Re \cong 50$, on the other hand, the skin friction decreases with increasing Mach number, in agreement with the slip effect. Skin friction in the slip flow regime is thus apparently affected by two interrelated phenomena of different sign, namely, the interaction of the thick boundary layer with the inviscid flow, and the slip at the surface. The former is more important at the higher densities, gradually giving way to the latter for densities extending into the transition regime.

The question of the degree of slip at the leading edge is of considerable interest, particularly at Mach numbers in the hypersonic range. Using a very small "free molecule flow" equilibrium temperature probe, Laurmann [*119*] has made a qualitative survey of the leading edge region, of the order of a hundred mean free paths square, at $M \cong 2$ and $M \cong 4$. His results indicate that the "shock wave" is as thick as the viscous

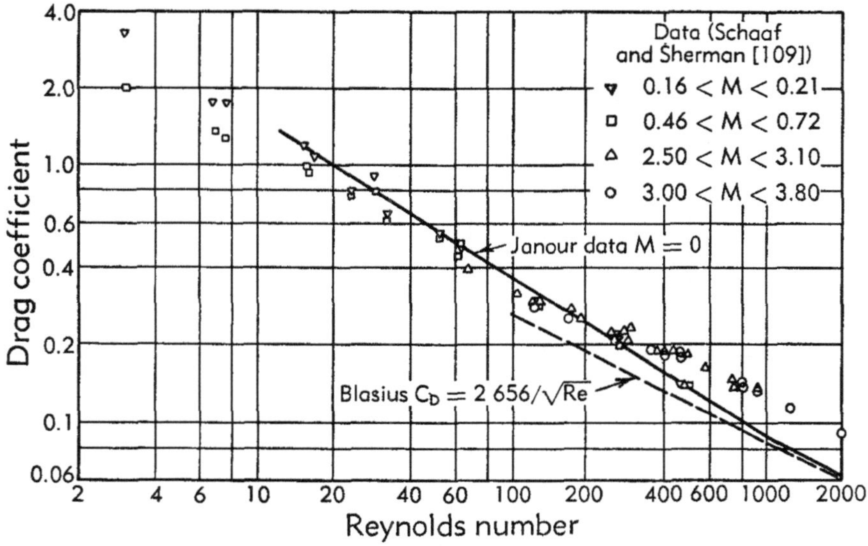

Fig. H,19. Skin friction in slip flow.

layer in this region, that there is a considerable region in which the shock wave coalesces with the viscous layer, and that there appears to be considerable slip on the plate at the higher Mach number.

Heat transfer for a flat plate with a temperature jump boundary condition has been investigated theoretically by Drake and Kane [*120*] and by Martino [*121*].

H,20. Cones. For an inviscid supersonic flow, the surface pressure on a cone at zero angle of attack is constant, provided that the Mach number is high enough to insure an attached shock wave at the cone vertex. In the slip and transition regimes, the boundary layer which forms on the cone surface is in general comparatively thick. This displaces the inviscid flow outwards and leads to a pressure distribution which rises rapidly toward the cone vertex. This induced pressure gradient interacts with the boundary layer to increase the skin friction, just as in the flat plate case. Estimating the boundary layer displacement thickness by no-slip flat plate theory and the Mangler transformation to conical flow [*122*], and using linearized supersonic flow theory to calculate the inviscid flow past a body consisting of the cone plus the boundary layer displacement thickness, Talbot [123] has obtained a result for the induced pressure in the form

$$\frac{p}{p_{\text{ideal}}} = 1 + \frac{f(M_1, \theta)}{\sqrt{Re_1}} \tag{20-1}$$

where θ is the cone semi-vertex angle, M_1 the Mach number, and Re_1 the Reynolds number based on conditions on the cone surface in inviscid

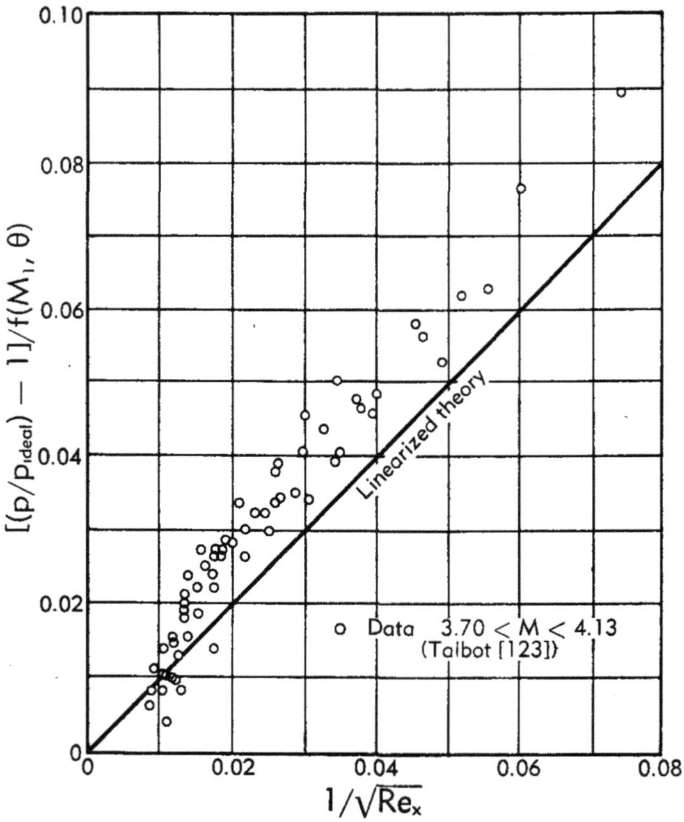

Fig. H,20a. Surface pressure distribution for cones with 5° semi-vertex angle.

flow and the slant distance x from the vertex. The quantity $f(M_1, \theta)$ is given by

$$f(M_1, \theta) = \frac{\gamma M_1^2 \left(3\theta \ln \frac{1}{\theta} - 4\theta\right)(1 + 0.277M_1^2)}{4[1 + \theta^2(1 + 2 \ln \theta)]} \tag{20-2}$$

Talbot has measured the surface pressure on slender (5° half angle) cones in the range $3.70 < M < 4.13$, $185 < Re < 12{,}000$. His results are indicated in Fig. H,20a, along with the theoretical result correspond-

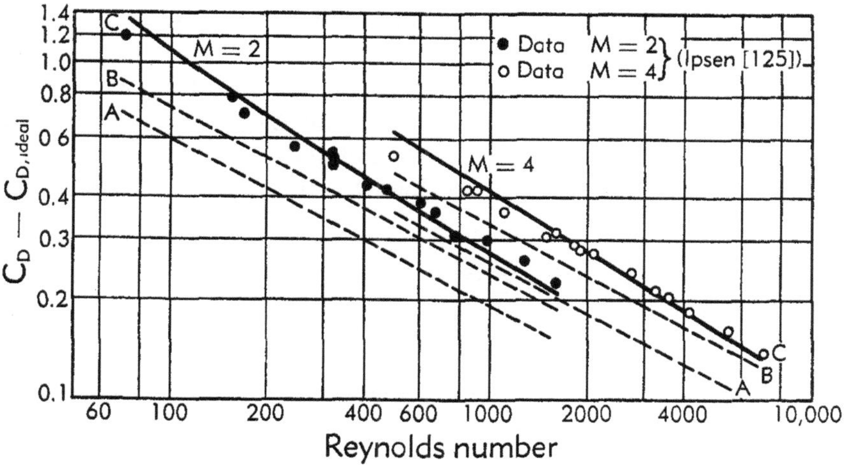

Fig. H,20b. Drag coefficients for cones with 15° semi-vertex angles.

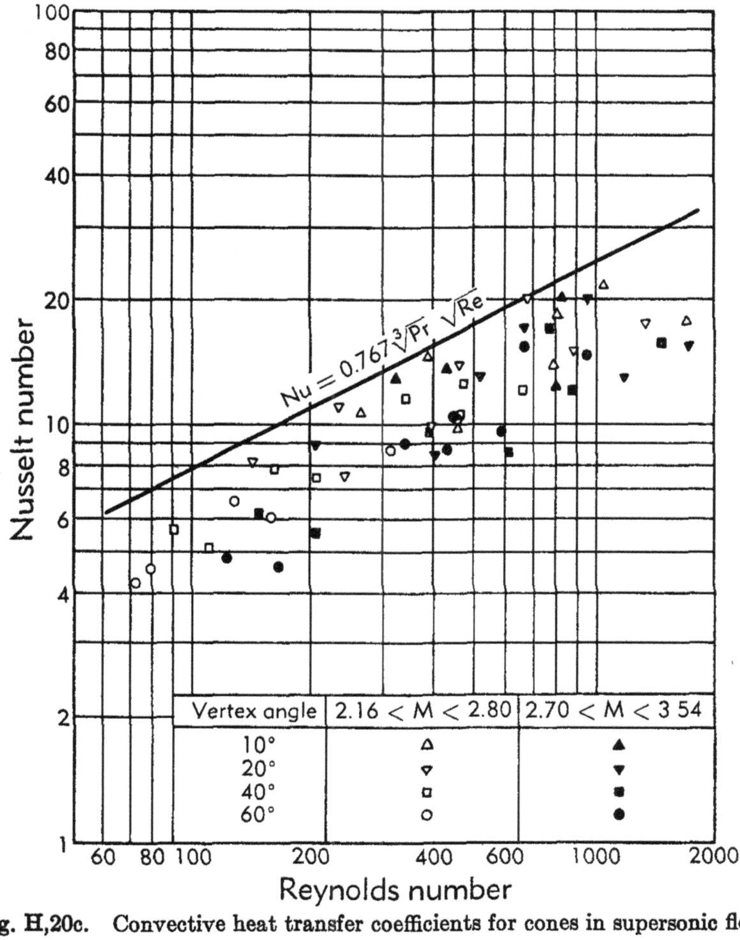

Fig. H,20c. Convective heat transfer coefficients for cones in supersonic flow ing to Eq. 20-1, confirming the general trend of the prediction. M and Re are based on free stream conditions.

The most complete theoretical estimate of the skin friction on cones in this range has been made by Probstein and Elliott [*124*], considering the interaction effect mentioned above and also the effect of transverse curvature on the boundary layer flow. Ipsen [*125*] has measured the

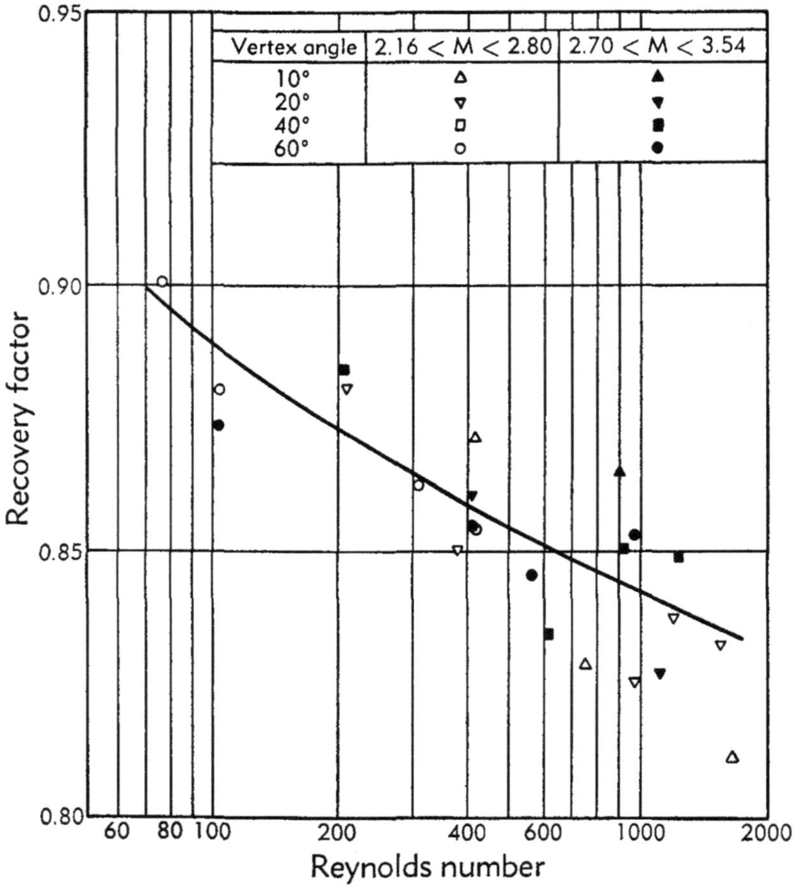

Fig. H,20d. Thermal recovery factors for cones in supersonic flow.

total drag on 15° half angle cones over the range $150 < Re < 1500$ at $M = 2$ and $1000 < Re < 7000$ at $M = 4$. His results are presented in Fig. H,20b. Here M and Re are based on the free stream conditions and the cone slant length. The quantity C_D is the drag coefficient based on cone surface area, while $C_{D,\,\text{ideal}}$ is the corresponding inviscid drag coefficient at the same Mach number. Curves A, B, and C are the calculated values of $C_D - C_{D,\,\text{ideal}}$ accounting for, respectively, the simple boundary layer skin friction, this plus the induced pressure, and finally this plus the transverse curvature correction to the skin friction. Again, none of these theories takes account of possible slip effects. The slight drop-off of the data for the lowest values of *Re* at each Mach number might possibly be due to such slip effects, although neither the experiments nor the theories are of sufficient accuracy to warrant such a conclusion.

The convection heat transfer characteristics of cones in the slip flow range have been determined by Drake and Maslach [*126*] over the range $78 < Re < 3270$, $2.16 < M < 3.54$ where M and Re are based on the free stream conditions and the cone slant length. The corresponding variation of the Nusselt number and the recovery factor with the Mach and Reynolds numbers are presented in Fig. H,20c and H,20d. The decrease in the heat transfer rate as well as the increase in the recovery factor with decreasing density are again observed.

H,21. Base Pressure. The base pressure p_b on a cone-cylinder configuration at zero angle of attack has been determined in the slip flow and transition regimes by Kavanau [*127,128*]. His results, together with a curve indicating the continuum results for geometrically similar models

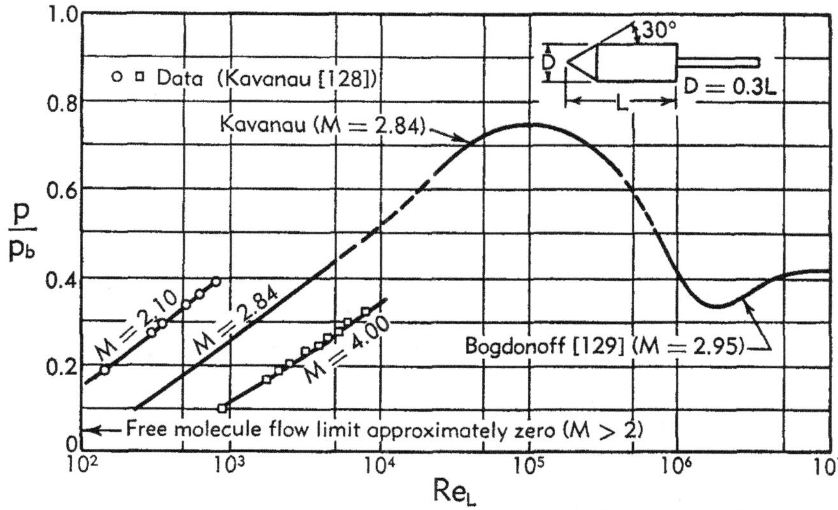

Fig. H,21. Base pressure coefficient for cone-cylinder configurations.

obtained by Bogdonoff [*129*], are given in Fig. H,21. The Reynolds number is based on the free stream and the model length, while p_∞ is the free stream static pressure. The peak in the base pressure curve near $Re = 10^5$ is interpreted as the point at which the transition point coincides with the critical point in the wake. An increase in Reynolds number introduces some turbulent mixing in the wake region, thus increasing the over-all mixing and hence decreasing the base pressure. On the other hand, a decrease in Reynolds number increases the laminar mixing and thus also decreases the base pressure. For the measurements in the lower Reynolds number range, Kavanau found that the pressure on the base varied considerably with radial position, being as much as four times as high at the center as at the periphery. The base pressures indicated in Fig. H,21 are area averages.

H,22. Cited References.

1. Maxwell, J. C. *The Scientific Papers of James Clerk Maxwell*, Vol. 2. Cambridge Univ. Press, 1890.
2. Kennard, E. H. *Kinetic Theory of Gases*. McGraw-Hill, 1938.
3. Loeb, L. B. *Kinetic Theory of Gases*, 2nd ed. McGraw-Hill, 1934.
4. Tsien, H. S. *J. Aeronaut. Sci. 13*, 653 (1946).
5. Tsien, H. S. *J. Aeronaut. Sci. 15*, 577 (1948).
6. Roberts, H. E. *Aeronaut. Eng. Rev. 8*, 19 (1949).
7. Donaldson, C. duP. An approximate method for estimating the incompressible laminar boundary layer characteristics on a flat plate in slipping flow. *NACA Research Mem. L9C02*, 1949.
8. Siegel, K. M. *J. Aeronaut. Sci. 17*, 191 (1950).
9. Van de Hulst, H. C. *The Atmospheres of the Earth and Planets*. Univ. Chicago Press, 1952.
10. Warfield, C. N. Tentative tables for the properties of the upper atmosphere. *NACA Tech. Note 1200*, 1947.
11. Cole, J., and Roshko, A. Heat transfer from wires at Reynolds numbers in the Oseen range. *Heat Transfer and Fluid Mech. Inst.*, 1954.
12. Epstein, P. S. *Phys. Rev. 23*, 710 (1924).
13. Zahm, A. F. *J. Franklin Inst. 217*, 153 (1934).
14. Grimminger, G., Williams, E. P., and Young, G. B. W. *J. Aeronaut. Sci. 17*, 241 (1950).
15. Stalder, J. R., Goodwin, G., and Creager, M. O. A comparison of theory and experiment for high speed free molecule flow. *NACA Tech. Note 2244*, 1950.
16. Stalder, J. R., Goodwin, G., and Creager, M. O. Heat transfer to bodies in a high speed rarefied gas stream. *NACA Tech. Note 2438*, 1951.
17. Sänger, E. Gaskinetik Sehr hoher Fluggeschwindigkeiten. *Deut. Luftfahrtforschung, Bericht 972*, 1938. Transl. as: The gas kinetics of very high altitude flight. *NACA Tech. Mem. 1270*, 1950.
18. Smoluchowski, M. V. *Wied. Ann. 64*, 101 (1898).
19. Knudsen, M. *Ann. Phys. 34*, 593 (1911).
20. *Hand und Jahrbuch der Chemischen Physik*, Vol. 3, Pt. II, Sec. IV. Akad. Verlagsgesellschaft, Leipzig, 1949.
21. Wiedmann, M. L. *Trans. Am. Soc. Mech. Engrs. 68*, 57 (1946).
22. Devienne, P. F. M. Condensation et adsorption des molécules sur une surface en atmosphère raréfiée. *Mém. sci. phys. 52*, 1952.
23. Devienne, P. F. M. Conduction thermique dans les gaz raréfiés coefficient d'accommodation. *Mém. sci. phys. 56*, 1953.
24. Hurlbut, F. C. *An Experimental Molecular Beam Investigation of the Scattering of Molecules from Surfaces. Ph.D. Thesis*, Univ. Calif., 1954.
25. Millikan, R. A. *Phys. Rev. 21*, 217 (1923).
26. Sänger, E., and Bredt, J. Uber einen Raketenantreib fur Fernbomber. *Deut. Luftfahrtforschung, Untersuch. u. Mitt. 3538*, 141 (1944).
27. Luntz, M. *Recherche aéronaut. Paris 7*, 17 (1949).
28. Stalder, J. R., and Jukoff, D. Heat transfer to bodies travelling at high speeds in the upper atmosphere. *NACA Rept. 944*, 1949.
29. Oppenheim, A. K. *J. Aeronaut. Sci*, 49 (1953).
30. Sauer, F. M. *J. Aeronaut. Sci. 18*, 353 (1951).
31. Hartree, D. R. *Mem. and Proc. Manchester Lit. and Phil. Soc. 80*, 85 (1935).
32. Heineman, M. *Commun. on Pure and Appl. Math. 1*, 259 (1948).
33. Snow, R. M. Aerodynamics in a highly attenuated atmosphere. *Appl. Phys. Lab., Johns Hopkins Univ. Rept. CM336*, 1946.

34. Ashley, H. *J. Aeronaut. Sci. 16*, 95 (1949).
35. Stalder, J. R., and Zurick, V. J. Theoretical aerodynamic characteristics of bodies in a free molecule flow field. *NACA Tech. Note 2423*, 1951.
36. Popov, S. G. *Vestnik Moskovskozo Universiteta 3*, 25 (1948). (*Am. Math. Soc. Transl. 36.*)
37. Garfunkel, I. M. Generalizations of applications of free molecule flow. *Univ. Mich. Rept. EMB58.*
38. Bell, S., and Schaaf, S. A. *J. Am. Rocket Soc. 23*, 314 (1953).
39. Bell, S., and Schaaf, S. A. *Jet Propulsion 25*, 168 (1955).
40. Sherman, F. S. A low density wind tunnel study of shock wave structure and relaxation phenomena in gases. *NACA Tech. Note 3298*, 1955.
41. Burgers, J. M. *Proc. Kon. Vlaamse Acad. voor Wet.*, 573 (1947).
42. Thomas, R. N., and Whipple, F. L. *J. Aeronaut. Sci. 18*, 636 (1951).
43. Sänger, E. *Schweiz. Arch. 16*, 43 (1950).
44. Jaffé, G. *Ann. Physik 6*, 195 (1930).
45. Keller, J. *Commun. on Pure and Appl. Math. 1*, 275 (1948).
46. Wang-Chang, C. S. Transport phenomena in very dilute gases. *Nav. Ord. Research and Develop. Rept. 7924-UMH3F*, 1950.
47. Wang-Chang, C. S., and Uhlenbeck, G. E. The heat transport between two parallel plates as functions of the Knudsen number. *Univ. Mich. Eng. Research Inst. Rept. M999*, 1953.
48. Szymanski, Z. *Arch. Mech. Stosowanej 8*, 1956.
49. Lunc, M. *Technika Lotnicza 5*, 39 (1950).
50. Lunc, M., and Lubonski, J. *Arch. Mech. Stosowanej 4*, 597 (1956).
51. Kryzwoblocki, M. Z., and Shinosaki, G. *Acta Phys. Austriaca 10*, 34–53 (1956).
52. Bowyer, J. M. *Drag on a Rotating Cylinder Apparatus near Conditions of Free Molecule Flow. Ph.D. Thesis*, Univ. Calif., 1956.
53. Lees, L., and Probstein, R. F. Hypersonic viscous flow over a flat plate. *Princeton Univ. Aeronaut. Eng. Dept. Rept. 195*, 1952.
54. Lees, L. *J. Aeronaut. Sci. 20*, 143–145 (1953).
55. Li, T. Y., and Nagamatsu, H. T. *J. Aeronaut. Sci. 20*, 345–351 (1953).
56. Grad, H. *Commun. on Pure and Appl. Math. 2*, 331 (1949).
57. Koga, T. *J. Chem. Phys. 22*, 1633–1646 (1954).
58. Enskog, D. *Arkiv Mat., Astron. Fysik Stockholm 21*, 13 (1928).
59. Burnett, D. *Proc. London Math. Soc. 40*, 382 (1935).
60. Chapman, S., and Cowling, T. G. *The Mathematical Theory of Non-Uniform Gases.* Cambridge Univ. Press, 1939.
61. *Heat Transfer.* Univ. Mich. Press, 1953.
62. Hirschfelder, J. O., Curtiss, C. F., and Bird, R. B. *Molecular Theory of Liquids.* Wiley, 1954.
63. Patterson, G. N. *Molecular Flow of Gases.* Wiley, 1956.
64. Greene, E. F., Cowan, G. R., and Hornig, D. F. *J. Chem. Phys. 19*, 427 (1951).
65. Greenspan, M. *J Acoust. Soc. Amer. 26*, 70 (1954).
66. Rosenhead, L., et al. A discussion on the first and second viscosity of fluids. *Proc. Roy. Soc. London A226*, 1–65 (1954).
67. Wang-Chang, C. S., and Uhlenbeck, G. E. Transport phenomena in polyatomic gases. *Univ. Mich. Eng. Research Inst. Rept. CM681*, July 1951.
68. Truesdell, C. *Proc. Natl. Acad. Sci. 34*, 342 (1948).
69. Truesdell, C. *J. Rat. Mech. and Anal. 1*, 125–300 (1952).
70. Mostov, P. M., Grad, H., and Borowitz, S. *Phys. Rev. 90*, 376 (1953).
71. Kohler, M. *Z. Physik 124*, 757 (1948).
72. Andersen, W. H., and Hornig, D. F. *J. Chem. Phys. 24*, 767–770 (1956).
73. Talbot, L., and Sherman, F. S. Structure of weak shock waves in a monatomic gas. *NACA Tech. Note.* (In press.)
74. Wang-Chang, C. S., and Uhlenbeck, G. E. On the behavior of a gas near a wall. *Univ. Mich. Eng. Research Inst. Rept. 2457-2-T*, Aug. 1956.
75. Ikenberry, E., and Truesdell, C. *J. Rat. Mech. and Anal. 5*, 1–54 (1956).

76. Truesdell, C. *J. Rat. Mech. and Anal. 5*, 55–128 (1956).
77. Schamberg, R. *The Fundamental Differential Equations and the Boundary Conditions for High Speed Slip Flow. Thesis*, Calif. Inst. Technol., 1947.
78. Smoluchowski, M. V. *Ann. Physik 33*, 1559 (1910).
79. Weber, S. *Det. Kgl. Danske Videnskabernes Selskab Math. fysiske Meddeleiser 16*, 9 (1939).
80. Kramers, H. A. *Nuovo Cimento 6, Suppl.*, 1949.
81. Welander, P. *Arkiv för Fysik 7*, 507–553 (1953).
82. Payne, H. *J. Chem. Phys. 21*, 2127–2132 (1953).
83. Van Dyke, K. S. *Phys. Rev. 21*, 250 (1923).
84. Lin, T. C., and Street, R. E. Effect of variable viscosity and thermal conductivity on high-speed slip flow between concentric rotating cylinders. *NACA Tech. Note 2895*, 1953.
85. Rose, M. *Phys. Rev. 91*, 469 (1953).
86. Chiang, S. F. *Drag Forces on Rotating Cylinders at Low Pressures. Ph. D. Thesis*, Univ. Calif., 1952.
87. Beams, J. W., Young, J. L., and Moore, J. W. *J. Appl. Phys. 17*, 886 (1946).
88. Kuhlthau, A. R. *J. Appl. Phys. 20*, 217 (1949).
89. Wang-Chang, C. S., and Uhlenbeck, G. E. Transport phenomena in very dilute gases. *Univ. Mich. Eng. Research Inst. Rept. CM579*, Nov. 1949.
90. Chambré, P. L. *Stability of Viscous Compressible Flow Between Rotating Cylinders. Ph.D. Thesis*, Univ. Calif., 1951.
91. Basset, A. B. *Hydrodynamics*. Deighton, Bell & Co., Cambridge, 1888.
92. Goldberg, R. *The Flow of a Rarefied Perfect Gas Past a Fixed Spherical Obstacle. Ph.D. Thesis*, New York Univ., 1954.
93. Millikan, R. A. *Phys. Rev. 22*, 1 (1923).
94. Kane, E. D. *J. Aeronaut. Sci. 18*, 259 (1951).
95. Drake, R. M., and Backer, G. H. *Trans. Am. Soc. Mech. Engrs. 74*, 1241–1250 (1952).
96. Eberley, D. K. Forced convection heat transfer from spheres to a rarefied gas. *Univ. Calif. Inst. Eng. Research Rept. HE-150-140*, July 1956.
97. Kavanau, L. L. *Trans. Am. Soc. Mech. Engrs. 77*, 617–624 (1955).
98. Mack, S. F., and Schaaf, S. A. Viscous effects on stagnation point temperatures. *Univ. Calif. Rept. HE-150-96*, 1951.
99. Homann, L. *Z. angew. Math. u. Mech. 16*, 153 (1936).
100. Lamb, H. *Z. agnew. Math. u. Mech. 16*, 615 (1936).
101. Finn, R. K. *J. Appl. Phys. 24*, 771 (1953).
102. Kovásznay, L. S. G., and Tormarck, S. I. A. Heat loss of hot wires in supersonic flow. *Johns Hopkins Univ. Aeronaut. Dept. Bumblebee Rept. 127*, Apr. 1950.
103. Laufer, I., and McClellan, R. Equilibrium temperature and heat transfer characteristics of hot wires in supersonic flow. *Guggenheim Aeronaut. Lab., Calif. Inst. Technol. External Rept. 315*, May 1956.
104. Sauer, F. M., and Drake, R. M. *J. Aeronaut. Sci. 20*, 175 (1953).
105. Lin, T. C., and Schaaf, S. A. The effect of slip on flow near a stagnation point and in a boundary layer. *NACA Tech. Note 2568*, 1951.
106. Mirels, H. Estimate of slip effect on compressible laminar boundary layer skin friction. *NACA Tech. Note 2609*, 1952.
107. Maslen, S. A. Second approximation to laminar boundary layer flow on flat plate in slip flow. *NACA Tech. Note 2818*, 1952.
108. Nonweiler, T. The laminar boundary layer in slip flow. *College of Aeronautics, Cranfield, Rept. 62*, Nov. 1952.
109. Schaaf, S. A., and Sherman, F. S. *J. Aeronaut. Sci. 21*, 85 (1954).
110. Alden, H. L. *J. Math. and Phys. 47*, 91 (1948).
111. Lewis, J. A., and Carrier, G. F. *Quart. Appl. Math. 7*, 228 (1949).
112. Kuo, Y. H. *J. Math. and Phys. 32*, 83 (1953).
113. Kaplan, S. *J. Appl. Math. and Phys. 5*, 111–136 (1954).
114. Latta, G. E. Investigations of the classical boundary layer associated with

Oseen's equation. *Guggenheim Aeronaut. Lab., Calif. Inst. Technol.* Unpublished manuscript.

115. Lagerstrom, P. A., and Cole, J. D. *J. Rat. Mech. and Anal. 4*, 817–882 (1955).
116. Goldstein, S. Summary interim report—Flow of an incompressible viscous fluid along a semi-infinite flat plate. *Univ. Calif. Inst. Eng. Research Rept. HE-150-144*, Dec. 1956.
117. Falkner, V. M., and Skan, S. W. Some approximate solutions of the boundary layer equations. *Brit. Aeronaut. Research Council Repts. and Mem. 1314*, Apr. 1930.
118. Janour, Z. Resistance of a plate in parallel flow at low Reynolds numbers. *NACA Tech. Mem. 1316*, 1951.
119. Laurmann, J. A. Experimental investigation of the flow about the leading edge of a flat plate. *Univ. Calif. Inst. Eng. Research Rept. HE-150-126*, Oct. 1954.
120. Drake, R. M., and Kane, E. D. A summary of the present status of heat transfer in rarefied gases. *Univ. Calif. Inst. Eng. Research Rept. HE-150-73*, Oct. 1950.
121. Martino, R. L. Heat transfer in slip flow. *Univ. Toronto Inst. Aerophys. Rept. 35*, 1955.
122. Mangler, W. *Z. angew. Math. u. Mech. 28*, 97 (1948).
123. Talbot, L. Viscosity corrections to cone probes in rarefied supersonic flow at a nominal Mach number of 4. *NACA Tech. Note 3219*, 1954.
124. Probstein, R. F., and Elliott, D. The transverse curvature effect in compressible axially-symmetric laminar boundary layer flow. *J. Aeronaut. Sci. 23*, 208–225 (1956).
125. Ipsen, D. C. *Cone Drag in a Rarefied Gas Flow. Ph.D. Thesis*, Univ. Calif., 1953.
126. Drake, R. M., and Maslach, G. J. Heat transfer from right circular cones to a rarefied gas in supersonic flow. *Univ. Calif. Inst. Eng. Research Rept. HE-150-91*, 1952.
127. Kavanau, L. L. *J. Aeronaut. Sci. 21*, 257 (1954).
128. Kavanau, L. L. Base pressure studies in rarefied supersonic flow. *J. Aeronaut. Sci. 23*, 193–208 (1956).
129. Bogdonoff, S. M. *J. Aeronaut. Sci. 19*, 201 (1952).

INDEX

GPSR Authorized Representative: Easy Access System Europe - Mustamäe tee 50, 10621 Tallinn, Estonia, gpsr.requests@easproject.com

www.ingramcontent.com/pod-product-compliance
Ingram Content Group UK Ltd.
Pitfield, Milton Keynes, MK11 3LW, UK
UKHW020615310326
469463UK00013B/304

* 9 7 8 0 6 9 1 6 2 6 4 9 9 *